Lecture Notes in Computer Science 10955

More information about this series at http://www.springer.com/series/7409

De-Shuang Huang · Kang-Hyun Jo
Xiao-Long Zhang (Eds.)

Intelligent Computing Theories and Application

14th International Conference, ICIC 2018
Wuhan, China, August 15–18, 2018
Proceedings, Part II

Editors
De-Shuang Huang
Tongji University
Shanghai
China

Kang-Hyun Jo
University of Ulsan
Ulsan
Korea (Republic of)

Xiao-Long Zhang
Wuhan University of Science and Technology
Wuhan City
China

ISSN 0302-9743 ISSN 1611-3349 (electronic)
Lecture Notes in Computer Science
ISBN 978-3-319-95932-0 ISBN 978-3-319-95933-7 (eBook)
https://doi.org/10.1007/978-3-319-95933-7

Library of Congress Control Number: 2018947576

LNCS Sublibrary: SL3 – Information Systems and Applications, incl. Internet/Web, and HCI

Printed on acid-free paper

This Springer imprint is published by the registered company Springer International Publishing AG
part of Springer Nature
The registered company address is: Gewerbestrasse 11, 6330 Cham, Switzerland

Preface

The International Conference on Intelligent Computing (ICIC) was started to provide an annual forum dedicated to the emerging and challenging topics in artificial intelligence, machine learning, pattern recognition, bioinformatics, and computational biology. It aims to bring together researchers and practitioners from both academia and industry to share ideas, problems, and solutions related to the multifaceted aspects of intelligent computing.

ICIC 2018, held in Wuhan, China, August 15–18, 2018, constituted the 14th International Conference on Intelligent Computing. It built upon the success of ICIC 2017, ICIC 2016, ICIC 2015, ICIC 2014, ICIC 2013, ICIC 2012, ICIC 2011, ICIC 2010, ICIC 2009, ICIC 2008, ICIC 2007, ICIC 2006, and ICIC 2005 that were held in Liverpool, UK, Lanzhou, Fuzhou, Taiyuan, Nanning, Huangshan, Zhengzhou, Changsha, China, Ulsan, Korea, Shanghai, Qingdao, Kunming, and Hefei, China, respectively.

This year, the conference concentrated mainly on the theories and methodologies as well as the emerging applications of intelligent computing. Its aim was to unify the picture of contemporary intelligent computing techniques as an integral concept that highlights the trends in advanced computational intelligence and bridges theoretical research with applications. Therefore, the theme for this conference was "Advanced Intelligent Computing Technology and Applications." Papers focused on this theme were solicited, addressing theories, methodologies, and applications in science and technology.

ICIC 2018 received 632 submissions from 19 countries and regions. All papers went through a rigorous peer-review procedure and each paper received at least three reviews. Based on the review reports, the Program Committee finally selected 275 high-quality papers for presentation at ICIC 2018, included in three volumes of proceedings published by Springer: two volumes of *Lecture Notes in Computer Science* (LNCS) and one volume of *Lecture Notes in Artificial Intelligence* (LNAI).

This volume of *Lecture Notes in Computer Science* (LNCS) includes 93 papers.

The organizers of ICIC 2018, including Tongji University, Wuhan University of Science and Technology, and Wuhan Institute of Technology, made an enormous effort to ensure the success of the conference. We hereby would like to thank the members of the Program Committee and the referees for their collective effort in reviewing and soliciting the papers. We would like to thank Alfred Hofmann, executive editor at Springer, for his frank and helpful advice and guidance throughout and for his continuous support in publishing the proceedings. In particular, we would like to thank all the authors for contributing their papers. Without the high-quality submissions from the authors, the success of the conference would not have been possible. Finally, we are especially grateful to the IEEE Computational Intelligence Society, the International Neural Network Society, and the National Science Foundation of China for their sponsorship.

May 2018

De-Shuang Huang
Kang-Hyun Jo
Xiao-Long Zhang

ICIC 2018 Organization

General Co-chairs

De-Shuang Huang, China
Huai-Yu Wu, China
Yanduo Zhang, China

Program Committee Co-chairs

Kang-Hyun Jo, Korea
Xiao-Long Zhang, China
Haihui Wang, China
Abir Hussain, UK

Organizing Committee Co-chairs

Hai-Dong Fu, China
Yuntao Wu, China
Bo Li, China

Award Committee Co-chairs

Juan Carlos Figueroa, Colombia
M. Michael Gromiha, India

Tutorial Chair

Vitoantonio Bevilacqua, Italy

Publication Co-chairs

Kyungsook Han, Korea
Phalguni Gupta, India

Workshop Co-chairs

Valeriya Gribova, Russia
Laurent Heutte, France
Xin Xu, China

Special Session Chair

Ling Wang, China

International Liaison Chair

Prashan Premaratne, Australia

Publicity Co-chairs

Hong Zhang, China
Michal Choras, Poland
Chun-Hou Zheng, China
Jair Cervantes Canales, Mexico

Sponsors and Exhibits Chair

Wenzheng Bao, Tongji University, China

Program Committee

Abir Hussain
Akhil Garg
Angelo Ciaramella
Ben Niu
Bin Liu
Bing Wang
Bingqiang Liu
Binhua Tang
Bo Li
Chunhou Zheng
Chunmei Liu
Chunyan Qiu
Dah-Jing Jwo
Daowen Qiu
Dong Wang
Dunwei Gong
Evi Syukur
Fanhuai Shi
Fei Han
Fei Luo
Fengfeng Zhou
Francesco Pappalardo
Gai-Ge Wang
Gaoxiang Ouyang
Haiying Ma
Han Zhang
Hao Lin
Hongbin Huang
Honghuang Lin
Hongjie Wu
Hongmin Cai
Hua Tang
Huiru Zheng
Jair Cervantes
Jian Huang
Jianbo Fan
Jiang Xie
Jiangning Song
Jianhua Xu
Jiansheng Wu
Jianyang Zeng
Jiawei Luo
Jing-Yan Wang
Jinwen Ma
Jin-Xing Liu
Ji-Xiang Du
José Alfredo Costa
Juan Carlos Figueroa-García
Junfeng Xia
Junhui Gao
Junqing Li
Ka-Chun Wong
Khalid Aamir
Kyungsook Han
Laurent Heutte
Le Zhang
Liang Gao
Lida Zhu
Ling Wang
Lining Xing
Lj Gong
Marzio Pennisi
Michael Gromiha Maria Siluvay
Michal Choras
Ming Li
Mohd Helmy Abd Wahab
Pei-Chann Chang

Ping Guo
Prashan Premaratne
Pu-Feng Du
Qi Zhao
Qingfeng Chen
Qinghua Jiang
Quan Zou
Rui Wang
Sabri Arik
Saiful Islam
Seeja K. R.
Shan Gao
Shanfeng Zhu
Shih-Hsin Chen
Shiliang Sun
Shitong Wang
Shuai Li Hong
Stefano Squartini
Sungshin Kim
Surya Prakash
Takashi Kuremoto
Tao Zeng
Tarık Veli Mumcu
Tianyong Hao
Valeriya Gribova
Vasily Aristarkhov
Vitoantonio Bevilacqua
Waqas Haider Khan Bangyal
Wei Chen
Wei Jiang
Wei Peng
Wei Wei
Wei-Chiang Hong
Weijia Jia
Weiwei Kong
Wen Zhang
Wenbin Liu
Wen-Sheng Chen
Wenyin Gong
Xiandong Meng
Xiaoheng Deng
Xiaoke Ma
Xiaolei Zhu
Xiaoping Liu
Xinguo Lu
Xingwen Liu
Xinyi Le
Xiwei Liu
Xuesong Wang
Xuesong Yan
Xu-Qing Tang
Yan Wu
Yan Zhang
Yi Xiong
Yong Wang
Yonggang Lu
Yongquan Zhou
Yoshinori Kuno
Young B. Park
Yuan-Nong Ye
Zhan-Li Sun
Zhao Liang
Zhendong Liu
Zhenran Jiang
Zhenyu Xuan
Zhihua Zhang

Additional Reviewers

Huijuan Zhu
Yizhong Zhou
Lixiang Hong
Yuan Wang
Mao Xiaodan
Ke Zeng
Xiongtao Zhang
Ning Lai
Shan Gao
Jia Liu
Ye Tang
Weiwei Cai
Yan Zhang
Zhang Yuanpeng
Han Zhu
Wei Jiang
Hong Peng
Wenyan Wang
Xiaodan Deng
Hongguan Liu
Hai-tao Li
Jialing Li
Kai Qian
Huichao Zhong
Huiyan Jiang
Lei Wang
Yuanyuan Wang
Biao Zhang
Ta Zhou
Wei Liao
Bin Qin
Jiazhou Chen
Mengze Du
Sheng Ding
Dongliang Qin
Syed Sadaf Ali
Zheng Chenc
Shang Xiang
Xia Lin
Yang Wu
Xiaoming Liu
Jing Lv
Lin Weizhong
Jun Li
Li Peng
Hongfei Bao
Zhaoqiang Chen
Ru Yang
Jiayao Wu
Dadong Dai
Guangdi Liu
Jiajia Miao
Xiuhong Yang
Xiwen Cai
Fan Li
Aysel Ersoy Yilmaz
Agata Giełczyk

Akila Ranjith
Xiao Yang
Cheng Liang
Alessio Ferone
José Alfredo Costa
Ambuj Srivastava
Mohamed Abdel-Basset
Angelo Ciaramella
Anthony Chefles
Antonino Staiano
Antonio Brunetti
Antonio Maratea
Antony Lam
Alfredo Pulvirenti
Areesha Anjum
Athar Ali Moinuddin
Mohd Ayyub Khan
Alfonso Zarco
Azis Ciayadi
Brendan Halloran
Bin Qian
Wenbin Song
Benjamin J. Lang
Bo Liu
Bin Liu
Bin Xin
Guanya Cai
Casey P. Shannon
Chao Dai
Chaowang Lan
Chaoyang Zhang
Zhang Chuanchao
Jair Cervantes
Bo Chen
Yueshan Cheng
Chen He
Zhen Chen
Chen Zhang
Li Cao
Claudio Loconsole
Cláudio R. M. Silva
Chunmei Liu
Yan Jiang
Claus Scholz
Yi Chen
Dhiya AL-Jumeily
Ling-Yun Dai
Dongbo Bu
Deming Lei
Deepak Ranjan Nayak
Dong Han
Xiaojun Ding
Domenico Buongiorno
Haizhou Wu
Pingjian Ding
Dongqing Wei
Yonghao Du
Yi Yao
Ekram Khan
Miao Jiajia
Ziqing Liu
Sergio Santos
Tomasz Andrysiak
Fengyi Song
Xiaomeng Fang
Farzana Bibi
Fatih Adıgüzel
Fang-Xiang Wu
Dongyi Fan
Chunmei Feng
Fengfeng Zhou
Pengmian Feng
Feng Wang
Feng Ye
Farid Garcia-Lamont
Frank Shi
Chien-Yuan Lai
Francesco Fontanella
Lei Shi
Francesca Nardone
Francesco Camastra
Francesco Pappalardo
Dongjie Fu
Fuhai Li
Hisato Fukuda
Fuyi Li
Gai-Ge Wang
Bo Gao
Fei Gao
Hongyun Gao
Jianzhao Gao
Gaoyuan Liang
Geethan Mendiz
Guanghui Li
Giacomo Donato Cascarano
Giorgio Valle
Giovanni Dimauro
Giulia Russo
Linting Guan
Ping Gong
Yanhui Gu
Gunjan Singh
Guohua Wu
Guohui Zhang
Guo-sheng Hao
Surendra M. Gupta
Sandesh Gupta
Gang Wang
Hafizul Fahri Hanafi
Haiming Tang
Fei Han
Hao Ge
Kai Zhao
Hangbin Wu
Hui Ding
Kan He
Bifang He
Xin He
Huajuan Huang
Jian Huang
Hao Lin
Ling Han
Qiu Xiao
Yefeng Li
Hongjie Wu
Hongjun Bai
Hongtao Lei
Haitao Zhang
Huakang Li
Jixia Huang
Pu Huang
Sheng-Jun Huang
Hailin Hu
Xuan Huo
Wan Hussain Wan Ishak
Haiying Wang
Il-Hwan Kim

Kamlesh Tiwari
M. Ikram Ullah Lali
Ilaria Bortone
H. M. Imran
Ingemar Bengtsson
Izharuddin Izharuddin
Jackson Gomes
Wu Zhang
Jiansheng Wu
Yu Hu
Jaya sudha
Jianbo Fan
Jiancheng Zhong
Enda Jiang
Jianfeng Pei
Jiao Zhang
Jie An
Jieyi Zhao
Jie Zhang
Jin Lu
Jing Li
Jingyu Hou
Joe Song
Jose Sergio Ruiz
Jiang Shu
Juntao Liu
Jiawen Lu
Jinzhi Lei
Kanoksak Wattanachote
Juanjuan Kang
Kunikazu Kobayashi
Takashi Komuro
Xiangzhen Kong
Kulandaisamy A.
Kunkun Peng
Vivek Kanhangad
Kang Xu
Kai Zheng
Kun Zhan
Wei Lan
Laura Yadira Domínguez Jalili
Xiangtao Chen
Leandro Pasa
Erchao Li
Guozheng Li
Liangfang Zhao
Jing Liang
Bo Li
Feng Li
Jianqiang Li
Lijun Quan
Junqing Li
Min Li
Liming Xie
Ping Li
Qingyang Li
Lisbeth Rodríguez
Shaohua Li
Shiyong Liu
Yang Li
Yixin Li
Zhe Li
Zepeng Li
Lulu Zuo
Fei Luo
Panpan Lu
Liangxu Liu
Weizhong Lu
Xiong Li
Junming Zhang
Shingo Mabu
Yasushi Mae
Malik Jahan Khan
Mansi Desai
Guoyong Mao
Marcial Guerra de Medeiros
Ma Wubin
Xiaomin Ma
Medha Pandey
Meng Ding
Muhammad Fahad
Haiying Ma
Mingzhang Yang
Wenwen Min
Mi-Xiao Hou
Mengjun Ming
Makoto Motoki
Naixia Mu
Marzio Pennisi
Yong Wang
Muhammad Asghar Nadeem
Nadir Subaşi
Nagarajan Raju
Davide Nardone
Nathan R. Cannon
Nicole Yunger Halpern
Ning Bao
Akio Nakamura
Zhichao Shi
Ruxin Zhao
Mohd Norzali Hj Mohd
Nor Surayahani Suriani
Wataru Ohyama
Kazunori Onoguchi
Aijia Ouyang
Paul Ross McWhirter
Jie Pan
Binbin Pan
Pengfei Cui
Pu-Feng Du
Iyyakutti Iyappan Ganapathi
Piyush Joshi
Prashan Premaratne
Peng Gang Sun
Puneet Gupta
Qinghua Jiang
Wangren Qiu
Qiuwei Li
Shi Qianqian
Zhi Xian Liu
Raghad AL-Shabandar
Rafał Kozik
Raffaele Montella
Woong-Hee Shin
Renjie Tan
Rodrigo A. Gutiérrez
Rozaida Ghazali
Prabakaran
Jue Ruan
Rui Wang
Ruoyao Ding
Ryuzo Okada
Kalpana Shankhwar
Liang Zhao

Sajjad Ahmed
Sakthivel Ramasamy
Shao-Lun Lee
Wei-Chiang Hong
Hongyan Sang
Jinhui Liu
Stephen Brierley
Haozhen Situ
Sonja Sonja
Jin-Xing Liu
Haoxiang Zhang
Sebastian Laskawiec
Shailendra Kumar
Junliang Shang
Guo Wei-Feng
Yu-Bo Sheng
Hongbo Shi
Nobutaka Shimada
Syeda Shira Moin
Xingjia Lu
Shoaib Malik
Feng Shu
Siqi Qiu
Boyu Zhou
Stefan Weigert
Sameena Naaz
Sobia Pervaiz
Somnath Dey
Sotanto Sotanto
Chao Wu
Yang Lei
Surya Prakash
Wei Su
Qi Li
Hotaka Takizawa
FuZhou Tang
Xiwei Tang
LiNa Chen
Yao Tuozhong
Qing Tian
Tianyi Zhou
Junbin Fang
Wei Xie
Shikui Tu
Umarani Jayaraman
Vahid Karimipour
Vasily Aristarkhov
Vitoantonio Bevilacqua
Valeriya Gribova
Guangchen Wang
Hong Wang
Haiyan Wang
Jingjing Wang
Ran Wang
Waqas Haider Bangyal
Pi-Jing Wei
Fangping Wan
Jue Wang
Minghua Wan
Qiaoyan Wen
Takashi Kuremoto
Chuge Wu
Jibing Wu
Jinglong Wu
Wei Wu
Xiuli Wu
Yahui Wu
Wenyin Gong
Zhanjun Wang
Xiaobing Tang
Xiangfu Zou
Xuefeng Cui
Lin Xia
Taihong Xiao
Xing Chen
Lining Xing
Jian Xiong
Yi Xiong
Xiaoke Ma
Guoliang Xu
Bingxiang Xu
Jianhua Xu
Xin Xu
Xuan Xiao
Takayoshi Yamashita
Atsushi Yamashita
Yang Yang
Zhengyu Yang
Ronggen Yang
Yaolai Wang
Yaping Yang
Yue Chen
Yongchun Zuo
Bei Ye
Yifei Qi
Yifei Sun
Yinglei Song
Ying Ling
Ying Shen
Yingying Qu
Lvjiang Yin
Yiping Liu
Wenjie Yi
Jianwei Yang
Yu-Jun Zheng
Yonggang Lu
Yan Li
Yuannong Ye
Yong Chen
Yongquan Zhou
Yong Zhang
Yuan Lin
Yuansheng Liu
Bin Yu
Fang Yu
Kumar Yugandhar
Liang Yu
Yumin Nie
Xu Yu
Yuyan Han
Yikuan Yu
Ying Wu
Ying Xu
Zhiyong Wang
Shaofei Zang
Chengxin Zhang
Zehui Cao
Tao Zeng
Shuaifang Zhang
Liye Zhang
Zhang Qinhu
Sai Zhang
Sen Zhang
Shan Zhang
Shao Ling Zhang
Wen Zhang
Wei Zhao
Bao Zhao

Zheng Tian
Zheng Sijia
Zhenyu Xuan
Fangqing Zhao
Zhipeng Cai
Xing Zhou
Xiong-Hui Zhou
Lida Zhu
Ping Zhu
Qi Zhu
Zhong-Yuan Zhang
Ziding Zhang
Junfei Zhao
Juan Zou
Quan Zou
Qian Zhu
Zunyan Xiong
Zeya Wang
Yatong Zhou
Shuyi Zhang
Zhongyi Zhou

Contents – Part II

Deep Convolutional Neural Network for Fog Detection

Jun Zhang[1(✉)], Hui Lu[1], Yi Xia[1], Ting-Ting Han[1], Kai-Chao Miao[2], Ye-Qing Yao[2], Cheng-Xiao Liu[2], Jian-Ping Zhou[2], Peng Chen[3], and Bing Wang[4]

[1] School of Electronic Engineering and Automation, Anhui University, Hefei 230601, Anhui, China
wwwzhangjun@163.com
[2] Anhui Meteorological Bureau, Hefei 230031, Anhui, China
[3] Institute of Health Sciences, Anhui University, Hefei 230601, Anhui, China
[4] School of Electrical and Information Engineering, Anhui University of Technology, Ma Anshan 243032, China

Abstract. Fog detection has becomes more and more important in recent years, real-time monitoring information is very beneficial for people to arrange production and life. In this paper, based on meterological satellite data (Himawari-8 standard data, HSD8), Covolutional Neural Network (CNN) is used to detect fog. Since HSD8 consists of 16 channels, the original CNN is extended to multiple channels for HSD8. Multiple Channels CNN (MCCNN) can make the full exploitation of spatial and spectral information effectively. A dataset is created from Anhui Area which consists of ground station data and grid data. Different image sizes and convolutional kernels are used to validate the proposed methods. The experimental results show that the proposed method achieves 91.87% accuracy.

Keywords: Fog · CNN · Detection · Himawari-8

1 Introduction

Fog has becomes one of the major catastrophic weather. The formation of fog is caused by a large number of tiny water droplets or ice crystals floating in the air near the ground. Therefore, if the horizontal visibility is less than 1 km, it can cause disastrous weather. In recent years, with the rapid development of socio-economic construction, especially land, sea and air transportation, fog detection the has become more and more important. Fog generally has the characteristics of rapid generation and development [1]. Traditional ground-based observation are limited to the local areas, it can not meet the requirements of large-scale and fast monitoring.

The earliest use methods of satellite remote sensing technology in fog monitoring is visible light image cloud identification and dissipation. In the 1960s, with the development of meteorological satellites, European and American countries started to use the satellite remote sensing technology for fog identification [2]. Usually, the methods of monitoring heavy fog mainly include two-channel method, principal component

D.-S. Huang et al. (Eds.): ICIC 2018, LNCS 10955, pp. 1–10, 2018.
https://doi.org/10.1007/978-3-319-95933-7_1

analysis method and texture processing method [3]. But now with the further research on satellite remote sensing, more scholars use different spectral characteristics to study the method of fog identification.

Although the previous method performs well, it still has some drawbacks. Most of these methods are based on manual feature learning, so the performance of these methods heavily depends on domain knowledge. Even satellite remote sensing data has rich spectral information, which can not be fully used by these traditional approaches [4]. However, deep learning is a effective method which can extract the features from mass of data automatically. In the last few years, deep learning has been a powerful machine learning technique for learning data-dependent and hierarchical feature representations from raw data. Convolutional Neural Network (CNN) is an algorithm in deep learning. CNN has been widely applied to image processing and computer vision problems, such as image classification,image segmentation [5], action recognition and object detection [6].

Himawari-8 Standard Data (HSD8) data has a very high spatial and spectral resolution. Every ten miniute, the satellite return 16 channel data. Such a large data set make CNN framework more suitable and more effective in handling classification [7, 8] problems. Deep learning has been widely used in hyperspectral image classification. Hyperspectral image data usually have hundreds of spectral image. Similar to hyperspectral image data, HSD8 consist by 16 channel.

In this research, we propose CNN framework that classifies HSD8 data into multiple classes. The experimental result show that the proposed method achieve a better classification performance.

2 First Section

2.1 CNN

Convolution Neural Network (CNN) is an efficient identification method that has been developed and attracted a great deal of attention in recent years. Generally, the basic structure of CNN includes two layers, one is a feature extraction layer, the input of each neuron is connected with a local acceptance field of the previous layer. Once the local feature is extracted, its location relationship with other features is also determined. The second is the feature mapping layer. The feature mapping structure uses the sigmoid function with small influence kernel as the activation function of the CNN, which makes the feature map invariant to displacement. In addition, due to the shared weights of neurons on a mapping surface, the number of free network parameters is reduced. The special structure of CNN sharing local weight has unique advantages in speech recognition [9] and image classification [10]. Weight sharing reduces the complexity of the network. In particular, the features of multi-dimensional input vector images can be directly input to the network avoid the complexity of data reconstruction in feature extraction.

CNN is a feed forward neural network, the structure of the model is shown in the Fig. 1. x is the input, h is the neuron output. When multiple cells are combined and have a hierarchical structure, a neural network model is established. Neural network connects many of these neurons into one network, the output of one neuron serves as input to another neuron. Neural networks can have a wide variety of topologies. One of the simplest is "multi-layer fully connected forward neural network." Its input is connected to every neuron on the first level of the network. The output of each neuron in the previous layer is connected to the input of each neuron in the next layer. The output of the last neuron is the output of the entire neural network.

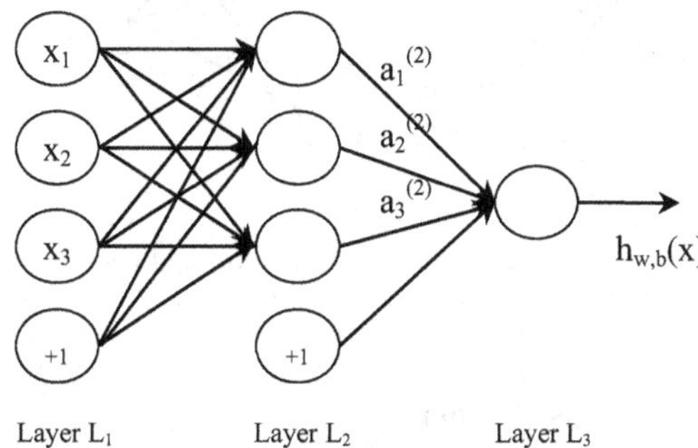

Fig. 1. Upper and lower neurons are all connected to the neural network

The order of CNN is the alternation of the convolutional layer with the pooling layer, which mimics the complex and simple cell nature of the mammalian visual cortex. Finally, the fully connected neural network ends. In ordinary deep neural networks, neurons connect to all the neurons in the next level. CNN Unlike ordinary neural networks, a typical CNN is as shown in Fig. 2. That is, each hidden activation is calculated by multiplying the entire input x by the weight W in that layer. However, in CNN, every hidden activation point h is obtained by inputting a small local weight W, so that the weight W can be shared in the entire input space. Neurons belonging to the same level share the same weight. The advantage of CNN weight sharing is that it helps to reduce the total number of training parameters and leads to more efficient training and more efficient models.

The pooling layer has two functions, the first effect is to retain the main features while reducing the parameters (similar to the PCA [11]) and the amount of calculation to prevent over-fitting and improve the model generalization ability. The second effect is introduced of invariance. The pooling layer has the max pooling and the average pooling, the typical pooling function is the max pooling. The max pooling divides the input data into a set of non-overlapping windows and outputs the maximum for each subregion, reducing the upper computational complexity and providing a form of translational invariance. The computational chain of the CNN terminates in fully connected network, the fully connected network contains the feature information extracted by the convolutional and pooling layers, we mainly use it for classification. In this article, we explore the appropriate architecture and strategy for CNN-based HSD8 classification.

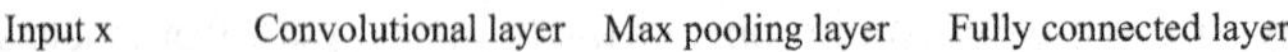

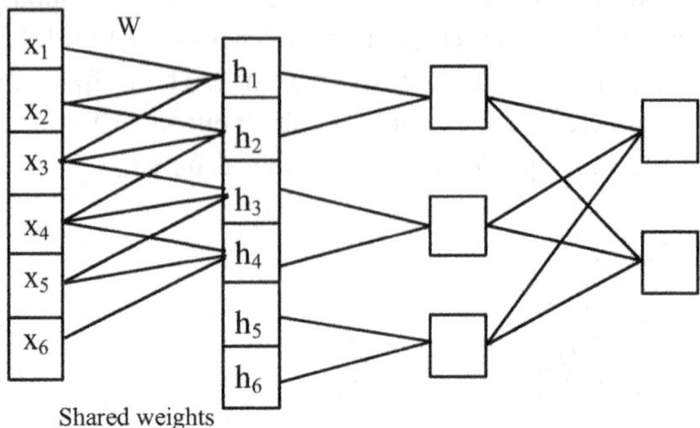

Fig. 2. A typical CNN architecture consisting of a convolutional layer, a max pooling layer, and a fully connected layer.

2.2 CNN-Based HSD8 Classification

Architecture of the Proposed CNN Classifier

We know that CNN can be more competitive and perform better than humans in certain visual issues and their ability to stimulate our research into the possibility of using CNN for classification of HSD8 using spectral features. In this paper, we adopt the CNN structure shown in Fig. 3. This network contains one input layer, two pairs of convolution layers and max pooling layers, two fully connected layers and one output layer in our network structure.

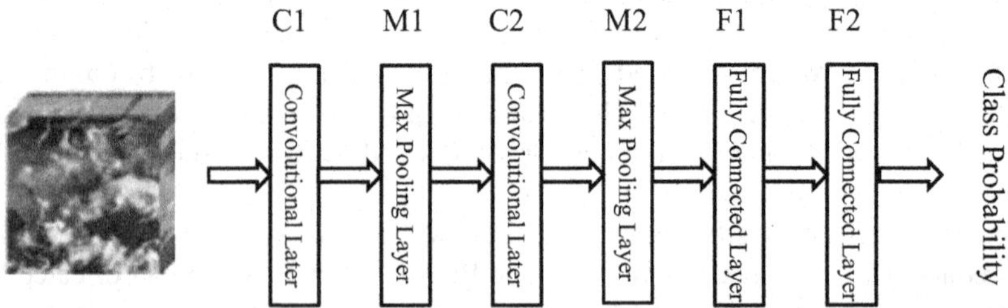

Fig. 3. The network structure of the CNN used as our discriminative classifier

Training Strategies

Here, we introduce how to learn the parameter space of the proposed CNN classifier. CNN's training process is mainly divided into two steps: forward propagation and back propagation [12]. The function of forward propagation is to calculate the actual classification result of the input data with the current parameters and the effect of back propagation is to update the trainable parameters [13], so that the difference between the actual classification output and the desired classification output as should be small as possible.

Forward Propagation

As shown in the Fig. 3, the complete network consists of six layers, including the input layer, the convolution layer C, the max pooling layer M, the full connection layer F. We first define the problem related notations used throughout the paper. We assume that θ is all the training parameters, $\theta = \{\theta i\}$ and i = 1, 2, 3, 4… where θi is the parameter set between the $(i-1)$ th and the i th layer. In this experiment we used a 6-layer convolution neural network. The output of the convolutional layer is as follows:

$$x_{i+1} = f_i(u_i) \tag{1}$$

$$u_i = \theta_i^T x_i + b_i \tag{2}$$

That x_i is the input of the i th layer, θ is weight. θ_i^T is a weight matrix of the i th layer acting on the input data, and b_i is an additive bias vector for the i th layer. $f()$ is the activation function. The activation function is to introduce nonlinear factors. The commonly used activation functions are Tanh, Sigmoid, Relu. In our designed architecture, we choose the hyperbolic tangent function Relu as the activation function. The maximum function max(u) is used in layer M.

We want to use the hypothetical function to estimate the probability value $p(y = j|x)$ for each category j. That is, we want to estimate the probability of each classification result for x occurring. Therefore, our hypothetical function will output a k-dimensional vector to represent the probabilities of these K estimates. Specifically, our hypothetical function takes the following form, the softmax regression model is defined as:

$$y_j^{(i)} = \begin{bmatrix} P(y^{(i)} = 1|\theta; x; b) \\ P(y^{(i)} = 2|\theta; x; b) \\ \cdots\cdot \quad \cdots\cdot \\ P(y^{(i)} = 2|\theta; x; b) \end{bmatrix} = \frac{1}{\sum_{j=1}^{k} e^{\theta_j^T x^{(i)} + b_j}} \begin{bmatrix} e^{\theta_1^T x^{(i)} + b_1} \\ e^{\theta_2^T x^{(i)} + b_2} \\ \cdots \\ e^{\theta_j^T x^{(i)} + b_j} \end{bmatrix} \tag{3}$$

Back Propagation

In the back propagation stage, the backpropagation algorithm gives an efficient way to use the gradient descent algorithm on all parameters, so that the loss function of the neural network model on the training data is as small as possible. Backpropagation algorithm is the core algorithm of training neural network, which can optimize the parameters of neural network according to the defined loss function, so that the loss function of neural network model on training data set reaches a smaller value. The optimization of the parameters in the neural network model directly determines the quality of the model, which is a very important step when using the neural network. The loss function used in this work is defined as:

$$J(\theta) = -\frac{1}{m}\sum_{i=1}^{m}\sum_{j=1}^{K} 1\left\{j = Y^{(i)}\right\} \log(y_j^{(i)}) \tag{4}$$

where m is the number of training samples. Y is the desired output. $y_j^{(i)}$ is the j th value of the actual output $y^{(i)}$ of the i th training sample and is a vector whose size is K. In the desired output $Y^{(i)}$ of the i th sample, the probability value of the labeled class is 1, and the probability values of other classes are $0.1\{j = Y^{(i)}\}$ means, if j is equal to the desired label of the i th training sample, its value is 1; otherwise, its value is 0.

In the training process, we need to get the prediction label by finding the maximum value in the output vector, which requires that we actually output is closer to the expectation. When the difference between them is small enough, the iteration stops, at which point the actual output is closer to the desired output. Experiments show that as the number of training iterations increases, the return value of the loss function becomes smaller, which can meet our needs. Finally, the trained CNN is ready for HSD8 classification. Training samples are first randomly divided into batches, each batch containing the same number of samples. The batch size has a certain impact on the experimental results. During training, we trained CNN using a stochastic gradient descent, each iteration only one batch was sent to the network for training. When the maximum number of iterations is reached, the training process will stop, then the training model will be output. In the testing process, we first input the test samples into the training network [14]. The parameters of the network are obtained by using the trained network parameters and then finding the maximum value in the output vector. The training and testing process is shown in Fig. 4.

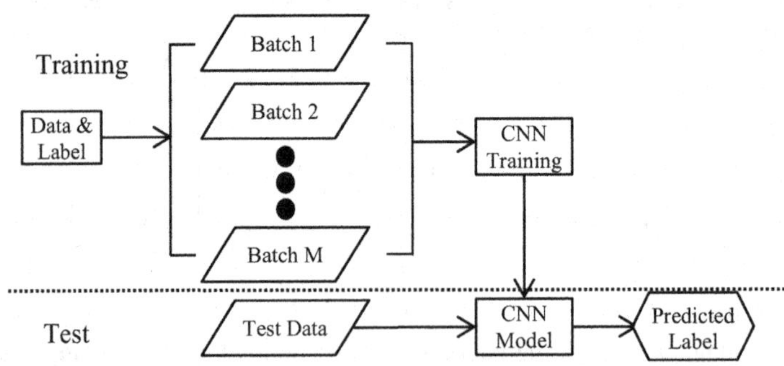

Fig. 4. Illustration of the parameter process

3 Experiments and Discussion

The Data Sets: This experiment using HSD8, total of 16 channels. When the visibility is greater than 1000 m, it is negative sample (without fog) and the other sample is positive sample (with fog).

Impact of parameter settings: The parameter selection is very important to the performance of the network. Such as the depth of the network, the size and number of convolution kernels, the batch size and the learning rate. In this experiment, we evaluated in detail the effect of different parameter settings on the sensitivity of the

CNN structure. In the following experiment, we will choose different parameters to adjust the network and select the appropriate network parameters.

Input the image sizes: The formation of fog is a certain range, when beyond a certain range may without fog, the test results have a negative impact. The size of the input image is actually the site as the center, to extract the grid data around them, in our experiments, set as 23 × 23,29 × 29,35 × 35,41 × 41,47 × 47. The experiment results in the following Fig. 5. For this the experiment, the best performance can be achieved by setting for the input size to 29, We will choose this parameter as the input size.

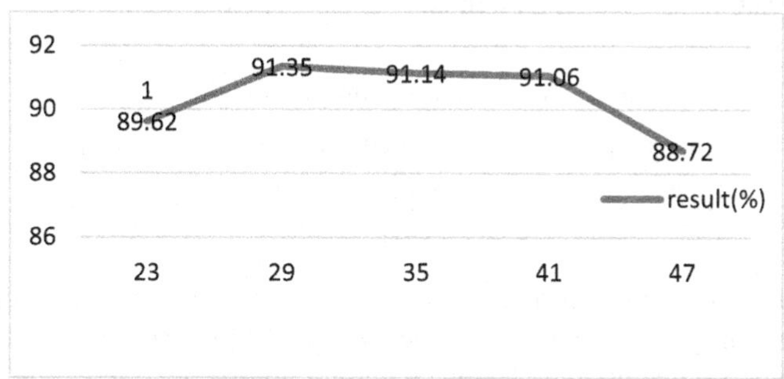

Fig. 5. Accuracy with different input sizes

Kernel Size: The larger the size of the convolution kernel, the larger the visual range, that is, the more information after processing, but it will increase the parameters of the network so that the network speed is slow. If the matrix is small, the visual range is small but The network is fast, so need pick the right size. In this experiment, we tested the effect of different kernel sizes. We fixed the second level kernel size to 3 and evaluated the performance by changing the kernel size in the first level. Generally, 1 × 1,2 × 2,3 × 3,4 × 4,5 × 5 these sizes. Figure 6 shows these results. From this table it can be seen that increasing kernel size does not always generate better result. For this the experiment, the best performance can be achieved by setting for the input size to 4.

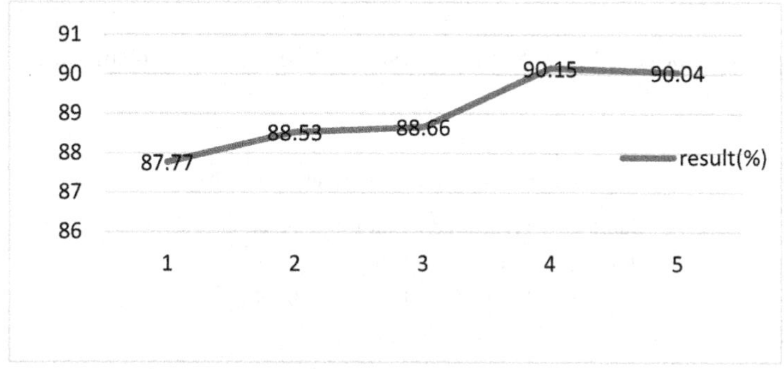

Fig. 6. Accuracy with different kernels

Learning rate: The smaller the learning rate, the more refined the learning, but at the same time the learning speed will also be commonly used to reduce. Many models are non-linear, like a curve. Gradient descent adopts many small linear iterations to approximate the nonlinear curve. Span too much (learning rate) will lose a lot of curve distortion information, local straight line too serious, too small to reach the end of the curve you need many steps, which requires more samples, The experimental learning rate setting from 0.00005 to 0.005, the result is as Table 1. From the experiment, it can be seen when the learning rate of 0.00005 choose the best, but too small will make the algorithm optimization is too slow, may be transferred to the local optimum. we choose learning rate 0.0001 as network paramater.

Table 1. Accuracy with different learning rate

Learning rate	0.00005	0.0001	0.0005	0.001	0.005
Result(%)	90.77	90.68	88.25	/	/

Network Depth: The depth of the network has a very big impact on the accuracy and speed of training. Therefore, we conduct comparative experiments to test the effect of different network depths by reducing or adding non-linear layers. We selected the depth of (3,4,5,6,7) the five networks for training and testing. The experimental results are summarized in Table 2. It can be seen from the experimental results that as the network depth increases, it can not produce better results, which may be the result of the gradient disappearance problem encountered by deep neural networks. From this table it can be seen that increasing the network depth does not always generate better result, which may be a result of the gradient vanishing problem encountered by deeper neural networks. For this experiment, From the experiment, we can seen that we choose a depth of 6 as our network depth.

Table 2. Accuracy with different network depth

Depth	3	4	5	6	7
Result(%)	89.33	89.87	90.53	90.67	90.01

Through the above CNN structural parameters set, the final results are as follow Table 3.

Table 3. The final classification results

Samples	Train total	Test total	Accuracy(%)
Positive	3642	1000	91.87
Negative	3642	1000	91.43

4 Conclusion

In this paper, we propose a new CNN-based fog detection method. By using a deep learning fog detection method, the information contained in the raw data can be fully utilized and even a small number of training samples can achieve higher accuracy. This work is the use of CNN-based fog detection preliminary exploration and achieved good results. Since we have fewer training samples, we did not choose a deeper network layer. In future work, we will expand the data set and use deep network training, such as the VGG16, VGG19 and Resnet series networks.

Acknowledgments. This work was supported by Jiangsu Province Meteorological Bureau Bei Ji Ge grant Nos. BJG201707, Anhui Province Meteorological Bureau meteorologist special grant Nos. KY201704, Anhui Provincial Natural Science Foundation (grant number 1608085MF136).

References

1. Linping, S., Chi, X.: The analysis and forecast of fog in North China Plain. Meteorol. Monthly (1995)
2. Anderson, R.K., Farr, G.R., Ashman, J.P., Smith, A.H., Ritter, L.F.: Application of meteorological satellite data in analysis and forecasting. J. Pharm. Pharmacol. **38**, 107–112 (1974)
3. Zhou, B., Du, J., Gultepe, I., Dimego, G.: Forecast of low visibility and fog from NCEP: current status and efforts. Pure. Appl. Geophys. **169**, 895–909 (2012)
4. Chen, Y., Zhao, X., Jia, X.: Spectral–spatial classification of hyperspectral data based on deep belief network. IEEE J. Sel. Top. Appl. Earth Obs. Remote Sens. **8**, 2381–2392 (2015)
5. Chen, L.C., Papandreou, G., Kokkinos, I., Murphy, K., Yuille, A.L.: Semantic image segmentation with deep convolutional nets and fully connected CRFs. Computer Science, pp. 357–361 (2014)
6. Szegedy, C., Toshev, A., Erhan, D.: Deep neural networks for object detection. Adv. Neural. Inf. Process. Syst. **26**, 2553–2561 (2013)
7. Bandos, T.V., Bruzzone, L., Camps-Valls, G.: Classification of hyperspectral images with regularized linear discriminant analysis. IEEE Trans. Geosci. Remote Sens. **47**, 862–873 (2009)
8. Krizhevsky, A., Sutskever, I., Hinton, G.E.: ImageNet classification with deep convolutional neural networks. In: International Conference on Neural Information Processing Systems, pp. 1097–1105 (2012)
9. Hinton, G., Deng, L., Yu, D., Dahl, G.E., Mohamed, A.R., Jaitly, N., et al.: Deep neural networks for acoustic modeling in speech recognition: the shared views of four research groups. IEEE Signal Process. Mag. **29**, 82–97 (2012)
10. Wang, J.,Yang, Y., Mao, J., Huang, Z., Huang, C., Xu, W.: CNN-RNN: a unified framework for multi-label image classification, pp. 2285–2294 (2016)
11. Agarwal, A., El-Ghazawi, T., El-Askary, H., Le-Moigne, J.: Efficient hierarchical-PCA dimension reduction for hyperspectral imagery. In: IEEE International Symposium on Signal Processing and Information Technology, pp. 353–356 (2008)
12. Lecun, Y., Boser, B., Denker, J.S., Henderson, D., Howard, R.E., Hubbard, W., et al.: Backpropagation applied to handwritten zip code recognition. Neural Comput. **1**, 541–551 (2014)

13. LeCun, Y., Bottou, L., Orr, G.B., Müller, Klaus -Robert: Efficient BackProp. In: Orr, G.B., Müller, K.R. (eds.) Neural Networks: Tricks of the Trade. LNCS, vol. 1524, pp. 9–50. Springer, Heidelberg (1998). https://doi.org/10.1007/3-540-49430-8_2
14. Yu, S., Jia, S., Xu, C.: Convolutional neural networks for hyperspectral image classification. Neurocomputing **219**, 88–98 (2016)

Prediction of Crop Pests and Diseases in Cotton by Long Short Term Memory Network

Qingxin Xiao[1], Weilu Li[1], Peng Chen[1(✉)], and Bing Wang[2(✉)]

[1] Institute of Physical Science and Information Technology, Anhui University, Hefei 230601, Anhui, China
pchen.ustc10@yahoo.com

[2] School of Electrical and Information Engineering, Anhui University of Technology, Ma'anshan 243032, Anhui, China

Abstract. This paper aims to predict the occurrence of pests and diseases for cotton based on long short term memory (LSTM) network. First, the problem of occurrence of pests and diseases was formulated as time series prediction. Then LSTM was adopted to solve the problem. LSTM is a special kind of recurrent neutral network (RNN), which introduces gate mechanism to prevent the vanished or exploding gradient problem. It has been shown good performance in solving time series problem and can handle the long-term dependency problem, as mentioned in many literatures. The experimental results showed that LSTM performed good on the prediction of occurrence of pests and diseases in cotton fields, and yielded an Area Under the Curve (AUC) of 0.97. The paper further verified that the weather factors indeed have strong impact on the occurrence of pests and diseases, and the LSTM network has great advantage on solving the long-term dependency problem.

Keywords: Long short term memory · Weather factors · Recurrent neural network · Occurrence of pests and diseases

1 Introduction

Recently, the occurrence frequency of regional cotton pests and diseases has increased rapidly, causing huge losses in agricultural production. There are many factors to affect its growth, of which the most significant one is abnormal climate change, which resulted in the continuous evolution of pests and further made them adaptive to the environment. All of that seriously influenced the yield and quality and made it very difficult to control the pests and diseases [1]. Many methods have been developed to control pest occurrence. One type was based on biochemical perspectives to suppress the occurrence of pests and diseases, i.e., pesticide screening [2], biological control [3]. The other type was based on historical data and tried to predict future occurrence trend of pests [4].

In recent years, deep learning has been widely used in many fields [5–10]. Long Short Term Memory (LSTM) is a deep learning model. It is a special kind of recurrent neutral network (RNN), which introduces gate mechanism into vanilla RNN to prevent

D.-S. Huang et al. (Eds.): ICIC 2018, LNCS 10955, pp. 11–16, 2018.
https://doi.org/10.1007/978-3-319-95933-7_2

the vanished or exploding gradient problem. LSTM has achieved good results in different fields. Li *et al.* adopted an LSTM auto-encoder with generating coherent text units from neural models to preserve and reconstruct multi-sentence paragraphs [8]. Gao *et al.* presented an mQA model, which contained a LSTM and a Convolutional Neural Network (CNN), to answer questions about the content of an image [9]. Theis and Bethge introduced a recurrent image model based on multi-dimensional LSTM units, which are particularly suited for image modeling [10].

In this paper, we propose a LSTM network based method to predict the occurrence of pests and diseases of cotton, with the use of weather factors. Results showed that our LSTM based model outperformed other traditional prediction models.

2 Methodology

2.1 Material and Problem Formulation

To investigate the impact of weather factors on the occurrences of pests and diseases, the datasets from Crop Pest Decision Support System (http://www.crida.in:8080/naip/AccessData.jsp) were used, which recorded cotton documents weekly (15, 375) for 10 insect pests and diseases in cotton along with corresponding weather conditions across 6 important locations in India. The weather features consist of Maximum Temperature (MaxT(°C)), Minimum Temperature (MinT(°C)), Relative Humidity in the morning (RH1(%)), Relative Humidity in the evening (RH2(%)), Rainfall (RF(mm)), Wind Speed (WS(kmph)), Sunshine Hour (SSH(hrs)) and Evaporation (EVP(mm)). Our aim is to predict the occurrences of pests and diseases under different weather conditions. Bollworm is the main target of biological control. So, bollworm records were used to build weather-pest forecasting model. Suppose X be the vector set of weather-pest records, $Y = \{0, 1\}$, from one single area along the whole recorded time, which is a time series set. The prediction problem can be then converted into predicting the occurrence ($Y_i = 0$) or non-occurrence ($Yi = 1$) of pests and diseases based on the feature vector X_i, $i = 1 \ldots N$, where N is the number of feature vectors. A LSTM based model was designed to capture the relationship of data (X_i, Y_i), $i = 1 \ldots N$, to predict future occurrence of pests and diseases under the weather features.

2.2 Long Short Term Memory

Like most RNNs, LSTM contains a memory function that can handle time series problems, while unlike traditional RNNs, LSTM is well-suited for long-term dependency problems because it solves the problem of gradient vanish and gradient explosion. There are three doors in LSTM. The input gate decides the input x_i entering into the current cell, the forget gate decides if and how much information be forgotten for the previous memory, and the output one controls the information outputting from the current cell. The gating operations ultimately determine which information is forgot and which information is entered into the neural network as useful information. For the weather-pest forecasting issue, it processes a series of temporal dependency inputs x_t at time t and the hidden vector h_{t-1} from the last time step then get the predicted h_t. The basic structure of LSTM cells can be seen in Fig. 1 and the related formulas are shown as below.

$$\begin{aligned} i_t &= \sigma\left(W^i \cdot [h_{t-1}, x_t] + b^i\right) \\ f_t &= \sigma\left(W^f \cdot [h_{t-1}, x_t] + b^f\right) \\ o_t &= \sigma(W^o \cdot [h_{t-1}, x_t] + b^o) \\ \mathrm{c}_t &= \tanh(W^c \cdot [h_{t-1}, x_t] + b^c) \\ C_t &= f_t \cdot C_{t-1} + i_t \cdot c_t \\ \mathrm{h}_t &= o_t \cdot \tanh(C_t) \end{aligned} \tag{1}$$

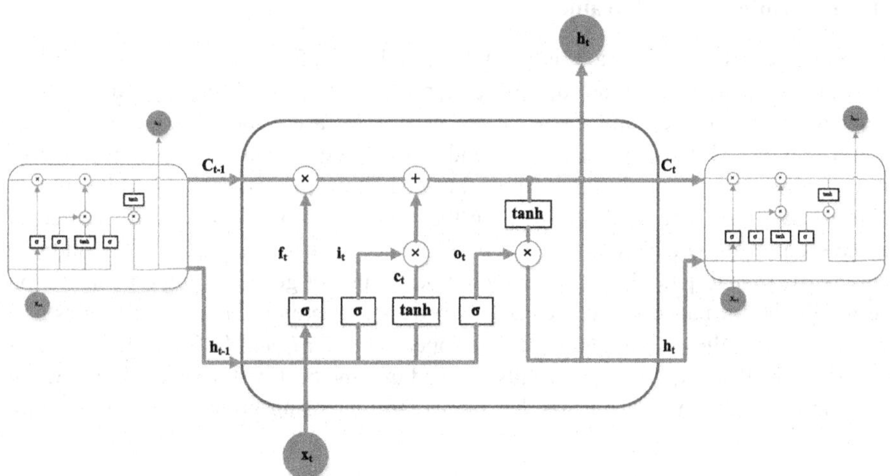

Fig. 1. Structure of LSTM cells.

where σ is the sigmoid function; *tanh* (*) is a nonlinear activation function; W is the recurrent weight matrix; b is the corresponding bias vector; i, f and o are the outputs of the input, forget, and output gates, respectively; and C and h are the memory vector and out vector of the cell, respectively.

According to the previous work [11], the output, (h_t, C_t), of a cell can be represented as a whole function *LSTM* (*):

$$(h_t, C_t) = LSTM([h_{t-1}, x_t], C_{t-1}, W) \tag{2}$$

where W concatenates the four weight matrices W^i, W^f, W^o and W^c.

2.3 Architecture of the Used LSTM Network

The prediction problem is converted into a time series problem, which uses the past weather-pest records to identify whether pests and diseases will occur in the future. The first thing that should be determined for the LSTM network is how long the historical observations is used for the prediction. Of course the longer the historical data is, the better the prediction will be, however the LSTM requires more computation. Here 'timesteps' is set as 4, i.e., four samples of weather-pest data are input together into the LSTM. Three other important parameters for the whole structure of the network should also be determined: the number of layers for LSTM layer l_r, the full-connected layer lfc and the corresponding number of hidden units denoted by *units_r*. In addition, some critical parameters have to be determined, i.e., the Adam optimization method [12] is adopted, and the learning rate and dropout are set as 0.001 and 0.1, respectively.

3 Experiment and Results

3.1 Determination of Parameters

Five top size datasets, denoted as p1, p2, p3, p4 and p5, were selected to train the LSTM network and determine the parameters. In this work, Accuracy (*ACC*), Area Under the Curve (*AUC*) and *F1-score* are used to measure the effectiveness of prediction methods. First, supposed l_r, l_{fc} and *units_fc* be 1, and set a proper value of *units_r* from {4, 5, 6, 7}. Table 1 shows the predictions on five datasets with different values of *units_rs*. The boldface items in the table represent the best performance. It can be seen from the table that the best performance occurs when *units_r* = 5 on three datasets p1, p2 and p4. Although the model performs not good enough on datasets p3 and p5, the performance differences are not obvious. Therefore, *units_r* was set as 5 in this work. Then, the determination of the proper value for l_r and l_{fc} from {1, 2, 3} is in the same way. The experimental results showed that the best performance occurs when $l_r = 1$ and $l_{fc} = 2$. The reason may be due to the increasing number of weights with

Table 1. Predictions on five datasets in terms of *units_rs*.

Units_r	Metrics	P1	P2	P3	P4	P5
4	ACC	0.9241	0.8973	0.9111	0.9017	0.8742
	AUC	0.9712	0.9532	0.9687	0.9578	0.9465
	F1-score	0.8857	0.8258	0.8316	0.8737	0.7804
5	ACC	**0.9329**	**0.9169**	0.9176	**0.9136**	0.8903
	AUC	**0.9764**	**0.9674**	0.9663	**0.9704**	**0.9715**
	F1-score	**0.8949**	**0.8555**	0.8580	**0.8955**	0.7903
6	ACC	0.9281	0.9063	0.9098	0.8949	0.8968
	AUC	0.9737	0.9643	0.9529	0.9628	0.9649
	F1-score	0.8896	0.8450	0.8420	0.8680	**0.8234**
7	ACC	0.9276	0.9013	**0.9255**	0.9000	**0.9032**
	AUC	0.9710	0.9557	**0.9717**	0.9551	0.9636
	F1-score	0.8870	0.8205	**0.8584**	0.8763	0.8104

increasing network layers, which results in lacking of insufficient data to train LSTM with large amount of weights. So we set $l_r = 1$ and $l_{fc} = 2$ to build the basic framework of the LSTM. We hope that the model has good generalization and could be applied in different cotton pests and diseases, so other pests and diseases records, such as jassid, whitely, and leaf blight, are input into the model to show the prediction power. The performance comparison on different kinds of datasets with LSTM network is listed in Table 2. Our model not only performs well in pests prediction, but also in disease prediction.

Table 2. Predictions on different kinds of pests and diseases with LSTM network.

Metrics	Bollworm	Whitefly	Jassid	Leaf blight
ACC	0.9207	0.9244	0.9354	0.9557
AUC	0.9719	0.9687	0.9776	0.9868
F1-score	0.8749	0.9243	0.9161	0.9204

3.2 Prediction Comparison with Other Methods

The bollworm dataset "p1" was adopted to implement the prediction comparison of our proposed method with other classical machine learning methods KNN [13], SVC [14] and Random Forest [15]. Their parameters were set as follows. For LSTM network, the parameters of *units_rs*, l_r and l_{fc} were set as 5, 1 and 2, respectively; for KNN, *n_neighbors* was set as 3; for SVC, LinearSVC was adopted and C was set as 10; for Random Forest, *n_estimators* was set as 100. Table 3 lists the prediction results. It can be seen from the table that LSTM network achieved the best prediction performance.

Table 3. Performance comparison on dataset "p1" with different methods.

Methods	ACC	AUC	F1-score
KNN	0.8426	0.8246	0.7365
SVC	0.7400	0.6353	0.4609
Random Forest	0.8563	0.8197	0.7481
LSTM network	**0.9174**	**0.9690**	**0.8603**

4 Conclusion

In this paper, we convert the problem of cotton pests occurrence into a time series classification problem. This is the first time, to our knowledge, to use LSTM to solve this prediction problem. The model could predict the occurrence of cotton pests and diseases according to weather conditions in the future, so that people can take real time precaution and reduce crop economic losses. Then, we also investigated the model on different types of cotton records, and achieved good predictions. In addition, some traditional machine learning methods were implemented to show the prediction comparison with LSTM model. Results showed that LSTM has certain advantages in processing time-dependent problem. However, this paper only addresses the issue of the occurrence of cotton pests and diseases and predicts the occurrence with respect of weather factors. So, in the future, we will construct model to predict the hazard level of pests and diseases, so that prediction results are more responsive to data, making it easier for people to develop detailed pest control strategies.

Acknowledgement. This work was supported by the National Natural Science Foundation of China (Nos. 61672035, 61300058 and 61472282).

References

1. Wu, K.M., Lu, Y.H., Wang, Z.Y.: Advance in integrated pest management of crops in China. Chin. Bull. Entomol. **46**(6), 831–836 (2009)
2. Luo, J.Y., Zhang, S., Ren, X.L., et al.: Research progress of cotton insect pests in China in recent ten years. Cotton Sci. **29**, 100–112 (2017)
3. Satnam, S., Mridula, G., Suneet, P., et al.: Selection of housekeeping genes and demonstration of RNAi in cotton leafhopper. Amrasca biguttula biguttula. **13**(1), e0191116 (2018)
4. Zhang, W.Y., Jing, T.Z., Yan, S.C.: Studies on prediction models of Dendrolimus superans occurrence area based on machine learning. J. Beijing For. Univ. **39**(1), 85–93 (2017)
5. Huang, D.S.: Systematic Theory of Neural Networks for Pattern Recognition. Publishing House of Electronic Industry of China, May 1996
6. Hochreiter, S., Schmidhuber, J.: Long short-term memory. Neural Comput. **9**(8), 1735–1780 (1997)
7. Graves, A.: Generating sequences with recurrent neutral networks. Comput. Sci. (2013)
8. Li, J.W., Luong, M.T., Dan, J.: A hierarchical neural autoencoder for paragraphs and documents. In: ACL 2015, v2, 6 June 2015
9. Gao, H.Y., Mao, J.H., Zhou, J., et al.: Are you talking to a machine? Dataset and methods for multilingual image question answering. arXiv:1505.05612 (2015)
10. Theis, L., Bethge, M.: Generative image modeling using spatial LSTMs. arXiv:1506.03478 (2015)
11. Kalchbrenner, N., Danihelka, I., Graves, A.: Grid long short-term memory. Comput. Sci. (2015)
12. Kingma, D.P., Ba, J.: Adam: A Method for Stochastic Optimization. Computer Science. arXiv:1412.6980. (2015)
13. Coomans, D., Massart, D.L.: Alternative k-nearest neighbour rules in supervised pattern recognition: Part 1. k-Nearest neighbour classification by using alternative voting rules. Anal. Chim. Acta **136**, 15–27 (1982)
14. Cortes, C., Vapnik, V.: Support-vector networks. Mach. Learn. **20**, 273–297 (1995)
15. Ho, T.K.: The random subspace method for constructing decision forests. IEEE Trans. Pattern Anal. Mach. Intell. **20**(8), 832–844 (1998)

Fault Diagnosis and Control of a KYB MMP4 Electro-Hydraulic Actuator for LDVT Sensor Fault

Tan Van Nguyen[1] and Cheolkeun Ha[2(✉)]

[1] Graduated Student, Department of Mechanical and Aerospace Engineering, Ulsan University, Ulsan, South Korea
nvtan@hueic.edu.vn

[2] Department of Mechanical Engineering, University of Ulsan, Ulsan, Korea
cheolkeun@gmail.com

Abstract. The electro-hydraulic actuator (EHA) has been used for the precise position and the force con-trol applications. Many controllers have been applied to EHA such as sliding mode, and fuzzy PID controllers. However, the sensor fault problem in the EHA system is difficult to solve and challengeable in controller design. The system can lead to a damage if any sensor becomes faulty. Therefore, in this paper, a fault tolerant control (FTC) of a KYB MMP4 electro-hydraulic actuator is applied based on the mathematical modeling for tracking control in the sensor fault condition. Herein, the FTC for detecting and isolating the sensor fault signal is designed using the unknown input observer (UIO). In the sensor fault case, the fault detection and isolation (FDI) generates the feedback signal to the PID controller to drive the position of the actuator. The simulation and experimentations are performed to verify the performance of the FTC.

Keywords: Unknown input observer · Fault-tolerant control · PID

1 Introduction

As we all know, hydraulic piston cylinders have been used extensively and have become commonplace in modern industries. They are controlled through control valves [1]. In contrast, the disadvantage of this method is its low energy efficiency due to leakage through the drain valve of the hydraulic pump and reduced valve speed. Recently, an electrohydraulic pump-driven actuator system (EHA) has been introduced as an enhanced hydraulic system. The structure of this EHA is shown in Fig. 1. The position of the hydraulic cylinder is adjusted directly by the two-way pump. Most of the previous studies used a control valve to control the force or to control the position of a hydraulic cylinder based on linear and linearized control techniques without concerning the system uncertainties. Many controllers applied and developed as PID controller [3], fuzzy PID controller [4], Sliding mode controller [5]. However, these controllers depend on experience or face chattering problems, which can damage the experimental instruments.

D.-S. Huang et al. (Eds.): ICIC 2018, LNCS 10955, pp. 17–25, 2018.
https://doi.org/10.1007/978-3-319-95933-7_3

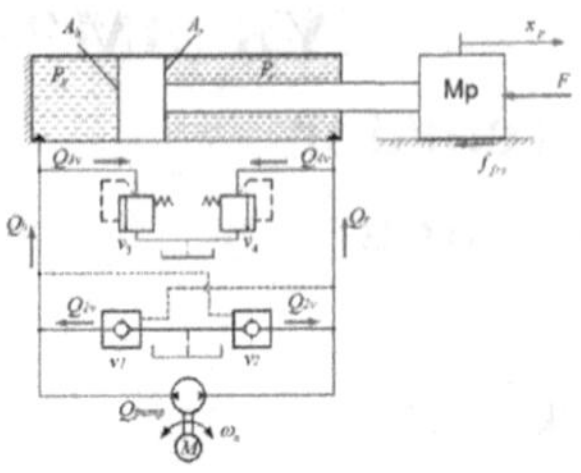

a) Modeling of the EHA system

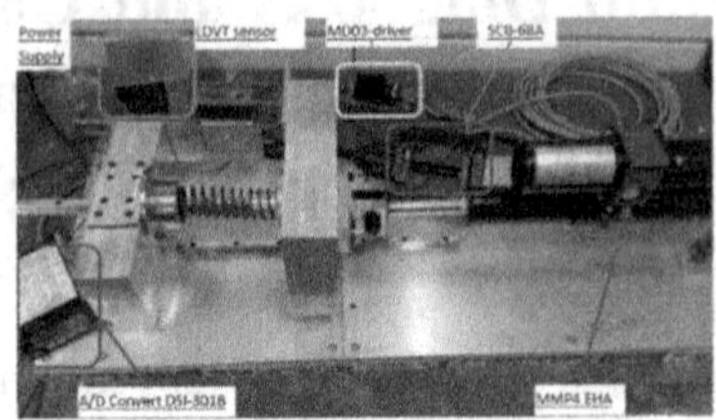

b) Modeling of the EHA for test bed

Fig. 1. The EHA pump controlled system

Nowadays, the control issues of a system are not only concerned with the control accuracy, but also the safety, reliability, maintainability, survivability, and sustainability of the system. These issues are motives to design the FTC controller with the ability to the tolerating system, endurances preventing loss of life. FTC is also expected to maintain desirable, robust performance and stability properties in the case of malfunctions in actuator, sensor or other system components. Generally, the FTC is a control method that ensures operating safety under the acceptable limit of a system when errors occur in the FDI system. FDI consists of two submodules as binary decision making and finding time error behavior. If the location, the time of the fault start and the severity of the fault are determined from the residual signal of the FDI or the estimated use of the error, the appropriate action can be taken switch or reconfigure the control system either using online or offline of the corresponding rules for many different error scenarios. the error estimating based on system observers has been widely used in various application [6, 7]. Several application studies of EHA have made with fault diagnosis based on the observer. Where Unknown Input Observer (UIO) error detection is one of the prominent methods in the FTC [8].

In this paper, the structure of the FTC is designed by two modules: FDI module and main control module. Fault is designed in order to diagnosis based on residual signals from unknown input observer, proposed as the FDI module. This module is involved in estimating and isolating the state of the system and the unknown input from sensor fault. Residual signals are filtered by a lowpass filter before they used for decision making. The part of fault information probably is also filtered out. So, response position error is reduced. Conversely, It can delay the FDI.

2 Modeling of EHA System

Modeling of the EHA system is shown in Fig. 1, the dynamics of the piston position may be written as [2]

$$m_p\ddot{x}_p + B_v\dot{x}_p + F + f_{frc} = (A_hP_h - A_rP_r) \tag{1}$$

Where m_p is the equivalent mass, $x_p, \dot{x}_p$ and $\ddot{x}_p$ are the position, the velocity and the acceleration of the piston, respectively, P_h, P_r and A_h, A_r are pressure and the area of the piston in two chamfers, respectively, F and f_{frc} are external load and the friction force on the cylinder, B_v is the viscosity damping coefficient.

F external load force can be computed as

$$F = K_{sp}x_p \tag{2}$$

Where K_{sp} is stiffness of the spring

The hydraulic continuity equations for the EHA system can be expressed as

$$\frac{dP_h}{dt} = \frac{\beta_e}{V_{ch} + x_pA_h}\left(Q_h - Q_i - \dot{x}_pA_h\right) \tag{3}$$

$$\frac{dP_r}{dt} = \frac{\beta_e}{V_{cr} - x_pA_r}\left(Q_r + Q_i + \dot{x}_pA_r\right) \tag{4}$$

Where β_e is the effective bulk modulus in each chamber and Q_i is the flow rate of the internal leakage of the cylinder

$$Q_i = C_{leak}(P_h - P_r) \tag{5}$$

C_{leak} is the coefficient of the internal leakage of the cylinder

$$Q_h = Q_{pump} + Q_{1v} - Q_{3v} \tag{6}$$

$$Q_r = -Q_{pump} + Q_{2v} - Q_{4v} \tag{7}$$

Q_{1v}, Q_{2v} are the flow rate through the pilot operated check valve

Where Q_{pump} is the pump flow rate

$$Q_{pump} = D_p\omega \tag{8}$$

Here D_p is displacement of the pump; ω is the speed of the servo pump system

Q_{1v}, Q_{2v} are the flow rate through the pilot operated check valve

According to the system dynamic Eqs. (1) to (8), the system can be represented by a state vector $\begin{bmatrix} x_1 & x_2 & x_3 & x_4 \end{bmatrix}^T = \begin{bmatrix} x_p & v_p & P_h & P_r \end{bmatrix}^T$ and can be expressed

$$\begin{bmatrix} \dot{x}_1 \\ \dot{x}_2 \\ \dot{x}_3 \\ \dot{x}_4 \end{bmatrix} = \begin{bmatrix} x_2 \\ \frac{1}{m_p}\left[(A_hx_3 - A_rx_4) - B_vx_2 - K_{sp}x_1 - f_{frc}\right] \\ -\frac{\beta_eA_h}{V_{ch} + x_1A_h}x_2 + \frac{\beta_e}{V_{ch} + x_1A_h}u_1 \\ \frac{\beta_eA_r}{V_{cr} - x_1A_r}x_2 + \frac{\beta_e}{V_{cr} - x_1A_r}u_2 \end{bmatrix} \tag{9}$$

Where $u_1 = Q_h - Q_i$; $u_2 = Q_r + Q_i$

3 Fault Tolerant Control for Sensor Faults in MMP4 EHA System

In this section, FTC is designed in order to diagnosis the sensor position faults of MMP4 EHA system, then determine appropriate signal according to the allowable tolerance. The PID main controller will operate conventional close-loop trajectory control. The signal from sensor was compared with the estimated position signal output of the UIO via an operator defined as a flash switch.

3.1 Unknown Inputs Observer (UIO) Design

Considering the design of observers for a class of linear dynamic systems in which system uncertainty can be modeled as an additive unknown disturbance term in the dynamic equation [6].

$$\begin{aligned} \dot{x}(t) &= Ax(t) + Bu(t) + Ed(t) + w(t) \\ y(t) &= Cx(t) + v(t) \end{aligned} \tag{10}$$

Where $X(t) \in R^n$ is the state vector, $y(t) \in R^m$ is the outputs vector and $d(t) \in R^q$ is the unknown input or disturbance vector, E is disturbance matrix, w_k, v_k are zero mean white noise.

The structure for a full order UIO is given by the dynamic system

$$\begin{aligned} \dot{z}(t) &= \Phi z(t) + \Upsilon Bu(t) + Ky(t) \\ \hat{x}(t) &= z(t) + \Gamma y(t) \end{aligned} \tag{11}$$

Where $\hat{x} \in R^n, z \in R^n$ are estimated state vector and states of observer. The observer matrixes Φ, Υ , Γ and K should be designed according to the state estimation error vector.

The state estimation error vector is defined as

$$e(t) = x(t) - \hat{x}(t) \tag{12}$$

The error term can be derived as

$$\begin{aligned} \dot{e}(t) = {} & \Phi e(t) + [(I - \Gamma C)A - \Phi - K_1 C]x(t) + [\Phi\Gamma - K_2]y(t) \\ & - [\Upsilon - (I - \Gamma C)]Bu(t) + (I - \Gamma C)Ed(t) + (I - \Gamma C)w(t) - \Gamma\dot{v}(t) - K_1 v(t) \end{aligned} \tag{13}$$

Where $K = K_1 + K_2$

It is easy to see that in order to make the estimation error a function of $\Phi\ e(t)$

$$\dot{e}(t) = \Phi e(t) \tag{14}$$

If this satisfied the conditions

$$(\Gamma C - I)E = 0; (I - \Gamma C) = \Upsilon; (A - \Gamma CA - K_1 C) = \Phi; \Phi\Gamma = K_2$$

The necessary and sufficient conditions for the existence of observers of the UIO equation are very classical and will not be discussed here. That condition is

(i) $rank(CE) = rank(E)$
(ii) (A_1, C) is detectable pair,

Where $A_1 = A - E\left[(CE)^T(CE)\right]^{-1}(CE)^T CA$; $\Gamma CE = E; \Gamma = E(CE)^+$
And $(CE)^+ = \left[(CE)^T(CE)\right]^{-1}(CE)^T$
So, the state estimation error can be written as

$$\dot{e}(t) = \Phi e(t) - K_1 v(t) + \Upsilon w(t) - \Gamma\dot{v}(t) \tag{15}$$

The disturbance estimation can be expressed from Eq. (14) as

$$\dot{y}(t) = C\dot{x}(t) \ = C[Ax(t) + Bu(t) + \mathrm{w}(t)] + CEd(t)$$

$$\hat{d}(t) = (CE)^+[\dot{y}(t) - C(Ax(t) + Bu(t) + \mathrm{w}(t))] \tag{16}$$

3.2 Fault Detection and Isolation (FDI)

In model based FDI techniques of system is used to decide about the occurrence of fault. In this paper, the UIO is evaluated for sensor fault detection based on estimated error information $\hat{d}^s$. The decision is executed as a Boolean indicator and defined as a *fault*$\in\{0,1\}$. $\hat{d}^s$ Should be zero when the sensor is not faulty and the *fault* is set to zero. Conversely, *fault* when $\hat{d}^s$ is not zero. This mean that the fault tolerant control of the system is determined by the coefficient, the value k to set the *fault* as below:

$$\begin{cases} fault = 1 & \text{if } \left|\hat{d}^s\right| \geq k \\ fault = 0 & otherwise \end{cases}$$

4 Simulation Results for Sensor Fault Detection

4.1 Results for Sensor Fault Detection

Assuming that the system only has the sensor fault when the linear faulty system will be written as

$$y(t) \ = Cx(t) + Ef_s(t)$$

where $C = diag([1 \quad 0_{1x3}])$ $and\, E = diag([1 \quad 0_{1x3}])$, and $f_s(t)$ is a sensor fault function

$$f_s(t) = \begin{bmatrix} g(t) \\ 0_{3x1} \end{bmatrix}; \quad \text{where } g(t) = \begin{cases} 0.005 & \text{if } 2.5 \leq t \leq 3 \\ -0.005 & \text{if } 9.5 \leq t \leq 10.5 \\ 0 \; otherwise & \end{cases}$$

Scheme of EHA pump controlled system is designed in Simulink expressed as Fig. 2.

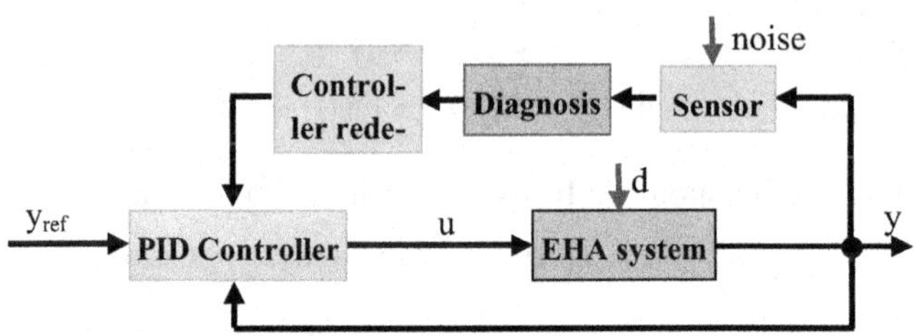

Fig. 2. The structure of EHA pump controlled system for control position using sensor FTC

Simulation with the parameters k = 0.1, results shown as in Fig. 3.

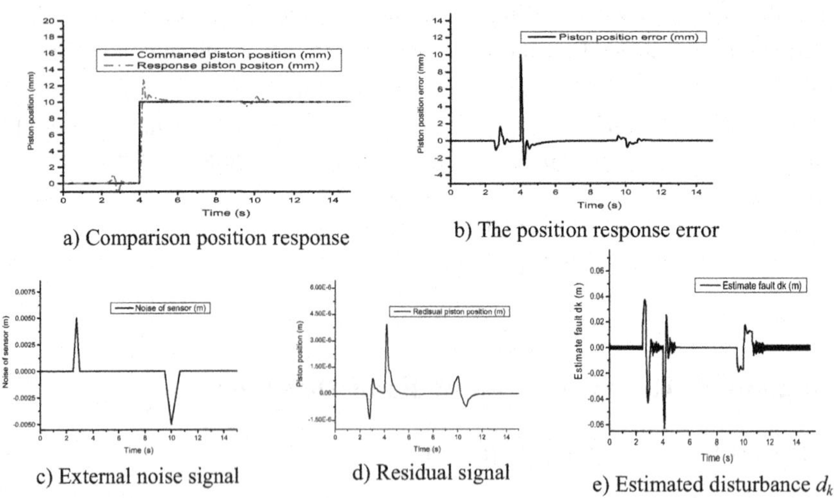

a) Comparison position response

b) The position response error

c) External noise signal

d) Residual signal

e) Estimated disturbance d_k

Fig. 3. The position response using PID controller

Simulation with the parameters k = 0.0001, comparison response piston position between PID without FTC and PDI using FTC.

From Figs. 3, 4. We can see that the accuracy of control signal depend on estimated disturbance *k*. The control signals of PID controller using FTC are better than the PID controller without FTC.

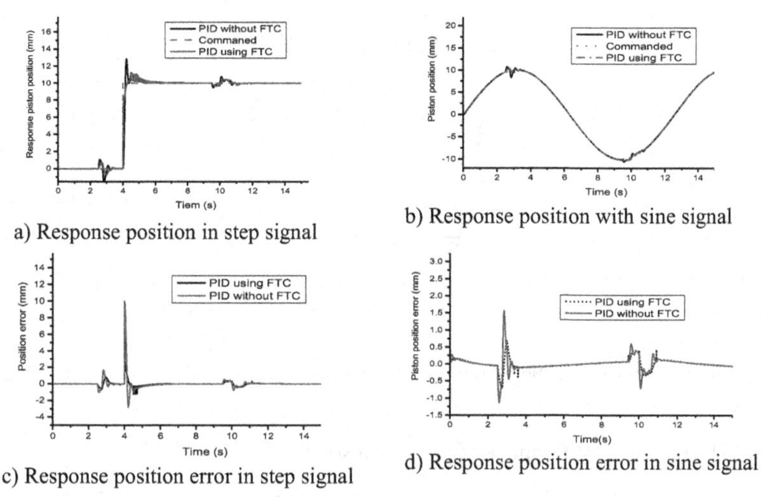

a) Response position in step signal

b) Response position with sine signal

c) Response position error in step signal

d) Response position error in sine signal

Fig. 4. Comparison of the responses of position and error between sine and step signal

4.2 Experiment Setup

The test bed using PID controller and FTC of the EHA system is shown in Fig. 5.

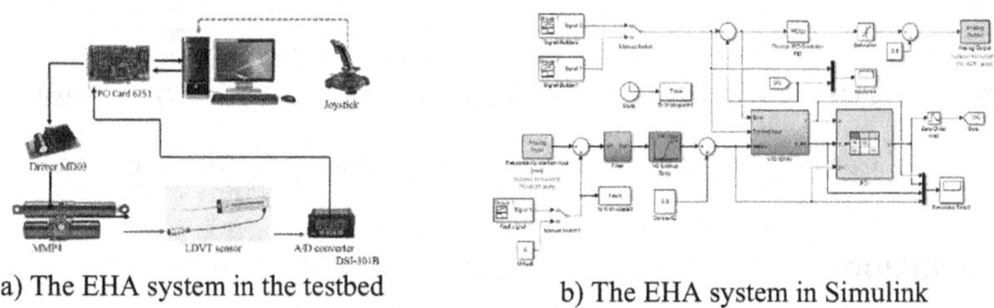

a) The EHA system in the testbed

b) The EHA system in Simulink

Fig. 5. The EHA system

A. The test bed using PID control based FCT
   The results for test bed with step and sine wave signal are shown as Fig. 6.

- In the case without FTC
- In the case using FTC

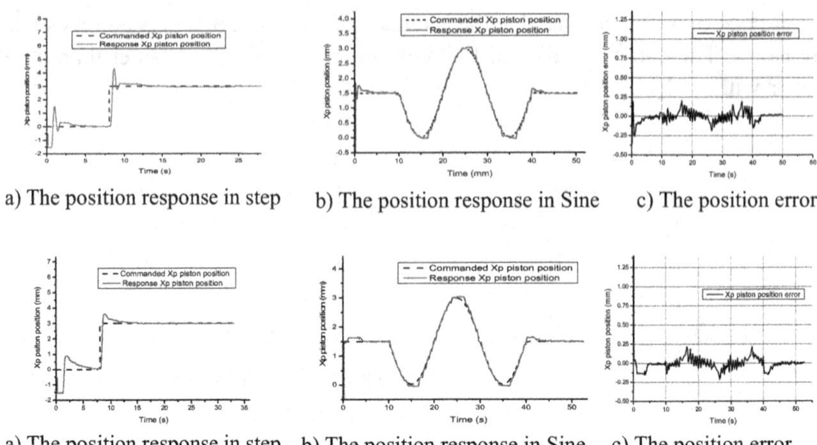

a) The position response in step b) The position response in Sine c) The position error

a) The position response in step b) The position response in Sine c) The position error

Fig. 6. Comparison of the responses of position and error to sine and step signal

B. The test bed using PID and FTC with sensor fault

Assuming that, the sensor fault is added in analog input signal of NI PCI 6251 as shown Fig. 5b and the signal is shown in Fig. 7c.

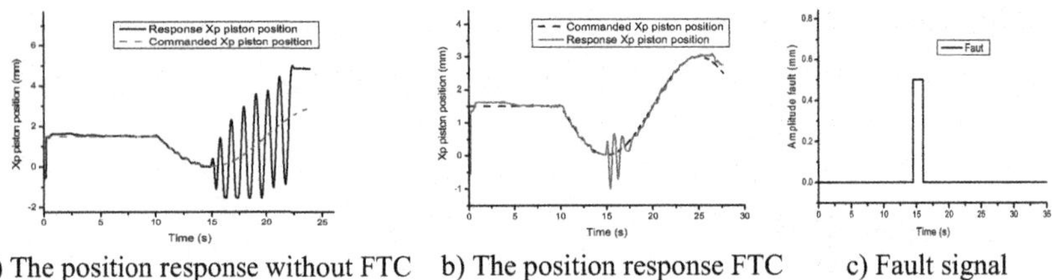

a) The position response without FTC b) The position response FTC c) Fault signal

Fig. 7. Comparison of the position responses in sine signal with sensor fault

5 Conclusions

In this paper, a mathematical modeling study of the KYB MMP4 electrohydraulic actuator was performed. The FTC is successfully designed for the PID main controller. The estimated disturbance signals d_k from unknown input observer are filter by a lowpass filter before they are used for decision making (FDI). The PID main controller will receive the estimated output signal, once the sensor fault occurred. Thus, the system is kept stability and safety. Furthermore, the control accuracy of the system is improved.

References

1. Ali, S.Y., Selim, S.: Nonlinear adaptive control of semi-active MR damper suspension with uncertainties in model parameters. Nonlinear Dyn. **79**, 2753–2766 (2015)
2. Moonumca, P., Depaiwa, N.: The force tracking control of electro-hydraulic system based on particle swarm optimization. Int. J. Innov. Comput. Inf. Control **12**(3), 809–821 (2016)
3. Yao, B., Bu, F., Chiu, G.T.C.: Adaptive robust motion control of single-rod hydraulic actuators: theory and experiments. IEEE Trans. Mechatron. **5**(1), 79–91 (2000)
4. Ahn, K.K., Truong, D.Q.: Online tuning fuzzy PID controller using robust extended Kalman filter. J. Process Control **19**, 1011–1023 (2009)
5. Perron, M., Lafontaine, J., Desjardins, Y.: Sliding-mode control of a servomotor-pump in a position control application. In: Proceedings of IEEE Conference on Electrical and Computer Engineering, Saskatoon, SK, Canada, pp. 1287–1291 (2005)
6. Odgaard, P.F., Stoustrup, J.: Unknown input observer based detection of sensor faults in a wind turbine. In: Proceedings of IEEE International Conference on Control Applications, pp. 310–315 (2010)
7. Fu, X., Liu, B., Zhang, Y., Lian, L.: Fault diagnosis of hydraulic system in large forging hydraulic press. Measurement **49**, 390–396 (2014)
8. Zarei, J., Poshtan, J.: Sensor fault detection and diagnosis of a process using unknown input observer. Math. Comput. Appl. **16**(4), 1010–1021 (2011)

A Mask R-CNN Model with Improved Region Proposal Network for Medical Ultrasound Image

Jun Liu[1(✉)] and PengFei Li[2]

[1] College of Computer Science and Technology, Wuhan University of Science and Technology, Wuhan, China
ljwhcn@qq.com

[2] Hubei Province Key Laboratory of Intelligent Information Processing and Real-Time Industrial System, Wuhan, China
807657578@qq.com

Abstract. In medical ultrasound image processing, it is often necessary to select the ROI before segmentation to obtain better segmentation accuracy. With the development of deep learning, the technology of object detection can well implement the function of automatically selecting ROI. The combination of object detection and image segmentation has also been proposed, such as Mask R-CNN, an end-to-end image segmentation model. However, the ROI selection by the algorithm above cannot meet the needs of medical image segmentation. Because its RPN layer is inherited from Faster R-CNN, a target classification framework. What we need is a region that can cover the whole object area with the details of edge. This information has an important influence for the further segmentation. Therefore, this paper improves the selection criteria of the anchor in the RPN layer, making the improved RPN layer more suitable for image segmentation tasks. Finally, the experimental results show that the improved model can achieve higher segmentation accuracy with the appropriate parameters selected.

Keywords: Image segmentation · Deep learning · Mask R-CNN Ultrasound image · Machine learning

1 Introduction

Image segmentation is an important step in image analysis and processing. The process of image segmentation is actually the process of clustering pixels in an image. With the requirement of higher accuracy in image segmentation, the segmentation ability of traditional image segmentation algorithms is unsatisfactory, such as threshold segmentation, edge segmentation and level set segmentation. Therefore, in order to get better clustering results, some clustering theories in machine learning, such as k-means clustering [1], Markov random fields [2] and support vector machines [3], are used in image segmentation. For some of the more complex images, in order to get better segmentation results, the above machine learning methods often need to manually select the ROI (region of interest) to remove most of the interference information.

D.-S. Huang et al. (Eds.): ICIC 2018, LNCS 10955, pp. 26–33, 2018.
https://doi.org/10.1007/978-3-319-95933-7_4

The object detection method can replace the artificial selection of ROI to realize the automatic selection of ROI. In recent years, deep learning has made great breakthroughs in the field of object detection. And the Mask R-CNN [4] model combines object detection with image segmentation.

The Mask R-CNN model is an instance segmentation model proposed by KaiMing He in 2016 combining object detection with image segmentation. This model combines Faster R-CNN [5] as object detection network and FCN [6] as image segmentation network. ResNeXt [7] and FPN [8] are used as feature extraction networks, and ROIAlign is used to replace the ROI pooling layer in Faster R-CNN.Faseter R-CNN is an end-to-end neural network proposed by Ross Girshick that combines object detection and classification. It has two earlier versions, R-CNN [9] and Fast R-CNN [10]. FCN is a network for image segmentation proposed by Jonathan Long et al. The network replaces the full-connected layer in the traditional image segmentation network with a convolutional layer, breaking through the stereotyped pattern at the pixel level in the traditional CNN [11]. ResNeXt is an upgraded version of the ResNet [12] network proposed by Saining Xie et al. This network can increase the accuracy without increasing the complexity of the parameters and reduce the number of hyperparameters. FPN is a multi-scale object detection algorithm proposed by Tsung-Yi Lin et al. It differs from traditional object detection in that prediction is performed independently at different feature layers. Since the RPN layer in the Mask R-CNN is still the same as the RPN layer in the Faster R-CNN, and the Faster R-CNN is a network for object detection and classification, the RPN layer in the Faster R-CNN selects the candidate frame selection criteria.In some aspects it is not suitable for image segmentation tasks. Sometimes the result is not satisfactory because Mask R-CNN directly uses RPN layer.

Aiming at the task of ultrasonic image segmentation of the follicles of cattle in this study, a task of two classifications, a new selection rule for anchor in the RPN layer is proposed. Under the new rule, an anchor that can contain ground truth will have a higher priority, making the training candidate box more suitable for image segmentation tasks.

2 Background

2.1 Mask R-CNN

Mask R-CNN is a neural network model combining object detection and image segmentation. It is based on Faster R-CNN's idea of combining object detection and classification. Connect the FCN directly after the ROI pooling layer in Faster R-CNN to perform image segmentation. However, compared with Faster R-CNN, Mask R-CNN also used ResNeXt and FPN instead of the CNN to extract the feature map, then replaced the ROI pooling layer with a more refined ROI Alignment layer, and finally used a side-by-side FCN layer. Also called the mask layer, not the single FCN. The overall network structure of Mask R-CNN is shown in Fig. 1:

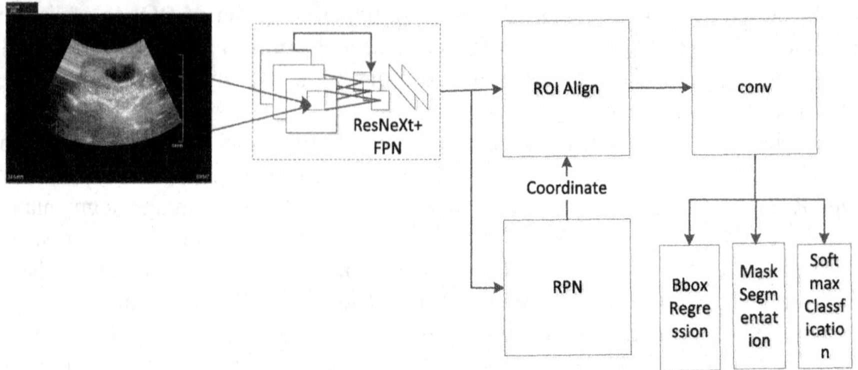

Fig. 1. The structure of Mask R-CNN

2.2 ROI Align

ROI Align is an improvement to the ROI pooling layer in Faster R-CNN. Since the ratio and size of the anchors in the RPN layer are not the same, after upgrading to the proposal according to certain rules, since the full connection layer behind needs the input of the same size, the ROI pooling layer is designed to normalize the size of the proposal. Since the RoIpooling layer uses nearest-neighbor interpolation, that is, when resize, rough coordinates are used for the case where the scaled coordinates cannot be exactly integers, which is equivalent to selecting the closest point to the target point. Kaiming He declared in his paper Mask RCNN that doing so would lose some degree of spatial symmetry, so he replaced the nearest neighbor interpolation with a bilinear interpolation. The improved ROI Pooling layer is called the ROI Align layer.

In one-dimensional space, the nearest point interpolation [13] is equivalent to rounding and rounding. In a two-dimensional image, the coordinates of the pixel are all integers. The method is to select the point closest to the target point, which will lose the spatial symmetry to some extent.

Bilinear interpolation [14] performs linear interpolation in both directions, respectively. In image processing, the position of the target pixel in the original image is first calculated according to the following formula (1):

$$\begin{aligned} srcX &= dstX * (srcWidth/dstWidth) \\ srcY &= dstY * (srcHeight/dstHeight) \end{aligned} \tag{1}$$

Where *dstX* and *dstY* are the coordinates of the pixel in the current proposal, *srcWidth* and *srcHeight* are the width and height of the original image, and *dstWidth* and *dstHeight* are the width and height of the proposal. And *srcX* and *srcY* are the coordinates of the target pixel in the original image. This pixel is usually a floating-point number. For example, pixel $f(1.2, 3.4)$, first find the four pixels adjacent to it: (1,3), (2,3), (1,4), (2,4). Write as the form $f(i+u, j+v)$, then $u = 0.2$, $v = 0.4$, $i = 1$, $j = 3$. Then there is formula (2):

$$f(i+u,j+v) = (1-u)(1-v)f(i,j) + (1-u)vf(i,j+1) + u(1-v)f(i+1,j) + uvf(i+1,j+1) \tag{2}$$

Doing so ensures the spatial symmetry of the proposal after the ROI Align layer. After the ROI Align layer resizes the feature map to the same size, the feature map is sent to the subsequent FCN for segmentation.

3 Mask R-CNN with Improved RPN

Before segmenting the image, the ROI is often manually selected. This is because some images are more complex and it is difficult to obtain good segmentation results by directly segmenting the original image. A good candidate box can greatly reduce the difficulty of segmentation.

In addition, this paper believes that the selection of the candidate box will affect the accuracy of the target segmentation.

The RPN layer selection candidate frame is selected based on the IOU between the anchor and ground truth, and the RPN layer was originally used in the target detection task in the Faster R-CNN. Therefore, the selection criteria for the candidate boxes in the RPN layer is not suitable for image segmentation tasks in some aspects. Therefore, Mask R-CNN directly uses RPN for image segmentation. Sometimes the obtained results are not satisfactory. The following will point out the deficiencies of the original RPN layer in the image segmentation task. First look at Fig. 2:

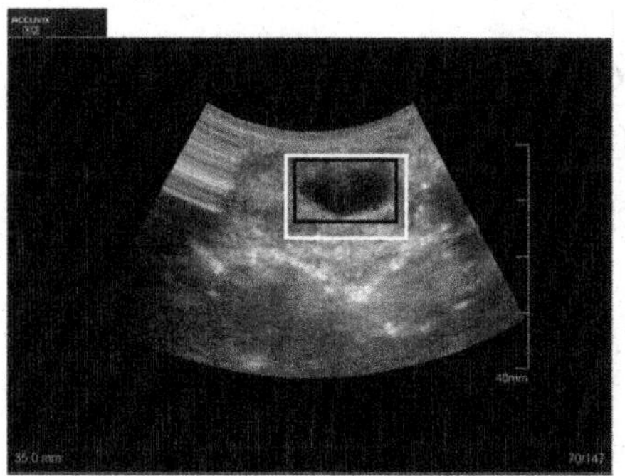

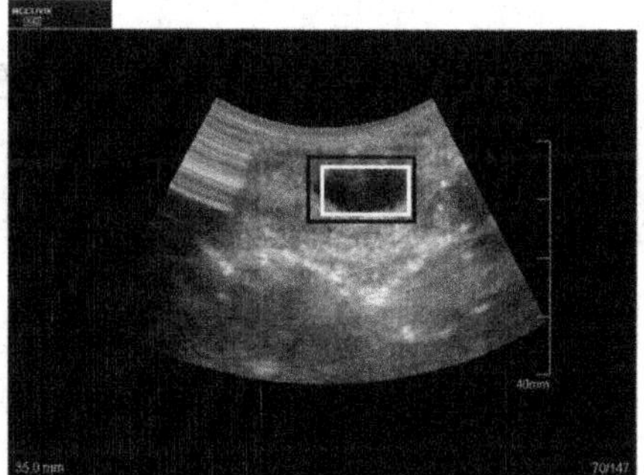

Fig. 2. Two different anchors under the same IOU in the same image

As shown in the above figure, the two images are the relationship between two different anchors and ground truth of the same ultrasound follicle image. The white box represents the anchor and the black box represents the ground truth. If the IOUs between the anchor and the ground truth of the two image are equal, which anchor is more suitable for the image segmentation task? Obviously the left image, because in the right image, some parts of the target area has exceeded the scope of the anchor, then in the image segmentation, the anchor of the right image will lose the information of the target area. Even if the RPN layer has a boundary regression correction, there is no guarantee that the modified bounding box can include the target area. And when doing image segmentation, it is often desirable that the borders retain a certain amount of background area information for classification, rather than tightly adhere to the target area. However, if the IOU of the right image is larger than the IOU of the left image at this time, the anchor's criterion is selected according to the RPN, and the anchor of the right image will have a higher priority than the left image.

First define the areas after the anchor and ground truth overlap. The specific definitions are as the following formula (3):

$$\begin{gathered} S_1 = area(C) - area(C) \cap area(G) \\ S_2 = area(G) - area(C) \cap area(G) \\ S_3 = area(c) \cap area(G) \end{gathered} \tag{3}$$

Therefore, this paper improves the criterion of selecting anchor for RPN layer in Mask R-CNN. Under the premise of guaranteeing a certain IOU, it is hoped that the selected anchor can retain the image information of the target area. So this article defines a new anchor selection criteria:

(1) In order to ensure a certain IOU, IOU is marked as a positive sample greater than 0.7, IOU less than 0.3 is recorded as a negative sample.

(2) In order to ensure that the extracted anchor does not lose the information of the target area, i.e. the acquired anchor can cover the entire ground truth, all anchors with positive samples are sorted by the size of η. Where η represents the ratio of the overlap between anchor and ground truth in ground truth, and the specific formula is as the following formula (4):

$$\eta = \frac{area(C) \cap area(G)}{area(G)} \tag{4}$$

(3) For all anchors with η being 1, sort by λ values. The definition of λ is as the following formula (5):

$$\lambda = |IOU - c| \tag{5}$$

4 Experimental Results

The software environment for the experiment code running in this paper is tensorflow 1.3.0, python 3.5. The hardware environment is 8 GB RAM and intel i5-6300HQ CPU and NVDIA GTX 960 GPU.Trained a total of 20,000 iteration times and it took 4 h 32 min.

In order to regress and correct the coordinate position of the prediction frame, it is usually necessary to extract about 200 positive samples and negative samples. Usually, $\eta = 1$ is extracted according to priority, and λ is a positive sample in an anchor with a certain range of values. Due to the need to limit the scope of λ, so that the priority of the extracted anchor box to meet a certain IOU to ensure the accuracy of the border coordinates regression. Therefore, for the case of insufficient positive samples, according to the original RPN layer processing sample does not do the situation, from the positive sample extracted in the first step, according to the size of the IOU to make up. By changing the value of c and the range of the different anchor selection criteria can be obtained, and then change the parameters of the frame regression, resulting in different prediction frame position coordinates. In order to verify the hypothesis that the choice of the candidate box will affect the segmentation accuracy, Mask R-CNN and the improved Mask R-CNN were used to segment the follicular image collection of cattle. Fixed modified Mask R-CNN $\lambda \leq 0.1$, Fig. 3 compares the average split accuracy of the improved model with that of the original model when c takes different values.

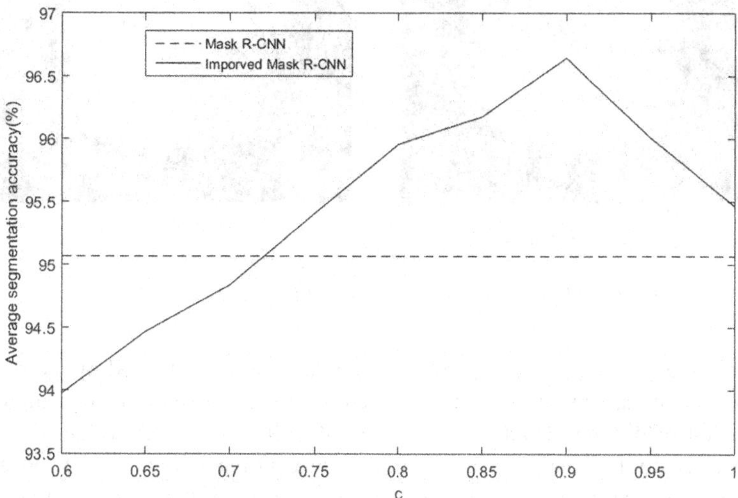

Fig. 3. Line chart of the average segmentation accuracy of the improved model and the original model with different value of c

When the range of λ is fixed, changing the value of the parameter c will affect the priority of the anchor box, thereby affecting the parameter that the RPN layer corrects the coordinate position of the candidate box, and changing the coordinate position of the prediction box generated by the RPN, thus Candidate boxes with different coordinate positions have been obtained. It can be known from the experimental results that the generation of different candidate boxes in the Mask R-CNN has a certain influence on the final segmentation result, and the improved Mask R-CNN model can be obtained if the appropriate parameters are selected. Higher average segmentation accuracy than the original Mask R-CNN model. Fig. 4 shows the segmentation results for the same ultrasound follicle image with the modified Mask R-CNN model under different value of c and the segmentation results for the original Mask R-CNN model when $\lambda \leq 0.1$ is fixed. The accuracy of the evaluation criteria is based on the Dice [15] similarity coefficient.

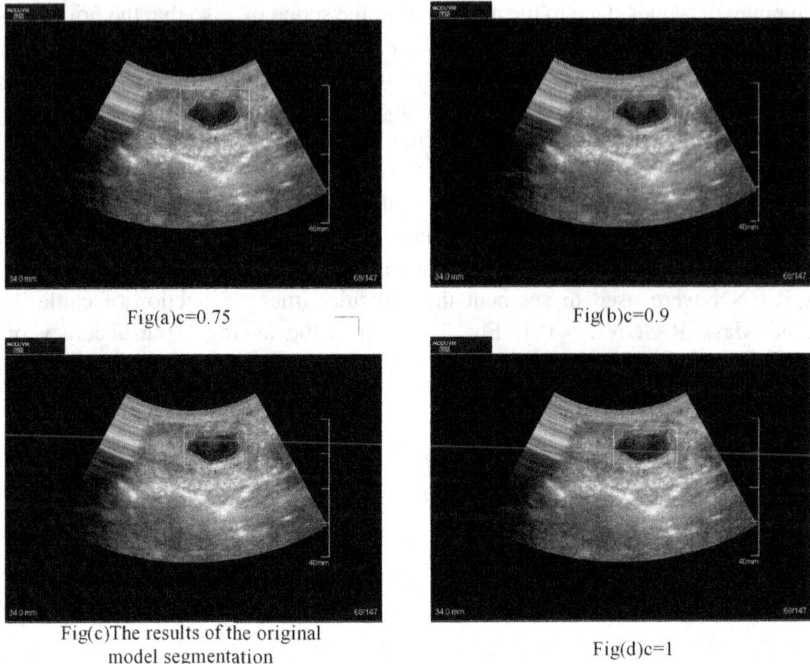

Fig(a)c=0.75
Fig(b)c=0.9
Fig(c)The results of the original model segmentation
Fig(d)c=1

Fig. 4. Segmentation results for various Mask R-CNN models

It can be seen from the Fig. 4 that the candidate frames extracted by the model for the same picture are different under different c values, and the results of the final segmentation of different candidate boxes are also different. If the value of c approaches too close to 1, it will cause the regression candidate frame to be too close to ground truth, so that the final candidate frame will be missing part of the target area. If the value of c is too small, the candidate frame will be too large, including too much interference information, and the segmentation accuracy will be reduced.

5 Conclusions

This paper improves the selection criteria of Anchor for RPN layer in Mask R-CNN, making the selected anchor more suitable for image segmentation task. The new anchor improves the regression parameters of the candidate boxes, and the new candidate box can increase the accuracy of the subsequent image segmentation.

The improvement of Mask R-CNN's RPN layer is based on human experience. Although it does improve the segmentation accuracy, it is not necessarily optimal. In order to avoid the limitations of human experience, we can learn the idea of deep learning, and use the convolutional layer to extract higher-level and deeper features of the object and try to avoid artificial design features. Researching and designing this kind of network structure is the direction that can be further studied.

Acknowledgement. This work was supported by the National Natural Science Foundation of China (Grant No.31201121, No.61373109 and No.61403287), the Natural Science Foundation of Hubei Province (Grant No.2014CFB288) and Open foundation of Hubei Province Key Laboratory of Intelligent Information Processing and Real-time Industrial System (Grant Nos. ZNSS2013A0001 and ZNSS2013A004).

References

1. Samundeeswari, E.S., Saranya, P.K., Manavalan, R.: Segmentation of breast ultrasound image using regularized K-means (ReKM) clustering. In: International Conference on Wireless Communications, Signal Processing and NETWORKING, pp. 1379–1383. IEEE (2016)
2. Li, L., Lin, J., Li, D., Wang, T.: Segmentation of medical ultrasound image based on Markov random field. In: The International Conference on Bioinformatics and Biomedical Engineering, pp. 968–971. IEEE Xplore (2007)
3. Nguyen, T.D., Sang, H.K., Kim, N.C.: Surface extraction using SVM-based texture classification for 3D fetal ultrasound imaging. In: International Conference on Communications and Electronics, pp. 285–290. IEEE (2007)
4. He, K., Gkioxari, G., Dollár, P., Girshick, R.: Mask R-CNN (2017)
5. Ren, S., He, K., Girshick, R., Sun, J.: Faster R-CNN: towards real-time object detection with region proposal networks. In: International Conference on Neural Information Processing Systems, vol. 39, pp. 91–99. MIT Press (2015)
6. Long, J., Shelhamer, E., Darrell, T.: Fully convolutional networks for semantic segmentation. In: Computer Vision and Pattern Recognition, vol. 79, pp. 3431–3440. IEEE (2015)
7. Xie, S., Girshick, R., Dollar, P., Tu, Z., He, K.: Aggregated residual transformations for deep neural networks, pp. 5987–5995 (2016)
8. Lin, T.Y., Dollar, P., Girshick, R., He, K., Hariharan, B., Belongie, S.: Feature pyramid networks for object detection, pp. 936–944 (2016)
9. Hariharan, B., Arbeláez, P., Girshick, R., Malik, J.: Simultaneous detection and segmentation. In: Fleet, D., Pajdla, T., Schiele, Bernt, Tuytelaars, T. (eds.) ECCV 2014. LNCS, vol. 8695, pp. 297–312. Springer, Cham (2014). https://doi.org/10.1007/978-3-319-10584-0_20
10. Girshick, R.: Fast R-CNN. Computer Science (2015)
11. Liu, F., Lin, G., Shen, C.: CRF learning with CNN features for image segmentation. Elsevier Science Inc. (2015)
12. He, K., Zhang, X., Ren, S., Sun, J.: Deep residual learning for image recognition, pp. 770–778 (2015)
13. Jiang, N., Wang, L.: Quantum image scaling using nearest neighbor interpolation. Quantum Inf. Process. **14**(5), 1559–1571 (2015)
14. Kirkland, E.J.: Bilinear interpolation. In: Kirkland, E.J. (ed.) Advanced Computing in Electron Microscopy. Springer, Boston (2010). https://doi.org/10.1007/978-1-4419-6533-2_12
15. Andrews, S., Hamarneh, G.: Multi-region probabilistic dice similarity coefficient using the Aitchison distance and bipartite graph matching. Computer Science (2015)

An Improved Double Hidden-Layer Variable Length Incremental Extreme Learning Machine Based on Particle Swarm Optimization

Qiuwei Li[1], Fei Han[1(✉)], and Qinghua Ling[2]

[1] School of Computer Science and Communication Engineering, Jiangsu University, Zhenjiang, Jiangsu, China
hanfei@ujs.edu.com

[2] School of Computer Science and Engineering, Jiangsu University of Science and Technology, Zhenjiang 212003, Jiangsu, China

Abstract. Extreme learning machine (ELM) has been widely used in diverse domains. With the development of deep learning, integrating ELM with some deep learning method has become a new perspective method for extracting and classifications. However, it may require a large number of hidden nodes and lead to the ill-condition problem for its random generation. In this paper, an effective hybrid approach based on Variable-length Incremental ELM and Particle Swarm Optimization (PSO) algorithm (PSO-VIELM) is proposed which can be used to regulate weights and extract features. In the new method, we build two hidden layers to establish a structure which is compact with a better generalization performance. In the first hidden layer named extraction layer, we make the feature learning to the raw data, and make dynamic updates for hidden layer nodes, and use the fitting error as the fitness function to update the weights corresponding to the hidden nodes with the method of PSO. In the second hidden layer named classification layer, we make a classification for the processed data from extraction layer and use cross-entropy as the fitness function to update the weights in the net. In order to find the appropriate number of hidden layer nodes, all hidden nodes will no longer grow in the case of a rebound in the fitness function on the validation set. The result in some datasets shows that PSO-VIELM has a better generalization performance than other constructive ELMs.

Keywords: Extreme Learning Machine · Particle Swarm Optimization
Feature extraction · Auto-encoder

1 Introduction

The chief cause of the widespread popularity of neural network (NN) is to be found its ability in data approximation and classification [1]. It is easily seen that NN has a strong and stable learning ability, and it can even address quite a few problems which traditional methods unable to do.

Based on the study of Single Hidden Layer Feedforward Network (SLFN), extreme learning machine (ELM) was proposed in 2004 [2, 3]. As opposed to routine SLFN,

D.-S. Huang et al. (Eds.): ICIC 2018, LNCS 10955, pp. 34–43, 2018.
https://doi.org/10.1007/978-3-319-95933-7_5

ELM algorithm randomly chooses its input weights and bias matrix then calculates the output weights by Moore-Penrose generalized pseudo-inverse afterward. The major advantage of ELM is that the process of training without iteration which let it learn faster than others. However, the number of hidden nodes in ELM should be selected at first, and the parameters of input matrix and bias will remain in the course of training. Huang et al. pointed out in [4] that ELM tends to require more hidden nodes rather than conventional tuning-based algorithms in many cases.

Seen from the viewpoint of ELM architectures, an open question on ELMs is how to determine a proper structure. To solve this problem, some strategies were proposed. An important concept about Incremental Extreme Learning Machine (I-ELM) which was mentioned by Huang et al. [5], proved that the incremental learning framework work well in generalization. Therefore, to better performance of the network framework and solve the ill-condition problem, among some evolutionary algorithms, a technique named particle swarm optimization (PSO) [6] is also an effective global search algorithm with its ability in searching and convergence. Therefore, the hybrids of PSO and ELM should be promising for training feedforward neural networks. Some research such as PSO-ELM [7] use PSO to optimize the input weights. However, the number of hidden nodes still predetermine before the model established.

The generalization feature about the network refers to the popularization capability of the model [8]. Many factors have influenced the feature of generalization, such as the structure of the network, learning algorithm and the characters of samples. In recent years, some experiments such as [9] proved that with some hidden nodes added superiorly, prediction error fell even faster. Although the parameters selected randomly could gradually converge, the speed of convergence may still vary. That is, how to select the proper parameters play an crucial role. In order to address this problem, Yang et al. [10] proposed parallel chaos which based incremental extreme learning machine (PC-ELM) to build a more compact network architecture.

In order to solve the problem for better generalization performance, with the influence by deep learning method, some feature learning method such as auto-encoder [11] was proposed. It has been a prevailing technique which replace traditional feature into deeper expression. Based on this method, some researches combined ELM and auto-encoder like ELM-AE [12] make an effective feature learning and obvious effect on the performance than other models. Nevertheless, this method is not combined with feature learning.

The traditional classification algorithm usually considers the feature learning stage and the classification stage. Although it is easy to implement on the algorithm, it cannot combine both to produce better combination results. Although deep learning based feature extraction method can achieve results well at present, it often needs to build multilayer network, which greatly increases time overhead.

In this paper, to extended the work in generalization and well-conditioned, we extend I-ELM with the aid of deep learning method, we proposed an improved double hidden-layer Variable Length Incremental Extreme Learning Machine based on Particle Swarm Optimization (in short, as PSO-VIELM). Similar to IELM, in PSO-VIELM, the hidden nodes, input weights and bias matrix will be automatically determined. In the optimization process of the first hidden layer, the number of step size of hidden nodes is added along with the error which contains not only the root mean squared error (RMSE) of training data but also the mean absolute percent error (MAPE) from validation set. Also, it is considered to selected the proper input weights and hidden bias, which may establish an effective and stability network rather than traditional SLFN. Similarly, in the second hidden layer, the step size is determined by the classification cross-entropy about validation set. Finally, the experiment results on six datasets to verify the performance of this method.

The rest of this paper is organized as follows. In Sect. 2, we introduce some preliminaries of IELM, PSO and PSO-ELM. Section 3 presents our algorithm which is integrating with some deep learning thoughts. Simulations are carried out and results are analyzed in Sect. 4. Conclusions are drawn in Sect. 5.

2 Preliminaries

2.1 Incremental Method on Extreme Learning Machine(I-ELM)

The Extreme Learning Machine which combined with incremental method is equally a kind of SLFN. As a variation of ELM, the input weights and hidden bias are randomly selected, then the output weights would be calculated with the number of hidden nodes confirmed. In traditional ELM classifier, many kinds of activity function can be used like sigmoidal, sine, tangent and relu. While in the process of training, given the input training datasets $X = \{x_i\}_{i=1}^{N}$ which accompany with the targets $T = \{t_i\}_{i=1}^{N}$, where N is the total number of the samples.

Assume that with L hidden nodes, ELM is modeled by

$$\boldsymbol{H}\beta = \boldsymbol{T} \tag{1}$$

where T represents the target matrix, H contains $g(x)$ which identified as the output weights of the hidden layer.

Different to traditional gradient descent method, in order to minimize the error $\sum_{i=1}^{N} ||o_i - t_i||$, where o_i means the predictor of ELM model, to be the equal of minimize the $||\boldsymbol{H}\beta - \boldsymbol{T}||$. With the input weights and hidden bias randomly selected, the output weights will be calculated with Moore-Penrose inverse of $\boldsymbol{H}$

$$\beta = \boldsymbol{H}^{+}\boldsymbol{T} \tag{2}$$

where $\boldsymbol{H}^{+}$ is the Moore-Penrose inverse of $\boldsymbol{H}$.

As already mentioned, the ELM models tend to have problems when hidden nodes added in the training and testing model. As an illustration of this, we consider an ELM classifier for the datasets which got from UCI (University of California Irvine) Machine Learning Repository. The datasets named Musk which contained 476 samples, 167 features and 2 types, then divided them into training set which contains 200 samples randomly and testing set which contains 276 samples. The hidden nodes of the ELM classifier which we used are set from 2 to 200, then repeat the process of classification in 100 times.

Figure 1(a) shows the accuracy in ELM training model and Fig. 1(b) shows the accuracy in testing model, and the average of the training time in the course of experiments under the hidden nodes altered is shown in Fig. 1(c). From those pictures, we could easily find that with the increasing of the hidden nodes, training accuracy rate has sustained growth. On the other hand, testing accuracy rate increased at first, however, with the number of hidden nodes growth, the accuracy rate has become decreased. Still and all, training time sustained growth.

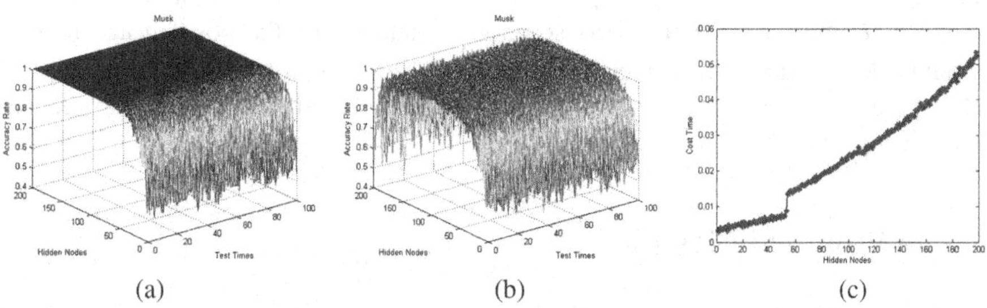

Fig. 1. (a) Example of a training result by ELM, (b) Example of a testing result by ELM, (c) Cost Time

In the Incremental Learning Model, the output weights relate to the errors

$$\boldsymbol{\beta}_L = \frac{\langle e_{L-1}, \boldsymbol{H}_L \rangle}{||\boldsymbol{H}_L||^2} \tag{3}$$

where e expresses error between predictors and targets.

While building a system about neural network of Incremental Extreme Learning Machine, the input weights and hidden bias were firstly irrevocably determined with $L-1$ hidden nodes, then the weights and bias are randomly selected for L-th node. has proved that as the number of hidden nodes increased, training error and prediction error consistently decreased. The output weights of the hidden nodes are based on least square method and it is calculated along with the hidden nodes increase which make it convergencing faster than ELM. At the same time, I-ELM algorithm and its structure are more compact.

2.2 Particle Swarm Optimization

Particle Swarm Optimization is an optimization algorithm, inspired by social learning, which was proposed by Eberhart and Kennedy. PSO works by initializing a flock of birds over the searching area randomly, where each bird named a particle or a solution. Normally, in each iteration, a fitness function was evaluated for every particle in swarm. Then all the birds rely on the previous best position and the global best position. Assume that the dimension of searching area is D, the total number of swarm is N, then the original PSO can be described as

$$v_{id}(t+1) = v_{id}(t) + c_1 \cdot R_1 \cdot [p_{ibd}(t) - x_{id}(t)] + c_2 \cdot R_2 \cdot [p_{gd}(t) - x_{id}(t)] \tag{4}$$

$$x_{id}(t+1) = x_{id}(t) + v_{id}(t+1), 1 \leq i \leq N, 1 \leq d \leq D \tag{5}$$

where $v_i(t)$ and $x_i(t)$ denote the velocity vector and the position of the i-th particle in PSO algorithm. Remarkably, the $p_{ib}(t)$ means the previous best solution of i-th particle and the $p_g(t)$ means the global best solution. c_1 and c_2 are the positive acceleration constants; R_i is a random number in order to add some randomness which between 0 and 1. Due to its strong search ability, it is generally used in the global search and searching for optimal solution.

3 The Proposed Method

In order to determine the proper number of hidden nodes and fine-tuned network structure is a crucial issue in Incremental ELM, the two hidden layers are important for extracting and classification. Moreover, a compact and well generalization framework is to enhance the performance of it. In this section, we introduce our proposed feature learning method which mainly based on incremental model. Draw support from deep learning method to enhance the feature extraction. This method uses auto-encoder and early stopping to build a proper network structure for feature extraction and the weights optimized by PSO. The proposed method is referred to as PSO-VIELM.

The detailed steps are presented as follows:

Given a set of raw data $\{(x_i, t_i)\}_{i=1}^{N} \in R^d \times R$, the proper number of the Tolerance Tol_{max}, Encouragement Enc_{max}, Growth Rate Gr, Drop Rate Dr, the maximum number of the hidden nodes L_{max} and the iteration *iter*, the expected $L = 1$:

Step 1: Randomly divided raw data into training set, validation set and testing set.
Step 2: Initializing the Network by training set with ELM.
Step 3: Determined the step L_{s1} for next iteration.
Step 4: Use PSO to select the proper weights:

- ***Substep 4.1:*** Randomly generate the swarm within the range of [–1, 1]. Each particle contains the input matrix and bias of the first hidden layer which both transformed into one dimensional vector like: $p_{si} = (w_{si}, b_{si})$
- ***Substep 4.2:*** Calculate the fitness of each particle. With the calculation among all the particles, the p_{ib} of each particle and the g_b are computed respectively.
- ***Substep 4.3:*** Each particle updates its own position with the influence among the best position it has visited, the global best position which the swarm has visited and itself. Moreover, all components in each particle should be limited into [–1, 1].
- ***Substep 4.4:*** Confirm whether reach the iteration or not. The Steps 3, 4 will be repeated if it is not.

Step 5: Retain the feature learning structure above and make feature extraction on the training set.

Step 6: Determined the step in the second hidden layer L_{s2} with the classification cross-entropy in validation set.
Step 7: Confirm and retain the best structure about the classification framework. Specific details and rules will be carried out in the following subsections.

3.1 Feature Extraction Layer

The traditional ELM classification method is based on the simple data pre-processing of the original data, and then the model is built directly. However, the original characteristics of the data are not conducive to the construction of the classification model and may even greatly affect the accuracy and generalization performance of the model after classification. The effective feature extraction can improve the learning ability of the model appropriately, on the other hand, it can improve the generalization performance of the model, so that it can perform well in the learning process of the whole sample set.

Based on the feature learning (FL) method mentioned above, we compared the data after feature learning and the raw data by inner and outer distance. This formula can be described as:

$$dis = \frac{\sum_{i=1}^{n}\sum_{k=1}^{n_i}\left(y_{ik} - \overline{y_i}\right)^2}{\sum_{i=1}^{n}\sum_{j=1}^{n}\left(\overline{y_i} - \overline{y_j}\right)^2} \tag{6}$$

where $\overline{y_i}$ is the center of every class. The top of the fraction indicates the inner distance in class i, the part which below the fraction means the sum of intra distance. In order to make the data have better performance after the feature learning, we should make the sum of the intra class distance of the sample as small as possible, and the distance between the samples should be as large as possible, so that the classifier can easily distinguish the different categories of samples. Therefore, the lower *dis* is, the better the classification performance are.

3.2 Weight Adjustment Criteria

Redundant feature and the noise of feature may cause the lower classification accuracy rate. In the light of original sample feature extraction, the effect of classification can be effectively reduced. Potential features and relation came from data-mining, especially on the datasets which are high dimension. In order to effectively measure the feature learning process in the first hidden layer which is using for feature learning in current network structure, with the idea of auto-encoder, the features of datasets will be reconstructed and try to reproduce the original data. Then with the aid of comparison between output and original data, the index of error size can be calculated with

$$E = \sqrt{\frac{\sum_{i=1}^{n}\left(y_{out,i} - y_{ori,i}\right)^2}{n}} + \sum_{i=1}^{n}\frac{\left|y_{out,i} - y_{ori,i}\right|}{y_{ori,i}} \tag{7}$$

where the y_{ori} means the original data, y_{out} means the prediction data output. The whole construct of this formula will be elaborated in Sect. 4.

Based on the method of calculation about the error index above, the fitness function of PSO can be confirmed. With the change of the number of hidden nodes, there will be a corresponding dimension of the weight need to be determined. Excellent weights are the key which could build a reasonable feature learning network.

4 Experiment Results and Discussion

All the experiments are carried out in MATLAB 7.14 (R2012a) environment running on a PC with Intel (R) Core (TM) i5-6500 CPU 3.20 GHz, and 4.00 GB RAM on Win 7 operation system. To verify the effectiveness and the performance about the algorithm, the PSO-VIELM is compared with ELM, I-ELM and PSO-ELM. In order to validate our proposed method, here, we present results on six classic classification issues from UCI machine repository, the specification of these datasets is listed in Table 1.

Table 1. Specification of six data sets

Datasets	Attributes	Classes	Training set	Validation set	Testing set
Heart	6.4025	3.9114	5.8893	3.7618	3.3632
Musk	3.5196	3.5741	5.7995	2.2533	2.8273
Wine	13	2	118	30	30
Iris	4	3	100	25	25
Segmentation	19	7	1500	405	405
Satellite	36	7	4435	1000	1000

In order to compare the different performance between the effect by original data and the data which have feature extracted by ELM. As the Table 2 shown below, we could find that data after feature extracted by ELM (FE-ELM) do better than raw data. This comparison shows from the side that adding the hidden layer for feature learning before classification can effectively improve the classification ability of the original sample.

Table 2. *dis* in Raw data and data after FE-ELM

Datasets	Heart	Musk	Wine	Iris	Segmentation	Satellite
dis in Raw data	6.4025	3.9114	5.8893	3.7618	3.3632	11.6663
dis in data after FE-ELM	3.5196	3.5741	5.7995	2.2533	2.8273	11.7250

The parameters used in the PSO algorithm are: the population size $m = 50$, the maximum iteration criterion $iter_{max} = 50$, the inertia weight $w = 0.7298$, the coefficient of self-recognition and social component $c_1 = 1.4962$, $c_2 = 1.4962$. The parameters used in the PSO-ELM and I-ELM are set the same as the PSO. For the ELM and PSO-ELM classifiers, the number of hidden nodes is randomly choosing in different datasets. The input weights and hidden biases are set randomly within [–1, +1], and the velocity of the particles are limited in [–0.5, +0.5].

In order to compare merits and demerits among the algorithms above, we compute several statistical performance metrics which include training time, mean and standard deviation of training and testing accuracy. And the performance comparison is shown in Table 3.

Table 3. Results of the classification problems

Problem	Algorithm	Training time(s)	Testing accuracy (%)	Hidden nodes
Heart	ELM	0.0034	81.80 ± 0.0147	[2,100]
	I-ELM	0.0523	75.36 ± 0.0071	80
	PSO-ELM	–	–	–
	PSO-VIELM	89.963	82.43 ± 0.0038	16(5)
Musk	ELM	0.0037	79.54 ± 0.0567	[2,100]
	I-ELM	0.2135	80.95 ± 0.0029	200
	PSO-ELM	–	–	–
	PSO-VIELM	85.228	85.43 ± 0.0451	108(17)
Wine	ELM	0.0042	90.78 ± 0.0544	[2,100]
	I-ELM	–	–	–
	PSO-ELM	9.1189	86.73 ± 0.0615	[2,100]
	PSO-VIELM	57.136	96.83 ± 0.0476	24(10)
Iris	ELM	0.0018	92.33 ± 0.0238	[2,100]
	I-ELM	0.0375	82.61 ± 0.1312	80
	PSO-ELM	6.3321	92.60 ± 0.057	[2,100]
	PSO-VIELM	14.384	98.07 ± 0.0372	8(17)
Segmentation	ELM	0.0222	92.82 ± 0.0073	[2,300]
	I-ELM	–	–	–
	PSO-ELM	187.5428	92.96 ± 0.077	[2,300]
	PSO-VIELM	163.187	87.23 ± 0.0837	57(35)
Satellite	ELM	0.0321	83.83 ± 0.066	[2,300]
	I-ELM	–	–	–
	PSO-ELM	311.639	84.96 ± 0.077	[2,300]
	PSO-VIELM	288.702	85.69 ± 0.0083	66(22)

It is easily can be seen in Table 3 that the comparison among the algorithms which were mentioned above mainly focus on mean, standard deviation, CPU computing time and corresponding hidden nodes. From Table 3, the number of hidden nodes is the least in all cases among four ELM, while the accuracy rate in testing set mainly the highest among them. While it is important to note that the first number of hidden nodes in PSO-VIELM is the amount of the node in feature extraction layer, and the second number is the amount of classification layer. Among all the datasets, PSO-VIELM has better performance in classification accuracy, while as a cost, it also has the highest CPU computing time. The other three algorithms use the best results in the existing papers.

5 Conclusions

In this paper, we proposed a new learning method Particle Swarm Optimization which optimize the Variable-length Incremental ELM feature extraction and classification. PSO-VIELM approach is proposed to determine the proper hidden nodes and corresponding number in input weights and hidden bias. With the improvement of algorithm, the effect of classifier accuracy rate has increased. It is easily can be seen from the experimental result that on various datasets, PSO-VIELM algorithm has better performance than ELM, IELM and PSO-ELM, and the structure can become stabilize. Furthermore, PSO-VIELM algorithm has stronger generalization performance than others. However, as we have discussed before, many parameters are lack of rigorous explanation in mathematics. The experiment effect determined the size of these parameters. In the future research works, we will add more derivations and arguments in order to make it more convincing.

Acknowledgements. This work was supported by the National Natural Science Foundation of China [Nos. 61572241 and 61271385], the National Key R&D Program of China [No. 2017 YFC0806600], the Foundation of the Peak of Six Talents of Jiangsu Province [No. 2015-DZXX-024], the Fifth 333 High Level Talented Person Cultivating Project of Jiangsu Province [No. (2016) III-0845], and the Research Innovation Program for College Graduates of Jiangsu Province [1291170030].

References

1. Miche, Y., Sorjamaa, A., Bas, P., Simula, O., Jutten, C., Lendasse, A.: Brief papers OP-ELM: optimally pruned extreme learning machine. IEEE Trans. Neural Netw. **21**(1), 158–162 (2010)
2. Huang, G.B., Zhu, Q.Y., Siew, C.K.: Extreme learning machine: a new learning scheme of feedforward neural networks. In: IEEE International Joint Conference on Neural Networks, vol. 2, pp. 985–990 (2004)
3. Huang, G.B., Zhu, Q.Y., Siew, C.K.: Extreme learning machine: theory and applications. Neurocomputing **70**(1–3), 489–501 (2006)
4. Huang, G.B., Siew, C.K.: Extreme learning machine: RBF network case. In: Control, Automation, Robotics and Vision Conference, pp. 1029–1036 (2004)
5. Huang, G.B., Chen, L., Siew, C.K.: Universal approximation using incremental constructive feedforward networks with random hidden nodes. IEEE Trans. Neural Netw. **17**(4), 879 (2006)
6. Kennedy, J., Eberhart, R.: Particle swarm optimization. In: IEEE International Conference on Neural Networks, vol. 4, pp. 1942–1948 (2002)

7. Xu, Y., Shu, Y.: Evolutionary extreme learning machine – based on particle swarm optimization. In: Wang, J., Yi, Z., Zurada, J.M., Lu, B.-L., Yin, H. (eds.) ISNN 2006. LNCS, vol. 3971, pp. 644–652. Springer, Heidelberg (2006). https://doi.org/10.1007/11759966_95
8. Zhao, G., Shen, Z., Miao, C., Gay, R.: Enhanced extreme learning machine with stacked generalization. In: IEEE International Joint Conference on Neural Networks, pp. 1191–1198 (2008)
9. Zhang, R., Lan, Y., Huang, G.B., Xu, Z.B.: Universal approximation of extreme learning machine with adaptive growth of hidden nodes. IEEE Trans. Neural Netw. Learn. Syst. **23** (2), 365 (2012)
10. Yang, Y., Wang, Y., Yuan, X.: Parallel chaos search based incremental extreme learning machine. Neural Process. Lett. **37**(3), 277–301 (2013)
11. Bastien, F., Lamblin, P., Pascanu, R., Bergstra, J., Goodfellow, I., Bergeron, A., Bouchard, N., Wardefarley, D., Bengio, Y.: Theano: new features and speed improvements. Computer Science (2012)
12. Sun, K., Zhang, J., Zhang, C., Hu, J.: Generalized extreme learning machine autoencoder and a new deep neural network. Neurocomputing **230**, 374–381 (2016)

Breast Cancer Medical Image Analysis Based on Transfer Learning Model

Yi Liu[1,2(✉)] and Xiaolong Zhang[1,2]

[1] School of Computer Science and Technology, Wuhan University of Science and Technology, Wuhan 430065, China
2420044571@qq.com, xiaolong.zhang@wust.edu.cn
[2] Intelligent Information Processing and Real-Time Industrial Systems Hubei Province Key Laboratory, Wuhan 430065, China

Abstract. Breast cancer becomes one of the common cancer among women. The computer-aided diagnosis is a very available technology, and it becomes an inevitable trend of modern medicine. There is litter data in the medical field which is not enough to support training model. The paper aims to solve the problem in the case of insufficient data volume, and a transfer learning model combined with Convolutional Neural Network (CNN) is proposed to achieve the goal. This model has three innovations. The first one takes Xavier method to initialize parameter, which can make the training process more stable. The second innovation takes dropout method to discard some network nodes randomly, which can reduce overfitting problem. The third one adds two convolutional layers and one max pooling layer before final fully connected layer. The experimental results have shown that this strategy is suitable for the problem in this paper. The paper indicates that the transfer learning model is an effective method with small-scale data, and it can be combined with deep learning algorithms.

Keywords: Breast cancer · Transfer learning · Convolutional Neural Network

1 Introduction

Breast cancer becomes one of the most common cancer among women. Early diagnosis can improve the cure rate and greatly reduce the mortality of breast cancer [1]. Computer-aided diagnosis method is useful for early diagnosis, and it can overcome the poor medical conditions [2].

Early detection of breast cancer is based on the analysis and classification of breast cancer X-ray medical images. For this kind of medical image, data collection and labeling are quite time-consuming. Moreover, breast cancer medical imaging data are still not enough to support the modeling. Nowadays, there are some common data sets like DDSM [3] and MIAS [4], and researchers mainly used machine learning algorithms to training these data. Wei compared several relatively advanced machine learning methods [5], such as Support Vector Machine (SVM), Kernel Fisher Discriminant (KFD), correlation vector Machine (RVM) and Board Machines (Set averages and AdaBoost). The accuracies of these traditional machine learning methods are

D.-S. Huang et al. (Eds.): ICIC 2018, LNCS 10955, pp. 44–53, 2018.
https://doi.org/10.1007/978-3-319-95933-7_6

mostly between in 80% and 90%. More recently researchers use deep learning methods to classify breast cancer. Ren [6] proposed an improved neural network method and acquired 85% accuracy rate on the DDSM. A multi-mode deep neural networks [7] is proposed and compare with three widely used methods in prediction of breast cancer, where the accuracy rate is 82.6% on DDSM. Deep learning and neural network methods can theoretically achieve better results, but they require large amount of data.

Early detection of breast cancer is very important, Shen [8] developed a set of shape factors to measure the roughness of contours in mammograms and used k-Nearest Neighbor (KNN) classifier to classify breast cancer, which obtained 73% accuracy on DDSM. A new metric was proposed to derive the micro-calcified roughness from the normalized distance markers [9], where a threshold classifier with fewer feature performed, this classification on DDSM acquires 80% accuracy.

This paper proposes a transfer learning model to classify the medical images of breast cancer. The model not only overcomes the poor classify performance of traditional machine learning methods, but also reduces the amount of data required. The transfer learning refers to extract knowledge and experience from one or more source tasks, and then applied them to a related target area [10]. As we know, it is difficult to capture and annotate images in specific fields. Usually, the training of deep learning requires a large number of labeled data as training samples [11], otherwise it is difficult to achieve the desired effect. In fact, the transfer learning could well address the problem of insufficient data volume in some special areas.

In this paper, the transfer learning is based on Google's Inception-V3 model [12]. Inception-V3 model has 46 layers and 96 convolutional layers. Training such a complex model requires a large amount of annotation data. By means of transfer learning with a small amount of data, we can still obtain better results, and greatly reduce training time.

2 Main Learning Algorithm

2.1 Convolutional Neural Network

Convolutional neural network (CNN) is a further development of neural network [13]. With the feature of parameter sharing and sparse connections, the CNN algorithm has competitive performance in image classification. CNN is multi-layer formed by the convolutional layer, the pooling layer, and the fully connected layer.

2.2 Medical Image Analysis

Medical image analysis is a common practice for medical diagnosis. Computer-aided diagnosis in medical image analysis can help a doctor to give the targeted treatment plan [14]. Medical image analysis such as noise reduction, segmentation, and classification can help doctors to do the diagnosis better.

This paper mainly analyzes the X-ray images of breast cancer. First, the medical image can convert into a pixel matrix. Second, the features of the matrix are extracted and analyzed by using our transfer learning model. Finally, the classification is performed using these features. The result can determine whether one is sick or not, which assists doctor in the diagnosis. It can reduce the burden on the doctor and improve the accuracy of diagnosis.

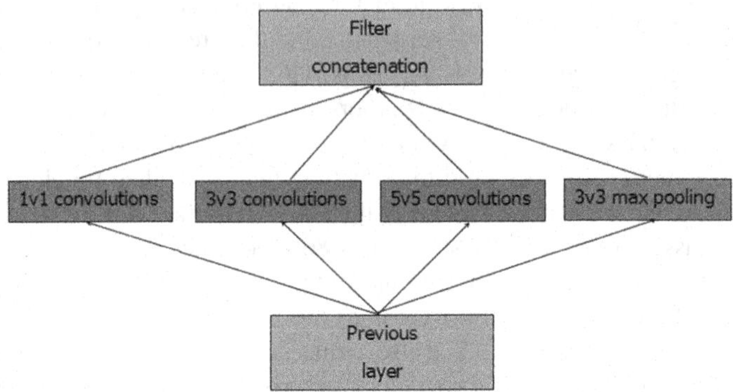

Fig. 1. Inception structure proposed by google which connected convolutional layers in parallel.

2.3 CNN Models

With the development of CNN, researchers have proposed many different models, such as inception-V3 model [12] and VGG model [16]. Google's inception-V3 model is a representative neural networks model which have great performance on ImageNet [15]. One of the more innovative aspects of the Inception-V3 model is to combine different convolutional layers in parallel by using inception structure as described in Fig. 1. For the purpose of keeping the size of the result, different convolutional layer with all zero padding. We can combine them in the depth dimension and extract more characteristic features. The Inception-V3 model had achieved high classification accuracy on the ImageNet database. In this paper, we choose it as the model of CNN for the transfer learning.

Inception-V3 model consists of 11 inception modules, the depth of the model up to 46 layers. We can arbitrarily change the size of the convolution filter, construct different inception structure, the methods can extract more different features. Due to its deep depth and the large number of parameters, high performance requires a large amount of training data, otherwise it is easy to cause over-fitting.

VGG is a deep convolutional neural network that explores the relationship between depth and performance in neural network. The main idea of VGG is to iterate small convolutional kernels and pooling layers to improve the performance of the networks.

2.4 Transfer Learning

A major assumption in many machine learning is that the training and future data must be in the same feature space and have the same distribution. However, in many real-world applications, this assumption may not hold. For example, we sometimes have a classification task in one domain of interest, but we only have sufficient training data in another domain of interest, where the latter data may be in a different feature space or follow a different data distribution. In such cases, knowledge transfer, if done successfully, would greatly improve the performance of learning by avoiding much expensive data-labeling efforts. Transfer learning is a one-size-fits-all ability for human, but for computers, it means to use the features extracted by one database to another database. For example, when the breast cancer X-ray medical images are not enough, we use another medical image to train our model, which can overcome the shortcomings. There are several cases of transfer learning: the source data domain is the same type as the target data domain; the source data domain is different type from the target data domain. This paper refers the latter case.

Given an image, the extracted features can be considered that composed of edges and corners. The elements of the image are similar. This phenomenon means that we can use other images features to help the image that we need process.

2.5 Transfer Learning Model Based on Convolutional Neural Network

Breast cancer X-ray image data is one kind of medical images data. The characteristics of medical image data are small data volume, labeling difficult and fuzzy imaging. CNN as a deep learning method has excellent performance for extracting these features. Moreover, transfer learning can make use of the features of the source domain training to improve the training although the target is insufficient. A transfer learning model is proposed in this paper based on convolutional neural networks.

For the transfer learning model based on VGG, we use a source data on VGG model firstly, where saving the parameters of CNN. Then, the target data is trained on the CNN model.

The steps of the transfer learning model based on inception-v3 are as follows. First, the source data is processed by Inception-V3 networks, meanwhile the parameters of each layer are saving [17]. Then, we modify the connection layer to accommodate classification tasks.

This transfer learning model of this paper has made some appropriate changes on the original one based on Inception-v3. The abbreviation of this algorithm is TL-AVC. For specific medical data, the changes are follows:

1. The parameter initialization is used as the Xavier initialization method [18]. As the experimental results fluctuate greatly, the initialization of parameters has an impact on the experimental results. Appropriate initialization parameters can avoid the training falling into local extreme points. Xavier method assumes the parameters of input dimension is n, the output dimension is m. Xavier initialization makes the parameters in the interval $\left[-\sqrt{\frac{6}{m+n}}, \sqrt{\frac{6}{m+n}}\right]$ uniformly distributed.
2. We use dropout method during the training phase [19]. Because the data is small, it is easy to over-fitting. Dropout refers to discard some of the nerve codes with a random probability. The method makes the model to study the more universal features. Practice has proved that dropout method could reduce over-fitting condition.

3. More convolutional and pooling layer are used. Before the final fully connected layer, adding two convolutional layer and a pooling layer. The kernel size of convolutional is 3×3. The features of medical image are not obvious, more convolutional and pool layer after Inception-v3 model are necessary. It has been proved by the experiments that adding two convolutional layers is the best choice.

Algorithm 1. TL-AVC

Input: Breast cancer medical image, the sample labels.
Output: Classification result of sample.
Initialize: i=1

1. Train the model on ImageNet and save parameter.
2. Divide the data into training, testing and validation data.
3. Add convolutional and pooling layer before the fully connection layer.
4. Use the Xavier method to initialize the neural networks parameters.
5. **Repeat**
6. Get the batch of training data.
7. Use the model in this paper to extract the feature.
8. Classify and calculate the loss.
9. Use gradient descent method to minimize the loss.
10. Calculate accuracy on verification data at regular intervals.
11. **Until** i =steps.
12. Calculate accuracy on testing data at last.

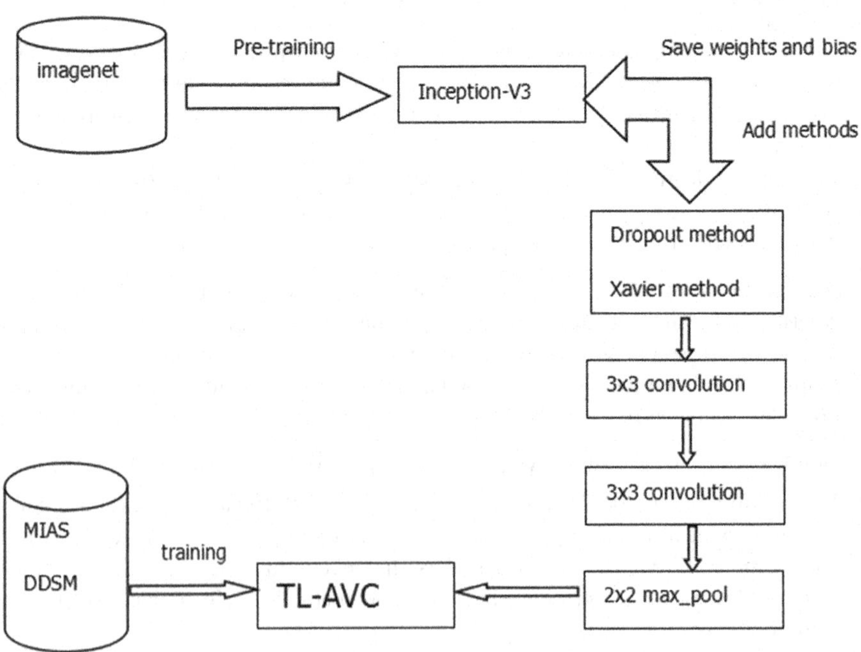

Fig. 2. The transfer learning model of this paper's system.

This paper proposes the TL_AVC algorithm, Fig. 2 is the transfer learning model of the system, Algorithm 1 describes the steps of the proposed algorithm. TL_AVC makes three improvements of the transfer learning model based on Inception-v3. For parameter initialization, the Xavier method make the parameters in the interval $[-\sqrt{\frac{6}{m+n}},\sqrt{\frac{6}{m+n}}]$ uniformly distributed. It can make the training results do not fall into the local minimum and achieve the overall optimal results. The convolutional and pooling layers have been added to the algorithm, and the dropout method has been used to reduce the over-fitting. By numerical experiments, it can be found that the modified model is more suitable for this problem.

3 Experimental Steps

3.1 Data Processing

The paper uses MIAS [4] and DDSM [3] database. The data is distributed as Table 1. Before training, we processed the data. We take noise-reduce and normalize operation before the feature extraction. Moreover, dividing the data into three parts: training data, test data and verification data. The training data is accounted for 60% of the total sample, verification is accounted for 20% of the total sample, and testing is accounted for 20% of the total sample too. The test sample did not cross the training sample and the validation sample. The result in this paper is the average of 10 results.

Table 1. Sample distribution of DDSM and MIAS.

	Lesions	Normal
MIAS	114	208
DDSM	286	327

3.2 Model Training

This paper uses three models: VGG transfer learning model, Inception-V3 transfer learning model and TL-AVA. The training procedures of the three transfer learning models in this paper are generally similar. First, the model is trained on the source domain while retaining the parameters, and then the target domain is input into the model for training and classification.

The input of the CNN is the pixel matrix processed in the data preparation phase. CNN is a multi-layer convolutional stack structure, x^{l-1} is the input of *l–level* convolutional layer, x^1 represents the output of *l-level* layer, w^l and b^l represent the *l-level* layer's weight and biases, f is the activation function of the convolutional layer, the commonly activation functions are sigmoid function and relu function. Then the forward propagation process is:

$$x^l = f\left(w^l * x^{l-1} b^l\right) \quad (1)$$

When wrong classification occurs, the cost of error classification is accumulated. By minimizing the cumulative cost, the accuracy of the model will more satisfactory. The cost function in this experiment is cross-entropy cost function. M means the number of the samples, $\left(x^{(i)}, y^i\right)$ is the data and its corresponding label. $h_{w,b}()$ is the hypothesis function. The cross-entropy loss function is:

$$J(w,b) = -\frac{1}{m}\sum\nolimits_{i=1}^{m} y^{(i)} log\left(h_{w,b}\left(x^{(i)}\right)\right) + \left(1 - y^{(i)}\right) log\left(1 - h_{w,b}\left(x^{(i)}\right)\right) \quad (2)$$

This paper uses the gradient descent method to adjust parameters, making the cost function minimum. $w_{ij}^{(l)}$ is the weights of the *l-level*, $b_{ij}^{(l)}$ is the biases of the *l-level* and a is the learning rate. The update formula for gradient descent sum as follows.

$$w_{ij}^{(l)} = w_{ij}^{(l)} - a\frac{\partial}{\partial w_{ij}^{l}} J(w,b) \qquad b_{ij}^{(l)} = b_{ij}^{(l)} - a\frac{\partial}{\partial b_{ij}^{l}} J(w,b) \quad (3)$$

After all kinds of transfer learning models are trained in the source domain, the values of the parameters in each layer will be saved. Then changing the final fully connected layer and classification layer, and make it applies to the source domain. Finally, we take the average of multiple results as the final results.

4 Results and Analysis

4.1 Normal and Lesions Classification

We use MIAS and DDSM databases in this paper. Transfer learning models include VGG transfer learning model, Inception-V3 transfer learning model and TL-AVC model.

Table 2. The classification accuracy on DDSM and MIAS using three kinds of transfer learning model.

Algorithm	DDSM	MIAS
VGG transfer learning model	78.23%	67.28%
Inception_V3 transfer learning model	95.81%	90.1%
TL-AVC (our method)	96.73%	93.42%

This paper divides the data into normal mammograms and lesions mammograms and uses three different transfer learning models to train the data. The classification results are shown in Table 2. The results in the table are the accuracy of the final test data, which are the results of multiple test averages, and each of the comparative tests repeated 10 times. For DDSM database, the accuracy rate of VGG transfer learning model is only 78.23%, while the accuracy of Inception-v3 transfer learning model is 95.81%. The method in this paper works best on the DDSM database with an accuracy of 96.73%. From above table, it can be found that TL-AVC has the best result. For MIAS database, the precision of VGG transfer learning model is only 67.28%, but it reached 90.1% accuracy by Inception-v3 transfer learning method. Finally, the accuracy of TL-AVC is 93.42%, which is still the best of the three methods. From another point of view, different convolutional neural networks transfer learning results are also very different. In this paper, we compare the VGG convolutional neural networks with the Inception-V3 convolutional neural networks, the results of different data all indicate the superiority of Inception-V3. Because of the Xavier method, when look at the intermediate results, we found the TL-AVC is more stable. Because of the dropout method and the new add layers, the accuracy of the result is more better.

Table 3. Compare the classification accuracy of TL-AVC with the results of other.

Algorithm	MIAS	DDSM
Shen [8]	70%	73%
Ren [6]	85%	85%
Ma [9]	62%	80%
TL-AVC (our method)	93.42%	96.73%

Breast cancer medical image classification is a very traditional issue, many researchers had greatly contribution in this work. Table 3 shows the related research results. Shen [8] developed a set of shape factors to measure the roughness of contours of calcification in mammograms and use KNN classifier to classify breast cancer, the algorithm obtained 70% accuracy on MIAS and 73% accuracy on DDSM. Ren [6] introduced equalization learning to neural networks and proposed an improved neural network classifier, then used ANN or SVM method to the classification problem, it reached 85% accuracy on two data sets. Ma [9] proposed a new metric to derive the micro-calcified roughness from the normalized distance maker using a threshold classifier, the accuracy rate was 62% on MIAS, and the accuracy rate on DDSM was 80%. The results of the transfer learning model in this paper are better than the above three methods on the two data sets. The method in this paper can obviously achieve the better results with a small amount of data.

4.2 Calcification and Mass Classification of Abnormal Data

This paper mainly solves the classification of normal and abnormal breast cancer images. From the above summary, TL-AVC has good performance on the classification of normal and abnormal breast cancer images problem. However, the breast lesions can divide into several types, after 4.1 experiments, this paper also classifies the different types of breast lesions. For the DDSM database, it includes 287 calcification sample and 675 mass sample. Table 4 shows the classification results of the three transfer learning models on abnormal samples of DDSM, each of the test results are the average of 10 times. The classification accuracy of VGG transfer learning model obtained

70.24%, the accuracy of Inception-v3 transfer learning is 69.98%. And it reached 79.93 accuracy by using TL-AVC, which is the best of the three methods. When classifying lesion types, the superiority of the method proposed in this paper can be found more obviously.

Table 4. The accuracy of mass and calcification data using three transfer learning models.

Algorithm	Mass and calcification data of DDSM
VGG transfer learning model	70.24%
Inception_V3 transfer learning model	69.98%
TL-AVC (our method)	79.93%

5 Conclusion

This paper proposes a transfer learning model combines with CNN. The model is pre-trained on ImageNet at first, then adjusts the network, uses the Xavier initialization method, dropout method and adds more convolutional and pooling layers to adapt to this data, we finally achieve satisfactory results. The performance of CNN combined with transfer learning model can be optimized by adjusting the parameter initialization mode and adding the dropout method.

Different CNN models have different influences for the performance of algorithm. A more suitable CNN model is the problem that we should consider in the future. Moreover, we would like to find a better way to combine transfer learning and CNN model.

Acknowledgment. This work was supported in part by National Natural Science Foundation of China (61273225, 61373109, 61702381).

References

1. Veronesi, U., Boyle, P., Goldhirsch, A., et al.: Breast cancer. Lancet **365**(9472), 1727–1741 (2015)
2. Kelsey, J.L., Hornross, P.L.: Breast cancer: magnitude of the problem and descriptive epidemiology. Epidemiol. Rev. **15**(1), 7 (1993)
3. Heath, M., Bowyer, K., Kopans, D., et al.: The digital database for screening mammography. In: Digital Mammography, pp. 457–460. Springer, Netherlands (2001)
4. Sucking, J., Boggis, C.R.M., Hutt, I., et al.: The Mini-MIAS database of mammograms. In: International Congress Series, vol. 1069, pp. 375–378 (1994)
5. Wei, L., Yang, Y., Nishikawa, R.M., et al.: A study on several Machine-learning methods for classification of malignant and benign clustered microcalcifications. IEEE Trans. Med. Imaging **24**(3), 371–380 (2005)
6. Ren, J., Wang, D., Jiang, J.: Effective recognition of MCCs in mammograms using an improved neural classifier. Eng. Appl. Artif. Intell. **24**(4), 638–645 (2011)

7. Sun, D., Wang, M., Li, A.: A multimodal deep neural network for human breast cancer prognosis prediction by integrating multi-dimensional data. IEEE/ACM Trans. Comput. Biol. Bioinform. 1 (2018)
8. Shen, L., Rangayyan, R.M., Desautels, J.E.L.: Application of shape analysis to mammographic calcifications. IEEE Trans. Med. Imaging **13**(2), 263–274 (1994)
9. Ma, Y., Tay, P.C., Adams, R.D., et al.: A novel shape feature to classify microcalcifications. In: IEEE International Conference on Image Processing, Western Carolina University, pp. 2265–2268. IEEE (2010)
10. Pan, S.J., Yang, Q.: A survey on transfer learning. IEEE Trans. Knowl. Data Eng. **22**(10), 1345–1359 (2010)
11. Lecun, Y., Bengio, Y., Hinton, G.: Deep learning. Nature **521**(7553), 436 (2015)
12. Szegedy, C., Liu, W., Jia, Y., et al.: Going Deeper with Convolutions, pp. 1–9 (2014)
13. Krizhevsky, A., Sutskever, I., Hinton, G.E.: ImageNet classification with deep convolutional neural networks. In: International Conference on Neural Information Processing Systems, pp. 1097–1105. Curran Associates Inc. (2012)
14. Doi, K.: Computer-aided diagnosis in medical imaging: historical review, current status and future potential. Comput. Med. Imaging Graph. **31**(4–5), 198–211 (2007)
15. Deng, J., Dong, W., Socher, R., et al.: ImageNet: a large-scale hierarchical image database. In: IEEE Conference on Computer Vision and Pattern Recognition, CVPR 2009, pp. 248–255. IEEE (2009)
16. Simonyan, K., Zisserman, A.: Very deep convolutional networks for large-scale image recognition. Computer Science, pp. 1–15 (2014)
17. Russakovsky, O., Deng, J., Su, H., et al.: ImageNet large scale visual recognition challenge. Int. J. Comput. Vis. **115**(3), 211–252 (2015)
18. Glorot, X., Bengio, Y.: Understanding the difficulty of training deep feedforward neural networks. J. Mach. Learn. Res. **9**, 249–256 (2010)
19. Srivastava, N., Hinton, G., Krizhevsky, A., et al.: Dropout: a simple way to prevent neural networks from overfitting. J. Mach. Learn. Res. **15**(1), 1929–1958 (2014)

A Fast Algorithm for Image Segmentation Based on Local Chan Vese Model

Le Zou[1,2,3], Liang-Tu Song[1,2], Xiao-Feng Wang[3(✉)], Yan-Ping Chen[3], Qiong Zhou[1,2,4], Chen Zhang[3], and Xue-Fei Li[1,2]

[1] Hefei Institute of Intelligent Machines, Hefei Institutes of Physical Science, Chinese Academy of Sciences, P.O. Box 1130, Hefei 230031, Anhui, China
[2] University of Science and Technology of China, Hefei 230027, Anhui, China
[3] Key Lab of Network and Intelligent Information Processing, Department of Computer Science and Technology, Hefei University, Hefei 230601, China
xfwang@hfuu.edu.cn
[4] School of Information and Computer, Anhui Agricultural University, Hefei 230036, Anhui, China

Abstract. Image segmentation plays a very important pole in image processing and computer vision field. Most of the energy minimization of level set methods are based on the steepest descent method and finite difference scheme. In this paper, we propose a sweeping algorithm to minimize Local Chan Vese (LCV) model. We calculate the energy change when a pixel is moved from the outside region to the inside region of evolving curves and vice versa, instead of directly solving the Euler-Lagrange equation. The algorithm is fast and robust to initial level set contour and can avoid solving partial differential equation. There is no need for the re-initialization step, any stability conditions and the distance regularization term. The experiments have shown the effectiveness of the proposed algorithm.

Keywords: Image segmentation · Region-based model · Level set Sweeping algorithm

1 Introduction

Image segmentation is a fundamental and key problem in image processing and computer vision fields. Over the past decades, thousands of image segmentation methods have attracted many scholars' attentions, in which level set method (LSMs) are very popular. The LSMs can be classified into edge-based methods [1], region-based methods [2] and hybrid methods [3]. The Chan-Vese (CV) model [2] is the most famous region-based method. However, the CV model can't segment the images with intensity inhomogeneity, and is sensitive to the placement of initial contour and setting of initial parameters. Hence, it converges slowly and apt to be trapped into local minima. Some improved models contain the local Chan-Vese (LCV) model [4], the region-scalable fitting (LBF) model [5], etc.

In almost all region based level set methods, the minimization of the energy is performed by the steepest descent (SD) method. In this process, one must solve the

D.-S. Huang et al. (Eds.): ICIC 2018, LNCS 10955, pp. 54–60, 2018.
https://doi.org/10.1007/978-3-319-95933-7_7

partial differential equation (PDE) until reaching the minimum. As a result, local minima may be reached and then wrong image segmentation is got. To ensure the stable evolution, Courant Friedrichs Lew (CFL) condition must be satisfied. The temporal step must be very small so that the evolution is time consuming. Some method are studied in literatures to overcome the shortcomings of SD method, such as implicit difference [1], Hermite differential operator [6] and so on. Recently, many scholars construct some new methods to get better image segmentation results. In [7, 8], the authors use the sweeping principle algorithm to get the optimization of the energy functional of the region based models. Inspired by ideas in [7, 8], we present a new algorithm for the optimization of the LCV model for image segmentation. The algorithm scans all the pixels of image and then checks the energy change for every pixel when the pixel is moved from the inside to the outside of the evolving curve and vice versa. Thus, it does not need to solve any PDE and has no numerical stability conditions. Besides, there is no need for the distance regularization term.

The remaining of this paper is as follows: in Sect. 2, a fast algorithm based on sweeping principle algorithm for LCV model is presented. In Sect. 3, some examples are given to show the efficiency of the proposed algorithm. Finally, some conclusions are provided in Sect. 4.

2 Fast Algorithm of Minimizing Local Chan Vese Model

The LCV model [4] is one of authors' previous works which can segment the intensity inhomogeneous images efficiently. It is based on the steepest descent method. The contour evolution is solved by the finite difference scheme which has some drawbacks, such as lager approximation error, time-cost consuming. The most direct results is the failed segmentation. To avoid this situation, inspired by the work of Song [7] and Boutiche [8], we propose a sweeping principle algorithm to minimize the LCV energy functional.

In the LCV energy functional, the curve evolution is dominated by the fidelity to data term which is written as follows:

$$\begin{aligned} F^{LCV} &= \alpha \cdot (|I(x) - c_1|^2 - |I(x) - c_2|^2) \\ &\quad + \beta \cdot (|g_k * I(x) - I(x) - d_1|^2 - |g_k * I(x) - I(x) - d_2|^2) \end{aligned} \tag{1}$$

where α and β are two small nonnegative constants. c_1, c_2, d_1, d_2 are updated at each iteration. g_k computes the averaging convolution in a $k \times k$ size window. More details of LCV model can be found in [4]. It is worth noting that the first term shows the global region fitting energy and the second one presents the local statistical information fitting energy. Here, the inside and outside energy was deducted as follows:

$$\begin{cases} \alpha \cdot |I(x) - c_1|^2 + \beta \cdot |g_k * I(x) - I(x) - d_1|^2 & \text{if } \phi(x) > 0 \\ \alpha \cdot |I(x) - c_2|^2 + \beta \cdot |g_k * I(x) - I(x) - d_2|^2 & \text{if } \phi(x) < 0 \end{cases} \tag{2}$$

In order to compute the energy change of each pixel when the pixel is moved from the outside of the curve to inside, we define the energy change as the following:

$$\begin{aligned}\Delta F_{21} = & [\alpha \cdot (|I(x) - c_1|^2 + \beta \cdot (|g_k * I(x) - I(x) - d_1|^2)\frac{m}{m+1}] \\ & - [\alpha \cdot |I(x) - c_2|^2) + \beta \cdot |g_k * I(x) - I(x) - d_2|^2)\frac{n}{n-1}]\end{aligned} \quad (3)$$

Similarly, we can get the Eq. (4),

$$\begin{aligned}\Delta F_{12} = & -[\alpha \cdot (|I(x) - c_1|^2 + \beta \cdot (|g_k * I(x) - I(x) - d_1|^2)\frac{m}{m-1}] \\ & + [\alpha \cdot |I(x) - c_2|^2) + \beta \cdot |g_k * I(x) - I(x) - d_2|^2)\frac{n}{n+1}]\end{aligned} \quad (4)$$

where m and n are areas of inside and outside parts of the evolving contour, respectively. v is a small nonnegative constant.

The main steps of the proposed sweeping principle algorithm are given as follows:

Step 1: Place the initial contour on the given image. Initialize the level set function $\phi = 1$ for inside part and $\phi = -1$ for outside part.
Step 2: Initialize parameters k, α, β. Set the total energy F to 99.
Step 3: Compute c_1, c_2, d_1, d_2 according to the update scheme in [4], calculate m, n.
For each pixel of ϕ do:
If $\phi = 1$ then compute the difference between the new and old energies ΔF_{12} according to Eq. (4),
If $\Delta F_{12} < 0$ then change ϕ from +1 to -1.
If $\phi = -1$ then compute the difference between the new and old energies ΔF_{21} according to Eq. (3),
If $\Delta F_{21} < 0$ then change ϕ from -1 to +1.
$F = \Delta F_{21} + \Delta F_{12}$,
Step 4: Repeat Step 3 until the total energy F remains unchanged.

It can be seen that the algorithm does not need derivative calculations, PDE and any numerical stability conditions. Besides, there is no need for F to be differentiable, and no need for the distance regularization term. Furthermore, the binary level set function can be used to avoid negative effects and speed up the optimization process.

3 Experimental Results

In this section, we give the experiments of the proposed algorithm on several synthetic and real images. The proposed algorithm was implemented by MatlabR2016a on a computer with Intel Core (TM) i7-6700 2.6 GHz CPU, 16G RAM, and 64 bit Win10 OS. For all experiments, we used the same parameters $\alpha = 10, \beta = 1, k = 3$. The level set function is initialized to a binary function. Here, we use green lines to represent the initial contours and the red lines for final evolution results.

Firstly, we give an example to show that the proposed algorithm is robust to initial setting. For most of the existing methods, the initial contour and parameters are usually set according to empirical experience. Only user is familiar with the image structure, the optimized segmentation result can be obtained. The improper initial contour and parameters setting may lead to failed segmentation. In this experiment, all the parameters are fixed. The initial contours are placed at four corners and center. Just as Fig. 1 shows, although different initial contour placement and parameters are given, the proposed algorithm successfully captured the object's boundaries.

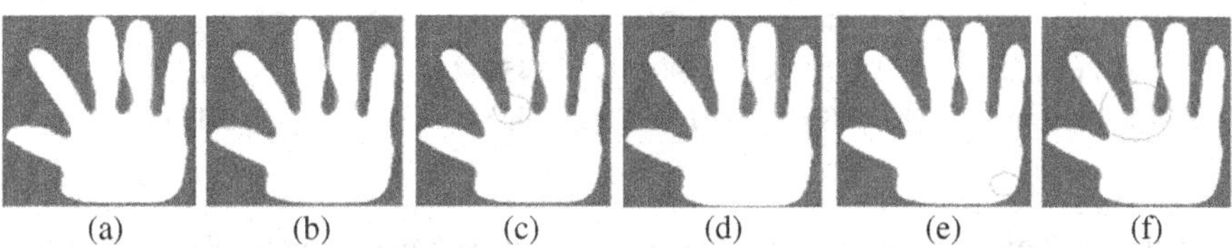

(a) (b) (c) (d) (e) (f)

Fig. 1. Demonstration of the robustness to initial contour placement and parameter of our algorithm. Image size: 108×130.

To make a fair comparison, we also used the LCV model [4], CV model [2], sweeping algorithm of CV model [7] and our algorithm based on LCV model to segment the synthetic images. All the methods used the same initialization. Figure 2 shows the final segment results and Table 1 gives the computational costs of four models. All the methods got satisfying results. From Table 2, we can see that the sweeping algorithm based LCV model consumes the smallest time than other three methods.

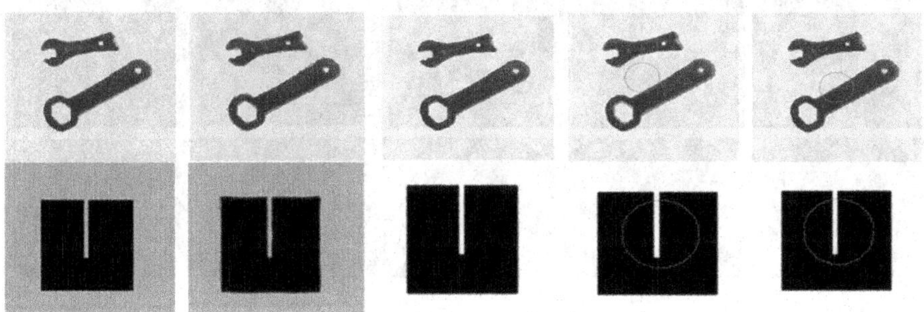

Fig. 2. Segmentation comparisons of CV model, LCV model, sweeping algorithm of CV model and the proposed algorithm on synthetic images. The first column: initial images. The second column: Final segmentation results of LCV model. The third column: Final segmentation results of CV model. The fourth column: Final segmentation results of sweeping algorithm of CV model. The last column: Final segmentation results of our method. Image size: 200×200, 105×108.

Table 1. Comparison of the computational cost of LCV model, CV model, the sweeping algorithm in [7] and the proposed algorithm in Fig. 2.

	LCV [4]	CV [2]	Algorithm of CV in [7]	Ours
The first image	6.804117 s	1.130172 s	0.189986 s	**0.156583 s**
The third image	3.998121 s	0.640465 s	0.127530 s	**0.106957 s**

Table 2. Comparison of the computational cost of LCV model, CV model, the sweeping algorithm in [7] and the proposed algorithm in Fig. 3

	LCV [4]	CV [2]	Algorithm of CV in [7]	Ours
The first image	12.185280 s	7.569005 s	**0.145644 s**	0.182315 s
The third image	38.175802 s	3.410595 s	0.199542 s	**0.154394 s**

In the third experiment, we give some real image segmentation results of the above four models. The iteration times for LCV and CV model are set to 400. We can see that the LCV and CV model cannot obtain right segment results. A perfect segmentation result is got by the proposed algorithm with a fast convergence speed. Just as shown in Table 2, for the first image in Fig. 3, the time of sweeping algorithm of CV model is smaller but can't get satisfy image segmentation. The proposed algorithm can give better image segment performance and use almost as the same time as the CV model. In order to see more clearly, we give a detailed comparison. Figure 4(a) and (b) show the segmentation result of the sweeping algorithm of CV model [7] and its zoom image in a local window. Figure 4(c) and (d) are corresponding results of our algorithm. It can be seen that our algorithm gives the slightly better result.

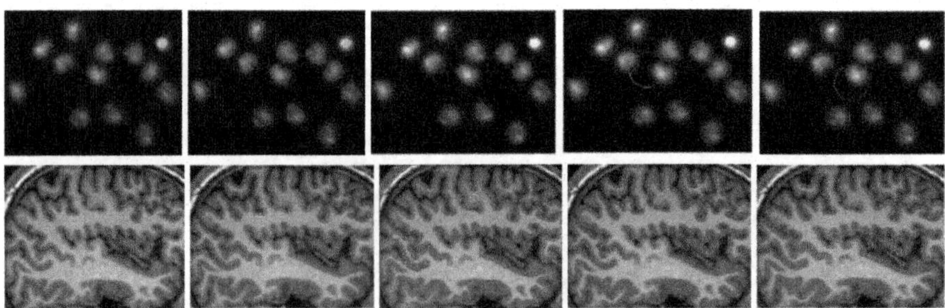

Fig. 3. Segmentation comparisons of LCV model, CV model, sweeping algorithm of CV model and the proposed algorithm. The first column: initial images. The second column: Final segmentation results of LCV model. The third column: Final segmentation results of CV model. The fourth column: Final segmentation results of sweeping algorithm of CV model. The last column: Final segmentation results of our method. Image size: 217×161, 600×479.

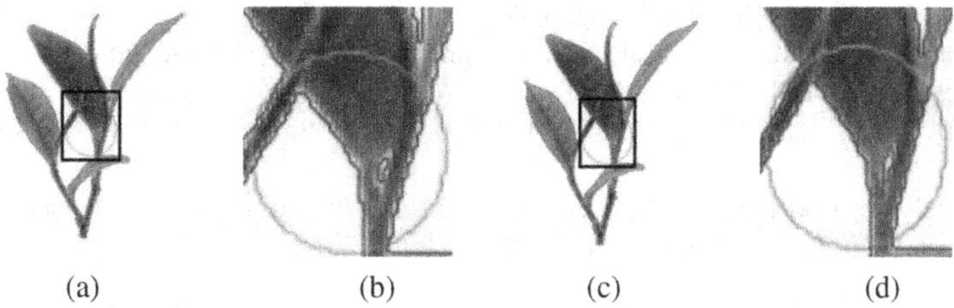

Fig. 4. Comparison of the sweeping algorithm in [7] and the proposed algorithm. (a) Segmentation results of [7]. (b) A zoom window of (a). (c) Segmentation results of the proposed algorithm. (d) A zoom window of the (c).

4 Conclusion

In this paper, we present a sweeping algorithm for LCV model which avoids solving PDE. Besides, the distance regularization term is not included so as to speed up the energy minimization. It should be noted that the sweeping algorithm of LCV model still has some shortcomings. For example, it can't segment image with strong noise, texture structure and serious intensity inhomogeneity. In future works, we will combine the proposed algorithm with some new methods and try to incorporate high-order image features to improve the performance for complex images.

Acknowledgements. The authors would like to express their thanks to Dr. Y. Boutiche, the author of reference [8], for discussing the algorithm of sweeping principle of CV model. This work was supported by the grant of the National Natural Science Foundation of China, No. 61672204, the Project of National Science and Technology Support Plan of China, No. 2015BAD18B05, the grant of Major Science and Technology Project of Anhui Province, No. 17030901026, the grant of the key Scientific Research Foundation of Education Department of Anhui Province, Nos. KJ2018A0555, KJ2016A603, KJ2017A152, KJ2017A542, the grant of Key Constructive Discipline Project of Hefei University, No. 2016xk05, Excellent Talents Training Funded Project of Universities of Anhui Province, No. gxfx2017099.

References

1. Malladi, R., Sethian, J.A., Vemuri, B.C.: Shape modeling with front propagation: a level set approach. IEEE Trans. Pattern Anal. Mach. Intell. **17**(2), 158–175 (1995)
2. Chan, T.F., Vese, L.A.: Active contours without edges. IEEE Trans. Image Process. **10**(2), 266–277 (2001)
3. Wang, X.F., Min, H., Zou, L., Zhang, Y.G.: A novel level set method for image segmentation by incorporating local statistical analysis and global similarity measurement. Pattern Recognit. **48**(1), 189–204 (2015)
4. Wang, X.F., Huang, D.S., Xu, H.: An efficient local Chan-Vese model for image segmentation. Pattern Recognit. **43**(3), 603–618 (2010)
5. Li, C.M., Kao, C.Y., Gore, J.C., Ding, Z.H.: Minimization of region-scalable fitting energy for image segmentation. IEEE Trans. Image Process. **17**, 1940–1949 (2008)

6. Wang, X.F., Min, H., Zou, L., Zhang, Y.G., Tang, Y.Y., Philip Chen, C.L.: An efficient level set method based on multi-scale image segmentation and Hermite differential operator. Neurocomputing **188**, 90–101 (2016)
7. Song, B., Chan, T.: A fast algorithm for level set based optimization. CAM-UCLA **68**, 02–68 (2002)
8. Boutiche, Y., Abdesselam, A.: Fast algorithm for hybrid region-based active contours optimization. IET Image Process **11**(3), 200–209 (2017)

Multi-modal Plant Leaf Recognition Based on Centroid-Contour Distance and Local Discriminant Canonical Correlation Analysis

Shanwen Zhang, Zhen Wang(✉), and Yun Shi

Department of Information Engineering, Xijing University, Xi'an 710123, China
wjdw716@163.com

Abstract. Leaf based plant species recognition plays an important research, but it is a challenging work because of the complexity and diversity of plant leaves. A multi-modal plant leaf recognition method is proposed based on centroid-contour distance (CCD) and local discriminant canonical correlation analysis (LDCCA). First, the CCD feature vector is extracted from each leaf image. Second, the extracted feature vectors of any two within-class leaves are integrated by LDCCA. Final, K-nearest neighbor classifier is applied to plant recognition. The experiment results on a public dataset validated the effectiveness of the proposed method.

Keywords: Plant recognition · Centroid-contour distance
Local discriminant canonical correlation analysis (LDCCA) · Feature extraction

1 Introduction

Plants are essential resources for nature, people lives, environmental protection and ecological balance. Plant recognition provides valuable information for plant research and development, and has great impact on environmental protection and exploration. Plant species identification is one of the most important research branches of botanical science. Many plant recognition methods and systems have been proposed, and achieve better results. Sabu et al. [1] gave a survey on different leaf recognition methods and classifications. Wäldchen et al. [2] completely reviewed a lot of papers with the aim of a thorough analysis and comparison of primary studies on computer vision approaches for plant species identification. Hu et al. [3] proposed a contour-based shape descriptor for plant recognition, called the multiscale distance matrix, to capture the shape geometry while being invariant to translation, rotation, scaling, and bilateral symmetry. The descriptor is further combined with a dimensionality reduction to improve its discriminative power. The proposed method avoids the time-consuming pointwise matching encountered in most of the previously used shape recognition algorithms. Mouine et al. [4] proposed an original method based on the leaf observation for plant species recognition, by considering two sources of information: the leaf margin and the leaf salient points. Du et al. [5] extracted leaf image feature and computed the distance between two manifolds modeled by leaf images, and applied a clustering procedure in order to express a manifold by a collection of local linear models, then measured the

D.-S. Huang et al. (Eds.): ICIC 2018, LNCS 10955, pp. 61–66, 2018.
https://doi.org/10.1007/978-3-319-95933-7_8

distance between local models which come from different manifolds that constructed above. Jyotismita et al. [6] proposed a methodology of characterizing and recognizing plant leaves using a combination of texture and shape features. In the method, leaf texture is modeled using Gabor filter and gray level co-occurrence matrix (GLCM) while leaf shape is captured using a set of curvelet transforms coefficients together with invariant moments. Zhang et al. [7] proposed a wavelet fractal feature based plant leaf image recognition approach. In the method, the preprocessed leaf images are pyramid decomposed with 5/3 lifting wavelet transform and sub images are obtained, and then fractal dimensions of each sub images are calculated to be the wavelet fractal feature of leaf images, finally back propagation artificial neural network is used to classify plant leaf images. Zeng et al. [8] proposed a shape descriptor, namely Periodic Wavelet Descriptor (PWD) for plant shape, and recognized plant species by a Back Propagation Neural Network (BPNN). Leaf shape is one of important visual feature of an image and used to describe plant species content. The Centroid contour distance (CCD) is formed by measuring distance between center and boundary of object [9–12]. The above methods are effective for plant recognition, but they are single-modal and their recognition rates are lower because plant leaves are complex and various. Local discriminant canonical correlation analysis (LDCCA) can integrate the different features from multimodal images [13]. Based on CCD and LDCCA, a plant recognition method is proposed and implemented on a leaf image dataset.

2 Related Works

In this section, we simply introduce centroid-contour distance and Local discriminant CCA (LDCCA).

2.1 Centroid-Contour Distance

Given a leaf image G, its p contour points $\{(x_1, y_1), (x_2, y_2), \ldots, (x_p, y_p)\}$ can be detected by the Harris edge detection operator, (x_i, y_i) is the ith edge point coordinate. CCD is defined as follows,

$$D = \{d_i | d_i = \sqrt{(x_i - x_0)^2 + (y_i - y_0)^2}\} \tag{1}$$

where (x_0, y_0) is the contour center point coordinate, $x_0 = \frac{1}{p}\sum_{i=1}^{p} x_i, y_0 = \frac{1}{p}\sum_{i=1}^{p} y_i$.

It is easy to reason that CCDV is related to the start point of contour and is sensitive to noise contour points. In order to extract the robust classifying features and reduce the feature dimensionality, we conduct discrete Fourier transform for D by,

$$F(k) = \frac{1}{K}\sum_{i=0}^{K-1} d_i e^{-\frac{j2\pi ik}{K}}, \quad k = 0, 1, \ldots, K-1 \tag{2}$$

Then the normalized Fourier descriptor can be calculated by,

$$F_V = \left\{ \frac{|F(2)|}{|F(1)|}, \frac{|F(3)|}{|F(1)|}, \ldots, \frac{|F(K-1)|}{|F(1)|} \right\} \tag{3}$$

It is generally not necessary to represent leaf shape with very high precision in practical applications, so the value of K is not necessary to be too big. Experimental test indicated that 32 or 64 of K can be enough to express the detail of leaf shape accurately. The value of K is set to 32 in the following experiments.

2.2 Local Discriminant CCA (LDCCA)

Suppose two different characteristic sets of n samples: $X = [x_1, x_2, \ldots, x_n] \in R^{p \times n}$ and $Y = [y_1, y_2, \ldots, y_n] \in R^{q \times n}$, CCA is to find two projection matrices $w_x \in R^{p \times d}$ and $w_y \in R^{q \times d}$ to maximize the correlation coefficient between X and Y. The corresponding objective function of LDCCA can be denoted as a linear programming problem as follows,

$$\max_{w_x, w_y} \frac{w_x^T S_{xy} w_y}{\sqrt{w_x^T S_{xx} w_x \cdot w_y^T S_{yy} w_y}} \tag{4}$$

where $S_{xx} = XX^T, S_{yy} = YY^T$ are the covariance of X and Y, respectively, $S_{xy} = S_w - \lambda S_b$ is the cross covariance of X and Y, λ is a adjustment factor to make a trade-off between S_w and S_b, S_w and S_b are local within-class covariance matrix and local between-class covariance matrix, defined respectively as follows,

$$\begin{aligned} S_w &= \sum_{i=1}^{n} \sum_{x_j \in N^w(x_i) \cap y_j \in N^w(y_i)}^{n} (x_i y_j^T + x_j y_i^T) \\ S_b &= \sum_{i=1}^{n} \sum_{x_j \in N^b(x_i) \cap y_j \in N^b(y_i)}^{n} (x_i y_j^T + x_j y_i^T) \end{aligned} \tag{5}$$

where $N^w(x_i)$ and $N^b(x_i)$ are k nearest neighborhoods of the within-class and between-class of xi.

Equation (5) can be solved by Lagrangian multiplier method [13], and obtain a generalized eigenvalue decomposition problem as follows,

$$\begin{pmatrix} & S_{xy} \\ S_{xy} & \end{pmatrix} \begin{pmatrix} w_x \\ w_y \end{pmatrix} = \lambda \begin{pmatrix} S_{xx} & \\ & S_{yy} \end{pmatrix} \begin{pmatrix} w_x \\ w_y \end{pmatrix} \tag{6}$$

Solving Eq. (6), we obtain d eigenvectors corresponding to the largest d $(d \leqslant n)$ eigenvalues, which form the projection matrices w_x and w_y corresponding the former d generalized eigenvalues, then, any two-view feature vector (x, y) can be projected into a low dimensional fusion feature vector $\begin{pmatrix} W_x^T x \\ W_y^T y \end{pmatrix}$.

3 Plant Recognition Method

Combining CCD and LDCCA to recognize plant species, as follows,

Suppose n training within-class leaf image pairs $(g_{1i}, g_{2i})_{i=1}^{n}$ and a test within-class leaf image pair (x, y).

Step 1. Extract CCD from each leaf image, respectively, and reduce the extracted features by PCA, and denote x_{1i} and x_{2i} are CCD features of g_{1i} and g_{2i}, respectively, and express $X = [x_{11}, x_{12}, \ldots, x_{1n}], Y = [x_{21}, x_{22}, \ldots, x_{2n}]$.
Step 2. Calculate $S_{xx} = XX^T, S_{yy} = YY^T$, S_w and S_b by Eq. (5).
Step 3. Perform SVD decomposition on Eq. (6).
Step 4. Select d eigenvectors corresponding to the largest d eigenvalues, and form the projection matrices W_x and W_y.
Step 5. Project each (x_i, y_i) and (x, y) into a low dimensional fusion feature vector $\begin{pmatrix} W_x^T x_i \\ W_y^T y_i \end{pmatrix}$ and $\begin{pmatrix} W_x^T x \\ W_y^T y \end{pmatrix}$.
Step 6. Utilize $\begin{pmatrix} W_x^T x_i \\ W_y^T y_i \end{pmatrix}_{i=1}^{n}$ to train K-nearest neighbor classifier and classify (x, y) by the classifier.

4 Experiments and Analysis

The proposed method is validated on ICL dataset and compared with three existing methods: Leaf image recognition based on wavelet and fractal dimension (WFD) [6], Leaf Recognition Based on Leaf Tip and Leaf Base Using Centroid Contour Gradient (CCG) [10], Leaf Shape Recognition using Centroid Contour Distance (CCD) [11]. The dataset has over 36,000 plant leaf images of 360 species. Some leaf examples are shown in Fig. 1.

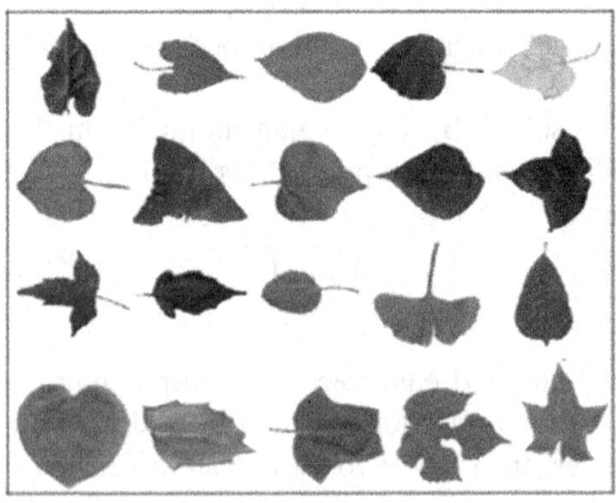

Fig. 1. 20 kinds of plant leaf

We randomly select a subset, including 1000 images from 20 kinds of plants, 50 leaf images from each kind of plant, and then combine 500 within-class leaf image pairs to conduct plant recognition experiments by 10-fold cross validation approach. Figure 2 is 10 leaf pairs. The image preprocessing processes are displayed in Fig. 3. All experiments are conducted using MATLAB implementations to solve the eigenvalue decomposition problem on the 1.8 GHz computer with 2 GB RAM. In the experiments, we experimentally set the parameters $\lambda = 0.4$, $d = 28$. We repeat these experiments 50 times and compute the average results. The average recognition and standard deviations in comparison with others are shown in Table 1.

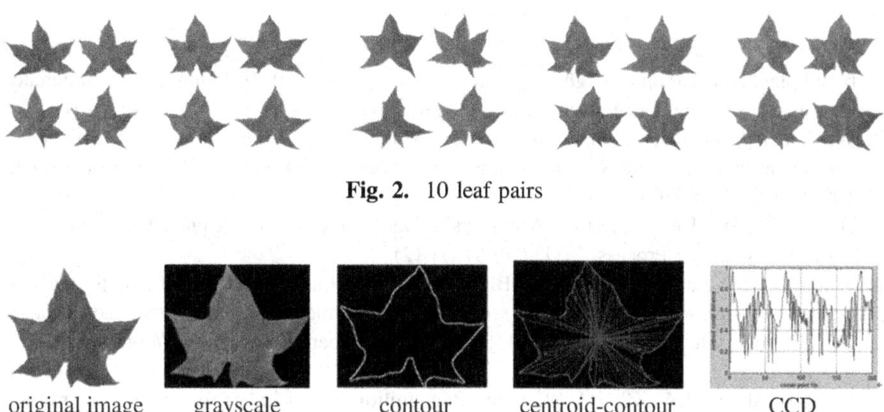

Fig. 2. 10 leaf pairs

Fig. 3. Leaf image processing

Table 1. Average identification results of WFD, CCG, CCD and our method.

Method	WFD	CCG	CCD	Ours
Results	0.8541 ± 0.135	0.8617 ± 0.182	0.8934 ± 0.154	0.9125 ± 0.132

From Table 1, it is seen that the proposed method outperforms other methods. The reason may be the method is two-modal view plant recognition method, which considers the correlation between two within-class leaves when extract the projection matrices.

5 Conclusions

In the paper, we proposed a plant recognition method based on CCD and LDCCA and implemented on ICL dataset. We extracted CCD from each leaf image and construct the objection function and then obtain two projection matrices to integrate two CCDs of leaf image pairs. Finally, we recognize plant species by 1-nearest neighbor classifier. The experiment results confirmed that the recognition rate of the proposed method was better than that of the existing leaf recognition methods. In future work, we will improve the proposed method and further improve its recognition performance. In addition, we will perform investigation in order to find a correct leaf contour extraction method in the complex backgrounds.

Acknowledgments. This work is supported by the China National Natural Science Foundation under grant Nos. 61473237, key research and development projects (2017ZDXM-NY-088), Key project (2016GY-141) of Shaanxi Department of Science and Technology. The authors would like to thank all the editors and anonymous reviewers for their constructive advice.

References

1. Sabu, A., Sreekumar, K.: Literature review of image features and classifiers used in leaf based plant recognition through image analysis approach. In: IEEE International Conference on Inventive Communication and Computational Technologies, pp. 145–149 (2017)
2. Wäldchen, J., Mäder, P.: Plant species identification using computer vision techniques: a systematic literature review. Arch. Comput. Methods Eng. 1–37 (2017). https://doi.org/10.1007/s11831-016-9206
3. Hu, R., Jia, W., Ling, H., et al.: Multiscale distance matrix for fast plant leaf recognition. IEEE Trans. Image Process. **21**(11), 4667 (2012)
4. Mouine, S., Yahiaoui, I., Verroust-Blondet, A.: Combining leaf salient points and leaf contour descriptions for plant species recognition. In: Kamel, M., Campilho, A. (eds.) ICIAR 2013. LNCS, vol. 7950, pp. 205–214. Springer, Heidelberg (2013). https://doi.org/10.1007/978-3-642-39094-4_24
5. Du, J.X., Shao, M.W., Zhai, C.M., et al.: Recognition of leaf image set based on manifold–manifold distance. Neurocomputing **188**, 131–138 (2014)
6. Jyotismita, C., Ranjan, P., Samar, B.: Plant leaf recognition using texture and shape features with neural classifiers. Pattern Recogn. Lett. **58**(1), 61–68 (2015)
7. Zhang, H., Tao, X.: Leaf image recognition based on wavelet and fractal dimension. J. Comput. Inf. Syst. **11**(1), 141–148 (2015)
8. Zeng, Q., Zhu, T., Zhuang, X., et al.: Using the periodic wavelet descriptor of plant leaf to identify plant species. Multimed. Tools Appl. **76**(17), 1–18 (2017)
9. Kohei, A., Cahya, R.: Content based image retrieval by using multi layer centroid contour distance. Int. J. Adv. Res. Artif. Intell. **2**(3), 16–20 (2013)
10. Fern, B.M., Sulong, G.B., Rahim, M.S.M.: Leaf recognition based on leaf tip and leaf base using centroid contour gradient. J. Comput. Theor. Nanosci. **20**(1), 209–212 (2014)
11. Hasim, A., Herdiyeni, Y., Douady, S.: Leaf shape recognition using centroid contour distance. IOP Conf. Ser. Earth Environ. Sci. **31**, (2016). https://doi.org/10.1088/1755-1315/31/1/012002. 012002
12. Khmag, A., Al-Haddad, S.A.R., Kamarudin, N.: Recognition system for leaf images based on its leaf contour and centroid. In: Student Conference on Research and Development, pp. 467–472. IEEE (2017)
13. Huang, X.Y., Zhang, B., Qiao, H., et al.: Local discriminant canonical correlation analysis for supervised PolSAR image classification. IEEE Geosci. Remote Sens. Lett. **14**(11), 2102–2106 (2017)

Fine-Grained Recognition of Vegetable Images Based on Multi-scale Convolution Neural Network

Xiu-Hong Yang, Ji-Xiang Du(✉), Hong-Bo Zhang, and Wen-Tao Fan

Department of Computer Science and Technology, Huaqiao University, Xiamen 361021, China
{yxh, jxdu}@hqu.edu.cn

Abstract. In recent years, deep learning has been widely used in various computer vision tasks. Because the task of solving vegetable pictures are different in the local critical areas, and solving the classification of vegetable categories to meet the needs of users has become an urgent problem to be solved. In this paper, we propose fine-grained image recognition based on Vegetable Dataset, and uses multi-scale iteration to extract critical area characteristics, which the learning at each scale consists of a classification subnetwork and the critical area. In addition, the multi-scale neural network is optimized by two loss functions, to learn accurate critical area and fine-grained feature. Finally, we further prove its scalability and effectiveness in comparing different datasets and different training methods, we get satisfactory results on Vegetable Dataset.

Keywords: Deep learning · Multi-scale · Classification and ranking Critical area

1 Introduction

Image classification is a classic research topic in the field of computer vision. The traditional image classification mainly deal with semantic level and instance level, and the fine-grained image classification is located between the two. Compared with object classification tasks (such as face recognition), the intra-class differences of fine-grained images are more significant. There are many uncertain factors such as pose, illumination, occlusion, background interference and so on.

Fine-grained image classification has extensive research needs and applications in both industry and academic. Relevant research topics include the identification of different species of birds [1], dogs [2], flowers [3], cars [4], aircraft [5], etc. Take the bird dataset as an example. The differences between these terns are very subtle and it is difficult to find critical information. Even for bird experts, it is not easy to fully identify these different species of birds. In recent years, deep learning has made great breakthroughs in various computer vision tasks. This paper mainly use the multi-scale neural network to classify the fine-grained image of vegetables.

D.-S. Huang et al. (Eds.): ICIC 2018, LNCS 10955, pp. 67–76, 2018.
https://doi.org/10.1007/978-3-319-95933-7_9

2 Relative Work

The difficulties in fine-grained image classification mainly include two aspects: (1) finding the key areas that contain the distinguishing information; (2) extracting the fine-grained characteristics of the local region. On the one hand, we should be able to accuracy locate the key areas in the image, and on the other hand, we should be able to extract valid information from key area.

2.1 Detection of Critical Areas

From the point of data, the research of fine-grained image classification has entered from strong supervision to the way of weak supervision. In early studies, the classification algorithm for fine-grained images relied on strong supervised learning. Its work mainly focuses on the use of bounding boxes and partial annotations to locate important regions in fine-grained recognition tasks [6–11]. In terms of detecting images, R-CNN can work very well; in the key point detection, it uses the pre-trained DPM (Deformable Part Model) [12, 13] algorithm to complete the key point detection, which can give a predefined the coordinates of the location point, and whether the point is visible and other information, and use these key points for posture alignment operations. However, the manual annotation of key areas seriously restricts the practicability of these algorithms. Therefore, more and more algorithms tend to no longer rely on manually annotated information in key areas, hoping to use only category labels and perform fine-grained classification tasks in a weakly supervised manner [14, 15]. Research on fine-grained image classification based on weak supervision depends only on the category label to complete classification task. Two Level Attention [16] algorithm proposed in this direction has achieved good classification effect, but it is still all the candidate region using convolution neural network of Selective Search [17] were selected, which need the large amount of calculation. Based on the idea of local area image representation, Zhang et al. [14] proposed an algorithm that can select local features with resolving power from convolution features. Compared with the traditional algorithm, it reduces the amount of computation which required to produce local areas, but still need to use the Selective Search [17] to generate the object region candidate for the input image.

2.2 Multi-scale Feature Extraction

The extraction of features is also a key factor to determine the accuracy of the image classification [18–20]. Traditional classification algorithms based on artificial features often face great limitations. With the success of deep features, which have significantly improved compared with hand-made features [21–25]. In order to learn more stronger feature representation, deep residual network [26] scales up CNN to 152 layers by optimizing residual functions, which reduces the error rate to 3.75% on ImageNet test set [24]. In order to better model the subtle differences in fine-grained categories, a bilinear structure [16] has been recently proposed to calculate the difference between two independent CNN's pairwise features to capture the local, which has achieved the state-of-the-art results in bird classification [9]. Besides, another method [27] proposes to unify CNN with spatially weighted representation by Fisher Vector [28], which shows superior results on both CUB-200-2011 [1] and Standford Dogs [2].

3 Proposed Approach

In this section, we will mainly introduce a Multi-scale convolution neural network for fine-grained image recognition. We consider the network with three scales, the stacking is performed in a similar fashion at a finer scale. The input is a complete image, which iteratively extracting critical areas twice by fine-grained differentiation.

Figure 1 shows the framework of Multi-scale convolution neural network. Input a complete coarse-grained image, and detect the key areas through the network to achieve the fine-grained image classification results. The single network structure achieves two tasks of classification and critical areas detection, and there are alternatively optimized by classification losses L_{cls} and pairwise ranking losses L_{rank}.

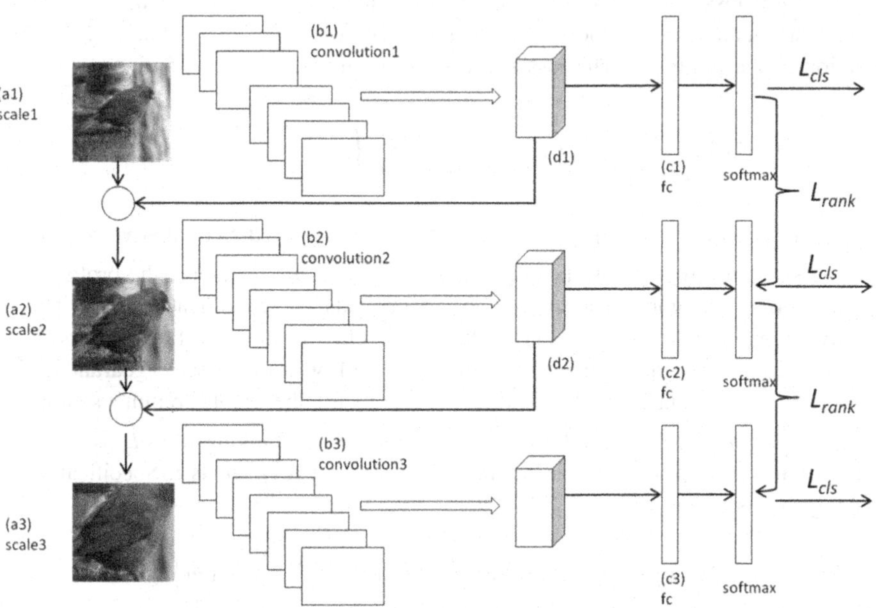

Fig. 1. The framework of multi-scale convolution neural network.

3.1 Focus on Location Critical Areas

The traditional fine-grained image recognition framework for subtle differences does not use neural network depth training to complete image localization and recognition. Inspired by the success of recent Region Proposal Network [29] and Two Level Attention Network, a method focused on extracting nuanced critical area was proposed, and the extraction and calculation of this critical area had almost no cost and was able to complete trained end-to-end.

Given an input image X, we first extract region-based depth features are denoted $W_c * X$, where * denotes a set of operations for convolution, pooling and activation, and W_c denotes the overall parameters. Because the single network has two tasks to accomplish classification and critical area detection, two outputs can be obtained. The first task is to generate a probability distribution p of fine-grained classes, as follows:

$$p(X) = f(W_c * X), \tag{1}$$

The second task is to predict the critical areas with a classification feature in a set of images. The critical areas that are detected to amplified to the fixed scale, which input to second networks to extract more finer features through higher resolution, and to local the key classification features.

3.2 Loss Function

The proposed Multi-scale convolution neural network is optimized by intra-scale classification loss and inter-scale ranking loss, for alternatively generating accurate critical area and learning more finer features. In particular, we minimized the loss function of the multitask. The loss function of the image sample is defined as:

$$L(X) = \sum_{s=1}^{2} \left\{ L_{cls}\left(Y^{(s)}, Y^{*}\right) \right\} + \sum_{s=1}^{2} \left\{ L_{rank}\left(p_t^{(s)}, p_t^{(s+1)}\right) \right\}, \tag{2}$$

where s represents each scale, $Y^{(s)}$ and Y^{*} of denote the predicted label vector from a specific size and ground truth label prediction, $p_t^{(s)}$ and $p_t^{(s+1)}$ denote the probabilities on the correct category, respectively. L_{cls} represents the classification loss. This is mainly the optimization of Fig. 1 convolution and classification level parameters (*b*1 to *b*3, *c*1 to *c*3), which has enough discriminating ability in every scale guarantee. The training is carried out through a softmax function for the whole training sample by fitting a category label. In addition, the $p_t^{(s)}$ from pairwise rankings loss L_{rank} indicates that the prediction probability is lost in the correct category labels *t*. Specifically, the ranking loss is given by:

$$L_{rank}(p_t^{(s)}, p_t^{(s+1)}) = \max\left\{0, p_t^{(s)} - p_t^{(s+1)} + margin\right\}, \tag{3}$$

which enforce $p_t^{(s+1)} > p_t^{(s)} + margin$ in training. Such a design can make the network from the coarse scale prediction as a reference, and gradually carry out a more detailed scale network to produce higher accuracy prediction near the most discriminating area.

3.3 Multi-scale Feature Fusion

The Multi-Scale Network has been trained on three scales, we can get a multi-scale concern from full-size images, from coarse to fine areas. In particular, the image X can be represented by a set of F_i, and F_i represents the feature descriptor on a specific scale generated by the complete connection layer (c1 to c3 in Fig. 1) in a classified network.

The feature of different network layers is extracted from the scale images obtained by the critical area detection network, and the feature fusion of the convolution layer with better effect is obtained, and the accuracy is improved.

4 Experimental Results

In this section, we will experiment with our own set of vegetable dataset and compare them with the experimental results obtained with different datasets and different network methods.

4.1 Datasets

We conduct experiments on three challenging fine-grained image recognition datasets, including Caltech-UCSD Birds(CUB-200-2011) [1], Standford Dogs [2] and Standford Cars [4]. Simultaneously, we conduct experiments on self-created dataset. The detailed statistics with category number and data splits are summarized in Table 1.

Table 1. The statistics of fine-grained datasets used in this paper.

Datasets	Category	Training set	Testing set
CUB-200-2011	200	5994	5794
Standford Dogs	120	12000	8580
Standford Cars	196	8144	8580
Vegetable	170	96900	10860

4.2 Comparison with the State of the Art

Multi-scale Localization: In Fig. 2, we can observe that these localizing regions on second and third scales are discriminatory to the corresponding categories, and are more easily classified than the first scale. The result is consistent with the human opinion, and it is helpful to carefully observe the fine-grained classification.

We put the results of the second-scale network (denoted as the Multi-scale convolution neural network (scale 2)), where the presence of the region at this scale can preserve both global bird structure and local visual cues, as shown in Fig. 2 and shown the results in Table 2. At the same time, the key areas of birds are located as shown in Fig. 3.

Fine-Grained Image Recognition: We use the open source deep learning framework to build the network and train it in our experiment. In order to verify the efficiency of the algorithm, use the following weak supervision method to compare with ours. All the methods are listed as follows:

- **DVAN** [30]**:** diverse attention network attends object from coarse to fine by multiple region proposals.
- **FCAN** [31]**:** fully convolutional attention network adaptively selects multiple task-driven visual attention by reinforcement learning.
- **B-CNN** [32]**:** bilinear-CNN proposes to capture pairwise feature interactions for classification.

Fig. 2. Five bird examples of the learned critical areas at different scales.

Table 2. Comparison of critical area location in the classification accuracy of CUB-200-2011 dataset.

Approach	Accuracy
FCAN	76.1
Multi-scale convolution neural network (scale 2)	82.4

And we use VGG-19 (pre-trained on ImageNet) for bird and car datasets, and VGG-16 for dogs. We can observe comparable results with the method in Table 3. In the dataset of CUB-200-2011, FCAN and DVAN propose similar ideas to zoom into critical area for classification, we can achieve better accuracy with 3.3% and 6.3% relative improvement because of the mutual reinforcement framework for critical area localization and region-based feature learning. The scale 2 in VGG-19 compares with scale 1 can gains 4.6%. Besides, the results of scale 3 slightly drop than scale 2, because of the missing of structural information existed in global bird images. The combination of triple single-scale network with different initial parameters only achieves 78.0%, 83.5%, 82.0% for the first, second and third scale, respectively. Besides, we lose the same important background information when we only focus on the characteristics of the most critical areas, which will have an impact on classification.

The classification accuracy on Standford Dogs dataset are also summarized in Table 3. The VGG-16 at the first scale takes the original images as input and achieves 76.7% recognition accuracy. Relying on the extracted critical regional features, the second scale of network achieves a significantly improved the recognition accuracy of 85.9%, with a relative gain of 9.3%. By combining the characteristics of two scales and three scales, we can increase the performance to 86.7% and 87.3%. Comparing with the two most relevant methods, the accuracy of DVAN and FCAN increased by 5.8% and 3.1%.

Fig. 3. Critical areas at the third scale for birds.

Table 3. Accuracy comparison of different datasets with different methods.

Approach	CUB-200-2011	Standford Dogs	Standford Cars
DVAN	79.0	81.5	87.1
FCAN	82.0	84.2	89.1
B-CNN	84.1		91.3
VGG-16		76.7	
VGG-19	77.8		84.9
Multi-Scale (scale2)	82.4	85.9	90.0
Multi-Scale (scale3)	81.2	85.0	89.2
Multi-Scale (scale1+2)	84.7	86.7	91.8
Multi-Scale (scale1+2+3)	85.3	87.3	92.5

The classification accuracy on Standford Cars are also summarized in Table 3. Although VGG-19 achieves only 84.9% accuracy at scale 1, the performance can increase to 90.0% after distinguishing regions from finer scales. By using the power of the feature set, the highest recognition accuracy of 92.5% is obtained from the integration of the original image, the whole vehicle and the front or rear regions. Compared with the state-of-art methods, the scale 1+2+3 of full model Multi-Scale Network exceeds DVAN and FCAN for large margins (5.4% and 3.4% relative gain) under the same settings. We also get better results than the high-dimensional B-CNN, and even achieve comparable performance with Multi-Scale Network, which depending bounding box defined by humans.

4.3 Experiments on the Dataset of Vegetable

The main purpose of fine-grained image classification is to identify the fine differences between vegetables of the same large category and subcategories, as shown in Fig. 4. The vegetable dataset contains a total of 170 species, with a total of 107760, with an average of 200–1000 pictures in each category. The dataset is collected in the market, and it faces many problems such as light, posture and placement angle, etc. It needs to use fine-grained image recognition method to achieve more accurate recognition results. Based on the experiments done before, we can draw a conclusion that Multi-Scale Network can achieve better recognition effect for fine-grained image recognition tasks. Based on the training method of the Multi-Scale Network, the vegetable data set is trained and tested at the same time. The results are compared with other methods, as shown in Table 4.

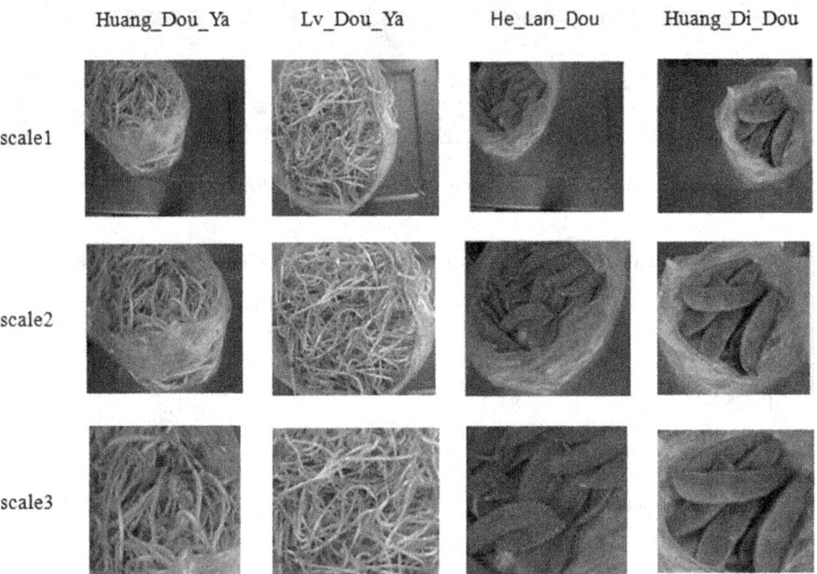

Fig. 4. Four vegetable examples of the learned critical areas at different scales.

Table 4. Accuracy comparison of vegetable dataset in different methods.

Approach	Accuracy
GoogLeNet	90.4
VGG-16	92.1
Multi-Scale (scale2)	90.1
Multi-Scale (scale3)	89.0
Multi-Scale (scale1+2)	91.5
Multi-Scale (scale1+2+3)	93.0

5 Conclusion

In this paper, we use the method of fine-grained image recognition to classify vegetable dataset, and achieve good result, partly because vegetable dataset do not need fine-grained classification to achieve better results. There is no doubt that multi-scale feature iterative extraction can effectively improve the accuracy of recognition of fine-grained images. At the same time, multi-scale iterative extraction, when the number of iterations is more, the less global information in scales, the more information of all iterated integration features, so the accuracy of training is higher.

Acknowledgement. This paper is supported by the Grant of the National Science Foundation of China (No. 61673186, 61502182, 61502183).

References

1. Wah, C., Branson, S., Welinder, P., Perona, P., Belongie, S.: The Caltech-UCSD Birds-200-2011 dataset. Technical report CNS-TR-2011-001, California Institute of Technology, USA (2011)
2. Aditya, K., Nityananda, J., Yao, B., Li, F.F.: Novel dataset for fine-grained image categorization. In: Proceedings of the IEEE Conference on Computer Vision and Pattern Recognition on First Workshop on Fine-Grained Visual Categorization. IEEE, Springs, USA (2011)
3. Nilsback, M.E., Zisserman, A.: Automated flower classification over a large number of classes. In: Proceedings of the 6th Indian Conference on Computer Vision, Graphics and Image Processing, pp. 722–729. IEEE, Bhubaneswar, India (2008)
4. Krause, J., Stark, M., Deng, J., Li, F.F.: 3D object representations for fine-grained categorization. In: Proceedings of the 4th IEEE Workshop on 3D Representation and Recognition. IEEE International Conference on Computer Vision, pp. 554–555. IEEE, Sydney, Australia (2013)
5. Maji, S., Kannala, J., Rahtu, E., Blaschko, M., Vedaldi, A.: Fine-grained visual classification of aircraft, 21 June 2013. https://arxiv.org/abs/1306.5151
6. Huang, S., Xu, Z., Tao, D., Zhang, Y.: Part-stacked CNN for fine-grained visual categorization. In: CVPR, pp. 1173–1182 (2016)
7. Lin, D., Shen, X., Lu, C., Jia, J.: Deep LAC: deep localization, alignment and classification for fine-grained recognition. In: CVPR, pp. 1666–1674 (2015)
8. Parkhi, O.M., Vedaldi, A., Jawajar, C., Zisserman, A.: The truth about cats and dogs. In: ICCV, pp. 1427–1434 (2011)
9. Welinder, P., Branson, S., Mita, T., Wah, C., Schroff, F., Belongie, S., Perona, P.: Caltech-UCSD birds 200. Technical report CNS-TR-2010-001, California Institute of Technology (2010)
10. Zhang, H., Xu, T., Elhoseiny, M., Huang, X., Zhang, S., Elgammal, A., Metaxas, D.: SPDA-CNN: unifying semantic part detection and abstraction for fine-grained recognition. In: CVPR, pp. 1143–1152 (2016)
11. Zhang, N., Donahue, J., Girshick, R.B., Darrell, T.: Part-based R-CNNs for fine-grained category detection. In: ECCV, pp. 1173–1182 (2014)
12. Branson, S., Van Horn, G., Belongie, S., Perona, P.: Bird species categorization using pose normalized deep convolutional nets, 11 June 2014. https://arxiv.org/abs/1406.2952

13. Branson, S., Beijbom, O., Belongie, S.: Efficient large-scale structured learning. In: Proceedings of the IEEE Conference on Computer Vision and Pattern Recognition, pp. 1806−1813. IEEE, Portland, USA (2013)
14. Zhang, Y., Wei, X.S., Wu, J., Cai, J., Lu, J., Nguyen, V.A., Do, M.N.: Weakly supervised fine-grained categorization with part-based image representation. IEEE Trans. Image Process. **25**(4), 1713–1725 (2016)
15. Lin, T.Y., RoyChowdhury, A., Maji, S.: Bilinear CNN models for fine-grained visual recognition. In: Proceedings of the 15th IEEE International Conference on Computer Vision, pp. 1449−1457. IEEE, Santiago, Chile (2015)
16. Xiao, T., Xu, Y., Yang, K., Zhang, J., Peng, Y., Zhang, Z.: The application of two-level attention models in deep convolutional neural network for fine-grained image classification. In: Proceedings of the IEEE Conference on Computer Vision and Pattern Recognition, pp. 842−850. IEEE, Boston, USA (2015)
17. Uijlings, J.R., van de Sande, K.E., Gevers, T., Smeulders, A.W.: Selective search for object recognition. Int. J. Comput. Vision **104**(2), 154–171 (2013)
18. Lin-Bo, Z., Chun-Heng, W., Bai-Hua, X., Yun-Xue, S.: Image representation using bag-of-phrases. Acta Automatica Sinica **38**(1), 46–54 (2012)
19. Wang-Sheng, Yu., Xiao-Hua, T., Zhi-Qiang, H.: A new image feature descriptor based on region edge statistical. Chin. J. Comput. **37**(6), 1298–1410 (2014)
20. Xue-Jun, Y., Chun-Xia, Z., Xia, Y.: 2DPCA-SIFT: an efficient local feature descriptor. Acta Automatica Sinica **40**(4), 675–682 (2014)
21. Fu, J., Mei, T., Yang, K., Lu, H., Rui, Y.: Tagging personal photos with transfer deep learning. In: WWW, pp. 344–354 (2015)
22. Fu, J., Wang, J., Rui, Y., Wang, X.-J., Mei, T., Lu, H.: Image tag refinement with view-dependent concept representations. IEEE T-CSVT **25**(28), 1409–1422 (2015)
23. Fu, J., Wu, Y., Mei, T., Wang, J., Lu, H., Rui, Y.: Relaxing from vocabulary: robust weakly-supervised deep learning for vocabulary-free image tagging. In: ICCV (2015)
24. Krizhevsky, A., Sutskever, I., Hinton, G.E.: Imagenet classification with deep convolutional neural networks. In: NIPS, pp. 1106–1114 (2012)
25. Wang, J., Fu, J., Mei, T., Xu, Y.: Beyond object recognition: visual sentiment analysis with deep coupled adjective and noun neural networks. In: IJCAI (2016)
26. He, K., Zhang, X., Ren, S., Sun, J.: Deep residual learning for image recognition. In: CVPR, pp. 770–778 (2016)
27. Zhang, X., Xiong, H., Zhou, W., Lin, W., Tian, Q.: Picking deep filter responses for fine-grained image recognition. In: CVPR, pp. 1134–1142 (2016)
28. Perronnin, F., Larlus, D.: Fisher vectors meet neural networks: a hybrid classification architecture. In: CVPR, pp. 3743–3752 (2015)
29. Girshick, R.B.: Fast R-CNN. In: ICCV, pp. 1440–1448 (2015)
30. Zhao, B., Wu, X., Feng, J., Peng, Q., Yan, S.: Diversified visual attention networks for fine-grained object classification. CoRR, abs/1606.08572 (2016)
31. Liu, X., Xia, T., Wang, J., Lin, Y.: Fully convolutional attention localization networks: efficient attention localization for fine-grained recognition. CoRR, abs/1603.06765 (2016)
32. Lin, T.-Y., RoyChowdhury, A., Maji, S.: Bilinear CNN models for fine-grained visual recognition. In: ICCV, pp. 1449–1457 (2015)

Feature Analysis and Risk Assessment of Android Group Based on Clustering

Zhijie Xiao[1(✉)], Tao Li[1,2(✉)], and Yuqiao Wang[1(✉)]

[1] College of Computer Science and Technology, Wuhan University of Science and Technology, Wuhan 430065, Hubei, China
544247884@qq.com, litaowust@163.com, leowon212@gmail.com

[2] Hubei Province Key Laboratory of Intelligent Information Processing and Real-Time Industrial System, Wuhan 430065, Hubei, China

Abstract. The security risk assessment of Android applications is an uncertain problem. It is difficult to determine whether the application authority is reasonable from a single application point of view, and whether some privacy rights overstep the functional scope of the application itself. To solve this problem, a group based method for feature analysis and risk assessment of large-scale Android applications is proposed. The permission that app applies is an important object for security analysis and evaluation. The same type of application has similar functions, so the required system privileges are similar. By comparing the application class of the same functional type to the population, the two layers model of group feature analysis and mass population clustering is used to evaluate the relative malicious program in the population, which shows the effectiveness and adaptability of the method.

Keywords: Android permission · Clustering · Assessment · Feature analysis

1 Introduction

The report shows [1] that in 2016 mobile Internet malware mainly aimed at the Android platform, with a total of 2053450, accounting for more than 99.9%, ranking first. The Google Bouncer [2] can detect malicious programs, but this detection is short of immediacy, and malicious applications have been heavily downloaded before they are detected [3]. The third party application market is more open and free, and at the same time, there is a lack of timely detection, resulting in the emergence of malicious software. These malicious applications will steal user privacy, consume resources maliciously, and maliciously deduct fees, which seriously violate user interests [4]. With the development of mobile payment in full swing, Android malicious applications are bound to increase rapidly, and the security of user privacy data is urgently needed. This paper proposes a method of large-scale feature analysis and risk assessment of Android application based on Clustering. Android applications of multiple functional types are collected from the application market for experiment. Statistical analysis of permissions and K-means algorithm in data mining are used for risk assessment of applications of different functional types.

D.-S. Huang et al. (Eds.): ICIC 2018, LNCS 10955, pp. 77–84, 2018.
https://doi.org/10.1007/978-3-319-95933-7_10

2 Related Work

The current Android permission mechanism does not achieve good results in data security protection [5]. Permission management follows the "minimum privilege principle" [6]. The application wants to access other files, data and resources, which must be declared in the AndroidManifest.xml file, which makes the Android system access to be an important factor of security evaluation.

Barrera [7] analyzes the application of the 22 kinds of applications in the Android application market and the first 50 of each class. It finds that most applications only apply for permission in a small set. Felt [8] and others have found that applications in the market use only a few parts of the entire set of permissions. These research work shows that the privilege mechanism of Android operation system can reflect the ability of application to access specific resources, which shows that the Android privilege mechanism can provide users with security reference for installing applications. But on the other hand, research shows that only 17% of users will be aware of the related permissions warning when installing the application software, while only 2.6% of users can correctly understand the permission specification provided by [9]. This shows that the lack of permission usage instructions can lead to the application list of Android applications that are presented to users at installation time, which can not effectively help users make security decisions, and weaken the design purpose of Android privilege mechanism. Users may still agree with applications to grant some unnecessary privileges and sensitive permissions. Therefore, it is not enough to rely solely on the Android system authority mechanism to provide protection for user privacy and data security.

The traditional application of static detection of [10] and dynamic detection of [11] is mainly from the perspective of individual to determine the safety of an application individual, and it is difficult to meet the needs of large-scale user detection. Document [12] has proposed an evaluation method for the effectiveness of privacy protection mechanism in Android system, monitoring the path that may be exploited by malicious programs, but it is not very suitable for dealing with large numbers of applications at the same time. The author [13] proposed a large-scale Android application similarity detection method, but this method can only identify the application of known malicious code. AndroidProtect [14] uses static analysis to mine massive application eigenvalues, and adjusts evaluation accuracy by dynamic target program behavior monitoring. The paper [15] proposes a similarity calculation method, which combines European distance to evaluate the dangerous trend of Android application. By calculating the distance between each application threshold and the minimum permissions set as a security threshold, it indicates the dangerous trend of application. However, the minimum privilege set does not exist in every type of application. The application similarity calculation based on Euclidean distance can not reflect the difference of the applied group's characteristics adaptively. And it is difficult to provide an intuitive risk assessment result.

3 Clustering-Based Risk Assessment System Framework

3.1 Safety Assessment System

This paper presents a system which based on the application of population and have access to the system permissions to make group features analysis and risk assessment, can automatically evaluate the unknown application, extraction application permission list and give the security reference for users to download applications. The framework and process of the system are shown in Fig. 1.

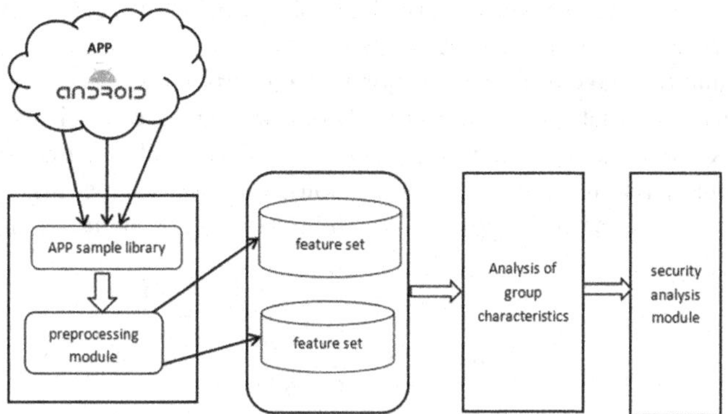

Fig. 1. System process and framework

The application with similar function is analogous to a population, through the authority feature preprocessing module, forming a set of privileges for each application. All obtained data are stored in the cloud to form a app sample library. Statistical method is used to make statistics for the privilege feature set, calculating the risk index of each permissions, and get the threat value of each app in the population, analysis the lightweight privilege feature for the group. The clustering operation module takes advantage of the K-means algorithm in data mining to cluster the extracted application permission feature set data, and clustering the Android application of the same population into multiple clusters. In the security analysis module, the population risk assessment is carried out by the data processing results of the group feature analysis module. The corresponding permissions for each group application are obtained according to the application category, and stored in the local and cloud database.

3.2 Method of Population Risk Assessment Based on Clustering

Definition 1. Permissions of APP: Permissions = $\{pi|pi \in$ permissions of Android$\}$. It presents the permissions of APP is all from Android.

The application of the same functional type as a population is divided into x classes. The category of population will change according to the total number of App.

Definition 2. Category label: Class $= \{C1, C2, C3\ldots Cx\}$. Class represents a set of different population. Cx is a category label for each population. Such as flashlights, cameras, players, social chat and other categories.

Definition 3. Population $=$ (Cx, Author, BackInfor, PermissionMatrix). Cx is a category label for each population. Author is the author's information of application. BackInfor represents the functional description of the application. PermissionMatrix represents the permissions information matrix for each application permissions. I'll use these terms throughout the rest of this article, based on the following definitions:

Definition 4. Permission matrix: PermissionMatrix $= \{\text{Pij} \mid \text{i} = 1, 2, \ldots, \text{m}; \text{j} = 1, 2, \ldots, \text{n}\}$. i represents the App in the population Cx numbered i, and if Appi has the privilege j, then Pij will have a value, otherwise Pij is 0.
Different populations have different functional characteristics, so the same permissions or permissions combinations have different security threat coefficients in different populations. A risk index PValue is set for each permissions. The higher risk index is, the higher security risk represents the permissions. For a certain permissions Pi in the population Cx, we define the privilege probability of a single permissions as:

Definition 5

$$P(Pi, Cx) = \frac{Pi\_Num}{Cx\_Sum}$$

The total number of permissions Pi appears in the population Cx is Pi_Num, and the total number of application in Cx is Cx_Sum. After giving the application permissions risk index and the single permissions probability, the application threat value definition is further given.

Definition 6. The Privilege vector of APPi is $(\text{p1}, \text{p2}, \ldots, \text{pn})$, The rights risk index vector of the APPi in population Cx is

PWightx $=$ (PValue1, PValue2, ..., PValuen). The method of calculating the threat value of Appi is

$$PSum_i = \sum_j P_{ij} PWight_x$$

In document [15], the author calculates the minimum permissions set of the population, that is, the set of permissions that all the applications of the population have to meet the basic function of the population. Then the minimum permissions set is used as the base threshold, and the Euclidean distance in the similarity calculation method is used to divide the dangerous trend of the application in the population. However, because of the diversity of rights and the existence of the public class library of Android, the minimum permissions set can't be found for some species.

Therefore, this paper uses the K-means algorithm in the similarity calculation method and the static statistics method to apply the population characteristics analysis and risk assessment.

The application of the same population {PSum1, PSum2, PSum3..., PSumn} as the input vector of the clustering algorithm, the result will be divided into K clusters.

$$\mu^{(0)} = \mu_1^{(0)}, \mu_2^{(0)} \ldots, \mu_k^{(0)} \tag{1}$$

$$\mathrm{C}^{(t)}(j) \leftarrow \arg\min_i \left\|\mu_i - PSum_j\right\|^2 \tag{2}$$

$$\mu_{\mathrm{i}}^{(t+1)} \leftarrow \arg\min_{\mu} \sum_{j:C(j)=i} \left\|\mu_i - PSum_j\right\|^2 \tag{3}$$

The formula (1), formula (2) and formula (3) introduce the clustering algorithm flow. The algorithm first initializes the K clustering centers $\mu^{(0)}$, and then calculates the distance from the center to each data $\mathrm{PSum_j}$ in the dataset, and updates the center point continuously until the center point does not change.

Finally, find out some of the nearest samples from each center to form a cluster. Clustering algorithm divides the application of population into K clusters, each application has its own threat value.

The distance measurement method adopted by the objective function is Euclidean distance method. The larger the distance is, the greater the difference between individuals is applied. Finally, all applications will converge to K cluster. After that, we can get the population vector P = (Cx,Cluster1,Cluster2..., ClusterK) for the population Cx.

Clusteri represents the K cluster of the population. For each population, the weights of K clusters and the size of cluster centers are counted. Combined with Euclidean distance, a Level value is set for each cluster, which is expressed as Level = (Cx, L1, L2...LK).

The value of the L1 to the LK level is increased in turn, and the security risk of the corresponding rank is also increased in turn. That is, the L1 level app threat range is the least, the corresponding security level is the highest, and the LK level application is the most risky.

4 Experiment and Analysis

The experimental operating environment is: Windows 10 operating system, 3.2 GHz four core processor, 8 GB memory. This paper chooses 360 application market as the data source and a crawler program is written in the python language. At present, 62 types of Android applications have been crawled in the 360 application market. Based on the design idea of the experiment, two kinds of commonly used app, camera and flashlight, are selected as the main research objects. Among them, 496 are flashlights and 485 are cameras.

For the following reasons, we chose two groups of cameras and flashlights as experimental objects. Firstly, the two types of applications have clear functional boundaries. For a app, it is easier to distinguish whether it belongs to a flashlight or camera category from its main permission statement and the application description that is filled in when it is uploaded. Secondly, flashlights and cameras are widely used in our daily life. Almost every user will install a flashlight or a camera for personal needs. If a feature rich and easy to use app is added to malicious code by illegal elements and uploaded again after shell processing, which will affect a large number of users (Fig. 2).

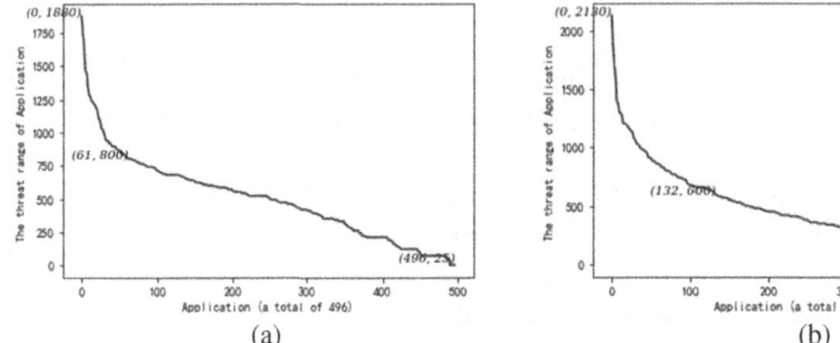

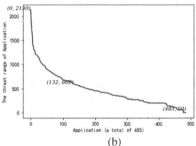

Fig. 2. The threat value distribution map of flashlight (a) and camera (b).

In the flashlight group, 87.72% of the app is concentrated between the range of 0 and 800 of the threat. Because the functional characteristics of this group are relatively simple, the right to apply for it is also relative to a set, which reflects the weight will be more concentrated, so the remaining 12.28% applications are likely to be malicious. Camera type application 3/4 is centrally distributed within the range of 0 to 600.

The probability of the same permissions is different for different populations. If the probability of a permission is high in a population, it shows that the permissions are the common permissions of the population, and the threat value will be relatively low. On the contrary, if the probability of permissions is low, then its threat will be multiplied. Since the minimum permissions set may be empty, so the permission probability became the important standard of measuring the threat value of authority (Fig. 3).

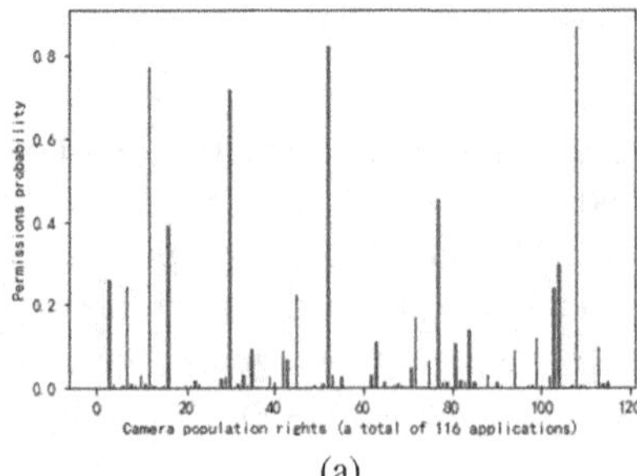

(a)

Permissions	probability
WRITE_EXTERNAL_STORAGE	0.87
INTERNET	0.82
ACCESS_NETWORK_STATE	0.77
CAMERA	0.72
WRITE_EXTERNAL_STORAGE	0.50
READ_PHONE_STATE	0.39

(b)

Fig. 3. (a) and (b) are the probability distribution of the privilege and the high frequency authority of the camera population.

In the experiment, two kinds of Android applications of flashlights and cameras are clustered with K-means algorithm. After analyzing the scale of data and the range of weight distribution, we tried to compare the results of clustering by multiple classification attempts, and finally clustered data into 4 clusters, that is, corresponding 4 security levels. The higher the level, the risk coefficient is high, the possibility of privacy will be higher, in the application of high grade, the probability of occurrence of malicious software is also bigger.

According to the definition, we calculate the permissions probability of the camera and flashlight population respectively, and calculate the threat value according to the single permissions risk index of each group. Finally, we use the clustering algorithm to evaluate the risk of cameras and flashlights.

As can be seen from Table 1, most of the applications are distributed in areas that are more secure or have a certain threat. In the flashlight population, 54 applications of the L4 level have the authority to accept and send SMS (Table 2).

Table 1. The result of camera population risk assessment.

Threat value	Level	Safety advice	Application number
[0, 260]	L1	Recommended installation	150
[270, 555]	L2	Allow installation	190
[565, 1060]	L3	Careful installation	117
[1075, 2130]	L4	Do not install	28

Table 2. The result of flashlight population risk assessment.

Threat value	Level	Safety advice	Application number
[0, 370]	L1	Recommended installation	179
[375, 560]	L2	Allow installation	143
[570, 825]	L3	Careful installation	120
[835, 1880]	L4	Do not install	54

Experiments show that there are certain differences in population distribution among different groups. Clustering method can well adapt to the distribution characteristics of each group, and calculate the range of threat applied in the population, so as to provide security risk assessment for different functional types. In the actual scenario, when a user needs to download an application, it can refer to the proposed installation recommendations to make a reasonable choice.

5 Conclusion

The experiments carried out permissions analysis and risk assessment for multiple population applications. The result shows that the clustering based method can provide risk assessment for different types of applications, and it has a scalability. The results and conclusions laid a good theoretical and data foundation for future design of

Android malware identifier for different function types. Next, we will consider permissions combination, code block similarity, monitoring application runtime and so on, and build an Android application malware recognizer based on population detection.

Acknowledgement. Authors are partially supported by Major projects of the Hubei Provincial Education Department (No. 17ZD014) and Hubei college students' innovation and entrepreneurship training program project (No. 201610488020).

References

1. The development of the China Mobile Internet and its security report (2017) [EB/OL]. http://www.isc.org.cn/zxzx/xhdt/listinfo-35398.html. Accessed 08 Mar 2018
2. Google Play [EB/OL], 14 December 2017. https://play.google.com/store
3. Peng, H., Gates, C., Sarma, B., et al.: Using probabilistic generative models for ranking risks of Android apps. In: ACM Conference on Computer and Communications Security, pp. 241–252. ACM (2012)
4. Zhou, Y., Jiang, X.: Dissecting Android malware: characterization and evolution. In: IEEE Symposium on Security and Privacy, pp. 95–109. IEEE (2012)
5. Jiawei, Z.H.U., Liangwen, Y.U., Zhi, G.U.A.N., et al.: A review of the security research of Android authority mechanism. Appl. Res. Comput. **32**(10), 2881–2885 (2015)
6. Sihan, Q.: Progress in research on Android security. J. Softw. **27**(1), 45–71 (2016)
7. Barrera, D., Oorschot, P.C.V., Somayaji, A.: A methodology for empirical analysis of permission-based security models and its application to Android. In: ACM Conference on Computer and Communications Security, pp. 73–84. ACM (2010)
8. Felt, A.P., Greenwood, K., Wagner, D.: The effectiveness of install-time permission systems for third-party applications (2010)
9. Felt, A.P., Ha, E., Egelman, S., et al.: Android permissions: user attention, comprehension, and behavior. In: Proceedings of the Eighth Symposium on Usable Privacy and Security, pp. 1–14. ACM (2012)
10. Burguera, I., Zurutuza, U., Nadjm-Tehrani, S.: Crowdroid: behavior-based malware detection system for Android. In: Proceedings of the 1st ACM Workshop on Security and Privacy in Smartphones and Mobile Devices, pp. 15–26. ACM, New York (2011)
11. Schmidt, A.D., Bye, R., Schmidt, H.G., et al.: Static analysis of executable for collaborative malware detection on Android. In: Proceedings of the 2009 IEEE International Conference on Communications, Piscataway, pp. 631–635. IEEE Press (2009)
12. Shuke, Z., Yang, Z., Liang, C., et al.: A Android system for privacy protection effectiveness evaluation method. J. Univ. Sci. Technol. China **44**(10), 853–861 (2014)
13. Sibei, J.I.A.O., Lingyun, Y.I.N.G., Yi, Y.A.N.G., et al.: An anti obfuscation method of similarity detection for large scale Android applications. J. Comput. Res. Dev. **51**(7), 1446–1457 (2014)
14. Zhang, T., Li, T., Wang, H., Xiao, Z.: AndroidProtect: Android apps security analysis system. In: Wang, S., Zhou, A. (eds.) CollaborateCom 2016. LNICST, vol. 201, pp. 583–594. Springer, Cham (2017). https://doi.org/10.1007/978-3-319-59288-6_58
15. Wang, H., Li, T., Zhang, T., Wang, J.: Android apps security evaluation system in the cloud. In: Guo, S., Liao, X., Liu, F., Zhu, Y. (eds.) CollaborateCom 2015. LNICST, vol. 163, pp. 151–160. Springer, Cham (2016). https://doi.org/10.1007/978-3-319-28910-6_14

Find the 'Lost' Cursor: A Comparative Experiment of Visually Enhanced Cursor Techniques

Chuanyi Liu(✉) and Rui Zhao

School of Information Science & Engineering,
Lanzhou University, Lanzhou, China
liuchuanyi96@hotmail.com, 18293508302@163.com

Abstract. We proposed eight visually enhanced cursor techniques (VECTs) to help users locate a cursor more easily and quickly. A thorough comparative experiment was conducted to evaluate the proposed and other existing prominent VECTs. The experimental results showed that the usefulness and performance of these techniques were dependent on their applied backgrounds and their cursor shapes. All the VECTs performed faster than the baseline in a common or a complex background. And the cursor-shape deforming technique totally outperformed all the others concerning mean task completion time in any kind of background.

Keywords: Lost cursor · Visually enhanced cursor techniques
Cursor shape

1 Introduction

Cursor is an essential widget in Graphic User Interfaces (GUIs). But it is not uncommon for us losing tracking the cursor in an interface. It often occurs after we pause the manipulation of the cursor for a moment, especially when the cursor is tiny. For example, the cursor is a small dot for a pen tool when we show slides with PowerPoint, and it is rather difficult to catch sight of the cursor and keep tracking it.

Most of the existing literature concerning 'lost' cursors only aim at some special population, e.g., the people with low vision [3–5, 7]; however, the problem exists universally no matter for those special population or for normal people. Furthermore, most of the literature proposed one or two more conspicuous effects to the cursor, e.g., appending red rings [4] or a flashing star [1] to the cursor; there are few studies to thoroughly explore possible techniques for finding the 'lost' cursor under different conditions. But we believe that such a technique is related to more than one factor, e.g., the background and the shape of a pointer.

In this paper, we conducted a thorough experiment to compare fourteen visually enhanced cursor techniques (VECTs) together with the default cursor (the baseline technique). These fourteen VECTs included all existing prominent techniques and eight novel ones (i.e., flashing, diffusing rings, appending halos, diffusing halos, shrinking halos, appending arrows, rotating arrows, and sparkling arrows). All the

D.-S. Huang et al. (Eds.): ICIC 2018, LNCS 10955, pp. 85–92, 2018.
https://doi.org/10.1007/978-3-319-95933-7_11

fifteen techniques had been compared under various conditions and measured by objective and subjective evaluations. The experimental results showed that VECT, background, and cursor shape all had significant effects on the subjects' performance regarding speed and accuracy, and subjective comments. Besides the aforementioned, an algorithm to activate the VECTs was proposed. The algorithm activates a VECT when necessary with minimum distractions. We believe that our work has valuable implications for VECT design.

2 Related Work

VECTs include techniques appending more conspicuous effects directly to a cursor (direct VECTs, D-VECTs) and outside a cursor (indirect VECTs, I-VECTs). Almost all the effects outside a cursor are fixed to it except "ColorEyes" [6], which are two red eyes keeping stationary in the window and looking at the mouse pointer, through eyes' cue we can faster locate the mouse pointer. But the separation between the mouse pointer and the red eyes distracts our attention from the work location.

D-VECTs highlight a cursor by adjusting its color [3, 7], size [7], or shape [3, 7]; while I-VECTs typically by appending some certain shapes (e.g., red rings [4], a ring [2], or a flashing star [1]) around it.

There are few researches to compare different VECTs except the work [7] of Wang et al. But they only compared five techniques (i.e., color adjustment, cursor enlargement, appearance change, appending an extra ring, and the default cursor) using a subjective questionnaire.

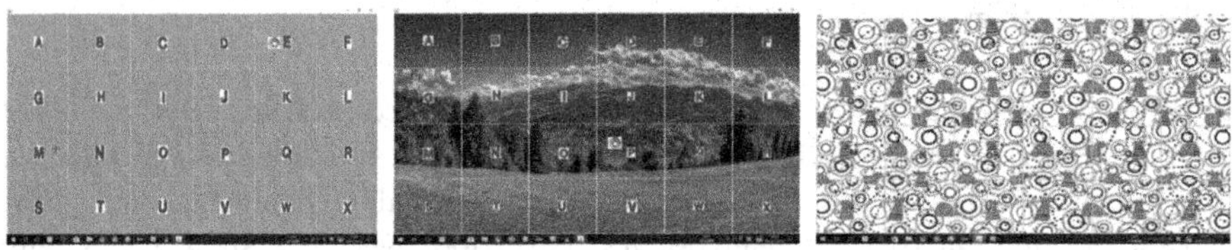

Fig. 1. Three types of backgrounds: simple (left), common (middle), and complex (right).

Fig. 2. Five types of cursor shapes, from left to right: E-Resize, arrow, I-Beam, CrossHair, and Dot.

Fig. 3. Baseline.

Fig. 4. Trailing.

2.1 Pilot Study

We need a time threshold to determine the longest time of each trail so as to conduct the formal experiment in a controlled way. A preliminary experiment was done to get the threshold. Nine participants (5 females) took part in the experiment. The experimental interface was displayed in a full screen mode. The screen was divided into 24 equal sections (see Fig. 1), each section was labeled with a capital letter. The cursor with one of the 5 different shapes (Fig. 2) was used to identify the target section, which randomly changed among the 24 sections. The participants clicked a key in the keyboard to select the target, if the letter on the key was the same as the target section's label, the trial was conducted correctly; otherwise, wrong. Each participant was requested to complete three blocks of trials. In a block, there were 3 types of backgrounds (Fig. 1) and 5 types of cursor shapes (see Fig. 2). Under each experimental condition, there were 12 target sections. Totally, the experiment includes 9 (participants) × 3 (blocks) × 3 (backgrounds) × 5 (cursor shapes) × 12 (selections) = 4860 selections. The performance time of each trial was recorded. After the experiment, we eliminated outliers and those of incorrect trials from the experimental data, and then sorted the left in ascending order. The number was chosen at 98% point of the data from each participant, then we got 9 numbers. The mean of the 9 numbers was utilized as the time threshold (T_m).

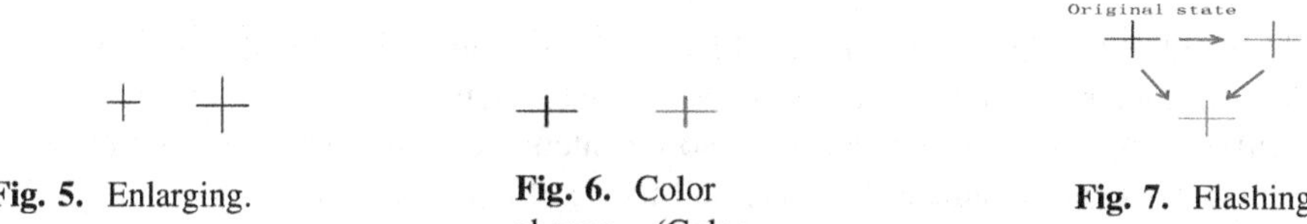

Fig. 5. Enlarging.

Fig. 6. Color change. (Color figure online)

Fig. 7. Flashing.

3 Experimental Method

3.1 Technique Design

There are 15 techniques: 14 VECTs together with the baseline (the default cursor). We introduce all these techniques taking the shape CrossHair as an example.

- Baseline: a normal cursor without any visual enhanced effect (see Fig. 3).
- Trailing: a series of cursors display in the motion trajectory of the original (Fig. 4).
- Enlarging: the cursor becomes larger (Fig. 5), it is different from an enlarged cursor of MS Windows for the enlarging is dynamic (activated automatically when necessary) not static as in MS Windows.
- Color change: the color of the cursor is changed from black to red (Fig. 6).
- Flashing: three colors alternate in a cycle with an interval of 200 ms (Fig. 7).

Fig. 8. Deforming.

- Deforming: shape of the cursor changes from CrossHair to hand (Fig. 8).
- Appending rings: a series of rings (whose centers are all at the crossing point of the CrossHair cursor) all appear simultaneously (Fig. 9 left).
- Diffusing rings: a series of rings (whose centers are all at the crossing point of the cursor) appear dynamically from small to big (Fig. 9 middle).
- Shrinking rings: a series of rings (whose centers are all at the crossing point of the cursor) appear dynamically from big to small (Fig. 9 right).

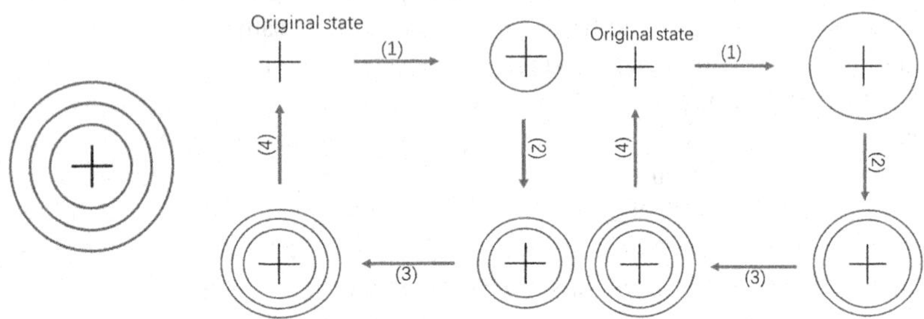

Fig. 9. A cursor with three kinds of appending rings: static (left), diffusing (middle), and shrinking (right) rings.

- Appending halos: a series of static halos (whose centers are all at the crossing point of the cursor) all appear simultaneously (Fig. 10 left).
- Diffusing halos: a series of halos (whose centers are all at the crossing point of the cursor) appear dynamically from small to big (Fig. 10 middle).
- Shrinking halos: a series of halos whose centers are all at the crossing point of the cursor appear dynamically from big to small (Fig. 10 right).

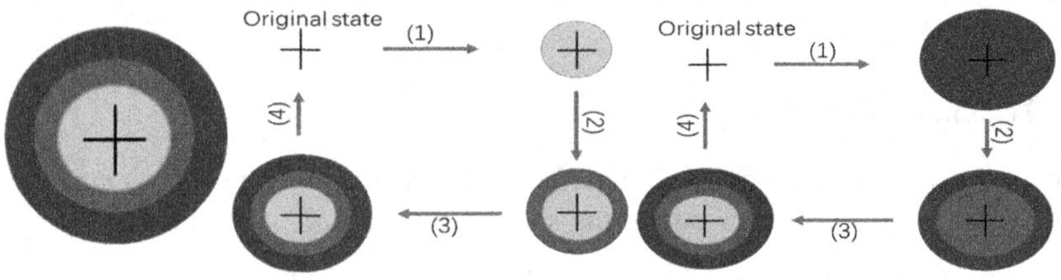

Fig. 10. A cursor with three kinds of appending halos: static (left), diffusing (middle), and shrinking (right) halos.

- Appending arrows: arrows simultaneously display in four directions (east, south, west, and north), all of them point to the crossing point of the cursor (Fig. 11 left).
- Rotating arrows: arrows dynamically and clockwise display in four directions (north, east, south, and west) with an interval of 200 ms (Fig. 11 middle).
- Sparkling arrows: the cursor alternates between its original state and state of appending arrows with an interval of 200 ms (Fig. 11 right).

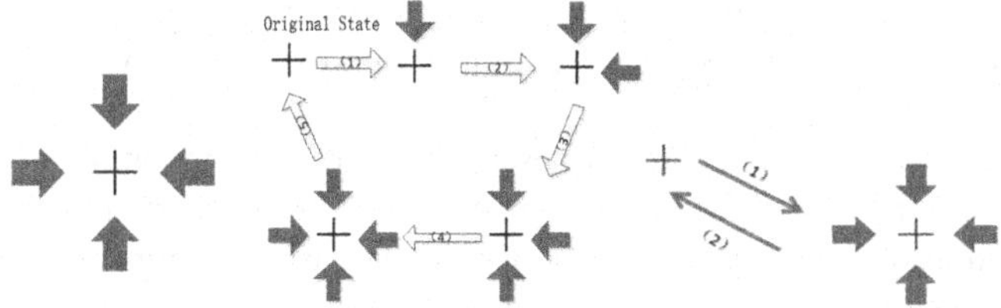

Fig. 11. A cursor with three kinds of appending arrows: static (left), rotating (middle), and sparkling (right) arrows.

3.2 Task and Stimuli

The task is the same as that of the pilot study, except that a VECT is activated (except the baseline) automatically when the given condition is met. A cursor with a VECT has two states, i.e., original state (without any visually enhanced effect, State0, Fig. 12) and visually enhanced state (with the given visually enhanced effect, State1, Fig. 12). After the mouse pointer keeps stationary for a time over the given threshold, T_O, and the motion distance of the cursor, S, is less than the given threshold, S_O, then the activation condition is met; and then the cursor changes from State0 to State1, i.e., a certain visually enhanced effect displays. If the mouse pointer stops moving or its motion distance exceeds S_O, the cursor changes from State1 to State0 and keeps the state. The activation algorithm ensures a given visually enhanced effect appears only when necessary and can minimize visual distractions of the enhanced effect.

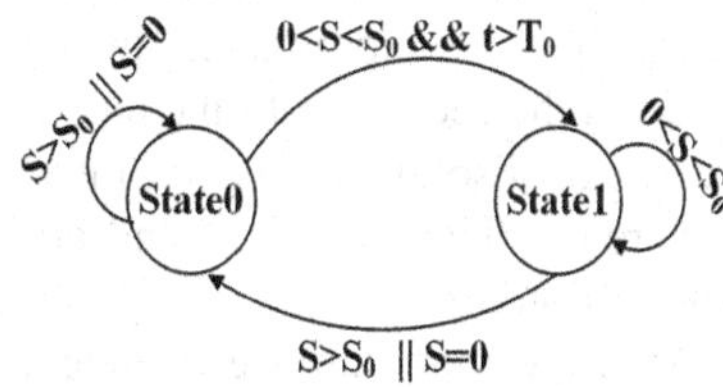

Fig. 12. State-transition diagram of a cursor with a VECT.

3.3 Subjects and Apparatus

Fifteen volunteers (8 females, aged between 23 and 32) participated in the study. All of them were experienced computer users, used a computer at least every other day for no less than one year. None of them had ever used the eight aforementioned novel VECTs.

The experiment was conducted on a PC with AMD Athlon II X2. 20 GHz processor running MS Windows 10. 1. A 19 in. monitor with a screen resolution of 1440 × 900 pixels, a HID-compliant optical mouse, and a HID keyboard were used in the experiment.

3.4 Design and Procedure

The experiment used a full-factor within-subjects design. The experimental independent variables were *Technique* (with 15 levels), *Block* (with 3 levels), *Background* (with 3 levels, the same as those of the pilot study), and *Cursor Shape* (with 5 levels, the same as those of the pilot study). The experimental dependent variables were task time and accuracy. There were two types of task errors: locating error (occurs when a section without the cursor was selected) and timeout error (occurs when the cursor had not been found within T_m).

The subjects were given a detailed description and explanation about the features and operation of each technique. They were conducted to have ten minutes' warm-up practice to familiarize themselves with the operation. During the formal experiment, the subjects were asked to complete the experimental tasks as quickly and accurately as possible. The experiment was conducted in three backgrounds and with five cursor shapes. Under each condition, subjects were required to complete eleven selections. The subjects performed a total of 15 (subjects) × 3(blocks) × 3 (backgrounds) × 5 (cursor shapes) × 15 (techniques) × 11 (selections) = 111375 selections. A Latin square was used to counterbalance the order efforts of the fifteen techniques. The background and cursor shape changed alternately between subjects.

During the experiment, the cursor with different shapes and techniques pseudo-randomly appeared in one of the 24 sections, a VECT was activated automatically if its condition was met. The subjects were allowed to have a rest between blocks. A trial began when the experimental interface was displayed and finished after a subject had identified the section with the cursor or a trial performance time exceeded T_m. A black screen was displayed between interfaces of any two sequent trials to eliminate visual interference. The task time, and error type and number were recorded by the experimental program.

After the quantitative experiment, the subjects were investigated for their subjective comments on the fifteen techniques. The questionnaire consisted of the ease of use, hand and eye fatigue and subjective preference (each factor has 7 rating levels, where 1 is the worst and 7 is the best).

4 Results

We conduct a 15(Technique) × 3(Block) × 5(Cursor Shape) × 3(Background) RM-ANOVA on task time and error rates.

4.1 Time

Mauchly's test indicated that the assumption of sphericity had been violated for the main effects of the factor of *Technique* ($\chi2(104) = 207.40$, $p < .001$), *Cursor shape* (($\chi2(9) = 31.81$, $p < .001$), and *Background* ($\chi2(2) = 8.62$, $p < .05$). Therefore, degrees of freedom were corrected using Greenhouse–Geisser ($\varepsilon = .30$ for *Technique*, .53 for *Cursor Shape*, and .67 for *Background*).

Technique

There was a significant effect of *Technique* ($F(4.15, 58.10) = 149.61$, $p < .001$) on mean task time. The ANOVA results showed that Deforming was the fastest technique, it was significantly faster than the default cursor ($F(1, 14) = 204.52$, $p < .001$). All techniques (except diffusing rings, shrinking rings, and diffusing halos) performed faster than the baseline. It indicates that most of the VECTs had positive effects on finding the 'lost' cursor.

Block

There was also a significant effect of *Block* on mean task time ($F(2, 28) = 11.10$, $p < .001$). This indicates that there was a significant learning effect for the subjects across the three blocks. The ANOVA results showed that mean time in the third block was approximately 7% lower than that in the first block.

Cursor Shape

Cursor shape significantly affected mean task time ($F(2.1, 29.9) = 603.49$, $p < .001$). Tests of within-subjects contrasts showed that differences of mean completion time between an Arrow cursor and an I-Beam cursor ($F(1,14) = 12.75$, $p < .005$), an E-Resize cursor and a CrossHair cursor ($F(1,14) = 22.33$, $p < .001$), a CrossHair cursor and a Dot cursor ($F(1,14) = 852.37$, $p < .001$) were all significant. Pairwise comparisons displayed that mean completion time with a Dot cursor was significantly longer than those with all the other cursors. It took the subjects approximately double time to complete a trial with a Dot cursor than with the other cursor shapes.

Background

Background had a significant main effect on mean completion time ($F(1.35, 18.86) = 36.98$, $p < .001$). Tests of within-subjects contrasts showed that mean completion time of a trial in a common background was significantly different from that in a simple one ($F(1,14) = 34.81$, $p < .001$) and that in a complex one ($F(1,14) = 23.56$, $p < .001$). The analytical results showed that the completion time increased together with the increase of a background's complexity.

4.2 Accuracy

Locating Error Rates

Mauchly's test indicated that the assumption of sphericity had been violated for all interaction effects except the main effect of *Background*; and each epsilon of all the other effects was less than 0.75, thus, Greenhouse–Geisser was used to correct the degrees for all of them.

The results of tests of within-subjects effects presented that only main effects of *Background* ($F(2, 28) = 3.46$, $p < .05$) and *Technique* ($F(1.82, 25.46) = 3.59$, $p < .05$) were significant on cursor locating accuracy. The analytical results showed that locating error rates of all the types of backgrounds were rather low (0.6%, the highest was of a complex background), and locating error rates varied across different levels of *Technique*, but all of them were minute (1%, the highest was of diffusing rings).

Timeout Error Rates

We only analyze two main effects, i.e., the effects of *Cursor shape* and *Technique*. Both *Cursor shape* ($\varepsilon = .25$) and *Technique* ($\varepsilon = .09$) violated the assumption of sphericity, we adjust their degrees by Greenhouse-Geisser.

Cursor shape ($F(1.0, 14.0) = 13.0$, $p < .005$) and *Technique* ($F(1.28, 17.88) = 7.93$, $p < .01$) both had significant effects on mean timeout error rates.

The analytical results showed that the timeout error rates of different cursor shapes were all approximately 0, except that (0.7%) of a Dot cursor; timeout error rates were all approximately 0 of different levels of *Technique* except the *Baseline* (0.6%), *Enlarging* (0.1%), *Color change* (0.6%), *Flashing* (0.1%), and *Trailing* (0.6%).

5 Subjective Comments

Ten out of the fifteen subjects thought that the techniques were not necessary in the simple background but were valuable in the common background or in the complex one. Most subjects believed that these techniques were necessary and valuable when they performed interaction tasks in a complex background or with a dot-shaped cursor.

6 Conclusion

In this paper, eight visually enhanced cursor techniques (VECTS) were proposed. The proposed and other existing VECTs were evaluated through a comparative experiment. The experimental results displayed that the value and the performance of these techniques were related with both the applied backgrounds and the shape of a cursor. Among these techniques, those appending enhanced effects directly to a cursor itself were more suitable to a simple context. All the VECTs outperformed the baseline concerning mean task completion time in either a common or a complex background. And these VECTs were valuable for a tiny-shaped cursor (e.g., a tiny dot-shaped cursor) in any type of background. The error rates of all the techniques in any background were minute. The experimental results indicate that context-aware and shape-related design is more suitable for VECTs.

Acknowledgment. Supported by the Fundamental Research Funds for the Central Universities Grants No. lzujbky-2016-k07.

References

1. Ashdown, M., Oka, K., Sato, Y.: Combining head tracking and mouse input for a GUI on multiple monitors. In: CHI 2005 Extended Abstracts on Human Factors in Computing Systems, pp. 1188–1191. ACM, Portland (2005)
2. Forlines, C., Vogel, D., Balakrishnan, R.: HybridPointing: fluid switching between absolute and relative pointing with a direct input device. In: ACM Symposium on User Interface Software and Technology, pp. 211–220 (2006)
3. Fraser, J., Gutwin, C.: A framework of assistive pointers for low vision users. In: International ACM Conference on Assistive Technologies, pp. 9–16 (2000)
4. Hollinworth, N.: Helping older adults locate 'lost' cursors using FieldMouse. In: International ACM Sigaccess Conference on Computers and Accessibility, pp. 315–316 (2010)
5. Hollinworth, N., Hwang, F.: Cursor relocation techniques to help older adults find 'lost' cursors. In: Proceedings of the 2011 Annual Conference on Human Factors in Computing Systems, pp. 863–866. ACM, Vancouver (2011)
6. Kline, R.L., Glinert, E.P.: Improving GUI accessibility for people with low vision. In: SIGCHI Conference on Human Factors in Computing Systems, pp. 114–121 (1995)
7. Wang, N.-T., Chen, Y.-L., Hsu, Y.-C.: Assistive design of cursors for low vision users. Int. J. Commun. Media Stud. (IJCMS) **6**, 1–6 (2016)

An Enhanced HAL-Based Pseudo Relevance Feedback Model in Clinical Decision Support Retrieval

Min Pan[1], Yue Zhang[2(✉)], Tingting He[2], and Xingpeng Jiang[2]

[1] National Engineering Research Center for E-Learning, Central China Normal University, Wuhan 430079, China
[2] School of Computer Science, Central China Normal University, Wuhan 430079, China
zhangyue.vip@mails.ccnu.edu.cn, tth@mail.ccnu.edu.cn

Abstract. In an actual electronic health record (EHR), patient notes are written with terse language and clinical jargons. However, most Pseudo Relevance Feedback (PRF) technique methods do not take into account the significant degree of candidate term in feedback documents and the co-occurrence relationship between a candidate term and a query term simultaneously. In this paper, we study how to incorporate proximity information into the Rocchio's model, and propose a HAL-based Rocchio's model, called HRoc. A new concept of term proximity feedback weight is introduced to model in the query expansion. Then, we propose three normalization methods to incorporate proximity information. Experimental results on 2016 TREC Clinical Support Medicine collections show that our proposed models are effective and generally superior to the state-of-the-art relevance feedback models.

Keywords: Clinical retrieval · Term proximity · Pseudo Relevance Feedback

1 Introduction and Motivation

The real notes (i.e., hospital records) are given as queries and are extracted by clinicians, which contain a lot of abbreviations and other language styles.[1] It is unreasonable to expect a busy clinician to rewrite his or her note in a manner similar to a case report simply.

PRF is a well-studied query expansion technique by making use of the feedback information [1]. Traditional PRF can still fail in some IR tasks. Some of the feedback documents are irrelevant to the query when the expansion terms generally select according to the candidate term frequency in the feedback documents, or the term distributions in the feedback documents and whole document collection.

Term proximity is an effective measure for term associations, which has been studied extensively in the past few years. The Hyperspace Analogue to Language

[1] http://www.trec-cds.org/.

D.-S. Huang et al. (Eds.): ICIC 2018, LNCS 10955, pp. 93–99, 2018.
https://doi.org/10.1007/978-3-319-95933-7_12

(HAL) [2] is a computational modeling of psychological theory of word meaning by considering context only as the words that immediately surround the given word. Most of these studies focus on the term proximity within the original query and adapt this in ranking documents [3]. However, the traditional term proximity methods ignores the significance of term frequency.

The main contributions of this paper are as follows: we adapt to the Rocchio's model [4] for proximity information, and integrate a term's co-occurrence proximity relationship into Rocchio's model. Introducing three normalization methods for a new concept of proximity-based term weighting. TREC clinical collections demonstrate the effectiveness of our proposed model.

2 Related Work

The Clinical Decision Support (CDS) track designed to complement previous biomedical inspired TREC tasks [5]. The CDS track has been heavily inspired by the TREC genomics [6] and the medical case-based retrieval track of Image-CLEF [7]. The TREC 2014 and 2015 in [8] proposed the use of short case reports, such as those published in biomedical articles, idealized representations of actual medical records. The 2016 CDS track used de-identified notes for actual patients, which focuses on topic modeling for query expansion [9].

PRF method is a common but effective technique for achieving better retrieval performance. For instance, Robertson et al. proposed a well-known automatic PRF algorithm in the Okapi system [10]. Miao et al. proposed a proximity-based feedback framework (called PRoc), which includes three different proximity measures (PRoc1, PRoc2 and PRoc3) to estimate relevance and importance of the candidate terms [11]. Ye and Huang proposed a uniform model (TF-PRF) to capture the local saliency of a candidate term associated in feedback documents [12]. In addition, many other methods have obtained significant performance in improving retrieval effectiveness [13].

In this paper, we propose a HAL-based co-occurrence PRF model, integrating a term's proximity weight information into a traditional PRF models: Rocchio's model. In our method, we estimate the weight of candidate expansion terms by taking their distance from query terms into account.

3 Our Proposed HAL -Based Proximity Weight Model

3.1 A HAL-Based PRF Model

Term proximity is effective to discriminate against these types of documents. HAL involves constructing a high-dimensional vector for each word by considering context simply as the words that immediately surround the given word [14]. We segment the document into a list of sliding windows, each window has a fixed window size *D*.

We adapt the original HAL model similar as in [14]. Then, a term can be represented by a semantic vector, in which each dimension is the weight for this term and other terms as follows:

$$HAL(t,q) = \sum_{l=1}^{|D|} w(l)p(t,l,q) \tag{1}$$

Where l is the distance from query term q to term t, $p(t,l,q)$ is the co-occurrence frequency within the sliding windows when the distance equals l, then $w(l) = D - l + 1$ denotes the strength.

In this paper, we also take into account the distinction factor of different query terms. Then the proximity-based term weighting $W_{HAL}(t,q)$ in the method is computed as follow:

$$W_{HAL}(t,q) = \sum_{i=1}^{|Q|} HAL(t,q_i)IDF(q_i) \tag{2}$$

The weighted HAL model completely includes the information of term distances and co-occurrence frequencies.

Let $Q_0 = \{q_1, q_2, \ldots, q_i\}$ represents the original query given by the user, when the corresponding term co-occurrence weight is integrated into classic Rocchio's model, a new query Q' will be generated by a method HRoc as follows:

$$Q' = (1-\alpha) * Q_0 + \alpha * ((1-\beta) * \sum\nolimits_{r \in R} \frac{r}{|R|} + \beta * \sum\nolimits_{r' \in R} \frac{r'}{|R|}) \tag{3}$$

where α and β are tuning coefficients of 0 to 1. $|R|$ is the number of feedback documents. r and r' are the vector of expansion term weight computed with BM25 and proximity co-occurrence weighting respectively. In order to better computing, the representation of query term weight as follows:

$$Q' = (1-\alpha) * Q_0 + \alpha * ((1-\beta) * \sum\nolimits_{r \in R} \frac{r}{|R|} + \beta * \sum\nolimits_{r' \in R} \frac{W_{HAL}(t,q)}{|R|}) \tag{4}$$

We would normalize term frequency weight, and the normalization methods are presented in detail in Sect. 3.2.

3.2 Normalization Methods

Normalization is a statistic method which transforms result data into number and maps them to the range of 0–1. It is convenient for data processing.

Then we linearly modified the formula as follows:

$$Q_1' = (1-\alpha) * Q_0 + \alpha * ((1-\beta) * Norm(\sum\nolimits_{r \in R} \frac{r}{|R|}) + \beta * Norm(\sum\nolimits_{r' \in R} \frac{W_{HAL}(t,q)}{|R|})) \quad (5)$$

Three different transformation methods In Table 1.

Table 1. The three kinds of normalization method

$Norm(t)$		
$norm_1(t)$	$norm_2(t)$	$norm_3(t)$
$t - \min(t)/(\max(t) - \min(t))$	$t/\sqrt{Sum(t^2)}$	$t/\max(t)$

In this paper, we take method of $norm_1(t)$, $norm_2(t)$ and $norm_3(t)$ to process data respectively, t represents the different weight value in Eq. (5). It also has the effect of reducing the influence of extreme values or outliers in the data without removing them from the data set.

4 Experimental Settings

4.1 Test Collections and Evaluation Metrics

We conduct a series of experiments on the standard 2016 TREC CDS Track collections, which contains articles published by PubMed Central (PMC).[2] We use pmc-00 and pmc-01 as test collections, pmc-02 and pmc-03 as experimental collections respectively. For the collections, the article numbers are 1.25 million, the topic numbers are 30, and each topic contains three fields (title, description and narrative).

The Mean Average Precision (MAP) performance measure for the top 1000 documents is used as evaluation metric. In addition, $P@k$ is used for evaluation so that we can emphasize the top-ranking documents, and we set $k \in \{5, 10\}$. F Value is the harmonic average of the Recall (R) rate and the Precision (P) rate. F Value can be represented $2PR/(P+R)$. Statistically significant evaluation metrics values based on the two-tailed paired p_value are computed at a 95% confidence level.

4.2 Parameter Setting

There are several controlling parameters tuning in our model and our experiments. First, in BM25, we sweep the values of b from 0 to 1.0 with an interval of 0.1, set $k1$, $k3$ to 1.2 and 8. Second, for medicine collection generally does not allow many candidate expansion words in other traditional medicine information retrieval research. We sweep the number of feedback documents and the feedback term from $\{10, 20, \ldots, 50\}$ respectively. Finally, we selected the HAL parameter D from

[2] https://www.ncbi.nlm.nih.gov/pmc/tools/openftlist/.

$\{0, 100, \ldots, 2500\}$, and the interpolation parameter $\alpha, \beta \in \{0.0, 0.1, \ldots, 1.0\}$. To evaluate the baselines and the proposed approaches, we use 2-fold cross-validation, in which the TREC queries are partitioned into two sets by the parity of their numbers on each collection. Then, the parameters learned on the training set are applied to the test set for evaluation purpose.

5 Results and Analysis

5.1 Comparison with the Recent Progress

We compared HRoc model with the strong baseline models (BM25 + Rocchio, RM3) and the state-of-the-art PRF models (PRoc2, TF-PRF) with different evaluation metric in Table 2. In particular, "a, b, c, d" respectively indicate a significant improvement over BM25 + Rocchio, RM3, PRoc2 and TF-PRF (Wilcoxon signed-rank test with $p < 0.05$). The bold phase style in a row means that it is the best result.

Table 2. Summary of comparison with BM25+Rocchio, RM3, PRoc2 and TF-PRF

	BM25+ Rocchio	RM3	PRoc2	TF-PRF	HRoc1	HRoc2	HRoc3
MAP	0.0490	0.0540	0.0600	0.0580	**0.0651**^{abcd} (32.86%, 20.56%, 8.50%, 12.24%)	0.0647^{abcd} (32.04%, 19.8%, 7.83%, 11.55%)	0.0642^{abcd} (31.02%,18.89%, 7.0%, 10.69%)
p@5	0.2733	0.2600	0.2867	0.2600	**0.2933**^{abcd} (7.32%, 12.81%, 2.30%, 12.81%)	0.2733^{bd} (0.0%, 5.12%, −4.67%, 5.12%)	0.2733^{bd} (0.0%, 5.12%, −4.67%, 5.12%)
p@10	0.2533	0.2467	0.2533	0.2467	**0.2733**^{abcd} (7.90%, 10.78%, 7.90%, 10.78%)	0.2667^{abcd} (5.29%, 8.11%, 5.29%, 8.11%)	0.2567^{bd} (1.34%, 4.05%, 1.34%, 4.05%)
F1	0.0853	0.0932	0.1031	0.1012	**0.1112**^{abcd} (30.36%, 19.31%, 7.86%, 9.88%)	0.1108^{abcd} (29.89%, 18.89%, 7.47%, 9.49%)	0.1096^{abcd} (28.5%, 17.6%, 6.30%, 8.30%)

In Table 2, we can clearly see that HRoc1 achieves better retrieval performance than HRoc2 and HRoc3. HRoc1 is markedly superior to PRoc2, and TF-PRF by up to 7.90% improvement in terms of P@10, which is more significant than on MAP.

5.2 Discuss and Analysis

D Value is key tuning parameter in the HRoc model. In our experiment, we set *D* value from 10 to 2500, and use HRoc1, HRoc2 and HRoc3 to measure the best result. According to the result in MAP metrics, we get a Precision trend value as following.

From Fig. 1, it is clear to see that HRoc1 and HRoc2 get optimal Precision value in $D = 1500$, and the two sides of the optimal value show a steady and slow decline trend. Experiments on TREC medicine collections demonstrate the effectiveness of the HRoc model. Furthermore, it is clear that the proposed proximity-based function outperforms the well-known two constraints (PRoc2, TF-PRF). One reason is that positional semantic information is important in clinical records and articles. Another reason is that significant degree of candidate term in feedback documents are not only decided by term frequency, because some clinical jargon only occur one time.

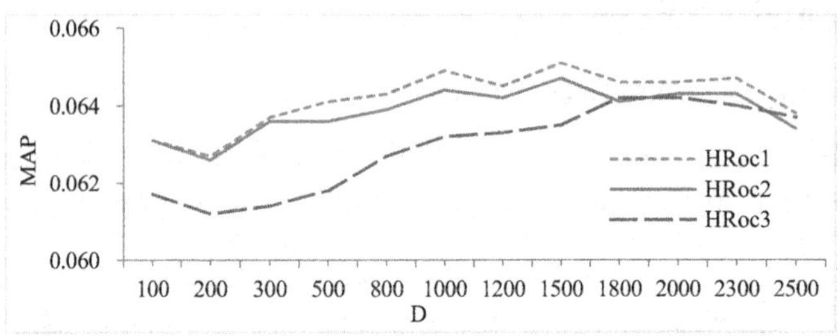

Fig. 1. The performance of HRoc with different D value in MAP

6 Conclusions and Future Work

In this paper, we proposed an enhanced HAL-based Rocchio's method. Then three normalization methods are integrated and proposed HRoc1, HRoc2 and HRoc3. Experiment results on CDS TREC clinical medicine data sets show that the proposed HRoc method is effective. Additionally, we carefully analyze the *D* value of our proposed three HRoc methods and get a trend picture in MAP.

There are several interesting future research directions for us to explore. First, the relationship between *D* value and average document length, we will make experiment on other category collections, and study features that are more suitable. Second, we use fixed window size, and it makes waste of running resource. We consider that window size varies as document length changes.

Acknowledgement. The National Natural Science Foundation of China (61532008), the National Key Research and Development Program of China (2017YFC0909502) support this research.

References

1. Ksentini, N., Tmar, M., Gargouri, F.: The impact of term statistical relationships on Rocchio's model parameters for pseudo relevance feedback. CISIM **8**, 135–144 (2016)
2. Rohde, D.L.T., Gonnerman, L.M., Plaut, D.C.: An improved model of semantic similarity based on lexical co-occurrence. Commun. ACM **8**, 627–633 (2005)
3. Qiao, Y.N., Du, Q., Wan, D.F.: A study on query terms proximity embedding for information retrieval. Int. J. Distrib. Sens. Netw. **13**(2) (2017)
4. Rocchio, J.J.: Relevance feedback in information retrieval. SMART Retr. Syst. Exp. Autom. Doc. Process. 313–323 (2007)

5. Kirk, R., Dina, D.F., Voorhees, E.M., William, R.H.: Overview of the TREC 2016 clinical decision support track. In: Proceedings of the 15th Text Retrieval Conference (2016)
6. Hersh, W., Voorhees, E.: TREC genomics special issue overview. Inf. Retr. **12**, 1–15 (2009)
7. Herrera, A.G.S.D., Kalpathy-Cramer, J., Demner-Fushman, D., Antani, S., Muller, H.: Overview of the image CLEF 2013 medical tasks. In: CLEF Working Notes (2013)
8. Roberts, K., Simpson, M.S., Voorhees, E., Hersh, W.: Overview of the TREC 2015 clinical decision support track. In: Proceedings of the 2015 Text Retrieval Conference (2015)
9. Johnson, A.E., Pollard, T.J., Shen, L., Wei, H., Lehman, L., Feng, M., Ghassemi, M., Mark, R.G.: MIMIC-III, a freely accessible critical care database. Sci. Data **3**, 160035 (2016)
10. Robertson, S.E., Walker, S., Beaulieu, M., Gatford, M., Payne, A.: Okapi at TREC-4. NIST Special Publications, pp. 73–96 (1996)
11. Miao, J., Huang, J.X., Ye, Z.: Proximity-based Rocchio's model for pseudo relevance. In: SIGIR 2012, pp. 535–544 (2012)
12. Ye, Z., Huang, J.X.: A simple term frequency transformation model for effective pseudo relevance feedback. In: SIGIR 2014, pp. 323–332 (2014)
13. Colace, F., Santo, M.D., Greco, L., Napoletano, P.: Improving relevance feedback based query expansion by the use of a weighted word pairs approach. JASIST **66**(11), 2223–2234 (2015)
14. Lund, K., Burgess, C., Atchley, R.A.: Semantic and associative priming in high-dimensional semantic space. In: Proceedings of the 17th Annual Conference of the Cognitive Science Society, pp. 660–665 (1995)

An RGBD Tracker Based on KCF Adaptively Handling Long-Term Occlusion

Xue-Fei Zhang[1,2], Ai-Ping Zeng[2], Shan Huang[2], Ming Qing[1], and Yi Zhou[1(✉)]

[1] Chaoying Technology, Co., Ltd., Chengdu, Sichuan 610041, China
zhouyi@chaoyingtec.com

[2] College of Electrical Engineering and Information Technology, Sichuan University, Chengdu 610065, China

Abstract. Since occlusion still be a challenge for object tracking in RGB data. In this paper, we propose an RGBD single-object tracker that built upon the well-known base KCF tracker and exploit how the depth information fusing to handle partial and long-term occlusion. To divides tracking model into parts, the proposed tracker could detect and handle occlusion of each part separately. Despite the robustness in tracking with long-term occlusion, our part-based tracker provides an adaptively updating learning matrix. Experimental results are conducted on our dataset, which demonstrate that our tracker contains stability in long-term tracking.

Keywords: RGBD tracking · Kernel · Long-term occlusion · RealSense

1 Introduction

One of the most fundamental and challenge areas in computer vision is object tracking, it is demonstrated in recent reviews [1–5]. A typical scenario of visual tracking is to track an unknown object which initialized by detecting automatically or selecting manually, consequently returning a bounding box as the tracked object in subsequent frames. In the last decade, several algorithms have evolved much in both accuracy and efficiency [6, 7], but there still contain several problems which make tracking as a challenging task in computer vision. One of them occurs in the online-learning step, which contains a constant learning rate for updating the model with whole tracked patch regard of some mistakes contained. Besides, in real scenarios, occlusion occurs quite often, creating tracking drift and failure. There are plenty of state-of-art algorithms [8, 9] provide powerful and robust trackers in hope for handling these issues.

Unlike traditional tracking algorithms, which working in the 2D data, we would like to advance them by introducing 3D dataset which contains both RGB and depth data. With digital sensor developing, many accurate depth sensors, such like Microsoft Kinect and Intel RealSense, could provide stable and accurate depth map. Equipping with this additional and useful information, our algorithm could handle previous problems in a quite easy way. Our approach builds on an observation based on prior work. That's the depth of our choosing target change a little, since the time interval is short, and we don't handle object moved extremely fast. According to this observation,

D.-S. Huang et al. (Eds.): ICIC 2018, LNCS 10955, pp. 100–107, 2018.
https://doi.org/10.1007/978-3-319-95933-7_13

our approach decomposes the tracking procedure into translation estimation of target, detection of the partial occlusion state with the depth map and adaptively on-line learning for model updating.

Our purpose is handling occlusion and adaptive updating part-based model. Firstly, we need to build a depth model in a way of the KCF model in [4, 5], instead we choose depth value to represent it. Accordingly, when talking about detecting, we mean it operators separately for each part. Furthermore, fusing the depth information in 2D RGB image, adaptively updating model to avoid long-term occlusion are the main contributions we make. The whole procedure is shown in Fig. 1.

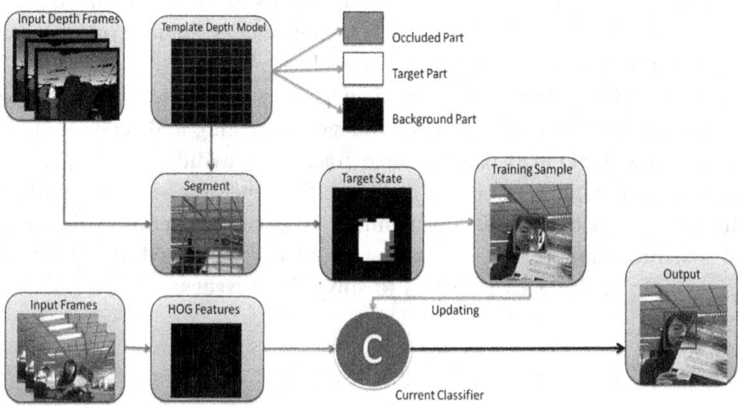

Fig. 1. Flowchart of our tracking approach. The input RGBD frames are divide in parts equally and combine depth segment and translation estimate for occlusion handling. After that adaptively updating the tracking model

The paper is structured as follow. In Sect. 2, several state-of-the-art methods focused on handling occlusion and recently published algorithms based on the fusion of color and depth information are reviewed. In Sect. 3, the base algorithm KCF [5] which we used is described briefly. In Sect. 4, we describe our proposed approach detailly. Experiment results are reported in Sect. 5. And conclusions are in the last section.

2 Related Works

Due to the availability of accurate and fast depth sensor, recently, computer vision has paid more attention on tracking algorithms which fuse the color and depth information. Robust performance in handling occlusion and deformation demonstrated in relatively few published algorithms. Contrast to traditional trackers that evolved from mean shift [10], particle filter [11] to robust correlation filter [7], the useful additional information opens a new area for object tracking and tackles traditional tracking tasks in remarkable ways.

One of the significant contribution published in [12], which provided a large dataset (over 100 RGBD videos), compensating the lack of benchmark evaluation dataset in RGBD Tracking area. Besides its author provides several baseline algorithms combining traditional feature and depth signal, as calculating HOG (histogram of gradient) in both color and depth map and extracting depth features for training novel models. Above all, it's author Song. constructs an important dataset that demonstrated by subsequent state-of-the-art RGBD algorithms, [13–15], all test in it.

Considering the high accuracy and processing speed, the KCF [5] tracker has been served as the baseline algorithm for many robust trackers. For example, the depth information has been fused in KCF tracker in [13], which ranking the fourth position in Princeton tracking benchmark. In [12], it's author using depth in diversified ways, treating the depth distribution as a single Gaussian model, identifying the occluding part as the region which not belong to the model.

Recently, the RGBD tracker presented in [14], which based on a particle filter framework [11], fusing the color and depth image by creating a 3D cuboid model. This algorithm provides a 3D part-based sparse tracker to avoid partial occlusion, and represents the target, which performs a robust performance for outranking all the other trackers in online Princeton RGBD benchmark [12]. Simultaneously it's author considers the synchronization and registration problems contained in RGBD dataset, providing a simple and effective method to solve these issues.

3 The Kernel Correlation Filters Tracker

The so-called KCF tracker presented by Henriques et al. [5], which creates an effective tracking framework, is chosen as our baselines algorithm. Besides, refer to further works presented in [6, 7, 16–18] for a comprehensive overview of the robust performance of KCF. In this section, we will briefly describe the KCF tracker with focus on aspects related to our proposed extensions.

Correlation tracker focuses on training a filter through finding a linear regression function: $\mathrm{f}(z) = w^T z$, that minimizes the squared error over the image patch and its regression output: $\min_{W} \sum_{i} (f(x_i) - y_i)^2 + \lambda \|W\|^2$. where the λ is a regularization parameter that controls overfitting, and the y_i stands for regression targets. Furthermore, the KCF use the 'kernel trick' to extend correlation filters for very fast RGB tracking, which changes the linear regression to ridge regression: $W = \sum_{i} \alpha_i \phi(x_i)$, where the variables α_i need to be estimated instead of W, and the ϕ is the function that maps feature into non-linear space. That the ridge regression function can be written as $\hat{\alpha} = (K + \lambda)^{-1} \hat{y}$, where the K is the circulant kernel matrix with elements $\hat{k}^{xx}$ corresponding to $\gamma(x_i, x_j)$ and the γ is the selected kernel function. Through the whole processing pipeline in KCF tracking that is consisted of training, tracking and updating, there always contains an $m \times m$ circular matrix $C(x)$ which constructed from an $m \times 1$ vector by applying a cyclic shift operator for impressive computation reduction. Since the circular matrix could be made diagonal by the Discrete Fourier Transform (DFT), regardless of generating vector $x[]$. With this useful property, as in [5]. the ridge regression could be solved as:

$$\hat{\alpha}=\mathcal{F}^{-1}((\mathcal{F}^*(\gamma^{xx})+\lambda)^{-1}\mathcal{F}(y)) \tag{1}$$

and after getting the coefficient α the detect process effectively with same circular matrix properties with calculating response of the classifier $f(z)$,

$$f(z)=\mathcal{F}^{-1}(\mathcal{F}^*(k^{xz})\odot\mathcal{F}(\alpha)) \tag{2}$$

The operation preformed in Fourier domain, and the new position is centered at the maximum response of $f(z)$.

4 Proposed Approach

In this paper, we propose a part-based updating tracker, where occlusion and deformation detecting by a depth model. We also provide a data-driven learning rate matrix replacing the constant one to adaptively update both depth model and tracking model. Considering both training, detecting and retraining working in element-wise operation, which implicitly densely sampling the candidate patches, we divide candidate patch into parts equally, and connect each part with correlated element in learning rate matrix.

4.1 Depth Segment

Recently several novel RGBD Tracking algorithms [15, 19–21] have being developed since the highly developing of reliable depth sensors. In our proposed approach, we choose the RealSense from Intel, which has high accuracy as binocular camera and process fast without adding extra calculate cost.

To extract target from region of interest(ROI), we assume the target has relatively significant share of the pixels in ROI. As shown in Fig. 2, the specific approach is described as follow: firstly, traversal the ROI, create a depth histogram; secondly define a slider which contains a fix constant rouge of depth value, use this slider to traversal the whole depth axis of histogram to find which one gather the most number of pixels; then extract target as the largest continuum whose pixels belong to the slider. As shown above, the pixels painting with red color represent the target extracted, which demonstrates our application works well.

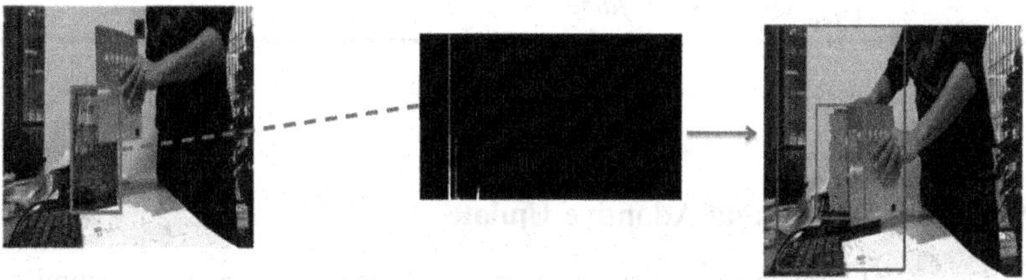

Fig. 2. The depth segments

4.2 Occlusion Detection

The work proposed in the Princeton [12], using depth cue as an identify signal to discriminate the tracked object from occluding one, it assumes the target has closest distance in the bounding box, which treating some part in front of target as new occluder. And the DSKCF [13] models the depth distribution as a single Gaussian and identifies candidate which don't belong to this model as an occluding part, meanwhile initials another tracker for the occluded part, which increase much computation cost.

In this paper, considering the depth continuity of the tracked object in adjoining patches, we can quickly figure out each part whether its depth distribution have changed suddenly in the candidate patch. After using translation tracker estimating the tracked position, we calculate difference of depth value between candidate patch and the tracked part in depth map. As said, we assume the object moved not particularly fast, so the depth difference of tracked object changed in a contrast little rouge. Compare to RGB tracker, which suffer from the long-term occlusion problem, this paper provides a simple but effective scheme to handle it, which described in Table 1.

Table 1. Occlusion detections at one part p^{ij}

Input:	Frame t. Depth patch D_t, depth template model D_{t-1}, target patch vector V, target state of part $Occl\_s_{t-1}^{ij}$,occlusion state $Tar\_s_{t-1}^{ij}$, size of part S^{ij}
Output:	Current occlusion state of part $Occl\_s_t^{ij}$

	Calculate depth difference sub^{ij} by $sub^{ij} = D_t - D_{t-1}$, census the number of changing pixels in part N^{ij}
	Acquire the current target state $Tar\_s_t^{ij}$ by depth patch segment
	If $Occl\_s_{t-1}^{ij} = true$ then
	  If $abs(sub^{ij}) < 200mm$, $Occl\_s_t^{ij} = true$
	Else
	  If $Tar\_s_{t-1}^{ij} = ture$ and $Tar\_s_t^{ij} = false$
	    If $sub^{ij} > 200mm$, $Occl\_s_t^{ij} = false$
	    Else if $sub^{ij} < -200mm$, $Occl\_s_t^{ij} = true$
	  Else $Occl\_s_t^{ij} = false$

4.3 Occlusion Handle and Adaptive Update

Above all, we could precisely divide each part into three categories: background, target, and foreground. According to their states, we could update the partial depth model separately and create an adaptive update learning matrix L_{IJ} which replacing the original constant learning rate: updating the unobstructed part with the current depth image while retaining its content of the occluded part for subsequent occlusion detection; after that, we translate the state of each part into a separate learning rate, which set according to the target state of correlated parts. Since a good compromise between retaining and updating, each element l_{ij} of the learning matrix is set as shown in Table 2.

Table 2. Setting value of learning rate matrix

	Background	Target	Foreground
l_{ij}	0.5	0.08	0

Differ from the work proposed in [22], which handles occlusion by creating a short-term classifier-pool to store the last K classifiers without occlusion, and select an optimal classifier form this pool for tracking when heavy occlusion occurs. As shown in Fig. 3, our proposed approach performs well for the frames with occlusion, even retains stability when completely occlusion occurs. With time of occlusion increasing, the classifier contains updating content of unobstructed parts and saved content of occlude parts can still retain its robust performance.

Fig. 3. Occlusion handling by retain the content of occluded parts

5 Experiment Result

This paper proposes a novel approach to handle long-term occlusion in object tracking. To evaluate the performance of our proposed tracker, we record a dataset consisting of 39 video clips with heavy occlusion for long time, which we use the Realsense to create both RGB and depth data, manually label the ground truth. The dataset contains many challenges including partial and full occlusion, long-term and short-term occlusion, fast motion and shape deformation.

To evaluate the performance of tracker, we consider two metrics which are performed in the Princeton [12], DSKCF [13] and 3D part-based Tracker: center position error(CPE) and the share of overlap between the output and ground truth. Meanwhile, to understand how much the tracker performance has been improved in occlusion handling with depth data, we also perform KCF [5], and DSKCF [13] trackers in our dataset. The performance measured by CPE and overlap ratios are shown in Fig. 4. As seen, our tracker significantly improves the baseline KCF tracker with increment in the average success rate from 0.571 to 0.677. And furthermore, in overlap threshold error plot, our proposed tracker outperforms the KCF tracker rapidly.

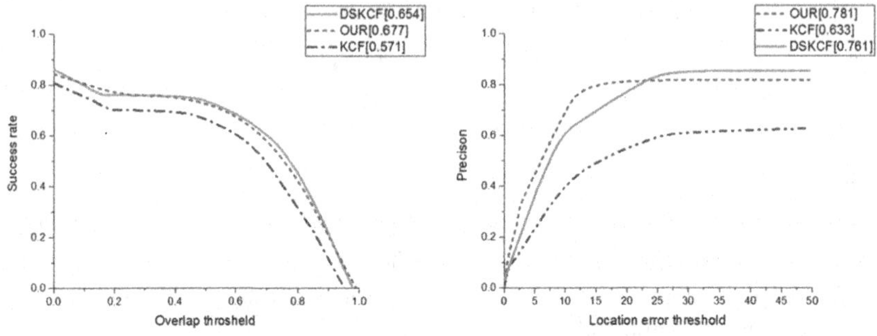

Fig. 4. Plots of CPE and OPE on all39 sequences

6 Conclusion

In this paper, we have proposed an effective occlusion handling tracker for real-time visual tracking, which adaptively updating unobstructed parts and retaining occluded parts. The translation is estimated by modeling a kernelized circulant structure. And the scale is estimated creating a bounding box around the segmented target. We further using depth data to detect and handle heavy occlusion, meanwhile adaptively updating each part of tracking model separately. Extensive experiment results show that the proposed approach performs favorably against the KCF and DSKCF in terms of robustness in long-term occlusion.

Acknowledgment. At the point of finishing this paper, we are grateful for the support by the ChaoYing Technology, Co, Ltd, Sichuan, China.

References

1. Ma, C., Huang, J.B., Yang, X., et al.: Adaptive correlation filters with long-term and short-term memory for object tracking. Int. J. Comput. Vis. 1–26 (2017)
2. Bibi, A., Mueller, M., Ghanem, B.: Target response adaptation for correlation filter tracking. In: Leibe, B., Matas, J., Sebe, N., Welling, M. (eds.) ECCV 2016. LNCS, vol. 9910, pp. 419–433. Springer, Cham (2016). https://doi.org/10.1007/978-3-319-46466-4_25
3. Danelljan, M., Bhat, G., Khan, F.S., et al.: ECO: Efficient Convolution Operators for Tracking. pp. 6931–6939 (2016)

4. Henriques, João F., Caseiro, R., Martins, P., Batista, J.: Exploiting the circulant structure of tracking-by-detection with kernels. In: Fitzgibbon, A., Lazebnik, S., Perona, P., Sato, Y., Schmid, C. (eds.) ECCV 2012. LNCS, vol. 7575, pp. 702–715. Springer, Heidelberg (2012). https://doi.org/10.1007/978-3-642-33765-9_50
5. Henriques, J.F., Rui, C., Martins, P., et al.: High-speed tracking with kernelized correlation filters. IEEE Trans. Pattern Anal. Mach. Intell. **37**(3), 583–596 (2014)
6. Danelljan, M., Hager, G., Khan, F.S., et al.: Discriminative scale space tracking. IEEE Trans. Pattern Anal. Mach. Intell. **39**(8), 1561–1575 (2017)
7. Dong, X., Shen, J., Wang, W., et al.: Occlusion-aware real-time object tracking by integrated circulant structure kernels classifier. IEEE Trans. Multimedia (2016)
8. Ma, C., Yang, X., Zhang, C., et al.: Long-term correlation tracking. In: Computer Vision and Pattern Recognition. pp. 5388–5396. IEEE (2015)
9. Liu, T., Wang, G., Yang, Q.: Real-time part-based visual tracking via adaptive correlation filters. In: IEEE Conference on Computer Vision and Pattern Recognition, pp. 4902–4912. IEEE Computer Society (2015)
10. Comaniciu, D., Meer, P.: Mean shift: a robust approach toward feature space analysis. IEEE Trans. Pattern Anal. Mach. Intell. **24**(5), 603–619 (2002)
11. Nummiaro, K., Koller-Meier, E., Gool, L.V.: An adaptive color-based particle filter. Image Vis. Comput. **21**(1), 99–110 (2003)
12. Song, S., Xiao, J.: Tracking revisited using RGBD camera: unified benchmark and baselines. In: IEEE International Conference on Computer Vision, pp. 233–240. IEEE (2014)
13. Hannuna, S., Camplani, M., Hall, J., et al.: DS-KCF: a real-time tracker for RGB-D data. J. Real-Time Image Process. 1–20 (2016)
14. Bibi, A., Zhang, T., Ghanem, B.: 3D part-based sparse tracker with automatic synchronization and registration. In: Computer Vision and Pattern Recognition, pp. 1439–1448. IEEE (2016)
15. Tang, F., Harville, M., Tao, H., et al.: Fusion of local appearance with stereo depth for object tracking. In: IEEE Computer Society Conference on, pp. 1–8. IEEE (2008)
16. Danelljan, M., Häger, G., Khan, F.S.: Accurate scale estimation for robust visual tracking. In: British Machine Vision Conference. pp. 65.1–65.11 (2014)
17. Bolme, D.S., Beveridge, J.R., Draper, B.A., et al.: Visual object tracking using adaptive correlation filters. In: Computer Vision and Pattern Recognition, pp. 2544–2550. IEEE (2010)
18. Tang, M., Feng, J.: Multi-kernel correlation filter for visual tracking. In: IEEE International Conference on Computer Vision, pp. 3038–3046. IEEE (2016)
19. Luber, M., Spinello, L., Kai, O.A.: People tracking in RGB-D data with on-line boosted target models. In: International Conference on Intelligent Robots and Systems, pp. 3844–3849. IEEE (2011)
20. Camplani, M., Paiement, A., Mirmehdi, M., et al.: Multiple Human Tracking in RGB-D Data: A Survey (2016)
21. Lim, H., Sinha, S.N.: Monocular localization of a moving person onboard a quadrotor MAV. In: IEEE International Conference on Robotics and Automation, pp. 2182–2189. IEEE (2015)
22. Dong, X., Shen, J., Yu, D., et al.: Occlusion-aware real-time object tracking. IEEE Trans. Multimedia **19**(4), 763–771 (2017)

Background Subtraction with Superpixel and k-means

Yu-Qiu Chen[1,2], Zhan-Li Sun[1(✉)], Nan Wang[1], and Xin-Yuan Bao[1]

[1] School of Electrical Engineering and Automation, Anhui University, Hefei, China
zhlsun2006@126.com

[2] Hefei Sunwin Intelligence Co., LTD, Hefei, China

Abstract. In this paper, we presents a background subtraction approach with superpixel and k-means that aims to use less memory to establish a background model and less computation time for moving object detection. We use superpixels to divide similar pixels into the same area, K-mean is used to obtain the main color values of the superpixel. The mean and variance of superpixels and changes in the number of previous attractions are used as the discriminative features. The main contribution of this paper is to propose features suitable for superpixel-based moving object detection. We test this method in different videos demonstrate that this method demonstrated equal or better segmentation than the other techniques and proved capable of processing 320 * 240 video at 114 fps, including post-processing, faster than most existing algorithms.

Keywords: Background subtraction · Superpixel · K-Means

1 Introduction

Background subtraction is the first step in event detection, object tracking and intelligent video surveillance. As a basic research topic of video processing, background subtraction has received more and more attention.

MOG [1] based on density estimation uses multiple single Gaussian models to describe the background model so that the background can be switched. Kim et al. proposed CodeBook [2] they introduced a layered codebook to simulate multiple backgrounds to obtain a time-series model for each pixel. Liang et al. proposed cp3 [3] use co-occurring pixels to accommodate sudden lighting changes and popping motion. ViBe [4] is one of the milestone algorithms for background subtraction widely used in industry because of its rapid operation and easy understanding. but these algorithms all need a lot of memory to store the background model.

Superpixels can significantly reduce memory requirements. In [5], they track superpixel in order to segment objects. This algorithm reduces the memory burden but is very computationally expensive. In order to reduce the operating time, Chen et al. proposed SuperBE [6] they treat each superpixel as a pixel and draws on ViBe to create a background model, calculate the mean and color covariance matrix of each superpixel to segment the foreground object. However, the SuperBE algorithm cannot accurately segment moving objects, and it is easy to miss small moving objects.

D.-S. Huang et al. (Eds.): ICIC 2018, LNCS 10955, pp. 108–112, 2018.
https://doi.org/10.1007/978-3-319-95933-7_14

The remainder of the paper is organized as follows. The proposed method is derived in Sect. 2, followed by experiments in Sect. 3. Finally, conclusions are made in Sect. 4.

2 Method

As mentioned above, we propose a background subtraction method based on superpixel and k-means. The algorithm consists of three parts: background model initialization, moving object detection and background model updating.

2.1 Background Model Initialization

Use SLIC [7] algorithm to find *N* similar pixels share the same background model can effectively reduce the memory burden. We only perform a superpixel clustering on the first frame, the segmentation result is stored and used for all future frames.

Next initialize the background model of each superpixel, calculate the following characteristics (Table 1):

Table 1. Superpixel background model

u_{bi}	$\mu_{bi} = \frac{1}{N_i}\sum_{j=1}^{N_i} x_{ij}$	The mean of superpixel
σ_{bi}	$\sigma_{bi} = \sqrt{\frac{1}{N_i}\sum_{j=1}^{N_i} (x_{ij} - \mu_{bi})}$	The standard deviation of superpixel
B_i	$\{b_1, b_2, b_3 \ldots b_{10}\}$	The main color value of superpixel
n		The number of noise points of superpixel

where x_{ij} *is* the jth member of the superpixel *i*, and N_i is the number of members of superpixel *i*. Use k-means and set the initial classification category $Ci = 4$, then fill the clustering results into B_i (the main color value of superpixel). An important feature n_{bi} can be calculated by the following formula:

$$t_{ij} = t_{ij} + 1 \quad if \ \left|x_{ij} - b_{ik}\right| < 20 \quad k = \{1, 2, 3 \ldots 0\} \tag{1}$$

$$n_{bi} = \begin{cases} n_{bi} & if \quad t_{ij} > 0 \\ n_{bi} + 1 & else \end{cases} \tag{2}$$

2.2 Moving Object Detection

The next step is to detect moving objects through the background model. Use formula (1) to calculate t_{ij}, then calculate the following features (Table 2):

Table 2. Current superpixel features

p_{ij}	$p_{ij} = \begin{cases} 0 & if\ t_{ij} > 0 \\ 255 & else \end{cases}$	Whether the pixel is a foreground point
u_{si}	$\mu_{si} = \frac{1}{N_i}\sum_{j=1}^{N_i} x_{ij}$	The mean of superpixel
σ_{si}	$\sigma_{si} = \sqrt{\frac{1}{N_i}\sum_{j=1}^{N_i} (x_{ij} - \mu_{si})}$	The standard deviation of superpixel
n_{si}	$n_{si} = \begin{cases} n_{si} & if \quad p_{ij} = 0 \\ n_{si}+1 & else \end{cases}$	The number of noise points of superpixel

when $|n_{pi} - n_{bi}| > 10$ and $|u_{pi} - u_{bi}| > 5$ often means that there is a moving object in the superpixel or the current background cannot express the superpixel accurately. In order to distinguish between these two cases we have established a weight function:

$$D_i = |\sigma_{si} - \sigma_{bi}| + 0.5 * |n_{si} - n_{bi}| \tag{3}$$

if $D_i < 20$, the superpixel is marked as being updated, otherwise it is marked as foreground. Then we need some post processing, first, zero out all p_{ij} of superpixel that are not marked as foreground, and then use closed and open operation filling the hole inside the moving object.

2.3 Background Model Updating

The superpixel that are marked for update need to be initialized again and the unmarked superpixel need to update their number of noise points n_{bi}:

$$n_{bi} = 0.8 * n_{bi} + 0.2 * n_{pi} \tag{4}$$

3 Experiments

In this section, to assess the performance of the proposed method, used the CDW2014 dataset [8], and the metrics shown in Table 3 to validate our algorithm. Partial comparison experiment results screenshots shown in Fig. 1, Tables 4 and 5.

Table 3. Metrics: TP: number of true positives; FP: number of false positives; FN: number of false negatives.

Recall	$\mathrm{Re} = \frac{TP}{TP+FN}$	Foreground classification accuracy
Precision	$\mathrm{Pr} = \frac{TP}{TP+FP}$	Overall accuracy measure
F-Measure	$F1 = \frac{2*\mathrm{Pr}*\mathrm{Re}}{\mathrm{Pr}+\mathrm{Re}}$	Overall accuracy measure

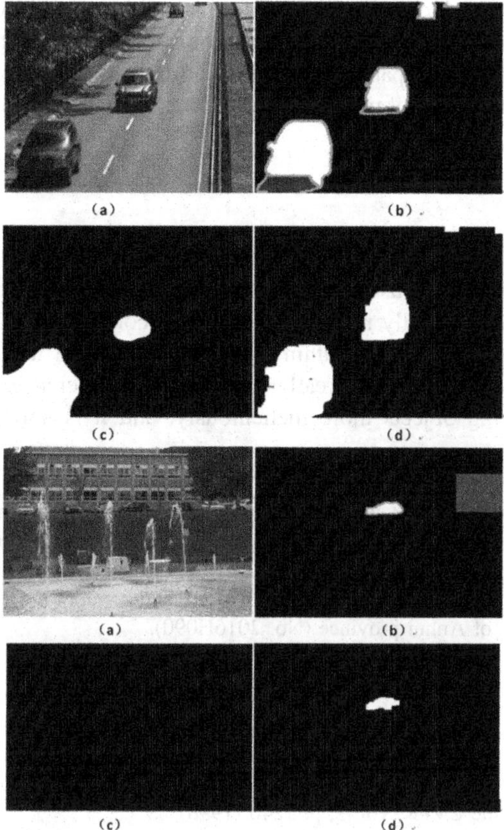

Fig. 1. (a) The current image; (b) The ground truth; (c) SuperBE; (d) Our algorithm;

Table 4. Good performance in some datasets

Video	Algorithm	Re	Pr	F1
Baseline	SuperBE	0.3458	0.8615	0.4135
	Our algorithm	0.8.303	0.8384	0.8331
cameraJitter	SuperBE	0.4801	0.8626	0.5851
	Our algorithm	0.6770	0.6215	0.6434

Table 5. Average time on CDW2014 (320 * 240)

Methods	Our algorithm	SuperBE
Time(ms)	8.77	13.31
Fps	114	75

Experiments show that our algorithm performs well in most videos. In comparative experiments we found that our algorithm is very resistant to foreground erosion and scene noise. Moreover, compared with the SuperBE algorithm, our algorithm can express the edge information of moving objects more meticulously, and it is not easy to lose small objects.

4 Conclusion

In this work, a new algorithm uses superpixel and k-means algorithms, the combination of these two algorithms greatly reduces the memory consumption of the background model, at the same time, our algorithm came up with some features that apply to superpixels so the algorithm can greatly resist noise interference, express the edge information of moving objects more meticulously, and it is not easy to lose small objects.

Acknowledgement. The work was supported by a grant from National Natural Science Foundation of China (No. 61370109), a key project of support program for outstanding young talents of Anhui province university (No. gxyqZD2016013), a grant of science and technology program to strengthen police force (No. 1604d0802019), and a grant for academic and technical leaders and candidates of Anhui province (No. 2016H090).

References

1. Stauffer, C., Grimson, W.E.L.: Adaptive background mixture models for real-time tracking. In: Proceedings of the CVPR, vol. 2, p. 2246 (1998)
2. Kim, K., Chalidabhongse, T.H., Harwood, D., et al.: Real-time foreground-background segmentation using codebook model. Real-Time Imaging **11**(3), 172–185 (2005)
3. Dong, L., Kaneko, S., Hashimoto, M., et al.: Co-occurrence probability-based pixel pairs background model for robust object detection in dynamic scenes. Pattern Recogn. **48**(4), 1374–1390 (2015)
4. Barnich, O., Droogenbroeck, M.V.: ViBe: a universal background subtraction algorithm for video sequences. IEEE Trans. Image Process. **20**(6), 1709 (2011)
5. Giordano, D., Murabito, F., Palazzo, S., et al.: Superpixel-based video object segmentation using perceptual organization and location prior. In: Computer Vision and Pattern Recognition, pp. 4814–4822 (2015)
6. Chen, T.Y., Biglari-Abhari, M., Wang, I.K.: SuperBE: computationally light background estimation with superpixels. J. Real-Time Image Process. **11**, 1–17 (2018)
7. Achanta, R., Shaji, A., Smith, K., et al.: SLIC superpixels compared to state-of-the-art superpixel methods. IEEE Trans. Pattern Anal. Mach. Intell. **34**(11), 2274–2282 (2012)
8. Wang, Y., Jodoin, P.M., Porikli, F., et al.: CDnet 2014: an expanded change detection benchmark dataset. In: Computer Vision and Pattern Recognition Workshops, pp. 393–400 (2014)

Application of Embedded Database SQLite in Engine Fault Diagnosis System

Li Huang[1(✉)] and Hui Huang[2]

[1] City College of Wuhan University of Science and Technology, Wuhan College of Foreign Languages and Foreign Affairs, Wuhan 430083, China
240113921@qq.com

[2] Central Southern China Electric Power Design Institute, China Power Engineering Consulting Group Corporation, Wuhan 430070, China

Abstract. In order to improve the performance of engine fault diagnosis and reduce the cost, a Embedded fault diagnosis system has been designed. This paper discusses the design and implementation of embedded database based on SQLite combined with the analysis method of fault tree.

Keywords: SQLite · Embedded database · Fault diagnosis

1 Introduction

At present, in order to meet the requirements of fast fault location and fault diagnosis, vehicle fault diagnosis system is developing in the direction of intellectualization and miniaturization. In this kind of system, it is not easy to use large database to manage the data. SQLite embedded database has many advantages, such as small volume, fast speed, convenient maintenance and abundant interfaces [1]. It has been applied to the engine fault diagnosis system, which greatly improves the efficiency of data management in the fault diagnosis system.

2 Technical Features of SQLite Database

SQLite is an open source file type database [2], which is widely used in embedded systems. It has the basic features of a general relational database, such as SQL syntax, transaction processing, data table management, and index. As an embedded database, it has the characteristics of small capacity, high reliability, simple management and convenient maintenance.

3 Design of Fault Diagnosis System

3.1 System Overall Design

The system combines fault tree analysis and production rule diagnosis expert system. With QT and embedded SQLite database as development platform, the engine fault

D.-S. Huang et al. (Eds.): ICIC 2018, LNCS 10955, pp. 113–119, 2018.
https://doi.org/10.1007/978-3-319-95933-7_15

diagnosis function based on embedded database is realized. The frame structure of the system is shown in Fig. 1.

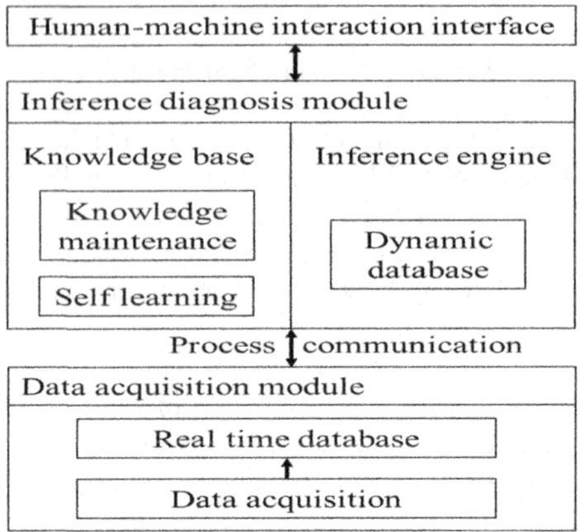

Fig. 1. Frame structure diagram of fault diagnosis system

The system is mainly composed of three parts. The data sampling module completes the collection of sensor information. It saves the information through preprocessing and conversion to real time database. The information includes fuel pressure, vacuum of the intake pipe, pressure in the fuel pipe, air flow, speed and so on.

The diagnostic reasoning module is the core part of the diagnosis system. It completes the diagnosis task and improve the ability of diagnosis system. And it completes the operation of increasing, modifying, deleting, and querying the data of the knowledge base. The module mainly includes knowledge management and maintenance of knowledge base, self learning function, inference engine and dynamic database.

The human-machine interaction module is the control and coordination mechanism of the system. It obtains information from the user through the interface, guides the user to participate in the diagnosis process, answers the questions raised by the system and completes the related operation instructions of the system. It mainly displays the diagnostic results in a graphical way.

3.2 Establishment of Fault Tree

The system constructs the fault tree model according to the causality between the fault source and the fault phenomenon. As an example of abnormal vibration of engine starting, a fault tree is set up as shown in Fig. 2.

In the above example, the engine start abnormal vibration is the top event of the fault tree. By retrieving the rule table in the database, three failure causes can be obtained. First, air leakage of the intake pipe. Second, abnormal pressure of fuel pressure. Third, the air flow sensor has trouble. According to the rules of positive reasoning, the system gives the diagnostic advice.

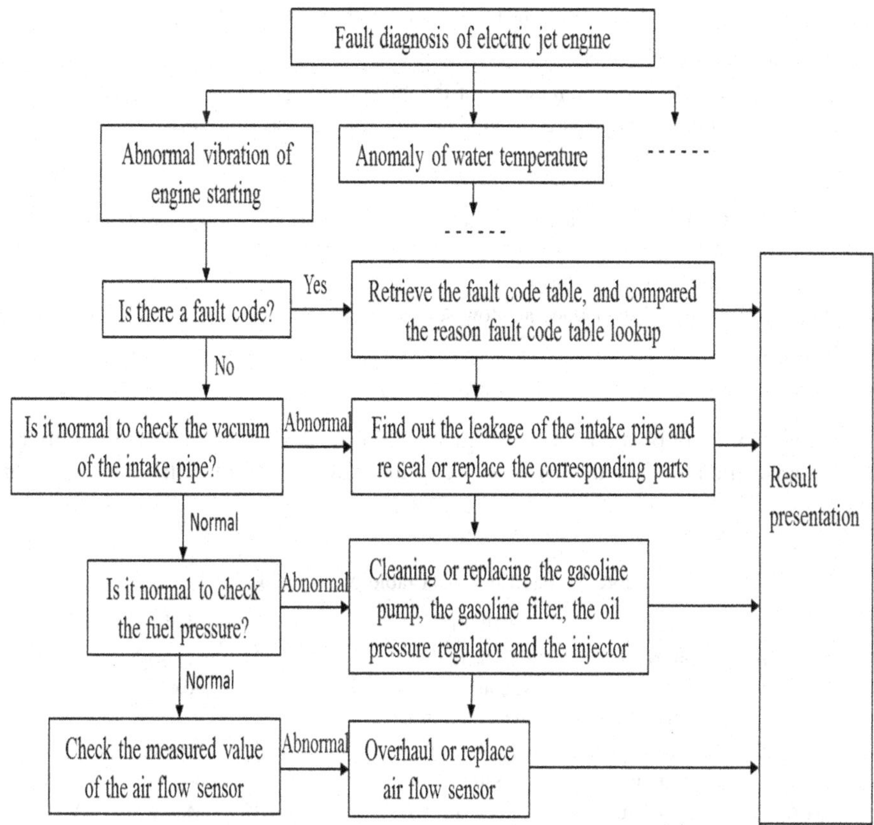

Fig. 2. Fault tree diagnosis of abnormal vibration of engine

In this way, the fault analysis branch is formed from the top event of the fault tree to every bottom event. These analytical branches are implemented by a Bi-directional chain list in database [3]. It is organically combined with a rule-based diagnostic system through fault tree analysis.

3.3 Database Design Examples

The database table file structure of abnormal vibration of engine starting is designed as follows.

- Basic fact database

The basic fact database stores the natural description forms of the facts, such as the rules, the conclusions, the intermediate conclusions and the final conclusions in the process of diagnosis and reasoning [4]. The design structure of the basic fact database table is shown in Table 1.

Table 1. Basic fact data table of abnvibration.db

ID	FactID	Fault behaviors
1	F1	Check the vacuum degree of the intake pipe
2	F2	Find out the leakage of the intake pipe and re seal or replace the corresponding parts
3	F3	Check fuel pressure
4	F4	Cleaning or replacing gasoline pump, gasoline filter, oil pressure regulator and injector
5	F5	Check air flow sensor
6	F6	Overhaul or replace air flow sensor

·Fault tree database

The common fault phenomenon of engine is described by fault tree according to the fault diagnosis method. The design structure of the fault tree database table is shown in Table 2.

Table 2. Fault tree data table of faltree.db

ID	FTAID	Entrance	TestItem	Ask	Method	FactID
1	FTA101	Abnormal vibration of engine starting	Check the vacuum degree of the intake pipe	Normal?	Message box	F1
2	FTA102	Normal intake pipe line	Check fuel pressure	Normal?	Message box	F3
3	FTA103	Normal fuel pressure	Check air flow sensor	Normal?	Message box	F5

·Rule precursor database

The rule precursor database stores all the premises of the rule and the system reasoning state information. The rule precursor database table design structure is shown in Table 3. In the table, the RuleID storage rules are encoded and the FactID represents the fact encoding of the rule precursor.

Table 3. Rule precursor data table of forule.db

ID	RuleID	FactID
1	R1	F1
2	R2	F3
3	R3	F5

·Rule post database

The conclusion of rule and the diagnosis reasoning state information are stored in the rule post database. For the convenience of the system, the latter fact of the rule encoding is encoded. The design of rule post database is shown in Table 4.

Table 4. Rule post database of bkrule.db

ID	RuleID	FactID
1	R1	F2
2	R2	F4
3	R3	F6

4 Application of SQLite in Fault Diagnosis System

4.1 Main Points of Database Operation

- Selection of memory database

The system use memory database because the operation of the process is faster in the memory database [5]. The main purpose is to improve the operation efficiency with memory database. Test data show that the running speed of the memory process is two times faster than that in NandFlash.

- Selection of NandFlash database

In order to avoid data loss in database when SQLite3 is in power down, the system uses NandFlash as storage medium. If the NandFlash is frequently erasable, its service life will be shortened. So large capacity NandFlash is used, and the equalization algorithm of the file system is used, averaging the whole nandfalsh's erasing, in order to ensure the service life of NandFlash.

- Sharing of memory databases between different processes

SQLite implements independent transaction processing through the exclusive and shared locking on the database level. This means that when multiple processes and threads can read data from the same database at the same time, only one can write to the data. Multiple processes or thread sharing databases are implemented through file locks.

4.2 Diagnostic System Flow Chart

The flow chart of the fault diagnosis system based on SQLite embedded database is shown in Fig. 3. The database implementation mainly includes two processes: data acquisition process and fault diagnosis process. The two processes are shared through the memory database.

In the process of data acquisition, the data acquisition driver module is completed in the Linux kernel. Then the collected data are written to the real time database. Since two processes share a memory database file, in order to avoid writing data conflicts in

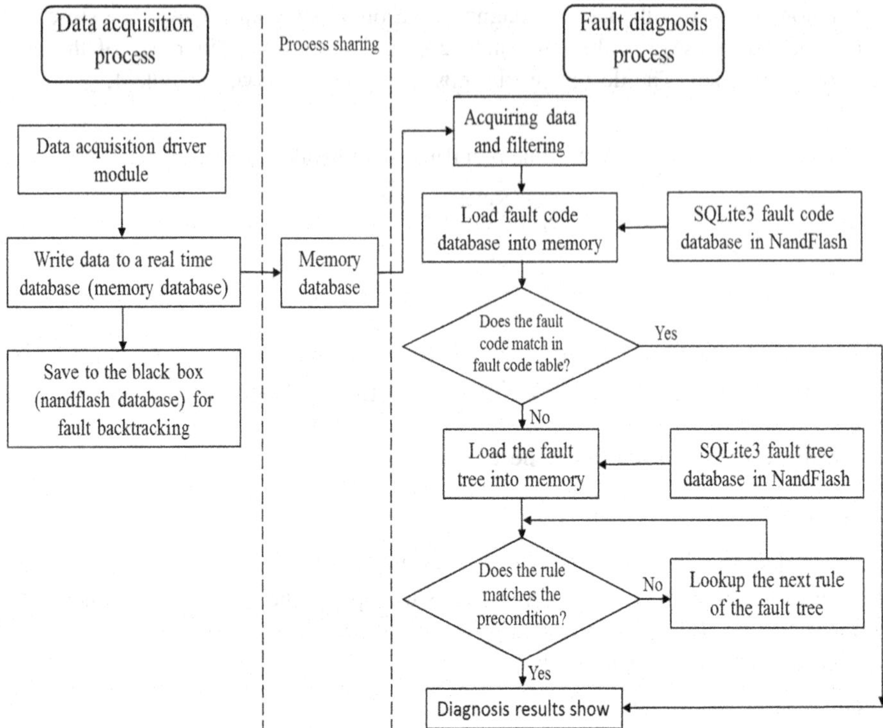

Fig. 3. Flow chart of fault diagnosis

the two processes, it must be locked when accessing the database file. At the same time, the collected data is stored in the NandFlash database for fault backtracking.

In the process of fault diagnosis, the collected data should be filtered to reduce the load in the system diagnosis process. The NandFlash fault code in the database of fault code table is loaded into memory, in order to improve the retrieval efficiency, and speed up the execution of the increasing, deleting, modifying, and searching of data [6]. If the matched fault code matching is found, the diagnosis results are displayed by the Human Machine Interface. If there is no matching fault code, the fault tree in the NandFlash fault tree database is loaded into the memory. Finally, through the search algorithm based on fault tree, we find out the cause of the failure and give the failure results and diagnosis advice. The main purpose of the use of NandFlash is to trace back to the engine's fault signs when the diagnostic system is powered down.

5 Conclusions

In this paper, the fault tree and the generation rule are combined to analyze the fault, and the rules are extracted and expressed by the knowledge representation method. Based on the advantages of embedded system, the fault diagnosis model of embedded system is set up. The engine fault diagnosis system based on SQLite embedded database is designed and implemented. The design plays the advantage of real-time diagnosis of embedded system. The system improves the efficiency of system diagnosis, and provides a reference for the design of other embedded fault diagnosis systems.

Acknowledgement. This paper is supported by Hubei Provincial Department of Education Science and Technology Research Foundation of B2017589; College Teachers Guiding Students' innovation and entrepreneurship training program project of 2017.

References

1. Lv, J., Xu, S., Li, Y.: Application research of embedded database SQLite. Int. Forum Inf. Technol. Appl. **2**, 539–543 (2009)
2. Yue, K., Jiang, L., Yang, L., Pang, H.: Research of embedded database SQLite application in intelligent remote monitoring system. Int. Forum Inf. Technol. Appl. **2**, 96–100 (2010)
3. Wang, F.C., Shen, P., Yan, X., Wang, L.D.: Study on Application of SQLite for Locomotive Fault Diagnosis System. Railway Locomotive & Car (2012)
4. Assaf, T., Bechta Dugan, J.: Diagnostic expert systems from dynamic fault trees. In: Reliability & Maintainability, Symposium-RAMS, pp. 444–450 (2004)
5. Adamson, M.S., Roberge, P.R.: The development of a deep knowledge diagnostic expert system using fault tree analysis information. Can. J. Chem. Eng. **69**(1), 76–80 (2010)
6. Ouarnoughi, H., Boukhobza, J., Olivier, P., Plassart, L., Bellatreche, L.: Performance analysis and modeling of SQLite embedded databases on flash file systems. Des. Autom. Embed. Syst. **17**(3–4), 507–542 (2013)

Research on Stock Forecasting Based on GPU and Complex-Valued Neural Network

Lina Jia, Bin Yang(✉), and Wei Zhang

School of Information Science and Engineering, Zaozhuang University, Zaozhuang 277160, China
batsi@126.com

Abstract. Accurately and rapidly forecasting stock index is a difficult and hot research in the economic field. In this paper, complex-valued neural network (CVNN) model is proposed to predict stock price. In order to improve the time of training CVNN model, a parallel particle swarm optimization (PSO) based on graphics processing unit (GPU) is proposed to optimize the complex-valued parameters of CVNN model. Shanghai stock exchange composite index is selected to demonstrate the performance of CVNN model with parallel PSO algorithm. The experiment results reveal that our proposed method could improve stock index accurately and reduce training time sharply.

Keywords: Stock forecasting · Neural network · Complex-valued Graphics processing unit

1 Introduction

Stock market plays a leading and crucial role in the market mechanism, which connects the savers and investors. The operating mechanism of the stock market reflects the situation of national economy and is recognized as the signal system of the national economy [1, 2]. Because of some uncontrollable factors, the prediction of stock market index is considered to be a difficult job.

Artificial neural network (ANN) model is a mathematical model, which simulates biology neural network. ANN model has very strong self-learning function and associative storage function, and could search the optimal solution quickly. Recently ANN model have been used for time series prediction, computer vision, image recognition, market analysis, tracking radar, decision optimization, battlefield management, and spacecraft attitude control. Pang et al. proposed a deep long short-term memory neural network (LSTM) to forecast the stock market [3]. Guan et al. proposed High-order-fuzzy-fluctuation-Trends-based Back Propagation (HTBP) Neural Network model to forecast TAIEX dataset of Taiwan stock exchange and the Shanghai stock exchange composite index [4].

But ANN model has the huge amount of parameters and training process is very slow. Many parallel techniques have been proposed to improve the training time of NN model. Cao et al. proposed a parallel particle swarm optimization (PSO) based on MapReduce framework to optimize BP neural network algorithm in order to improve classification accuracy of big data [5]. Mei et al. proposed a multichannel structure of

D.-S. Huang et al. (Eds.): ICIC 2018, LNCS 10955, pp. 120–128, 2018.
https://doi.org/10.1007/978-3-319-95933-7_16

Hopfield neural network (MHNN) based on graphics processing unit (GPU) for the spectral mixture unmixing of hyperspectral images [6]. Beyeler et al. proposed a cortical neural network model to guide robot navigation on GPU platform [7].

Complex-valued versions of classical ANN models have been proposed to solve the complex problems. The experiments results revealed complex-valued neural network (CVNN) is more flexible and functional than real-valued NN model. In this paper, CVNN model is proposed to predict stock index. In order to reduce the training runtime of CVNN model, a parallel particle swarm optimization based on GPU framework is proposed to optimize the complex-valued parameters of CVNN model.

2 Method

2.1 Complex-Valued Neural Network

An example of three-layer complex-valued neural network is described in Fig. 1, which has m input nodes, n hidden nodes and one output node. In a complex-valued neural network, input variables, weights, threshold values and output variable are all complex numbers. Suppose that input vector $[z_1, z_2, \cdots, z_m]$. The output of i-th hidden node is computed as followed.

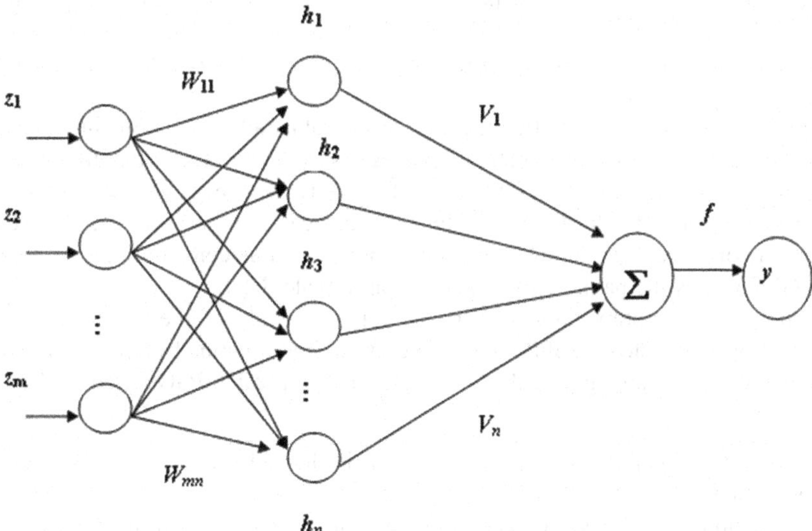

Fig. 1. An example of three-layer complex-valued neural network.

$$h_i = f(W_i + W_{i1}z_1 + W_{i2}z_2 + \ldots + W_{im}z_m). \tag{1}$$

Where W_i is threshold value of i-th hidden node and $W_{i1}, W_{i2}, \ldots W_{im}$ are the complex-valued weights between input layer and hidden layer. $f(\cdot)$ is complex-valued Sigmoid activation function, which is described as followed.

$$f(z) = \frac{1}{1+e^{-\mathrm{Re}(z)}} + j\frac{1}{1+e^{-\mathrm{Im}(z)}}. \tag{2}$$

Where $\mathrm{Re}(z)$ is the real part of complex value z and $\mathrm{Im}(z)$ is the imaginary part of complex value z.

The output layer is computed by

$$y = f(h_1V_1 + h_2V_2 + \ldots + h_nV_n). \tag{3}$$

Where $V_1, V_2, \ldots V_n$ are the complex-valued weights between hidden layer and output layer.

2.2 Parallel Optimization of Complex-Valued Parameters

Complex-valued parameters in a CVNN need to be optimized, containing weights (W_{ij} and V_i) and threshold value (W_i). For a CVNN m-n-1 in Fig. 1, the number of complex-valued parameters is $mn + n$. Each complex value has the real part and the imaginary part, so the number of parameters optimized is $2mn + 2n$. The parameter vector is $[\mathrm{Re}(W_{ij}), \mathrm{Im}(W_{ij}), \mathrm{Re}(W_i), \mathrm{Im}(W_i), \mathrm{Re}(V_i), \mathrm{Im}(V_i)]$. In order to find the optimal parameter vector, a parallel particle swarm optimization (PSO) based on GPU framework is proposed.

Graphics processing unit (GPU) has a more special parallel computing hierarchy, in which there are a large number of stream processors [8, 9]. General computing uses the cooperative processing mode of CPU and GPU. GPU (device) is generally used as a coprocessor of the CPU (host), and the host and device have their own memory. CPU deals with complex logic and the serial operations, such as complex logic operators, while GPU makes the parallel processing of large-scale data.

Particle swarm optimization is derived from the study of the behavior of bird predator [10]. The basic idea of PSO is to find the optimal solution by collaboration and information sharing among particles in the group. The parallel PSO based on GPU is described as followed.

(1) Initialize the population ($x_1, x_2, \ldots, x_N$, N is the number of particles) and the parameters.
(2) Population, velocity vector $(v_1, v_2, \ldots, v_N)$ and parameter information are converted into GPU data structure and stored in GPU memory. Population in GPU device is $[dev\_x_1, dev\_x_2, \ldots, dev\_x_N]$ and velocity vector in GPU device is $[dev\_v_1, dev\_v_2, \ldots, dev\_v_N]$.
(3) Population is divided into many subsets. The kernel function <Fitness_Kernel> is used to calculate the fitness of the particles in each subset.
(4) The kernel function <Update_Fitness_Kernel> is used to update the current optimal fitness value and position of each particle.

(5) The kernel function <Getbest_Kernel> is used to search the global optimal fitness value and position, which is stored in variable *dev_Gbest*.

(6) The kernel function <update_VP_Kernel> is used to update the velocity and position of each particle. A new velocity for particle i is updated by

$$\begin{aligned} dev\_v_i(t+1) &= dev\_v_i(t) + c_1 r_1 (dev\_Pbest_i - dev\_x_i(t)) \\ &+ c_2 r_2 (dev\_Gbest(t) - dev\_x_i(t)). \end{aligned} \tag{4}$$

Where c_1 and c_2 are positive constants, r_1 and r_2 are uniformly distributed random number in $[0, 1]$, and dev_Pbest_i is the current optimal solution of particle i.

According to the updated velocity dev_v_i, the position of particle i is updated by the following equation.

$$dev\_x_i(t+1) = dev\_x_i(t) + dev\_v_i(t+1). \tag{5}$$

(7) If maximum generation is reached, go to (8); otherwise go to (3).

(8) The optimal particle and the fitness value are copied to the CPU memory as the optimization result.

2.3 The Flowchart of Stock Prediction by Our Method

The flowchart of stock index prediction based on CVNN model and PSO on GPU is described in Fig. 2.

3 Experiment

Shanghai stock exchange composite index (Shanghai index) reflects the stock price changes in the Shanghai Stock Exchange, and was officially released since July 15, 1991. The experiment selects stock data from 1 January, 2011 to 1 January, 2015 for testing our method. The root mean squared error (*RMSE*) and S*peedUp* are proposed to evaluate the performance of our proposed method.

$$RMSE = \sqrt{\frac{1}{T}\sum_{i=1}^{T}\left(y^i_{target} - y^i_{forecast}\right)^2} \tag{6}$$

$$SpeedUp = \frac{Runtime(CPU)}{Runtime(GPU)} \tag{7}$$

Where T is the number of stock sample points, y^i_{target} is the actual stock index in the $i-th$ day and $y^i_{forecast}$ is the forecasting stock index in the $i-th$ day.

The experiments are executed on CPU and GPU, respectively. CPU environment has Intel Xeon E5-1620 3.5 GHz processor and 16 GB of RAM memory. GPU environment has NVIDIA Quadro K2200 graphic board. The used CVNN model has 5

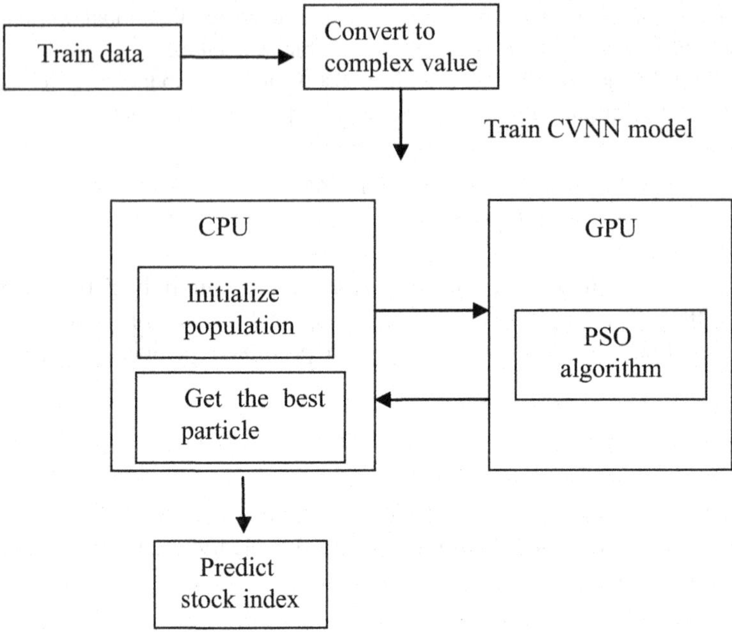

Fig. 2. The flowchart of stock prediction.

input variable, 6 hidden points and one output point. In PSO algorithm, the number of particles is set to 200 and maximum generation is set to 1000. 500 data points are used to train CVNN model and 244 samples are used for testing. The predicting result of CVNN model on GPU is described in Fig. 3. From Fig. 3, we can see that our method could predict stock index accurately. Real-valued neural network (RVNN), wavelet neural network (WNN), ordinary differential equation (ODE) and CVNN model on CPU are also used to predict the same stock data. The results are listed in Table 1. From Table 1, it can be clearly seen that CVNN model has smaller RMSE than state-of-the-art methods. CVNN model based on GPU has the similar performance with the model in CPU, which reveals that GPU does not affect the prediction accuracy of CVNN model.

In order to test the performance of parallel PSO algorithm based on GPU, PSO algorithms with different numbers of particles and different maximum generations are executed on CPU and GPU, respectively. The running times of PSO with different numbers of particles are listed in Table 2. When the number of particles is equal to 100, the runtime of our method on GPU is more than that on CPU. This is due to that when the amount of data is relatively small, GPU improvement time is less than the wasted time because of transmitting data between CPU and GPU. As the number of particles is much more, GPU can obviously improve the execution time of the algorithm. S*peedUp* performance is depicted in Fig. 4. From Fig. 4, it can be seen that as the number of particles increases, the amplitude of improved runtime increases. When the amount of parallel processing data is too large, the amplitude of the improvement tends to be

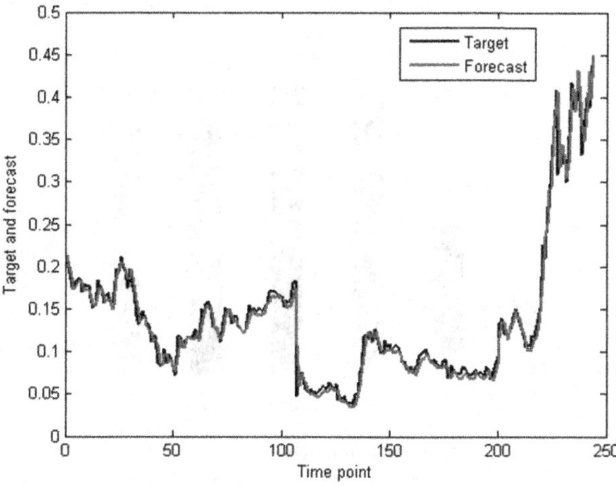

Fig. 3. The prediction result of the Shanghai index using CVNN model on GPU.

Table 1. RMSE results of five different methods for Shanghai index.

Methods	RMSE
RVNN-PSO [11]	0.01325
WNN-PSO [11]	0.01385
ODE [12]	0.01302
CVNN+CPU	0.01278
CVNN+GPU	0.01273

stable. The running time of PSO with different maximum generations is described in Fig. 5, which reveals that GPU obviously improves the training time of CVNN model. Figure 6 shows S*peedUp* performance of PSO with different maximum generations. S*peedUp* is about 8 times and very stable, due to that iteration optimization is serial.

Table 2. Runtime of PSO with different number of particles on CPU and GPU for Shanghai index.

Number of particles	Runtime (CPU)	Runtime (GPU)
100	16.19	39.329
500	75.728	39.602
1000	146.182	41.282
2000	293.907	46.616
3000	442.053	62.692
4000	577.215	96.074
5000	759.068	127.609

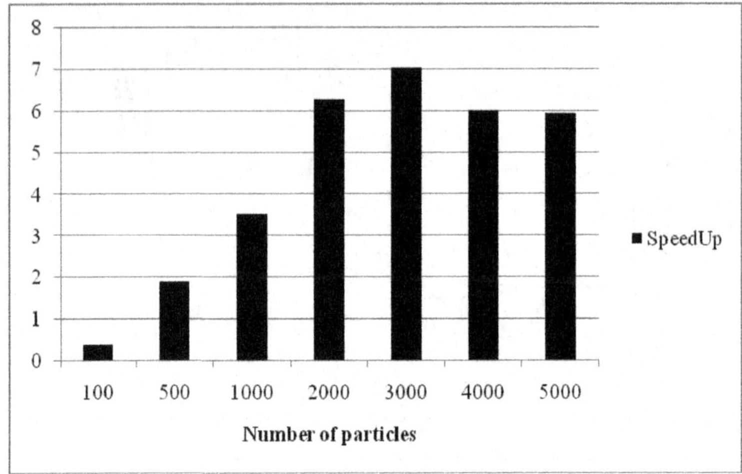

Fig. 4. SpeedUp of PSO with different number of particles for Shanghai index.

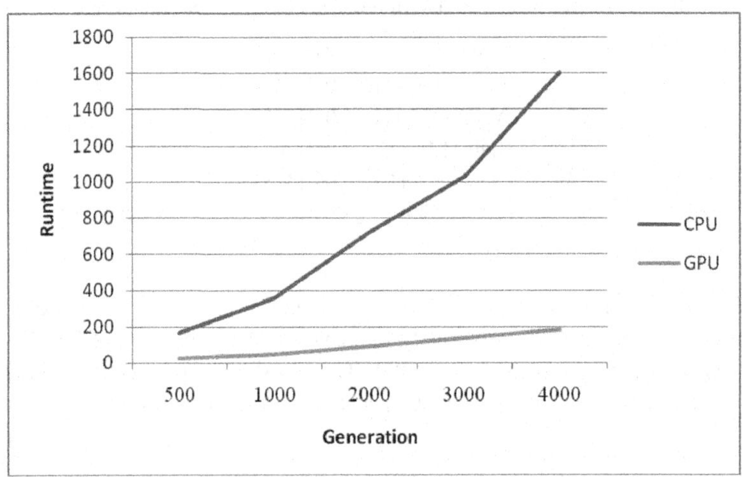

Fig. 5. Runtime of PSO with different maximum generations on CPU and GPU for Shanghai index.

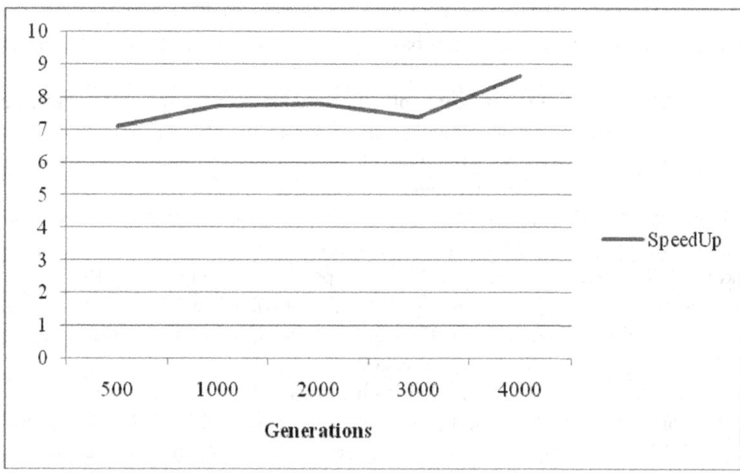

Fig. 6. *SpeedUp* performance of PSO with different maximum generations on CPU and GPU for Shanghai index.

4 Conclusion

In order to improve the accuracy and time of forecasting stock index, this paper proposed a novel prediction method based on complex-valued neural network model and parallel particle swarm optimization. Parallel particle swarm optimization based on GPU platform is proposed to evolve the complex-valued parameters of CVNN model. Our proposed method is used to predict Shanghai index. Our method could predict stock index accurately. CVNN model has smaller RMSE than RVNN model and GPU does not affect the prediction accuracy of CVNN model. PSO based on GPU could reduce training time of CVNN model sharply.

Acknowledgments. This work was supported by the Natural Science Foundation of China (No. 61702445), the PhD research startup foundation of Zaozhuang University (No.2014BS13), Zaozhuang University Foundation (No. 2015YY02), and Shandong Provincial Natural Science Foundation, China (No. ZR2015PF007).

References

1. Fama, E.F.: The behavior of stock-market prices. J. Bus. **38**(1), 34–105 (1965)
2. Chen, N.F., Roll, R., Ross, S.A.: Economic forces and the stock market. J. Bus. **59**(3), 383–403 (1986)
3. Pang, X., Zhou, Y., Wang, P., Lin, W., Chang, V.: An innovative neural network approach for stock market prediction. J. Supercomput. **1**, 1–21 (2018)
4. Guan, H., Dai, Z., Zhao, A., He, J.: A novel stock forecasting model based on High-order-fuzzy-fluctuation trends and back propagation neural network. PLoS ONE **13**(2), e0192366 (2018)

5. Cao, J., Cui, H., Shi, H., Jiao, L.: Big data: a parallel particle swarm optimization-back-propagation neural network algorithm based on MapReduce. PLoS ONE **11**(6), e0157551 (2016)
6. Mei, S., He, M., Shen, Z.: Optimizing hopfield neural network for spectral mixture unmixing on GPU platform. IEEE Geosci. Remote Sens. Lett. **11**(4), 818–822 (2013)
7. Beyeler, M., Oros, N., Dutt, N., Krichmar, J.L.: A GPU-accelerated cortical neural network model for visually guided robot navigation. Neural Netw. **72**, 75–87 (2015)
8. Shiraki, A., Masuda, N., Tanaka, T., Sugie, T., Ito, T.: Computer generated holography using a graphics processing unit. Opt. Express **14**(2), 603–608 (2006)
9. Brodtkorb, A.R., Hagen, T.R., Sætra, M.L.: Graphics processing unit (GPU) programming strategies and trends in GPU computing. J. Parall. Distrib. Comput. **73**(1), 4–13 (2013)
10. Kennedy, J.: Particle swarm optimization. In: Sammut, C., Webb, G.I. (eds.) Encyclopedia of Machine Learning. Springer, Boston (2002)
11. Chen, Y.H., Abraham, A.: Hybrid learning methods forstock index modeling. In: Kamruzzaman, J., Begg, R.K., Sarker, R.A. (eds.) Artificial Neural Networks in Finance, Health and Manufacturing: Potential and Challenges, IdeaGroup Inc. Publishers, USA (2006)
12. Chen, Y., Yang, B., Meng, Q., Zhao, Y., Abraham, A.: Time-series forecasting using a system of ordinary differential equations. Inf. Sci. **181**(1), 106–114 (2011)

Development of Remote Monitoring System of Communication Base Station Using IoT and Particle Filter Technology

Yang-Weon Lee(✉)

Honam University, 417, Eodeung-daero, Gwangsan-gu, Gwangju 62399, Republic of Korea
ywlee@honam.ac.kr

Abstract. This research is carried out mainly through the method of IoT and particle filter applications. We collected the information by looking through the homepages of the world famous IoT appliances brands on the internet. And we combined the collected information in the literature investigating how to combine to particle filters for monitoring base communication station remotely.

The developed system collects environmental data of wireless communication base station using several sensors, analyzes the collected data using particle filters, and controls the remotely base station. As a wireless remote control method, we implemented Bluetooth, Ethernet and Wi-Fi. Finally, it is designed for users to enable remote control and monitoring when the user is not in the base station.

Keywords: Particle filter · Smart control system · Wireless sensor network

1 Introduction

The fourth industrial revolution has become a hot topic in recent years [1]. The basic idea of the fourth industrial revolution is as shown in Fig. 1, as the super connection, super intelligence, and large fusion, and the industrial ecosystem generates vast data through IoT and IoP (Internet of People). AI can be understood as providing the super intelligent product production/service by making appropriate judgment and autonomous control based on Deep Learning of Big Data and leading the industrial revolution.

The IoT refers to the Internet environment in which people, objects, and data are all connected to a wired/wireless network to generate, collect, share, and utilize information. On the other hand, applying the Internet of Things to real life requires the integrated implementation of the underlying technologies. The technologies required are largely the result of the middleware software that is used to store and analyze sensor and network hardware technologies, such as controller, wireless chip, and data reconstruction, and to represent the data representation [2] (Fig. 2).

In recent years, radio base stations have become so widespread in large areas that they are helping to ensure seamless communication. Furthermore, these base stations contain islands and mountainous areas, so it is very difficult to maintain them.

D.-S. Huang et al. (Eds.): ICIC 2018, LNCS 10955, pp. 129–135, 2018.
https://doi.org/10.1007/978-3-319-95933-7_17

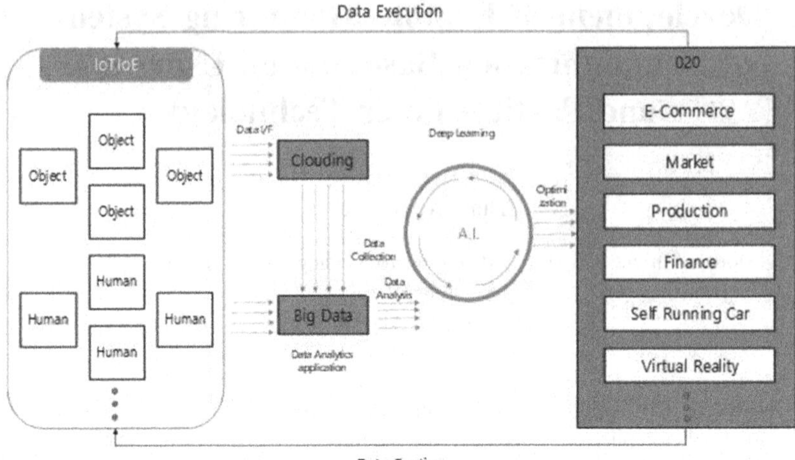

Fig. 1. Mechanism of 4th industrial revolution

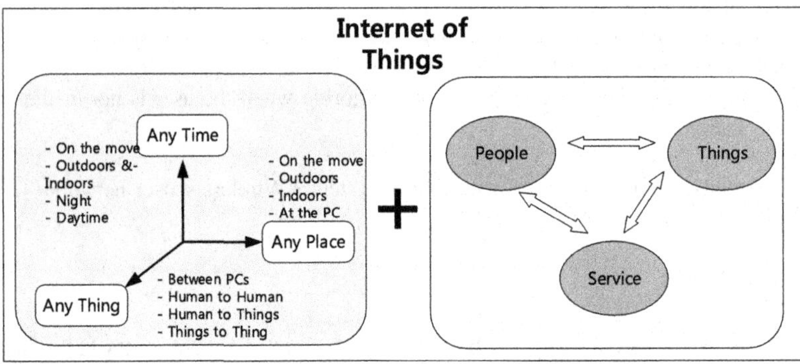

Fig. 2. Communication environmental of IoT

In particular, in recent years, it has become increasingly difficult to maintain the system, such as to intentionally destroy base station facilities or to attempt to prevent arson on the ground that it harms EM threats to electromagnetic humans and animals and the surrounding beauties. Consequently, the need for remote monitoring in real time is increasing [3].

In this study, to solve these problems, a system that is based on cloud services is constructed, and a wireless base station scattered across the country is connected to the Internet, and IoT sensor (fire, temperature and humidity) is detected at each base station (Fig. 3).

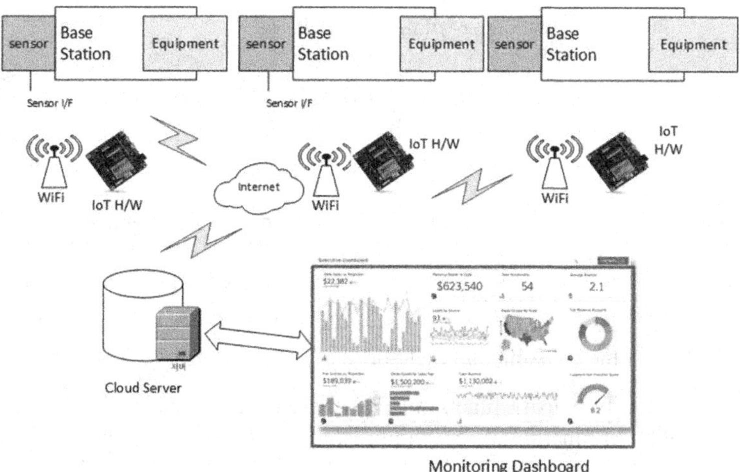

Fig. 3. Overall configuration of remote monitoring system of base station

2 Modeling of Particle Filter for Remote Monitoring System Analysis

The general particle filter approach to manage the sensor data of base communication station, also known as the condensation algorithm [4] and Monte Carlo localization [5, 6], uses a large number of particles to explore the state space. Each particle represents a hypothesized sensor location in state space. Initially the particles are uniformly randomly distributed across the state space, and each subsequent frame the algorithm cycles through the steps illustrated in Fig. 4:

1. To initialize, draw *N* particles from an *a priori* distribution $p(x_0)$. Set all weights to be normalized and equal.

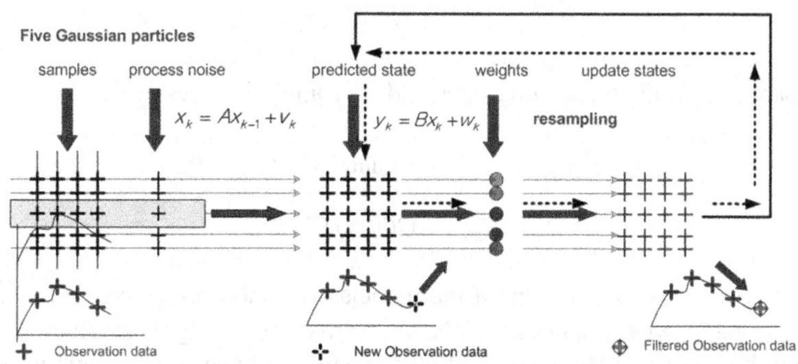

Fig. 4. Particle filter calculation process

$$x_0^i \sim p(x_0), \quad i = 1 : N \tag{1}$$

$$m_0^i = \frac{1}{N}, i = 1 : N \tag{2}$$

2. Propagate the particle set according to a process model of target dynamics.

$$x_k^i = f\left(x_{k-1}^i\right) + w_{k-1}^i, i = 1 : N \tag{3}$$

Here $f(x_{k-1}^i)$ is the dynamics model and w_{k-1}^i is the noise.

3. Update probability density function (PDF) based on the received measurement z_k and the likelihood function of receiving that measurement given the current particle value.

$$m_k^i = m_{k-1}^i p\left(z_k | x_k^i\right), i = 1 : N \tag{4}$$

4. Normalize the weights.

$$m_k^i = \frac{m_k^i}{\sum_{i=1}^N m_k^i} \tag{5}$$

5. Compute the effective sample size. The bounds on the effective sample size are given by $1 \le N_{eff} \le N$.

$$N_{eff} = \frac{1}{\sum_{i=1}^N (m_k^i)^2} \tag{6}$$

6. Compare N_{eff} to the resampling threshold. Resample if necessary.

$$N_{eff} < N_{thr} \rightarrow \text{Resample and GO TO STEP 2}$$

$$N_{eff} \ge N_{thr} \rightarrow \text{GO TO STEP 2} \tag{7}$$

Here, N_{eff} is used to determine if the particles are unevenly distributed. Also if a few particles have large weighting and the other have small weighting values, resulting in N_{eff} will be close to 1. In contrast, if all the particles have equal weighting values, N_{eff} will be equal to N.

At any time step, the value of the estimate is as follows:

$$\bar{x}_k = \sum\nolimits_{i=1}^{N} \left(x_k^i - \bar{x}_k\right)^2 m_k^i \tag{8}$$

The covariance can be computes using

$$P_k = \sum\nolimits_{i=1}^{N} \left(x_k^i - \bar{x}_k\right)^2 m_k^i \tag{9}$$

This results in particles congregating in regions of high probability and dispersing from other regions, thus the particle density indicates the most likely sensor node states. See [6] for a comprehensive discussion of this method. The key strengths of the particle filter approach to localization and tracking are its scalability (computational requirement varies linearly with the number of particles), and its ability to deal with multiple hypotheses (and thus more readily recover from tracking errors). However, the particle filter was applied here for several additional reasons [8–10]:

- It provides an efficient means of searching for a sensor in a multi-dimensional state space.
- Reduces the search problem to a verification problem, i.e. is a given hypothesis face-like according to the sensor information?
- Allows fusion of cues running at different frequencies.

3 Implementation of Remote Control Base Station System

Figure 5 shows the functional diagram of the remote monitoring system of the cloud service of based wireless communication base station to be developed in this research. The flow of the main functions is as follows: First, each wireless base station detects intrusion detection, temperature, humidity, and flame by using four sensors, and transmits the transmission frequency status and power status in the base station equipment to remote logging information (transmission frequency, transmission output, transmission bandwidth range, power information), preprocess in IoT Device H/W, and connect to Internet AP using Wi-Fi. Information connected to the Internet is transmitted to the cloud system in real time. The cloud system is implemented as Software as a Service (SaaS), so that the actual company does not need to buy a separate cloud server, and it is implemented in a way that only the monthly fee is paid.

The wireless base station manager monitors the status information of each base station displayed on the dashboard, and automatically sets alarm ranges and sends alarm characters to the system.

The sensor and base station information are stored in real time in the database and are designed to accumulate data in a round-robin manner, so that the desired time zone can be shortened or enlarged.

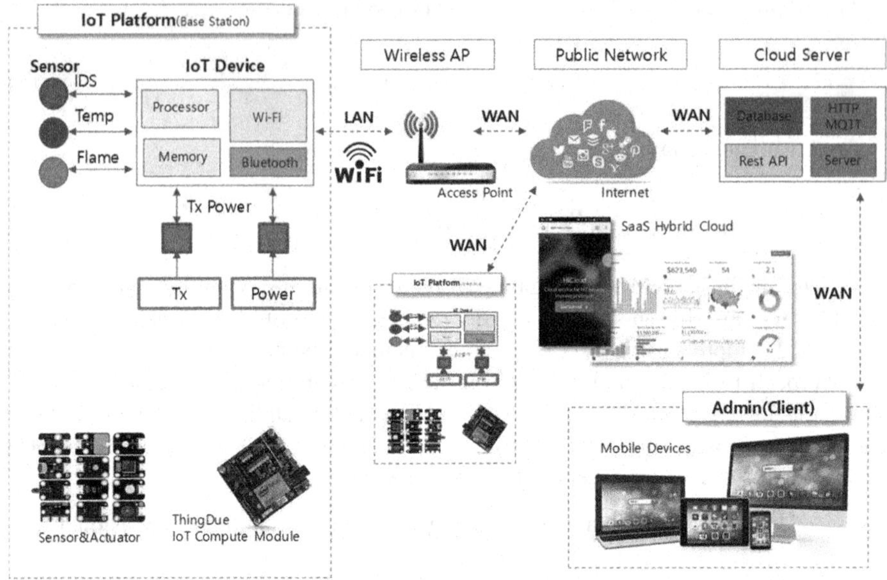

Fig. 5. System operation flow diagram of S/W structure

4 Conclusion and Future Work

In this study, we developed the hardware and software that applies IoT and cloud service technologies, which are key keywords in the fourth industrial revolution.

The IoT platform and node-RED proposed in this paper are used to provide services that are more capable of providing faster maintenance and also to drastically reduce maintenance costs.

References

1. Marcelo, T.O.: IOT and industry 4.0: the industrial new revolution. In: International Conference on Management and Information Systems, pp. 75–82 (2017)
2. Atzori, L., Iera, A., Morabito, G.: The internet of things: a survey. Comput. Netw. **54**(15), 2787–2805 (2010)
3. Carpenter, D.O.: Human health effects of nonionizing electromagnetic fields. In: Bingham, E., Cohressen, B. (eds.) Patty's Toxicology, 6th edn., vol. 6, pp. 109–32. John Wiley & Sons, Inc., Hoboken. (2012). Chapter 100
4. Fortmann, T.E., Bar-Shalom, Y., Scheffe, M.: Sonar tracking of multiple targets using joint probabilistic data association. IEEE J. Oceanic Eng. **8**(3), 173–184 (1983)
5. Gordon, N., Salmond, D., Smith, A.: Novel approach to nonlinear/non-Gaussian Bayesian state estimation. In: IEE Proceedings .F, Radar and signal processing, pp. 107–113 (1993)
6. Isard, M., Blake, A.: Condensation. Conditional density propagation for visual tracking. Int. J. Comput. Vis. 5–28 (1998)

7. Lee, Y.-W.: Design of smart garden system using particle filter for monitoring and controlling the plant cultivation. In: Huang, D.-S., Hussain, A., Han, K., Gromiha, M. Michael (eds.) ICIC 2017. LNCS (LNAI), vol. 10363, pp. 461–466. Springer, Cham (2017). https://doi.org/10.1007/978-3-319-63315-2_40
8. Lee, Y.W.: Implementation of mutual localization of multi-robot using particle filter. In: Huang, D.-S., Jiang, C., Bevilacqua, V., Figueroa, J.C. (eds.) ICIC 2012. LNCS, vol. 7389, pp. 87–94. Springer, Heidelberg (2012). https://doi.org/10.1007/978-3-642-31588-6_12
9. Lee, Y.: Optimization of moving objects trajectory using particle filter. In: Huang, D.-S., Bevilacqua, V., Premaratne, P. (eds.) ICIC 2014. LNCS, vol. 8588, pp. 55–60. Springer, Cham (2014). https://doi.org/10.1007/978-3-319-09333-8_7
10. Lee, Y.-W.: Implementation of interactive interview system using hand gesture recognition. Neurocomputing **116**, 272–279 (2012)

Prediction of Dissolved Oxygen Concentration in Sewage Using Support Vector Regression Based on Fuzzy C-means Clustering

Xing-Liang Shi[1], Jian Zhou[1], Xiao-Feng Wang[2(✉)], and Le Zou[2]

[1] Department of Environmental Engineering, Hefei University, Hefei 230601, Anhui, China

[2] Key Lab of Network and Intelligent Information Processing, Department of Computer Science and Technology, Hefei University, Hefei 230601, China

xfwang@hfuu.edu.cn

Abstract. In order to solve the problem of real-time measurement of dissolved oxygen in wastewater treatment process, a support vector regression algorithm based on fuzzy C-means clustering is proposed to predict the content of dissolved oxygen (DO) in sewage. Firstly, the whole samples are divided into many sub-samples by fuzzy C-mean clustering. Then, a support vector regression model is established on each sub sample. Compared with other prediction methods, the proposed model has good comprehensive prediction performance. It can satisfy the actual demand prediction of DO dissolved oxygen in sewage.

Keywords: Fuzzy clustering · Support vector regression · Dissolved oxygen DO prediction

1 Introduction

Dissolved oxygen (DO) is one of the important indexes to evaluate water quality. The deficiency or excess of DO will affect the water quality. Accurate prediction of DO is of great significance. However, many of the current measurements are aimed at the measurement of the current DO concentration, mathematical models [1] should be established to predict future concentration values.

In 1940s, foreign scholars began to study sewage treatment models. In 1980s, domestic and foreign scholars introduced microbial growth kinetics into the field of wastewater treatment. The International Water Quality Association (IAWQ) developed the activated sludge Model No. 1 (ASM1A) [2] on the basis of previous studies. In 90s, support vector machine (SVM) appeared. The algorithm is simple and robust. Song [3] used SVM to predict effluent TP. Cheng [4] used least square support vector machine to predict COD. However, the training sample size of the above models are all small, the measurement of influent parameters did not take the lag into account. Hence, there might be under-fitting for the massive data, which leads to the poor prediction accuracy [5].

D.-S. Huang et al. (Eds.): ICIC 2018, LNCS 10955, pp. 136–142, 2018.
https://doi.org/10.1007/978-3-319-95933-7_18

In this paper, a support vector regression model based on fuzzy C-means clustering (FCM-SVR) is proposed to predict the concentration of dissolved oxygen. The FCM-SVR can improve the accuracy of the dissolved oxygen prediction.

2 Related Methods

2.1 Fuzzy C-Means Clustering Algorithm

Fuzzy C-means clustering (FCM) algorithm [6, 7] is a partition-based clustering algorithm. FCM takes n vectors x_i (i = 1, 2,… n) are divided into c fuzzy groups, and find the cluster center of each group.

FCM allows each given data point to determine the degree of belonging to each group with 0 ~ 1. The sum of the membership degrees of a data set is equal to 1.

FCM's value function (or objective function) such as Formula 1:

$$J(U, c_1, \ldots, c_c) = \sum_{i=1}^{c} \sum_{j=1}^{n} u_{ij}^m d_{ij}^2 \tag{1}$$

Where u_{ij} is between 0 and 1, c_i is the cluster center of fuzzy group I, $d_{ij} = \|x_j - c_i\|$ denotes the distance from the sample point x_j to cluster center c_i.

In order to minimize the objective function, the Lagrange multiplier method is used to establish the objective optimization function such as Formula 2.

$$J(U, c_1, \ldots, c_c) = \sum_{i=1}^{c} \sum_{j=1}^{n} u_{ij}^m d_{ij}^2 + \sum_{j=1}^{n} \lambda_j \left(\sum_{i=1}^{c} u_{ij} - 1 \right) \tag{2}$$

$\lambda_j, j = 1, \ldots, n$ is the Lagrange multipliers of the n constraints of formula 1, in order to obtain the derivation of all input parameters so that Formula 2 reaches the minimum necessary condition such as Formulas 3 and 4.

$$C_i = \frac{\sum_{j=1}^{n} u_{ij}^m x_j}{\sum_{j=1}^{n} u_{ij}^m}, \; i = 1, 2, \ldots, c \tag{3}$$

$$u_{ij} = \frac{1}{\sum_{k=1}^{c} \left(\frac{d_{ij}}{d_{kj}} \right)^{\frac{2}{m-1}}}, i = 1, 2, \ldots, c; j = 1, 2, \ldots, n \tag{4}$$

As mentioned above, the steps of the FCM algorithm are as follows:

1. Initialize the membership matrix u_{ij} with a random number between 0 and 1.
2. Calculation of c cluster centers c_i by Formula 3, i = 1,…, c.
3. The target Function 1 is calculated according to Formula 2. If it is less than a certain threshold, the algorithm stops.
4. Calculation of New Matrix u_{ij} by Formula 4, then return to step 2.

2.2 Support Vector Regression Algorithm

The basic idea of SVR [8] is to find an optimal classification surface to minimize the error of all training samples of the optimal classification surface. Suppose the regression function is shown in Eq. (5):

$$f(x) = w\phi(x) + b \tag{5}$$

Φ(x) is a non-linear mapping function, w is the weight of each sample, b is the bias factor. The objective function is shown in Eq. (6):

$$L = min\frac{1}{2}\|w\|^2 + C\sum_{i=1}^{n}(\xi_i + \xi_i^*) \tag{6}$$

C is the penalty factor which is given. ξ_i, ξ_i^*, respectively, the relaxation factor of the upper and lower boundaries. The introduction of Lagrange multipliers α, α^*, β, β^*. The Eq. (6) is converted to dual form is shown in Eq. (7):

$$\begin{aligned} \mathrm{L} = \frac{1}{2}\|w\|^2 + C\sum\nolimits_{i=1}^{n}(\xi_i + \xi_i^*) - \sum\nolimits_{i=1}^{n}\alpha(\varepsilon + \xi_i - y(i) + w\phi(x) + b) \\ - \sum_{i=1}^{n}\alpha^*(\varepsilon + \xi_i^* + y(i) - w\phi(x) - b) - \sum_{i=1}^{n}(\beta\xi_i + \beta^*\xi_i^*) \end{aligned} \tag{7}$$

Find the partial derivatives of parameters *w, b,* ξ_i, *and* ξ_i^* and make them 0, then substitute the formula (5) to obtain the formula (8) as follows:

$$\mathrm{f}(x) = \sum\nolimits_{i=1}^{n}(\alpha_i - \alpha_i^*)(x_i, x) + b \tag{8}$$

α_i, α_i^* is the solution of the dual problem, b is the optimal solution of the threshold.

3 Prediction of Dissolved Oxygen Concentration by Support Vector Regression Based on Fuzzy Clustering

3.1 Data Correlation Analysis

In order to exclude some irrelevant indicators, we do correlation analysis on data samples. The factors affecting the DO concentration are as follows: PH at *t-1* moment, MLSS, ORP, inlet NH4N, effluent COD, effluent TP, inflow cumulative flow and effluent cumulative flow. The output is DO concentration at *t* moment.

The correlation coefficient formula is as follows:

$$R(X,Y) = \frac{\sum_{i=1}^{n}(x_i - \bar{x})(y_i - \bar{y})}{\sqrt{\sum_{i=1}^{n}(x_i - \bar{x})^2}\sqrt{\sum_{i=1}^{n}(y_i - \bar{y})^2}} \tag{9}$$

In this equation, x are factors, If $R(X, Y) > 0$, X is positively related to Y and its value ranges in [0,1]. If $R(X, Y) < 0$, there is a negative correlation between X and Y and its value ranges in [−1, 0]. If $R(X, Y) = 0$, there is no correlation between X and Y.

3.2 Construction of FCM-SVR Model

Multiple DO related index can be obtained by correlation analysis. The data is pre-treated based on fuzzy clustering algorithm. We can get the sample data of C sub-classes. Then, we use the SVR algorithm to model the C subclasses and standardize each type of samples. Its standardized formula (10) is shown as follows:

$$\bar{x} = \frac{x - mean(x)}{std(x)} \tag{10}$$

where x is the training sample. mean(x) and std (x) are the mean and standard deviation of the training samples.

The data samples of each category are divided into training data and test data according to the proportion of 8:2. The RBF function is chosen as the kernel function of the SVR model which is shown as follows:

$$\mathrm{k}(x, y) = exp\left(\frac{-\|x - y\|^2}{2g^2}\right) \tag{11}$$

The penalty coefficient V for the training data of C classes and the best super parameters of the kernel parameter are selected by cross validation. The principle of cross validation is to randomly divide the data sets into k groups, each of which takes one of the groups as test data and the remaining k-1 groups as the training data. The best parameters obtained are used to train the training data of C classes to obtain different support vector regression models. The test samples are used to predict and verify the established model. In the same way, we first classify the prediction samples by fuzzy C-means clustering, then classify them into the corresponding SVR models, and predict them. The prediction results are compared with the actual values. The root mean square error and the mean absolute error are used to verify the advantages and disadvantages of the model as follows:

$$\mathrm{RMSE} = \left(\frac{1}{n}\sum\nolimits_{i=1}^{n} (y_i - y_{testi})^2\right)^{\frac{1}{2}} \tag{12}$$

$$\mathrm{MAE} = \frac{1}{n}\sum\nolimits_{i=1}^{n} |y_i - y_{testi}| \tag{13}$$

where y_i is the predictive value, y_{testi} is the test data, n is the number of samples.

4 Results and Analysis

The used data were collected from a sewage treatment plant in Hefei, Anhui Province in 2017 which contains 46000 groups of samples. The water quality parameters are PH, MLSS, ORP, inlet NH4N, effluent COD, effluent TP, the accumulative flow of water and the cumulative discharge of the effluent.

The correlation between the concentration of dissolved oxygen DO and other indexes was calculated by Eq. (9). The results are shown in Table 1.

Table 1. The correlation between dissolved oxygen DO and other indexes

Correlation	PH	MLSS	ORP	Inlet NH4N	Effluent COD	Effluent TP	Inflow cumulative flow	Effluent cumulative flow
DO	0.65	−0.03	0.45	0.76	0.53	0.71	0.75	0.74

It can be seen from Table 1 that PH, ORP, inlet NH4N, effluent COD, effluent TP, inflow accumulative flow and effluent cumulative flow is positively correlated with DO. The correlation degree is relatively high. MLSS is negatively correlated with DO, and the correlation is low. Therefore, the sample data of clustering are PH, ORP, inlet NH4N, effluent COD, effluent TP, inflow accumulative flow, and effluent cumulative flow. The variation curve of the original data of each indicator over time is shown in Fig. 1.

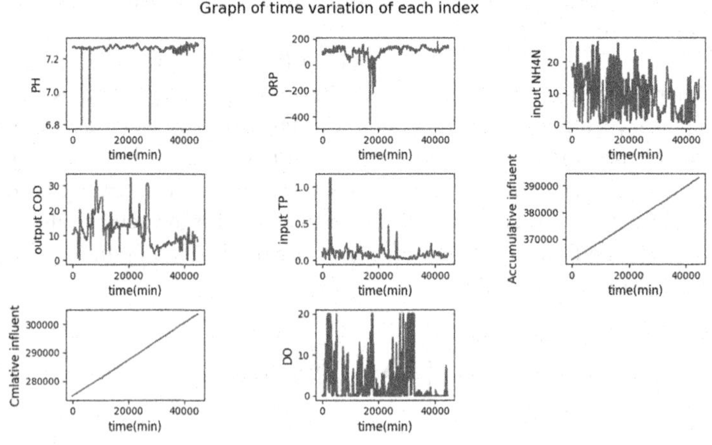

Fig. 1. Graph of time variation of each index

We use FCM-SVR to predict the dissolved oxygen DO and compare it with a single support vector regression model. The result is shown in Fig. 2. The abscissa is the annual test data sequences in 2017, 80% of it are selected as training data, and 20%

remaining as test data. The ordinate is the concentration of dissolved oxygen DO. The curves in Fig. 2 show the result of the actual value and the predicted value.

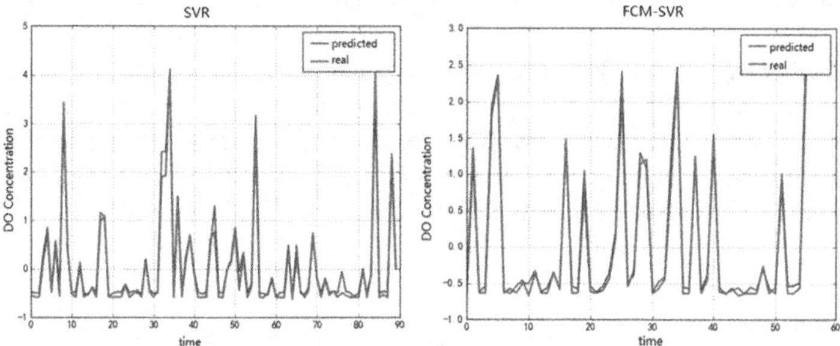

Fig. 2. The diagram comparison of SVR and FCM-SVR

In order to verify the results of the model, we use the formulas (12) and (13) to measure the prediction effect. The results are shown in Table 2.

Table 2. Error analysis of SVR and FCM-SVR

Quarter	SVR		FCM-SVR	
	MAE%	RMSE%	MAE%	RMSE%
First	0.14	0.24	0.14	0.22
Second	0.12	0.21	0.12	0.18
Third	0.15	0.24	0.13	0.23
Fourth	0.11	0.18	0.11	0.16

From Fig. 2 and Table 2, we can see that the FCM-SVR model can better match the complex nonlinear relationship between the indicators of the sewage and the dissolved oxygen concentration. The prediction result is obviously better than the single SVR model. According to Table 2, the FCM-SVR algorithm compared with the SVR algorithm, the MAE did not change significantly in the first quarter to the fourth quarter, and RMSE decreased by 2%, 3%, 1% and 2% respectively. Through the comparison of the four groups of experiments, we can see that the FCM-SVR model has a better prediction accuracy than the SVR model. Compared with SVR model, it has good comprehensive performance, which can meet the actual needs of sewage quality prediction.

5 Conclusion

In this paper, we used sewage water quality index as input and constructed the prediction model of dissolved oxygen concentration based on FCM-SVR. Under the same conditions, the simulation results show that the proposed FCM-SVR model is superior to the SVR model. The proposed algorithm not only reduces the time complexity but also has higher prediction accuracy. This proposed model is superior to other time series prediction methods in comprehensive performance. More importantly, it provides an effective solution for predicting water quality quickly and accurately.

Acknowledgements. This work was supported by the grant of the National Natural Science Foundation of China, No.61672204, the grant of Major Science and Technology Project of Anhui Province, No.17030901026, the grant of Key Constructive Discipline Project of Hefei University, No. 2016xk05, the grant of the key Scientific Research Foundation of Education Department of Anhui Province, No. KJ2018A0555, KJ2017A542.

References

1. Liu, T.: Modeling and parameter optimization analysis of activated sludge system (A2/O). Harbin Institute of Technology (2017)
2. Henzm, M.: Activated sludge models ASM1, ASM2, ASM2d and ASM3, pp. 13–15. IW-A Publishing, London (2000)
3. Song, X.M.: Study on soft sensor model of wastewater measuring based on SVM. Nanchang University (2007)
4. Cheng, C.: Mixed multi-modeling soft measurement research on biological aerated filter sewage treatment. Anhui University of Technology (2016)
5. Huang, W.: Research on the engineering application of high concentration activated sludge process treating wastewater. Harbin Institute of Technology (2012)
6. Balafar, M.A.: Fuzzy C-mean based brain MRI segmentation algorithms. Artif. Intell. Rev. **413**, 441–449 (2014)
7. You, J.C., Mao, H.H.: Classification of seabed sediment using the fuzzy C mean clustering and support vector machine methods. Mar. Sci. **11**, 122–130 (2014)
8. Ding, S.F., Qi, B.J., Tan, H.Y.: An overview on theory and algorithm of support vector machines. J. Univ. Electron. Sci. Technol. China **40**(1), 2–7 (2011)

A Deep Reinforcement Learning Method for Self-driving

Yong Fang(✉) and Jianfeng Gu(✉)

Key Laboratory for Specialty Fiber Optics and Optical Access Networks, Shanghai Institute for Advanced Communication and Data Science, Shanghai University, Shanghai 200444, China
yfang@staff.shu.edu.cn, jianfeng@shu.edu.cn

Abstract. Self-driving technology is an important issue of artificial intelligence. Basing on the end-to-end architecture, deep reinforcement learning has been applied to research for self-driving. However, self-driving environment yields sparse rewards when using deep reinforcement learning, resulting in local optimum to network training. As a result, the self-driving vehicle does not obtain correct actions from outputs of neural network. This paper proposes a deep reinforcement learning method for self-driving. According to the classification threshold value that is dynamically adjusted by reward distributions, the sparse rewards is divided into three groups. The experience information for different rewards is fully utilized and the local optimum problem in the network training process is avoided. By comparing with the traditional method, simulation results show that the proposed method significantly reduces the training time of network.

Keywords: Self-driving · Deep reinforcement learning · Sparse rewards
Reward classification

1 Introduction

In the early decades of the 21th century, the pace of innovation is speeding up and the industry is on the brink of a new technological revolution: "self-driving" vehicles [1].

Deep Q-Network (DQN) which combines reinforcement learning and neural networks was proposed by Mnih et al. [2, 3] in 2013. The DQN is a convolutional neural network, trained with a variant of Q-learning [4], whose input is RGB images and whose output is a value function estimating future rewards. DQN has achieved great success at the Atari 2600 games, even surpassing the level of human top players in some games. Afterwards Lillicrap et al. [5] presented an actor-critic [6], model-free algorithm based the deterministic policy gradient (DPG) [7] that can operate over continuous action spaces, to solve the problem of "dimension disasters" with DQN.

Supported by the National Natural Science Foundation of China under Grants 61673253 and 612 71213, and the Ph.D. Programs Foundation of Ministry of Education of China under Grant 20133 108110014.

D.-S. Huang et al. (Eds.): ICIC 2018, LNCS 10955, pp. 143–152, 2018.
https://doi.org/10.1007/978-3-319-95933-7_19

These deep reinforcement learning algorithms provide a new idea to solve the problem of agent control.

Basing on the end-to-end architecture, deep reinforcement learning has some applications in self-driving tasks. Literature [8] proposed a reinforcement learning approach using DQN to steer a vehicle in a 3D physics simulation. Relying solely on camera image input the approach directly learns steering the vehicle in an end-to end manner. In [9], the authors used DQN algorithm to solve disadvantages of automated driving systems (ADS) in the scenario of the on-ramp merging. In reality, the ADS will not work in the scenario of the on-ramp merging until the car reaches the main road of highway. When entering the main road from the bypass road, the car still has to be manually operated, causing some troubles for the drivers. Thus, authors employed DQN algorithm to control car replacing driver. Moreover, they used the Long Short-Term Memory (LSTM) [10] to send useful historical information for network training. The deep reinforcement learning method for continuous control is based on historical data, which would make unpredicted decisions in unfamiliar scenarios. In other words, the generalization ability of the neural networks are so weak that it may causes some traffic accidents. In order to handle this issue, Xiong [11] developed a new method which combines the artificial potential field, path tracking and DDPG algorithm. Sallab [12] proposed a self-driving framework that used deep reinforcement learning algorithm. The authors used Recurrent Neural Networks (RNN) to handle partially observable scenes. Furthermore, the "glimpse" and action networks of the self-driving framework can guide the convolution kernel to the area associated with the driving scene and ignore the influence of unrelated scenes.

In this paper, we propose a deep reinforcement method to handle the problem of sparse rewards in self-driving. Firstly, we give the reasons which lead to sparse rewards in self-driving task. Then, we divide sparse rewards into three groups according to the classification threshold which is dynamically adjusted by reward distributions, to avoid local optimum in network training. Finally we compare our method with the raw DDPG algorithm in self-driving scenarios and discuss the impact of reward classification for networks training.

2 Related Work

2.1 Deep Q-Network

We consider a standard reinforcement learning where an agent interacts with the ε (Environment) over a number of discrete time steps. At each time step t, the agent receives state s_t and selects an action from the action space according to the policy π, where π is a mapping from states s_t to actions a_t. After taking an action a_t, the agent will receive the next state s_{t+1} and reward r_t from the simulation system. This process continues until it encounters an end state. For example, the car hits the fence or gets out of the road. The agent aims to find a policy π, which maximizes the expected rewards from each state s_t. We can express the expected rewards from time step t as $R_t = \sum_{\tau=t}^{\infty} \gamma^{\tau-t} r_t$, where γ represents the discount factor. If it equals to 0, the system only considers instant rewards. In contrast, the system will focus on future rewards when the γ equals 1. If we know action a and state s, we can express the expected rewards as

$$Q^{\pi}(s,a) = \mathbb{E}[R_t|s_t = s, a] \tag{1}$$

Formula (1) is also called the action value function. And the optimal action function is $Q^{*}(s,a) = \max_{\pi} Q^{\pi}(s,a)$ that gives the maximum action value for state s and action a achievable by any policy.

In the model-based reinforcement learning methods, the action value function is generally represented by a function approximator (deep neural networks). If we lead into weights θ, the action value function can be expressed as $Q(s, a; \theta)$. The neural network inputs are states s that often are images or other low-dimensional data. The outputs correspond to the predicted Q-value of individual action. In one-step Q-Learning, the parameters θ are updated by iteratively minimizing the loss functions, which are shown in the (2).

$$L_i(\theta_i) = \mathbb{E}(r + \gamma \max_{a'}(s', a'; \theta_{i-1}) - Q(s, a; \theta_i))^2 \tag{2}$$

where s' indicates the next state and a' is next action. Corresponding to the neural networks, $r + \gamma \max_{a'} Q(s', a'; \theta_{i-1})$ is equivalent to target. Because of the discount factor γ $(0 \sim 1)$, the neural network outputs will tend to a maximum along with the number of iterations increasing. Furthermore, DQN adopted ε-greedy algorithm [13] to explore the environment. Using this method will enhance the neural networks generalization ability.

2.2 Deep Deterministic Policy Gradient

The neural network outputs are discrete in DQN algorithm. So if we use DQN algorithm to handle physical control tasks having high dimensional and continuous action spaces such as controlling 7-joint manipulator, it will inevitably result in the "dimensional disaster" of the action value. To solve this difficulty, Lillicrap proposed DDPG algorithm in the 2016 ICLR. Their work was based on the deterministic policy gradient (DPG) algorithm. By using policy-based algorithm, they extended deep reinforcement learning methods to continuous action domain, making it possible to control agent with high number of degrees of freedom.

Figure 1 show the actor-critic network architecture that is used by DDPG algorithm. The two networks are connected by action value. For the critic networks, we can express the network weights as θ^{Q}. As for the actor networks, we express the network weights as θ^{μ}. The loss function of critic networks is defined as

$$L = \frac{1}{N}\sum\nolimits_i (y_i - Q(s_i, a_i|\theta^{Q}))^2 \tag{3}$$

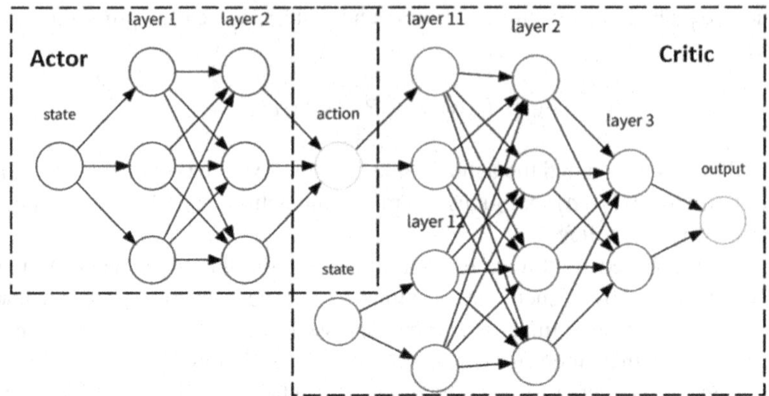

Fig. 1. Actor-critic network architecture used by DDPG

where N is the number of states that we need to process each iteration, and y_i is the target of the neural network. The y_i can be expressed by

$$y_i = r_i + \gamma Q'(s_{i+1}, \mu'(s_{i+1}|\theta^{\mu'})|\theta^{Q'}) \tag{4}$$

where r_i is the reward from environment after taken our current action. The $\mu'(s_{i+1}|\theta^{\mu'})$ represents our action a_{i+1}, which is neural network output.

Different from the deep Q learning loss function, the Q value here is not chosen the maximum because the output value of the current neural network is only 1×1 vector, resulting in difficulty for neural network convergence. The actor networks update the weights along the gradient direction returned by the critic networks. Its policy gradient is shown as follows:

$$\nabla_{\theta^{\mu}} J \approx \frac{1}{N} \sum_i \nabla_a Q(s, a|\theta^Q)|_{s=s_i, a=\mu(s_i)} \nabla_{\theta^{\mu}} \mu(s|\theta^{\mu})|_{s_i} \tag{5}$$

The expression (5) represent gradient for actor networks training, so the outputs of actor networks are equivalent to the weights of whole actor-critic networks. Because actor networks output is continuous, this method can be applied to control agent with high number of degrees of freedom. Note that, the two networks are divided in training process. First, DDPG uses a method similar to deep Q learning for the critic networks weights update. Then the gradient of critic networks is transmitted to actor networks. Finally, the actor networks update the weights using gradient ascent approach. The common goal of both networks is to make the Q value of critic network output as large as possible.

3 Reward Classification

In this section, we give the reward distributions for different stages in self-driving environment. Then, we classify reward into three groups based on this distribution. When the rewards were classified, the network training maybe meet the overfitting problem, so we also present a way to avoid it.

3.1 Sparse Rewards

In the process of our growing up, through communication and learning with parents, teachers and friends, we have accumulated a lot of prior knowledge about the living environment. These prior knowledge will enable us to live well, solve problems encountered, and make effective choices [14]. However, unlike humans, the agent in deep reinforcement learning does not possess any prior knowledge. It can only perceive the surroundings through constantly making mistakes. In some sparse rewards circumstances such as self-driving, using this perceptive method will waste much time. Thus, we need classify the rewards to guide agent apperceiving environment quickly.

In deep reinforcement learning algorithm, experience replay [15, 16] is mainly used to training neural network. Through using this algorithm, it becomes possible to break the temporal correlations by mixing more and less recent experience for the updates. Therefore, experience replay algorithm can improve the generalization capability of network. However, in environment with sparse rewards, there are maybe a large number of states with no rewards, resulting in the agent does not know how to optimize their actions to get more rewards.

In this paper, we set up the reward function as follows:

$$R = V\cos(\theta) - V\sin(\theta) - V|\mathrm{trackPos}| \tag{6}$$

where R is the reward obtained by the car, V is speedX (Table 1) and θ is the deflection angle of car. The trackPos is shown in Table 1. From (6), we can see that our goal is to maximize the car speed along the road and punishes the other direction speed.

Table 1. State information during driving

Name	Range (unit)	Description
Angle	$[-\pi, +\pi]$ (rad)	The angle between the direction of the car and the track axis
Track	(0,200) (m)	Vector of 19 range finder sensors
Trackpos	$(-1, 1)$	Distance between the car and the track axis
speedX	$(-\infty, +\infty)$ (km/h)	Speed of the car along the longitudinal axis of the car
speedY	$(-\infty, +\infty)$ (km/h)	Speed of the car along the transverse axis of the car
speedZ	$(-\infty, +\infty)$ (km/h)	Speed of the car along the Z axis of the car
wheelSpinVel	$[0, +\infty)$ (rad/s)	4 wheels rotation speed
Rpm	$[0, +\infty)$ (rpm)	Number of rotation per minute of the car engine

Figure 2 show the initial stage of training. The speed is low in the car startup phase, so the rewards are low. As time goes on, the car speed will increase resulting in rewards increasing. However, due to using the ε-greedy algorithm, the agent does not have the ability to avoid obstacles. The probability of car collision increase after the speed improved, causing the car get negative rewards.

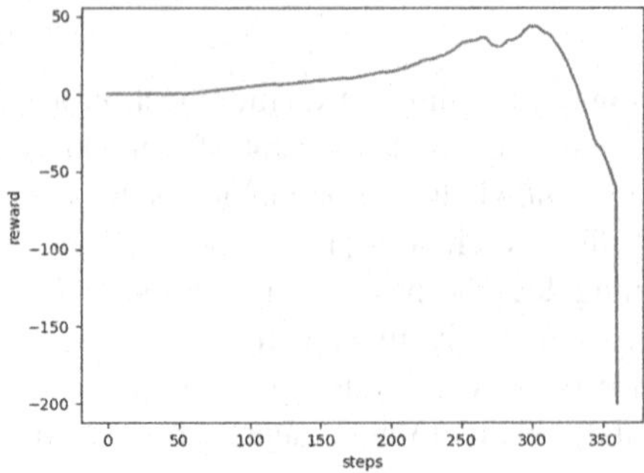

Fig. 2. The reward distributions in the initial stage of training

From the Fig. 2, we know that there are small number of states for the high and low rewards. If we adopt experience replay method to training, it will cause the agent to spend more time in dealing with the common states, not to perceive the rewards of the system. It is likely to fall into a local optimum affecting the results of self-driving.

In the self-driving environment, there is a problem of sparse rewards. To handle it, authors in [17] pointed out that we can divided our returns into negative and positive category, then randomly selected state-action pairs from the two states buckets. Based on this, we propose a deep reinforcement method. According to the reward distributions, we divide neural network training into three stages including initial stage, middle stage and final stage. The later stages are shown in the Fig. 3.

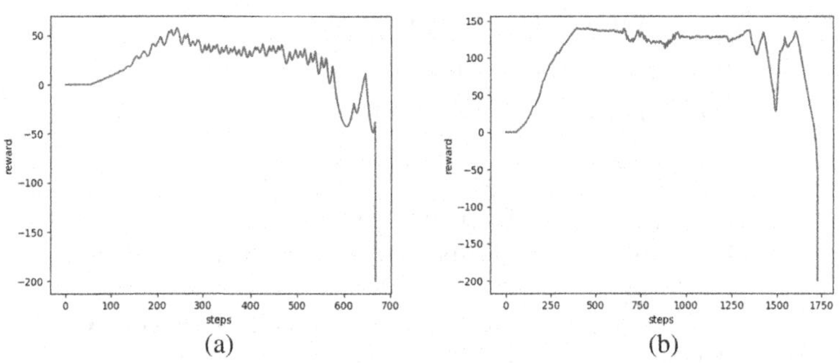

Fig. 3. The reward distributions for different training stage. (a) Middle stage. (b) Final stage

In the middle stage, the car decreases training time for exploring environment instead of increasing training time for getting high speed, which lead to the reward distributions change. For this change, we increase classification threshold compared with initial training stage. With training time raising, the speed of car will become larger and tend to stable. The reward distributions of final stage is not as severe as the before stages. In this situation, we reuse the raw experience replay to training networks.

3.2 Overfitting

In the initial stage of training, the quantities of different rewards encountered by the car is shown in the Fig. 4. In the figure, the normal rewards have a huge advantage in quantity compared to the high rewards and low rewards. Due to the neural networks randomly sample from three reward buckets, this will inevitably lead to the neural networks reuse the data with the high rewards and low rewards many times. In reality, if the rewards are classified into three groups, the network training will be likely to encounter the overfitting problem.

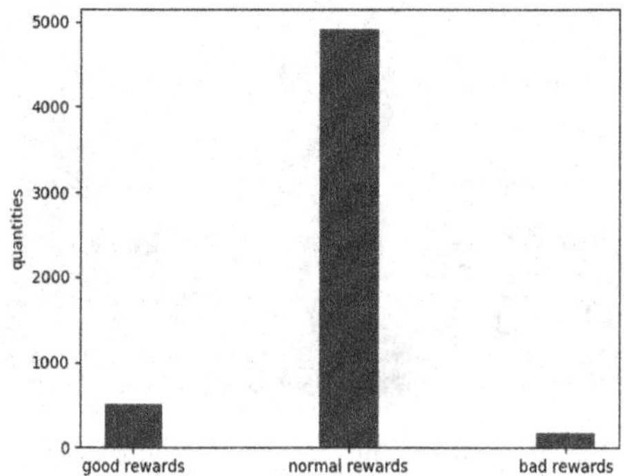

Fig. 4. The number of different kinds of rewards under ten cycles

We estimate whether the neural networks encounters the overfitting problem according to the total rewards returned by system. If the total rewards of system is no apparent change, the system maybe has the overfitting problem. When meeting this overfitting problem, the system will automatically adjust the sampling amount for different buckets of rewards. For example, if the total rewards always stay low level, the system will increase the sampling amount for normal rewards and decreasing sampling amount for good rewards and bad rewards until the total awards raising.

4 The Experimental Results and Analysis

In this section, self-driving simulation environment is introduced. The states information we used is also provided. At the end of section, we give the results for our method compared with raw DDPG algorithm.

4.1 Self-driving Simulation Environment

The self-driving experiment is based on Torcs simulation environment [18] which is an excellent open source simulation software for 3D racing. It provides some practical factors that need to be considered during driving, such as automotive dynamics, aerodynamics, and fuel consumption. Torcs system communicates with self-driving car through the UDP protocol and provides real-time information during driving, such as the angle between the direction of the car and road. The simulation platform is shown in the Fig. 5. It can be set to split-screen for display, enabling us to monitor the states of self-driving car and other vehicles in real time. The simulation platform is similar to the highway environment.

Fig. 5. Self-driving simulation environment

Torcs is open source and provides program control interfaces, so most of the deep learning algorithms adopted it for simulation. In self-driving tasks, the inputs of neural networks are low-dimensional states information (such as angle and position), and outputs are the control signals which include Steering, Acceleration, and Brake. Compared to real life self-driving, low-dimensional states information are similar to those collected by on-board sensors. Therefore, the neural network adopt the low-dimensional states as input conforming to the self-driving of real-world situations.

4.2 State Information of Self-driving

In the Torcs simulation environment, the self-driving car receives data from system through the UDP protocol. Table 1 show part of self-driving car information. These state information ensure the car driving safely. Remarkably, the Track state represent a vector of outputs of 19 range finder sensor, one of which returns the distance between the car and road edge within 200 m. By default, the sensors sample the space in front of the car every 10°, spanning clock from −90° up to +90° with respect to the car axis. The speedX is the speed of the car along the axis of the road. The purpose of our method is to maximize the speedX.

From Table 1, we found that angle, trackpos and track mainly reflect the position information of the vehicle, while speedX, speedY, wheelSpinVel and rpm reflect the speed information of the vehicle. In order to reduce the complexity of neural network, we integrate these states and use the neural networks to learn the relevance of each state.

4.3 Comparison with DDPG

The simulation racetrack used in this paper is shown in the Fig. 6(a). The road length is 2,057.56 m and the width is 15 m. The road is relatively simple and can clearly reflect the different performance between the two algorithms. To ensure the fairness, the two algorithms do not contain any priori knowledge. They can only obtain the states of the car and send control signals to the car's angle, acceleration and braking according to the outputs of neural networks with same number of neurons. Both car rely on the rewards of system to complete the perception of the surrounding environment.

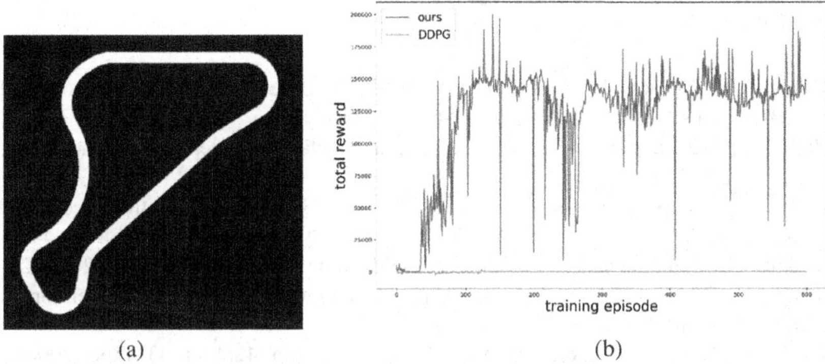

(a) (b)

Fig. 6. Self-driving test environment and results for our method compare with traditional DDPG. (a) Test environment. (b) Comparison result.

The simulation results for comparison are shown in the Fig. 6(b). At the beginning of the system, self-driving cars adopted respectively our method and raw DDPG algorithm do not achieve running the whole racetrack. It is very easy to collide and end an episode. With the progressing of the neural network training, it will be easy to fall into a local optimum in raw DDPG algorithm because the networks not fully utilize states with different rewards. While the networks in our method make full use of states with different rewards, having the ability to overcome the local optimum problem. From the results, it can be seen that the self-driving car used our method completed racetrack running. All in all, after solving the problem of sparse rewards, the performance of the self-driving car has been greatly improved.

5 Conclusion

In this paper, a deep reinforcement learning method for self-driving was proposed. The reason and harm of sparse rewards in self-driving environment was presented. Through classified the reward based on dynamic threshold, the local optimum was avoided in network training process. The overfitting problem was also considered. The self-driving simulation environment and the states information that ensure the car driving safely was introduced. We gave experiment results for our method compared with raw DDPG algorithm. From the results, we can see that our method has obvious advantages to maintain self-driving car stable operation.

However, due to the complex environment, there are many factors that need to be considered in self-driving tasks. If just relying on deep reinforcement learning algorithms, the safety of the vehicle may be not guaranteed. Deep reinforcement learning is still at the research stage. There are many problems needed to study, such as low adaptability to the environment and weak perception of rewards. In the future research, deep reinforcement learning should be combined with other self-driving algorithms. Using deep reinforcement learning algorithms evaluates whether the current decisions is right or wrong, to promote the self-driving development.

References

1. Silberg, G., Wallace, R.: Self-driving cars: the next revolution. Cent. Automot. Res. 36 (2012). https://doi.org/10.1007/978-3-642-21381-6-16
2. Mnih, V., Kavukcuoglu, K., Silver, D., et al.: Playing atari with deep reinforcement learning. arXiv: 1312.5602 (2013)
3. Mnih, V., Kavukcuoglu, K., Silver, D., et al.: Human-level control through deep reinforcement learning. Nature **518**(7540), 529–533 (2015)
4. Watkins, C.J.C.H., Dayan, P.: Q-learning. In: Machine Learning, pp. 279–292 (1992)
5. Lillicrap, T.P., Hunt, J.J., Pritzel, A., et al.: Continuous control with deep reinforcement learning. Comput. Sci. **8**(6), A187 (2015)
6. Konda, V.: Actor-critic algorithms. Siam J. Control Optim. **42**(4), 1143–1166 (2006)
7. Silver, D., Lever, G., Heess, N., Degris, T., Wierstra, D., Riedmiller, M.: Deterministic policy gradient algorithms. In: ICML (2014)
8. Wolf, P., et al.: Learning how to drive in a real world simulation with deep Q-Networks. In: 2017 IEEE Intelligent Vehicles Symposium (IV), pp. 244–250. Los Angeles, CA (2017). https://doi.org/10.1109/ivs.2017.7995727
9. Wang, P., Chan, C.Y.: Formulation of Deep Reinforcement Learning Architecture Toward Autonomous Driving for On-Ramp Merge (2017)
10. Hochreiter, S., Schmidhuber, J.: Long short-term memory. Neural Comput. **9**(8), 1735–1780 (1997)
11. Xiong, X., Wang, J., Zhang, F., et al.: Combining Deep Reinforcement Learning and Safety Based Control for Autonomous Driving. (2016). arXiv preprint arXiv:1612.00147
12. Sallab, A.E., Abdou, M., Perot, E., et al.: Deep reinforcement learning framework for autonomous driving. Electron. Imaging **2017**(19), 70–76 (2017)
13. Gomes, E.R., Kowalczyk, R.: Dynamic analysis of multiagent Q -learning with ε-greedy exploration. In: International Conference on Machine Learning. pp. 369–376. ACM (2009)
14. Dubey, R., Agrawal, P., et al.: Investigating Human Priors for Playing Video Games. eprintarXiv: 1803.05262 (2018)
15. Lin, L.-J.: Self-improving reactive agents based on reinforcement learning, planning and teaching. Mach. Learn. **8**(3–4), 293–321 (1992)
16. Schaul, T., Quan, J., Antonoglou, I., et al.: Prioritized experience replay. (2015). arXiv preprint arXiv:1511.05952
17. Narasimhan, K., Kulkarni, T., Barzilay, R.: Language understanding for textbased games using deep reinforcement learning. In: Conference on Empirical Methods in Natural Language Processing (EMNLP) (2015)
18. Loiacono, D., Cardamone, L., Lanzi, P.L.: Simulated car racing championship: Competition software manual. (2013). arXiv preprint arXiv:1304.1672

Image Classification Based on Improved Spatial Pyramid Matching Model

Li Feng[1,2(✉)], Xiaofeng Wang[1,2], and Dongfang Chen[1,2]

[1] Computer Science and Technology, Wuhan University of Science and Technology, Wuhan 430065, China
1473357859@qq.com

[2] Hubei Province Key Laboratory of Intelligent Information Processing and Real-Time Industrial, Wuhan 430065, China

Abstract. Spatial pyramid matching (SPM) uses the statistics of local features in an image sub region as a global feature. It shows good performance in terms of generic image recognition. However, the disadvantages of this method are that the constructed visual dictionary is easy to fall into a local optimal solution due to the randomness of the initial centroid of k-means and it ignores the spatial distribution of salient object in images. In this research, we propose a new clustering method that using black hole algorithm to determine the initial center of k-means when constructing a visual dictionary and making the result have globally optimal solution and less computational costs. To better distinguish the target and background in the image, we propose discriminative SPM, which is a new representation that forms the image feature as a weighted sum of features over all pyramid levels. The weights are selected by the spatial distribution of salient objects in images. The resulting feature is compact and preserves high discriminative power. Thus reducing the effect of image background on classification. As documented in the experimental results, the proposed schemes can improve the classification accuracy of image compared to the other existing methods.

Keywords: K-means · Black hole algorithm · Spatial pyramid
Initial center · Salient object · Discriminative SPM

1 Introduction

In recent years, image classification is receiving increasingly significant attention owing to its great potential in both industry applications and classification problems The methods [1, 2] based on bag of features (BOF) have achieved good results for image classification. BOF method treats an image as a collection of unordered appearance descriptors extracted from local patches, quantizes them into discrete "visual words", and then computes a compact histogram representation for semantic image classification. Although this framework has demonstrated to be simple and efficient, it ignores spatial information of local features which has been observed very helpful in improving classification accuracy.

D.-S. Huang et al. (Eds.): ICIC 2018, LNCS 10955, pp. 153–164, 2018.
https://doi.org/10.1007/978-3-319-95933-7_20

To improve classification accuracy, many approaches have been proposed. Lazenik [3] proposed Spatial Pyramid Matching (SPM) method. The method partitions an image into $2^l * 2^l$ segments in different scales $l = 0, 1, 2$, computes the BoF histogram within each of the 21 segments, and finally concatenates all the histograms to form a vector representation of the image. In case where only the scale $l = 0$ is used, SPM reduces to BoF. But there are also many difficulties in constructing visual dictionaries and feature merging. The main manifestations are as follows:(1) when using K-means to construct visual dictionary, the randomness of initial center selection leads to easy falling into local optimal solution. (2) when the spatial features are combined, it is impossible to distinguish target object region and the background region in the image, and it is difficult to select the appropriate eigenvector organization method so that both the calculation and the image loss of information can be reduced.

In the process of K-means clustering, document [6] proposed a double word bag model to construct a visual word histogram that is more capable of representing images, it reduces the interference caused by the instability of the K-means algorithm to the experimental results, but it also causes a great deal of time consumption. Document [7] got better performance through replacing the K-means with sparse coding SC, to a certain extent that improves the construction of the visual dictionary. Document [19] proposed an evaluation of different representations obtained from the bag of features approach to classify histopathology images. But the algorithms of document [7] and document [19] take too much time.

After obtaining the visual dictionaries, they need to be organized to represent the image. The original SPM model has a fixed weight for each pyramid level and does not distinguish the weights for different grid cells. Document [4] proposed a method to learn the pyramid weights in space. The method used cross-validation to find the optimal weights. Although the pyramid level weights were studied, the optimization of the weight of each grid cell was not considered. In document [5], random forest was used to classify the images and randomly select weights for each pyramid level at the nodes. This randomness makes it difficult to choose suitable weights for different levels of the pyramid. In addition, it does not consider the weight of different grid cells.

In this paper, an image classification method is proposed to optimize K-means clustering and to construct discriminative grid weights. Firstly, we propose a BH-Kmeans clustering algorithm that combines the black hole (BH) algorithm and the K-means algorithm when constructing the visual dictionary. That is, BH [10] algorithm is used to determine the initial center of K-means clustering, the number of iterations in the clustering process is reduced, and the clustering result is stable and accurate. After the generation of the visual dictionary, the images located at different levels and different spatial locations are made histogram construction based on their corresponding spatial visual dictionaries. After obtaining the quantified features, this paper proposed a method to construct a discriminative grid weight. Firstly, the salient object region of the image is calculated and different weights are given according to the proportion of the salient object region in different grids of the pyramid. Then, all the weighted quantization feature vectors are connected to form a final image description, so that the optimal weight is generated for each grid area, thus avoiding the interference of the background region to the image classification.

The proposed method optimizes the K-means clustering and considers the contribution of the salient features of the image to the image classification when blending different levels of information. The second section discusses the proposed method; the third section verifies the effectiveness and robustness of the proposed method by experimenting on two standard data sets; and finally concludes the conclusion.

2 Improved SPM Model for Image Classification

In this paper, there are two main innovations in image classification based on SPM model. Firstly, BH-Kmeans clustering algorithm is proposed that using black hole algorithm to determine the initial center of k-means and the algorithm is used to generate visual dictionary, so that the constructed visual dictionary has the global optimal solution and the clustering results are stable and accurate. Secondly, a method is proposed to construct the distinctive grid weight by using the salient object of the image. Different weights are assigned to different grids to form the final image representation, so that each grid area produces the best weight and avoids the background area interference on the image classification.

2.1 BH-Kmeans Clustering Algorithm

K-means clustering algorithm is widely used in various fields due to its advantages of fast clustering and strong ability of the local search ability. However, it also has some limitations, such as the dependence on the initial cluster centers, the poor global search ability, easy to fall into local optimum, the poor clustering accuracy and low efficiency. Black-hole algorithm [10] is a heuristic optimization algorithm based on the black hole phenomenon in nature. It has been applied to the field of power flow calculation [15], image processing [16, 18], parameter optimization [17] and so on. It has the advantage of high precision and is easy to achieve global optimal solution. In the process of optimizing the cluster centers, the black hole is always kept as the global optimal solution. Using the black hole clustering algorithm can solve the problem that k-means algorithm is sensitive to the initial center that easy to fall into the local optimal. Therefore, this paper uses the BH-Kmeans clustering algorithm to construct a visual dictionary. According to the specific situation of the clustering problem, the black hole coding method and the fitness function are designed to effectively overcome the disadvantages of the K-means algorithm. The method has the advantages of fast convergence speed, strong stability and high clustering precision, and obtains good clustering effects. BH-Kmeans clustering algorithm structure shown in Fig. 1.

Firstly, dense sift features of training images are extracted to form a large feature set. Then we design the coding method of BH-Kmeans clustering algorithm. When the black hole algorithm is used to optimize the K-means initial clustering center, the key is how to represent each of the stars and determine the fitness function. Since each of the stars in the black hole algorithm is poly the candidate solution of the class result, so each of the stars (candidate solution) is shown in Fig. 2.

The encoding method of BH-Kmeans clustering algorithm is based on the clustering center, and the dimension of C cluster center is K, then the position of the star is

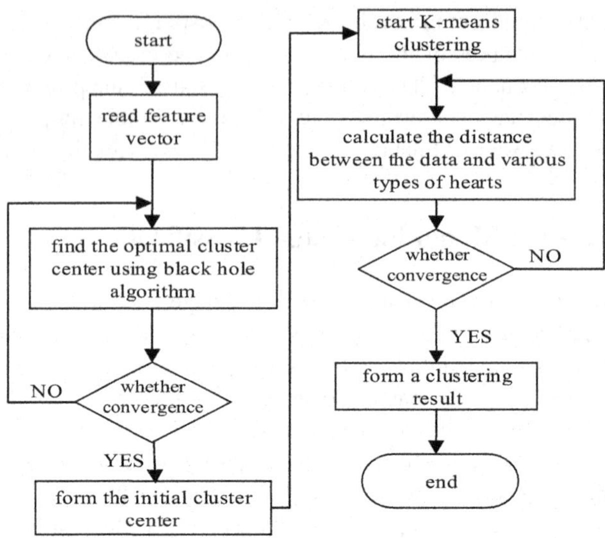

Fig. 1. The flow chart of BH-Kmeans clustering algorithm

$Z_{11}Z_{12}Z_{13}\ldots Z_{1K}$	$Z_{21}Z_{22}Z_{23}\ldots Z_{2K}$	......	$Z_{C1}Z_{C2}Z_{C3}\ldots Z_{CK}$

Fig. 2. Coding structure of the stars

a dimensional variable is K, that is, the position of each star is composed of the C cluster center. In Fig. 2, Z_{CK} represents the K dimension of the C center in optimal initial cluster center.

After designing the coding mode, the black hole algorithm is used to find the initial clustering center of K-means. The specific BH-Kmeans algorithm flow is as follows:

(1) Initialization parameters: the size of the dictionary is the cluster number of clusters C, the maximum number of iterations is max_*iteration*, the black hole is defined as BH, the number of stars is defined as *planetnum*.
(2) Calculate the value of the fitness of the stars, the fitness function is as shown in Eq. (1). Select the star with the best fitness value as the black hole.

$$fit = \sum_{i=1}^{T} \sum_{j=1}^{C} w_{ij} \cdot \left\| x_i - z_j \right\| \tag{1}$$

In Eq. (1) $X = (x_1, x_2, \ldots\ldots, x_i, x_T)$ is the collection of image feature vectors, x_i is the i feature vector. $Z = (z_1, z_2, \ldots\ldots z_j, \ldots, z_C)$,$z_j$ is the j cluster center; T is the number of feature vector. C is the number of cluster. if the i feature vector belongs the j cluster, $w_{ij} = 1$, else $w_{ij} = 0$.

(3) According to Eq. (2) to move the location of the various stars

$$l(t+1) = l_i(t) + rand \cdot (l_{BH} - l_i(t)) \tag{2}$$

The $l_i(t)$ indicates the position of the$i$ star at the time of the t search. l_{BH} represents the position of the black hole. *rand* is a random number between 0 and 1.

(4) Recalculate the fitness values of all the stars. If the fitness value of a star is better than the fitness value of a black hole, make the star a new black hole.
(5) Calculate the radius according to Eq. (3). If there is a distance between a star and a black hole less than the radius, delete the star and randomly generate a new star.

$$r = \frac{fit_{BH}}{\sum_{i=1}^{planetnum} fit_i} \tag{3}$$

In Eq. (3), fit_{BH} and fit_i respectively represent the fitness value of the black hole and the fitness value of the i star; *planetnum* is the number of star.

(6) If the algorithm has reached the maximum number of iterations or has converged, the implementation process ends; otherwise, step (3) is performed.
(7) K-means clustering is performed by using the clustering center obtained in the previous step as K-means initial centroid to calculate the distance between the feature and the centroid of the image until the algorithm converges to form the final clustering result.

2.2 Salient Object Map and Calculation of Differentiated Grid Weight

In the SPM model, firstly extract the original image of the global features, then divide the image horizontally into smaller and smaller grids in each pyramid. The features are extracted from each grid at each pyramid level and connected into a large feature vector. However, the same weight is assigned to each grid area in the SPM model without considering the different contribution of different features of the image to the image classification. In order to solve this problem, we proposed using the salient object of the image to construct the differentiated grid weights. Give each grid of each layer a different weight according to the different contribution. Then the weight of each grid per layer is weighted in series to form the final image description.

Image saliency has been widely used in many fields such as target recognition [11], image segmentation [12] and image retrieval [13]. In an image, human visual attention will first be drawn to the salient features of the image and then get the different information together. Therefore, in addition to the salient object in the image, the rest usually have a small effect on image classification. Therefore, the spatial distribution of salient object has a significant effect on image classification [9].

In this paper, the calculation of salient object map adopts the method in document [8] which is a salient object detection based on clustering. The advantage of this method is that there is no arduous learning process, simple, common and effective.

We use a single picture of a significant target detection method: First, extract the CIE Lab image color features and Gabor features, and then use the BH-Kmeans clustering algorithm to cluster the image features. In the same way, the problem of generating local optimal solution using K-means clustering in document [8] is avoided. Then calculate the feature contrast of each class with other classes and the center offset measure of each cluster, combining the feature contrast and the center offset to obtain a significant probability image of the image. Finally, calculate the significant probability of each pixel to form the salient map of the image and obtain the salient object in the image. Through the calculation of salient regions in document [8], the salient object in an image are effectively analyzed. For convenience of calculation, after the salient map is calculated, the salient map is binarized processing that uses the Otsu algorithm [14], the image is divided into obvious black and white area. After this method, the pixel value of the salient object area of the image is 1, and the pixel value of other area is 0. The salient object of the final image is shown in Fig. 3.

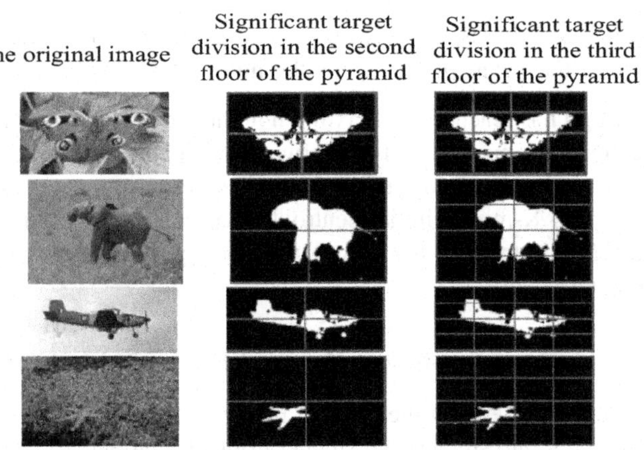

Fig. 3. Salient object distribution of image

After the image's salient object is extracted, we use the salient object to build differentiated grid weights that we give more weight to salient object in the image and give less weight to background area. this method may improve the image classification better. To take advantage of the salient object of the image to construct a distinguishing grid, we design the following two principles: (1) Fine-grained images take more weight than coarse-grained ones, that is, the fine-grained grid can better characterize the image information. The weight of each level of the pyramid is given a different weight according to the proportion of salient regions in each grid, significantly larger proportion of the target is given greater weight, and small proportion of the significant target is given as smaller weight.

In order to facilitate the calculation, the image is first normalized to $M * M$, image segmentation level l, $l \in \{0, 1, \ldots, L\}$, here the image was divided into three levels. The number of grids divided by the image at the l level is $G(l) = 4^l$.Defined the expression vector of the k grid at the l level is q_k^l, the final feature expression vector of the k grid at the l level is defined as f_k^l:

$$f_k^l = w_k^l * q_k^l \tag{4}$$

The image final expression of the pyramid representation with distinguishing weight is defined as Eq. (5):

$$F_{image} = \sum_{l=0}^{L} \sum_{k=1}^{4^l} w_k^l * q_k^l \tag{5}$$

The size of w_k^l is decided by δ_k^l,δ_k^l is determined by the proportion of significant objects in the k grid in the pyramid. As shown in Fig. 4, the proportion of significant targets in each grid is different. In SPM model, the 0th image of pyramid is not meshed, we set the same weight with the document [3], that is $w_1^0 = 1/4$. The proportion of the k grid of the significant target in l level is defined as Eq. (6):

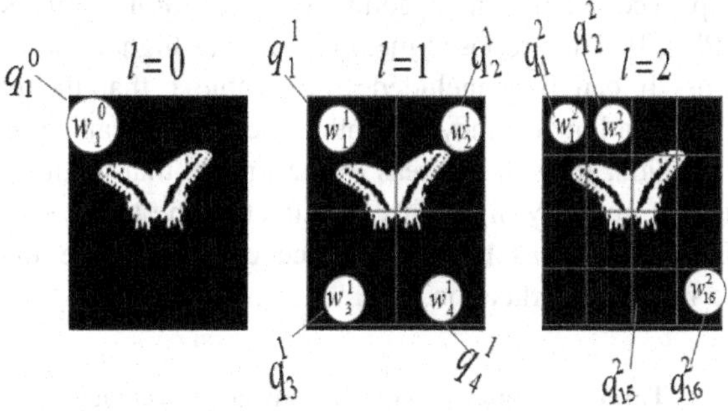

Fig. 4. Spatial pyramid of discriminant weight

$$\delta_k^l = \frac{p_k^l}{T_l} \tag{6}$$

The T_l represents the total number of pixels in each grid at the l level of the pyramid that $T_l = (M * M)/4^l.p_k^l$ is the sum of the values of the pixels in the k grid in the l level. At the level $l = 1$ and $l = 2$, set w_k^l different weight according to δ_k^l. The bigger the δ_k^l value, the greater the w_k^l weight. The smaller the δ_k^l value, the smaller the w_k^l weight. Finally, the final expression vector f_k^l of each grid is connected in order to get the final expression vector of the image F_{image}.

3 Experimental Results and Analysis

3.1 Image Dataset Introduction and Selection

In order to verify the performance of the classification method in this paper, we experiment with two standard image libraries, caltech101 [21] and caltech256 [22]. The Caltech 101 dataset consists of a total of 9146 images, split between 101 different object categories, as well as an additional background/clutter category. Each object category contains between 40 and 800 images on average. The Caltech 256 dataset consists of a total of 30607 images, split between 256 different object categories.

In this experiment, we randomly select five types of images from the caltech101 and caltech256 data sets, respectively.

3.2 Algorithm Comparison

In this experiment, each experiment randomly selected 50 images of each class as training images from the two datasets, and 30 of the remaining images were selected as test images, setting the visual dictionary size to 500. We use LIBSVM to classify the experiment and repeat the experiment 10 times to obtain the average classification accuracy. The improved algorithm is compared with BOF [20], KernelBOF [19], ScSPM [7] and SPM [3]. The proposed method achieves higher classification accuracy than other methods. It can be concluded from Table 1 that the proposed method achieves higher classification accuracy than other methods. It can be seen from this that it is necessary to consider the salient areas of the image, that is, the salient targets for image classification. It not only highlights the effect of the target object on the image classification, but also weakens the interference caused by the image background information to the image classification.

Table 1. Comparison of classification accuracy

Algorithm	Caltech101	Caltech256
	Accuracy/%	Accuracy/%
BOF [20]	85.2 ± 0.8	68.72 ± 1.1
KernelBOF [19]	88.73 ± 1.4	74.26 ± 2.4
ScSPM [7]	94.5 ± 0.9	78.67 ± 1.4
SPM [3]	94.33 ± 1.2	78.13 ± 1.9
Algorithm of this paper	95.46 ± 1.2	81.33 ± 0.7

3.3 The Influence of Weights on Classification Accuracy

In this experiment, we analyzed the impact of weights on the accuracy of image classification, and analyzed the classification performance under four kinds of weights: (1) the same weight value; (2) weight values set according to the inverse ratio of the number of grids; (3) weight values set randomly; (4) weight values set according to the proportion of the significant goal in each grid, as follows:

(1) The same weight value is denoted as the weight $w(1)$, that is $w(1) = [1, 1, 1]$, the feature vectors in the three levels of the spatial pyramid are given the same weight.
(2) weight values set according to the inverse of each grid number, denoted as $w(2)$. As we can see from the second section, the spatial pyramid partitioning method is evenly divided, the blocks in each level are the same size and the number of blocks in the first level is 0. According to the proportion of the number of Chinese blocks in each level Weight. Choose a set of weights inversely proportional to the number of blocks in each level, that is $w(2) = 1/G(l) = [1, 1/4, 1/16]$.
(3) weight values set randomly, that is $w(3)$ and $w(4)$.randomly choose some weights of arbitrary size to carry on the experiment, choose the weight that coincides with the coefficient of the nuclear matrix in document [3], that is $w(3) = [1/4, 1/4, 1/2]$. And then randomly select a group of pyramid level in the number of independent weight that is $w(4) = [1/8, 1, 1/6]$.
(4) weight values set according to the proportion of the significant goal in each grid

In this paper, we set the different weights according to the proportion of significant objects in each grid. In this article, we normalize the goal as $200 * 200$. When $l = 0$, $w_k^l = 1/4$; when $l = 1$, the weights in different grids are set as follows: if $\delta_k^l \geq 0.2$, $w_k^l = 5/8$, if $\delta_k^l < 0.2$, $w_k^l = 1/4$ when $l = 2$, the weights in different grids are set as follows: if $0.2 \leq \delta_k^l \leq 0.3$, $w_k^l = 3/4$; if $0.3 < \delta_k^l < 0.5$, $w_k^l = 7/8$; if $0.5 < \delta_k^l$, $w_k^l = 9/8$; if $\delta_k^l < 0.5$, $w_k^l = 1/2$.

The above five kinds of weight settings on the Caltech101 and Caltech256 data set classification performance is shown in Figs. 5 and 6. As can be seen from the figure: In this paper, according to the significant target of the image, the classification accuracy of different weights in each grid is obviously higher than the weight of weight and the weight of the inverse ratio of the grid number. From this, we can see that this paper makes effective use of the spatial information according to the spatial distribution of the significant objects of the image and obtains a relatively high classification accuracy.

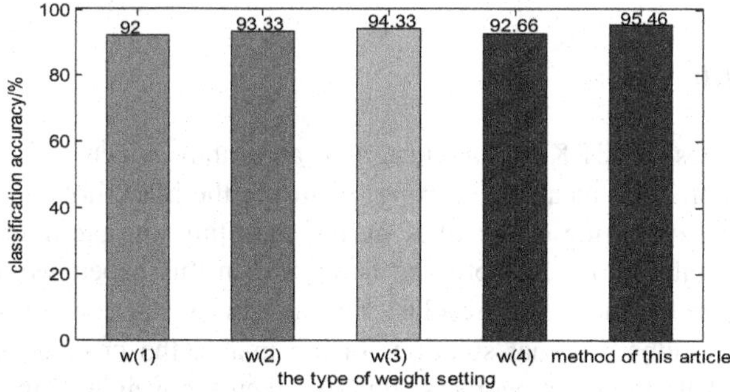

Fig. 5. Classification accuracy for different weight settings on Caltech 101

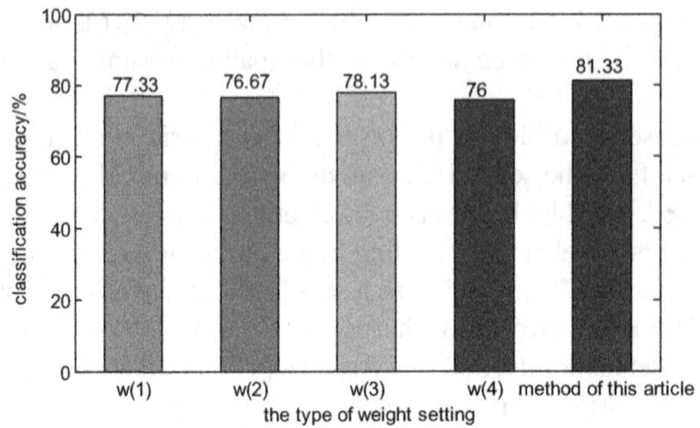

Fig. 6. Classification accuracy for different weight settings on Caltech 256

3.4 Comparing the Time Efficiency of Building a Dictionary

We compare the time efficiency and classification accuracy of the three methods to regenerate the dictionary. The weight is set according to the original weight of the document [3]. (1) In the BH-Kmeans algorithm, we set the black hole algorithm to iterate 3 times, with 8 initial stars and 10 K-means iterations. (2) The K-means algorithm is set to iterate 100 times in document [3]. (3) Sparse coding in document [7]. The results are shown in Table 2: As can be seen from Table 2, this method has obvious advantages in terms of time and classification accuracy.

Table 2. Comparing the time to build a dictionary

Algorithm	Caltech101	Caltech256
	Time (s)/Accuracy (%)	Time (s)/Accuracy (%)
K-means	467.90/93.33	471.21/78
SC	530/94	540/79
BH-Kmeans	387.86/94.67	389.68/81

4 Conclusion

This paper proposed BH-Kmeans clustering algorithm which using Black Hole Algorithm to Improve K-means Clustering. By using the black hole algorithm to find the optimal initial clustering center of K-means algorithm, the problem that K-means algorithm easily falls into local optimum is solved. In this paper, we apply the BH-Kmeans algorithm to image feature clustering and make the constructed visual dictionary have the global optimal solution. In the image, the prominent target in the image is more likely to attract people's attention, then the weight of this area is large, indicating that the characteristics of this area are more important. Because the background area in the image is likely to cause interference to the image classification, the weight of this block is relatively small, indicating that the feature of this region has a relatively small effect on the entire image. Therefore, this paper proposes a method to construct distinguishing grid weights by using the salient object of the image, and gives the best weight for each image area according to the proportion of each target grid in each grid. Incorporating the saliency mechanism of this kind of image into the SPM model increases the weight of the significant target area in the image and avoids the interference of the background area in the image. Experiments show that the proposed method is effective and the accuracy of image classification is obviously improved.

References

1. Zheng, Z., Zhang, Y., Yan, L.: Global and local exploitation for saliency using bag-of-words. IET Comput. Vis. **8**(4), 299–304 (2014)
2. Wu, L., Hoi, S.C., Yu, N.: Semantics-preserving bag-of-words models and applications. IEEE Trans. Image Process. Publ. IEEE Sig. Process. Soc. **19**(7), 1908–1920 (2010)
3. Lazebnik, S., Schmid, C., Ponce, J.: Beyond bags of features: Spatial pyramid matching for recognizing natural scene categories. In: CVPR (2006)
4. Bosch, A., Zisserman, A., Munoz, X.: Representing shape with a spatial pyramid kernel. In: ACM International Conference on Image and Video Retrieval, pp. 401–408. ACM (2007)
5. Bosch, A., Zisserman, A., Munoz, X.: Image classification using random forests and ferns. In: ICCV (2007)
6. Yu, Q.: Optimization of initial clustering center selection using K-means algorithm. Appl. Comput. **5**, 170–174 (2017)
7. Yang, J., Yu, K., Gong, Y., Huang, T.: Linear spatial pyramid matching using sparse coding for image classification, pp. 1794–1801 (2009)
8. Fu, H., Cao, X., Tu, Z.: Cluster-based co-saliency detection. IEEE Trans. Image Process. Publ. IEEE Sig. Process. Soc. **22**(10), 3766 (2013)
9. Underwood, G., Templeman, E., Lamming, L., Foulsham, T.: Is attention necessary for object identification? Evidence from eye movements during the inspection of real-world scenes. Conscious. Cogn. **17**(1), 159–170 (2008)
10. Hatamlou, A.: Black hole: a new heuristic optimization approach for data clustering. Inf. Sci. **222**(3), 175–184 (2013)
11. Huo, S., Zhou, Y., Lei, J., Ling, N., Hou, C.: Linear feedback control system based salient object detection. IEEE Trans. Multimed. **1**(1), 1 (2017)
12. Goferman, S., Zelnikmanor, L., Tal, A.: Context-aware saliency detection. IEEE Trans. Pattern Anal. Mach. Intell. **34**(10), 1915–1926 (2012)
13. Alamdar, F., Keyvanpour, M.R.: Effective browsing of image search results via diversified visual summarization by clustering and refining clusters. Sig. Image Video Process. **8**(4), 699–721 (2014)
14. Xing, H., Huang, C.: Feature extraction of raindrops based on ostu algorithm. Meteorological Science & Technology (2017)
15. Bouchekara, H.R.E.H.: Optimal power flow using black-hole-based optimization approach. Appl. Soft Comput. **24**, 879–888 (2014)
16. Yaghoobi, S., Hemayat, S.,Mojallali, H.: Image gray-level enhancement using Black Hole algorithm. In: International Conference on Pattern Recognition and Image Analysis, pp. 1–5. IEEE (2015)
17. Tong, W., Gao, X.W., Jiang, Z.J.: Parameters optimizing of lssvm based on black hole algorithm. J. Northeast. Univ. **35**(2), 170–174 (2014)
18. Pashaei, E., Aydin, N.: Binary black hole algorithm for feature selection and classification on biological data. Appl. Soft Comput. **56**, 94–106 (2017)

19. Caicedo, Juan C., Cruz, Angel, Gonzalez, Fabio A.: Histopathology image classification using bag of features and kernel functions. In: Combi, Carlo, Shahar, Yuval, Abu-Hanna, Ameen (eds.) AIME 2009. LNCS (LNAI), vol. 5651, pp. 126–135. Springer, Heidelberg (2009). https://doi.org/10.1007/978-3-642-02976-9_17
20. Deselaers, T., Pimenidis, L., Ney, H.: Bag-of-visual-words models for adult image classification and filtering. In: International Conference on Pattern Recognition, pp. 1–4. IEEE (2008)
21. Li, F.F., Fergus, R., Perona, P.: Learning generative visual models from few training examples: an incremental bayesian approach tested on 101 object categories. Comput. Vis. Image Underst. **106**(1), 178 (2007). http://www.vision.caltech.edu/Image_Datasets/Caltech101
22. Griffin, G., Holub, A., Perona, P.: Caltech-256 object category dataset. California Institute of Technology (2007). http://www.vision.caltech.edu/Image_Datasets/Caltech256/

A Novel ECOC Algorithm with Centroid Distance Based Soft Coding Scheme

Kaijie Feng, Kunhong Liu(✉), and Beizhan Wang

Software School of Xiamen University, Xiamen, Fujian, China
fengkaijie1995@foxmail.com, {lkhqz,wangbz}@xmu.edu.cn

Abstract. In ECOC framework, the ternary coding strategy is widely deployed in coding process. It relabels classes with $\{-1, 0, 1\}$, where $-1/1$ means to assign the corresponding classes to the negative/positive group, and label 0 leads to ignore the corresponding classes in the training process. However, the application of hard labels may lose some information about the tendency of class distributions. Instead, we propose a Centroid distance-based Soft coding scheme to indicate such tendency, named as CSECOC. In our algorithm, Sequential Forward Floating Selection (SFFS) is applied to search an optimal class assignment by maximizing the ratio of inter-group and intra-group distance. In this way, a hard coding matrix is generated initially. Then we propose a measure, named as coverage, to describe the probability of a sample in a class falling to a correct group. The coverage of a class in a group replace the corresponding hard element, so as to form a soft coding matrix. Compared with the hard ones, such soft elements can reflect the tendency of a class belonging to positive or negative group. Instead of classifiers, regressors are used as base learners in this algorithm. To the best of our knowledge, it is the first time that soft coding scheme has been proposed. The results on five UCI datasets show that compared with some state-of-art ECOC algorithms, our algorithm can produce comparable or better classification accuracy with small scale ensembles.

Keywords: Error correcting output codes · Multiclass · Coverage
Soft codeword

1 Introduction

Nowadays, the multi-class classification problem has been a significant issue in the field of the pattern recognition and machine learning [1]. Usually, there are two popular solutions: the first is the application of a classifier capable of dealing with multi-class problem directly; another is to decompose a multi-class problem into multiple binary-class problems. As it is found that a single classifier can't guarantee high performances for some hard problems, the latter is a more feasible way.

Currently, there are some widely used approaches, such as One-versus-one (OVO), one-versus-all (OVA) and error correcting output codes (ECOC). In detail, OVO combines two classes to form a binary class problem and ignores all other classes in turn [2]; OVA considers one class as the positive group and all other classes as the negative group [3, 4]. Both methods deploy the majority voting scheme to decide the

D.-S. Huang et al. (Eds.): ICIC 2018, LNCS 10955, pp. 165–173, 2018.
https://doi.org/10.1007/978-3-319-95933-7_21

final labels. However, it was proved that ECOC can reduce bias and variance errors produced by the binary classifiers more effectively compared with OVA and OVO.

ECOC is a more general framework, allowing classes to be relabeled according to a coding strategy [5, 6]. With the widely deployed ternary coding strategy, an ECOC algorithm mainly includes two steps: encoding and decoding. In the encoding phase, a $N_c * N$ coding matrix is created with elements taking value $\{-1, 0, 1\}$, where N_c represents the number of classes and N represents the number of base learners. For the coding matrix, each column represents a class partition scheme, matching a base learner. The classes labeled 1/−1 are assigned to positive/negative group, and those labeled 0 are ignored during training process. In the decoding phase, N base learners produce N labels for an unknown sample. And the vector consisting of predicted labels is compared with each row in the coding matrix. The class with the least loss, such as Hamming distance and Euclidean distance, is set as the final result [7].

There are mainly two categories of ECOC algorithms: problem-independent and problem-dependent. Dense Random ECOC (DRECOC) and Sparse Random ECOC (SRECOC) algorithms are problem-independent designs, as they do not take data distribution into account when generating the coding matrix. On the contrary, problem-dependent algorithms, such as DECOC [8, 9], Forest-ECOC [10] and ECOC-ONE [11], take the intrinsic characteristics of data into the consideration. In most cases, it is found that problem-dependent algorithms are superior to the former.

However, the ternary coding scheme neglects the probability of each class belonging to positive group or negative group [12, 13]. And up to now, there is no solution for it in both problem dependent and independent designs. In this paper, we propose a Centroid distance based Soft coding ECOC algorithm, named as CSECOC. It aims to utilize the probability to improve the performance. In this algorithm, classes are allocated to one group at first. Then, based on the distances among the centroids of classes and groups, some classes are assigned to another group by maximizing the ratio of inter-group distance to intra-group distance with a Sequential Forward Floating Selection (SFFS) algorithm. After the class assignment scheme is settled, a measure is defined to describe data distribution within each group, named as coverage. It is calculated by the proportion of samples whose nearest centroids matching the correct group. Then coverage is deployed as the membership of a class to its group, taking values in the range of [−1, 1]. By setting the elements in coding matrix with the corresponding coverages, the coding matrix consists of real values depicting the probability of a class belonging to a group. Our ECOC algorithm aims to fit data better, so as to improve the generalization ability.

This paper is organized as follow. Section 2 presents the detail of CSECOC. Section 3 presents the experiment results along with some discussions, and Sect. 4 concludes this paper.

2 The Framework of CSECOC

This section gives the detailed steps of CSECOC in Fig. 1. SFFS algorithm [14] is employed to search the best binary partitions by maximizing the ratio of the inter-group distance over intra-group distance as criteria.

Algorithm CSECOC

Input: $\{C_1, C_2, \cdots, C_{N_c}\}$
Output: Coding matrix M
Initialization:
$G_1^0 = \emptyset; G_2^0 = G; G = \{C_1, C_2, \cdots, C_{N_c}\}; k = 0; col = 1.$

Step 1. Search the k-th binary partition G_1^k/G_2^k of G:
1.1. Inclusion:
$C^+ = argmax_{C_i \in G_2^k} E(G_1^k + C_i, G_2^k - C_i)$
$G_1^{k+1} = G_1^k + C^+$
$G_2^{k+1} = G_2^k - C^+$
$k = k + 1$
1.2. Conditional exclusion:
$C^- = argmax_{C_i \in G_1^k} E(G_1^k - C_i, G_2^k + C_i)$
If $E(G_1^k - C^-, G_2^k + C^-) > E(G_1^k, G_2^k)$ **Then**
$G_1^{k+1} = G_1^k - C^-$
$G_2^{k+1} = G_2^k + C^-$
$k = k + 1$
go to **Step1.2**
Else If $E(G_1^k, G_2^k) \approx E(G_1^{k-1}, G_2^{k-1})$ **Then**
go to **Step 2**
Else
go to **Step 1.1**
Step 2. Calculate the col-th column in coding matrix M:
For $i \in \{1,2, \cdots, N_c\}$
$M_{i,col} = coverage_{i,col}$
$col = col + 1$
Step 3. Coding the G_1^k and G_2^k
If $|G_i^k| > 1$ $i \in [1,2]$ **Then**
$G = G_i^k$
go to **Step 1**

Fig. 1. The detail of CSECOC algorithm.

Assume for training data set, there are m samples $X = \{x_1 \cdots x_m\}$ with N_c classes, and $C = \{C_1, C_2, \cdots, C_{N_c}\}$ represents the class label set. Let L be the total number of features, and x_i^l represent the l-th feature of x_i, where $l \in [1, \ldots L]$. Assume the i-th class contains N_{C_i} samples. The original class set G would be divided into a binary partition G_1^k/G_2^k, including T_1^k/T_2^k classes respectively, where T_1^k/T_2^k represent the number of classes in G_1^k/G_2^k and k denotes the times of iteration. Let $center_1^k/center_2^k$ represents the centroid of G_1^k/G_2^k, and $center_{C_i}$ stands for the centroid of class C_i. In the calculation, these centroids are the mean values of overall samples in a group or class.

$$d(x_i, x_j) = \sqrt{\sum_{l=1}^{L} \left(x_i^l - x_j^l\right)^2} \tag{1}$$

$$S(G_i^k) = \begin{cases} \frac{2}{T_i^{k2} - T_i^k} \sum_{p \neq q; C_p, C_q \in G_i} d\left(center_{C_p}, center_{C_q}\right), & T_i^k \neq 1 \, and \, T_i^k \neq N_c - 1 \\ 0, & otherwise \end{cases} \tag{2}$$

$$E(G_1^k, G_2^k) = \begin{cases} \frac{d\left(center_1^k, center_2^k\right)}{S(G_1^k) + S(G_2^k)}, & T_1^k \neq 1 \, and \, T_2^k \neq N_c - 1 \\ 0, & otherwise \end{cases} \tag{3}$$

$$I(x_i) = \begin{cases} 1, & d\left(x_i, center_1^k\right) \leq d\left(x_i, center_2^k\right) \\ 0, & d\left(x_i, center_1^k\right) > d\left(x_i, center_2^k\right) \end{cases} \tag{4}$$

$$coverage_{r,l} = \begin{cases} \frac{\sum_{x_j \in C_r} I(x_j)}{N_{C_i}}, & \text{if } C_r \in G_1 \\ -\frac{\sum_{x_j \in C_r} 1 - I(x_j)}{N_{C_i}}, & \text{if } C_r \in G_2 \\ 0, & otherwise \end{cases} \tag{5}$$

The intra-group distance and inter-group distance are proposed here to describe two relationships: (1) the relationship among classes in one group; (2) the relationship of classes between two groups. It is intuitive that a large inter-group distance suggests a wide margin between two groups, facilitating the learning task of each base learner. A large inter-group distance paves the way for the next iteration. The intra-group distance is calculated based on the centroids of all classes in a group, while inter-group refers to the distance between centroids of both groups. $S(G_i^k)$ in formula (2) evaluates the intra-group distance for group G_i^k. There could be $T_i^k(T_i^k - 1)/2$ possible ways for T_i^k classes in G_i^k. And $E(G_1^k, G_2^k)$ in formula (3) calculates the ratio of the inter-group distance and intra-group distance. As a larger $E(G_1^k, G_2^k)$ offers a large margin between groups, it is used as the optimal objective for the searching algorithm.

Here, a measure, coverage, is defined to describe the probability of a sample falling into a true group. Coverage is the proportion of samples in a class whose nearest centroid is the correct group. And formula (5) calculates the coverage of class r. It is obvious that if a large proportion of samples in a class can be assigned to a group the class truly belonging to, then the reliability of this assignment scheme is higher. On the contrary, a small coverage reveals an unreasonable class assignment to a group. $I(x_i)$ is used to record to which group sample x_i belongs. By setting the $coverage_{r,l}$ of the r-th class for the l-th classifier as an element $M_{r,l}$, the final coding matrix contains the membership of every class. The details of CSECOC are shown in Fig. 1.

In Fig. 1, step 1 employs SFFS to find the best binary partition G_1^0/G_2^0 for the original group G. The initialized partition is G_1^0 and G_2^0, where G_1^0 is empty and G_2^0 contains all classes in the training set. At step1.1, the algorithm tries to remove a class C^+ from G_2^0 and add it to G_1^0 to maximize the evaluation function $E(G_1^0, G_2^0)$.

At step1.2, the worst class C^{-} is moved from G_1^0 to G_2^0 to keep criterion increasing. If the new partition is better than the original partition, then new partitions would be accepted. Otherwise our algorithm goes back to step1.1. Step 1 is terminated when $E(G_1^0, G_2^0)$ cannot be increased anymore.

Step 2 aims to fill the coding matrix by means of estimating the probability of each class belonging to positive or negative group. For each class, Euclidean distance measurement is used to calculate the distance to two group centroids for each sample. After this step, both G_1^0 and G_2^0 are checked. If one of them contains more than one class, the division process will continue. The algorithm stops only when each group consists of one class. In general, for a N_c class problem, the algorithm repeats this splitting process N_c-1 times, and each iteration contributes a new column to the coding matrix. So, our algorithm produces a coding matrix with N_c-1 columns.

The class partition process can be mapped as a binary tree, and the creation of a coding matrix is illustrated as Fig. 2. Here the left child node represents positive group and the right child node represents negative group. The root node N1 represents six classes, {0, 1, 2, 3, 4, 5}. Assume that $E(G_1^0, G_2^0)$ is maximized by assigning {0, 4} to the positive group and {1, 2, 3, 5} to the negative group, then the coverage of each class is calculated. The classes assigned to positive group will obtain positive coverage values, and the classes belonging to negative group can only receive negative values. This class decomposition process repeats and produces a coding matrix with five columns, as shown in Fig. 3.

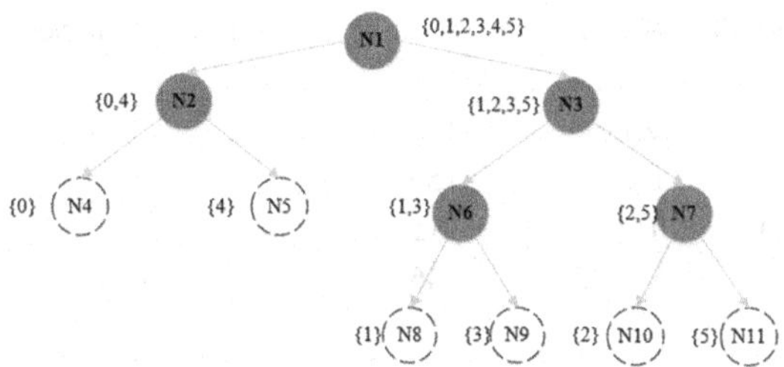

Fig. 2. Process of creating coding matrix.

In Fig. 3, $\{C_0, C_1, \ldots, C_5\}$ represent different classes and $\{H_0, H_1, \ldots, H_4\}$ represent different base learners for each column. As each element show the membership of corresponding class belonging to a group, from the first column, it is found that the samples of C_1, C_2, C_3 belong to the negative group at 100%. Only 91% samples in C5 belong to the negative group, and the remaining 9% samples close to the positive group instead. In another hand, only 73% samples in C_4 close to the centroid of positive group. So, such soft coding scheme provides us more information about data distributions in different classes.

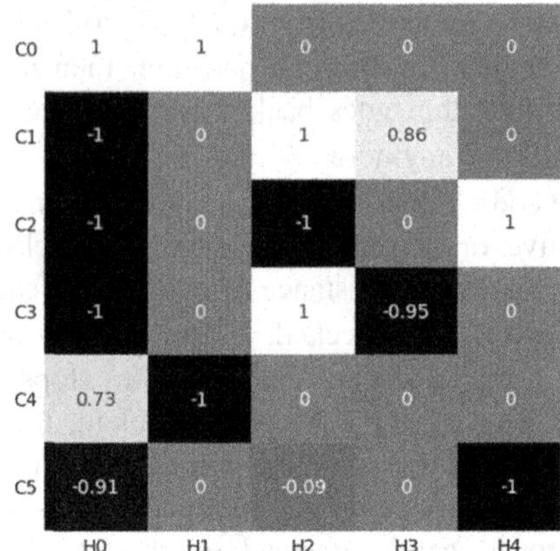

Fig. 3. Coding Matrix based on Fig. 2.

3 Experiments and Discussions

In the experiment, five datasets from UCI repository are deployed to validate our approach, and the details are listed in Table 1.

Table 1. Description of the data sets used in experiments

#Index	#Name	#Samples	#Features	#Classes
A	Dermatology	358	34	6
B	Wine	178	13	3
C	Iris	150	4	3
D	Thyroid	215	5	3
E	Vehicle	846	18	4

We compare our approach with OVO, OVA, DRECOC, SRECOC and DECOC. Because of different working principle, different ECOC algorithms produce different coding matrices with various sizes, as shown in Table 2, where N_c represents the number of classes. DECOC uses the formula (3) as evaluation function to create hierarchical structure. In our algorithm, the targets are real values instead of class labels. So, two regressors are deployed as base learners, SVR with RBF kernel [15, 16] and KNN (K = 5) based regressor. As other ECOC algorithms require classifiers as base learners, SVM and KNN are applied instead. These base learners are picked from Scikit-learn toolbox with default settings [17].

Table 2. Ensemble sizes of different algorithms

OVO	DRECOC	SRECOC	OVA, DECOC, CL- ECOC
$(N_c - 1) \times N_c$	$10logN_c$	$15logN_c$	N_c-1

In experiments, stratified 10-fold cross validation is applied ten times with random splitting. So, the mean accuracies and standard deviation results are listed in Tables 3 and 4, and the highest accuracies are marked in bold font. To simplify our discussions, all features are used in the whole process.

Table 3. Results of different methods using SVM

Measures	Datasets	OVA	OVO	DRECOC	SRECOC	DECOC	CSECOC
Accuracy	A	95.2 ± 1.70	97.2 ± 1.49	96.3 ± 1.37	89.9 ± 7.04	**97.5 ± 0.93**	97.4 ± 0.81
	B	96.9 ± 1.45	97.4 ± 1.89	**98.5 ± 0.74**	98.1 ± 1.43	98.1 ± 1.17	97.8 ± 1.39
	C	95.8 ± 2.32	95.5 ± 2.81	96.0 ± 2.40	95.8 ± 2.71	**96.8 ± 1.78**	96.2 ± 2.23
	D	95.2 ± 2.96	95.5 ± 2.62	94.5 ± 2.50	94.2 ± 2.98	95.3 ± 1.95	**96.9 ± 1.69**
	E	61.6 ± 2.26	76.0 ± 1.51	72.8 ± 4.21	75.9 ± 2.80	75.1 ± 2.53	**78.7 ± 2.59**
	mean	88.9 ± 2.14	92.3 ± 2.06	91.6 ± 2.24	90.8 ± 3.39	92.6 ± 1.67	**93.4 ± 1.74**
F-score	A	95.0 ± 1.83	97.2 ± 1.41	96.3 ± 1.43	88.2 ± 8.30	**98.2 ± 1.42**	97.4 ± 0.81
	B	96.8 ± 1.50	97.4 ± 1.90	**98.5 ± 0.71**	98.1 ± 1.42	98.3 ± 1.20	97.8 ± 1.42
	C	95.7 ± 2.35	95.6 ± 2.80	96.0 ± 2.33	95.8 ± 27.11	**97.1 ± 2.61**	96.3 ± 2.23
	D	95.0 ± 3.27	95.3 ± 2.82	94.2 ± 2.71	93.8 ± 3.23	95.0 ± 2.35	**96.8 ± 1.88**
	E	58.5 ± 2.30	74.9 ± 1.93	70.8 ± 5.32	74.8 ± 3.02	75.3 ± 3.26	**78.4 ± 2.81**
	mean	88.2 ± 2.69	92.1 ± 2.17	91.2 ± 2.50	90.1 ± 8.62	92.8 ± 2.17	**93.3 ± 1.83**

Table 4. Results of different methods using KNN

Measures	Datasets	OVA	OVO	DRECOC	SRECOC	DECOC	CSECOC
Accuracy	A	94.4 ± 1.71	94.9 ± 1.86	**95.6 ± 1.16**	87.7 ± 10.03	94.4 ± 1.71	93.5 ± 3.70
	B	97.2 ± 1.71	97.2 ± 1.71	95.7 ± 1.67	96.1 ± 1.93	97.2 ± 1.71	**98.1 ± 0.83**
	C	94.7 ± 2.47	94.7 ± 2.47	**95.3 ± 2.71**	94.9 ± 2.00	94.6 ± 3.01	94.7 ± 2.47
	D	92.2 ± 3.11	92.2 ± 3.11	94.0 ± 2.96	94.0 ± 4.32	92.0 ± 3.14	**94.2 ± 3.88**
	E	68.4 ± 2.05	71.5 ± 2.45	70.6 ± 1.39	70.4 ± 3.36	71.1 ± 2.31	**71.5 ± 2.72**
	mean	89.4 ± 2.21	90.1 ± 2.32	90.2 ± 2.30	88.6 ±5.12	89.9 ± 2.38	**90.4 ± 2.72**
F-score	A	94.5 ± 1.71	95.0 ± 1.89	**95.7 ± 1.11**	86.5 ± 11.0	95.1 ± 2.31	93.1 ± 5.01
	B	97.2 ± 1.72	97.2 ± 1.74	95.7 ± 1.72	96.1 ± 2.01	97.5 ± 2.22	**98.2 ± 0.81**
	C	94.7 ± 2.53	94.7 ± 2.51	**95.4 ± 2.71**	94.9 ± 2.03	95.3 ± 3.13	94.7 ± 2.53
	D	91.6 ± 3.49	91.6 ± 3.42	93.7 ± 3.28	93.6 ± 4.76	91.2 ± 3.62	**94.0 ± 4.12**
	E	66.7 ± 2.53	70.1 ± 2.81	69.9 ± 1.89	69.4 ± 3.74	70.2 ± 3.71	**70.3 ± 3.11**
	mean	88.9 ± 2.40	89.7 ± 2.47	90.1 ± 2.14	88.1 ± 4.71	90.0 ± 2.99	**90.1 ± 3.11**

F-score and accuracy are two widely used measures for the evaluation of different algorithms performances. The original F-score and accuracy are designed for binary problems. When applied in a multiclass problem, the average F-score and accuracy among classes are used. That is, for the i-th binary problem, the i-th class is regard as the positive class, and others are labelled as the negative class, so positive rate (P_i), negative rate (N_i), true positive (TP_i), true negative (TN_i), false positive (FP_i) and false negative (FN_i) are calculated as those in a binary problem. The final score is the average of all binary problems, as shown by formulas (6–9), and β is set to 1 to get balanced results.

From Tables 3 and 4, it is found in general, problem-dependent ECOC algorithms can beat problem-independent ones in most cases. That is, DECOC and CSECOC algorithm can achieve high accuracies and F-scores in most cases with the smallest ensemble size. As DRECOC and SRECOC are based on random coding algorithms, their performances are not so stable, and SRECOC never wins in experiments.

$$\text{Accuracy} = \text{avg}\left(\sum\nolimits_{i=1}^{R} \frac{TP_i + TN_i}{P_i + N_i}\right) \tag{6}$$

$$\text{Precision} = \text{avg}\left(\sum\nolimits_{i=1}^{R} \frac{TP_i}{TP_i + FP_i}\right) \tag{7}$$

$$\text{Recall} = \text{avg}\left(\sum\nolimits_{i=1}^{R} \frac{TP_i}{P_i}\right) \tag{8}$$

$$\text{Fscore} = \text{avg}\left(\sum\nolimits_{i=1}^{R} \frac{(\beta^2 + 1) * Presicion_i * Recall_i}{\beta^2 * Presicion_i + Recall_i}\right) \tag{9}$$

With SVM/KNN as base learner, our algorithm wins two/three out of five cases based on average accuracies and achieves the highest average mean results. Even though OVO employs much more base learners compared with our algorithm, its performance slightly worse. The same conclusions can be drawn from the results of F-scores. So, the advantage of our algorithm is obvious. And it should be noted that because our algorithm requires less learners, it can be trained and tested faster.

4 Conclusion

In this paper, we introduced a new ECOC based on the evaluation of centroid loss. It split classes into two partitions by maximizing the ratio of inter-group and intra-group. After the class assignment is settled, the algorithm evaluates the probability of each class belonging to positive group or negative group, which is deployed as elements in the coding matrix. In this way, our algorithm produces soft coding matrix, and each element takes value within $[-1, 1]$. Comparing with other methods, our algorithm can achieve the best average accuracy and F-scores. It also obtains the highest accuracy and F-scores with SVM and KNN in most cases.

It is obvious that there are still some more topics concerning with this algorithm, such as some new manners to define the soft codes, in ECOC algorithm. It is our future research direction.

Acknowledgement. This work is supported by National Key Technology Research and Development Program of the Ministry of Science and Technology of China (2015BAH55F05); Natural Science Foundation of Fujian Province (No. 2016J01320 and 2015J05129), and National Natural Science Foundation of China (Grant No. 61502402 and 61772023).

References

1. Crammer, K., Gentile, C.: Multiclass classification with bandit feedback using adaptive regularization. Mach. Learn. **90**(3), 347–383 (2013)
2. Xue, A., Wang, X., Song, Y., Lei, L.: Discriminant error correcting output codes based on spectral clustering. Pattern Anal. Appl. **20**(3), 653–671 (2017)
3. Zhou, J.D., Zhou, H.J., Zhou, H.J., Jia, N., Jia, N.: Decoding design based on posterior probabilities in ternary error-correcting output codes. Pattern Recogn. **45**(4), 1802–1818 (2012)
4. Procaccia, A.D., Shah, N., Zick, Y.: Voting rules as error-correcting codes. Artif. Intell. **231**(2), 1–16 (2016)
5. Dietterich, T.G., Bakiri, G.: Solving multiclass learning problems via error-correcting output codes. J. Artif. Intell. Res. **2**(1), 263–286 (2012)
6. Liu, K.H., Li, B., Zhang, J., Du, J.X.: Ensemble component selection for improving ICA based microarray data prediction models. Pattern Recogn. **42**(7), 1274–1283 (2009)
7. Escalera, S., Pujol, O., Radeva, P.: On the decoding process in ternary error-correcting output codes. IEEE Trans. Pattern Anal. Mach. Intell. **32**(1), 120–134 (2009)
8. Pujol, O., Radeva, P., Vitrià, J.: Discriminant ECOC: a heuristic method for application dependent design of error correcting output codes. IEEE Trans. Pattern Anal. Mach. Intell. **28**(6), 1007–1012 (2006)
9. Liu, K.H., Zeng, Z.H., Ng, V.T.Y.: A hierarchical ensemble of ECOC for cancer classification based on multi-class microarray data. Inf. Sci. **349**, 102–118 (2016)
10. Escalera, S., Pujol, O., Radeva, P.: Boosted landmarks of contextual descriptors and Forest-ECOC: a novel framework to detect and classify objects in cluttered scenes. Pattern Recogn. Lett. **28**(10), 1759–1768 (2007)
11. Escalera, S., Pujol, O., Radeva, P.: ECOC-ONE: a novel coding and decoding strategy. In: 18th International Conference on Pattern Recognition, ICPR 2006, pp. 578–581. Institute of Electrical and Electronics Engineers Inc, Hong Kong (2006)
12. Khowaja, S.A., Yahya, B.N., Lee, S.L.: Hierarchical classification method based on selective learning of slacked hierarchy for activity recognition systems. Expert Syst. Appl. **88**(11), 165–177 (2017)
13. Japkowicz, N., Barnabe-Lortie, V., Horvatic, S., Zhou, J.: Multi-class learning using data driven ECOC with deep search and re-balancing. In: IEEE International Conference on Data Science and Advanced Analytics, DSAA 2015, pp. 1–10. Institute of Electrical and Electronics Engineers Inc., Pairs (2015)
14. Pudil, P., Novovičová, J.Kittler: J.: Floating search methods in feature-selection. Pattern Recogn. Lett. **15**(11), 1119–1125 (1994)
15. Yang, Q., Qian, Z., Zheng, G., Wei, C., Xie, L., Zhu, Y., Li, Y.: The combination approach of SVM and ECOC for powerful identification and classification of transcription factor. BMC Bioinform. **9**(1), 1–8 (2008)
16. Tong, M., Liu, K.H., Xu, C., Ju, W.: An ensemble of SVM classifiers based on gene pairs. Comput. Biol. Med. **43**(7), 729–737 (2013)
17. Pedregosa, F., Gramfort, A., Michel, V., Thirion, B., Grisel, O., Blondel, M., Prettenhofer, P., Weiss, R., Dubourg, V., Vanderplas, J.: Scikit-learn: machine learning in python. J. Mach. Learn. Res. **12**(10), 2825–2830 (2012)

A Compound Algorithm for Parameter Estimation of Frequency Hopping Signal Based on STFT and Morlet Wavelet Transform

Bao-Lin Zhang[1], Jun Lv[1](✉), and Jia-Rui Li[2]

[1] Department of Information and Communication, Army Academy of Armored Forces, Beijing 100072, China
15600786982@163.com

[2] Beijing University of Posts and Telecommunications, Beijing 100876, China

Abstract. When the frequency hopping signals are analyzed in time and frequency domain, the time of arithmetic operation by Short-time Fourier Transform (STFT) is short, but the estimation accuracy of hopping time is not high; Morlet wavelet transform algorithm is an accurate estimation algorithm but its complexity is so high that it spends a lot of time in calculating. Therefore, based on the study of the two algorithms, this paper proposes a composite algorithm for parameter estimation of hopping frequency signal. Firstly, the STFT algorithm is used to search the suspected region of the hopping time, and then the Morlet wavelet transform is used to find the time accurately. So, it can get the accurate estimation of the hopping time and frequency. Theoretical analysis and simulation results show the feasibility and effectiveness of the proposed algorithm.

Keywords: STFT algorithm · Morlet wavelet transform
Hopping time estimation · Frequency estimation

1 Introduction

In view of the current shortage of spectrum resources, the use of time-frequency analysis technology to obtain the exact signal in the time and frequency domain and ultimately realize the cognitive radio [1] technology has become a hot research topic. On the contrary, the bandwidth of frequency hopping signal [2] is usually wide, which is very unfavorable to the utilization of spectrum resources.

At present, there are many kinds of methods for the time-frequency analysis [3, 4], such as short-time Fourier transform, continuous wavelet transform, Wigner-Ville distribution, Cohen distribution and so on. Many scholars do a lot of useful explorations and researches in the aspect. In references [5], Sun Ji proposed a way for frequency hopping signal time-frequency analysis based on STFT, which has great advantages in computation, but not both time domain and frequency domain it has better resolution; Li in [6] thinked that wavelet transform can overcome the shortcomings of STFT after comparing the linear time frequency analysis methods, and achieve the high frequency resolution in low frequency and high frequency respectively, while time resolution decreasing in the high frequency; in the literature [7] wavelet transform is used to analyze non-stationary

D.-S. Huang et al. (Eds.): ICIC 2018, LNCS 10955, pp. 174–181, 2018.
https://doi.org/10.1007/978-3-319-95933-7_22

signal, and characteristics obtained in the spectrum analysis results are clearer and more accurate compared to STFT. In addition, its anti-noise performance is better. For the time-frequency analysis of the frequency hopping signals, using wavelet transform [8], owing to the region the hopping time may appears unknowns, needs to make calculation on the whole bandwidth. It will be useless and occupy a large amount of computation time in non-value region, and can not accurately estimate the hopping time.

Therefore, on the basic research of the two algorithms, this paper proposes a compound algorithm for parameter estimation of frequency hopping signal based on STFT and Morlet wavelet transform. Firstly, the STFT algorithm is used to roughly search the suspected region containing the hopping moment, and then the Morlet wavelet transform is used to find the exact region so as to obtain the accurate estimation value of the hopping time and frequency.

2 Mathematical Model of Frequency Hopping Signal

Figure 1 is a mathematical model of the generation and transmission of frequency hopping signals, and $s_1(t)$ is set to the transmitting frequency hopping signals, which is expressed as:

$$s_1(t) = m(t)\cos[(w_0 + nw_\Delta)t + \varphi_n] \tag{1}$$

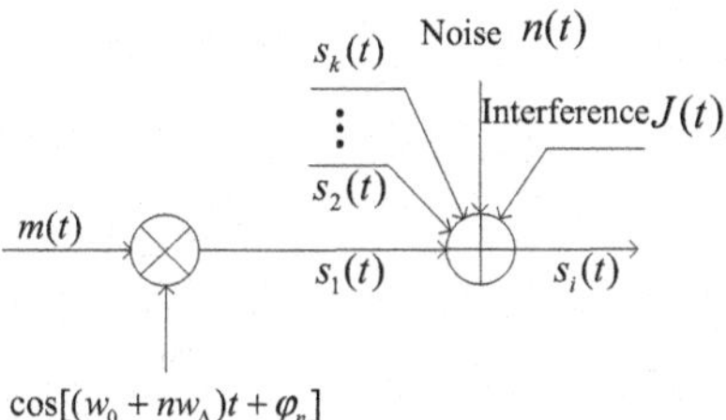

Fig. 1. Frequency hopping signal model received by receiver in frequency hopping communication system.

Among them: $n = 0, 1, 2, \ldots, N - 1$; w_Δ is hopping interval of FH synthesizer, each hop duration is T, usually $w_\Delta = 2\pi/T$; $m(t)$ is the information flow to be sent; φ_n is the initial phase.

$s_1(t)$ in the channel combining with other address signal $s_j(t)$, noise $n(t)$ and interference $J(t)$ enters the receiver, which is called $s_i(t)$:

$$s_i(t) = s_1(t) + \sum_{j=2}^{k} s_j(t) + n(t) + J(t) \tag{2}$$

Where $s_j(t)$ is another address signal $(j = 2, \ldots, k)$.

3 Composite Algorithm

STFT is the earliest and most commonly method for time-frequency analysis, and it is a natural extension of Fourier transform. Due to the influence of fixed window function, the time-frequency aggregation performance of STFT is bad so that the performance of estimating hopping time is poor, so STFT is only suitable for the condition of requiring low estimation accuracy.

Morlet wavelet is widely used for the analysis and processing of frequency hopping signals because of its good time-frequency aggregation performance [9]. Therefore, the Morlet wavelet is used as mother wavelet to analyze and process the simulation frequency hopping signals. The approximate discretization form of Morlet wavelet transform is:

$$W_f(a,k) = \frac{1}{\sqrt{a\sqrt{\pi}}}\left[\sum_{n=0}^{N-1} f(nT_s)e^{j\omega_0\frac{(k-n)T_s}{a}}e^{-\frac{1}{2}\left[\frac{(k-n)T_s}{a}\right]^2}\right] \tag{3}$$

The increase of a, while the time window is elongated, shortens the bandwidth, bandwidth becoming narrow, the center frequency of $\omega = \omega_0/a$ decreasing, and the frequency resolution getting better; instead, the bandwidth will increase, the center frequency becomes larger, the time resolution and frequency resolution get smaller. Combination of Two Algorithms.

Searching the entire bandwidth by STFT, the region the hopping time may exist in can largely be determined. Then the Morlet wavelet transform is used in the region hopping time may exist in, accurate estimation of hopping position can be obtained, as shown in Fig. 2.

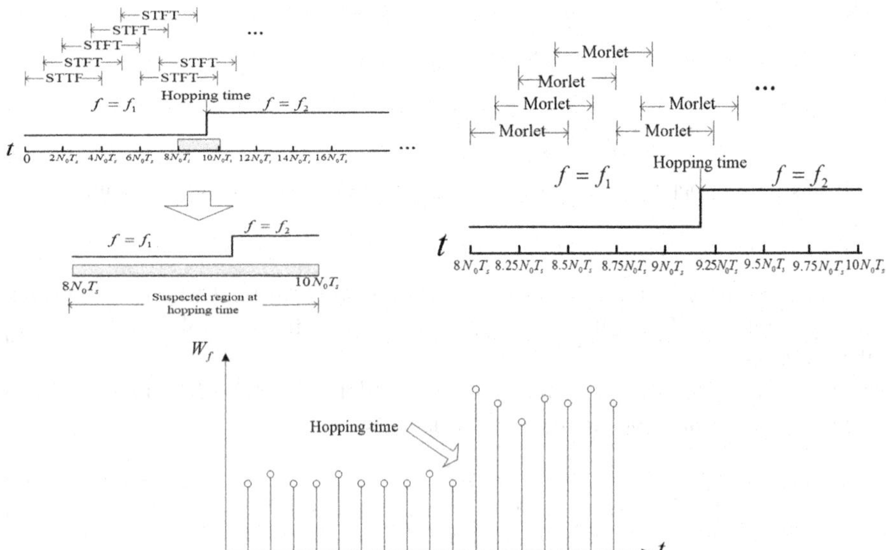

Fig. 2. Accurate estimation of frequency hopping signals

3.1 Smoothing Filtering

Considering the interference of noise, some large peak of pulses may appear in the time-frequency ridges of frequency hopping signals, as shown in Fig. 3(a). Therefore, the peak value of these pulses should be removed by smoothing filter. Specific method is: first in the time-frequency ridges intercepting a length of Δm signal, if values of the signal's both endpoints are same and there is point's value greater than a threshold value of p among the signal, the value of the point is modified to endpoints'; if there is no point such that, continue to detect the next signal whose length is Δm. The smoothed time-frequency ridges are shown in Fig. 3(b).

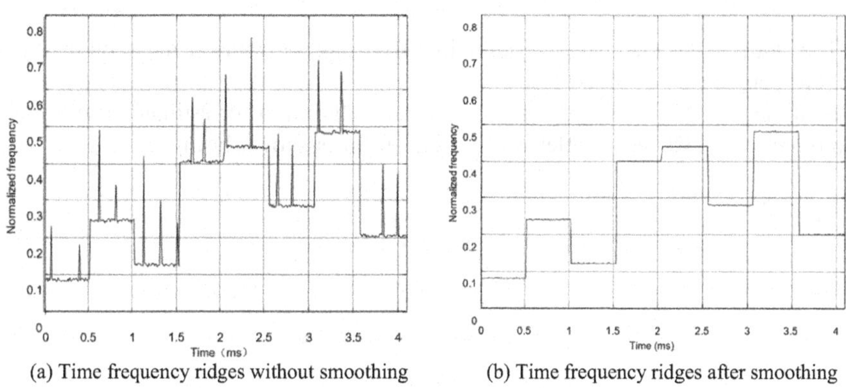

(a) Time frequency ridges without smoothing (b) Time frequency ridges after smoothing

Fig. 3. Comparison of no smooth and smooth time-frequency ridges

3.2 Accurate Hop Hop Duration Obtained by Least Squares Fitting

Select the K peaks obtained by differential operation as $p(i), i = 1, 2, 3, \ldots, K$; convert each one into spherical coordinate as (x_i, y_i), where $x_i = i$, representing the number of slot; $y_i = p(i)$, representing the hopping time obtained by fine estimation. The K coordinate values are substituted into the formula (4) to the formula (6) of the least squares method to obtain the straight line (Fig. 4).

$$k = \frac{K\sum_{i=1}^{K} x_i y_i - \sum_{i=1}^{K} x_i \sum_{i=1}^{K} y_i}{K\sum_{i=1}^{K} x_i^2 - \left(\sum_{i=1}^{K} x_i\right)^2} \tag{4}$$

$$b = \frac{1}{K}\left(\sum_{i=1}^{K} y_i - k\sum_{i=1}^{K} x_i\right) \tag{5}$$

$$y = kx + b \tag{6}$$

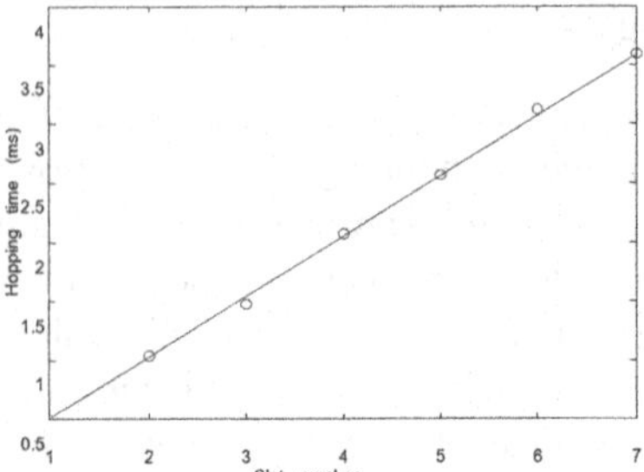

Fig. 4. Least squares fitting of hop time

The slope of the line is the hop duration. By using the value of the x_i on the fitting line, the hopping time can be further corrected.

In summary, the flow chart of the composite algorithm for estimation of hopping time based on STFT and Morlet wavelet transform is shown in Fig. 5.

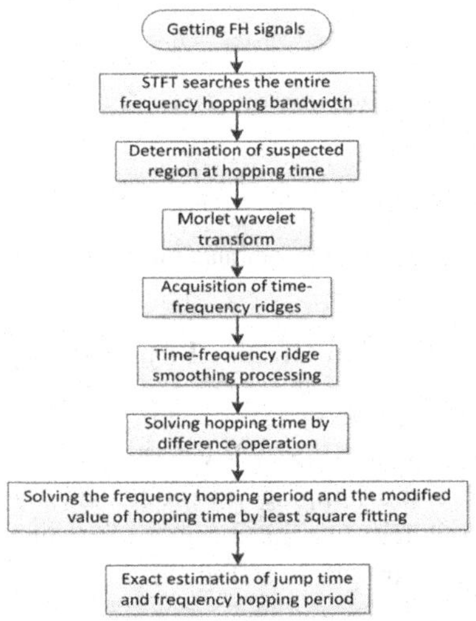

Fig. 5. Flow chart of composite algorithm

4 Simulation and Result Analysis

The general form of frequency hopping signals can be expressed as:

$$s(t) = A\cos(2\pi f_j t + \theta_j) + w(t), 0 < t \leq T \tag{7}$$

In the formula: A is the signal amplitude, f_j and θ_j are the frequency and phase of frequency hopping signal component whose order is the j at $jT \leq t < (j+1)T$, $w(t)$ is Gauss white noise, and T is the hop duration. In order to more easily simulate, the hopping frequency range of the frequency hopping signal is [1MHz,12MHz], the frequency interval is 1MHz, and the hop duration-$T = 0.512$ ms is used. Therefore, the frequency hopping signal for simulation contains a total of 12 hopping frequencies, which can be expressed as $f_i = 1 + i\Delta f MHz$, $i = 0, 1, \cdots, 11$.

The frequency hopping sequences of the simulation are expressed as $f_i = \{f_1, f_5, f_2, f_9, f_{10}, f_6, f_{11}, f_4\}$, and the hopping time as t_i = {0.512, 1.024, 1.536, 2.048, 2.56, 3.072, 3.584}. And the sampling frequency is 25 MHz, so each hop of the frequency hopping signal contains 12800 sample points. STFT algorithm, Morlet wavelet transform and composite estimation algorithm based on STFT and Morlet wavelet transform are used to estimate the hopping time of frequency hopping signals, respectively. The estimated results are shown in Tables 1, 2 and 3.

Table 1. Estimation error of hopping time of STFT algorithm.

Hopping time	t_1	t_2	t_3	t_4	t_5	t_6	t_7
True values (ms)	0.512	1.024	1.536	2.048	2.56	3.072	3.584
The estimated values (ms)	0.587	1.106	1.589	2.114	2.619	3.134	3.633
The estimation error (ms)	0.055	0.042	0.053	0.051	0.045	0.047	0.049

Table 2. Hopping time estimation error of Morlet wavelet transform.

Hopping time	t_1	t_2	t_3	t_4	t_5	t_6	t_7
True values (ms)	0.512	1.024	1.536	2.048	2.56	3.072	3.584
The estimated values (ms)	0.535	1.042	1.551	2.061	2.577	3.08	3.596
The estimation error (ms)	0.023	0.018	0.015	0.013	0.017	0.008	0.012

Table 3. Hopping time estimation error of composite algorithm.

Hopping time	t_1	t_2	t_3	t_4	t_5	t_6	t_7
True values (ms)	0.512	1.024	1.536	2.048	2.560	3.072	3.584
The estimated values (ms)	0.532	1.043	1.554	2.066	2.577	3.088	3.600
The estimation error (ms)	0.020	0.019	0.019	0.018	0.017	0.016	0.016

Seen from Table 1, 2 and 3, due to the influence of fixed time-frequency window function, STFT aggregation performance is poor, resulting in the hop time estimation performance is relatively poor, so estimation error of hop time is relatively large by STFT algorithm; owing to the multi-scale characteristic of Morlet wavelet transform, its time-frequency properties are better than STFT, so the hopping time estimation performance is relatively high and the estimation error of hop time is relatively smaller.

In order to compare the computational complexity of each algorithm, the time spent in the simulation is shown in Table 4. It can be seen that the compute of STTF algorithm

Table 4. Calculation time of each algorithm.

Algorithm	STFT	Morlet wavelet transform	Composite algorithm
Computational time (ms)	1.586	7.232	1.839

is very small, and the Morlet wavelet transform has a large amount of computation due to a large number of complex exponential operations. Therefore, the Morlet wavelet transform takes more time than STFT, and can not respond quickly to the time-frequency characteristics of the signal, and the real-time performance is poor.

In the composite algorithm, a good estimation performance of Morlet wavelet transform retains the advantage of high precision, estimation error is similar to Morlet wavelet transform's, but the calculation time of Morlet wavelet transform is greatly reduced, so the ability of responsing time and frequency domain is greatly enhanced in real time.

In order to test the anti-noise performance of the algorithm, the signal-to-noise ratio is $-5 \sim 5$ dB, STFT algorithm, Morlet wavelet transform algorithm and the compound algorithm proposed in this paper are used for hop time estimation about the simulation of frequency hopping signal, the estimation error curve is shown in Fig. 6.

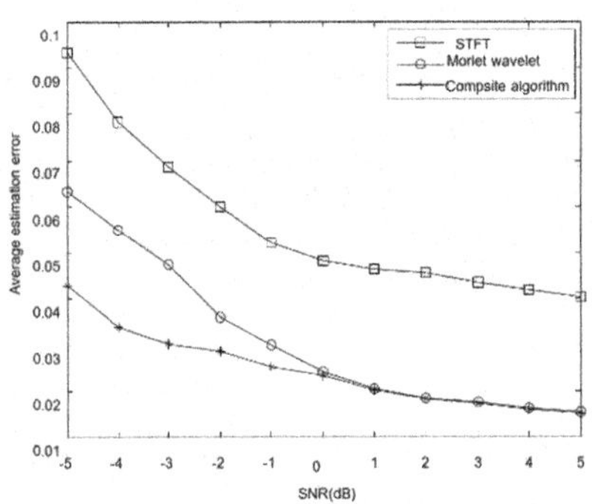

Fig. 6. Error curve of hopping time estimation for various algorithms

From Fig. 6, the anti-noise performance of composite algorithm is proposed in this paper is better than the other two algorithms, before the composite algorithm estimates the hop time, time-frequency ridge of FH signals are smoothed, reducing the noise influence on the performance of the algorithm.

5 Conclusion

The comprehensive performance of the proposed algorithm based on STFT and Morlet wavelet transform is obviously better than that of STFT algorithm and Morlet wavelet transform method. The simulation results show that it improves the signal time domain and frequency domain resolution compared with STFT algorithm, and accelerates the estimation speed relative to the wavelet transform, the computation time is also greatly reduced, the utilization of spectrum is fed back timely to provide support for the spectrum access technology, so as to increase the utilization ratio of spectrum.

References

1. Chen Bing, H., Feng, Z.K.: Advances in cognitive radio re-search. J. Data Acquis. Process. **31**(3), 440–451 (2016)
2. Zhi-chao, S., Zhang-meng, L., Zhi-tao, H., et al.: Online hop timing detection and frequency estimation of multiple FH signals. ETRI J. **35**(5), 748–756 (2013)
3. Boashash, B.: Time Frequency Signal Analysis and Processing: a Comprehensive Reference. Elsevier Science Ltd., Amsterdam (2003)
4. Cohen, L.: Time-Frequency Analysis. Prentice-Hall, Englewood Cliffs (1995)
5. Sun, J., Yang, Z., Li, L.: Frequency hopping signal analysis method based on short time Fourier transform. Commun. Countermeas. **4**(4), 17–21 (2015)
6. Li, Z., Diao, R., Han, W., et al.: A review of linear time frequency analysis methods. Reserv. Eval. Dev. **33**(4), 239–246 (2010)
7. Han-bo, Z., Yi, Y., Kui-xi, Y.: Analysis and processing of non-stationary signals based on wavelet transform. J. Nanjing Norm. Univ. (Eng. Tech. Edit.) **14**(1), 63–69 (2014)
8. Zhang, X., Wang, X., Du, X.: Blind estimation of frequency hopping signal parameters based on wavelet transform. J. Circuits Syst. **14**(4), 60–65 (2009)
9. Zheng, Y., Chen, X., Zhu, R., et al.: Based on wavelet decomposition and Hilbert Huang transform frequency hopping signal detection. Transducer Microsyst. Technol. **12**(9), 132–135 (2017)

Recognition of Comparative Sentences from Online Reviews Based on Multi-feature Item Combinations

Jie Zhang[1], Liping Zheng[1], Lijuan Zheng[2(✉)], and Junyan Ge[3]

[1] School of Computer Science, Liaocheng University, Liaocheng 252000, Shandong, China

[2] School of Business, Liaocheng University, Liaocheng 252000, Shandong, China
zhangjieliaoc@163.com

[3] School of Data Science and Software Engineer, Qingdao University, Qingdao 266000, Shandong, China

Abstract. At present, comparative sentences in online reviews are a common and convincing expression. In the autonomous recognition of Chinese comparative sentences, the selection of feature items plays a important role. The previous research mainly adopt the pattern recognition methods. This paper focuses on the recognition of comparative sentences for multi-feature item combinations in online reviews and use the text classification algorithm in machine learning to achieve. First, analyze the influence of the number of different feature items in comparative sentence recognition about the classification performance, and select the number of feature items with the highest mean of classification accuracy, make a combination of different feature items. Then use the document frequency method to reduce the dimension of feature items and select the Boolean weights to construct feature vector. Finally, using SVM classifier to discern comparative sentences. Based on the online reviews of mobile phone, This paper studies the recognition of comparative sentences for thirty feature items.

Keywords: Feature items · Compare elements · Word frequency
Accuracy

1 Introduction

In recent years, thanks to the increasing popularity of social media, consumers have a broader platform for their opinions and experiences on various products. Many users post online reviews on social media such as weibo, post bar, BBS, and blogs. The online reviews contain a lot of subjective information. If these subjective information are excavated, people can quickly obtain information about products and other products of the same type, and more effectively understand people's true views. It has important practical significance and research significance for the business field and people's daily life. In the past, online reviews were limited to the advantages and disadvantages of things, and the variety of goods was relatively unitary. Most of the researches are based

D.-S. Huang et al. (Eds.): ICIC 2018, LNCS 10955, pp. 182–193, 2018.
https://doi.org/10.1007/978-3-319-95933-7_23

on the non-comparative sentences of online reviews, and analyze the tendency of users' comments. Scholars study the extraction of emotional information, the classification of emotional information, and the selection of reference words of praise and demerit, and determine the propensity through lexical similarity [1].

The comparative sentence occupies an increasingly important position in the current online comments. Comparison influences people's daily life in all aspects, people can learn new things by comparison and judge the advantages and disadvantages of different kinds of things. But, there are many ways to recognition the sentences at present, some are based on keywords, some are based on comparison marks and comparison results, and others are based on comparison elements to identify comparative sentences. It is not conducive to the comprehensive analysis of the accuracy of each feature item. In order to make a list of all the possible features, and effectively determine the differences and connections between them. In this paper, different combinations of five comparative elements are used as feature items to study the recognition of comparative sentences.

2 Comparative Sentence Identification Related Work

2.1 Related Research Review

At present, the recognition of Chinese comparative sentences has been studied in terms of keywords, comparison objects, comparison elements, comparison marks, and comparison results. In the subject of automatic recognition of comparative sentences, The topic of automatic identification of comparative sentences was first proposed by Jindal and Liu [2, 3], To classify every English sentence by adopting a supervised learning method, in the process of model training, category sequence rule characteristics are introduced. For English comparative sentences, foreign scholar Doran [4] lists the types of comparative sentences, such as Nominal Comparatives, Adjectival Comparatives, Adverbial Comparatives, Adjectival Superlatives, and Adverbial Superlatives. Moreover, Jindal and Liu divide the comparative sentences into the following subcategories: (Table 1).

Table 1. The subcategory of comparison

		Semantics	Example sentences
Nominal comparatives		Compare object attributes with the same or similar properties	Hongmi mobile phone is as good as Glory
Different contrast	Sequential contrast	There is a sequential difference between the two	Hongmi is more durable than Glory
	Different	There are differences between the two	Hongmi's interface is different from the iPhone
	Extreme contrast	The extreme value between multiple items	Hongmi is the most cost-effective

According to the Chinese comparison of sentence meanings, construct a comparative sentence pattern library. Zhou et al. [5] took into account the "symbiotic relationship" between comparative sentences and comparative elements, and constructed a comparative feature dictionary system. Bai [6] proposed a method that compares the comparison mark and the comparison result to the recognition of the comparison sentence, to sum up the categories of Chinese comparison sentences. Wu et al. [7] adopted a method of recognition the number of comparison objects. When there are two comparison object numbers, it is judged whether it is a comparative sentence. Wang [8] extracts candidate comparison sentences containing at least one keyword to form a candidate comparison sentence set, and then uses a multi-feature fusion classification method to classify candidate comparison sentence sets. Zheng [9] focuses on sentences and paragraphs as samples,the effects of various factors on sentiment classification accuracy in chinese online reviews are discussed. Zheng [10] chose n-char-grams and n-pos-grams as potential emotional features. The improved document frequency method is used to select the feature subset. It also conducted chi-square test and explored the influence of feature selection on Chinese online comment analysis. Lu [23] proposed a sentence similarity calculation method based on a deep learning model Word2Vector combined with edit distance algorithm.

2.2 Comparative Factors

The comparison element refers to the basic unit that constitutes the comparative sentence, which consists of the comparison subject, comparison mark, comparison result, comparison object and comparison point.

The Comparing subjects is a very important part of the comparison sentence, which is the topic and focus of the speaker discusses in comparing several subjects. At the same time, the comparison subject can sometimes be missing.

The comparison mark is the most important part of the comparison sentence, and it is the most important factor to judge whether a sentence is a comparative sentence. Most scholars classify different sentences according to the different marks. Similarly, the study of this paper is based on the category of comparison marks [6]. The commonly used comparisons are as follows: compare, Less than, have, haven't, most, more, like, same as, almost, not as good as. Some sentences that contain comparison marks but have no comparison results cannot be regarded as comparative sentences.

The comparison result is also a more important part of comparative sentences, which is the focus of comparative relation extraction. The reason for the different sentence patterns in comparison sentences is the difference in comparison results. The identification of comparative results is significant for sentiment analysis, and the comparison result is generally the majority of adjectives.

The comparative object is also a very important part of comparative sentence, and it is the reference object in which the comparison subject is evaluated. At the same time, the object of comparison can also be missing.

The comparison point is the factor that compares the comparative subject with the comparative subject. For example: Hongmi is a very cost-effective mobile phone.

2.3 The Text Representation of Comparison Sentence

The text representation of comparative sentences using vector space model. First, the text is divided into word processing. Then, generate a sequence of feature items based on the training sample set. According to the generated sequence, Boolean assignment is performed on the documents in the training sample set and the test sample set to generate a set of matrix vectors. The basic process includes the selection of feature items, the dimension reduction of feature items, and the weight calculation of feature items.

The Selection of Feature Items. The selection of feature items, that is, selecting a semantic unit as a feature item. The selection of feature is based on a certain evaluation criterion to choose the term frequency ranking independently of the selected words, and select some feature items with higher frequency from them, filter out the other low frequency feature items. This paper uses the term frequency method to select the feature items.

The Dimension Reduction of Feature Items. The methods of the dimension reduction of feature items are: Document Frequency, Information Gain, Chi-square Statistic, Mutual Information, etc. Comparing the research on the dimension reduction methods of existing feature items. Hwee et al. [13] results show that the dimension reduction methods of DF, IG and CHI are better than those of MI and TS. At the same time, Liu [14] showed that the reduction effect of DF was better than CHI and IG.

The Weight Calculation of Feature Items. The methods for calculating the weight of feature items are Boolean weight method, inverted document frequency, absolute term frequency, term frequency-inverse document frequency. Pang [15] used Boolean weight method to calculate the weight of feature items, and the accuracy of sentence classification was 82.9%. In most studies, only the feature items appear, regardless of the number of occurrences, and the Boolean weight method meets the requirements.

2.4 The Selection of Classifiers

The commonly used text classifiers are Support Vector Machines, Naïve Bayes, Maximum Entropy, etc. The study found that all features used in the SVM model use the L1 regular method to help mitigate overfitting. The parameter adjustment of SVM model mainly focuses on the penalty coefficient C, which is the tolerance of error. The higher the penalty coefficient, the less tolerance for error. After parameter search, the optimal parameter combination is obtained. The study found that in the case of limited training data, compared to many classifiers, the SVM is better for classification. For this reason, SVM has also become the preferred classifier for comparative sentence recognition. Among them, Xia [16] uses SVM for classification. With the increase of the number of comments, the accuracy of the classification of SVM is also increased. Li [17] used 144 comments as a training set and a test set. Using the SVM classifier for classification, the accuracy rate was as high as 85.4%. Shi [18] used SVM for classification. The results show that SVM is more excellent in Chinese sentiment analysis than in English sentiment classification. Pang [15] manually tagged feature words in film reviews, using the frequency of feature items appearing in the text as a classification basis, using NB, ME, and SVM three classifiers to perform experiments. The results show that the SVM classification effect is best.

3 Recognition of Comparative Sentences Based on Multiple Feature Item Combinations

3.1 The Basic Flow of Comparison Sentence Recognition

In this paper, the different combination of the comparison elements is used as the feature item to deal with the collected training corpus data. First, manually annotate the feature items contained in the training set. Then statistics are made on different combinations of elements to form a set of feature items. Next, according to frequency of use arrange feature items. Finally, according to the number of special words to select the required feature items [11, 12]. The process of comparing sentence recognition is shown in Fig. 1.

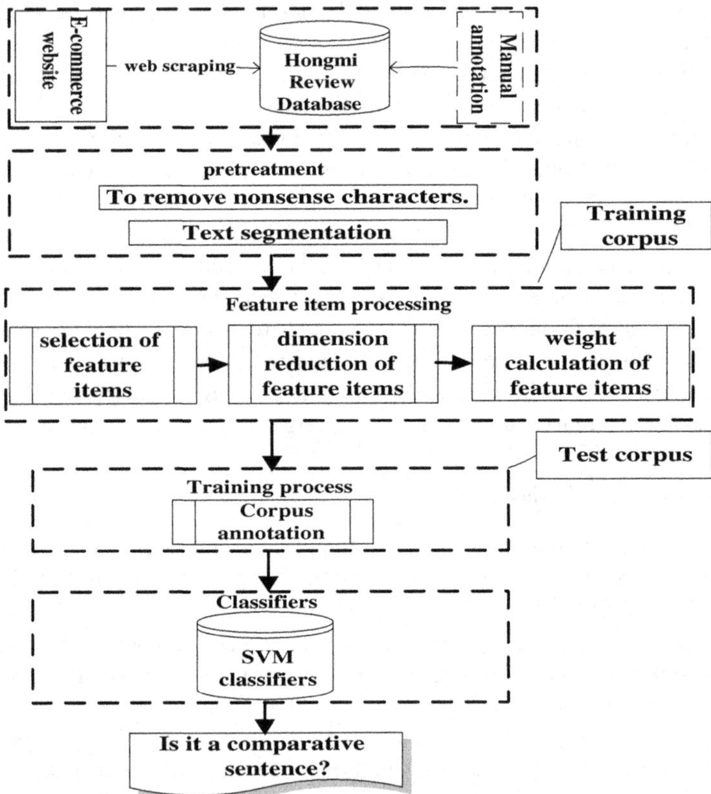

Fig. 1. Comparison sentence recognition flow chart

3.2 Data Corpus

The experiment uses the Hongmi and the honors mobile phone evaluation data provided by jingdong website as the training set and test set. All data are obtained through the web scraping program. This article takes the reviews of the Hongmi mobile phone as the training set and honors's reviews as the test set.

Considering the effect of the number of feature items on the accuracy of comparative sentence recognition. If the quantity of corpus is not enough, the classifier will be difficult to classify the comparative sentence effectively, resulting in errors in classification accuracy, and too much training corpus will cause the experiment process to be too cumbersome and affect the efficiency and progress of the experiment. Therefore, it is necessary to choose the appropriate quantity of corpus. This paper selects 600 training corpus and 600 test corpus respectively to conduct experiments.

3.3 Data Preprocessing Process

In the classification of comparative sentences, the data preprocessing process includes removing noise, removing stop words, removing meaningless characters and chinese word segmentation, etc. Stop words refer to words that are not helpful to the classification of comparative sentences, such as pronouns, conjunctions, prepositions, and other high-frequency pronouns, usually requires the maintenance of a particular task's stopwatch list, and the words that appear in the stop table are deleted during feature extraction. Different from English with delimiters, the accuracy of Chinese word segmentation will directly affect the final classification result.

3.4 The Extraction of Feature Items

Based on the comparison subject, comparison object, comparison point, comparison mark and comparison results, this paper uses the different free combinations of the five elements as the feature item of the comparative sentence. Among them, the comparison subject is represented by "Z" letters, the comparison object is represented by the "K" letter. The Compare points with "D" letters, the comparison marks are represented by "B" letters, the comparison results are represented by "J" letters. The combination is a group of B(B $\leq$ A) elements from A different element, which is called a combination of B elements from A different element. The number of combinations of B(B $\leq$ A) elements taken from A different element is called the number of combinations of B elements taken from A different element. Denoted by the symbol c(a,b). c(a,b) = p(a, b)/b! = a!/((a−b)!*b!). When the type of feature item is one, the number of combinations is $C_5^1 = 5$,which is Z,B,D,K,J; When the type of feature item is two,the number of combinations is $C_5^2 = 10$, which is ZK,ZD,ZB,ZJ,KD,KB,KJ,DB,DJ,BJ; When the type of feature item is three, the number of combinations is $C_5^3 = 10$, which is ZKD, ZKB,ZKJ,ZDB,ZDJ,ZBJ,KDB,KBJ,DBJ,KDJ; When the type of feature item is four, the number of combinations is $C_5^4 = 5$,which is ZKDJ,ZKDB,ZKBJ,ZDBJ,KDBJ; When the type of feature item is five, the number of combinations is $C_5^5 = 1$, which is ZKDBJ. Therefore, the number of feature items in this experiment is $31(C_5^1 + C_5^2 + C_5^3 + C_5^4 + C_5^5 = 31)$. As shown in Table 2:

Table 2. Feature List

Combination	Ab.	Combination of future item	combination	Ab.	Combination of future item
Comb1	Z	comparison subject	Comb17	ZKB	subject+object+mark
Comb2	B	comparison mark	Comb18	ZKJ	subject+object+result
Comb3	D	comparison point	Comb19	ZDB	subject+point+mark
Comb4	K	comparison object	Comb20	ZDJ	subject+point+result
Comb5	J	comparison result	Comb21	ZBJ	subject+mark+result
Comb6	ZK	subject+object	Comb22	KDB	object+point+mark
Comb7	ZD	subject+point	Comb23	KBJ	object+mark+result
Comb8	ZB	subject+mark	Comb24	DBJ	point+mark+result
Comb9	ZJ	subject+result	Comb25	KDJ	object+point+result
Comb10	KD	object+point	Comb26	ZKDJ	subj+obj+point+result
Comb11	KB	object+mark	Comb27	ZKDB	sub+obj+point+mark
Comb12	KJ	object+result	Comb28	ZKBJ	subj+obj+mark+result
Comb13	DB	point+mark	Comb29	ZDBJ	sub+point+mark+result
Comb14	DJ	point+result	Comb30	KDBJ	obj+point+mark+result
Comb15	BJ	mark+result	Comb31	ZKDBJ	sub+obj+point+mark +result
Comb16	ZKD	subject+object+point			

In these thirty-one combinations, the comparison subject is the topic and focus of the discussion. Whether the comment statement is a comparative sentence or not, the comparison subject must exist, so the Z factor itself cannot be used as the basis for discriminating comparison sentences. Therefore, remove the Z combination and only consider the remaining thirty features.

3.5 The Selection of Classifiers

As described above, SVM classifier has better effect on emotion classification. In the first piace, this article uses word frequency method to select feature items, Selecting Method Based on Document Frequency for Feature Word Dimensionality Reduction. Finally use Boolean weight method to calculate the weight of feature items and using SVM as a classifier. Therefore, SVM is used as the classifier. The SVM classifier is affected by the type of kernel function [21, 22]. The commonly used kernel functions are Radius Basis Function, Polynomial Kernel Function and Sigmoid Kernel Function. The most widely used kernel function is RBF. Li [19] proposed that RBF is a universal kernel function for any sample. Bian [20] experimentally concludes that RBF has the advantage of solving small sample problems. Therefore, the kernel function used in this paper is RBF. The operation steps are shown in Fig. 2.

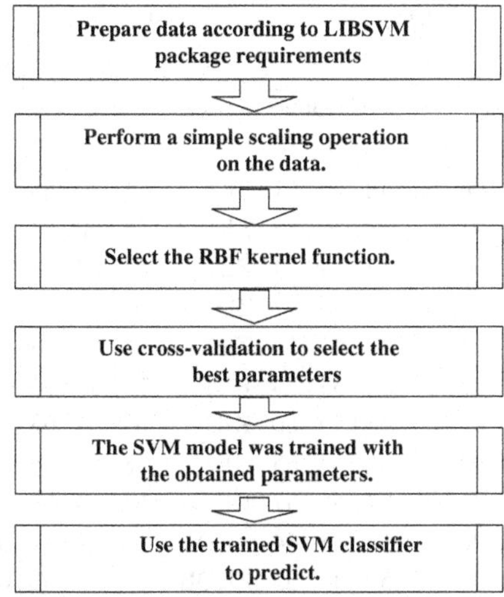

Fig. 2. SVM operation steps

4 Experimental Results and Analysis

This paper uses DF method to select five, ten, twenty, thirty and fifty semantic units as feature items to conduct experiments.

4.1 The Experimental Results

Figure 3 shows the line graphs of representative experimental data of six groups with higher accuracy when the number of thirty feature items is five, ten, twenty, thirty and fifty. Table 3 shows the recognition accuracy mean. Figure 4 is the accuracy of the comparative sentence recognition of feature item combinations in the experiment.

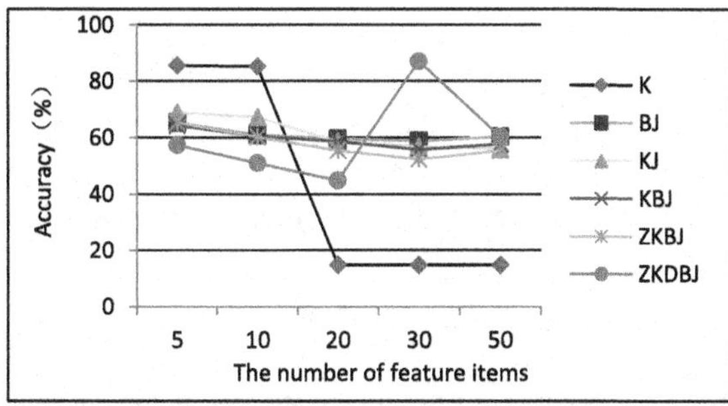

Fig. 3. The number of feature items affects the classification performance

Table 3. Accuracy of recognition

FI	A										
	The number of feature items										
	5	10	20	30	50	FI	5	10	20	30	50
B	76.06	66	48	61	61	ZKB	71	70	65	56	56
D	72.91	67	58	56	55	ZKJ	63	62	61	61	60
K	85.6	85	14	14	14	ZDB	63	62	62	60	60
J	68.92	67	56	53	47	ZDJ	64	64	63	62	61
ZK	85.26	66	14	41	14	ZBJ	65	64	63	62	64
ZD	72.91	67	55	54	54	KDB	67	66	66	66	65
ZB	71.92	66	63	59	60	KBJ	65	64	63	63	63
ZJ	68.92	66	66	65	65	DBJ	58	56	56	55	55
KD	72.91	71	54	55	57	KDJ	64	64	63	63	63
KB	75.9	65	64	65	64	ZKDJ	64	64	63	63	63
KJ	68.92	67	50	45	35	ZKDB	63	63	62	62	61
DB	67.53	65	65	64	63	ZKBJ	65	65	64	64	63
DJ	65.73	64	63	65	63	ZDBJ	60	58	58	57	57
BJ	65.34	60	59	58	60	KDBJ	57	57	56	56	55
ZKD	73.48	67	54	55	56	ZKDBJ	57	50	44	87	60
AM	68.34	65	56	58	56	AM	68	65	56	58	56

Ab. A: Accuracy%, FI: Feature Items, AM: Accuracy Mean

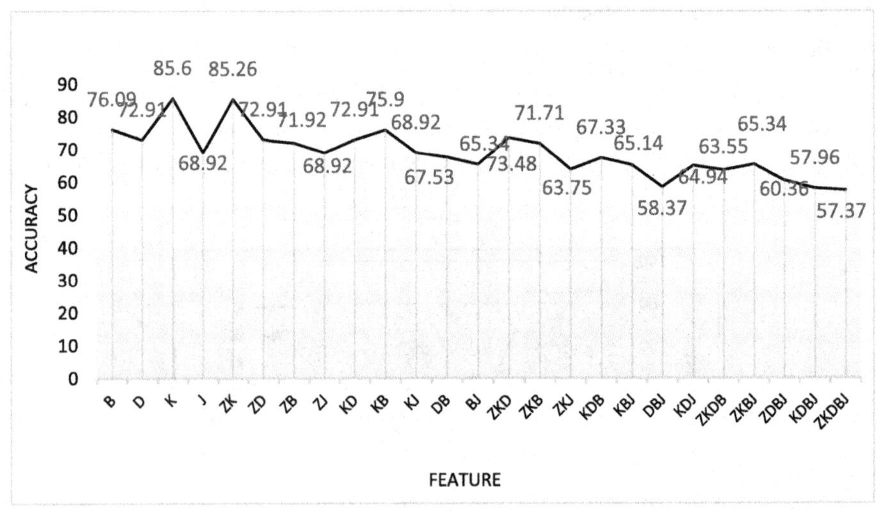

Fig. 4. The accuracy of the combination of multiple features

4.2 Analysis of Experimental Results

The Comparison of Different Feature Item Selection Methods. Figure shows the accuracy of different feature items in the recognition of Chinese comparative sentences. Experimental analysis of 600 corpora concluded K feature item in the recognition of comparative sentences with higher accuracy, the maximum can reach 85.6%. It's 20% more accurate than BJ feature item. Moreover, K features more balanced efficiency and accuracy.

The Influence of The Number of Feature Items on The Classification Performance. This paper adopts DF method to reduce dimension treatment, the number of feature items has a greater influence on the classification accuracy. When the number of feature items is small, the classification accuracy is higher. With the increase of the number of feature items, the classification accuracy is decreasing. Some classification accuracy rates will rise under the overall downward trend.
The ZKDBJ combination has a low accuracy rate when the number of special words is 5. When the number of feature items is 30, the recognition accuracy rate is the highest among all experimental data. It shows that the number of feature items with the highest accuracy of each combination is different. Therefore, it is necessary to analyze the recognition accuracy of the number of different characteristic words when choosing feature items.

Training time increases with the number of feature items. Therefore, it is necessary to balance efficiency and accuracy when choosing the number of feature items. Experimental results show that the highest average accuracy rate is 5. Therefore, the experiment selected five feature items as experimental features.

The Comparison of The Accuracy of Different Feature Items. Among them, K combination has higher accuracy in the identification of comparative sentences. When the Z factor is increased, the recognition accuracy of the ZK combination shows a short downward trend, and the recognition rate at this time is not greatly affected. Therefore, the ZK combination can be used as a feature items of comparative sentence recognition; When adding B elements, the accuracy of KB composition also shows a downward trend. This phenomenon indicates that when adding one feature word, the accuracy rate is not reduced. When the ZD elements are added, the recognition accuracy of the ZKD combination shows a clear downward trend. The reason why KD elements have a greater impact on the recognition accuracy of comparative sentences is that D elements are expressed in a paradoxical manner in the identification of comparative sentences, and lack formalized marks. At the same time, when adding two elements, the accuracy rate drops more.

The J element has low accuracy in the identification of comparative sentences. In all combinations, the feature items that contain J elements are generally lower in the recognition of the comparison sentences, and the recognition accuracy of the feature items that do not contain J elements is relatively high. Through research and analysis, it is believed that the J elements are mostly in the form of adjectives. And the adjective cannot be used alone as the basis for judging whether the sentence is a comparative sentence. Therefore, feature items containing J elements are generally less accurate in the identification of comparative sentences.

5 Conclusions

The recognition of comparative sentences has always been a difficult problem in the field. This paper takes online reviews in chinese context as the research object, and using the combination of five elements as feature items to identify comparative sentences, which provides a new ideas for the text mining of comparative sentences. Then, Study the effect of the number of characteristic items on the accuracy of classification, And analyze the accuracy of comparative sentence recognition under multiple feature item combinations.

In order to improve the research value of the experiment, this paper chooses five feature items to carry out the recognition of comparative sentences, and the accuracy is higher. However, with the increase of the number of feature items, the classification accuracy has a downward trend instead. There are no professional conclusions about the causes of this phenomenon. The number of feature items with the highest accuracy for each combination is also different. And it is necessary to analyze the recognition accuracy of different feature items when choosing feature items. Therefore, the choice of the number of feature items in different training sets will be another question worth exploring.

Acknowledgement. This work is supported by the National Institute of Education Humanities and Social Sciences Research Youth Fund Project (16YJCZH159), Shandong Provincial Institute of Humanities and Social Sciences Research Project (J16YF25), Liaocheng University Scientific Research Project (31801140).

References

1. Xiong, D.L., Cheng, J.M., Tian, S.L.: Sentence orientation research based on How Net. Comput. Eng. Appl. **44**(22), 143–145 (2008)
2. Jindal, N., Liu, B.: Identifying comparative sentences in text documents. In: Proceedings of the 29th Annual International ACM SIGIR Conference on Research and Development in Information Retrieval, pp. 244–251. ACM (2006)
3. Jindal, N., Liu, B.: Mining comparative sentences and relations. In: Proceeding of AAAI, Palo Alto, pp. 1331–1336 (2006)
4. Doran, C., Egedi, D., Hockey, B.A., et al.: XTAG system: a wide coverage grammar for English. In: Proceedings of the 15th Conference on Computational Linguistics, vol. 2, pp. 922–928. Association for Computational Linguistics (1994)
5. Zhou, H., Hou, M., Hou, M., et al.: Chinese comparative sentences identification and comparative elements extraction based on semantic classification. J. Chin. Inf. Process. **28** (3), 136–141 (2014)
6. Bai, L., Hu, R., Liu, Z.: Recognition of comparative sentences based on syntactic and semantic rules-system. Acta Scientiarum Naturalium Universitatis Pekinensis **51**(2), 275–281 (2015)
7. Wu, C., Wei, X.F.: Opinion analysis and recognition of comparative sentences in user views. Comput. Sci. **43**(s1), 435–439 (2016)
8. Wang, W., Zhao, T., Bing, X.U., et al.: Automatic identify Chinese comparative sentences. Intell. Comput. Appl. **5**, 1–3 (2015)

9. Zheng, L.J., Wang, H.W.: Sentimental polarity and strength of online cellphone reviews based on sentiment ontology. J. Ind. Eng. Eng. Manag. **31**(2), 47–54 (2017)
10. Zheng, L.J., Wang, H.W., Gao, S.: Sentimental feature selection for sentiment analysis of Chinese online reviews. Int. J. Mach. Learn. Cybernet. **9**(1), 75–84 (2018)
11. Wang, H.W., Zheng, L.J., et al.: Sentiment feature selection from Chinese online reviews based on statistical machine learning. In: Academic Annual Conference of China Branch of Information Systems Association (2011)
12. Zheng, L.J., Wang, H.W., et al.: Sentiment intensity of online reviews based on fuzzy-statistics sentiment words. J. Syst. Manage. **32**(4), 376–384 (2013)
13. Hwee, T.N., Wei, B.G., Kok, L.L.: Feature selection, perceptron learning and a usability case study for text categorization. In: Proceedings of the 20th Annual International ACM SIGIR Conference on Research and Development in Information Retrieval, vol. 31, pp. 67–73 (1997)
14. Liu, X.: A study on affective polarity classification based on statistical natural language. Tongji university master's thesis (2011)
15. Pang, B., Lee, L., Vaithyanathan, S.: Sentiment classification using machine learning techniques. In: Proceedings of the Conference on Empirical Methods in Natural Language Processing, Philadelphia, US, pp. 79–86 (2002)
16. Xia, H.S., Peng, L.Y.: SVM-based comments classification and mining of virtual community: for case of sentiment classification of hotel reviews. In: Proceedings of the International Symposium on Intelligent Information Systems and Applications (IISA 2009), vol. 10, pp. 507–511 (2009)
17. Li, J.: An approach of sentiment classification using SVM for Chinese texts. In: Proceedings of 2006 International Conference on Artificial Intelligence - 50 Years' Achievements, Future Directions and Social Impacts, pp. 759–761 (2006)
18. Shi, W., Qi, G.Q., Meng, F.J.: Sentiment classification for book reviews based on SVM model. In: Proceedings of the 2005 International Conference on Management Science and Engineering, pp. 214–217 (2005)
19. Li, X.Y., Zhang, X.F., Shen, L.: A selection means on the paramrter of radius basis function. Acta Electronica Sinica **33**(B12), 2459–2463 (2005)
20. Bian, Z.Q.: Pattern Recognition. Tsinghua University Press, Beijing (1998)
21. Gao, S., Wang, H.W., Feng, G., et al.: Review of comparative opinions mining studies of online comments. New Technol. Libr. Inf. Serv. **32**(10), 1–12 (2016)
22. Wang, S., Zhao, C., Liu, H.: Comparison element ellipsis identification based on rules and sequence patterns. J. Shanxi Univ. **38**(1), 85–92 (2015)
23. Lu, Y.H.: A sentence similarity calculation method based on Word2Vector and edit distance. Comput. Knowl. Technol. **13**(5), 146–147 (2017)

Hierarchical Hybrid Code Networks for Task-Oriented Dialogue

Weiri Liang and Meng Yang(✉)

School of Data and Computer Science,
Sun Yat-sen University, Guangzhou, China
yangm6@mail.sysu.edu.cn

Abstract. Task-oriented dialog system is a research hotspot in natural language processing field. In recent years, the application of neural network (NN) has greatly improved the performance of dialog agent. However, there is still a big gap of performance between human beings and dialog agent, in which the domain knowledge and semantic analysis are not well exploited. In this paper we propose a model of Hierarchical Hybrid Code Networks (HHCNs), in which a word-character RNN for semantic representation and a NN-based selection for domain knowledge are integrated. Thus the proposed HHCNs can effectively conduct semantic analysis (e.g., identify proper nouns and misspelling word) and select meaningful responses for the dialog. The experimental results on the dataset of Dialog State Tracking Challenge 2 (DSTC2) have shown a superior performance of HHCNs.

Keywords: Task-oriented dialogue · Hybrid Code Network · Dialog systems

1 Introduction

Task-oriented dialogue system is a new way of communication between human and machine, receiving great attention from the industrial world and the academic world. Unlike open-domain chatbot, task-oriented dialogue system mainly focuses on a specific task such as shopping, fault resolution, ticket booking, etc. In a dialog, agent should guide users to complete the task and provide needed information for the task, instead of only chitchat.

Traditional methods consider task-oriented dialogue system as a pipeline of four components [1–8]: natural language understanding, dialogue state tracking, dialogue policy learning, and natural language generation, where each component implements independently. This kind of agent not only suffers from poor migration capabilities, but also transfers errors from one component to another. Besides, labeling data for each component respectively also consumes lots of time.

A more popular way of task-oriented system considers the entire task as a whole, using an end-to-end model to train the agent. This agent can directly build a model from raw corpus and learn the dialogue policy which could be used in practical application. In recent years, deep learning [9–19] becomes a prominent method in building an end-to-end dialog system, which achieved promising performance. In Memory Networks (MemN2N) [20], they attempt to use reading comprehension model

D.-S. Huang et al. (Eds.): ICIC 2018, LNCS 10955, pp. 194–204, 2018.
https://doi.org/10.1007/978-3-319-95933-7_24

to address dialog task. As its advanced version, Gated End-to-End Memory Networks (GMemN2N), make significant progress [21] via building gated connections between the memory access layers and the controller stack of a MemN2N. The direct application of the seq2seq model with copy [22, 23] and attention mechanisms [24, 25] has also achieved quite advanced performance, indicating the effectiveness of the seq2seq model on dialogue system [26]. In addition, reinforcement learning has been proved to possess a good prospect in the dialogue system [27], in which utterance-level LSTM and dialog-level LSTM are used to build agent and virtual user. Recently, Query-Reduction Network (QRN) [28] addresses the long-term dependency problem by simplifying the recurrent update, surpassing other pure neural networks in question answering and dialog tasks. Hierarchical Neural Network [9] is proposed to take full account of the role of context in dialogue. Although above deep learning models have achieved promising performance on task-oriented dialogue, they still have the following disadvantages. First, the behaviors of the task-oriented dialogue can be strange when the answer is completely generated by the network. Second, the human domain knowledge can not be fully used to build a good model.

Hybrid Code Networks(HCNs) are the first model trying to combine codes (e.g., entity extraction and database query program) and neural networks [29], in which an RNN is used to accumulate dialog state and choose correct actions. One character of HCNs is that they use developer-provided action templates, which can contain entity references, such as "<city>, right?". However, HCNs are too simple in network structure. Bag of words and average word embedding model can not capture the location information of sentence. Moreover, it can't identify misspelling and proper nouns that may appear in practical dialog. Last but not least, there is still room for improvement in the combination of networks and codes.

In order to solve the issues above, we proposed the Hierarchical Hybrid Code Networks (HHCNs) to improve the performance of task-oriented system. In HHCNs, the codes and neural networks are more effectively exploited together. In order to make a powerful semantic representation, we propose a word-character RNN to encode sentence, with another RNN to process each sentence vector. Meanwhile, a neural network is applied to select meaningful response from the domain knowledge. Thus the proposed HHCNs model can simultaneously get the advantages of the codes and neural networks to further improve the performance of the dialog system. The experimental results on the dataset of Dialog State Tracking Challenge 2 (DSTC2) [30] clearly show a superior performance of HHCNs.

The rest of this paper is organized as follows. Section 2 presents a brief review of the related works. Section 3 gives the proposed hierarchical hybrid code network. Section 4 conducts the experiments and Sect. 5 concludes the paper.

2 Related Work

Query-Reduction Network (QRN) [28] is a variant of Recurrent Neural Network (RNN). It can effectively handle both short-term and long-term sequential dependencies to reason over multiple facts. It considers the context sentences as a sequence of state-changing triggers, and reduces the original query to a more informed query as it observes each trigger (context sentence) through time. As a pure neural network method, it's one of the best model in question answering and dialog tasks. However, the responses of the dialog agent may not well match with human being since they are completely generated by the network and human domain knowledge is ignored.

Hybrid Code Networks(HCNs) [29] tried to combine codes and neural networks, in which domain knowledge is used to write codes, and recurrent neural network (RNN) is used to construct dialogue system. As shown in Fig. 1, the cycle begins when the user provides an utterance. The sentence will be encoded in several ways. HCNs use bag of words and average of word2vec [31] embedding to encode the utterance. In addition, domain knowledge is used to acquire context features about entity and other information (such as the states of database). All these vectors will be concatenated and pass to RNN. After that, the dense and softmax layer will output the probability of all 68 designed templates. HCNs choose the template with the highest probability and fill entity into the slot of template, which will be translated to the phase of language generation. Finally, the action taken is provided as a feature (encoded into a size 68 onehot vector) to the RNN in the next timestep.

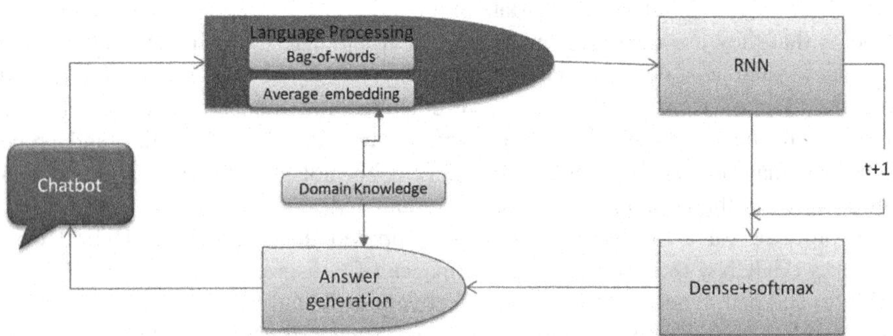

Fig. 1. Flow chart of HCNs. It uses a RNN to model the whole process of dialogue, with the help of domain knowledge.

However, there are still several drawbacks in HCNs. For instance, bag of words and average word embedding model can not well do semantic representation and the combination of networks and codes can be further improved. In this paper, we propose HHCNs to effectively integrate codes and neural network via introducing a hierarchical processing: a word-character RNN for semantic representation and a neural network (NN) based selection for action templates. Compared to HCNs, our model use the word-character RNN to better capture in-formation of sentences, and acquire the ability to identify unknown word. What's more, we add neural network selection to help the agent select the best target action in dialogue.

3 Hierarchical Hybrid Code Networks (HHCNs)

3.1 Overview of HHCNs

The HHCN model is summarized in Fig. 2. The whole flowchart of HHCNs is similar to that of HCNs but with two additional hierarchical processes: word-character RNN and neural network selection.

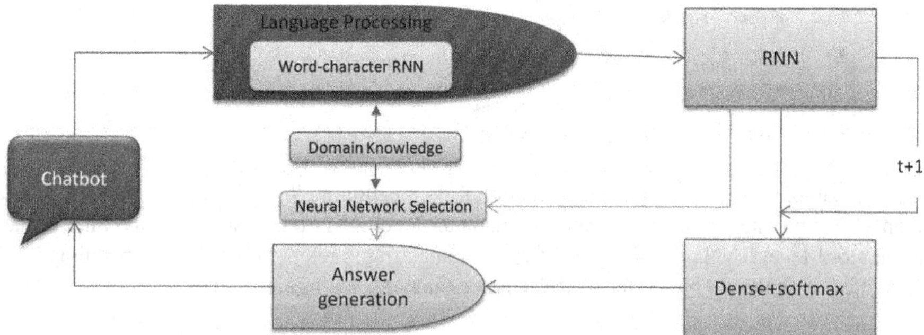

Fig. 2. Flow chart of HHCNs. Note that orange frames refer to the new elements introduced in our model. Word-character RNN and RNN are both trainable modules.

For the word-character RNN, the inputs are the utterance and the latest response of the dialog agent. After processing them, word-character RNN will output the semantic representation vector of the utterance and the vector of the latest answer. With the word-character RNN, the agent can effectively get the information of the current dialog and identify proper nouns and misspelling word.

For the neural network selection, the input is the hidden state of RNN in the latest step and the domain knowledge. The neural network will output selections of actions, e.g., back to previous recommend, keeping current recommend and proposing next recommend. The advantage of neural network selection is that the agent learning is simplified with domain knowledge and the powerful selection ability of neural network is effectively used.

3.2 Word-Character RNN

In most cases, Hybrid Code Networks use a pre-trained word2vec model to convert words into vectors. However, under the condition of practical task, we often come across proper nouns like names of shop, names of brand, etc. Sometimes, user may input sentence with misspelling. As a result, there can be some words that could not directly be found in pre-trained word2vec model. In order to tackle these problems, we introduce a character-level RNN to encode these unknown words [32]. The word-character RNN is displayed in Fig. 3.

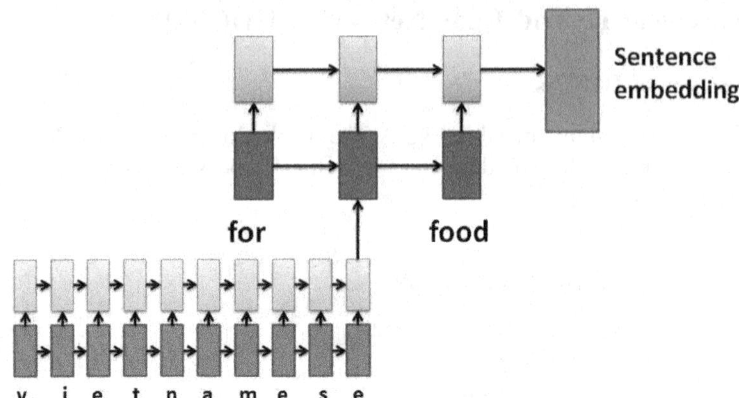

Fig. 3. Example of a word-character model for embedding "for vietnamese food?" into a sentence vector. The green frames refer to character-level RNN processing, and the blue frames refer to word-level RNN processing. Lastly, the orange frame refers to sentence embedding. Note that frames of shallow color refer to RNN processing. (Color figure online)

Suppose we have a sentence, which has n tokens.

$$S = (w_1, w_2, \ldots, w_i, \ldots, wn) \tag{1}$$

where S is the given sentence composed of n tokens, and w_i is the embedding of the ith token. We want to convert tokens of word to a series of word embedding.

If the word is in our word2vec model, we use the embedding in word2vec to fully utilize language model knowledge.

If the word, denoted by w_i (suppose the length of the word is k), is not in word2vec, we use the unidirectional character-level RNN to process the word independently.

$$w_i = (c_1, c_2, \ldots, c_j, \ldots, c_k) \tag{2}$$

where c_j is the onehot vector of the j character in the encoded word.

Denote the hidden units of character-level RNN as h. We encode each character within the word as a onehot vector c_j, and pass vectors to the character-level RNN.

$$h_t = \text{RNN}(c_t, h_t - 1) \tag{3}$$

where h_t is the hidden states of character-level RNN at step t of handling the tth character.

To unify the shape of w_i with other word2vec embedding, h_k should multiply a trainable matrix W, with the size of u-by-a. Here u is the dimension of hidden layer, and a is the dimension of word2vec. We take the last hidden states of character-level RNN h_k as the represents of word embedding.

$$w_i = h_k W \quad (4)$$

where W is a matrix in feature dense combination layer in the network. It is used to recombine h_k, extending its dimension from u to a.

In this way, the character-level RNN will automatically learn how to convert a unknown word into an embedding, which could be perfectly integrated with the word2vec model. Then we input each word embedding into a bidirectional word-level RNN, and use the last hidden states of the word-RNN as the sentence embedding.

When user inputs an utterance such as text, the utterance will be encoded into feature. We encode a sentence into vector with a Word-character RNN, which could be either a long short-term memory (LSTM) [33] or gated recurrent unit (GRU) [34], specifically for utterance sentence embedding. The last hidden states of the Word-character RNN will be used as utterance embedding. Both the current utterance and the last response of dialog agent will be convert into utterance embedding.

3.3 Domain Knowledge with Neural Network Selection

In this section, we illustrate in detail the domain knowledge introduced in HHCNs, and how to combine these domain knowledge with neural network through codes. Domain knowledge is applied in the following aspects: Key entities detection, Action templates summarization and Database results recommend.

Take the dialogue listed in Table 2 as an example. There are four kinds of entity that are important to dialog task: cuisine, unregistered cuisine, location and price. Firstly, we distinguished the registered cuisine from the unregistered cuisine by using domain knowledge. We wrote a simple string matching software, to extract entity from sentence. So the existence of each entity could be indicated by a binary context vector. Secondly, we used the entity extractor to summarize templates from training set. For example, "pipasha restaurant is a nice place in the east of town and the prices are expensive" could map into template "<restaurant> is a nice place in the <location> of town and the prices are <price>". This resulted in 77 templates in our task. This part is similar to HCNs.

What's more, we prepared a database searching API for the DSTC2 dataset, which could find suitable restaurant according to the request of user. If many of restaurants suit for the request, they are sorted by rating. Unlike HCNs, our model recommend restaurant through both neural networks and codes. HHCNs use software to create recommended list by sorting, and RNN uses another dense and softmax layer to output a probability of actions and to choose how to move on the list. There are three actions can be chosen: back to previous recommend, keeping current recommend and proposing next recommend. The state of API result is indicated by the context feature. The detail of context feature is shown in Table 1.

Table 1. Context features about entity and database state.

Feature	num
Presence of each entity in dialog state	4
Presence of each entity in current utterance	4
Whether DB has been queried yet	1
Whether DB results are empty	1
Whether DB results are non-empty	1
Whether any DB results have been presented	1
Whether all DB results have been presented	1
Whether any DB results are available to present	1
Whether current query yields no results in training set	1
Total context features	15

4 Experimental Results

4.1 Dataset

We prepare the data in our study based the corpus from the second Dialog State Tracking Challenge (DSTC2) [30]. This dataset includes an end-to-end dialog learning task in the restaurant domain, which is a practical dataset used in human-computer dialog. In the dialog, the agent should recommend a restaurant based on the preferring location/cuisine/price of user. Besides, agent should be able to answer question about address or phone number of restaurant, and recommend another restaurant when the user rejects the recommendation. Note that DSTC2 intentionally used different dialog policies in the training and test sets. To make the evaluation closer to real cases, we redistribute the total dataset to training set, development set and test set to unify their style. Table 2 shows the statistics of the dataset used in our experiments.

Table 2. Statistics of the dataset.

Train/dev/test dialog num	1735/500/1000
Average dialog length	9.2
Action template	77
Location/cuisine/price	5/77/3

4.2 Training Procedure

We trained the HHCNs on the training set. The HHCNs were specified using pytorch version 0.4.0a0+c65bd66. We selected the LSTM as the recurrent layer for both RNN and Word-character RNN, using the Adam optimizer [35]. The dimension of word2vec was 300. We used the development set to tune the number of hidden units, and the number of epochs. Hidden layer size of RNN for dialog modeling was set as 96. In the Word-character RNN, the hidden units of sentence-level RNN for utterance encoding were set as 128, while character-level RNN was set as 64. All of them were made up of

2 layers. Initial learning rate was set as 0.001. Dropout [36] (p = 0.5) was applied during model training to prevent model from over- fitting. Then we trained for 25 epochs for our model and took the one with best performance in development set during the 25 epochs. The word embeddings were static during training. In training, each dialog formed one minibatch. The training loss was the sum of categorical cross-entropy of action template and database operator.

4.3 Results and Analysis

We report average turn accuracy and dialog accuracy of each model. Following past work, we suppose the previous dialogue is consistent with the dataset when predicting next action. The turn is correct if the string matches the reference exactly. The dialog accuracy indicates if all turns in a dialog are correct.

We compare our proposed HHCNs with two latest approaches, Query-reduction Network (QRN) and HCNs. QRN is a prominent model aiming at address question answering problem, which also show good performance on task-oriented dialog in its paper [28]. Hybrid Code Networks are the first approaches trying to combine codes and networks, which achieving state-of-the-art performance in dialog task. And our HHCNs could also be considered as an advanced version of HCNs. As QRN has shared its code on github, we can directly evaluate its performance. However, the author of HCNs doesn't share his code, so we have no choice but use our own reproduce HCNs.

For the QRN, we used the reset gate, setting the dimension of the hidden state to 50 with 2 layers, while not using match mode. All other parameters used the default setting in its paper. We trained 500 epochs on training set, and picked the one achieving best performance in dev dataset. We repeated 10 times and selected the highest accuracy of QRN model. For the HCNs, we set the hidden units to 128 as it did in its paper, but used the adam optimizer and set learning rate to 0.001 as we found this would perform better. Other setting was consistent with the HCNs paper. Results are shown in Table 3.

Table 3. Results on DSTC2 dataset.

Model	DSTC2	
	Turn Acc	Dialog Acc
QRN	53.0%	–
HCN	63.5%	7.6%
HHCN	66.1%	7.9%

As can be seen in Table 3, HHCNs surpass existed model and achieving state-of-the-art result in both turn accuracy and dialog accuracy. Nonetheless, these two models are the best approaches in task-oriented dialog we are aware of. We attribute HHCNs beyond HCNs to its better ability to extract sentence information, as well as the ability to identify unknown words. In HCNs, bag-of-words and average embedding lack the ability to identify the position information of words in sentence, which word-character RNN could capture.

5 Discussion and Conclusion

This paper has introduced a Hierarchical Hybrid Code Network for task-oriented dialog systems. Based on the proposed hierarchical processes of a word-character RNN for semantic representation and a NN-based selection for domain knowledge, the codes and neural networks are more effectively exploited together. The word-character RNN effectively encodes last response of agent to acquired complete information of the dialog. And we successfully address the problem of misspelling as well as proper noun, which makes our model behave more robust in practical task. What's more, HHCNs implement a neural network selection, which is a smarter way to combine networks with codes. The experimental results on the dataset of Dialog State Tracking Challenge 2 (DSTC2) [30] also clearly show a better performance of HHCNs compared to the latest QRN [28] and HCNs [29].

The appeal of our model comes from effectiveness and simplicity. We could use human knowledge to improve and achieve promising performance, using a rather simple and easy model. As in most cases, we have lots of human knowledge to reuse and avoid to build a too complicated model. It is expected that this simple and effective model can be a good baseline in our future research on task-oriented dialog.

Acknowledgments. This work is partially supported by the National Natural Science Foundation of China (Grant no. 61772568), Guangzhou Science and Technology Program (Grant no. 201804010288), and Shenzhen Scientific Research and Development Funding Program (Grant no. JCYJ20170302153827712).

References

1. Levin, E., Pieraccini, R., Eckertm, W.: A stochastic model of human-machine interaction for learning dialog strategies. IEEE Trans. Speech Audio Process. **8**(1), 11–23 (2000)
2. Singh, S., Litman, D., Kearns, M., Walker, M.: Optimizing dialogue management with reinforcement learning: experiments with the NJFun system. J. Artif. Intell. Res. **16**(1), 105–133 (2011)
3. Williams, J.D., Young, S.: Partially Observable Markov Decision Processes for Spoken Dialog Systems. Academic Press Ltd., London (2007)
4. Hori, C., Ohtake, K., Misu, T., Kashioka, H., Nakamura, S.: Statistical dialog management applied to WFST-based dialog systems. In: IEEE International Conference on Acoustics, Speech and Signal Processing, pp. 4793–4796 (2009)
5. Lee, C., Jung, S., Kim, S., Lee, G.G.: Example-based dialog modeling for practical multi-domain dialog system. Speech Commun. **51**(5), 466–484 (2009)
6. Griol, D., Hurtado, L.F., Segarra, E., Sanchis, E.: A statistical approach to spoken dialog systems design and evaluation. Speech Commun. **50**(8–9), 666–682 (2008)
7. Young, S., Gai, M., Thomson, B., Williams, J.D.: Pomdp-based statistical spoken dialog systems: a review. Proc. IEEE **101**(5), 1160–1179 (2013)
8. Li, L., He, H., Williams, J.D.: Temporal supervised learning for inferring a dialog policy from example conversations. In: Spoken Language Technology Workshop, pp. 312–317 (2014)

9. Serban, I.V., Sordoni, A., Bengio, Y., Courville, A., Pineau, J.: Building end-to-end dialogue systems using generative hierarchical neural network models. In: Thirtieth AAAI Conference on Artificial Intelligence, pp. 3776–3783 (2016)
10. Sordoni, A., Galley, M., Auli, M., Brockett, C., Ji, Y., Mitchell, M., Nie, J.Y., Gao, J., Dolan, B.: A neural network approach to context-sensitive generation of conversational responses (2015)
11. Shang, L., Lu, Z., Li, H.: Neural responding machine for short-text conversation, pp. 52–58 (2015)
12. Vinyals, O., Le, Q.: A neural conversational model. Computer Science (2015)
13. Yao, K., Zweig, G., Peng, B.: Attention with intention for a neural network conversation model. Computer Science (2015)
14. Li, J., Galley, M., Brockett, C., Spithourakis, G., Gao, J., Dolan, B.: A persona-based neural conversation model. Meeting of the Association for Computational Linguistics, pp. 994–1003 (2016)
15. Luan, Y., Ji, Y., Ostendorf, M.: LSTM based conversation models (2016)
16. Xu, Z., Liu, B., Wang, B., Sun, C., and Wang, X.: In-corporating loose-structured knowledge into LSTM with recall gate for conversation modeling. pp. 3506–3513 (2016)
17. Li, J., Galley, M., Brockett, C., Gao, J., Dolan, B.: A diversity-promoting objective function for neural conversation models. Computer Science (2015)
18. Lowe, R.T., Pow, N., Serban, I.V., Charlin, L., Liu, C.-W., Pineau, J.: Training end-to-end dialogue systems with the ubuntu dialogue corpus. Dialogue Discourse **8**(1), 31–65 (2017)
19. Serban, I.V., Sordoni, A., Lowe, R., Charlin, L., Pineau, J., Courville, A., Bengio, Y.: A hierarchical latent variable encoder-decoder model for generating dialogues (2016)
20. Sukhbaatar, S., Szlam, A., Weston, J., Fergus, R.: End-to-end memory networks. Computer Science (2015)
21. Perez, J., Liu, F.: Gated end-to-end memory networks (2016)
22. Gu, J., Lu, Z., Li, H., Li, V.O.K.: Incorporating copying mechanism in sequence-to-sequence learning. pp. 1631–1640 (2016)
23. Gulcehre, C., Ahn, S., Nallapati, R., Zhou, B., Bengio, Y.: Pointing the unknown words. pp. 140–149 (2016)
24. Bahdanau, D., Cho, K., Bengio, Y.: Neural machine translation by jointly learning to align and translate. Computer Science (2014)
25. Luong, M.T., Pham, H., Manning, C.D.: Effective approaches to attention-based neural machine translation. Computer Science (2015)
26. Eric, M., Manning, C.D.: A copy-augmented sequence-to-sequence architecture gives good performance on task-oriented dialogue. pp. 468–473 (2017)
27. Liu, B., Lane, I.: Iterative policy learning in end-to-end trainable task-oriented neural dialog models (2017)
28. Seo, M., Min, S., Farhadi, A., Hajishirzi, H.: Query-reduction networks for question answering (2016)
29. Williams, J.D., Asadi, K., Zweig, G.: Hybrid code networks: practical and efficient end-to-end dialog control with supervised and reinforcement learning. pp. 665–677 (2017)
30. Henderson, M., Thomson, B., Williams, J.D.: The second dialog state tracking challenge. In: Proceedings of the 15th Annual Meeting of the Special Interest Group on Discourse and Dialogue (SIGDIAL), pp. 263–272 (2014)
31. Mikolov, T., Sutskever, I., Chen, K., Corrado, G., Dean, J.: Distributed representations of words and phrases and their compositionality. **26**, 3111–3119 (2013)
32. Luong, M.T., Manning, C.D.: Achieving open vocabulary neural machine translation with hybrid word-character models. pp. 1054–1063 (2016)

33. Hochreiter, S., Schmidhuber, J.: Long short-term memory. Neural Comput. **9**(8), 1735–1780 (1997)
34. Chung, J., Gulcehre, C., Cho, K.H., Bengio, Y.: Empirical evaluation of gated recurrent neural networks on sequence modeling. Eprint Arxiv (2014)
35. Kingma, D., Ba, J.: Adam: a method for stochastic optimization. Computer Science (2014)
36. Srivastava, N., Hinton, G., Krizhevsky, A., Sutskever, I., Salakhutdinov, R.: Dropout: a simple way to prevent neural networks from overfitting. J. Mach. Learn. Res. **15**(1), 1929–1958 (2014)

Characterization of Radiotherapy Sensitivity Genes by Comparative Gene Set Enrichment Analysis

Min Zhu[3(✉)], Xiaolai Li[3], Shujie Wang[1,2], Wei Guo[1,2], and Xueling Li[1,2(✉)]

[1] Anhui Province Key Laboratory of Medical Physics and Technology, Center of Medical Physics and Technology, Hefei Institutes of Physical Science, Chinese Academy of Sciences, 350 Shushanhu Road, Hefei, Anhui 230031, People's Republic of China
xuelingli16@foxmail.com, xlli@cmpt.ac.cn

[2] Cancer Hospital, Chinese Academy of Sciences, 350 Shushan Road, Hefei 230031, Anhui, People's Republic of China

[3] Hefei Institute of Intelligent Machines, Chinese Academy of Sciences, 350 Shushanhu Road, Hefei 230031, Anhui, People's Republic of China
mzhu17@foxmail.com

Abstract. Postoperative and preoperative radiotherapy has been widely applied to kill the local cancer cells, prevent metastasis and lessen cancer burden in the treatment of cancers. However, the response to radiotherapy varies among cancer patients. In this study we mine and characterize the radiotherapy efficacy associated genes, radiosensitivity genes, in rectal cancer gene expression profile (GSE3493) from a previous study by gene set enrichment analysis. 381 genes were identified by comparing the gene expression profiles of responder and nonresponder rectal cancer patients who underwent preoperative radiotherapy. The top radiotherapy sensitive genes include MCF2, WHAMMP2, PCDHGA8, SHOX2, FAS, X81001, HAVCR1, PLXDC2, OPRM1 and PWAR5. We performed enrichment analysis of transcription factor, chromosome position, and gene sets reported in literatures by comparing this gene set with reported functional or structural gene sets. We find that the gene set has significant overlap with radiotherapy response, irradiation response, inflammation, XRCC3, ATM and BRCA1 related gene sets in different cancers from previous reports. Enriched chromosome positions include 16q13 and 17q21. The top enriched transcription factors with most number of radiotherapy response target genes include FOXP1, TP63, AR, STAT3, SOX2, SMAD4, BACH1, SMAD2, SMAD3, ZNF217 and RELA. The present study suggested the potential molecular mechanism behind the radiotherapy responders and non-responders, where both inflammatory and immune response and DNA damage response are very likely to control the radiotherapy sensitivity. The results may provide insights into the development of novel therapeutic approaches.

M. Zhu and X. Li—Contribute equally to the work.

D.-S. Huang et al. (Eds.): ICIC 2018, LNCS 10955, pp. 205–216, 2018.
https://doi.org/10.1007/978-3-319-95933-7_25

Keywords: Radiotherapy sensitivity genes · Radiogenomics
Gene set enrichment analysis · Transcription factors · Bioinformatics

1 Introduction

Radiotherapy (RT) is one of important cancer treatment options besides operation and chemotherapy. Above 50% cancer patients have experienced radiotherapy for treatment [1], and has been estimated to have a 16% increase from 2012 [2] to 2025 [3] for Europe. However, response to radiotherapy differs among patients receiving RT at similar dose pre or post operation. While local cancer for some patients may be under control without progression and metastasis, the efficacy is limited for other patients. Some patients even experience acute or late adverse treatment effects, including radiation injury and secondary cancer [4]. This interpatient variability in efficiency may be largely due to the difference in the genetic and genomics background of both the cancer and normal tissue, including DNA repair efficiency and pathways, especially DNA double-strand break repair (DSBR). The role(s) of inflammatory and immune response may be interconnected with DNA damage response and affect patient radiotoxicity [5]. Prediction of radiotherapy responder or non-responder and prediction of dose based on identified radiosensitive genes have been reported [6–8]. The goal of the radiotherapy dose selection is to enhance radiation response in cancer tissue, and meanwhile to decrease its toxicity on normal tissue [9]. However, the molecular mechanism behind the difference in the RT efficacy of cancer patients is still not fully understood. Transcriptional regulatory network as one of important molecular mechanisms of biological systems [10–13] can be constructed by comparative gene set enrichment analysis. In the present study, we have firstly obtained differentially expressed genes by comparing two patient groups, i.e., responders versus non-responders. We then have characterized RT sensitivity genes by comparative gene set enrichment analysis with transcription factor target gene sets identified by ChIP-Seq from literature reports and UCSC Genome Browser database. The result may shed light on the molecular network mechanism including transcriptional regulatory network leading to the radiosensitivity in term of clinical RT efficacy. The results may provide insights into the development of novel sensitization or combination therapeutic approaches.

2 Methods

2.1 DEGs of Noresponder Versus Responder

DEGs of Noresponder versus Responder were identified by analyzing the GSE3493 [14] downloaded from Gene Expression Omnibus (GEO) with non-paired t-test by using t-test2 function of Matlab Version 2016a with equal variance.

2.2 Gene Set Enrichment Analysis

Method of Transcription Factor Enrichment Analysis can be found in our previous published paper [11]. Briefly, ChEA ChIP-X data were downloaded from the website http://amp.pharm.mssm.edu/lib/chea.jsp of 2014 version. The ChIP-X data consist of 345 TF binding experiments performed on human, mouse and rat, among which 148 were performed on human. For the transcription factor analysis, we also downloaded the uniform histone and TF binding peaks from Encyclopedia of DNA Elements (ENCODE) Consortium website: (http://ftp.ebi.ac.uk/pub/databases/ensembl/encode/integration_data_jan2011/byDataType/signal/jan2011/bigwig/). There were total 1169 distinct ChIP-Seq experiments. The target genes (TGs) corresponding to the TF binding peaks were identified according to PAVIS criteria [15]. The enrichment for each transcription factor regulated target gene (TG) was calculated by P-value = 1-hygecdf ($k-1$, N, K, n) using the Matlab build-in function hygecdf. Here, the hyper-geometric probability distribution was used as our background distribution. k represents the number of overlapping genes of the DEGs and the target genes for the specific transcription factor, N is the number of genome-wide genes: 12,834, K, the number of the TGs of the TF in the ChIP-data, and n, the number of DEGs: 381. For multiple testing corrections, a Benjamini-Hotchberg and Bonferroni correction was also performed to estimate the false discover rate (FDR) Q-value. Enrichment fold ratio was calculated with k/n/(K/N). If the Q-value is smaller than 0.01 and fold ratio is greater than 1.5, the corresponding transcription factor was considered significantly enriched. Chromosome locus mapping, and comparative gene set enrichment [16] were performed with default parameters based on GSEA online webserver [17].

2.3 Visualization of the Protein Interaction Network of the Radiosensitivity Genes and the Enriched Transcription Factors

We have obtained the list of binding protein pairs from the Protein Interactions and Network Analysis (PIANA) database for the proteins with $p < 0.05$ and fold ratio change greater than 1.4. with both proteins for each interacting pair were differentially expressed between RT responders and nonresponders. 109 protein pairs of those RT sensitive genes were obtained and visualized in Cytoscape. For the protein interaction within the enriched transcription factors, the protein-protein interaction network was also extracted from pinar and visualized in Cytoscape. For detail, see our previous work published in [12, 13].

2.4 Radiosensitivity Index Calculation

Radiosensitivity index (RSI) was calculated by using the ten reported marker genes [6], i.e. AR, JUN, STAT1, PRKCB, RELA, ABL1, SUMO1, CDK1, HDAC1, IRF1. Briefly, the 10 genes were ranked again according to their rank and assigned to 1 to 10 values according to their ranks. When multiple probesets are available for a gene, we chose the probeset with the highest expression level.

3 Results and Discussion

3.1 Differentially Expressed Genes of Nonresponders Versus Responders

At a threshold of P-Value of 0.05, 395 probesets corresponding to 381 genes were identified as differentially expressed genes with not paired *t*-test and equal variance. The top genes were listed in Table 1 in the order of descending fold ratio of the mean expression of each gene from nonresponders versus responders. Most top genes with highest fold ration of the mean expression of nonresponder versus responder are associated with RT prognostics, which include MCF2, MCF.2 cell line derived transforming sequence [18], WHAMMP2, WAS protein homolog associated with actin, golgi membranes and microtubules pseudogene 2, PCDHGA8, protocadherin gamma subfamily A, 8 [19], SHOX2, short stature homeobox 2 [20], FAS, Fas cell surface death receptor [21], X81001, uncharacterized LOC105379655. Protein interaction of the RT Sensitivity genes visualized by Cytoscape revealed the hub genes, which include FAS, OPRM1, MCF2, PLXDC2 and SHOX2 (Fig. 1). Pathway enrichment analysis with Cytoscape plug-in Reactome FI revealed the top pathways, which include Proteoglycans in cancer(K), Pathways in cancer(K), Apoptosis(K), Focal adhesion(K), MAPK signaling pathway(K), HIV-1 Nef: Negative effector of Fas and TNF-alpha(N), Hepatitis B(K), Signalling by NGF(R) (Table 2). The main enriched pathways are concordant with previous reports.

Table 1. The top RT reponsive genes sorted with descending fold ratio of the mean expression of each gene of nonresponders over responders.

Gene Name	Gene Description	Fold ratio	P-Value
MCF2	MCF.2 cell line derived transforming sequence	2.0	0.010
WHAMMP2	WAS protein homolog associated with actin, golgi membranes and microtubules pseudogene 2///WAS protein homolog associated with actin, golgi membranes and microtubules pseudogene 3	1.7	0.004
PCDHGA8	protocadherin gamma subfamily A, 8	1.6	0.030
SHOX2	short stature homeobox 2	1.5	0.024
FAS	Fas cell surface death receptor	1.5	0.006
X81001	uncharacterized LOC105379655	1.5	0.011
HAVCR1	hepatitis A virus cellular receptor 1	1.5	0.000
PLXDC2	plexin domain containing 2	1.5	0.021
OPRM1	opioid receptor mu 1	1.4	0.037
PWAR5	Prader Willi/Angelman region RNA 5	1.4	0.004
IL33	interleukin 33	1.4	0.018
FOXA1	forkhead box A1	1.4	0.011
NFKB2	nuclear factor kappa B subunit 2	0.8	0.016
NTRK2	neurotrophic receptor tyrosine kinase 2	0.8	0.042
PLA2G4A	phospholipase A2 group IVA	0.8	0.008
WHAMMP2///WHAMMP3	WAS protein homolog associated with actin, golgi membranes and microtubules pseudogene 2///WAS protein homolog associated with actin, golgi membranes and microtubules pseudogene 3	0.8	0.001

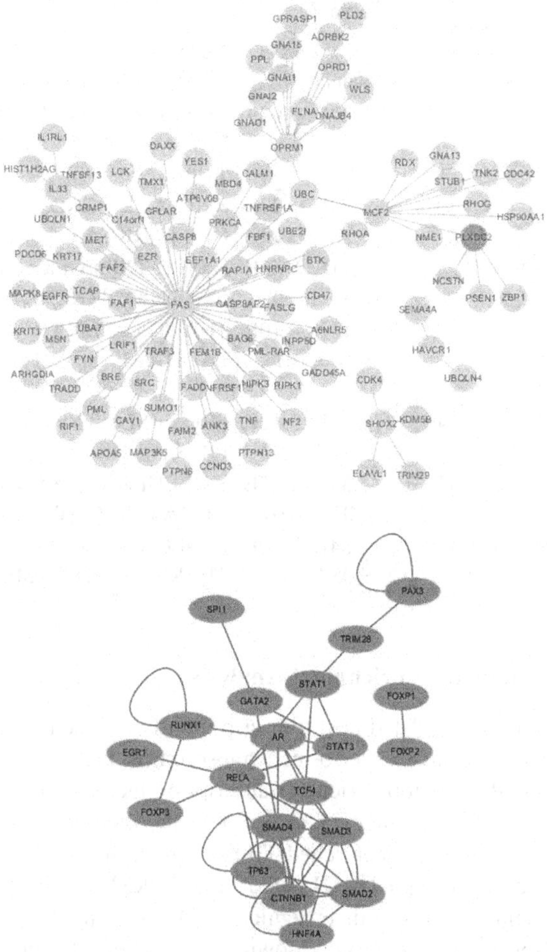

Fig. 1. Protein interaction network visualization for the genes with $p < 0.05$ and mean fold ratio change greater than 1.4 of radiotherapy responders over nonreponders (top), and enriched transcription factors (bottom), respectively.

Table 2. Top enriched pathways sorted with descending number of RT sensitive genes involved

GeneSet	*K*	*k*	P-value	FDR	Nodes
Proteoglycans in cancer (K)	205	12	3.23E−13	9.27E−11	SRC,FASLG,TNF,EGFR,CAV1,MSN, PRKCA,RHOA,FAS,PTPN6,EZR,MET
Pathways in cancer(K)	397	11	9.78E−09	3.12E−07	FASLG,EGFR,GNAI2,MAPK8,FADD, HSP90AA1,PRKCA,RHOA,CDK4, FAS,MET

(continued)

Table 2. (*continued*)

GeneSet	*K*	*k*	P-value	FDR	Nodes
Apoptosis(K)	140	10	6.27E −12	8.97E −10	TRADD,FASLG,TNF,MAPK8,RIPK1, FADD,GADD45A,PTPN13, TNFRSF1A,FAS
Focal adhesion(K)	201	9	4.83E −09	2.27E −07	SRC,EGFR,CCND3,MAPK8,RAP1A, CAV1,PRKCA,RHOA,MET
MAPK signaling pathway (K)	255	9	3.71E −08	8.53E −07	FASLG,TNF,EGFR,MAPK8,RAP1A, GADD45A,PRKCA,TNFRSF1A,FAS
HIV-1 Nef: Negative effector of Fas and TNF-alpha(N)	33	8	7.13E −14	4.10E −11	TRADD,FASLG,TNF,MAPK8,RIPK1, FADD,TNFRSF1A,FAS
Hepatitis B(K)	146	8	8.37E −09	3.12E −07	SRC,FASLG,TNF,MAPK8,FADD, PRKCA,CDK4,FAS
Signalling by NGF(R)	421	8	2.31E −05	1.85E −04	SRC,EGFR,MAPK8,RAP1A,PRKCA, RHOA,LCK,MET
Signaling by Interleukins (R)	460	8	4.34E −05	2.80E −04	FASLG,TNF,EGFR,IL33,HSP90AA1, TNFRSF1A,PTPN6,MET

3.2 Transcription Factor Enrichment Analysis

Gene expression may be regulated post transcription through activation of TF protein. Taking it into consideration, we performed transcription factor enrichment analysis. Tables 3 and 4 show that the top enriched transcription factors (TF) from transcription Factor Enrichment Analysis. Three of six of TFs reported in radiosenstivity index (RSI) [7] can be found in our top enriched TF list, which includes AR, STAT1, RelA. Among the left three TFs, cJun, HDAC1, IRF1, used in RSI calculation, TFs from their same family or protein complex or their regulatory TF were enriched, including JUND, YY1, and IRF8 from mouse (data not shown), respectively. Other enriched TFs are involved in DNA repair pathways, such as TP63, SMAD4, SMAD3, SMAD2 and inflammation and immunity, such as STAT3 (Table 4), NFKB, FOXP1, and SOX2.

Analysis of the protein-protein interaction among the enriched TFs shows that that AR, STAT1, and RelA, three of the six TFs included in radiosensitivity index formula, are the hubs of the TF protein-protein interaction network (Fig. 1), although there is no overlap between the significant RT sensitivity genes identified here and the previous reported 10 gene signature of RSI [7]. We calculated the RSI of the RT responders and nonresponders by using previous reported 10-gene signature [6, 7]. Our results demonstrated there is significant RSI difference (Fig. 2) between the RT responders and nonresponders. The enriched TFs can be used as clinical radosensitivity TF candidates for further study.

Table 3. Transcription factor enrichment analysis based on CHIA human data with 12625 genome-wide genes and 395 radiotherapy sensitivity Genes

TF	PMID	#genome-wide	#this gene set	Q-value	Fold ratio	Cell line
TP63	23658742	1654	82	0.0001	1.58	EP156T
STAT3	23295773	1447	76	0.0001	1.68	U87
BACH1	22875853	868	51	0.0001	1.88	HELA-AND-SCP4
FOXP1	21924763	1780	85	0.0003	1.53	HESC
SMAD4	21799915	1142	60	0.0003	1.68	A2780
SOX2	20726797	1197	61	0.0005	1.63	SW620
SMAD3	18955504	935	50	0.0005	1.71	HaCaT
AR	19668381	1631	77	0.0006	1.51	PC3
SMAD2	18955504	935	50	0.0006	1.71	HaCaT
ZNF217	24962896	792	43	0.0009	1.74	MCF7
PADI4	21655091	514	31	0.001	1.93	MCF7
EST1	17652178	355	24	0.001	2.16	JURKAT
PAX3	20663909	532	31	0.0016	1.86	RHABDOMYOSARCOMA
DROSHA	22980978	222	17	0.0016	2.45	HELA
RELA	24523406	791	41	0.002	1.66	SAEC
NR1H3	23393188	344	22	0.0022	2.04	ATHEROSCLEROTIC-FOAM
FOXP2	23625967	447	26	0.0029	1.86	PFSK-1 AND SK-N-MC
EGR1	19374776	57	7	0.0029	3.93	THP-1
SPI1	20517297	528	29	0.0034	1.76	HL60
TTF2	22483619	734	37	0.0039	1.61	HELA
TRIM28	17542650	464	26	0.004	1.79	NTERA2
TCF4	18268006	226	15	0.0054	2.12	LS174T
FOXP3	21729870	660	33	0.0055	1.6	TREG
CLOCK	20551151	296	18	0.0056	1.94	293T
STAT1	20625510	276	17	0.006	1.97	HELA

Table 4. Transcription factor enrichment analysis based on ENCODE ChIP-Seq Data with 12625 genome-wide genes and 395 radiotherapy sensitivity Genes

TF	Cell line	#genome-wide	#this gene set	P-value	Q-value (BH[a])	Q-value (B[b])	Fold ratio
ZNF263	K562b	704	45	9.6E−07	3.8E−06	4.4E−03	2.04
Max	H1-hESC	2167	122	1.0E−11	7.6E−12	8.8E−09	1.80
TCF12	H1-hESC	1381	76	1.7E−07	5.3E−07	6.2E−04	1.76
H3K9me3	GM12878	2435	130	5.8E−11	4.9E−11	5.7E−08	1.71
Nrf1	H1-hESC	1635	87	9.7E−08	2.8E−07	3.3E−04	1.70
Nrf1	H1-hESC	1756	93	4.8E−08	1.2E−07	1.4E−04	1.69
Nrf1	K562	1705	90	8.4E−08	2.4E−07	2.8E−04	1.69
Max	H1-hESC	2506	130	3.5E−10	3.6E−10	4.2E−07	1.66

(*continued*)

Table 4. (*continued*)

TF	Cell line	#genome-wide	#this gene set	P-value	Q-value (BH[a])	Q-value (B[b])	Fold ratio
MEF2C_(SC-13268)	GM12878	1534	79	1.0E−06	4.2E−06	4.9E−03	1.65
NANOG_(SC-33759)	H1-hESC	1908	98	7.7E−08	2.2E−07	2.5E−04	1.64
SETDB1	U2OS	1964	99	1.4E−07	4.5E−07	5.3E−04	1.61
SETDB1	U2OS	2212	111	3.6E−08	8.3E−08	9.7E−05	1.60
BCL11A	GM12878	2065	103	1.3E−07	4.0E−07	4.7E−04	1.59
STAT3	MCF10A-Er-Src	2134	105	1.8E−07	5.8E−07	6.8E−04	1.57
Nrf1	GM12878	2363	116	4.9E−08	1.3E−07	1.5E−04	1.57
Nrf1	GM12878	2327	114	7.0E−08	1.9E−07	2.2E−04	1.57
JunD	HepG2	3343	163	7.6E−11	6.6E−11	7.7E−08	1.56
SRF	GM12878	3200	156	2.6E−10	2.5E−10	2.9E−07	1.56
GATA2_(CG2-96)	K562	2083	101	5.6E−07	2.1E−06	2.4E−03	1.55
c-Myc	K562	2525	122	4.9E−08	1.2E−07	1.4E−04	1.54
NFKB	GM19193	1989	96	1.2E−06	5.0E−06	5.9E−03	1.54
NRSF	HTB-11	1954	94	1.8E−06	7.5E−06	8.8E−03	1.54
YY1_(C-20)	H1-hESC	2887	138	1.2E−08	2.1E−08	2.5E−05	1.53
STAT3	MCF10A-Er-Src	3580	171	9.6E−11	8.5E−11	1.0E−07	1.53
MafK_(SC-477)	HepG2	3780	180	2.7E−11	2.1E−11	2.4E−08	1.52
FOXA2_(SC-6554)	HepG2	3390	161	7.0E−10	8.1E−10	9.4E−07	1.52
MafK_(SC-477)	HepG2	4306	204	4.8E−13	2.6E−13	3.1E−10	1.51
SRF	GM12878	2620	124	1.0E−07	3.0E−07	3.5E−04	1.51
BATF	GM12878	2879	136	2.8E−08	6.1E−08	7.1E−05	1.51
BATF	GM12878	3198	151	4.3E−09	6.5E−09	7.5E−06	1.51
STAT3	MCF10A-Er-Src	3715	175	1.5E−10	1.4E−10	1.6E−07	1.51
H3K27me3	H7-hESC	2784	131	6.0E−08	1.6E−07	1.9E−04	1.50
TAL1_(SC-12984)	K562	3002	141	2.0E−08	4.2E−08	5.0E−05	1.50

[a]BH, Benjamini-Hotchberg correction; [b]B, Bonferroni correction.

3.3 Chromosome Position Enrichment of the Radiotherapy Sensitivity Genes

By using the GSEA online tool, we performed chromosome band enrichment analysis. The top enriched chromosome band includes 16q13 and 17q21, which may suggest that the radiosensitive chromosome or chromatin region at the nuclear or subcellular epigenetic level mechanism.

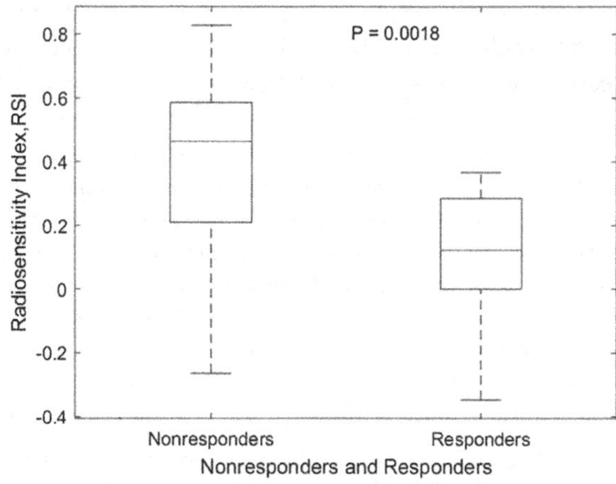

Fig. 2. Radiosensitivity index scores of nonresponders and responders

3.4 RT Sensitivity Gene Set Comparison with Gene Sets Reported in Literatures

Gene Set Enrichment Analysis (GSEA) of the radiotherapy sensitivity Genes against the curated C2 genesets showed enrichment in DNA repair, UV irradiation sensitivity genes. The radiotherapy sensitivity genes are most significantly enriched in the ATM-PCC network (PCC: Pearson Correlation Coefficient, 64/381 radiotherapy Sensitivity genes were present in the set of 1442 ATM-associated genes; p = 2.67E−30). This gene set comprises genes whose expression positively correlated (Pearson correlation coefficient, PCC > = 0.4) with that of ATM across a compendium of normal tissues. Other gene sets originating from this previous study (Pujana et al. [22]), including the BRCA1-PCC were also identified as highly enriched (da Costa et al. [23]). The results suggest an association between radiotherapy sensitivity genes and DNA damage repair factors. The radiotherapy sensitivity genes are also enriched in common down-regulated transcripts in fibroblasts expressing either XP/CS or TDD mutant forms of ERCC3 [GeneID = 2071], after UVC irradiation, suggesting common pathway, ERCC3 mediated DNA repair pathway, may be shared between radiotherapy and UV irradiation. The radiotherapy sensitivity genes are also enriched in KLF1 regulatory network, MAPK8, Alzheimer's disease and diabetes II related genes. This is concordant with the above protein network analysis, where reveals the FAS hub gene, and suggest important functional role of inflammation and metabolism in radiotherapy.

The GSEA C2 database gene set enrichment analysis demonstrates that DNA repair pathway, inflammatory and immune response may be involved in regulating radiation sensitivity. This is cannot be disregarded when it comes to RT outcome for patients. In a word, the identified radiosensitive genes are biologically relevant to radiotherapy. The result indicates potential molecular mechanism involved in the radiotherapy sensitivity and resistance.

One of the future study directions on the clinical radiosensitivity related genes may include cell composition deconvolution analysis, which will reveal the tissue cancer cell types and cancer related cell composition, including filtrated immune cells, basal cells in order to provide multilevel cellular microenvironment mechanism hypothesis. Another study direction is to infer the mutations, the genetic basis for clinical RT sensitivity and patient outcome.

Acknowledgements. This study was supported by the National Science Foundation of China, (Grant Nos. 31371340, 61673369, 61273324, and 81572948, and was also supported by a start-up program funded by Center of Medical Physics and Technology & Cancer Hospital, Chinese Academy of Sciences, Hefei, Program No. Y6BF0Q1391.

References

1. Torre, L.A., Bray, F., Siegel, R.L., Ferlay, J., Lortet-Tieulent, J., Jemal, A.: Global cancer statistics, 2012. CA Cancer J. Clin. **65**, 87–108 (2015)
2. Ferlay, J., Soerjomataram, I., Dikshit, R., Eser, S., Mathers, C., Rebelo, M., Parkin, D.M., Forman, D., Bray, F.: Cancer incidence and mortality worldwide: sources, methods and major patterns in GLOBOCAN 2012. Int. J. Cancer **136**, E359–E386 (2015)
3. Borras, J.M., Lievens, Y., Barton, M., Corral, J., Ferlay, J., Bray, F., Grau, C.: How many new cancer patients in Europe will require radiotherapy by 2025? an ESTRO-HERO analysis. Radiother. Oncol. **119**, 5–11 (2016)
4. Scaife, J.E., Barnett, G.C., Noble, D.J., Jena, R., Thomas, S.J., West, C.M., Burnet, N.G.: Exploiting biological and physical determinants of radiotherapy toxicity to individualize treatment. Br. J. Radiol. **88**, 20150172 (2015)
5. Georgakilas, A.G., Pavlopoulou, A., Louka, M., Nikitaki, Z., Vorgias, C.E., Bagos, P.G., Michalopoulos, I.: Emerging molecular networks common in ionizing radiation, immune and inflammatory responses by employing bioinformatics approaches. Cancer Lett. **368**, 164–172 (2015)
6. Scott, J.G., Berglund, A., Schell, M.J., Mihaylov, I., Fulp, W.J., Yue, B., Welsh, E., Caudell, J.J., Ahmed, K., Strom, T.S., Mellon, E., Venkat, P., Johnstone, P., Foekens, J., Lee, J., Moros, E., Dalton, W.S., Eschrich, S.A., McLeod, H., Harrison, L.B., Torres-Roca, J.F.: A genome-based model for adjusting radiotherapy dose (GARD): a retrospective, cohort-based study. Lancet Oncol **18**, 202–211 (2017)
7. Eschrich, S.A., Pramana, J., Zhang, H., Zhao, H., Boulware, D., Lee, J.H., Bloom, G., Rocha-Lima, C., Kelley, S., Calvin, D.P., Yeatman, T.J., Begg, A.C., Torres-Roca, J.F.: A gene expression model of intrinsic tumor radiosensitivity: prediction of response and prognosis after chemoradiation. Int. J. Radiat. Oncol. Biol. Phys. **75**, 489–496 (2009)
8. Eschrich, S., Zhang, H., Zhao, H., Boulware, D., Lee, J.H., Bloom, G., Torres-Roca, J.F.: Systems biology modeling of the radiation sensitivity network: a biomarker discovery platform. Int. J. Radiat. Oncol. Biol. Phys. **75**, 497–505 (2009)
9. Guo, Z., Shu, Y., Zhou, H., Zhang, W., Wang, H.: Radiogenomics helps to achieve personalized therapy by evaluating patient responses to radiation treatment. Carcinogenesis **36**, 307–317 (2015)
10. Yang, J., Zhao, Y., Kalita, M., Li, X., Jamaluddin, M., Tian, B., Edeh, C.B., Wiktorowicz, J. E., Kudlicki, A., Brasier, A.R.: Systematic Determination of Human Cyclin Dependent Kinase (CDK)-9 Interactome Identifies Novel Functions in RNA Splicing Mediated by the DEAD Box (DDX)-5/17 RNA Helicases. Mol. Cell. Proteomics **14**, 2701–2721 (2015)

11. Tian, B., Li, X., Kalita, M., Widen, S.G., Yang, J., Bhavnani, S.K., Dang, B., Kudlicki, A., Sinha, M., Kong, F., Wood, T.G., Luxon, B.A., Brasier, A.R.: Analysis of the TGFbeta-induced program in primary airway epithelial cells shows essential role of NF-kappaB/RelA signaling network in type II epithelial mesenchymal transition. BMC Genom. **16**, 529 (2015)
12. Li, X., Zhu, M., Brasier, A.R., Kudlicki, A.S.: Inferring genome-wide functional modulatory network: a case study on NF-kappaB/RelA transcription factor. J. Comput. Biol. **22**, 300–312 (2015)
13. Li, X., Zhao, Y., Tian, B., Jamaluddin, M., Mitra, A., Yang, J., Rowicka, M., Brasier, A.R., Kudlicki, A.: Modulation of gene expression regulated by the transcription factor NF-kappaB/RelA. J. Biol. Chem. **289**, 11927–11944 (2014)
14. Watanabe, T., Komuro, Y., Kiyomatsu, T., Kanazawa, T., Kazama, Y., Tanaka, J., Tanaka, T., Yamamoto, Y., Shirane, M., Muto, T., Nagawa, H.: Prediction of sensitivity of rectal cancer cells in response to preoperative radiotherapy by DNA microarray analysis of gene expression profiles. Cancer Res. **66**, 3370–3374 (2006)
15. Huang, W., Loganantharaj, R., Schroeder, B., Fargo, D., Li, L.: PAVIS: a tool for Peak Annotation and Visualization. Bioinformatics **29**, 3097–3099 (2013)
16. Liberzon, A., Birger, C., Thorvaldsdóttir, H., Ghandi, M., Mesirov, Jill, P., Tamayo, P.: The molecular signatures database hallmark gene set collection. Cell Syst. **1**, 417–425 (2015)
17. Subramanian, A., Tamayo, P., Mootha, V.K., Mukherjee, S., Ebert, B.L., Gillette, M.A., Paulovich, A., Pomeroy, S.L., Golub, T.R., Lander, E.S., Mesirov, J.P.: Gene set enrichment analysis: A knowledge-based approach for interpreting genome-wide expression profiles. Proc. Natl. Acad. Sci. **102**, 15545–15550 (2005)
18. Jang, E.R., Lee, J.H., Lim, D.-S., Lee, J.-S.: Analysis of ataxia-telangiectasia mutated (ATM)- and Nijmegen breakage syndrome (NBS)-regulated gene expression patterns. J. Cancer Res. Clin. Oncol. **130**, 225–234 (2004)
19. Paik, S.M.P., PA, US), Kim, S. (Gyeonggi-do, KR), Kang, W.K. (Seoul, KR), Lee, J.Y. (Seoul, KR), Bae, J.M. (Seoul, KR), Sohn, T.S. (Seoul, KR), Noh, J.H. (Seoul, KR), Choi, M.G. (Seoul, KR), Park, Y.S (Seoul, KR), Park, J.O. (Gyeonggi-do, KR), Park, S.H. (Seoul, KR), Lim, H.Y. (Gyeonggi-do, KR), Jung, S.H. (Chapel Hill, NC, US): Marker for predicting gastric cancer prognosis and method for predicting gastric cancer prognosis using the same. Samsung Life Public Welfare Foundation (Seoul, KR), United States (2016)
20. Schmidt, B., Beyer, J., Dietrich, D., Bork, I., Liebenberg, V., Fleischhacker, M.: Quantification of Cell-Free mSHOX2 Plasma DNA for Therapy Monitoring in Advanced Stage Non-Small Cell (NSCLC) and Small-Cell Lung Cancer (SCLC) Patients. PLoS ONE **10**, e0118195 (2015)
21. Ogawa, Y., Nishioka, A., Hamada, N., Terashima, M., Inomata, T., Yoshida, S., Seguchi, H., Kishimoto, S.: Expression of fas (CD95/APO-1) antigen induced by radiation therapy for diffuse B-cell lymphoma: immunohistochemical study. Clin. Cancer Res. **3**, 2211–2216 (1997)
22. Pujana, M.A., Han, J.D., Starita, L.M., Stevens, K.N., Tewari, M., Ahn, J.S., Rennert, G., Moreno, V., Kirchhoff, T., Gold, B., Assmann, V., Elshamy, W.M., Rual, J.F., Levine, D., Rozek, L.S., Gelman, R.S., Gunsalus, K.C., Greenberg, R.A., Sobhian, B., Bertin, N., Venkatesan, K., Ayivi-Guedehoussou, N., Sole, X., Hernandez, P., Lazaro, C., Nathanson, K.L., Weber, B.L., Cusick, M.E., Hill, D.E., Offit, K., Livingston, D.M., Gruber, S.B., Parvin, J.D., Vidal, M.: Network modeling links breast cancer susceptibility and centrosome dysfunction. Nat. Genet. **39**, 1338–1349 (2007)
23. da Costa, R.M., Riou, L., Paquola, A., Menck, C.F., Sarasin, A.: Transcriptional profiles of unirradiated or UV-irradiated human cells expressing either the cancer-prone XPB/CS allele or the noncancer-prone XPB/TTD allele. Oncogene **24**, 1359–1374 (2005)

24. Baelde, H.J., Eikmans, M., Doran, P.P., Lappin, D.W., de Heer, E., Bruijn, J.A.: Gene expression profiling in glomeruli from human kidneys with diabetic nephropathy. Am. J. Kidney Dis. **43**, 636–650 (2004)
25. Blalock, E.M., Geddes, J.W., Chen, K.C., Porter, N.M., Markesbery, W.R., Landfield, P.W.: Incipient Alzheimer's disease: microarray correlation analyses reveal major transcriptional and tumor suppressor responses. Proc. Natl. Acad. Sci. USA **101**, 2173–2178 (2004)
26. Pilon, A.M., Arcasoy, M.O., Dressman, H.K., Vayda, S.E., Maksimova, Y.D., Sangerman, J. I., Gallagher, P.G., Bodine, D.M.: Failure of terminal erythroid differentiation in EKLF-deficient mice is associated with cell cycle perturbation and reduced expression of E2F2. Mol. Cell. Biol. **28**, 7394–7401 (2008)
27. Yoshimura, K., Aoki, H., Ikeda, Y., Fujii, K., Akiyama, N., Furutani, A., Hoshii, Y., Tanaka, N., Ricci, R., Ishihara, T., Esato, K., Hamano, K., Matsuzaki, M.: Regression of abdominal aortic aneurysm by inhibition of c-Jun N-terminal kinase. Nat. Med. **11**, 1330–1338 (2005)

MiRNN: An Improved Prediction Model of MicroRNA Precursors Using Gated Recurrent Units

Meng Cao[1], Dancheng Li[1](✉), Zhitao Lin[1], Cheng Niu[1], and Chen Ding[2]

[1] Software College, Northeastern University, Shenyang, China
LDC@mail.neu.edu.cn
[2] College of Life and Health Science, Northeastern University, Shenyang, China
dingchen@mail.neu.edu.cn

Abstract. MicroRNAs (miRNAs) are small noncoding RNAs that derived from hairpin-forming miRNA precursors (pre-miRNAs) and regulating gene expression at the post-transcriptional level. Many sophisticated computational tools have been developed for miRNA prediction. However, all these existing approaches for predicting miRNA require large amounts of task-specific knowledge in the form of handcrafted features and data pre-processing. In this article, we introduce MiRNN (MiRNN is available at https://github.com/CadenC/MiRNN), a novel computational predictor based on bidirectional gated recurrent units (GRUs). Our system is truly end-to-end, requiring no feature engineering or data preprocessing, thus making it applicable to a wide range of sequence classification tasks. Its main purpose is to omit the procedure of feature extraction and to provide accurate prediction by using the high-level features extracted from the bidirectional recurrent neural network. The experimental results show that MiRNN can produce state-of-the-art performance on pre-miRNA prediction task. The overall prediction accuracy of our model on miRBase data sets is 93.70%. In addition, we trained our model on various clade specific dataset and obtained increased accuracy.

Keywords: MiRNN · MicroRNA prediction · Deep learning
Bidirectional RNN · GRUs · End-to-end model

1 Introduction

MicroRNAs (miRNAs) are about 22 nucleotides long, non-coding RNAs that regulate the translation of mRNAs at the post-transcription level. They are expressed in a wide variety of organisms including viruses, plants, and animals [1, 2]. The miRNAs biogenesis involves number of steps. First, primary transcripts of miRNA (pri-miRNA) are transcribed from genes that are several kilobases long. The pri-miRNAs are then clopped by Rnase-III enzyme Drosha into ~70 base pairs (bp) long hairpin-looped precursor miRNAs (pre-miRNAs). The exportin-5 protein transports pre-miRNAs hairpin into the cytoplasm through nuclear pore. In cytoplasm, pre-miRNAs are further cleaved by Rnase-III enzyme Dicer to produce a ~22 bp double stranded intermediate

D.-S. Huang et al. (Eds.): ICIC 2018, LNCS 10955, pp. 217–222, 2018.
https://doi.org/10.1007/978-3-319-95933-7_26

called miRNA/miRNA* duplex. A strand of the duplex with relative thermodynamic stability becomes a mature miRNA. The final products are incorporated into RISC complexes, which works through base pairing of the target mRNAs. The miRNAs play key roles in development, cell proliferation and cell death. Thus, their deregulation has been connected with neurode generative disease, cancer and metabolic disorders [3].

In the past few years, many experiments have found that miRNAs exist in many biological processes [4]. MicroRNAs are involved in the development of animals and plants [5], such as differentiation of embryonic [6], muscle [7], skeletal [8], hematopoietic [9] and many other types of cells. They are also known to control cell death [10] and proliferation [11], insulin secretion [12] or lipid metabolism [13]. Finally, several studies showed that organisms under various stress have a responsive miRNAs signature pattern, allowing resistance and adaptation [14, 15]. MiRNAs are even used by virus to infect hosts [16, 17].

Distinguishing real pre-miRNAs from other pseudo hairpins is a problem that can be considered as a binary classification problem. The general approach of existing models can be summarized as follows: First, conduct feature engineering on pre-miRNA sequence and design fixed-length feature vector. Second, build the model based on certain classification algorithm. Common pre-miRNA features include the secondary structure, nucleotide frequency, sequence length, bulges, etc. [18] These existing models have two severe limitations. First, they all require heavy work on extracting features from candidate hairpin sequences. Since there is not an explicit way to tell which features are useful for prediction, strenuous efforts might be taken for trying different features combinations. Second, the artificially extracted features contain only part of the information of the original sequence, thus limiting the potential of the model to improve the prediction accuracy. DeepMiRGene [19] is the only existing tool we know applies recurrent neural network.

In this study, we introduce a novel model using GRUs to distinguish real pre-miRNA from pseudo hairpins which have similar stem-loop features but do not contain mature miRNAs. In our approach, we make no use of any pre-extracted features including the secondary structure of the pre-miRNA hairpin. Our model is truly end-to-end, no feature engineering and data pre-processing are required. The experiment result shows that our model obtains state-of-the-art performance in pre-miRNA prediction in case of sensitivity and specificity.

2 Materials and Methods

2.1 Datasets

MiRNAs and pre-miRNAs sequences were downloaded from miRbase [20] release 21, which contains 28645 entries representing hairpin precursor miRNAs, expressing 35828 mature miRNA products, in 223 species. All the (almost) sequences are verified as true pre-miRNA by laboratory experiment. We retrieved pre-miRNA sequences of different species respectively.

For the purpose of training classifiers, negative sets of non-miRNAs were generated from different sources. Human pseudo pre-miRNA hairpins are obtained suggested by microPred [21]. We obtained 8494 non-redundant human pseudo hairpin sequences which have been previously used in triple-SVM, MiPred and miPred methods. For plant pseudo pre-miRNAs sequences, we have adopted the same method as Plant-MiRNAPred [22]. Negative datasets of mouse, nematoda, arthropoda are downloaded from the UCSC database [23]. The different filtering criteria, including non-overlapping sliding window, no multiple loops, lowest base pair number set to 18 and maximum of −15 kcal/mol free energy of the secondary structure are applied on these sequences to resemble the real pre-miRNA properties, and the pseudo pre-miRNAs hairpins are set to have the same length distribution as the real sequences.

2.2 Model Structure

In this section, we describe the implementation of our recurrent network model. More specifically, we employ gated recurrent units, which has been found to be very successful in a wide range of tasks related to sequence data [24]. Figure 1 shows the architecture of our model. The model receives a candidate hairpin sequence as input and predicts a probability of whether the input sequence will be a real pre-miRNA.

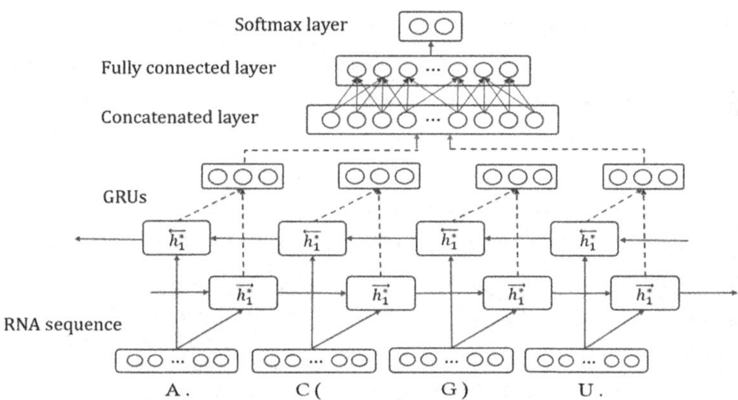

Fig. 1. Our neural network architecture. The dotted lines indicate vector concatenation.

The input candidate hairpin sequences are first mapped to a vector space. We apply one-hot encoding for each type of nucleotide and use an embedding layer to map it to a four-dimensional vector. Therefore, each hairpin sequence of length 200 can be vectorized as a 4 * 200 matrix with each column represents an encoded nucleotide. The nucleotide embeddings are randomly initialized at first, and the optimal value is achieved during neural network training using gradient descent algorithm.

In the RNA secondary structure, there are only two statuses for each nucleotide, paired or unpaired. In the dot-bracket notation, unpaired nucleotides are represented as "." and base-paired nucleotides are represented using as "("s or closing ")"s. To make use of secondary structure information, we apply another embedding layer for each status of nucleotide in the secondary structure. We only consider if the nucleotide is paired or not. Therefore, we use two two-dimensional vectors represent each pair status. And the nucleotide embedding is concatenated with the pair-status embedding as final vector representation.

3 Results and Discussion

3.1 Classification Performance

We first trained our classifier on a balanced data set consisting of 22915 pre-miRNA sequences (irrespective of species) and the same number of negative samples obtained using the methods mentioned above. We applied 5-fold cross-validation to get a more accurate estimation. To prevent the model from overfitting the data set, we used dropout training and set the dropout rate to be 0.5. We evaluate our trained model on test set which consists 5729 real and pseudo pre-miRNAs. The sensitivity, specificity and accuracy are 94.14%, 93.27%, 93.70% respectively, with a Matthews correlation coefficient (MCC) of 0.87. The ROC curve in Fig. 2 shows the performance of our classifier on different species.

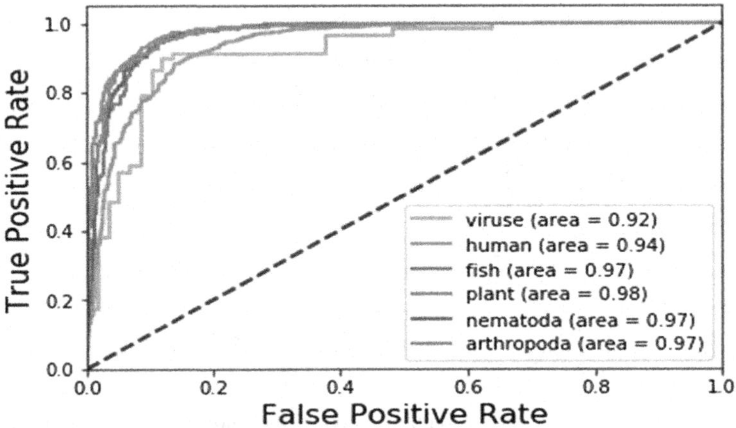

Fig. 2. The performance of our classifier on different species.

Since miRNAs properties are different between species, we trained and evaluated our model separately on each of five clades. The results are presented in Table 1. As seen from the table, the prediction accuracy is improved by training the classifier separately on different clades, which verify the difference of miRNAs between specifies.

3.2 Comparison with State-of-the-Art Tools

We compare our model with existing microRNA prediction tools, the results are shown in Table 2. We list the performances of our model in the human, plant, and

Table 1. Prediction accuracy of lineage-specific MiRNN predictors.

Classifier	Number of instances	Correctly classified	Sensitivity	Specificity	ACC	MCC
Mammals	10271	9461	0.918	0.939	0.928	0.860
Plants	6713	6327	0.981	0.956	0.969	0.937
Nematoda	1855	1791	0.922	0.972	0.946	0.893
Arthropods	3295	3113	0.971	0.955	0.963	0.926
Fish	1622	1515	0.950	0.939	0.944	0.889

cross-species precursor miRNA data sets. From Table 2, we can see MiRNN performs competitively compared to existing pre-miRNA prediction models. On plant pre-miRNAs dataset, our MiRNN model achieves new state-of-the-art performance.

Table 2. Comparison MiRNN with existing prediction tools.

Tool	Target	Classifier	Sensitivity (%)	Specificity (%)
MiRAlign	Human	Alignment	89.9	–
Triplet	Human	SVM	93.30	88.10
Mipred	Human	RF	89.35	93.21
mipred	Human	SVM	84.55	97.97
microPred	Human	SVM	90.02	97.82
MiRenSVM	Homo sapiens	SVM	93.05	96.50
PlantMiRNAPred	Plant	SVM	ACC > 90.0	
HeteroMirPred	Zebrafish	SVM, RF, KNN	94.8	98.3
deepMiRGene	Human	LSTM	89.0	99.0
deepMiRGene	Cross-species	LSTM	91.0	98.0
MiRNN	Human	GRU	93.81	92.47
MiRNN	Plant	GRU	98.14	95.63
MiRNN	Cross-species	GRU	94.14	93.27

4 Conclusions

In this paper, we proposed a recurrent neural network architecture for pre-microRNA prediction. Our model, which is based on bidirectional gated recurrent units, is truly end-to-end. Compared with previously computational tools, our model relies on no task-specific resources, feature engineering or data pre-processing. In the miRbase dataset, MiRNN can achieve 93.70% accuracy, and the performance is improved by training the model separately on different clades.

There are several potential directions for future work. First, our model can be further improved by combining more useful and correlated information. Another interesting direction is to apply our model to other biological fields.

References

1. Ambros, V.: A hierarchy of regulatory genes controls a larva-to-adult developmental switch in C. elegans. Cell **57**(1), 49–57 (1989)
2. Ruvkun, G.: Glimpses of a tiny RNA world. Science **294**(5543), 797–799 (2001)
3. Witkos, T.M., Koscianska, E., Krzyzosiak, W.J.: Practical aspects of microRNA target prediction. Curr. Mol. Med. **11**(2), 93–109 (2011)
4. Lim, L.P., Lau, N.C., Garrett-Engele, P., et al.: Microarray analysis shows that some microRNAs downregulate large numbers of target mRNAs. Nature **433**(7027), 769 (2005)
5. Carrington, J.C., Ambros, V.: Role of microRNAs in plant and animal development. Science **301**(5631), 336–338 (2003)
6. Suh, M.R., Lee, Y., Kim, J.Y., et al.: Human embryonic stem cells express a unique set of microRNAs. Dev. Biol. **270**(2), 488–498 (2004)
7. Williams, A.H., Liu, N., Van Rooij, E., Olson, E.N.: Microrna control of muscle development and disease. Curr. Opin. Cell Biol. **21**(3), 461–469 (2009)
8. Chen, J.F., Mandel, E.M., Thomson, J.M., et al.: The role of microRNA-1 and microRNA-133 in skeletal muscle proliferation and differentiation. Nat. Genet. **38**(2), 228 (2006)
9. Shivdasani, R.A.: MicroRNAs: regulators of gene expression and cell differentiation. Blood **108**(12), 3646–3653 (2006)
10. Ambros, V.: The functions of animal microRNAs. Nature **431**(7006), 350 (2004)
11. Brennecke, J., Hipfner, D.R., Stark, A., et al.: bantam encodes a developmentally regulated microRNA that controls cell proliferation and regulates the proapoptotic gene hid in Drosophila. Cell **113**(1), 25–36 (2003)
12. Poy, M.N., Eliasson, L., Krutzfeldt, J., et al.: A pancreatic islet-specific microRNA regulates insulin secretion. Nature **432**(7014), 226 (2004)
13. Wilfred, B.R., Wang, W.X., Nelson, P.T.: Energizing miRNA research: a review of the role of miRNAs in lipid metabolism, with a prediction that miR-103/107 regulates human metabolic pathways. Mol. Genet. Metab. **91**(3), 209–217 (2007)
14. Fujii, H., Chiou, T.J., Lin, S.I., et al.: A miRNA involved in phosphate-starvation response in Arabidopsis. Curr. Biol. **15**(22), 2038–2043 (2005)
15. Guy, C.L.: Cold acclimation and freezing stress tolerance: role of protein metabolism. Annu. Rev. Plant Biol. **41**(1), 187–223 (1990)
16. Pfeffer, S., Zavolan, M., Grässer, F.A., et al.: Identification of virus-encoded microRNAs. Science **304**(5671), 734–736 (2004)
17. Nelson, J.A.: Small RNAs and large DNA viruses. N. Engl. J. Med. **357**(25), 2630–2632 (2007)
18. Leclercq, M., Diallo, A.B., Blanchette, M.: Computational prediction of the localization of microRNAs within their pre-miRNA. Nucleic Acids Res. **41**(15), 7200–7211 (2013)
19. Park, S., Min, S., Choi, H., et al.: deepMiRGene: deep neural network based precursor microRNA prediction. arXiv preprint arXiv:1605.00017 (2016)
20. S, G.J.: miRBase. http://www.mirbase.org/
21. Batuwita, R., Palade, V.: microPred: effective classification of pre-miRNAs for human miRNA gene prediction. Bioinformatics **25**(8), 989–995 (2009)
22. Xuan, P., Guo, M., Liu, X., et al.: PlantMiRNAPred: efficient classification of real and pseudo plant pre-miRNAs. Bioinformatics **27**(10), 1368–1376 (2011)
23. Fujita, P.A., Rhead, B., Zweig, A.S., et al.: The UCSC genome browser database: update 2011. Nucleic Acids Res. **39**(suppl_1), D876–D882 (2010)
24. Hinton, G.E., Srivastava, N., Krizhevsky, A., et al.: Improving neural networks by preventing co-adaptation of feature detectors. arXiv preprint arXiv:1207.0580 (2012)

Using Deep Neural Network to Predict Drug Sensitivity of Cancer Cell Lines

Yake Wang[1], Min Li[1(✉)], Ruiqing Zheng[1], Xinghua Shi[2], Yaohang Li[3], Fangxiang Wu[4], and Jianxin Wang[1]

[1] School of Information Science and Engineering, Central South University, Changsha, China
limin@mail.csu.edu.cn

[2] Department of Bioinformatics and Genomics, College of Computing and Informatics, University of North Carolina at Charlotte, Charlotte, NC 28223, USA

[3] Department of Computer Science, Old Dominion University, Norfolk, VA 23529, USA

[4] Division of Biomedical Engineering, University of Saskatchewan, Saskatoon, SK S7N5A9, Canada

Abstract. High-throughput screening technology has provided a large amount of drug sensitivity data for hundreds of compounds on cancer cell lines. In this study, we have developed a deep learning architecture based on these data to improve the performance of drug sensitivity prediction. We used a five-layer deep neural network, named as DeepPredictor, that integrated both genomic features of cell lines and chemical information of compounds to predict the half maximal inhibitory concentration on the Cancer Cell Line Encyclopedia (CCLE) dataset. We demonstrated the performance of our deep model using 10-fold cross-validations and leave-one-out strategies and showed that our model outperformed existing approaches.

Keywords: Cancer cell lines · Drug sensitivity · DeepPredictor
Deep learning · Predictive models

1 Introduction

Cultured Cancer cell lines are fundamental materials to study the molecular basis of drug activity [1] and to discover novel anticancer drugs in cancer biology research. Several large-scale high-throughput screening efforts have catalogued genomic information of a panel of in vitro cell lines as well as their drug sensitivity profiles against hundreds of compounds.

These datasets have been studied in many aspects, such as novel anticancer drug discovery [2], and anticancer drug repositioning [3]. In the context of drug discovery and repositioning, computational approaches have been widely used to predict drug sensitivity data over cancer cell lines. Though initial method Quantitative Structure-Activity Relationship (QSAR) was widely adopted, it could not generalize across cancer cell lines since the properties of cell lines were not considered. Menden et al. [4]

D.-S. Huang et al. (Eds.): ICIC 2018, LNCS 10955, pp. 223–226, 2018.
https://doi.org/10.1007/978-3-319-95933-7_27

made the first effort to integrate cell line genomic features and drug chemical features to model drug sensitivity data using a three-layer neural network and random forests (RF). Ammad-ud-din et al. [5] proposed an extended QSAR model via integrating drug properties matrix and cell line properties matrix to predict drug response data using Kernelized Bayesian Matrix Factorization (KBMF). [6] built a dual-layer integrated cell line-drug network to deduce drug response data. Cortes-Ciriano et al. [7] integrated the transcript profiles of genes and compounds' Morgan fingerprints to predict the drug responses using random forests (RF) and support vector machine (SVM).

In this study, we propose a novel deep learning architecture to predict drug sensitivity of cancer cell lines by integrating gene expression data of cell lines and compound fingerprints. The experimental results show that our approach outperforms the state-of-the-art approaches on drug sensitivity prediction.

2 Method

2.1 Dataset

We extracted 504 cell lines with response data against 24 drugs from CCLE dataset. To reduce the input feature dimension, we chose the top-1000 genes that display the highest variance across cell lines and normalized their expression data to zero mean and unit variance. We calculated compounds' Morgan fingerprints and used them as compounds' feature representations. The response data IC50 s were converted to $\ln IC50(\mu M)$ to compare with previous study [7]. The final data matrix contains 491 cell lines and 23 drugs with 96.25% completeness.

2.2 DeepPredictor

Deep learning has recently achieved remarkable success in many areas including computer vision [8], speech recognition [9], and natural language processing [10]. A feedforward deep network is one of the classic deep learning models. It contains several stacked layers with each neural unit connecting to all the units at next layer. Given enough neural units or deep layers, a feedforward deep network can become a universal approximator [11]. We can use a deep network to model drug sensitivity prediction considering the existence of large amount of drug pharmacological profiling data in a classification task. We train a feedforward deep neural network to model the continuous drug response data, which showed that this regression-based deep network outperformed start-of-the-art work in this field.

Specifically, our newly developed deep model, namely DeepPredictor, works as follows as illustrated in Fig. 1. We integrated the genomic information of cancel cell lines and chemical structures of compounds as the feature vectors, which were taken as the inputs of a feedforward neural network to predict the sensitivity data of corresponding cell line-compound pairs.

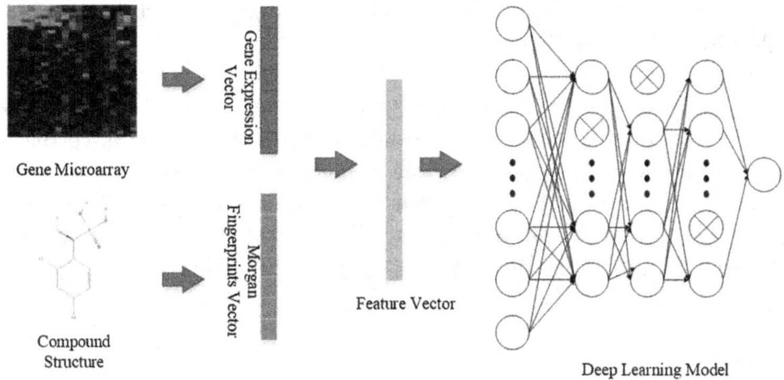

Fig. 1. The overall workflow of our DeepPredictor model.

3 Results

We performed 10-fold cross-validations to obtain the predictive performance measurement of our deep learning models. And we also left cell lines from each tissue out and each compound out to test the ability of the models trained on the rest data to predict sensitivity data on novel cell lines and compounds. The final RMSE and R^2 values were averaged.

We compared DeepPredictor with the state-of-the-art study [7], RF, on CCLE dataset. As summarized in Table 1, cross-validation results showed that DeepPredictor outperforms RF with lower RMSE value of 0.51 and higher R^2 value of 0.75. Leave-one-tissue-out and leave-one-compound-out experiments also showed that DeepPredictor outperformed RF on the CCLE dataset in predicting sensitivity data involving novel cell lines or compounds.

Table 1. Results on CCLE dataset.

Metrics	CV		LOTO		LOCO	
	RMSE	R^2	RMSE	R^2	RMSE	R^2
RF	1.02 ± 0.05	0.74 ± 0.03	0.97 ± 0.26	0.75 ± 0.12	1.62 ± 1.32	0.18 ± 0.15
DeepPredictor	0.51 ± 0.06	0.75 ± 0.04	0.60 ± 0.22	0.68 ± 0.13	1.36 ± 1.56	0.04 ± 0.08

4 Conclusion

In this paper, we proposed DeepPredictor, a deep learning model, to predict drug sensitivity of cancer cell lines. The model was trained on CCLE dataset. And the results showed DeepPredictor outperformed state of the art models with the lowest prediction errors and high coefficient determination. The cross-validation results show that DeepPredictor has the best interpolation ability to fill in missing drug sensitivity values, and the LOTO and LOCO results show that DeepPredictor has the lowest extrapolation error. We thus believe that DeepPredictor can benefit clinical cancer therapy and future study on drug sensitivity prediction.

Acknowledgement. We would like to thank Isidro Cortés Ciriano at Department of Biomedical Informatics, Harvard Medical School for providing the research data and discussing with us during research. This work is supported by the National Science Fund for Excellent Young Scholars under Grant No. 61622213, the National Natural Science Foundation of China under grant No. 61772552, and the Fundamental Research Funds for the Central Universities of Central South University under grant No. 2018zzts560 and No. 2018zzts028.

References

1. Weinstein, J.N.: Drug discovery: cell lines battle cancer. Nature **483**(7391), 544–545 (2012)
2. Wilding, J.L., Bodmer, W.F.: Cancer cell lines for drug discovery and development. Cancer Res. **74**(9), 2377–2384 (2014)
3. Jin, G., Wong, S.T.: Toward better drug repositioning: prioritizing and integrating existing methods into efficient pipelines. Drug Discov. Today **19**(5), 637–644 (2014)
4. Menden, M.P., Iorio, F., Garnett, M., et al.: Machine learning prediction of cancer cell sensitivity to drugs based on genomic and chemical properties. PLoS ONE **8**(4), e61318 (2013)
5. Ammad-ud-din, M., Georgii, E., Gonen, M., et al.: Integrative and personalized QSAR analysis in cancer by kernelized Bayesian matrix factorization. J. Chem. Inf. Model. **54**(8), 2347–2359 (2014)
6. Zhang, N., Wang, H., Fang, Y., et al.: Predicting anticancer drug responses using a dual-layer integrated cell line-drug network model. PLoS Comput. Biol. **11**(9), e1004498 (2015)
7. Cortes-Ciriano, I., van Westen, G.J., Bouvier, G., et al.: Improved large-scale prediction of growth inhibition patterns using the NCI60 cancer cell line panel. Bioinformatics **32**(1), 85–95 (2016)
8. Krizhevsky, A., Sutskever I., Hinton G.E.: Imagenet classification with deep convolutional neural networks. In: Advances in Neural Information Processing Systems, pp. 1097–1105 (2012)
9. Hinton, G., Deng, L., Yu, D., et al.: Deep neural networks for acoustic modeling in speech recognition: the shared views of four research groups. IEEE Sig. Process. Mag. **29**(6), 82–97 (2012)
10. Sutskever, I., Vinyals, O., Le, Q.V.: Sequence to sequence learning with neural networks. In: Advances in Neural Information Processing Systems, pp. 3104–3112 (2014)
11. Hornik, K., Stinchcombe, M., White, H.: Multilayer feedforward networks are universal approximators. Neural Netw. **2**(5), 359–366 (1989)

PCM: A Pairwise Correlation Mining Package for Biological Network Inference

Hao Liang[1], Feiyang Gu[2], Chaohua Sheng[1], Qiong Duan[1], Bo Tian[1], Jun Wu[3], Bo Xu[1(✉)], and Zengyou He[1(✉)]

[1] School of Software, Dalian University of Technology, Dalian, China
{BoXu, zyhe}@dlut.edu.cn

[2] Baidu Inc., Beijing, China

[3] School of Information Engineering, Zunyi Normal University, Zunyi, China

Abstract. One fundamental task in molecular biology is to understand the dependency among genes or proteins to model biological networks. One widely used method is to calculate the pairwise correlation or association scores between genes or proteins. To date, a software package supporting various types of correlation measures has been lacking. In this paper, we present a pairwise correlation mining package, termed PCM, which supports the commonly used marginal correlation measures, together with two algorithms enabling the estimation of conditional correlations. Two example data sets are used to illustrate how to use this package and demonstrate the importance of having an integrated software package that incorporates various correlation measures. The package and source codes of the implementations are available at https://github.com/FeiyangGu/PCM.

Keywords: Pairwise correlation · Network inference · Correlation mining

1 Background

To understand the relationships between DNA, RNA, proteins or other cellular molecules, it is necessary to infer biochemical networks from genomic data and proteomic data by exploring various types of computational and statistical methods. Such network inference or reverse engineering problem is quite fundamental in bioinformatics, which has drawn much attention during the past decades [1, 3, 5].

The key issue in the network inference procedure is to obtain the interaction relationships among the molecules such as genes and proteins. This issue can be tackled from various angles, leading to different types of computational and statistical network inference models. Probably the most straightforward and commonly used formulation is to cast the network inference problem as a pairwise correlation mining problem, i.e., calculating the correlation or association scores among each gene or protein pair. Although many correlation measures have been explored in different network inference applications, there is still no consensus on the best one even for one specific application such as the gene regulatory network inference [3]. Therefore, it is highly necessary to have a pairwise correlation mining package that supports various types of correlation measures.

D.-S. Huang et al. (Eds.): ICIC 2018, LNCS 10955, pp. 227–231, 2018.
https://doi.org/10.1007/978-3-319-95933-7_28

In this paper, we provide PCM, an open-source implementation of pairwise correlation mining algorithms. PCM enables the pairwise correlation mining with a series of marginal correlation measures under the same umbrella. In addition, it implements two low-order conditional correlation mining methods and provides the functionality of clustering-based approximate correlation mining. Examples are provided to illustrate how to use the package and performance evaluation on several real data sets are used to justify the rationale for developing such a package.

2 Correlation Measures and Algorithms

PCM was implemented in C++, which uses a matrix as the input. Columns and rows in the input matrix correspond to variables (e.g. genes, proteins) and samples (e.g. gene expression profiles), respectively. PCM consists of three main modules.

2.1 Marginal Correlation Measures

The Marginal Correlation module calculates the marginal pairwise correlation scores among all candidate pairs and returns a set of pairs whose correlation scores are above the user-specified threshold. Furthermore, it is further divided into two sub-modules: one is used for handling continuous variables and another one is designed to mine pairwise correlations among binary variables.

In the current version, we have implemented 9 correlation measures for continuous variables and 27 correlation measures for binary variables. The correlation measures for continuous variables can be applied to quantify the association strength between gene profiles in gene regulatory network inference, and those measures for binary variables can be used for network inference from qualitative affinity purification-mass spectrometry (AP-MS) data [5].

The names and detailed definitions of these correlation measures are presented in Tables 1 and 2 in the full version of this paper which is available at https://github.com/FeiyangGu/PCM.

2.2 Clustering-Based Approximate Correlation Mining

To conduct an exhaustive pairwise correlation mining, the time complexity is at least $O(n^2)$, where n is the number of variables. To handle the data sets with large number of variables, an alternative approach is to generate an approximate set of correlated pairs so as to reduce the time complexity. We also provide a fast algorithm for generating an approximate set of correlated pairs. This algorithm first uses clustering algorithms such as *k*-means to partition the variables into different clusters, then the correlation coefficients of all candidate pairs within each cluster are calculated. Finally, the mining results across all clusters are gathered to generate the final set of correlated pairs.

In the PCM package, we implement a clustering-based method for this purpose. This method has two steps: clustering and correlation mining. In the first step, it uses clustering algorithms to partition the variables into different groups. For continuous variables, the clustering algorithm employed is the well-known *k*-means method, which requires the number of clusters as an input parameter. For binary variables, the CLOPE algorithm [6] is utilized, which also has an user-specified parameter called repulsion for controlling the level of intra-cluster similarity. It is expected that highly correlated variables will be put into the same cluster so that it is sufficient to perform correlation mining within each cluster. Therefore, in the second step, the correlation coefficients of all candidate pairs within each group are calculated. And the mining results across all clusters are merged to generate the final set of correlated pairs.

Suppose the number of generated clusters is k (all clusters has the same size) and the clustering algorithm has a linear time complexity, then this method will reduce the time complexity of pairwise correlation mining from $O(n^2)$ to $O(n^2/k)$. Apparently, such performance gain in running efficiency is at the cost of missing some really correlated pairs of variables if these variables are not assigned into the same cluster. For instance, if we handle the example data set used in the experimental section when the number of cluster is set to be 2 and the threshold for the Pearson's correlation coefficient is 0.5, 9 of the 49 correlated pairs that are above the threshold will be missed.

2.3 Algorithms for Conditional Correlation Mining

Since the marginal correlation cannot distinguish direct and indirect associations (the induced associations due to other variables), the Conditional Correlation module implements two existing algorithms for mining low-order conditional pairwise correlations: LOPC [8] and CMI2NI [7]. The conditional correlation measures the correlation between two variables after their dependence on other variables is removed. Thus, these algorithms are able to remove pairs whose correlation relationship may mainly come from their mutual dependencies on other variables.

CMI2NI. For the CMI2NI method, we provide two implementations for different types of variables: CMI2NIC and CMI2NIB. The former implementation strictly follows [7], which can be used to handle data sets with continuous variables. While the latter implementation is designed for data sets with binary variables, which computes the conditional mutual information between two binary variables directly.

LOPC. The LOPC [8] algorithm is similar to that of CMI2NI. In this method, the zero-th, first and second order partial correlation coefficients for a candidate variable pair are calculated in an iterative manner. For instance, if the zero-th order partial correlation coefficient of one variable pair is less than the given threshold, then this pair would not be considered in the subsequent evaluation with respect to higher order partial correlations.

3 Results

Two example data sets were delivered within the PCM package. The first data set, "data_DREAM3.txt", is a DREAM3 data set about the Yeast knock-out gene expressions, which is composed of 100 continuous variables and 100 samples. The second data set is produced from the AP-MS data in the reference [2], which has 2761 binary variables (the presence/absence of each protein in a purification) and 2166 samples (each sample corresponds to a purification).

3.1 An Example Application

In this section, we use the first data set as the example data. The following commands can be used to find all gene pairs whose Pearson's correlation coefficients are no less than 0.2 from this data set and put the mining results into the file "out.txt":

PCM PearsonC data_DREAM3.txt out.txt 0.2.
To rapidly obtain a set of correlated pairs, we can use the following commands:
PCM PearsonC data_DREAM3.txt out.txt 0.2 1 5,

where "1" means that we use clustering methods to obtain an approximate mining results, and "5" is the number of clusters specified by the users.

Furthermore, the following commands will use the algorithm LOPC in [8] to retain only pairs whose conditional correlation coefficients (from 0th order up to 2nd order) are significant enough.

PCM LOPCC data_DREAM3.txt out.txt.

3.2 Comparison Results

In this section, we use the second data set, the AP-MS data set from Gavin et al. [2], as an example to illustrate the necessity of implementing a software package that are composed of many correlation measures and algorithms.

To validate network inference results from the AP-MS data, we use three reference sets of experimentally validated binary protein interactions for the performance assessment. These reference sets are denoted as Y2H, PCA, and BGS, respectively [4]. As shown in Fig. 1, different correlation measures may yield significantly different network inference results. Moreover, we do not know which measure will achieve the best performance for a specific data set. Therefore, it is necessary to have software package that can provide various correlation measures for the end users.

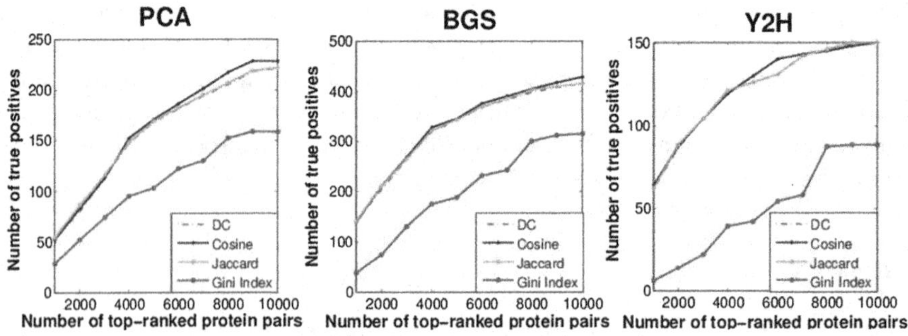

Fig. 1. The performance comparison on the Gavin data set when DC (Dice Coefficient), Cosine, Jaccard and Gini index are used as the correlation measures. For each reference set and correlation measure, we report a set of top-ranked interactions (x-axis) to check how many interactions are contained in the reference set (y-axis).

4 Conclusions

PCM enables automated pairwise correlation mining using various types of correlation measures. Different correlation mining methods can be applied in a single framework, enabling easy and fast comparison and selection of the most suitable correlation mining method for a biological network inference task.

Acknowledgement. This work was partially supported by the Natural Science Foundation of China (Nos.61572094, 61502071), the Fundamental Research Funds for the Central Universities (No.DUT2017TB02) and the Science-Technology Foundation for Youth of Guizhou Province (No.KY[2017]250).

References

1. De Smet, R., Marchal, K.: Advantages and limitations of current network inference methods. Nature Rev. Microbiol. **8**(10), 717–729 (2010)
2. Gavin, A.C., Aloy, P., Grandi, P., Krause, R., Boesche, M., Marzioch, M., Rau, C., Jensen, L. J., Bastuck, S., Dumpelfeld, B., et al.: Proteome survey reveals modularity of the yeast cell machinery. Nature **440**(7084), 631–636 (2006)
3. Kurt, Z., Aydin, N., Altay, G.: A comprehensive comparison of association estimators for gene network inference algorithms. Bioinformatics **30**(15), 2142–2149 (2014)
4. Schelhorn, S.E., Mestre, J., Albrecht, M., Zotenko, E.: Inferring physical protein contacts from large-scale purification data of protein complexes. Mol. Cell. Proteomics **10**(6), M110–004929 (2011)
5. Teng, B., Zhao, C., Liu, X., He, Z.: Network inference from AP-MS data: computational challenges and solutions. Brief. Bioinform. **16**(4), 658–674 (2015)
6. Yang, Y., Guan, X., You, J.: CLOPE: a fast and effective clustering algorithm for transactional data. In: Proceedings of the Eighth ACM SIGKDD International conference on Knowledge Discovery and Data Mining, pp. 682–687. ACM (2002)
7. Zhang, X., Zhao, J., Hao, J.K., Zhao, X.M., Chen, L.: Conditional mutual inclusive information enables accurate quantification of associations in gene regulatory networks. Nucleic Acids Res. **43**(5), e31 (2015)
8. Zuo, Y., Yu, G., Tadesse, M.G., Ressom, H.W.: Biological network inference using low order partial correlation. Methods **69**(3), 266–273 (2014)

RP-FIRF: Prediction of Self-interacting Proteins Using Random Projection Classifier Combining with Finite Impulse Response Filter

Zhan-Heng Chen[1,2], Zhu-Hong You[1,2(✉)], Li-Ping Li[1], Yan-Bin Wang[1,2], and Xiao Li[1]

[1] The Xinjiang Technical Institute of Physics and Chemistry, Chinese Academy of Sciences, Urumqi 830011, China
zhuhongyou@ms.xjb.ac.cn, zhuhongyou@gmail.com
[2] University of Chinese Academy of Sciences, Beijing 100049, China

Abstract. The self-interacting proteins (SIPs) plays a significant part in the organism and the regulation of cellular functions. Thence, we developed an effective algorithm to predict SIPs, named RP-FIRF, which merges the Random Projection (RP) classifier and Finite Impulse Response Filter (FIRF) together. More specifically, the Position Specific Scoring Matrix (PSSM) was firstly converted from protein sequence by exploiting Position Specific Iterated BLAST (PSI-BLAST). Then, we obtained the same size of matrix by implementing a valid matrix multiplication on PSSM, and applied FIRF approach to calculate the eigenvalues of each protein. The Principal Component Analysis (PCA) approach is used to extract the most relevant information. Finally, the performance of the proposed method is performed on *human* dataset. The results show that our model can achieve high average accuracies of 97.89% on *human* dataset using the 5-fold cross-validation, which demonstrate that our method is a useful tool for identifying SIPs.

Keywords: Self-interacting proteins · Random projection
Finite impulse response filter

1 Introduction

Protein is a significant component of all cells. No protein, no life. Most of proteins often work together with their partner or other proteins which is called protein-protein interactions (PPIs) [1–4]. Self-interacting proteins (SIPs) is a particular form of PPIs, where can interact in terms of duplicate their own genes. SIPs play an essential role in the evolution of protein interaction networks (PINs) [5]. The functionality of protein refers to that it could control the transport of small molecules and ions across cell membranes, relies on their homo-oligomers [6]. Ispolatov et al. discovered that the average homodimers of SIPs more than twice as much as that of non-SIPs in the PINs [7]. It is crucial for clarifying the function of SIPs to further understand the regulation of protein function, so as to better understand the effect of disease mechanism [8]. From the past years, many studies show that homo-oligomerization must play an important role in many biological processes, such as gene expression regulation, signal

D.-S. Huang et al. (Eds.): ICIC 2018, LNCS 10955, pp. 232–240, 2018.
https://doi.org/10.1007/978-3-319-95933-7_29

transduction, enzyme activation and immune response [9–13]. More and more computational methods were developed at the historic moment [14–26]. However, these methods could be applied to detect PPIs well [27], but they are not good enough to predict SIPs. Hence, we put forward a random projection (RP) bind with Finite Impulse Response Filter (FIRF) model for predicting SIPs from protein sequence information.

2 Results and Discussion

2.1 5-Fold Cross-Validation on *Human* Dataset

In view of the fairness and over-fitting problems, in the experiment, we repeated the technique five times on *human* dataset, termed 5-fold cross-validation. we cut the dataset into five non-overlapping pieces, and the training set was randomly chosen in 4/5 of characteristic values and selected the remaining values as independent test set. Then, we repeat it five times to test our RP-FIRF model on *human* dataset, and got the results are shown in Table 1. From the table, the average Accuracy(Acc), Sensitivity (Sen), Specificity(Sp), and Matthews correlation coefficient(MCC) of 97.89%, 74.46%, 100.00%, and 85.31% respectively.

Table 1. Measure the quality of RP-FIRF model and the other methods on *human* dataset

Model	Acc (%)	Sp (%)	Sen (%)	MCC (%)
SLIPPER [5]	91.10	95.06	47.26	41.97
DXECPPI [28]	30.90	25.83	87.08	8.25
PPIevo [29]	78.04	25.82	87.83	20.82
LocFuse [30]	80.66	80.50	50.83	20.26
CRS [31]	91.54	96.72	34.17	36.33
SPAR [31]	92.09	97.40	33.33	38.36
Proposed method	**97.89**	**100.00**	**74.46**	**85.31**

2.2 Measure Our RP-FIRF Model Against Other Existing Methods

In order to testify our model can achieve better results, we measured the quality of proposed model with other existing methods based on the *human* dataset. We listed a clear statement of account in Table 1, it is obvious that the overall results of our prediction approach is also significantly better than the other six methods on *human* dataset. The prediction accuracy of the overall experimental results can be improved. This fully illustrates that a good feature extraction method and a suitable classifier are very important for predicting model. It is further illustrated that the proposed method is superior to the other six approaches and quite suitable for predicting SIPs.

3 Conclusion

In our research, a RP-FIRF-based prediction model using protein sequence information was put forward to predict SIPs. In the experiment, the key focus and difficult problem is that the dataset used by the classifier are unbalanced. The mainly improvements for this model are attributable to the following three aspects: (1) Finite impulse response filter (FIRF) was regarded as an effective feature processing method which can capture the main useful information of the dataset. (2) PCA was treated as a reasonable dimensional-reduced method which was applied to capture the evolutionary information without the noisy features from the data. It can provide help to increase the accuracy of prediction system. (3) The RP classifier is strongly suitable for predicting SIPs. The experimental results measured by the presented model on *human* dataset revealed that the overall prediction performance of RP-FIRF model is significantly better than that of the SVM-based classifier and other exiting approaches. For the future research, there will be more and more effective feature extraction methods and deep learning techniques explored for detecting SIPs.

4 Materials and Methodology

4.1 Datasets

In our study, we constructed the datasets mainly derived from the UniProt database [32] which contains 20,199 curated *human* protein sequences. There are many different types of resources such as DIP [33], BioGRID [34], IntAct [35], InnateDB [36] and MatrixDB [37], we can get the related information from them and construct the dataset for the experiment by applying 2994 *human* SIPs sequences. Then, we need to single out the dataset for the experiment and assess the performance of our RP-FIRF model, which mainly includes three steps [31]: (1) We will remove the protein sequences which may be fragments, and retain the length of protein sequences between 50 residues and 5000 residues; (2) To build up the positive dataset of *human*, we formed a high-grade SIPs data which should meet one of the following conditions: (a) the self-interactions were revealed by at least one small-scale experiment or two sorts of large-scale experiments; (b) the protein has been announced as homo-oligomer (containing homodimer and homotrimer) in UniProt; (c) it has been reported by more than two publications for the self-interactions; (3) For the *human* negative dataset, we removed all the types of SIPs from the whole *human* proteome and SIPs detection in UniProt database. To sum it up, we obtained the ultimate *human* dataset for the experiment which was mainly composed of 1441 SIPs and 15,938 non-SIPs [31].

4.2 Assessment Tools

In our study, in the interest of size up the steadiness and effectiveness of our present method, we calculated the values of five parameters: Accuracy (Acc), Sensitivity (Sen), specificity (Sp), and Matthews's Correlation Coefficient (MCC), respectively. These parameters can be described as follows:

$$Acc = \frac{TP + TN}{TP + FP + TN + FN} \quad (1)$$

$$Sen = \frac{TP}{TP + FN} \quad (2)$$

$$Sp = \frac{TN}{FP + TN} \quad (3)$$

$$MCC = \frac{TP \times TN - FP \times FN}{\sqrt{(TP + FN) \times (TN + FP) \times (TP + FP) \times (TN + FN)}} \quad (4)$$

where, *TP* represents the number of true positives. *FP* represents the quantity of false positives. *TN* represents the count of true negatives. *FN* represents the quantity of false negatives.

4.3 Position Specific Scoring Matrix

In our achievements, Position Specific Scoring Matrix (PSSM) is a helpful tool which was applied to detect distantly related proteins [38–42]. Accordingly, a PSSM was converted from each protein sequence information by employing the Position Specific Iterated BLAST (PSI-BLAST) [43]. And then, a given protein sequence can be transformed into an $H \times 20$ PSSM which can be announced as follow:

$$M = \{M_{\alpha\beta}\ \alpha : 1 = 1 \cdots H, \beta = 1 \cdots 20\} \quad (5)$$

where the row H is the length of a protein sequence, and the column represents the number of amino acids. For the query protein sequence, the score $C_{\alpha\beta}$ represents the β-*th* amino acid in the position of α which can be distributed from a PSSM.

$$C_{\alpha\beta} = \sum\nolimits_{k=1}^{20} p(\alpha, k) \times q(\beta, k) \quad (6)$$

where $p(\alpha,k)$ denotes the appearing frequency value of the k-*th* amino acid at position of α with the probe, and $q(\beta,k)$ is the value of Dayhoff's mutation matrix between β-*th* and k-*th* amino acids.

Eventually, a strongly conservative position can achieve a greater score, and otherwise a lower degree denotes a weakly conservative position. To get a high degree and a wide range of homologous sequences, the *E*-value parameter of PSI-BLAST was set to be 0.001 which reported for a given result represents the number of two sequences' alignments and chose three iterations in this process.

4.4 Finite Impulse Response Filters

In our work, we applied FIRF to process the characteristics of protein sequences, which would be used to predict the SIPs. The main feature is that some characteristics of the problem can be fully highlighted by transformation. We design it by using Fourier series method in details as follows.

At first, the corresponding Frequency Response Function of FIRF transfer function can be defined as:

$$H(e^{jw}) = \sum_{n=0}^{N-1} h(n)e^{-jwn} \tag{7}$$

where, *h(n)* is the available impulse response sequence, and *N* represents the sample sizes of frequency response $H(e^{jw})$. Given the frequency response $H_d(e^{jw})$ of ideal filter, and let $H(e^{jw})$ approach $H_d(e^{jw})$ infinitely.

$$H_d(e^{jw}) = \sum_{n=-\infty}^{\infty} h_d(n)e^{-jwn} \tag{8}$$

where $h_d(n)$ is a finite length. If $h_d(n)$ is an infinite length, we can intercept $h_d(n)$ by applying a finite length of the windows function sequence *w(n)*.

$$h(n) = \frac{1}{2\pi} h_d(n)w(n) \tag{9}$$

In our study, we multiply the transpose of PSSM by PSSM to achieve 20×20 matrix. and then, we employ the FIRF method to obtain a feature vector. Afterwards, the feature values of each protein sequence can be calculated as a *400*-dimensional vector. Finally, each protein sequence from *human* dataset was converted into a *400*-dimensional vector by applying FIRF method.

4.5 Random Projection Classifier

There are lots of techniques for solving classification problems [44–55]. In our work, Random projection(RP) methods showed that *N* points in *N* dimensional space can almost always be projected onto a space of dimension *ClogN* with control on the ratio of distances and the error [56]. We introduce the RP algorithm as follow, Let

$$\Gamma = \{Ai\}_{i=1}^{N}, \quad Ai \in R^n \tag{10}$$

be the primitive high dimensional space dataset, where *n* is the high dimension and *N* is the number of the dataset. The goal of dimensionality reduction is embedding the vectors into a lower dimensional space R^q from a high dimension R^n, where $q << n$.

$$\tilde{\Gamma} = \{\tilde{A}_i\}_{i=1}^{N}, \quad \tilde{A}_i \in R^q \tag{11}$$

where q is close to the intrinsic dimensionality of Γ. Thus, the vectors in the set of Γ was regarded as embedding vectors.

In this work, the random projections were divided into non-overlapping blocks where $B1 = 10$ and each one carefully chosen from a block of size $B2 = 30$ that obtained the smallest estimate of the test error. We used the k-Nearest Neighbor (KNN) as base classifier and the leave-one-out test error estimate. where $k = seq\ (1, 30, by = 8)$. Our classifier integrates the results of taking advantage of the base classifier on the selected projection, with the data-driven voting threshold to confirm the final mission.

References

1. De Las Rivas, J., Fontanillo, C.: Protein–protein interactions essentials: key concepts to building and analyzing interactome networks. PLoS Comput. Biol. **6**(6) (2010)
2. Zhu, L., You, Z.H., Huang, D.S., Wang, B.: t-LSE: A novel robust geometric approach for modeling protein-protein interaction networks. PLoS ONE **8**(4) (2013)
3. Lei, Y.K., You, Z.H., Ji, Z., Zhu, L., Huang, D.S.: Assessing and predicting protein interactions by combining manifold embedding with multiple information integration. BMC Bioinform. **13**(7), S3 (2012)
4. Li, Z.W., You, Z.H., Chen, X., Li, L.P., Huang, D.S., Yan, G.Y., Nie, R., Huang, Y.A.: Accurate prediction of protein-protein interactions by integrating potential evolutionary information embedded in PSSM profile and discriminative vector machine classifier. Oncotarget **8**(14), 23638 (2017)
5. Liu, Z., Guo, F., Zhang, J., Wang, J., Lu, L., Li, D., He, F.: Proteome-wide prediction of self-interacting proteins based on multiple properties. Mol. Cell. Proteomics **12**(6), 1689–1700 (2013)
6. Marianayagam, N.J., Sunde, M., Matthews, J.M.: The power of two: protein dimerization in biology. Trends Biochem. Sci. **29**(11), 618–625 (2004)
7. Ispolatov, I., Yuryev, A., Mazo, I., Maslov, S.: Binding properties and evolution of homodimers in protein–protein interaction networks. Nucleic Acids Res. **33**(11), 3629–3635 (2005)
8. Wang, Y.-B., You, Z.-H., Li, L.-P., Huang, Y.-A., Yi, H.-C.: Detection of interactions between proteins by using legendre moments descriptor to extract discriminatory information embedded in PSSM. Molecules **22**(8), 1366 (2017)
9. Woodcock, J.M., Murphy, J., Stomski, F.C., Berndt, M.C., Lopez, A.F.: The dimeric versus monomeric status of 14-3-3ζ is controlled by phosphorylation of Ser58 at the dimer interface. J. Biol. Chem. **278**(38), 36323–36327 (2003)
10. Baisamy, L., Jurisch, N., Diviani, D.: Leucine zipper-mediated homo-oligomerization regulates the Rho-GEF activity of AKAP-Lbc. J. Biol. Chem. **280**(15), 15405–15412 (2005)
11. Katsamba, P., Carroll, K., Ahlsen, G., Bahna, F., Vendome, J., Posy, S., Rajebhosale, M., Price, S., Jessell, T., Ben-Shaul, A.: Linking molecular affinity and cellular specificity in cadherin-mediated adhesion. Proc. Nat. Acad. Sci. **106**(28), 11594–11599 (2009)
12. Koike, R., Kidera, A., Ota, M.: Alteration of oligomeric state and domain architecture is essential for functional transformation between transferase and hydrolase with the same scaffold. Protein Sci. **18**(10), 2060–2066 (2009)

13. You, Z.H., Huang, Z.A., Zhu, Z., Yan, G.Y., Li, Z.W., Wen, Z., Chen, X.: PBMDA: a novel and effective path-based computational model for miRNA-disease association prediction. PLoS Comput. Biol. **13**(3) (2017)
14. You, Z.H., Zhou, M.C., Xin, L., Shuai, L.: Highly efficient framework for predicting interactions between proteins. IEEE Trans. Cybern. **47**(3), 731–743 (2017)
15. You, Z.H., Lei, Y.K., Gui, J., Huang, D.S., Zhou, X.: Using manifold embedding for assessing and predicting protein interactions from high-throughput experimental data. Bioinformatics **26**(21), 2744–2751 (2010)
16. You, Z.H., Yin, Z., Han, K., Huang, D.S., Zhou, X.: A semi-supervised learning approach to predict synthetic genetic interactions by combining functional and topological properties of functional gene network. BMC Bioinform. **11**(1), 343 (2010)
17. An, J.Y., You, Z.H., Chen, X., Huang, D.S., Yan, G., Wang, D.F.: Robust and accurate prediction of protein self-interactions from amino acids sequence using evolutionary information. Mol. BioSyst. **12**(12), 3702–3710 (2016)
18. An, J.Y., You, Z.H., Chen, X., Huang, D.S., Li, Z.W., Liu, G., Wang, Y.: Identification of self-interacting proteins by exploring evolutionary information embedded in PSI-BLAST-constructed position specific scoring matrix. Oncotarget **7**(50), 82440–82449 (2016)
19. Huang, Y.A., Chen, X., You, Z.H., Huang, D.S., Chan, K.C.C.: ILNCSIM: improved lncRNA functional similarity calculation model. Oncotarget **7**(18), 25902–25914 (2016)
20. Zhu, L., You, Z.H., Huang, D.S.: Increasing the reliability of protein–protein interaction networks via non-convex semantic embedding. Neurocomputing **121**, 99–107 (2013)
21. Xia, J.F., You, Z.H., Wu, M., Wang, S.L., Zhao, X.M.: Improved method for predicting phi-turns in proteins using a two-stage classifier. Protein Pept. Lett. **17**(9), 1117–1122 (2010)
22. You, Z.H., Yu, J.Z., Zhu, L., Li, S., Wen, Z.K.: A MapReduce based parallel SVM for large-scale predicting protein–protein interactions. Neurocomputing **145**, 37–43 (2014)
23. Li, S., You, Z.H., Guo, H., Luo, X., Zhao, Z.Q.: Inverse-free extreme learning machine with optimal information updating. IEEE Trans. Cybern. **46**(5), 1229–1241 (2016)
24. Lei, W., You, Z.H., Xia, S.X., Feng, L., Xing, C., Xin, Y., Yong, Z.: Advancing the prediction accuracy of protein-protein interactions by utilizing evolutionary information from position-specific scoring matrix and ensemble classifier. J. Theor. Biol. **418**, 105–110 (2017)
25. You, Z.H., Li, S., Gao, X., Luo, X., Ji, Z.: Large-scale protein-protein interactions detection by integrating big biosensing data with computational model. Biomed. Res. Int. **2014**, 598129 (2014)
26. Gao, Z.-G., Wang, L., Xia, S.-X., You, Z.-H., Yan, X., Zhou, Y.: Ens-PPI: a novel ensemble classifier for predicting the interactions of proteins using autocovariance transformation from pssm. Biomed. Res. Int. **2016**, 8 (2016)
27. Wang, Y.-B., You, Z.-H., Li, X., Jiang, T.-H., Chen, X., Zhou, X., Wang, L.: Predicting protein–protein interactions from protein sequences by a stacked sparse autoencoder deep neural network. Mol. BioSyst. **13**(7), 1336–1344 (2017)
28. Du, X., Cheng, J., Zheng, T., Duan, Z., Qian, F.: A novel feature extraction scheme with ensemble coding for protein–protein interaction prediction. Int. J. Mol. Sci. **15**(7), 12731–12749 (2014)
29. Zahiri, J., Yaghoubi, O., Mohammad-Noori, M., Ebrahimpour, R., Masoudi-Nejad, A.: PPIevo: Protein–protein interaction prediction from PSSM based evolutionary information. Genomics **102**(4), 237–242 (2013)
30. Zahiri, J., Mohammad-Noori, M., Ebrahimpour, R., Saadat, S., Bozorgmehr, J.H., Goldberg, T., Masoudi-Nejad, A.: LocFuse: human protein–protein interaction prediction via classifier fusion using protein localization information. Genomics **104**(6), 496–503 (2014)

31. Liu, X., Yang, S., Li, C., Zhang, Z., Song, J.: SPAR: a random forest-based predictor for self-interacting proteins with fine-grained domain information. Amino Acids **48**(7), 1655–1665 (2016)
32. Consortium, U.: UniProt: a hub for protein information. Nucleic Acids Res. **43**(D1), D204–D212 (2014)
33. Salwinski, L., Miller, C.S., Smith, A.J., Pettit, F.K., Bowie, J.U., Eisenberg, D.: The Database of Interacting Proteins: 2004 update. Nucleic Acids Res. **32**, D449–D451 (2004)
34. Chatr-Aryamontri, A., Breitkreutz, B.-J., Oughtred, R., Boucher, L., Heinicke, S., Chen, D., Stark, C., Breitkreutz, A., Kolas, N., O'donnell, L.: The BioGRID interaction database: 2015 update. Nucleic Acids Res. **43**(D1), D470–D478 (2014)
35. Orchard, S., Ammari, M., Aranda, B., Breuza, L., Briganti, L., Broackes-Carter, F., Campbell, N.H., Chavali, G., Chen, C., Del-Toro, N.: The MIntAct project—IntAct as a common curation platform for 11 molecular interaction databases. Nucleic Acids Res. **42** (D1), D358–D363 (2013)
36. Breuer, K., Foroushani, A.K., Laird, M.R., Chen, C., Sribnaia, A., Lo, R., Winsor, G.L., Hancock, R.E., Brinkman, F.S., Lynn, D.J.: InnateDB: systems biology of innate immunity and beyond—recent updates and continuing curation. Nucleic Acids Res. **41**(D1), D1228–D1233 (2012)
37. Launay, G., Salza, R., Multedo, D., Thierry-Mieg, N., Ricard-Blum, S.: MatrixDB, the extracellular matrix interaction database: updated content, a new navigator and expanded functionalities. Nucleic Acids Res. **43**(D1), D321–D327 (2014)
38. Gribskov, M., McLachlan, A.D., Eisenberg, D.: Profile analysis: detection of distantly related proteins. Proc. Nat. Acad. Sci. **84**(13), 4355–4358 (1987)
39. Wang, Y., You, Z., Li, X., Chen, X., Jiang, T., Zhang, J.: PCVMZM: using the probabilistic classification vector machines model combined with a zernike moments descriptor to predict protein-protein interactions from protein sequences. Int. J. Mol. Sci. **18**(5), 1029 (2017)
40. You, Z.H., Li, J., Gao, X., He, Z., Zhu, L., Lei, Y.K., Ji, Z.: Detecting protein-protein interactions with a novel matrix-based protein sequence representation and support vector machines. BioMed Res. Int. **2015**, 1–9 (2015)
41. You, Z.H., Chan, K.C.C., Hu, P.: Predicting protein-protein interactions from primary protein sequences using a novel multi-scale local feature representation scheme and the random forest. PLoS ONE **10**(5) (2015)
42. Lei, Y.-K., You, Z.-H., Dong, T., Jiang, Y.-X., Yang, J.-A.: Increasing reliability of protein interactome by fast manifold embedding. Pattern Recogn. Lett. **34**(4), 372–379 (2013)
43. Altschul, S.F., Koonin, E.V.: Iterated profile searches with PSI-BLAST—a tool for discovery in protein databases. Trends Biochem. Sci. **23**(11), 444–447 (1998)
44. You, Z.-H., Li, X., Chan, K.C.: An improved sequence-based prediction protocol for protein-protein interactions using amino acids substitution matrix and rotation forest ensemble classifiers. Neurocomputing **228**, 277–282 (2017)
45. Wang, L., You, Z.-H., Chen, X., Li, J.-Q., Yan, X., Zhang, W., Huang, Y.-A.: An ensemble approach for large-scale identification of protein-protein interactions using the alignments of multiple sequences. Oncotarget **8**(3), 5149 (2017)
46. Li, J.-Q., You, Z.-H., Li, X., Ming, Z., Chen, X.: PSPEL. In silico prediction of self-interacting proteins from amino acids sequences using ensemble learning. IEEE/ACM Trans. Comput. Biol. Bioinform. **14**(5), 1165–1172 (2017)
47. Zhu, H.-J., You, Z.-H., Zhu, Z.-X., Shi, W.-L., Chen, X., Cheng, L.: DroidDet: effective and robust detection of android malware using static analysis along with rotation forest model. Neurocomputing **272**, 638–646 (2018)

48. Chen, X., Huang, Y.-A., You, Z.-H., Yan, G.-Y., Wang, X.-S.: A novel approach based on KATZ measure to predict associations of human microbiota with non-infectious diseases. Bioinformatics **33**(5), 733–739 (2016)
49. Huang, Y.-A., You, Z.-H., Chen, X., Yan, G.-Y.: Improved protein-protein interactions prediction via weighted sparse representation model combining continuous wavelet descriptor and PseAA composition. BMC Syst. Biol. **10**(4), 120 (2016)
50. Huang, Y.-A., You, Z.-H., Li, X., Chen, X., Hu, P., Li, S., Luo, X.: Construction of reliable protein–protein interaction networks using weighted sparse representation based classifier with pseudo substitution matrix representation features. Neurocomputing **218**, 131–138 (2016)
51. Chen, X., Huang, Y.-A., Wang, X.-S., You, Z.-H., Chan, K.C.: FMLNCSIM: fuzzy measure-based lncRNA functional similarity calculation model. Oncotarget **7**(29), 45948 (2016)
52. Chen, X., Yan, C.C., Zhang, X., You, Z.-H.: Long non-coding RNAs and complex diseases: from experimental results to computational models. Brief. Bioinform. **18**(4), 558–576 (2016)
53. Huang, Y.-A., You, Z.-H., Chen, X., Chan, K., Luo, X.: Sequence-based prediction of protein-protein interactions using weighted sparse representation model combined with global encoding. BMC Bioinform. **17**(1), 184 (2016)
54. Huang, Y.-A., You, Z.-H., Gao, X., Wong, L., Wang, L.: Using weighted sparse representation model combined with discrete cosine transformation to predict protein-protein interactions from protein sequence. BioMed Res. Int. **2015**, 10 (2015)
55. Wang, L., You, Z.-H., Xia, S.-X., Chen, X., Yan, X., Zhou, Y., Liu, F.: An improved efficient rotation forest algorithm to predict the interactions among proteins. Soft Comput. **22**, 1–9 (2017)
56. Schclar, A., Rokach, L.: Random projection ensemble classifiers. In: Filipe, J., Cordeiro, J. (eds.) Enterprise Information Systems. Lecture Notes in Business Information Processing, vol. 24, pp. 309–316. Springer, Heidelberg (2009). https://doi.org/10.1007/978-3-642-01347-8_26

NetCoffee2: A Novel Global Alignment Algorithm for Multiple PPI Networks Based on Graph Feature Vectors

Jialu Hu, Junhao He, Yiqun Gao, Yan Zheng, and Xuequn Shang(✉)

School of Computer Science, Northwestern Polytechnical University, 1 Dongxiang Road, Xi'an 710129, China
shang@nwpu.edu.cn

Abstract. Network alignment provides a fast and effective framework to automatically identify functionally conserved proteins in a systematic way. However, due to the fast growing biological data, there is an increasing demand for more accurate and efficient tools to deal with multiple PPI networks. Here, we present a novel global alignment algorithm NetCoffee2 to discover functionally conserved proteins. To test the algorithm performance, NetCoffee2 and several existing algorithms were applied on eight real biological datasets. Results show that NetCoffee2 is superior to IsoRankN, NetCoffee and multiMAGNA++ in terms of both coverage and consistency. The binary and source code are freely available at https://github.com/screamer/NetCoffee2.

Keywords: PPI network alignment · Simulated annealing
Functionally conserved proteins

1 Introduction

Protein function is a fundamental problem that attracts many researchers in the fields of both molecular function and evolution. Although many researchers have put a great of efforts to develop public protein annotation databases, such as Uniprot [1], the task of protein characterization is far to be completed. Thanks to the development of next generation sequencing, computational methods become a major strength for discovering the molecular function and phylogenetic [2].

Global network alignment provides an effective computational framework to systematically identify functionally conserved proteins from a global node map between two or more protein-protein interaction (PPI) networks [3]. These alignments of two networks are called pairwise network alignment [4]. These of more than two networks are termed as multiple network alignment [5, 6]. There are two types of node maps: one-to-one and multiple-to-multiple. With a global network alignment, one can easily predict function of unknown proteins by using "transferring annotation".

IsoRank was the first algorithm proposed to solve global network alignment, which takes advantage of a method analogous to Google's PageRank method [7]. An updated version IsoRankN was proposed to perform multiple network alignment based on spectral clustering on the induced graph of pairwise alignment score [8]. Intuitively

D.-S. Huang et al. (Eds.): ICIC 2018, LNCS 10955, pp. 241–246, 2018.
https://doi.org/10.1007/978-3-319-95933-7_30

guided by T-Coffee [9], a fast and accurate program NetCoffee [10] was developed to search for a global alignment by using a triplet approach. However, it cannot work on pairwise network alignment. To improve the edge conservation, a genetic algorithm MAGNA was proposed, which mimics the evolutionary process [6]. MAGNA++ speeds up the MAGNA algorithm by parallelizing it to automatically use all available resources [11]. A more advanced version multiMAGNA++ was applied to find alignment for multiple PPI networks [12]. However, there still exists a gap between network alignment and the prediction of unknown protein function in a systematical level.

Here, we present a novel network alignment algorithm NetCoffee2 based on graph feature vectors to identify functionally conserved proteins. Unlike NetCoffee, NetCoffee2 can perform tasks of both pairwise and multiple network alignments. Furthermore, it outperforms existing alignment tools in both coverage and consistency.

2 Methods

2.1 Definition and Notation

Network alignment is a problem to search for a global node mapping between two or more networks. Suppose there is a set of PPI networks $\{G_1, G_2, \ldots, G_k\}, \mathrm{k} \geq 2$, each network can be modeled as a graph $G_i = \{V_i, E_i\}$, where V_i and E_i represents proteins and interactions appearing in networks. A matchset consists of a subset of proteins from $\bigcup_{i=k}^{k} V_i$. A global network alignment is to find a set of mutually disjoint matchsets from a set of PPI networks. Note that, each protein can only appear in one matchset in a global alignment solution. Each matchset represents a functionally conserved group of proteins.

2.2 An Integrated Model

Sequence information is one of important factors in charactering biological function of genes, RNA and proteins. PPI network topology can provide complementary information for the prediction of protein function. As used in many other network aligners such as IsoRank, and Magna, both topology and sequence information are integrated in one similarity measure to search for functionally conserved proteins across species.

2.3 Sequence-Based Similarity

Intuitively guided by an assumption that structures determine functions, most of existing network aligners use both amino acid sequences and network topology to predict protein functions. Here, we performed an all-against-all sequence comparison using BLASTP on all protein sequences. These protein pairs with significant conserved regions are taken into consideration for further filtrations. Note that e-value is an input parameter to control the coverage of network alignment. Let Ω denote the candidates of homology proteins. Given a protein pair u and v, the sequence similarity $\mathrm{s(u, v)}$ can be calculated in the following formula, $s_h(u,v) = \frac{\epsilon(u,v) - \varepsilon_{min}(u,v)}{\Delta\varepsilon}$. Here, $\varepsilon(\mathrm{u, v})$ can be log (evalue) or bitscore of the protein pair u and v, and $\Delta\varepsilon$ is the largest difference between any two pairs of homolog in Ω, $\Delta\varepsilon = \varepsilon_{max}(u, v) - \varepsilon_{min}(u, v)$, which servers as a normalization factor. The most similar one is 1, the least 0.

2.4 Topology-Based Similarity

To find the topologically similar protein pairs, a similarity measure is necessary for evaluating the topological similarity. Our method works on a principle that if two nodes are aligned, then the local induced-subgraphs should be similar. Given a network $G = (V, E), V = (v_1, v_2 \ldots v_n), v_i \in V$, we design a 5-tuple-feature vector $(\gamma, \delta, \tau, \eta, \theta)$ for each node in V to represent the local connection of the node. Without loss of generality, we denote the adjacent matrix of G as $M_{n \times n}$. Since M is real and symmetric, there must exist a major normalized eigenvector $K = (k_1, k_2 \ldots k_n)$. In another words, K is the normalized eigenvector of the largest eigenvalue. Then, k_i represents the reputation of the node v_i. The greater the reputation is, the more important the node is. Therefore, we use k_i as the first element of the 5-tuple-feature vector (i.e. γ) to character the node v_i. Let us denote the neighbor of v as N_v. Then, we use $|N_v|$ as the second element of the 5-tuple-feature vector (i.e. δ), the sum of the reputation of these nodes $\sum_{x \in N_v} k_x$ as the third element (i.e. τ). Let us denote these nodes that are 2-step away from v as N_v^2. It notes that all nodes in N_v^2 are not directly connected to v. Then, we use $|N_v^2|$ as the fourth element (i.e. η). The last element θ is calculated by the formula $\frac{1}{2}\sum_{x \in N_v^2} k_x p_{xv}$. Here, we denote the number of the shortest paths from x to v as p_{xv}. As shown in Fig. 1(a), the 5-tuple-feature vector of a_1, a_2, a_3, a_4, a_5 in G_1 are $(1, 3, 2.63, 1, 0.16)$, $(0.88, 3, 2.33, 1, 0.75)$, $(0.33, 1, 0.88, 2, 1)$, $(0.75, 2, 2, 1, 0.88)$, $(1, 3, 2.63, 1, 0.16)$, respectively. The vector of each element of all nodes should be normalized in the following step as shown in Fig. 1(b). With the normalized 5-tuple-feature vector, the node similarity of any two nodes $s_t(u, v)$ can be calculated with the Gaussian function $s_t(\mathrm{u}, \mathrm{v}) = \exp(-\frac{1}{2}x^2)$, where x represents the Euclidean distance between the 5-tuple-feature vectors of node u and v.

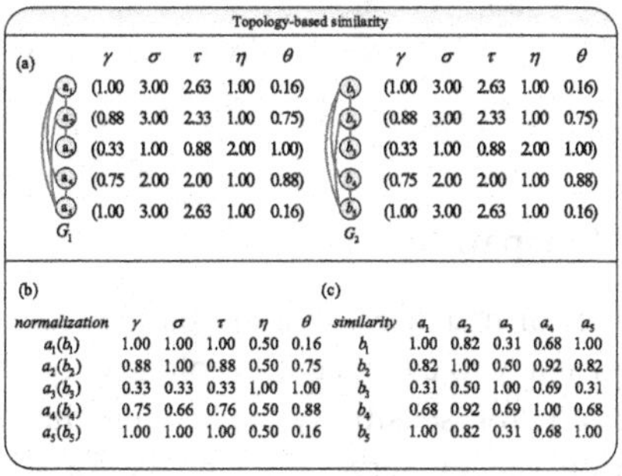

Topology-based similarity

(a)

G_1	γ	σ	τ	η	θ
a_1	(1.00	3.00	2.63	1.00	0.16)
a_2	(0.88	3.00	2.33	1.00	0.75)
a_3	(0.33	1.00	0.88	2.00	1.00)
a_4	(0.75	2.00	2.00	1.00	0.88)
a_5	(1.00	3.00	2.63	1.00	0.16)

G_2	γ	σ	τ	η	θ
b_1	(1.00	3.00	2.63	1.00	0.16)
b_2	(0.88	3.00	2.33	1.00	0.75)
b_3	(0.33	1.00	0.88	2.00	1.00)
b_4	(0.75	2.00	2.00	1.00	0.88)
b_5	(1.00	3.00	2.63	1.00	0.16)

(b)

normalization	γ	σ	τ	η	θ
$a_1(b_1)$	1.00	1.00	1.00	0.50	0.16
$a_2(b_2)$	0.88	1.00	0.88	0.50	0.75
$a_3(b_3)$	0.33	0.33	0.33	1.00	1.00
$a_4(b_4)$	0.75	0.66	0.76	0.50	0.88
$a_5(b_5)$	1.00	1.00	1.00	0.50	0.16

(c)

similarity	a_1	a_2	a_3	a_4	a_5
b_1	1.00	0.82	0.31	0.68	1.00
b_2	0.82	1.00	0.50	0.92	0.82
b_3	0.31	0.50	1.00	0.69	0.31
b_4	0.68	0.92	0.69	1.00	0.68
b_5	1.00	0.82	0.31	0.68	1.00

Fig. 1. The calculation of similarity matrix between two networks G_1 and G_2. (a) A 5-tuple-feature vector (γ, σ, τ, η, θ) was calculated on each node. (b) Vectors of σ, τ, η, θ were normalized by its maximal element. (c) The similarity matrix was calculated by a Gaussian-based similarity measure $s_t(u, v) = \exp(-\frac{1}{2}x^2)$.

2.5 Simulated Annealing

To find an optimal network alignment, the alignment score can be formulated as $f(\mathbb{A}) = \sum_{m \in \mathbb{A}} s_m$, where $\mathbb{A}$ and m is refer to a global alignment and a matchset, respectively. Suppose $m = (m_1, m_2, \cdots, m_v)$, the alignment score of the matchset is $s_m = \sum_{i=m_1}^{m_{v-1}} \sum_{j=i}^{m_v} \alpha s_h(i,j) + (1 - \alpha) s_t(i,j)$. Therefore, the problem of global network alignment can be modeled as an optimization problem, which is to search for an optimal alignment $\mathbb{A}^*$, such that $\mathbb{A}^* = \arg\max_{\mathbb{A}} f(\mathbb{A}) = \sum_{m \in \mathbb{A}} s_m$. To solve this problem, we used a simulated annealing algorithm to search for an approximately optimal solution.

3 Result and Discussion

3.1 Test Datasets and Experimental Setup

To test our method on real biological data, PPI network of five species were downloaded from IntAct (https://www.ebi.ac.uk/intact/). The five species include mus musculus (MM), saccharomyces cerevisiae (SC), drosophila melanogaster (DM), arabidopsis thaliana (AT) and homo sapiens (HS). To improve the data quality, these interactions of the spoke-expanded co-complexes are filtered out. We collected 41,043 proteins and 193,576 interactions as test datasets. All of our test datasets can be freely accessible at http://www.nwpu-bioinformatics.com/netcoffee2/. As seen in Table 1, eight datasets were generated as benchmark datasets. The number of PPI networks in eight benchmark datasets ranges from two to five.

Table 1. Algorithms performance were tested on eight datasets.

Species	D1	D2	D3	D4	D5	D6	D7	D8
MM		√	√		√		√	√
SC	√			√		√	√	√
DM		√	√	√		√	√	√
AT			√	√	√		√	√
HS	√				√	√		√

3.2 Performance and Comparison

Our goal is to identify a set of matchsets that are biologically meaningful. Therefore, we use coverage and consistency to evaluate the biological quality of alignment results. Coverage serves as a proxy for sensitivity, indicating the amount of proteins the alignment can explain. Consistency serves as a proxy for specificity, measuring the functional similarity of proteins in each match set. There is a trade-off between coverage and consistency.

Given an alignment solution, we used the percentage of aligned proteins as coverage. As NetCoffee is not applicable on pairwise network alignment, there is no

NetCoffee result for D1 and D2. The results show that NetCoffee2 stably found coverage of 76.7% on average for all the eight datasets. It is followed by multiMAGNA++, which found 70.4% proteins on average. Although the coverage of MultiMAGNA++ can be more than 80% on D3, D4 and D7, it rapidly fell to 50% on D1, D2 and D5. NetCoffee approximately identifies about 35% proteins on average, which is less than the coverage of NetCoffee2 and multiMAGNA++. IsoRankN found only an average of 9.6% proteins on eight datasets, which is obviously smaller than the coverage of the other competitor. Overall, the results show that NetCoffee2 is superior to multiMAGNA++, NetCoffee and IsoRankN in terms of coverage and it is more stable than all of its competitors.

Consistency is used to measure the biological quality of matchsets in alignment results. We employed two concepts to evaluate global alignment algorithms based on Gene Ontology (GO) terms: mean entropy (ME) and mean normalized entropy (MNE) [8, 10]. The mean normalized entropy (MNE) is the arithmetic mean of normalized entropy for all matchsets in a global alignment. It should be noted that these alignments with lower ME and MNE values are more functionally coherent. As can be seen in Table 2, NetCoffee2 has the best performance on D2, D7 and D8 in terms of ME, which are 0.73, 1.01 and 1.10, respectively. And mutliMAGNA++ obtains the best ME on D1 (0.94), D3 (0.91), D5 (0.98) and D6 (1.00). NetCoffee gets the best ME on D4 (0.85) and D6 (1.00). Overall, NetCoffee2 found the best ME (0.973) on average, which is followed by multiMAGNA++ (1.005), NetCoffee (1.022) and IsoRankN (1.144). Furthermore, NetCoffee2 obtains an average of 0.53 in terms of MNE, which is followed by NetCoffee (0.55), multiMAGNA++ (0.56) and IsoRankN (0.58). It outperforms it competitors on all the eight datasets in terms of MNE. Therefore, we can draw a conclusion that NetCoffee2 is superior to the existing algorithms multiMAGNA++, NetCoffee and IsoRankN in terms of both ME and MNE.

Table 2. Consistency was measured by mean entropy (ME) and mean normalized entropy (MNE).

Algorithm	Consistency	D1	D2	D3	D4	D5	D6	D7	D8	Average
IsoRankN	ME	1.09	1.07	1.15	1.07	1.18	1.20	1.19	1.20	1.144
	MNE	0.58	0.56	0.58	0.59	0.53	0.60	0.60	0.58	0.58
NetCoffee	ME	*	*	0.99	**0.85**	1.05	**1.00**	1.07	1.17	1.002
	MNE	*	*	0.54	0.54	0.53	0.55	0.58	0.57	0.55
multiMAGNA++	ME	**0.94**	0.94	**0.91**	0.93	**0.98**	**1.00**	1.16	1.18	1.005
	MNE	0.55	0.55	0.53	0.58	**0.52**	0.57	1.63	0.59	0.56
NetCoffee2	ME	1.04	**0.73**	0.94	0.87	1.04	1.05	**1.01**	**1.10**	**0.973**
	MNE	**0.54**	**0.46**	**0.52**	**0.54**	**0.52**	**0.55**	**0.56**	**0.55**	**0.53**

4 Conclusion

Network alignment is a very important computational framework for understanding molecular function and phylogenetic relationships. Here, we developed an efficient algorithm NetCoffee2 based on graph feature vectors. NetCoffee2 is a fast, accurate and scalable program for both pairwise and multiple network alignment problems. To evaluate the algorithm performance, NetCoffee2 and three existing algorithms have been performed on eight real biological datasets. Results show that NetCoffee2 is apparently superior to multiMAGNA++, NetCoffee and IsoRankN in terms of both coverage and consistency. It can be concluded that NetCoffee2 is a versatile and efficient computational tool that can be applied to both pairwise and multiple network alignments. Hopefully, its application can benefit the research community in the fields of molecular function and evolution.

Funding. This project has been funded by the National Natural Science Foundation of China (Grant No. 61332014 and 61702420); the China Postdoctoral Science Foundation (Grant No. 2017M613203); the Natural Science Foundation of Shaanxi Province (Grant No. 2017JQ6037); the Fundamental Research Funds for the Central Universities (Grant No. 3102018zy032).

References

1. U. P. Consortium: Uniprot: a hub for protein information. Nucleic Acids Res. **43**(Database issue), 204–212 (2015)
2. Marcotte, E., Pellegrini, M., Ng, H.L., Rice, D.W.: Detecting protein function and protein-protein interactions from genome sequences. Science **285**(5428), 751–753 (1999)
3. Klau, G.W.: A new graph-based method for pairwise global network alignment. BMC Bioinform. **10**(Suppl. 1), 1–9 (2009)
4. Narad, P., Chaurasia, A., Wadhwab, G., Upadhyayaa, K.C.: Net2align: analgorithm for pairwise global alignment of biological networks. Bioinformation **12**(12), 408 (2016)
5. Hu, J., Reinert, K.: LocalAli: an evolutionary-based local alignment approach to identify functionally conserved modules in multiple networks. Bioinformatics **31**(3), 363–372 (2015)
6. Saraph, V., Milenković, T.: MAGNA: Maximizing accuracy in global network alignment. Bioinformatics **30**(20), 2931 (2013)
7. Singh, R., Xu, J., Berger, B.: Global alignment of multiple protein interaction networks with application to functional orthology detection. Proc. Natl. Acad. Sci. U.S.A. **105**(35), 12763–12768 (2008)
8. Liao, C.S., Lu, K., Baym, M., Singh, R., Berger, B.: Isorankn: spectral methods for global alignment of multiple protein networks. Bioinformatics **25**(12), 253–258 (2009)
9. Notredame, C., Higgins, D.G., Heringa, J.: T-coffee: a novel method for fast and accurate multiple sequence alignment. J. Mol. Biol. **302**(1), 205–217 (2000)
10. Hu, J., Kehr, B., Reinert, K.: Netcoffee: a fast and accurate global alignment approach to identify functionally conserved proteins in multiple networks. Bioinformatics **30**(4), 540 (2014)
11. Vijayan, V., Saraph, V., Milenković, T.: MAGNA++: maximizing accuracy in global network alignment via both node and edge conservation. Bioinformatics **31**(14), 2409–2411 (2015)
12. Vijayan, V., Milenković, T.: Multiple network alignment via multiMAGNA++. IEEE/ACM Trans. Comput. Biol. Bioinform. **PP**(99), 1 (2017)

Dynamics of a Stochastic Virus Infection Model with Delayed Immune Response

Deshun Sun[1], Siyuan Chen[1], Fei Liu[1,2(✉)], and Jizhuang Fan[3(✉)]

[1] Control and Simulation Center, Harbin Institute of Technology, Harbin 150001, China
liufei@hit.edu.cn

[2] School of Software Engineering, South China University of Technology, Guangzhou 510006, China

[3] State Key Laboratory of Robotics and System, Harbin Institute of Technology, Harbin 150001, China
fanjizhuang@hit.edu.cn

Abstract. Patients are always affected by external environmental noises and random fluctuations inside bodies. Considering this, in this paper we propose a stochastic virus infection model with delayed immune response, which consists of a system of four-dimensional stochastic delayed equations. We verify that there is a unique global positive solution for this model with any positive initial value, and establish the sufficient conditions for extinction and persistence of the model. Further, we perform a couple of numerical simulations to illustrate our theoretical analysis results.

Keywords: Virus infection · Stochastic delayed model
Extinction, persistence

1 Introduction

Mathematical modeling and analysis play an important role in understanding the dynamics of infectious viruses, and offering insights on the development of treatment strategies and antiviral drug therapies [1]. Many discrete and continuous mathematical models have been proposed to describe and analyze virus infection [2–4], immune response [5–8], and anti-retroviral treatment [9], For example, Chiacchio proposed discrete approache based on Agent-based modeling. However, mostly using deterministic ordinary or delayed differential equations.

In fact, patients are always affected by stochastics, such as external environmental noises, and random fluctuations inside bodies. Consequently, many researchers have introduced stochastics into deterministic models [10–16]. For instance, Meng et al. [10] illustrated that the infectious disease, which would persist in a deterministic model, will, however, go extinct if stochastic perturbations are considered. To the best of our knowledge, there is little work on the research of stochastic virus infection models with time delay; see, e.g., [17–19]. Based on paper [7], we propose a stochastic virus infection model with delayed immune response:

D.-S. Huang et al. (Eds.): ICIC 2018, LNCS 10955, pp. 247–258, 2018.
https://doi.org/10.1007/978-3-319-95933-7_31

$$\begin{cases} dT = (\lambda - \beta TV - dT)dt + \delta_1 TdB_1(t), \\ dI = (\beta TV - aI)dt + \delta_2 IdB_2(t), \\ dV = (kI - uV - qBV)dt + \delta_2 VdB_3(t), \\ dB = (ge^{-\mu\tau}B(t-\tau)V(t-\tau) - cB)dt + \delta_4 BdB_4(t). \end{cases} \tag{1}$$

where T, I, V and B denote the numbers of uninfected cells, infected cells, viruses and B cells, respectively. Parameters λ, k and g represent the production rates of uninfected cells, viruses and B cells, respectively. β is the infection rate of uninfected cells. d, a, u and c are the death rates of uninfected cells, infected cells, viruses and B cells. q models the removal rate of viruses by B cells. τ is the time delay of immune response. $B_i(t)(i = 1, 2, 3, 4)$ are independent standard Brownian motions, and δ_i $(i = 1, 2, 3, 4)$ are the corresponding intensities of white noises.

In this paper, we prove the existence and uniqueness of the positive solution to the model, and establish the sufficient conditions for extinction and persistence of the model. Moreover, we perform a couple of numerical simulations to illustrate our theoretical analysis results and campare to corresponding deterministic model.

2 Existence and Uniqueness of the Positive Solution

Considering the actual situation, the solution $(T(t), I(t), V(t), B(t))$ to system (1) should be nonnegative and unique. In this section, we will verify that there is a unique global positive solution to system (1) for any positive initial value.

Theorem 1. For any initial value $T(0) > 0$, $I(0) > 0$, $V(\xi) \geq 0$, and $B(\xi) \geq 0$ for all $\xi \in [-\tau, 0]$ with $V(0) > 0$ and $B(0) > 0$, system (1) has a unique positive solution $(T(t), I(t), V(t), B(t))$ on $t > 0$ and the solution will remain in $\Re^4_+$ with probability 1. Here we define $\Re^4_+ = \{x \in \Re^4 : x_i > 0, 1 \leq i \leq 4\}$.

Proof. The beginning proof is similar to paper [18, 19], so we omit it. In the following, we are going to prove $\tau_\infty = \infty$ almost surely $(a.s.)$.

Assuming that $\tau_\infty = \infty\, a.s.$ is false, there exists a pair of constants $T_1 > 0$ and $\varepsilon_1 \in (0, 1)$ such that $P\{\tau_\infty \leq T_1\} > \varepsilon_1$. Thus, we know that there exists an integer $N_1 \geq N_0$ such that $P\{\tau_N \leq T_1\} \geq \varepsilon_1$, $\forall N > N_1$. Define

$$V_1(T, I, V, B) = (T - \frac{au}{k} - \frac{au}{k}\ln\frac{Tk}{au}) + (I - 1 - \ln I) + \frac{a}{k}(V - \frac{ce^{\mu\tau}}{g} - \frac{ce^{\mu\tau}}{g}\ln\frac{Vg}{ce^{\mu\tau}}) + \frac{aqe^{\mu\tau}}{kg}(B - 1 - \ln B) + \frac{aq}{k}\int_{t-\tau}^{t} B(\theta)V(\theta)d\theta. \tag{2}$$

Apparently, the function $V_1(T, I, V, B)$ is nonnegative. Applying Itô's formula [20], we obtain

$$
\begin{aligned}
&dV_1(T,I,V,B)\\
&= [(1-\frac{au}{kT})(\lambda-\beta TV-dT)+\frac{au}{2k}\delta_1^2+(1-\frac{1}{I})(\beta TV-aI)+\frac{1}{2}\delta_2^2\\
&+\frac{a}{k}(1-\frac{ce^{\mu\tau}}{gV})(kI-uV-qBV)+\frac{ace^{\mu\tau}}{2kg}\delta_3^2+\frac{aq}{k}BV-\frac{aq}{k}B(t-\tau)V(t-\tau)\\
&+\frac{aqe^{\mu\tau}}{kg}(1-\frac{1}{B})(ge^{-\mu\tau}B(t-\tau)V(t-\tau)-cB)+\frac{aqe^{\mu\tau}}{2kg}\delta_4^2]dt\\
&+(1-\frac{au}{kT})\delta_1 TdB_1(t)+(1-\frac{1}{I})\delta_2 IdB_2(t)+(1-\frac{ce^{\mu\tau}}{gV})\frac{a}{k}\delta_3 VdB_3(t)\\
&+(1-\frac{1}{B})\frac{aqe^{\mu\tau}}{kg}\delta_4 BdB_4(t).
\end{aligned}
$$

The equation above can be written as

$$
\begin{aligned}
dV_1(T,I,V,B) = LV_1(T,I,V,B)dt &+ (1-\frac{au}{kT})\delta_1 TdB_1(t)+(1-\frac{1}{I})\delta_2 IdB_2(t)\\
&+(1-\frac{ce^{\mu\tau}}{gV})\frac{a}{k}\delta_3 VdB_3(t)+(1-\frac{1}{B})\frac{aqe^{\mu\tau}}{kg}\delta_4 BdB_4(t).
\end{aligned}
$$

Further, we have

$$
LV_1(T,I,V,B) \le \lambda+\frac{au}{k}d+a+\frac{acue^{\mu\tau}}{gk}+\frac{acqe^{\mu\tau}}{gk}-dT \le K,
$$

where K is a positive constant. Hence, we obtain the following inequality:

$$
\begin{aligned}
dV_1(T,I,V,B) \le Kt &+ (1-\frac{au}{kT})\delta_1 TdB_1(t)+(1-\frac{1}{I})\delta_2 IdB_2(t)\\
&+(1-\frac{ce^{\mu\tau}}{gV})\frac{a}{k}\delta_3 VdB_3(t)+(1-\frac{1}{B})\frac{aqe^{\mu\tau}}{kg}\delta_4 BdB_4(t).
\end{aligned} \tag{3}
$$

Integrating both sides of inequality (3) from 0 to $t_N \wedge T_1$ (here $\wedge$ means the bigger one), and computing the expectations, we have

$$
\begin{aligned}
&EV_1[T(\tau_N\wedge T_1), I(\tau_N\wedge T_1), V(\tau_N\wedge T_1), B(\tau_N\\
&\wedge T_1)] \le V_1(T(0), I(0), V(0), B(0)) + KT_1.
\end{aligned} \tag{4}
$$

Set $\Omega_N = \{\tau_N \le T_1\}$ for $N \ge N_1$, and we can see that $P(\Omega_N) \ge \varepsilon_1$. For any $\omega \in \Omega_N$, at least one of $T(\tau_N,\omega), I(\tau_N,\omega), V(\tau_N,\omega)$ and $B(\tau_N,\omega)$ is equal to N or $\frac{1}{N}$. By the definition of V_1 and the inequality (4), we then have

$$
V_1(T(\tau_N,\omega), I(\tau_N,\omega), V(\tau_N,\omega), B(\tau_N,\omega)) \ge (N-1-\ln N)\wedge(\frac{1}{N}-1+\ln N).
$$

$$V_1(T(0),I(0),V(0),B(0))+KT_1 \geq E[I_{\Omega_N}V_1(T(\tau_N,\omega),I(\tau_N,\omega),V(\tau_N,\omega),B(\tau_N,\omega))]$$
$$\geq \varepsilon_1(N-1-\ln N)\wedge(\frac{1}{N}-1+\ln N), \tag{5}$$

where I_{Ω_N} is the indicator function of Ω_N. When $N \to \infty$, we have

$$\infty > V_1(T(0),I(0),V(0),B(0))+KT_1 = \infty,$$

which is contradictory, and then we obtain $\tau_\infty = \infty\, a.s.$ Hence, $T(t), I(t), V(t)$, and $B(t)$ will not explode in a finite time almost surely and Theorem 1 is proved.

3 Extinction

In this section, we investigate the case that the disease becomes extinct. Before drawing the main conclusion, we first give the following Lemmas 1 and 2.

Lemma 1. Let $(T(t),I(t),V(t),B(t))$ be the solution to system (1) with any initial value $T(0)>0$, $I(0)>0$, $V(\xi)\geq 0$ and $B(\xi)\geq 0$ for all $\xi\in[-\tau,0]$ with $V(0)>0$, $B(0)>0$. We then have
$\lim_{t\to\infty}\frac{T(t)}{t}=0\,a.s.$, $\lim_{t\to\infty}\frac{I(t)}{t}=0\,a.s.$, $\lim_{t\to\infty}\frac{V(t)}{t}=0\,a.s.$, $\lim_{t\to\infty}\frac{B(t)}{t}=0\,a.s.$, and $\lim_{t\to\infty}\frac{ge^{-\mu t}\int_{t-\tau}^{t}e^{\mu s}B(s)V(s)ds}{t}=0\,a.s.$

Proof. Define
$x(t)=T(t)+I(t)+V(t)+B(t)+ge^{-\mu t}\int_{t-\tau}^{t}e^{\mu s}B(s)V(s)ds$, and $W(x)=(1+x)^{\theta}$, where θ is a positive constant. Applying Itô's formula, we have

$$dW = LW(x)dt+\theta(1+x)^{\theta-1}[\delta_1TdB_1(t)+\delta_2IdB_2(t)+\delta_3VdB_3(t)+\delta_4BdB_4(t)],$$

$$LW(x)=\theta(1+x)^{\theta-1}[\lambda-dT-(a-k)I-uV-cB-g\mu e^{-\mu\tau}\int_{t-\tau}^{t}e^{-\mu s}B(s)V(s)ds$$
$$-(q-g)BV]+\frac{\theta(\theta-1)}{2}(1+x)^{\theta-2}(\delta_1^2T^2+\delta_2^2I^2+\delta_3^2V^2+\delta_4^2B^2).$$

When $a>k$ and $q>g$, we define
$A=\min\{d,(a-k),u,c,(q-g)\}$, and $\delta^*=\max\{\delta_1,\delta_2,\delta_3,\delta_4\}$.
Similar to the proof given in [18, 19, 21], we further have

$$\ln((1+x(t))^{\theta})\leq(\frac{1}{\theta}+v)\ln t \tag{6}$$

where $0<v<1-\frac{1}{\theta}$ and $\theta>1$.

From inequality (6), we have $\limsup\limits_{t\to\infty}\frac{x(t)}{t} \leq \limsup\limits_{t\to\infty}\frac{t^{\frac{1}{\theta}+v}}{t} \leq 0$. Therefore,

$$\limsup_{t\to\infty}\frac{x(t)}{t} = \lim_{t\to\infty}\frac{T(t)+I(t)+V(t)+B(t)+ge^{-\mu t}\int_{t-\tau}^{t} e^{\mu s}B(s)V(s)ds}{t} \leq 0\,a.s.$$

Because of the positivity of $(T(t), I(t), V(t), B(t))$, we have

$$\lim_{t\to\infty}\frac{T(t)+I(t)+V(t)+B(t)+ge^{-\mu t}\int_{t-\tau}^{t} e^{\mu s}B(s)V(s)ds}{t} = 0\,a.s..$$

Therefore, the Lemma 1 is proved.

Lemma 2. $\lim\limits_{t\to\infty}\frac{\int_0^t T(s)dB_1(s)}{t} = 0$, $\lim\limits_{t\to\infty}\frac{\int_0^t I(s)dB_2(s)}{t} = 0$, $\lim\limits_{t\to\infty}\frac{\int_0^t V(s)dB_3(s)}{t} = 0$, and $\lim\limits_{t\to\infty}\frac{\int_0^t B(s)dB_4(s)}{t} = 0\,a.s.$

Lemma 2 can be proved in a similar way as given in [16, 18, 19]. The proof is omitted.

Theorem 2. If $\tilde{R}_0 = \frac{\beta\frac{\lambda}{d}+\frac{1}{2}\delta_2^2+\frac{1}{2}\delta_3^2+\frac{1}{2}\delta_4^2}{\underline{A}} < 1$, where $\underline{A} = \min\{a-k, u, c\}$, then we have

$$\lim_{t\to\infty}\langle T\rangle = \frac{\lambda}{d}\,a.s.,\ \lim_{t\to\infty}\langle I\rangle = 0\,a.s.,\ \lim_{t\to\infty}\langle V\rangle = 0\,a.s.,\ \lim_{t\to\infty}\langle B\rangle = 0\,a.s.,\ \lim_{t\to\infty}\langle B, V\rangle = 0\,a.s.$$

Proof. Define the following operation

$$\langle x(t)\rangle = \frac{1}{t}\int_0^t x(s)ds,\quad \langle x(t), y(t)\rangle = \frac{1}{t}\int_0^t x(s)y(s)ds.$$

Applying Itô's formula to $T+I+V+B+ge^{-\mu\tau}\int_{t-\tau}^{t} B(s)V(s)ds$, we obtain

$$\begin{aligned} &d(T+I+V+B+ge^{-\mu\tau}\int_{t-\tau}^{t} B(s)V(s)ds) \\ &= \lambda - dT - aI + kI - uV - qBV - cB + ge^{-\mu\tau}BV + \delta_1 TdB_1(t) \\ &+ \delta_2 IdB_2(t) + \delta_3 VdB_3(t) + \delta_4 BdB_4(t). \end{aligned} \tag{7}$$

Integrating both sides of (7) from 0 to t, we have

$$\begin{aligned} &\frac{T+I+V+B+ge^{-\mu\tau}\int_{t-\tau}^{t} BVds}{t} - \frac{T(0)+I(0)+V(0)+B(0)+ge^{-\mu\tau}\int_{-\tau}^{0} BVds}{t} \\ &= \lambda - d\langle T\rangle - (a-k)\langle I\rangle - u\langle V\rangle - c\langle B\rangle - (q-ge^{-\mu\tau})\langle B, V\rangle + \frac{\delta_1}{t}\int_0^t TdB_1(s) \\ &+ \frac{\delta_2}{t}\int_0^t IdB_2(s) + \frac{\delta_3}{t}\int_0^t VdB_3(s) + \frac{\delta_4}{t}\int_0^t BdB_4(s). \end{aligned} \tag{8}$$

Let

$$\varphi(t)=\frac{T+I+V+B+ge^{-\mu\tau}\int_{t-\tau}^{t}BVds}{t}-\frac{T(0)+I(0)+V(0)+B(0)+ge^{-\mu\tau}\int_{-\tau}^{0}BVds}{t}$$
$$-\frac{\delta_1}{t}\int_0^t TdB_1(s)-\frac{\delta_2}{t}\int_0^t IdB_2(s)-\frac{\delta_3}{t}\int_0^t VdB_3(s)-\frac{\delta_4}{t}\int_0^t BdB_4(s).$$

Obviously, $\lim_{t\to\infty}\varphi(t)=0\,a.s.$

From Eq. (8), we obtain

$$\langle T\rangle=\frac{\lambda}{d}-\frac{(a-k)}{d}\langle I\rangle-\frac{u}{d}\langle V\rangle-\frac{c}{d}\langle B\rangle-\frac{(q-ge^{-\mu\tau})}{d}\langle B,V\rangle-\frac{\lambda}{d}\varphi(t).$$

On the other hand, applying Itô's formula to $\ln(I+V+B)$, we have

$$d\ln(I+V+B)=\frac{1}{I+V+B}[\beta TV-aI+kI-uV-qBV+ge^{-\mu\tau}B(t-\tau)V(t-\tau)$$
$$-cB+\frac{1}{2}\delta_2^2 I+\frac{1}{2}\delta_3^2 V+\frac{1}{2}\delta_4^2 B]+\frac{\delta_2 IdB_2+\delta_3 VdB_3+\delta_4 BdB_4}{I+V+B}. \tag{9}$$

Define $\underline{A}=\min\{a-k,u,c\}$. Because of the positivity of $(T(t),I(t),V(t),B(t))$, we get

$$d\ln(I+V+B)\le\beta T-\underline{A}-\frac{qBV}{I+V+B}+\frac{ge^{-\mu\tau}B(t-\tau)V(t-\tau)}{I+V+B}+\frac{1}{2}\delta_2^2$$
$$+\frac{1}{2}\delta_3^2+\frac{1}{2}\delta_4^2+\delta_2 dB_2+\delta_3 dB_3+\delta_4 dB_4. \tag{10}$$

Integrating both sides of (10) from 0 to t, we have

$$\frac{\ln(I+V+B)-\ln(I(0)+V(0)+B(0))}{t}$$
$$\le\beta\langle T\rangle-\underline{A}-\frac{1}{t}\int_0^t\left(\frac{qBV}{I+V+B}-\frac{ge^{-\mu\tau}B(t-\tau)V(t-\tau)}{I+V+B}\right)ds+\frac{1}{2}\delta_2^2+\frac{1}{2}\delta_3^2+\frac{1}{2}\delta_4^2$$
$$+\frac{\delta_2 B_2+\delta_3 B_3+\delta_4 B_4}{t},$$
$$\le\beta\frac{\lambda}{d}-\frac{\lambda}{d}\varphi(t)-\underline{A}+\frac{1}{2}\delta_2^2+\frac{1}{2}\delta_3^2+\frac{1}{2}\delta_4^2+\frac{\delta_2 B_2+\delta_3 B_3+\delta_4 B_4}{t}.$$

Therefore, we have

$$\frac{\ln(I+V+B)}{t} \leq \underline{A}(\tilde{R}_0 - 1) - \frac{\lambda}{d}\varphi(t) + \frac{\delta_2 B_2 + \delta_3 B_3 + \delta_4 B_4}{t} + \frac{\ln(I(0)+V(0)+B(0))}{t},$$

where $\tilde{R}_0 = \frac{\beta\frac{\lambda}{d} + \frac{1}{2}\delta_2^2 + \frac{1}{2}\delta_3^2 + \frac{1}{2}\delta_4^2}{\underline{A}}$.

Based on the Lemma 3 of [18, 19],

we have $\lim_{t\to\infty} \frac{\delta_2 B_2 + \delta_3 B_3 + \delta_4 B_4}{t} = 0\, a.s.$ and $\limsup_{t\to\infty} \frac{\ln(I+V+B)}{t} \leq \underline{A}(\tilde{R}_0 - 1)\, a.s.$ When $\tilde{R}_0 < 1$, $\limsup_{t\to\infty} \frac{\ln(I+V+B)}{t} < 0\, a.s.$, which implies $\lim_{t\to\infty}(I+V+B) = 0\, a.s.$ $\lim_{t\to\infty} I(t) = 0\, a.s.$, $\lim_{t\to\infty} V(t) = 0\, a.s.$, and $\lim_{t\to\infty} B(t) = 0\, a.s.$

Because

$$\langle T\rangle = \frac{\lambda}{d} - \frac{(a-k)}{d}\langle I\rangle - \frac{u}{d}\langle V\rangle - \frac{c}{d}\langle B\rangle - \frac{(q - ge^{-\mu\tau})}{d}\langle B, V\rangle - \frac{\lambda}{d}\varphi(t),$$

$\lim_{t\to\infty} I(t) = 0\, a.s.$, $\lim_{t\to\infty} V(t) = 0\, a.s.$, $\lim_{t\to\infty} B(t) = 0\, a.s.$, we further have $\lim_{t\to\infty}\langle I\rangle = 0$, $\lim_{t\to\infty}\langle V\rangle = 0$, $\lim_{t\to\infty}\langle B\rangle = 0$, $\lim_{t\to\infty}\langle B, V\rangle = 0$ and $\lim_{t\to\infty}\langle T\rangle = \frac{\lambda}{d}$ $a.s.$

4 Persistence

In this section, the persistence of the disease is studied. We will prove when the threshold $\tilde{R}_0 > 1$ and some conditions are satisfied, there exists a stationary distribution.

Theorem 2. If $\tilde{R}_0 > 1$, and $\beta\frac{\lambda}{d} + \frac{3}{2}\delta_*^2 + k > a + u + c$, where $\delta_* = \min\{\delta_1, \delta_2, \delta_3\}$, $\underline{A} = \min\{a-k, u, c\}$, $\bar{A} = \max\{a-k, u, c\}$, we have

$$\limsup_{t\to\infty}\langle(I+V+B)\rangle \leq \frac{\beta\lambda - \underline{A}d + \frac{1}{2}\delta_2^2 d + \frac{1}{2}\delta_3^2 d + \frac{1}{2}\delta_4^2 d}{\underline{A}} \quad a.s,$$

$$\liminf_{t\to\infty}\langle T\rangle \geq \frac{\lambda}{d} - \frac{\beta\lambda - \underline{A}d + \frac{1}{2}\delta_2^2 d + \frac{1}{2}\delta_3^2 d + \frac{1}{2}\delta_4^2 d}{d} \quad a.s,$$

$$\liminf_{t\to\infty}\langle(I+V+B)\rangle \geq \frac{\beta\lambda - ad + kd - ud - cd + \frac{3}{2}\delta_*^2 d}{\bar{A}} \quad a.s,$$

$$\limsup_{t\to\infty}\langle T\rangle \leq \frac{\lambda}{d} - \frac{\beta\lambda - ad + kd - ud - cd + \frac{3}{2}\delta_*^2 d}{d} \quad a.s..$$

Proof. Integrating both sides of (10) from 0 to t, we have

$$\ln(I+V+B) \leq (\beta\frac{\lambda}{d} - \underline{A} + \frac{1}{2}\delta_2^2 + \frac{1}{2}\delta_3^2 + \frac{1}{2}\delta_4^2)t - \frac{(a-k)}{d}\int_0^t Tds - \frac{u}{d}\int_0^t Vds - \frac{c}{d}\int_0^t Bds$$
$$- \frac{\lambda}{d}\varphi(t)t + \delta_2 B_2 + \delta_3 B_3 + \delta_4 B_4.$$

We further have

$$\ln(I+V+B) \leq (\beta\frac{\lambda}{d} - \underline{A} + \frac{1}{2}\delta_2^2 + \frac{1}{2}\delta_3^2 + \frac{1}{2}\delta_4^2)t - \frac{\underline{A}}{d}\int_0^t (I+V+B)ds$$
$$- \frac{\lambda}{d}\varphi(t)t + \delta_2 B_2 + \delta_3 B_3 + \delta_4 B_4.$$

When $\tilde{R}_0 > 1$, by the Lemma 5.2 of [16], we have

$$\limsup_{t\to\infty} \frac{1}{t}\int_0^t (I+V+B)ds = \limsup_{t\to\infty} \langle (I+V+B)\rangle \leq \frac{\beta\lambda - \underline{A}d + \frac{1}{2}\delta_2^2 d + \frac{1}{2}\delta_3^2 d + \frac{1}{2}\delta_4^2 d}{\underline{A}} \quad a.s$$

Given $\langle T\rangle = \frac{\lambda}{d} - \frac{(a-k)}{d}\langle I\rangle - \frac{u}{d}\langle V\rangle - \frac{c}{d}\langle B\rangle - \frac{(q-ge^{-\mu\tau})}{d}\langle B, V\rangle - \frac{\lambda}{d}\varphi(t)$,
we obtain

$$\liminf_{t\to\infty} \langle T\rangle \geq \frac{\lambda}{d} - \frac{\underline{A}}{d}\langle (I+V+B)\rangle = \frac{\lambda}{d} - \frac{\beta\lambda - \underline{A}d + \frac{1}{2}\delta_2^2 d + \frac{1}{2}\delta_3^2 d + \frac{1}{2}\delta_4^2 d}{d} \quad a.s.$$

Defining $\delta_* = \min\{\delta_1, \delta_2, \delta_3\}$, we have

$$d\ln(I+V+B) = \frac{1}{I+V+B}[\beta TV - aI + kI - uV - qBV + ge^{-\mu\tau}B(t-\tau)V(t-\tau) - cB$$
$$+ \frac{1}{2}\delta_2^2 I + \frac{1}{2}\delta_3^2 V + \frac{1}{2}\delta_4^2 B] + \frac{\delta_2 IdB_2 + \delta_3 VdB_3 + \delta_4 BdB_4}{I+V+B}.$$

Integrating both sides of the equation above from 0 to t, we have

$$\ln(I+V+B) \geq (\beta\frac{\lambda}{d} - a + k - u - c + \frac{3}{2}\delta_*^2)t - \frac{(a-k)}{d}\int_0^t Tds - \frac{u}{d}\int_0^t Vds - \frac{c}{d}\int_0^t Bds$$
$$- \frac{\lambda}{d}\varphi(t)t + \delta_2 B_2 + \delta_3 B_3 + \delta_4 B_4.$$

Defining $\bar{A} = \max\{a-k, u, c\}$, we have

$$\ln(I+V+B) \geq (\beta\frac{\lambda}{d} - a + k - u - c + \frac{3}{2}\delta_*^2)t - \frac{\bar{A}}{d}\int_0^t (I+V+B)ds - \frac{\lambda}{d}\varphi(t)t$$
$$+ \delta_2 B_2 + \delta_3 B_3 + \delta_4 B_4.$$

When

$$\tilde{R}_0 > 1 \text{ and } \beta\frac{\lambda}{d} + k + \frac{3}{2}\delta_*^2 > a + u + c \tag{11}$$

By using the Lemma 5.1 of [16], we have
$\liminf_{t\to\infty} \frac{1}{t}\int_0^t (I+V+B)ds = \liminf_{t\to\infty} \langle (I+V+B) \rangle \geq \frac{\beta\lambda - ad + kd - ud - cd + \frac{3}{2}\delta_*^2 d}{A}$ $a.s$,
$\limsup_{t\to\infty} \langle T \rangle \leq \frac{\lambda}{d} - \frac{\beta\lambda - ad + kd - ud - cd + \frac{3}{2}\delta_*^2 d}{d}$ $a.s.$ Thus, the proof is completed.

5 Numerical Simulations and Discussions

In this section, we will numerically illustrate the theoretical results obtained above, and discuss the effect of stochastic disturbances on the virus dynamics.

5.1 $\tilde{R}_0 < 1$ with Different Intensities of Noises

For the following simulations, we choose the values of parameters for system (1) as follows: $\lambda = 1.5026$, $\beta = 0.401$, $d = 2.0561$, $a = 6.5051$, $k = 3.7525$, $u = 0.3818$, $q = 3.8046$, $g = 0.9129$, $\mu = 0.7$, $c = 0.3752$, and $\tau = 0.8$. The initial values are: $T(0) = 1$, $I(0) = 0.1$, $V(\xi) = 0.1$, and $B(\xi) = 0.1$ when $\xi \in [-\tau, 0]$. When the intensities of the white noises are set to $\delta_1 = \delta_2 = \delta_3 = \delta_4 = 0.01$, the dynamics of the stochastic system and the corresponding deterministic system are shown in Fig. 1.

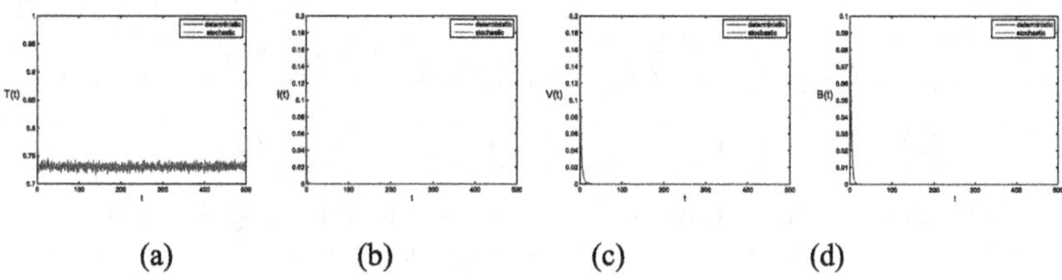

(a) (b) (c) (d)

Fig. 1. The dynamics of the stochastic system and the corresponding deterministic system, when $\tilde{R}_0 < 1$ with $\delta_1 = \delta_2 = \delta_3 = \delta_4 = 0.01$.

From Fig. 1, we can see when $\tilde{R}_0 < 1$ with $\delta_1 = \delta_2 = \delta_3 = \delta_4 = 0.01$, the dynamics of $I(t)$, $V(t)$ and $B(t)$ for the stochastic system is similar to deterministic system (see Fig. 1(b)–(d)), while $T(t)$ has small fluctuations around the value $\lim_{t\to\infty} \langle T \rangle = 0.7308$ (see Fig. 1 (d)). When $\tilde{R}_0 < 1$ with $\delta_i = 0.2$, $I(t)$, $V(t)$ and $B(t)$ also have small fluctuations, while $T(t)$ has large fluctuations. However, no matter whether the disturbance intensity is small or large, $T(t)$, $I(t)$, $V(t)$ and $B(t)$ all fluctuate around the free virus equilibria of the corresponding deterministic system.

5.2 Condition (11) Is Satisfied with Different Intensities of Noises

The parameter values for system (1) as follows: excepting $\lambda = 9.8026$, $d = 1.0561$, other parameters and the initial values are same to Fig. 1. When the intensities of white noises are $\delta_1 = \delta_2 = \delta_3 = \delta_4 = 0.01$, condition (11) is satisfied. In this case, Fig. 2. gives the dynamics of the stochastic system and the corresponding deterministic system.

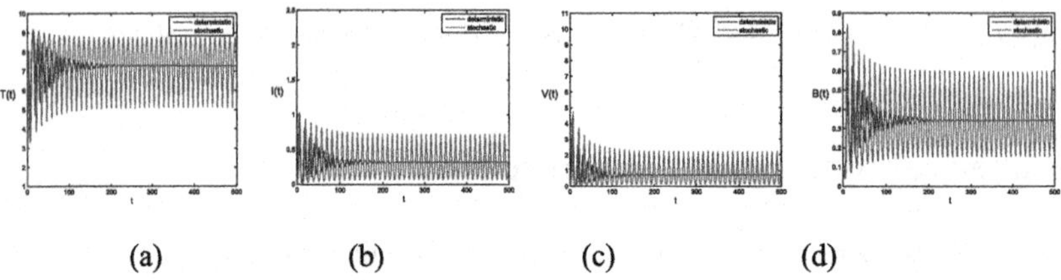

(a) (b) (c) (d)

Fig. 2. The dynamics of the stochastic system and the corresponding deterministic system, when condition (11) is satisfied with $\delta_1 = \delta_2 = \delta_3 = \delta_4 = 0.01$.

When the intensities of the white noises are set to $\delta_1 = \delta_2 = \delta_3 = \delta_4 = 0.1$, and condition (11) is satisfied. Figure 3. gives the dynamics of the stochastic system and the corresponding deterministic system.

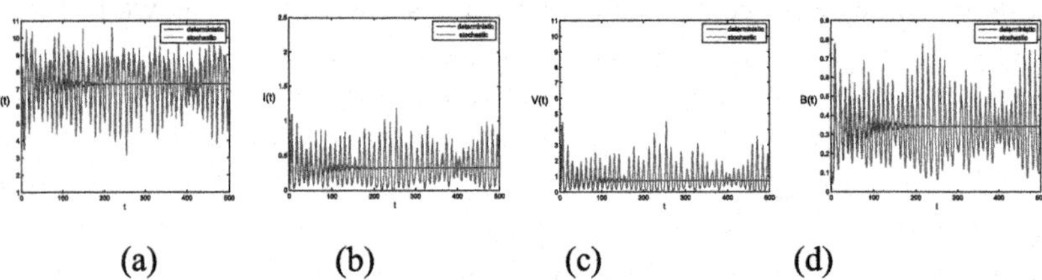

(a) (b) (c) (d)

Fig. 3. The dynamics of the stochastic system and the corresponding deterministic system, when condition (11) is satisfied with $\delta_1 = \delta_2 = \delta_3 = \delta_4 = 0.1$.

When condition (11) is satisfied with a small disturbance intensity (e.g., 0.01), the stochastic system will produce approximate periodic solutions; however, the deterministic system has asymptotically stable solutions (see Fig. 2(a)–(d)). When the disturbance intensity becomes large (e.g., 0.1), the periodic solutions will be destroyed, and the solutions become unstable (see Fig. 3(a)–(d)).

5.3 Condition (12) Is Satisfied with Stochastic Disturbances

Excepting $\lambda = 9.8026$, $\beta = 0.601$, $d = 2.0561$, parameters and the initial values are same to Fig. 1. When the intensities of the white noises set to $\delta_1 = \delta_2 = \delta_3 = \delta_4 = 0.2$, we have

$$R_0 > 1, \ \tilde{R}_0 > 1, \ \text{and} \ \beta\frac{\lambda}{d} + k + \frac{3}{2}\delta_*^2 < a + u + c. \tag{12}$$

Figure 4 shows when condition (12) is satisfied, the infected state will exist in the deterministic system, but in the stochastic system, the disease will become extinct due to the influence of white noises.

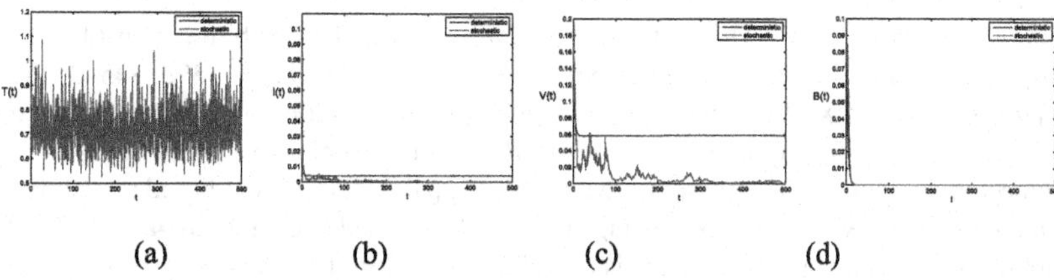

(a) (b) (c) (d)

Fig. 4. The dynamics of the stochastic system and the corresponding deterministic system, when condition (12) is satisfied with $\delta_1 = \delta_2 = \delta_3 = \delta_4 = 0.2$.

6 Conclusion

In this paper, a stochastic virus infection model with delayed immune response is proposed. We then verify that there is a unique global positive solution with any positive initial value. We also establish the sufficient conditions for extinction and persistence of the model. By simulation, we find that if $\tilde{R}_0 < 1$, no matter whether the disturbance intensity is small or large, the disease won't persist. When condition (11) is satisfied, small disturbance intensity can result in approximate periodic solutions; however, the solution of deterministic system is asymptotically stable. The interesting result is that when condition (12) is satisfied, in the stochastic system, the disease will become extinct due to the influence of white noises. However, in the deterministic system, the infected state is persistent. Hence, this may be an alternative way to consider the impact of the disturbance to control the evolution of virus infection.

Acknowledgments. We thank the support of the NSFC (51675124, 61273226) and Science and Technology Program of Guangzhou, China (201804010246)

References

1. Elaiw, A.M., AlShamrani, N.H.: Global stability of humoral immune virus dynamics models with nonlinear infection rate and removal. Nonlinear Anal. Real World Appl. **26**, 161–190 (2015)
2. Min, L.Q., Su, Y.M., Kuang, Y.: Mathematical analysis of a basic virus infection model with application to HBV infection. Rocky Mt. J. Math. **38**, 1573–1584 (2008)
3. Manna, K., Chakrabarty, S.P.: Chronic hepatitis B infection and HBV DNA-containing capsids: Modeling and analysis. Commun. Nonlinear Sci. Numer. Simul. **22**, 383–395 (2015)
4. Chen, Y.M., Zou, S.F., Yang, J.Y.: Global analysis of an SIR epidemic model with infection age and saturated incidence. Nonlinear Anal. Real World Appl. **30**, 16–31 (2016)
5. Su, Y.M., Sun, D.S., Zhao, L.: Global analysis of a humoral and cellular immune virus dynamics model with the Beddington–DeAngelis incidence rate. Math. Methods Appl. Sci. **38**, 2984–2993 (2015)
6. Song, X.Y., Wang, S.L., Dong, J.: Stability properties and Hopf bifurcation of a delayed viral infection model with lytic immune response. J. Math. Anal. Appl. **373**, 345–355 (2011)
7. Wang, T.L., Hu, Z.X., Liao, F.C.: Stability and Hopf bifurcation for a virus infection model with delayed humoral immune response. J. Math. Anal. Appl. **411**, 63–74 (2014)
8. Chiacchio, F., Pennisi, M., Russo, G., Motta, S., Pappalardo, F.: Agent-based modeling of the immune system: NetLogo, a promising framework. Biomed. Res. Int. **2014**, 907171 (2014)
9. Monica, C., Pitchaimani, M.: Analysis of stability and Hopf bifurcation for HIV-1 dynamics with PI and three intracellular delays. Nonlinear Anal. Real World Appl. **27**, 55–69 (2016)
10. Meng, X.Z., Zhao, S.N., Feng, T., Zhang, T.H.: Dynamics of a novel nonlinear stochastic SIS epidemic model with double epidemic hypothesis. J. Math. Anal. Appl. **433**, 227–242 (2016)
11. Jovanovic, M., Krstic, M.: Stochastically perturbed vector-borne disease models with direct transmission. Appl. Math. Model. **36**, 5214–5228 (2012)
12. Yu, J.J., Jiang, D.Q., Shi, N.Z.: Global stability of two-group SIR model with random perturbation. J. Math. Anal. Appl. **360**, 235–244 (2009)
13. Liu, Q., Chen, Q.M.: Dynamics of a stochastic SIR epidemic model with saturated incidence. Appl. Math. Comput. **282**, 155–166 (2016)
14. Li, D., Cui, J.A., Liu, M., Liu, S.Q.: The evolutionary dynamics of stochastic epidemic model with nonlinear incidence rate. Bull. Math. Biol. https://doi.org/10.1007/s11538-015-0101-9
15. Chen, C., Kang, Y.M.: The asymptotic behavior of a stochastic vaccination model with backward bifurcation. Appl. Math. Model. **40**, 1–18 (2016)
16. Ji, C.Y., Jiang, D.Q.: Threshold behaviour of a stochastic SIR model. Appl. Math. Model. **38**, 5067–5079 (2014)
17. Chen, G.T., Li, T.C.: Stability of stochastic of delayed SIR model. Stoch. Dyn. **9**, 231–252 (2009)
18. Liu, Q., Chen, Q.M., Jiang, D.Q.: The threshold of a stochastic delayed SIR epidemic model with temporary immune. Phys. A **450**, 115–125 (2016)
19. Appleby, J.A.D., Mao, X.R.: Stochastic stabilisation of functional differential equations. Syst. Control Lett. **54**, 1069–1081 (2005)
20. Has'minskij, R.Z.: Stochastic Stability of Differential Equations, Sijthoof & Noordhoof, Aplohen aan der Rijn, The Nederlands (1980)
21. Mao, X.: Stochastic Differential Equations and Applications. Horwood Publishing, Chichester (1997)

An Approach for Glaucoma Detection Based on the Features Representation in Radon Domain

Beiji Zou[1], Qilin Chen[1], Rongchang Zhao[1(✉)], Pingbo Ouyang[1,2], Chengzhang Zhu[1], and Xuanchu Duan[2]

[1] School of Information Science and Engineering, Central South University, Changsha 410083, China
zhaorc100@163.com
[2] The Second Xiangya Hospital of Central South University, Changsha 410011, China

Abstract. Glaucoma is a chronic and irreversible eye disease that leads to the structural changes of the Optic Nerve Head (ONH). In clinical practice, ONH assessment is one of the most significant measurements for glaucoma detection. However, the structural changes of ONH reveals complex mixture of visual patterns that are challenging to be represented. In this paper, a novel features representation approach in Radon domain is proposed to capture these complex patterns. In our method, fundus images are projected into Radon domain with Radon Transform (RT) in which the spatial radial variations of ONH are converted to a discrete signal for constraint optimization, feature enhancement and dimensionality reduction. Subsequently, the Discrete Wavelet Transform (DWT) is adopted to obtain subtle differences and quantize them. The experiments show that our approach achieves excellent detection results on RIMONE-r2 dataset with the accuracy and Area Under the Curve (AUC) of receiver operating characteristic curve at 86.154% and 0.906 respectively, much better than other algorithms. The results demonstrate that the proposed method can be used as an effective tool for glaucoma detection in the mass screening of fundus images.

Keywords: Computer-aided diagnosis · Glaucoma detection
Radon transform

1 Introduction

Glaucoma is a chronic and highly blinding eye disease, in which the vision is permanently damaged. It is predicted that the number of people suffered from glaucoma will break through 80 million by 2020 [1]. Generally, glaucoma detection is mainly based on the analysis of intraocular pressure, visual field test and optic nerve head assessment. However, the intraocular pressure measurement provides low accuracy in glaucoma detection and the visual field examination requires the special equipment usually presented in the hospital. Only ONH assessment can be accomplished by

D.-S. Huang et al. (Eds.): ICIC 2018, LNCS 10955, pp. 259–264, 2018.
https://doi.org/10.1007/978-3-319-95933-7_32

automatic algorithms based on the analysis of fundus images. Therefore, ONH assessment emerges as a preferred modality for large-scale glaucoma detection.

In fundus images, the ONH is also called the Optic Disk (OD), which can be divided into two distinct zones, namely, a central bright zone called the Optic Cup (OC) and a peripheral region called the neuroretinal rim. The structural changes of ONH caused by glaucoma are: the enlargement of the cup, neuroretinal rim loss and the changes of Peripapillary Atrophy (PPA). A series of strategies have been put forward for glaucoma detection based on these pathological features, including optic disc and cup segmentation [2], Cup to Disk Ratio (CDR) calculation [3], neuroretinal rim detection [4] and PPA [5]. Since all these measurements are usually focus on one aspect of glaucoma and heavily depend on the precise segmentation methods, effectively capturing the more comprehensive features to promote the glaucoma detection is our main interest.

In this paper, we propose a novel features representation in Radon domain for glaucoma detection, which avoids segmentation and outperforms previous methods on two major performance indicators, i.e., accuracy of classification and AUC. Our method adopt Radon transform and discrete wavelet transform for feature representation, which combines their respective advantages for capturing the complex pattern of glaucoma. Then Principal Component Analysis (PCA) is performed on these features to reduce the dimensions. Support Vector Machine (SVM) is finally developed for automatic detection on the challenging and public database (RIMONE-r2 and Drishti-GS) using 10-fold cross validation method. The progress are shown in Fig. 1.

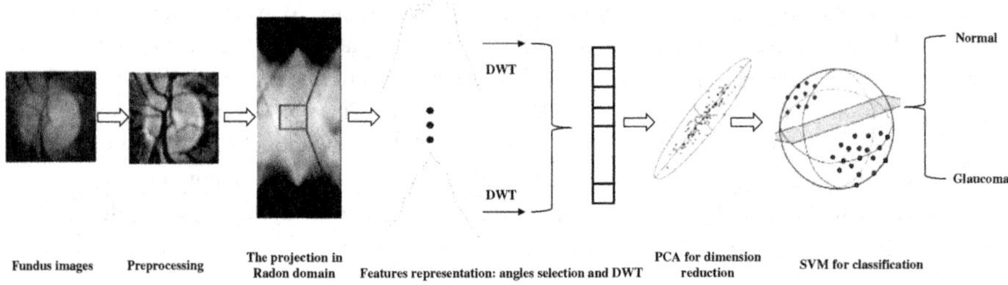

Fig. 1. The block diagram of the proposed method

2 Proposed Methodology

Since it is difficult to define an effective patterns to describe the radial variation of ONH in Cartesian coordinates, in this paper, we introduce a Radon transformation for characterizing the glaucoma better. The Radon transform provides a pixel-wise representation of the original image in the Radon domain, which has the following properties:

(1) Constraints Optimization: In the original fundus image, the CDR is calculated by the ratio of Vertical Cup Diameter (VCD) to Vertical Disc Diameter (VDD). The loss of neuroretinal rim is not consistent in different directions. And the location

of PPA occurs more frequently in the temporal region, as shown in Fig. 2(a). All of them reveal an obvious trend of radial variation and 2D spatial distribution relationship. Radon transform transfers these radial variation and spatial relationship to a 1D relationship, where the main differences are concentrated in the middle part, as shown in Fig. 2(b).

(2) Equivalent Features Enhancement: In order to improve the performance of glaucoma detection, most of existing methods adopt the combination of the CDR, the neuroretinal rim loss and the blood vessels to achieve the comprehensive representation of glaucoma. The result of Radon transform inherently integrates these radial variation, thus the features enhancement for glaucoma detection can be done during the Radon transformation.

(3) Dimensionality Reduction: Radon transform is a projection algorithm of the image, which can reduce the original data dimension and effectively avoid information loss. For an image size of 300 × 400 pixels, the feature dimension is only about 500 after Radon transform.

The specific process is as follows.

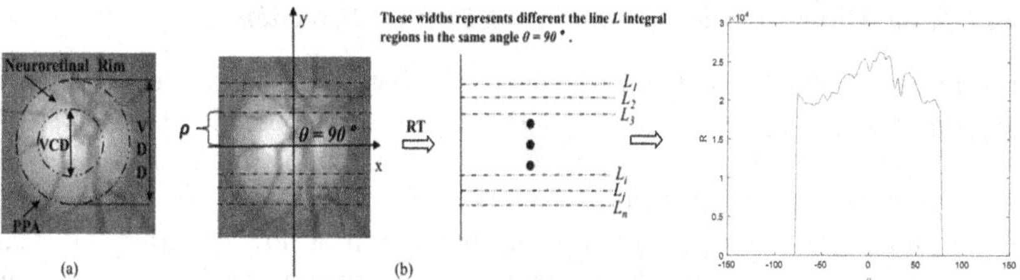

Fig. 2. (a) The main measurements of ONH (b) 90° Radon transform diagram

2.1 Preprocessing: Illumination Correction and Contrast Enhancement

The purpose of preprocessing is to eliminate the non-uniform illumination and contrast enhancement. Color fundus images are converted into gray-scale image and followed by Contrast Limited Adaptive Histogram Equalization (CLAHE). CLAHE is more suitable to improve the local contrast of the image and get more image details than common histogram equalization.

2.2 Radon Transform for Feature Projection

For a given image I, Radon transform is described as follows:

$$R(\rho, \theta) = \iint_D f(x, y)\delta(\rho - x\cos\theta - y\sin\theta)dxdy \qquad (1)$$
$$s.t. \quad \delta(t) = \begin{cases} 1, & t = 0 \\ 0, & t \neq 0 \end{cases}$$

where D stands for the whole image, $f(x, y)$ means the intensity of the image at a point (x, y), ρ is the distance of the straight line to the origin, θ describes as the angle from the horizontal, the characteristic function δ is the Kronecker delta function, limiting the projection along the straight line. In this paper, the combination of nine angles ($\theta = 20°, 40°, 60°, 80°, 100°, 120°, 140°, 160°, 180°$) yields the best performances.

2.3 Features Quantization with 1D DWT

The result of Radon transform is a discrete signal and varies heavily due to the different angles, the individual differences and the complex visual pattern of glaucoma. Wavelet transform inherits and develops the idea of short-time Fourier transform, which is an ideal tool for signal analysis and processing. Its main characteristic is the localization analysis in time-frequency domain, and realizes the multi-scale refinement of signal by stretching and translation. So in this paper, 1D DWT is employed to convert the differences of Radon transform into coefficients and quantitatively represent them accurately. The Biorthogonal (bior1.1) wavelet decomposition at level 1 is chosen for research by extensive experiments.

2.4 PCA for Dimension Reduction and Classifier Selection

The advantage of principal component analysis is that it will find a linear subspace in which projection data retains the variation as much as possible. For this reason, extracted features are fed to PCA for dimensionality reduction. In this paper, we adopt the 94 percent of cumulative contributions proportion.

Radon transform and wavelet quantization form a low dimensional feature descriptor, and there are only 455 samples in experimental database. SVM can be superior to these situations. SVM is a kind of excellent small sample learning method due to the theory of maximum margin hyper-plane, which can not only grasp the key samples but also has good robustness. In addition, SVM adopts kernel function to project the non-linear data into a high dimensional space to separate them effectively, which is suitable for the complex characteristics of glaucoma. Other classifiers like Random Forest (RF) are also taken into account for comparison.

3 Experimental Results

3.1 Database and Evaluation Criteria

The performance of the proposed method are evaluated through AUC of Receiver Operation Characteristic curve (ROC) in publicly available RIMONE-r2 and Drishti-GS. ROC curve shows the tradeoff between sensitivity and specificity, defined as sensitivity = TP/(TP + FN), specificity = TN/(FP + TN), where TP and TN are the number of true positives and true negatives, respectively, and FP and FN are the number of false positives and false negatives, respectively. The RIMONE-r2 is consisted of 200 glaucoma and 255 normal fundus images. The Drishti-GS contains 101 fundus images, and 70 images are glaucoma cases.

3.2 Results

Table 1 is the accuracy and AUC obtained from the experiments. For the RIMONE–r2, SVM receives the highest performance with the accuracy and AUC of 86.154% and 0.906, respectively. But for the Drishti-GS, SVM performance degrades due to the uneven distribution of data. In addition, RF does not work well under the circumstances for random sampling and inability to data projection, which are not suitable for glaucoma detection.

Table 1. Performance of the proposed method with ten-fold validation

Database	Classifier	Accuracy(%)	AUC
RIMONE-r2	SVM	86.154	0.906
RIMONE-r2	RF	77.100	0.769
Drishti-GS	SVM	74.000	0.732
Drishti-GS	RF	78.000	0.733

Table 2 shows the comparison between the proposed method and the others. In our work, we are able to detect the normal and glaucoma with the accuracy of 86.154% and the AUC of 0.906, which much better than existing methods. The current study indicates that the proposed method possesses better performance for glaucoma detection.

Table 2. Compared with other algorithms

Authors	Database	Images	Accuracy(%)	AUC
Bock et al. [6]	private	575	80.000	0.880
Cheng et al. [7]	private	650	-	0.830
Maheshwari et al. [8]	RIMONE-r2	455	81.320	-
Ours	RIMONE-r2	455	86.154	0.906

3.3 Effectiveness and Robustness: The Performance of Different Angles and Dimensions in Radon Domain

Feature dimensions after Radon transform are different because of various image resolutions. In this paper, we adopt 6, 9, 18 angles and 600, 800, 1000 dimensions for the experiments. The accuracy of all results is above 80%, which reveals that Radon transform can effectively capture and represent the characteristics of glaucoma. In this paper, the combination (9 even angles and 690 dimensions) is superior to the representation of original images and yields the best accuracy.

We also verify the classification performance with the different kernel functions of SVM such as linear, polynomial and radial basis functions. For the powerful capability of data processing and nonlinear projection, RBF ($\sigma = 6.727$) shows the best classification performance. In addition, after using CLAHE, the accuracy increases from 85.274% to 86.154%, and AUC is approximately the same. Features dimension drops from 6210 to 244 with PCA at 94% of cumulative contributions proportion, and accuracy also increases 0.88%, which greatly removes redundant information and improves the performance of method.

4 Conclusion

In this paper, we present a novel features representation method based on Radon domain for glaucoma detection, which is able to capture the discriminative features that better characterize the complex visual patterns related to glaucoma. Because Radon transform can effectively reduce the dimension of color fundus images, the whole method is very quick and efficient, providing favorable conditions for its application in the mass glaucoma detection. Moreover, the performance of AUC is up to 0.906, which makes it serve as an efficient tool for glaucoma diagnosis in clinical practice.

Our future work will focus on the improvement of our method to enable it to deal with various of image resolutions better for clinical applications. Moreover, we will explore the more exquisite angles combination of Radon transform, and try to establish a theory for between-class and within-class distance between normal and glaucoma images in Radon domain, which aim to promote the performance of glaucoma detection.

Acknowledgement. This work is supported by the National Natural Science of Foundation of China (No. 61573380) and Hunan Natural Science Foundation of China (No. 2017JJ3411).

References

1. Chen, X., Xu, Y., Yan, S., Wong, D.W.K., Wong, T.Y., Liu, J.: Automatic feature learning for glaucoma detection based on deep learning. In: Navab, N., Hornegger, J., Wells, W.M., Frangi, A.F. (eds.) MICCAI 2015. LNCS, vol. 9351, pp. 669–677. Springer, Cham (2015). https://doi.org/10.1007/978-3-319-24574-4_80
2. Fu, H., Cheng, J., Xu, Y., Wong, D.W.K., Jiang, L., Cao, X.: Joint optic disc and cup segmentation based on multi-label deep network and polar transformation. IEEE Trans. Med. Imaging **PP**(99), 1 (2018)
3. Cheng, J., Zhang, Z., Tao, D., Wong, D., Jiang, L., Baskaran, M.: Similarity regularized sparse group lasso for cup to disc ratio computation. Biomed. Opt. Express **8**(8), 3763 (2017)
4. Harizman, N., Oliveira, C., Chiang, A., Tello, C., Marmor, M., Ritch, R.: The ISNT rule and differentiation of normal from glaucomatous eyes. Arch. Ophthalmol. **124**(11), 1579–1583 (2006)
5. Teng, C.C., De Moraes, C.G., Prata, T.S., Tello, C., Ritch, R., Liebmann, J.M.: Beta-Zone parapapillary atrophy and the velocity of glaucoma progression. Ophthalmology **117**(5), 909–915 (2010)
6. Bock, R., Nyul, M.L.G., Hornegger, J., Michelson, G.: Glaucoma risk index: automated glaucoma detection from color fundus images. Med. Image Anal. **14**(3), 471–481 (2010)
7. Cheng, J., Yin, F., Wong, D.W.K., Tao, D., Jiang, L.: Sparse dissimilarity-constrained coding for glaucoma screening. IEEE Trans. Biomed. Eng. **62**(5), 1395–1403 (2015)
8. Maheshwari, S., Pachori, R.B., Acharya, U.R.: Automated diagnosis of glaucoma using empirical wavelet transform and correntropy features extracted from fundus images. IEEE J. Biomed. Health Inf. **21**(3), 803–813 (2017)

Screening of Pathological Gene in Breast Cancer Based on Logistic Regression

Yun Zhao[1] and Xu-Qing Tang[1,2(✉)]

[1] School of Science, Jiangnan University, Wuxi, China
txq5139@jiangnan.edu.cn
[2] Wuxi Engineering Research Center for Biocomputing, Wuxi, China

Abstract. Breast cancer has become the focus of the pathological gene screening research. In this paper, logistic regression and multiple hypothesis testing are used to screen the pathological gene based on the existing breast cancer genetic data. Then referencing the confidence level, the p-value of logistic regression is used to screen the pathological gene initially. Furthermore, by considering the Type I error, the multiple hypothesis testing is used to make the result accurate. In addition, SVM is used to test the reliability of this paper's methods. In order to illustrate the feasibility of this method, each gene which screened by this method is tested and verified by the literature of breast cancer.

Keywords: Logistic regression · BH testing · Bonferroni testing
Gene screening

1 Introduction

Human Genome Project [1] and gene sequencing technology have provided new directions and massive data for disease-related research, especially for cancer research. With the development of gene sequencing, data mining and other related technologies, breast cancer has gotten a huge help in the treatment. The common measures of the pathological gene screening are almost based on gene sequencing technology [2, 3] which rely on biological experiments. Different from the common measures, the problem that pathological gene screening is converted into a binary classification problem.

Logistic regression model [4] is a common method to solve the binary classification problem, and has a widely used in some researches, such as the health problems [5], analysis data and diagnose of breast cancer [6]. So we set up a logistic regression model according to the gene sequence data, and use the p-value of hypothesis testing [7] to screen pathological gene initially. Then, based on the preliminary screening, we use multiple hypothesis testing to obtain pathological gene.

According to the gene dataset of breast cancer, we verify the actual pathological connection between the pathogenic genes screened and the various stages of breast cancer, and verify the feasibility of the methods. As for the problem of pathological genes screening on breast cancer, the model has some limitations, because breast cancer is a complex disease. However, logistic regression, multiple hypothesis testing does not depend on a large number of gene sequencing experiments.

D.-S. Huang et al. (Eds.): ICIC 2018, LNCS 10955, pp. 265–271, 2018.
https://doi.org/10.1007/978-3-319-95933-7_33

2 Binary Classification Model Solving Method

According to the structure of dataset, we set two labels to solve the binary classification problem, and set up logistic regression model. Then the confidence level of the p-value, Bonferroni multiple hypothesis testing are used to screen the pathological genes.

2.1 Logistic Regression [4]

According to the definition of classification problem, we consider dichotomous dependent variables Y and N dimensional independent variables X satisfy the Eq. (1).

$$\text{logit}(Y^*) = \beta_0 + \sum_{i=1}^{N} \beta_i x_i \ , \ (x_i \in X) \tag{1}$$

In Eq. (1), variable Y is a dichotomous dependent variable. Because of the dichotomous dependent variable Y, the probability of the $Y = 1$ is more important than the value of the Y. From the nature of probability, we know that P(Y = 1) must be controlled within [0,1]. The value of the right side of the equation maybe over [0,1]. We use the odds of the $Y = 1$ to define the $Y^* = P(Y = 1)/(1-P(Y = 1))$. So we use the ln($Y^*$) to replace the Y^*, and record as logit (Y^*) = ln(Y^*).

In logistic regression, we use p-value of the current regression coefficients to determine whether A and B are related. Moreover, the p-value is smaller, the significance is higher, and the possibility is bigger.

2.2 Bonferroni Multiple Hypothetical Testing Based on FWER [8]

During the using of logistic regression, one of the main basis is the hypothetical testing of the model, and it is possible to make mistakes of Type I error [8]. So when the hypothetical testing is used continuously, the possibility of the Type I error is rose.

We define the total error rate of the whole models as FWER (Family-wise Error Rate), and the equation is FWER $= \Pr(V \geqslant 1)$, with V is the sum of the Type I error; Pr means probability.

Bonferroni multiple hypothesis testing can be constructed on the basis of FWER. The test principle is as follows: If the continuous inspections of the dataset have the n independent assumptions, and the significant level of every hypothetical test is α. So the significant level which is used for test is account for the significant level of one hypothetical test's $\alpha' = \alpha/n$.

2.3 Screening Accuracy Based on Score-Weight

In this paper's methods, the p-value is calculated for each probe to screen the genes, and the p-value is transformed into p' by using $-\log_{10}(\cdot)$. Therefore, p' can be used to define the score and evaluate the screening results. The equation of score is $S_i = p_i' = -\log_{10}(p_i)$, with $i = 1, 2, \cdots, M$; M is the total number of selected genes.

However, the results have the possibility to make error screening. So the main function of the selected genes and their relationship with breast cancer are found out to define the scoring weights ω. When $i \in nCo$, $\omega = 0$; when $i \in iCo$, $\omega = 0.5$; when $i \in Co$, $\omega = 1$. Where nCo denotes a gene set unrelated; iCo denotes a gene set indirectly related; Co denotes a gene set directly related.

The final score of the gene corresponding to each probe can be calculated from the score S and the score weight ω. In addition, we need the following Hypothesis:

Hypothesis: The genes screened by this method are the correct classification, which is directly related to breast cancer.

Through the Hypothesis, we can calculate the theoretical final score of the gene corresponding to each probe and quantify the evaluation difference by comparing the actual final score with the theoretical final score and define the screening accuracy rate ***Ac*** which is the effect of algorithm screening. ***Ac*** expression is shown as $\boldsymbol{AC} = \sum \omega S / \sum \omega' S$. Where ω' is the assumed weight, due to Hypothesis, $\omega' = 1$.

3 Experiment and Discussion

GEO (Gene Expression Omnibus) is the largest and most comprehensive public gene expression data resource nowadays. This paper refers to the Series GSE15852 [9] of breast cancer in GEO. Series GSE15852 rely on the GPL96 [HG-U133A] Affymetrix Human Genome U133A Array which has 22215 probes. Every probe has 86 samples, which has 43 abnormal samples and 43 abnormal samples (URL: https://www.ncbi.nlm.nih.gov/geo/query/acc.cgi?acc=GSE15852).

3.1 Result of Logistic Regression Model

According to Eq. (1), we set up the logistic regression model which has the disease sample ($Y = 1$) and the normal sample ($Y = 0$) on the Series GSE15852 dataset, and obtain 4308 significant probes under the significant level $\alpha = 0.05$. Because these results are shown densely, we transform the data interval is converted from (0, 1] to (0, 6.7) by using $p' = -\log_{10}(p)$, which the threshold α is changed to $\alpha' = -\log_{10}(\alpha)$, and the data visualization is optimized. 4308 probes are screened from 22,215 probes, and cover 19.39% of all the probes. This result of the confidence level is coarse and we can't find the key genes through the result. Therefore, we need to introduce the more accurate testing on the results.

3.2 Result of Bonferroni Multiple Hypothesis Testing

Based on the Subsect. 3.1, we make 22,215 hypotheses when the logistic regression model is set up, and each hypotheses has the possible to make mistakes of Type I error. According to the 22215 hypotheses and the significance level $\alpha = 0.05$, the Bonferroni multiple hypothesis testing shows that the significant level of each hypothesis after calibration is $\alpha_1 = \alpha/n = 0.05/22215$, so $\alpha'_1 = -\log_{10}(\alpha_1) = 5.64$.

Based on the Bonferroni multiple hypothesis testing, we screened 22 from 22,215 probes. According to the relationship between probe and Gene in Affymetrix genechip U133A Genomic Chip, 20 genes that correspond 22 probes are obtained, and they is respectively 201839_s_at-EPCAM, 202286_s_at-TACSTD2, 203548_s_at-LPL, 203980_at-FABP4, 204151_x_at/216594_x_at-AKR1C1, 204388_s_at-MAOA, 205382_s_at-CFD, 218087_s_at-SORBS1, 209699_x_at-AKR1C2, 209763_at-CHRDL1, 209771_x_at-CD24, 210201_x_at-BIN1, 212419_at-ZCCHC24, 213524_s_at-G0S2, 215695_s_at-GYG2, 207092_at-LEP, 207175_at-ADIPOQ, 209555_s_at-CD36, 205913_at-PLIN1, 43427_at/49452_at-ACACB.

Furthermore, we can analyze the major functions of the 20 genes based on NCBI Gene database. Through the analysis of the major functions, we sort out the genes that have a direct relationship with breast cancer, which these gene names and supporting literatures is respectively EPCAM [10], TACSTD2 [11], LPL [12], FABP4 [13], AKR1C1/AKR1C2 [14], CFD [15], LEP [16], ADIPOQ [17], CD36 [18], CD24 [19], BIN1 [20].

In addition, some of the 20 genes have an indirect relationship with breast cancer, they are G0S2 [21] and ACACB [22].

Among the 20 genes corresponding to 22 probes screened by Bonferroni multiple hypothesis testing, except for the genes related to breast cancer, there is no relationship between the other genes and breast cancer. So we can get the score and weight of 22 probes, and the result is shown in Table 1.

Table 1. The score and weight of probes

REF_ID	Score	Weight	REF_ID	Score	Weight
201839_s_at	6.243987264	1	209699_x_at	6.214244943	1
202286_s_at	5.968669823	1	209763_at	5.664361091	0
203548_s_at	6.12787437	1	209771_x_at	5.944417334	1
203980_at	6.228107604	1	210201_x_at	6.091431078	1
204151_x_at	5.911241838	1	212419_at	5.718824907	0
204388_s_at	5.720079881	0	213524_s_at	5.967609555	0.5
205382_s_at	5.817147434	1	215695_s_at	6.014857614	0
205913_at	5.863511649	0	216594_x_at	6.027867144	1
207092_at	6.213961063	1	218087_s_at	5.936872923	0
207175_at	5.693033469	1	43427_at	6.604120295	0.5
209555_s_at	6.132704573	1	49452_at	6.338055624	0.5

According to the definition of $\boldsymbol{Ac}$, we get the conclusion that the $\boldsymbol{Ac}$ of the Bonferroni multiple hypothesis testing based on FWER is 66.5%. Because there are 6 unrelated genes in the screening results and the scores of 6 genes are relatively high.

3.3 Comparison with SVM

In order to discuss the feasibility of logistic regression in pathological gene screening problem, we use SVM [23] to screen the same gene. Based on the screening results of logistic regression and the Bonferroni multiple hypothesis testing, we know that breast cancer is related to the $-\log_{10}(p)$. So we use the $-\log_{10}(p_{Co})$ which is related to breast cancer and the $-\log_{10}(p_{nCo})$ which is contrary to build a feature dataset $T = \{[-\log_{10}(p_{Co}), 1], [-\log_{10}(p_{nCo}), -1]\}$, and train SVM by T. Then, we use the result of the training to classify the 22215 probes, under the same p-value condition of logistic regression. The probes are 201839_s_at-EPCAM, 203548_s_at-LPL, 203980_at-FABP4, 207092_at-LEP, 209699_x_at-AKR1C2, 210201_x_at-BIN1, 216594_x_at-AKR1C1, 209555_s_at-CD36, 43427_at/49452_at -ACACB.

The genes screened of the results are all the results shown in Table 1. No new probes and genes are obtained based on the classification and screening of SVM, and the main results of the screening are similar. Through the comparison of the two methods, similar results are obtained, which shows that logistic regression is feasibility.

3.4 Discussion

First, we screen 4308 probes from 22215 probes by confidence level which is 19.39% of the total number of probes. So the result is not representative. Second, we screen 22 probes from 22215 probes by Bonferroni multiple hypothesis testing based on FWER. By comparing the database, we know that 22 probes correspond to 20 genes, of which 13 probes correspond to 12 genes are directly related to breast cancer, and 3 probes corresponding to 2 genes are indirectly related to breast cancer. So, the ***Ac*** of the result is 66.5% and result is representative. Last, we used SVM to verify the reliability of logistic regression.

According to the results, we know that the result of confidence level can't screen the pathological genes well, and the result of Bonferroni multiple hypothesis testing based on FWER can screen the pathological genes well and most of the probes are associated with breast cancer.

4 Conclusion

From the result in this paper, we can know the methods which are logistic regression model, Bonferroni multiple hypothesis testing based on FWER on screening pathological genes of breast cancer is very effective. Meanwhile, we don't use the specific parameters to make and solution the model. So the model in this paper to screen pathological genes can apply well in other diseases which the gene databases is known.

Acknowledgements. This work was supported by the National Natural Science Foundation of China (Grand No. 11371174 and 11271163).

References

1. Collins, F.S., Morgan, M., Patrinos, A.: The human genome project: lessons from large-scale biology. Science **300**(5617), 286 (2003)
2. Peng, X., Sun, L., Huo, H.: The effect of apelin-13 on breast MCF-7 cell proliferation and invasion via activate ERK1/2 signaling pathways. J. Northeast Normal Univ. (Nat. Sci.) **47** (3), 127–131 (2015)
3. Tang, X., Zhou, Y., Zhang, W.: Correlation between expression of TOP2A and HER2 signaling pathway in breast cancer. J. Xi‘an Jiaotong Univ. Med. Sci. **36**(4), 519–522 (2015)
4. Edition, S.: Applied logistic regression analysis. Technometrics **38**(2), 184–186 (2017)
5. Zhang, Y., Kwon, D., Pohl, K.M.: Computing group cardinality constraint solutions for logistic regression problems. Med. Image Anal. **35**, 58–69 (2016)
6. Zou, J., Liang, Q.: Progress in data mining techniques of diagnosis of breast cancer. J. Biomed. Eng. **29**(2), 375–378 (2012)
7. Jin, H., Zhou, L.: Reconsideration on hypothesis test and P value. J. Environ. Occup. Med. **34**(2), 95–98 (2017)
8. Hwang, Y.T., Lai, J.J., Ou, S.T.: Evaluations of FWER-controlling methods in multiple hypothesis testing. J. Appl. Stat. **37**(10), 1681–1694 (2010)
9. Pau Ni, I.B., Zakaria, Z., Muhammad, R., et al.: Gene expression patterns distinguish breast carcinomas from normal breast tissues: the Malaysian context. Pathol. Res. Pract. **206**(4), 223–228 (2010)
10. Spizzo, G., Went, P.S., Obrist, P., et al.: High Ep-CAM expression is associated with poor prognosis in node-positive breast cancer. Breast Cancer Res. Treat. **86**(3), 207–213 (2004)
11. Ambrogi, F., Fornili, M., Boracchi, P., et al.: Trop-2 is a determinant of breast cancer survival. PLoS ONE **9**(5), e96993 (2014)
12. Thomassen, M., Tan, Q., Kruse, T.A.: Gene expression meta-analysis identifies chromosomal regions and candidate genes involved in breast cancer metastasis. Breast Cancer Res. Treat. **113**(2), 239–249 (2009)
13. Wang, W., Yuan, P., Yu, D., et al.: A single-nucleotide polymorphism in the 3′-UTR region of the adipocyte fatty acid binding protein 4 gene is associated with prognosis of triple-negative breast cancer. Oncotarget **7**(14), 18984–18998 (2016)
14. Ji, Q., Aoyama, C., Nien, Y.D., et al.: Selective loss of AKR1C1 and AKR1C2 in breast cancer and their potential effect on progesterone signaling. Cancer Res. **64**(20), 7610 (2004)
15. Chen, L., Ye, C., Huang, Z., et al.: Differentially expressed genes and potential signaling pathway in Asian people with breast cancer by preliminary analysis of a large sample of the microarray data. J. South. Med. Univ. **34**(6), 807–812 (2014)
16. Gao, J., Zhang, J., Dong, Y.: Clinical significance of detection of serum Leptin, Insulin_like growth factor-1 and Tunor necrosis factor α in breast cancer patients. Lab. Med. **20**(1), 28–29 (2005)
17. Reddy, N.M., Kalyani, P., Kaiser, J.: Adiponectin and Leptin molecular actions and clinical significance in breast cancer. Int. J. Hematol.-Oncol. Stem Cell Res. **8**(1), 31 (2014)
18. Lai, G., Jiang, B.: Progress in the relationship between the differential expression of thrombospondin-1 and breast cancer. J. Gannan Med. Univ. **32**(3), 481–482 (2012)
19. Wang, Z., Wang, Q., Wang, Q., et al.: Prognostic significance of CD24 and CD44 in breast cancer: a meta-analysis. Int. J. Biol. Mark. **32**(1), e75–e82 (2016)
20. Ghaneie, A., Zembapalko, V., Itoh, H., et al.: Bin1 attenuation in breast cancer is correlated to nodal metastasis and reduced survival. Cancer Biol. Ther. **6**(2), 192–194 (2007)

21. Jones, J.E.C., Esler, W.P., Patel, R., et al.: Inhibition of acetyl-CoA carboxylase 1 (ACC1) and 2 (ACC2) reduces proliferation and de novo lipogenesis of EGFRvIII human glioblastoma cells. PLoS ONE **12**(1), e0169566 (2017)
22. Mukherjee, B., Mcellin, B., Camacho, C.V., et al.: EGFRvIII and DNA double-strand break repair: a molecular mechanism for radioresistance in glioblastoma. Can. Res. **69**(10), 4252–4259 (2009)
23. Nello, C., John, S.-T.: An Introduction to Support Vector Machines and Other Kernel-based Learning Methods, 1st edn. Cambridge University Press, Cambridge (2000)

Analysis of Disease Comorbidity Patterns in a Large-Scale China Population

Mengfei Guo[1], Yanan Yu[1], Tiancai Wen[2,6], Xiaoping Zhang[3], Baoyan Liu[3], Jin Zhang[4], Runshun Zhang[5], Yanning Zhang[6(✉)], and Xuezhong Zhou[1(✉)]

[1] School of Computer and Information Technology and Beijing Key Lab of Traffic Data Analysis and Mining, Beijing Jiaotong University, Beijing 100044, China
xzzhou@bjtu.edu.cn

[2] Institute of Basic Research in Clinical Medicine, China Academy of Chinese Medical Sciences, Beijing 100700, China

[3] China Academy of Chinese Medicine Sciences, Beijing 100070, China

[4] Data Center of Traditional Chinese Medicine, China Academy of Chinese Medical Sciences, Beijing 100700, China

[5] Guang'anmen Hospital, China Academy of Chinese Medical Sciences, Beijing 100053, China

[6] School of Computer Science, Northwestern Polytechnical University, Xi'an 710129, Shanxi, China
ynzhang@nwpu.edu.cn

Abstract. Background: Disease comorbidity is popular and has significant indications for disease progress and management. We aim to detect the general disease comorbidity patterns in Chinese populations using a large-scale clinical data set.

Materials and Methods: We extracted the diseases from a large-scale anonymized data set derived from 8,572,137 inpatients in 453 hospitals across China. We built a Disease Comorbidity Network (DCN) with significant disease co-occurrence and detected the topological patterns of disease comorbidity using both complex network and data mining methods.

Results: We obtained the DCN with 5702 nodes and 258,535 edges, which shows a power law distribution of the degree and weight. It indicated that there exists high heterogeneity of comorbidities for different diseases. Meanwhile, we found that the DCN is a hierarchical modular network with community structures. We further divided the network into 10 modules using community detection algorithm, which showed two types of modules exist in the DCN.

Conclusions: Our study indicates that disease comorbidity is significant and valuable to understand the disease incidences and their interactions in real-world populations, which will provide important insights for detection of the patterns of disease classification, diagnosis and prognosis.

Keywords: Disease comorbidity · Complex network · Network medicine

M. Guo, Y. Yu and T. Wen—These authors contribute equally to this work.

D.-S. Huang et al. (Eds.): ICIC 2018, LNCS 10955, pp. 272–278, 2018.
https://doi.org/10.1007/978-3-319-95933-7_34

1 Introduction

Disease comorbidity has become a major problem for chronic disease management [1]. For example, when a patient suffers from multiple diseases, the treating is particularly complicate [2] and the popular therapies with multiple drugs might cause serious side effects due to their interactions [3]. Unfortunately, the patterns and the underlying mechanisms of disease comorbidity are far from fully elucidated. Therefore, recently, it has become a hot research topic on disease comorbidity both from clinical observations and molecular network mechanisms, which in most cases is derived from the data in Europe and United States. Here, we utilized a large-scale clinical data and conducted our research across the full range of diseases in China population. We built a disease comorbidity network (DCN) with thousands of diseases and obtained the topological properties and their relationships by complex network measurements. The results have implications for the disease comorbidity patterns and would be helpful to medical research.

2 Methods

2.1 Data Source and Preprocessing

Our data sets were derived from the data center of the China Academy of Chinese Medical Sciences. We only use the clinical diagnostic information of patients to carry on our research, so our research strictly preserves the privacy of patients.

After removing of the records with missing values and inconsistent diagnosis codes, we obtained 8,572,137 high quality clinical cases. We used ICD10 (the 10th revision of the International statistical classification of diseases [4]) to represent the diagnosis and deal with them in the form of four-digit ICD10 codes for further analysis.

2.2 Data Analysis Methods

We applied the frequent pattern growth algorithm (FP-Growth) to obtain the co-occurrence disease pairs [5]. We used Relative Risk (RR) and Φ-correlation [6, 7] to measure the correlations between disease pairs. When two diseases d_i and d_j co-occur more frequently than expected by chance, we would have $RR_{ij} > 1$ and $\Phi_{ij} > 0$.

We constructed the DCN with those significant disease pairs with $RR > 1$ and $\Phi > 0$. Then we used four basic topological measurements, namely degree, betweenness centrality(BC), clustering coefficient(CC_1) and closeness centrality(CC_2) to evaluate the centrality of nodes in the network. In addition, we calculated the correlations between some topological measurements to identify the coupling and hierarchical patterns underlying the DCN.

3 Results

3.1 Basic Properties of the Disease Comorbidity Network

We constructed the DCN with diseases whose co-occurrence >5 and RR >1.0, Φ-correlation >0.0. The DCN has 5,702 nodes and 258,535 edges with average degree 90.717 (see Fig. 1.A for degree distribution) and average edge weight 12904.494 (see Fig. 1.B for weight distribution). In addition, the average path length is 2.528 and the average CC_1 is 0.629 (see Fig. 1.C for CC_1 distribution), which indicated that DCN is a highly clustering network, with the neighbors of a disease closely connected.

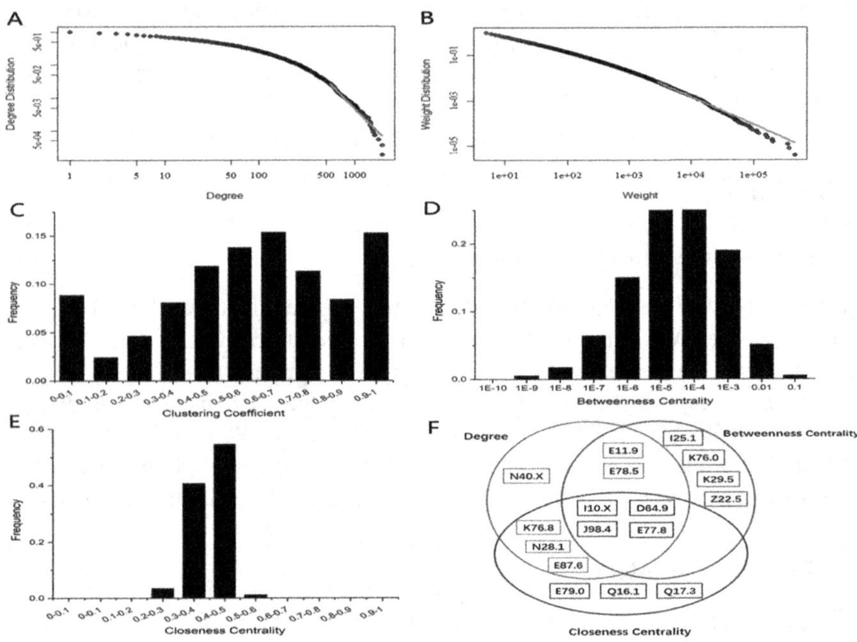

Fig. 1. Basic properties of the network A. Distribution of degree. B. Weight distribution of edges. C. Distribution of CC_1. D. Distribution of BC. E. Distribution of CC_2. F. The top 10 diseases with the highest degree, CC_2 and BC, respectively.

The power law distribution of degree and weight (Fig. 1.A and B) showed that DCN is a scale-free network [8], which means that some diseases (e.g. hypertension, atherosclerotic heart disease) have very high comorbidities in China population. We obtained the three disease lists, which are ranked as the top 10 diseases of degree, betweenness centrality and CC_1 (Fig. 1.F). It showed that hypertension, anaemia, other disorders of lung and other disorders of glycoprotein metabolism are the top 4 diseases included in all these rank lists.

3.2 Hierarchical Modular Structures of Disease Comorbidity Network

To identify the more elucidated patterns in the DCN, we calculated the correlations between several pairs of network topological measurements (Fig. 2A–F). We found that there exists negative correlation between degree and CC_1 (Pearson correlation coefficient (PCC) = −0.398199, see Fig. 2.A) in DCN, which indicated that DCN is a hierarchical modular network [9]. Furthermore, consistently, we found that there exists negative correlation between CC_1 and CC_2 (PCC = −0.1551085, see Fig. 2.B). These two results showed that in DCN, the neighbors of diseases located in the center of the network (easier to get to other nodes) have large diversity and diseases with less CC_2 tend to occur simultaneously with diseases in the same module.

Furthermore, the positive correlation between CC_2 and degree (PCC = 0.5964138, see Fig. 2.C) indicates that the data is reliable, because both the degree and close centrality reflect the centrality of a node.

The BC can reflect the diversity of disease connotation. There exists negative correlation between BC and CC_1 (PCC = −0.1814519, see Fig. 2.F), which shows that neighbors of the disease with large CC_1 are not connected closely as a hub node. For example, as a hub node in DCN, hypertension has high BC and degree (BC = 0.092988, degree = 1926), which reflects its diverse mechanisms and comorbid phenotypes. Also, the relationships between its neighbors are sparse (CC_1 = 0.05050615.), which indicate that there exist potential subtypes of hypertension disorder. For disorders of choroid (H31.8), its BC is 0. It has much fewer neighbors (degree = 12) but is more closely related to them than hypertension (CC_1 = 1). That is to say, the number of the comorbidity diseases of the disease is few, but their relationship between their comorbid diseases is strong.

3.3 Disease Comorbidity Communities

To identify the disease comorbidity groups from the DCN, we applied BGLL community detection method [10] to find the communities, which resulted in 10 communities with denser comorbidity links between the diseases other than random expectations (see Fig. 2.G–H). There are both homogeneous and heterogeneous comorbidity diseases in the same communities. Meanwhile, there exist branching relationships between categories. That is, a large number of disease comorbidities appear between different categories of diseases. For example, in the above mentioned communities, the most common diseases (157 diseases) belong to the eye related disease category, accounting for 74.8% and also contains 53(25.2%) diseases from other categories, most of which are caused by cataracts (H25–H26). Ocular comorbidity diseases are common in people with cataracts in real-world clinical settings [11]. This would be insightful for the refinement of disease classification.

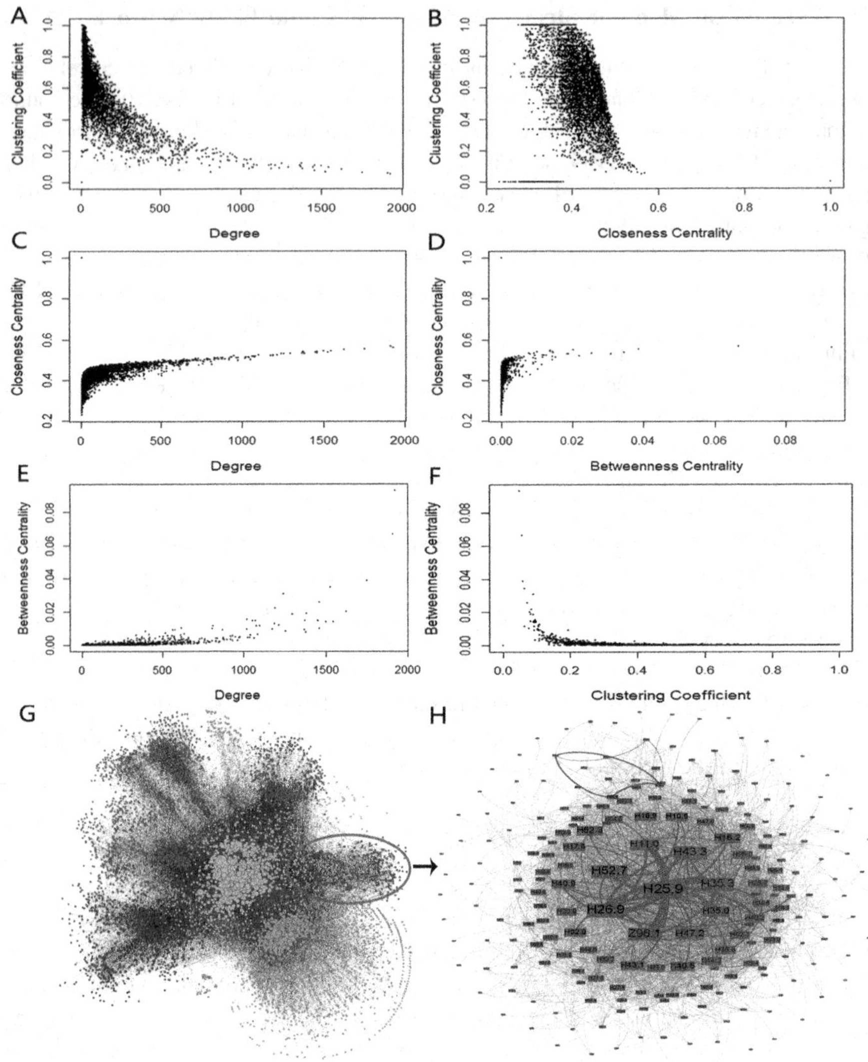

Fig. 2. The relationship between topological properties and the network structure. A. Degree and CC_1 B. CC_2 and CC_1 C. Degree and CC_2 D. BC and CC_2 E. Degree and BC F. CC_1 and BC G. Modules in the network H. One of the modules in the network

4 Discussion

Following the several pioneer studies on network disease comorbidities and their underlying molecular mechanisms [6, 7], it has been a hot research topic in both clinical and network medicine field [12], which were mainly derived from the clinical data in Europe and United States. Due to the influence from environment factors, ethnicity and social factors to disease patterns, it is important to investigate the disease comorbidity patterns in large-scale populations in China [13, 14].

Our research is carried out across 5,702 diseases in 22 categories and 8,572,137 patients of all ages. Therefore, the range of our study is more extensive in both data and study than before, which has great significance for the study of disease comorbidities. We focus on the DCN and analyzed the correlation of diseases in the network. Also, we pay attention to the topological characteristics of the network and the relationship between them, and extend it to medical meanings.

The major limitation of our research is that the recording of diseases in clinical data would prone to incomplete diagnoses because clinical practitioners would tend to record the diagnosis they primarily treated rather than all the diseases of patients.

Acknowledgement. This work is partially supported by the National Natural Science Foundation of China (Nos. 61105055 and 81230086), the National Basic Research Program of China (No. 2014CB542903), the Special Programs of Traditional Chinese Medicine (Nos. 201407001, JDZX2015168, JDZX2015171 and JDZX2015170), National Key R&D Project (2017YFC1 703506) and the National Key Technology R&D Programs (Nos. 2013BAI02B01 and 2013BAI 13B04).

References

1. Hu, J.X., Thomas, C.E., Brunak, S.: Network biology concepts in complex disease comorbidities. Nat. Rev. Genet. **17**(10), 615–629 (2016)
2. Gijsen, R., Hoeymans, N., Schellevis, F.G., et al.: Causes and consequences of comorbidity: a review. J. Clin. Epidemiol. **54**(7), 661–674 (2001)
3. Von Lueder, T.G., Atar, D.: Comorbidities and polypharmacy. Heart Fail. Clin. **10**, 367–372 (2014)
4. World Health Organization: ICD-10: international statistical classification of diseases and related health problems 10th rev. World Health Organ. **56**(3), 65 (1992)
5. Han, J., Pei, J., Yin, Y.: Mining frequent patterns without candidate generation. ACM SIGMOD Rec. **29**(2), 1–12 (2000)
6. Hidalgo, C.A., Blumm, N., Barabási, A., et al.: A dynamic network approach for the study of human phenotypes. PLoS Comput. Biol. **5**(4), e1000353 (2009)
7. Park, J., Lee, D., Christakis, N.A., et al.: The impact of cellular networks on disease comorbidity. Mol. Syst. Biol. **5**(1), 262 (2009)
8. Newman, M.E.J.: The structure and function of complex networks. SIAM Rev. **45**, 167–256 (2003)
9. Ravasz, E., Barabási, A.L.: Hierarchical organization in complex networks. Phys. Rev. E: Stat. Nonlin. Soft Matter Phys. **67**(2), 026112 (2003)
10. Chaturvedi, P., Dhara, M., Arora, D.: Community detection in complex network via BGLL algorithm. Int. J. Comput. Appl. **48**(1), 32–42 (2012)
11. Pham, T.Q., Wang, J.J., Rochtchina, E., et al.: Systemic and ocular comorbidity of cataract surgical patients in a western Sydney public hospital. Clin. Exp. Ophthalmol. **32**(4), 383–387 (2004)
12. Chen, Y., Xu, R.: Network analysis of human disease comorbidity patterns based on large-scale data mining. In: Basu, M., Pan, Y., Wang, J. (eds.) ISBRA 2014. LNCS, vol. 8492, pp. 243–254. Springer, Cham (2014). https://doi.org/10.1007/978-3-319-08171-7_22

13. Liu, J., Ma, J., Wang, J., et al.: Comorbidity analysis according to sex and age in hypertension patients in China. Int. J. Med. Sci. **13**(2), 99–107 (2016)
14. Chen, H., Zhang, Y., Wu, D., et al.: Comorbidity in adult patients hospitalized with type 2 diabetes in northeast China: an analysis of hospital discharge data from 2002 to 2013. Biomed. Res. Int. **2016**(11), 1–9 (2016)

Stacking Multiple Molecular Fingerprints for Improving Ligand-Based Virtual Screening

Yusuke Matsuyama[1,3] and Takashi Ishida[1,2,3(✉)]

[1] Department of Computer Science, Graduate School of Information Science and Engineering, Tokyo Institute of Technology, Tokyo, Japan
matsuyama@cb.cs.titech.ac.jp, ishida@c.titech.ac.jp
[2] Department of Computer Science, School of Computing, Tokyo Institute of Technology, Tokyo, Japan
[3] Education Academy of Computational Life Sciences, Tokyo Institute of Technology, Tokyo, Japan

Abstract. Currently, most of machine learning based virtual screening methods use a molecular fingerprint. There are numerous fingerprints proposed for various aims, and it is known that the best fingerprint is different for each target, and it is difficult to select the most suitable fingerprint. To overcome this problem, we propose a new technique for the use of multiple fingerprints for drug activity prediction. The method implies that each molecular fingerprint extracts different features of a compound, and prediction based on a different fingerprint returns different results. We applied the ensemble learning technique to integrate predictions based on multiple fingerprints. The method builds prediction models based on 8 different major molecular fingerprints, and then integrates multiple prediction results from those models. As a result of performance evaluation, the proposed method increased the predicted performance as compared to the prediction models involving a single molecular fingerprint.

Keywords: Ligand-based virtual screening · Molecular fingerprint
Ensemble learning

1 Background

Drug activity prediction based on the machine learning method is useful for ligand-based virtual screening. These days, many machine learning algorithms have been proposed and used for drug activity prediction, but all the modern machine learning models require a vector representation of fixed length as input. In image recognition, this is not a problem because we can easily generate a fixed length vector from images of any size by scaling them. In drug activity prediction, we also have to encode chemical-compound information into a fixed-length vector for application of a machine learning algorithm. Because the number of atoms and the numbers of bonds vary among compounds, they cannot be input into the model as it is unless accompanied by some characteristic quantization technique. A number of studies have been conducted for many years to convert the structure and chemical properties of a compound into a feature vector of a certain length [1–3]. The feature vectors are often called a molecular

D.-S. Huang et al. (Eds.): ICIC 2018, LNCS 10955, pp. 279–288, 2018.
https://doi.org/10.1007/978-3-319-95933-7_35

fingerprint of a compound, and the aim of generating such a feature vector for a compound was originally database search. Nonetheless, most of machine learning-based drug activity prediction methods have employed this vectorization technique for generating the input for the machine learning models.

Although many molecular fingerprints have been proposed, they can be roughly subdivided into five categories [4]. The first category is a method for determining the key structure and the address of the vector corresponding to the structure in advance by a dictionary method and set it to 1 if the compound has the structure, or 0 if not. MACCS key [2] is a representative approach in this category. The second category is a method focusing on the bonding of a compound. This category is based on the length and type of the bonds the compound has, the type of both ends. The atompair fingerprint [5] is a representative approach in this category. In the third category, local chemical structure within r atoms is first replaced with a unique numerical value by some method such as Morgan method [1]. Then, bits at the address corresponding to the number obtained through the hash function are set. The extended connectivity fingerprint (ECFP) [3] is a representative of the category. The fourth category is a method for listing candidates of structures in the compound that are considered to be important for binding to the target protein (pharmacophores) and for encoding this information. 2-Dimensional Pharmacophore Fingerprint (2DPF) [6] is a representative technique in this category.

Currently, there are many fingerprints with different features, and thus it is difficult to select appropriate fingerprint for constructing a machine learning model. As a result, in most of the existing drug activity prediction studies, the authors did not specify the reason why they had selected the specific fingerprint as the input for machine learning. For tackling this problem, there are several studies that benchmark fingerprints [4, 7]. Riniker et al. reported that ECFP4 offers the highest precision on average according to the database search by compound similarity based on the fingerprint, while the differences in prediction accuracy by difference of feature vectorization method were small. In addition, the best fingerprint is different for each target and task, and there is no feature vectorization method that always shows good performance on any target proteins. It can be said that each fingerprint is focused on different points of the compound. Thus, this observation suggests that it is important to understand the differences in the nature of each fingerprint, and its advantages and disadvantages. Nevertheless, we still do not have sufficient knowledge about fingerprints, and selection of the appropriate fingerprint for each specific problem is still a serious problem.

What will happen if we simultaneously use multiple fingerprints at hand that are not good for prediction? If a fingerprint used with another one can complement information in a compound from the other fingerprint, we can expect an improvement of prediction accuracy in comparison with a single fingerprint. Of course, when the fingerprints contain exactly the same information about a compound, then the improvement of prediction accuracy cannot be expected. Nonetheless, a combination of multiple fingerprints has already shown practically good performances. For instance, Ma et al., used a combination of the "atompair" fingerprint and "donor-acceptor pair" [8] fingerprints for their QSAR prediction [9]. Most of the previous studies have used a combination of two different fingerprints, but we can try more than two fingerprints for this purpose. One of the reason why previous researches did not use more than two fingerprints is that simple concatenation of multiple fingerprints may cause a problem. This is because the size of the input vector becomes too large and it will decrease prediction accuracy owing to collinearity and the high dimensional curse [10].

In this study, we proposed a prediction method based on stacking [11], which is one of the ensemble methods, for effective use of multiple fingerprints. We evaluated the accuracy of the proposed method and compared it with that of prediction based on a single fingerprint and accuracy of the prediction involving simple concatenation of multiple molecular fingerprints. As a result, the proposed method showed a significant performance improvement.

Table 1. An assay list used in this work.

PubChem AID	Target or goal	#Active	#Inactive
1915	Group A Streptokinase expression inhibition, streptkinase	2219	1017
2358	Protein phosphatase 1	1006	934
463213	Identification of tim10 yeast inhibitors	4141	3235
463215	Identify small molecule inhibitors of tim10 yeast	2941	1695
488912	Sentrin-specific protease 8	2491	3705
488915	Sentrin-specific protease 6	3568	2628
488917	Sentrin-specific protease 7	4283	1913
488918	Sentrin-specific proteases	3691	2505
492992	Two pole domain potassium channel	2094	2820
504607	Mdm2/MdmX interaction	4830	1412
624504	Mitochondrial permeability transition pore	3944	1090
651739	Inhibition of trypanosoma cruzi	4051	1324
651744	NIH, 3T3 toxicity	3102	2306
652065	Binding r(CAG) RNA repeats	2966	1287

2 Materials and Methods

2.1 The Dataset

First of all, we describe the datasets and molecular fingerprints used in this work. Assay datasets for 14 targets were downloaded from the PubChem database [12]. We used the same datasets that were used in the study by Dahl et al. [13]. The data include a chemical structural formula of a compound and the activity label whether it is active toward the targets or not. The assay targets and the numbers of active and inactive compounds in the datasets are shown in Table 1. As an example, some of the compounds belonging to AID1915 assay data are shown in Fig. 1. We used eight popular molecular fingerprints shown in Table 2. All of them were generated by RDKit [14]. The MACCS key used in this study has only 166 dimensions whose structure is open to the public.

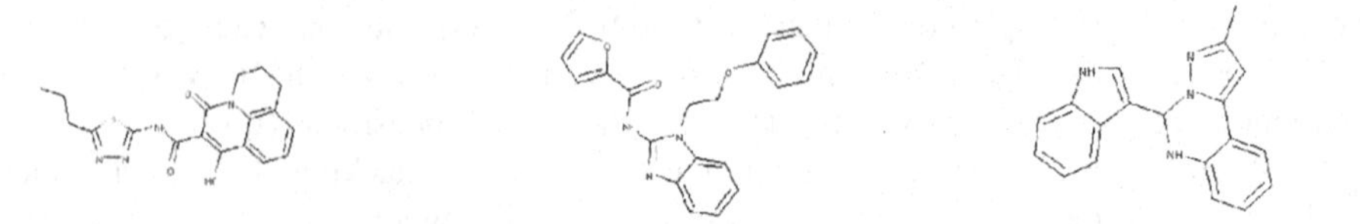

CCCc1nnc(NC(=O)c2c(O)c3cccc4c3n(c2=O)CCC4)s1
Positive

O=C(Nc1nc2ccccc2n1CCOc1ccccc1)c1ccco1
Positive

Cc1cc2n(n1)C(c1c[nH]c3ccccc13)Nc1ccccc1-2
Negative

Fig. 1. Example of compounds in AID1915 assay data.

Table 2. The list of molecular fingerprints and descriptor that were used in this work. Abbreviation of each fingerprint is shown in parentheses.

Name	Length
2 dimensional pharmacophore fingerprint (2dpf) [6]	3348
Atompair (atompair)	2048
Daylight (daylight)	2048
Extended connectivity fingerprint, radius = 4 (ecfp4) [3]	2048
Extended connectivity fingerprint, radius = 6 (ecfp6)	2048
MACCS keys (maccskey)	166
Topological Torsion (topo_tor)	2048
Chemical/Structure descriptors (chem_desc)	193

2.2 The Prediction Method Based on Stacking

Even in the recent research on drug activity prediction, only one or two molecular fingerprints have been used as the input of a prediction model. By concatenating a large number of molecular fingerprints horizontally, we can easily construct an input vector based on multiple fingerprints. Such an input vector contains more information than a single molecular fingerprint does, and may increases the prediction performance. On the other hand, the dimensionality of such a vector is too large for the size of datasets, and it often causes the curse of high dimensionality. It may be harmful to use many fingerprints at once because the dimensionality of input exceeds 10000. Therefore, we propose a method to incorporate a large number of molecular fingerprints into a prediction model reasonably by means of the stacking technique.

Stacking generalization [11] is one of the ensemble methods. The prediction results from the first-stage model are called a meta feature, and they are used as the input into the second-stage learning model. This kind of techniques are often called meta-learning methods, and such methods have been already used in the several bioinformatics studies [15, 16]. Nonetheless, such methods change the datasets and learning algorithms for each independent learner. In contrast, our method mainly changes the feature extraction or vectorization methods, i.e., the molecular fingerprint, for them. In our staking model, the first-stage learning models are constructed from different molecular fingerprints and learning algorithms with the same training data and learning hyper-parameters. Next, the prediction results of those models are collected as meta-features for second-stage.

The Fig. 2 shows the protocol for generation of a meta-feature. First, the dataset is divided into a training set and test set. The training set is subdivided into 4 parts. For each combination of learning algorithms, learning hyperparameters, and a molecular fingerprint, a meta-feature is generated as follows. A model is trained using 3 pieces of data, and then the drug activities of the remaining 1 piece of data are predicted using the trained model. The processes are repeated 4 times by changing pieces as in cross-validation, and the prediction results are concatenated to compile 1 meta-feature vector. For generating a meta-feature of test data, a new prediction model is trained with the same hyperparameters used above by means of the whole training data, and the model is then applied to the test data. The processes are repeated for all combinations of learning algorithm, learning hyperparameters, and a molecular fingerprint.

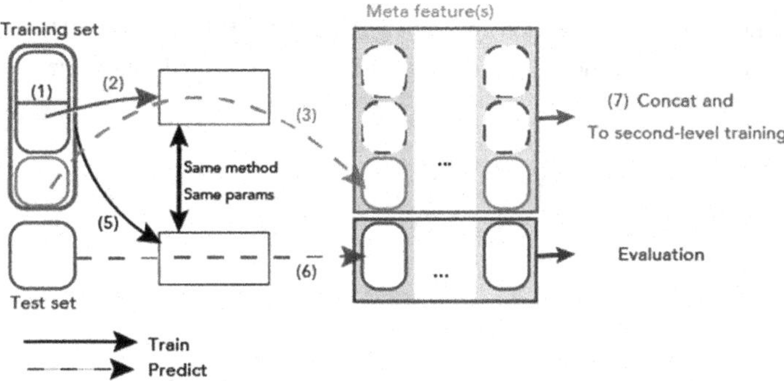

Fig. 2. An overview of meta-feature generation. Step 1: prepare learning models given certain parameters. Step 2: Training with 2/3 of training data. Step 3: Making prediction using the remaining 1/3 of the training data. Step 4: Perform steps 2–3 three times and then generate a meta feature for the entire training data. Step 5: Training model using whole training data. Step 6: Predict the test data with model of step 5, let output be meta feature for test data. Step 7: Concatenate for the meta feature obtained in steps 2–6 and make this an input to the second level training.

In this work, we used 8 fingerprints shown in Table 2, the support vector machine (SVM) [17] and random forest (RF) as the learning algorithms. We employed 20 combinations of hyperparameters for SVM, and 10 different combinations of hyperparameters for RF. Thus, the total dimensionality of the meta-feature was $240 (= (20 + 10) \times 8)$. List of all parameters used in each prediction model is shown in Table 3. Finally, a second-level prediction model was trained using the meta-features of the training data. As the learning model at this level, we compared random forest, logistic regression, and SVM with the linear kernel, and a simple averaging method that averages all meta-features. We mainly used scikit-learn which is a machine learning library of python for the implementation (Fig. 3).

Table 3. List of parameters applying grid search in each method and learning model.

Method/Level	Model	Hyperparameters for grid search
Baselines	Random forest	*num_estimators*: [50,100,150,200,250,300,350,400,450,500], *max_features*: [50,100,150,200,300,350,400]
Proposed/First	Random forest	*num_estimators*: [100,150,200,300], *max_features*: [100,150,200,300,400]
Proposed/First	SVM	*kernel*: ['linear','rbf'], *C*: [100,10,1,0.1,0.01]
Proposed/Second	Random forest	*num_estimators*: [50,100,150,200], *max_features*: [30,60,100,150]
Proposed/Second	SVM	*kernel*: ['linear'], *C*: [100,10,1,0.1,0.01,0.001]
Proposed/Second	Logistic regression	*penalty*: ['l1','l2'] *C*: [0.1,0.3,0.5,0.7,0.9]
Proposed/Second	Mean of meta feature	mean of meta feature as predicted value

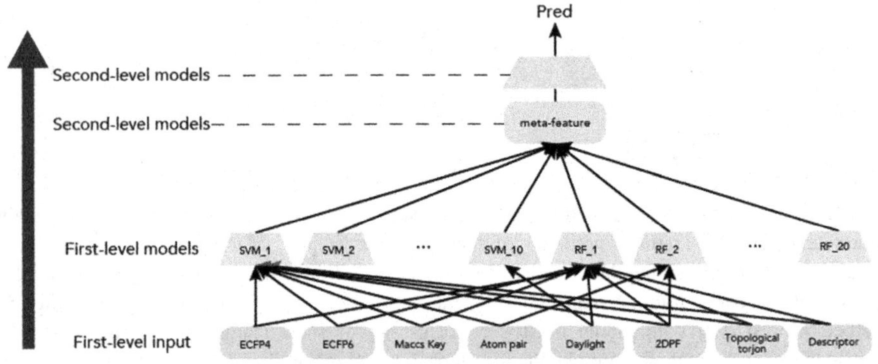

Fig. 3. An overview of stacking with a multiple fingerprint ensemble.

3 Results

For evaluating the performance of the proposed method, we performed 4 experiments by changing a part of a test set as in a 4-fold cross-validation test, and obtained the average of the results for the test data. We used the area under the receiver operating characteristics curve (AUC) as the performance metrics.

As the baseline of the prediction performance, we constructed prediction models using random forest with a single molecular fingerprint, and evaluated their performances using 4-fold cross-validation. In this evaluation, we searched optimal hyperparameters using grid search with following conditions; the number of estimators: 50, 100, 200, 300, 400, max features: 50, 100, 200, 300, 400. For each target, we selected the best model and used it as the baseline. Table 4 shows the performance of each prediction model with a single molecular fingerprint.

Table 4. The performance of the prediction model with a single fingerprint. The asterisk indicates the best performance.

PubChem AID	2dpf	atom-pair	Daylight	ecfp4	ecfp6	maccskey	topo tor	chem-_desc
1915	0.726	0.753	0.757	0.749	0.743	0.757	0.761	0.767*
2358	0.746	0.775	0.763	0.771	0.76	0.764	0.775	0.782*
463213	0.651	0.691	0.697	0.707*	0.698	0.689	0.696	0.687
463215	0.595	0.636	0.645	0.648*	0.64	0.628	0.644	0.624
488912	0.636	0.702*	0.678	0.691	0.683	0.673	0.694	0.692
488915	0.691	0.741*	0.722	0.739	0.738	0.703	0.728	0.733
488917	0.773	0.821*	0.805	0.806	0.809	0.79	0.817	0.808
488918	0.736	0.786*	0.768	0.784	0.784	0.748	0.785	0.779
492992	0.835	0.854	0.855	0.864*	0.856	0.836	0.86	0.843
504607	0.659	0.69	0.668	0.664	0.652	0.673	0.68	0.692*
624504	0.89	0.879	0.893	0.893	0.889	0.896*	0.896	0.895
651739	0.793	0.8	0.798	0.803	0.8	0.782	0.8	0.806*
651744	0.874	0.878	0.879	0.88	0.878	0.858	0.878	0.883*
652065	0.757	0.786	0.765	0.754	0.748	0.766	0.758	0.808*
Average	0.74	0.771	0.764	0.768	0.763	0.755	0.769	0.771
s.d.	0.086	0.071	0.075	0.074	0.076	0.074	0.074	0.075

For comparison, we also evaluated the prediction performance of learning models with a feature vector generated by concatenating multiple fingerprints horizontally. We used SVM and random forest as machine learning algorithms, and performed a grid search to obtain optimal hyperparameters.

3.1 Performance Evaluation

Table 5 shows the results of evaluation for 14 assay datasets. The performance of both prediction methods based on multiple fingerprints (concatenation and stacking) were better than that of the best model based on a single molecular fingerprint in all cases. In addition, in all 14 assays, the stacking model showed better prediction accuracy than the model based on simple concatenation of multiple molecular fingerprints did. To test statistical significance of the improvement by the proposed method, we applied the paired-sample t-tests. The p-value for single-molecular-fingerprint based model and the proposed method (stacking) was 1.53e−06, and that for the concatenation and the proposed method was 5.51e−06. Both p-values are sufficiently small ($p < 0.01$). The Cohen's d was 2.077 for single-molecular-fingerprint based model and the proposed method, and it also indicates that the improvement of prediction accuracy by the proposed method was significant.

Table 5. The prediction accuracy (AUC) of each method.

Pubchem AID	Single FP	Concatenation	Stacking (Averaging)
1915	0.767	0.785	0.794
2358	0.782	0.794	0.798
463213	0.707	0.709	0.718
463215	0.648	0.651	0.659
488912	0.702	0.712	0.717
488915	0.741	0.755	0.759
488917	0.821	0.832	0.840
488918	0.786	0.799	0.809
492992	0.864	0.872	0.879
504607	0.692	0.703	0.704
624504	0.896	0.913	0.918
651739	0.806	0.806	0.811
651744	0.883	0.884	0.891
652065	0.801	0.804	0.803
Avg.	0.779	0.787	0.793

3.2 Optimization of the Second Level Prediction

The proposed method showed better prediction performance even when using a simple averaging algorithm in the second-level prediction. Next, we applied a more sophisticated learning algorithm to the second-level learning for improving the prediction. As described in the Methods section, we tested random forest, logistic regression, and SVM with the linear kernel in addition to the simple averaging method. The hyper-parameters were optimized using a cross-validation test on the training set. Table 6 shows the results of performances evaluation. Logistic regression showed better performance than did the others. Random forest and SVM are more sophisticated and can often solve more complex problems as compared with logistic regression because they are nonlinear models. Nonetheless, this result is not surprising because linear models, e.g., blending and stacking, have often been employed in meta-learning studies [18]. This is because the first-level learning model maps the raw feature space to the meta-feature space, and the feature space is much simpler than the original space so that a nonlinear identification model is not required in the process.

3.3 Performance Evaluation

The proposed method needs more computing resources than single fingerprint based methods. In this study, we used a TSUBAME supercomputer, which is a cluster machine of Tokyo Institute of Technology. Each compute node of it has 2 sockets of Intel Xeon X5670. The second level training requires only a few minutes, and thus we ignore it. The first-level trainings were done with 10 parallel processes. At this time, the calculation time correlates with the number of data and the difficulty of training. For training in the 4-fold cross validation, AID463213 needs the longest calculation time

Table 6. A comparison of second-level classifiers; Random Forest (RF), Logistic Regression (LR), Linear kernel SVM (LSVM), and the mean of meta-features (Averaging). The results suggest that the linear classifier beats nonlinear classifiers by a wide margin.

Pubchem AID	RF	LR	LSVM	Averaging
1915	0.786	0.793	0.794	0.794
2358	0.782	0.797	0.800	0.798
463213	0.712	0.717	0.718	0.718
463215	0.642	0.656	0.634	0.659
488912	0.702	0.718	0.719	0.717
488915	0.746	0.760	0.759	0.759
488917	0.832	0.843	0.842	0.840
488918	0.804	0.813	0.812	0.809
492992	0.878	0.882	0.882	0.879
504607	0.678	0.707	0.644	0.704
624504	0.917	0.920	0.921	0.918
651739	0.811	0.817	0.815	0.811
651744	0.892	0.897	0.896	0.891
652065	0.803	0.816	0.814	0.803
Avg.	0.784	0.795	0.789	0.793

for the training (59,985 s.), and the median of calculation time is 27,832 s. The average calculation time is longer than 6 h. Nonetheless, the heavy computation cost of first level training is owing to a large number of independent training processes. Thus, we can accelerate the process by using more computing nodes.

4 Conclusion and Future Work

In this work, we proposed a new drug activity prediction method based on stacking for the effective use of multiple molecular fingerprints. We conducted performance evaluation tests and verified the significance of the improvement of prediction accuracy made by using multiple fingerprints and the ensemble-based method.

As future works, we are considering the following three objects of research. The first one is that, diversity of predicted compounds by ligand based prediction methods often is lower as compared with that by structure-based prediction methods such as a docking simulation. This property is often a disadvantage for practical applications. By contrast, our approach produces good variety in the predicted compounds at the first stage because the proposed method uses multiple molecular fingerprints with different features. Thus, we believe that it is possible to ensure the diversity of the predicted compounds by improving the meta learning method without degrading performance. Second, on the subject of increasing the type and amount of data to be incorporated into the learning model. In machine learning-based prediction, a data shortage or bias is often a major problem. One of the causes is that often several experimental methods and evaluation scales of compounds are often intermixed in public databases. This state of affairs means that the number of data that can be used simultaneously will decrease. In our approach, it is likely that multiple experiments and methods can be used for training without difficulty for constructing a prediction model by means of data based on a plurality of experimental information in the first stage-learning models. Finally, we also consider that it is necessary to make it public web service so that our method can be applied to user's own data set.

References

1. Morgan, H.L.: The generation of a unique machine description for chemical structures-a technique developed at chemical abstracts service. J. Chem. Doc. **5**(2), 107–113 (1965)
2. Durant, J.L., Leland, B.A., Henry, D.R., Nourse, J.G.: Reoptimization of MDL keys for use in drug discovery. J. Chem. Inf. Comput. Sci. **42**(6), 1273–1280 (2002)
3. Rogers, D., Hahn, M.: Extended-connectivity fingerprints. J. Chem. Inf. Model. **50**(5), 742–754 (2010)
4. Riniker, S., Landrum, G.A.: Open-source platform to benchmark fingerprints for ligand-based virtual screening. J. Cheminform. **5**(5), 26 (2013)
5. Carhart, R.E., Smith, D.H., Venkataraghavan, R.: Atom pairs as molecular features in structure-activity studies: definition and applications. J. Chem. Inf. Comput. Sci. **25**(2), 64–73 (1985)
6. McGregor, M.J., Muskal, S.M.: Pharmacophore fingerprinting. 2. Application to primary library design. J. Chem. Inf. Comput. Sci. **40**(1), 117–125 (1999)
7. O'Boyle, N.M., Sayle, R.A.: Comparing structural fingerprints using a literature based similarity benchmark. J. Cheminform. **8**(1), 1–14 (2016)
8. Kearsley, S.K., Sallamack, S., Fluder, E.M., Andose, J.D., Mosley, R.T., Sheridan, R.P.: Chemical similarity using physiochemical property descriptors. J. Chem. Inf. Comput. Sci. **36**(1), 118–127 (1996)
9. Ma, J., Sheridan, R.P., Liaw, A., Dahl, G.E., Svetnik, V.: Deep neural nets as a method for quantitative structure–activity relationships. J. Chem. Inf. Model. **55**(2), 263–274 (2015). PMID: 25635324
10. Bishop, C.M.: Pattern Recognition and Machine Learning. Springer, New York (2006)
11. Wolpert, D.H.: Stacked generalization. Neural Netw. **2**(505), 241–259 (1992)
12. Wang, Y., Bryant, S.H., Cheng, T., Wang, J., Gindulyte, A., Shoemaker, B.A., Thiessen, P.A., He, S., Zhang, J.: PubChem BioAssay: 2017 update. Nucleic Acids Res. (2017)
13. Dahl, G., Jaitly, N., Salakhutdinov, R.: Multi-task Neural Networks for QSAR Predictions. arXiv preprint arXiv:1406.1231, pp. 1–21 (2014)
14. Landrum, G.: RDKit: Open-source cheminformatics (2006)
15. Ishida, T., Kinoshita, K.: Prediction of disordered regions in proteins based on the meta approach. Bioinformatics **24**(11), 1344–1348 (2008)
16. Yuan, Q., Gao, J., Wu, D., Zhang, S., Mamitsuka, H., Zhu, S.: DrugE-Rank: improving drug – target interaction prediction of new candidate drugs or targets by ensemble learning to rank. Bioinformatics **32**, 18–27 (2016)
17. Steinwart, I., Christmann, A.: Support Vector Machines, 1st edn. Springer, New York (2008). https://doi.org/10.1007/978-0-387-77242-4
18. Sill, J., Takacs, G., Mackey, L., Lin, D.: Feature-Weighted Linear Stacking, pp. 1–17 (2009)

Combining mRNA, microRNA, Protein Expression Data and Driver Genes Information for Identifying Cancer-Related MicroRNAs

Jiawei Lei[1], Shu-Lin Wang[1(✉)], and Jianwen Fang[2]

[1] College of Computer Science and Electronics Engineering, Hunan University, Changsha 410082, Hunan, China
smartforesting@gmail.com

[2] Biometric Research Branch, Division of Cancer Treatment and Diagnosis, National Cancer Institute, Rockville, MD 20850, USA

Abstract. As is well-known, microRNAs (miRNAs), a short nor-coding RNA, play a vital role in important biological processes such as gene expression and transcriptional regulation. And it was reported that miRNAs have involved in the occurrence and development of various human cancer, which shows the potentiality of miRNAs in cancer treatment and diagnosis. However, it is a great challenge for the detection and prioritization of cancer-related miRNAs. In this paper, we proposed a novel approach which combines mRNA, miRNA, protein expression data by introducing dirver genes for identifying glioblastoma (GBM)-related miRNAs. And identified miRNAs were ranked by related scores. The performance of our method was evaluated by the proportion of the previously known miRNAs and the area under the receiver operating characteristic curves (AUC). A literature survey was also used to validate the detected results. A miRNA-gene regulatory module was constructed for understanding the biological function of ranked miRNAs in cancer.

Keywords: Cancer-related miRNA · Gene expression · Dirver genes

1 Introduction

MicroRNAs (miRNAs), a small non-coding RNA, approximately of 20–24 nucleotides in length, are involved in a wide variety of biological processes such as cell proliferation, development, and apoptosis. They regulate the expression of genes through mRNAs cleavage or translation inhibition by recognizing the complementary target site in the 3`-untranslated region of mRNA [1]. Recent studies have indicated that miRNAs play an important role in the process of various cancer. Differentially expressed miRNAs in various human cancer have been reported such as breast cancer, ovarian cancer and lung cancer, which shows miRNAs may be viewed as a marker in human cancer diagnosis and it provides a new direction in cancer treatment [2]. However, systematically and accurately identifying cancer-related miRNAs is a challenge with limited data of cancer-related miRNA. And the interaction between genes and mRNAs

D.-S. Huang et al. (Eds.): ICIC 2018, LNCS 10955, pp. 289–300, 2018.
https://doi.org/10.1007/978-3-319-95933-7_36

is complicated. Generally, each miRNA regulates more than one gene, and each gene may be regulated by various miRNAs.

So far, the identification of cancer-related miRNAs has been widely studied. Zhang et al. [3] proposed a method based on mRNAs and miRNAs expression, where abnormally expressed miRNAs and mRNAs were selected for analysis. Jin and Lee [4] adopted statistic methods to prioritize miRNAs by measuring the value of functional feature miRNAs and counting the number of paired genes. Above two methods only consider mRNAs expression data, and these two methods have not considered protein expression level. Zhao et al. [5] provided an approach for identifying cancer-related miRNAs based on mRNAs expression data, where the dysfunctional pathway of cancer was integrated into calculating a score for ranking miRNAs, but this method was time-consuming. Seo et al. [6] reported that integrated miRNA, miRNA, and protein expression data for identifying cancer-related miRNAs recently, but this approach collected genes from mRNA and protein expression data, which could not cover those genes that only show abnormal expression in mRNAs. Furthermore, none of these methods considers the information of dirver genes. It is well known that the dirver genes play a critical role in the cancer process, which decide the major reason of cancer [7]. Thus, miRNAs are related to cancer driven gene, which provides further evidence to confirm the relation between selected miRNAs and cancers.

Based on the above idea, a novel approach was provided to detect cancer-related miRNAs by introducing dirver genes to integrate mRNA expression data and protein expression, and three gene expression datasets were integrated. Then a related score was proposed to prioritize the detected miRNAs. Additionally, the percentage of known miRNAs, AUC, and literature survey were used for assessing the result.

2 Materials and Methods

Our approach includes five steps described in Fig. 1. In the first step, the expression value of each mRNAs was normalized by Z-score, and the two-sample t-tests was used to hunt for mRNAs with abnormal expression. Then we selected the common genes included in both the selected mRNAs expression datasets and dirver genes, and obtained the second common gene set where genes are included in both proteins expression datasets and dirver genes list. Two selected genes expression databases were obtained from these two common gene sets and gene expression data. In the third step, two correlation matrices, mRNA-miRNA correlation matrix and protein-miRNA correlation matrix, were calculated. And the gene-miRNA pairs with marked correlation were selected from two correlation matrices. In the last step, a related score was defined to prioritize the identified miRNAs.

2.1 Data Collection

We obtained mRNA, miRNA, and protein tumor expression data of GBM generated by Chin et al. [8] and Brennan et al. [9], and mRNA normal samples were extracted from The Cancer Genome Atlas (TCGA) data portal (http://cancergenome.nih.gov). After samples were paired, mRNA expression contains 124 tumor samples and 5 unmatched

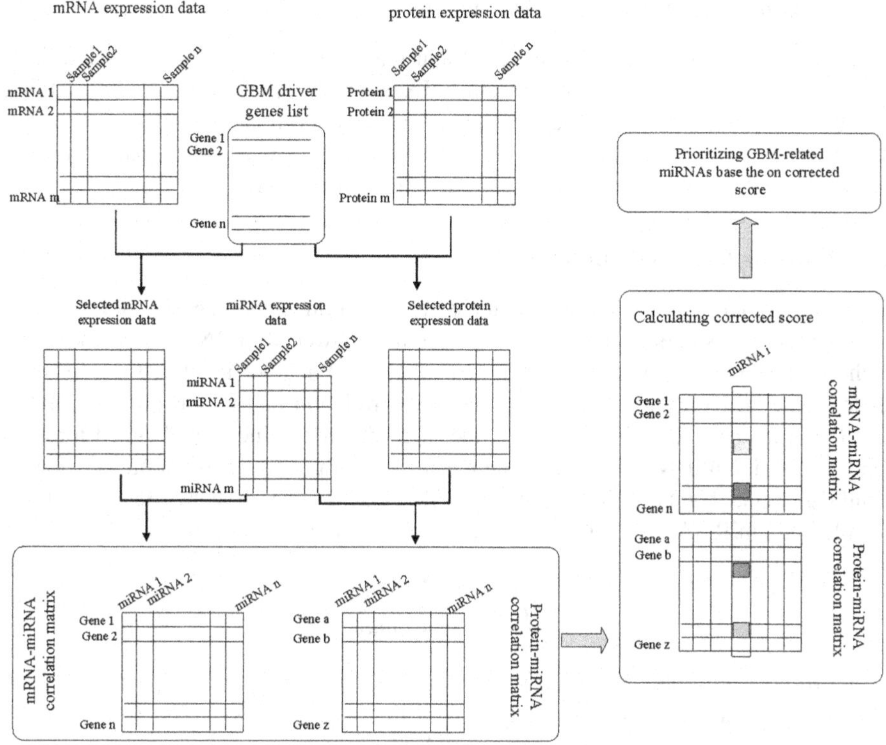

Fig. 1. Workflow for identifying and ranking GBM-related miRNAs

normal samples. The miRNAs expression data includes 124 tumor samples, and protein expression data has also 124 tumor samples. Finally, 10470 mRNAs, 405 miRNAs and 202 proteins were obtained. The known 229 dirver genes of GBM was obtained from Driverdbv2 [10]. The information of 118 GBM-related miRNA from the Human & Disease Database V2.0 (HMDD v2.0) [11, 12] shows that about 93 miRNAs were reported to have been related to GBM in 405 miRNAs.

2.2 Detecting Abnormal Expression mRNAs and Selecting Relevant Genes

Z-score was used to normalize the expression value of each mRNA, then we used the two-sample t-tests, Bonferroni corrected p-value less than 0.05, to hunt mRNAs with obviously different expression level between tumor samples and normal samples. Then a gene set included in both the detected mRNAs expression databases and the dirver genes list of GBM was selected. And we obtained second gene set included in both the proteins expression databases, proteins name was converted into genes name, and the dirver genes list of GBM. The two gene sets are combined to simulate two selected genes expression matrixes: mRNA expression matrixes and protein expression matrixes.

2.3 Obtaining the Correlation Matrices of Genes-miRNA

Based on above two selected gene expression matrixes and miRNA expression database, two correlation matrices, mRNAs-miRNA correlation matrix and protein-miRNA correlation matrix, were calculated by Spearman's rank correlation coefficients (SCC). According to the two correlation matrices, gene-miRNA pairs are identified, which are situated at top $\alpha\%$ absolute correlation values in both matrices.

2.4 Ranking Candidate miRNAs

In general, for each miRNA, it can regulate more than one gene in expression level, so a related score was defined to measure the regulated power of miRNAs, and it regarded as the foundation for prioritizing identified miRNAs. First, we identified significantly related gene-miRNA pairs in mRNAs-miRNA correlation matrix and protein-miRNA correlation matrix. Then the absolute value of correlation value was obtained for each pair. For given a miRNA, afterward, the mean value of positive correlation values with significant related gene-miRNA pairs was defined its related score. Base on above notable gene-miRNA binding data, candidate miRNAs were ranked.

$$Micors_i = \frac{\sum |X_{ij}|}{Count_i} \tag{1}$$

where *Micorsi* presents the related score of miRNA *i*, and X_{ij} is the Spearman's rank correlation coefficients value of miRNA *i* with related genes *j*. The $Count_i$ is the number of genes significantly related miRNA *i*.

2.5 Constructing miRNA-Gene Regulation Modules

Ranked miRNAs are clustered by hierarchical clustering analysis (HCA). HCA is divided into two types: agglomerative cluster and divisive cluster based on different strategies for hierarchical clustering. In this paper, we used previous one to cluster ranked miRNAs. It is composed of four basic steps [13]. The distance between each miRNA and each miRNA is their similarity, and miRNAs are clustered based on mRNA-miRNA correlation matrix.

After miRNAs clustering, 3519 mRNAs with marked abnormal expression are assigned to each miRNA cluster by calculating the ratio of miRNA-mRNA in each cluster and selecting a cluster with the maximum ratio. For each cluster, the ratio computed as follows:

$$As_{C_i j} = \frac{\sum S_{k,j}}{Count(C_i)} \tag{2}$$

where $As_{C_i j}$ denotes the ratio of mRNA *j* when mRNA *j* is assigned to miRNA cluster *Ci*. *k* is miRNA's ID belonging to miRNA cluster C_i. $S_{k,j}$ denotes the SCC of mRNA *j* and miRNA *k*. $Count(C_i)$ denotes the number of miRNAs in miRNA cluster C_i. The greater the $As_{C_i j}$ is, the greater the probability of mRNA associated with the miRNA cluster is.

Finally, proteins were assigned to above mRNA- miRNA clusters by calculating the averaged SCC of miRNA-protein in each cluster based on protein-miRNA correlation matrix with the same strategy.

3 Results

3.1 Prioritization of miRNA Markedly GBM-Related miRNAs

In the 10470 mRNAs of GBM, we obtained 3519 mRNAs that show markedly different expression level compared with the normal samples. About 59 genes were got from the dirver genes of GBM in protein and mRNAs expression databases with abnormal expression.

In mRNA-miRNA and protein-miRNA correlation matrices, the value of Spearman's correlation coefficient was calculated for each markedly genes-miRNA pair that among the top 1% or top 5% from two matrices. For a given miRNA, the related score was calculated by the value of positive Spearman's rank correlation coefficient in miRNA-gene pairs of this miRNA. Out of 405 miRNAs, 391 miRNAs have credibly related scores and were efficiently ranked, and we obtained two lists by setting thresholds, 1% or 5%.

3.2 Validation of the Ranked miRNAs

Based on the information of HMDDv2 and the data of gene-miRNA pairs, 93 known miRNAs related with GBM in all of 391 miRNAs that have credible related scores. In our ranked miRNA list, the 50% (10/20) miRNAs included in HMDD in the top 5% of all miRNAs, and 40% (16/40) miRNAs are known to have relationship with GBM from HMDD in the top 10% of all miRNAs with same threshold. When the threshold was set to 1%, 55% (11/20) miRNAs are shown in the top 5% of all miRNAs, and 42.5% (17/40) miRNAs are reported to have relationship with GBM from HMDD in the top 10% of all miRNAs.

As shown in Table 1, we collected the reports of 20 miRNAs in top 5% ranked list where the significant threshold is set as 5%. Out of all 20 miRNAs, 18 miRNAs were reported that were related to GBM, and 2 miRNAs were validated its related with other cancers. Then, AUC value of ROC curve was assessed, which shows the AUC value of 0.772 and 0.793 in 1% and 5% GBM ranking.

3.3 Comparison with Other Methods

We compared our method with the one from Seo et al. [6]. The percentage of known GBM-related miRNAs was evaluated the accuracy of methods. And the running time of algorithms also assess the performance of methods. As shown in the Fig. 2, our approach performs better than Seo's method when significant threshold is 5%. In top 40, the percentage of known GBM-related is more than 40%. Compared with Seo's method, our method achieves almost same performance among top 20 when the

Table 1. The studies of top 20 miRNAs are shown in the ranked list. Columns 1 and 2 present the rank and the name of miRNA. GBM-related miRNAs are labeled with 1 (included in HMDD), or 0 (not included in HMDD) in column 3. The columns 4 and 5 show the representative studies for the relationship of miRNAs with GBM, respectively. And the * represent that this miRNA has not been reported in GBM but shows it has relation with other cancer.

Rank	miRNA	HMDD	Evidence	Reference
1	hsa-miR-30a	1	miR-30a inhibits glioma progression and stem cell-like properties by repression of Wnt5a	Zhang et al. [14]
2	hsa-miR-33a	0	miR-33a promotes glioma-initiating cell self-renewal via PKA and NOTCH pathways	Wang et al. [15]
3	hsa-miR-204	0	RETRACTED: Downregulation of miR-204 expression correlates with poor clinical outcome of glioma patients	Ye et al. [16]
4	hsa-miR-124	1	RETRACTED ARTICLE: MiR-124 Functions as a Tumor Suppressor via Targeting hCLOCK1 in Glioblastoma	He et al. [17]
5	hsa-miR-9*	1	MiR-9 regulates the expression of CBX7 in human glioma	Chao et al. [18]
6	hsa-miR-200b	1	miR-200b as a prognostic factor targets multiple members of RAB family in glioma	Liu et al. [19]
7	hsa-miR-128	1	Micro-RNA-128 (miRNA-128) down-regulation in glioblastoma targets ARP5 (ANGPTL6), Bmi-1 and E2F-3a, key regulators of brain cell proliferation	Cui et al. [20]
8	hsa-miR-595	0	MiR-595 targeting regulation of SOX7 expression promoted cell proliferation of human glioblastoma	Hao et al. [21]
9	hsa-miR-181d	1	MiR-181d acts as a tumor suppressor in glioma by targeting K-ras and Bcl-2	Wang et al. [22]
10	hsa-miR-219-1-3p	0	*MicroRNA-219 is downregulated in non-small cell lung cancer and inhibits cell growth and metastasis by targeting HMGA2	Sun et al. [23]
11	hsa-miR-142-3p	1	Effect of miR-142-3p on the M2 macrophage and therapeutic efficacy against murine glioblastoma	Xu et al. [24]
12	hsa-miR-338-3p	0	Mir-338-3p Inhibits Malignant Biological Behaviors of Glioma Cells by Targeting MACC1 Gene	Shang et al. [25]
13	hsa-miR-101	1	miR-101 is down-regulated in glioblastoma resulting in EZH2-induced proliferation, migration, and angiogenesis	Smits et al. [26]

(*continued*)

Table 1. (*continued*)

Rank	miRNA	HMDD	Evidence	Reference
14	hsa-miR-195	1	MicroRNA-195 plays a tumor-suppressor role in human glioblastoma cells by targeting signaling pathways involved in cellular proliferation and invasion	Zhang et al. [27]
15	hsa-miR-582-3p	0	*MicroRNA-582 promotes tumorigenesis by targeting phosphatase and tensin homologue in colorectal cancer	Song et al. [28]
16	hsa-miR-601	0	*miR-601 is a prognostic marker and suppresses cell growth and invasion by targeting PTP4A1 in breast cancer	Hu et al. [29]
17	hsa-miR-200a	0	MiR-200a impairs glioma cell growth, migration, and invasion by targeting SIM2-s	Su et al. [30]
18	hsa-miR-34a	1	miR-34a attenuates glioma cells progression and chemoresistance via targeting PD-L1	Wang and Wang [31]
19	hsa-miR-106b	0	Down-regulation of miR-106b suppresses the growth of human glioma cells	Zhang et al. [32]
20	hsa-miR-153	1	MicroRNA-153 is tumor suppressive in glioblastoma stem cells	Zhao et al. [33]

significant threshold was set to 1%. This comparison result in running time was shown in Table 2, and our approach is superior to the method of Seo et al.

3.4 Assessment of miRNA-mRNA-Protein Regulatory Modules

In the ranked list, 391 miRNAs that have credible related scores were clustered to 20 sets by HCA. 3519 mRNAs and 201 proteins were assigned to these 20 clusters. Then Gene Ontology (GO) in Biological Processes (BP) and Kyoto Encyclopedia of Genes and Genomes (KEGG) pathways analysis are applied to assess the biological significance of the modules [34]. p-values (adj.) have been obtained through hypergeometric analysis corrected by FDR method.

As shown in Table 3, focal adhesion is significant in the results of KEGG pathways enrichment of module 3. Previous studies show focal adhesion pathway is involved in multiple tumors. For example, prostate cancer [35], hepatocellular carcinoma tumor [36], and lung cancer [37]. Module 3 also contains Jak-STAT pathways which further indicates cancer pathways enrichment. As well known, Jak-STAT can lead to abnormal growth of many malignancies including breast cancer [38].

As shown in Fig. 3, the top 15 GO biological process terms of module 3 are all enriched as p-value <0.05. The most significant biological processes in module 3 include immune response, signal transduction, innate immune response, positive regulation of gamma-delta T cell differentiation, positive regulation of interleukin-6 production.

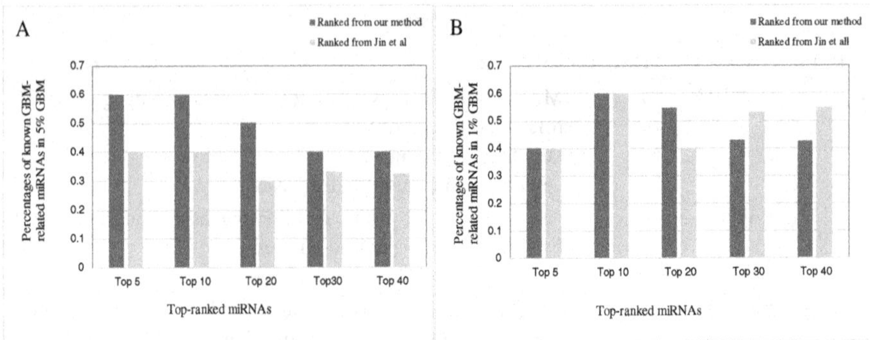

Fig. 2. Performance comparison with the method from Seo et al. These pictures show the result of comparison between our approach and method from Seo et al. The y-axis indicates the percentage of known miRNA in HMDD in the top-ranked miRNAs, and the x-axis represents top ranked miRNAs from two methods. Figure A and B show the results of the comparison in GBM top 5% and 1%, respectively.

Table 2. Comparison of running time between our method and method form Seo et al. [6].

Methods	1% GBM	5% GBM
Our method	40.96 s	42.62 s
Seo's method	169.24 s	174.90 s

Table 3. KEGG pathways enrichment of module 3

ID	Items	Details	p-value (adj.)	Genes
1	Kegg:05200	Pathways in cancer, Non-small cell lung cancer, Glioma	5.34E-04	PLCG2, EGF, PIK3CG, CDKN2A
2	Kegg:04145	Phagosome	9.17E-04	NCF1, FCGR2A, MSR1, NCF4, M6PR, TUBA1C
3	Kegg:05200	Toxoplasmosis, Pathways in cancer, Focal adhesion, Small cell lung cancer, Amoebiasis	9.21E-04	LAMB2, LAMA2, PIK3CG
4	Kegg:04810	Regulation of actin cytoskeleton, Focal adhesion, Prostate cancer, Melanoma	1.02E-03	PDGFD, EGF, PIK3CG
5	Kegg:04630	Jak-STAT signaling pathway	1.20E-03	IL7R, IL15, IL10RA, IL7, PIK3CG, SOCS2

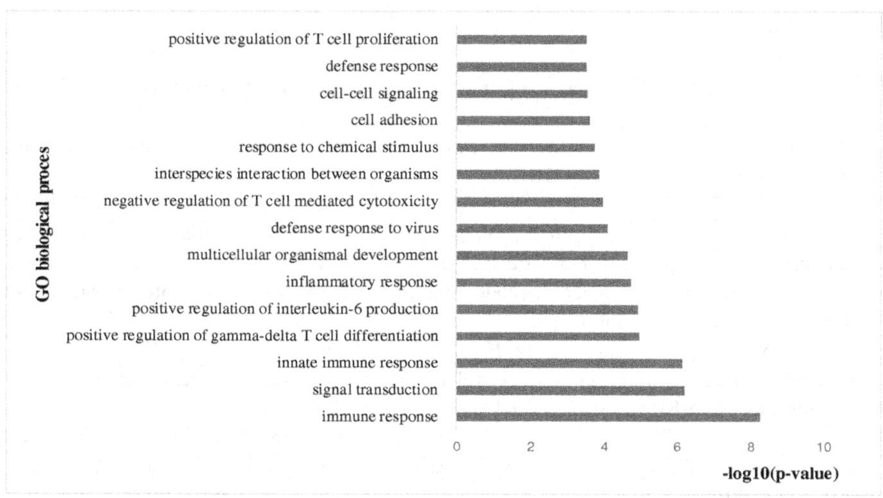

Fig. 3. The top 15 GO (BP) items statistical significance of module 3.

4 Conclusion and Discussion

It was reported by many studies that the miRNA plays a critical role in various biological processes, and they correspond to the occurrence and development of cancer. Consequently, identifying and prioritizing cancer-related miRNAs is a key step for the diagnosis and treatment of cancer.

In this paper, we proposed an approach for identifying and ranking GBM-related miRNAs by integrating the mRNA, miRNA, protein expression data and the driven gene information of GBM. Our method obtained two lists about GBM-related miRNAs, which show the degree of correlation between identified miRNAs with GBM. In our method, we introduced dirver genes to integrate the mRNA expression data and protein expression data. For evaluating our method, the percentage of known GBM-miRNA included in HMDD was calculated, and the evidence of identified and ranked miRNAs in existed research were retrieved. A miRNAs-mRNAs-protein regulatory module was constructed for understanding the biological senses of ranked miRNAs. Our method shows a better performance than the existing methods based on miRNA and protein expression in accuracy and running time.

Although our approach extends the methods of identifying and ranking cancer-related miRNAs, and the value of significant threshold is sensitive to results proportion of known miRNAs. Thus, we future work will focus on the development of marked pairs selection in our method.

Acknowledgement. This work was supported by the grants of the National Science Foundation of China (Grant Nos. 61472467, 61672011, and 61471169) and the Collaboration and Innovation Center for Digital Chinese Medicine of 2011 Project of Colleges and Universities in Hunan Province.

References

1. Bartel, D.P.: MicroRNAs: genomics, biogenesis, mechanism, and function. Cell **116**(2), 281–297 (2004)
2. Humeau, M., Torrisani, J., Cordelier, P.: miRNA in clinical practice: pancreatic cancer. Clin. Biochem. **46**(10–11), 933–936 (2013)
3. Zhang, W.Y., Zang, J., Jing, X.H., Sun, Z.D., Yan, W.Y., Yang, D.R., Guo, F., Shen, B.R.: Identification of candidate miRNA biomarkers from miRNA regulatory network with application to prostate cancer. J. Transl. Med. **12**, 66 (2014)
4. Jin, D., Lee, H.: Prioritizing cancer-related microRNA by integrating microRNA and mRNA datasets. Sci. Rep. **6**, 35350 (2016)
5. Zhao, X.M., Liu, K.Q., Zhu, G., He, F., Duval, B., Richer, J.M., Huang, D.S., Jiang, C.J., Hao, J.K., Chen, L.: Identifying cancer-related microRNAs based on gene expression data. Bioinformatics **31**(8), 1226–1234 (2015)
6. Seo, J., Jin, D., Choi, C.H., Lee, H.: Integration of microRNA, mRNA, and protein expression data for the identification of cancer-related microRNAs. PLoS ONE **12**(1), e0168412 (2017)
7. Tamborero, D., Gonzalez-Perez, A., Perez-Llamas, C., Deu-Pons, J., Kandoth, C., Reimand, J., Lawrence, M.S., Getz, G., Bader, G.D., Ding, L., Lopez-Bigas, N.: Comprehensive identification of mutational cancer driver genes across 12 tumor types. Sci. Rep. **3**, 2650 (2013)
8. Chin, L., Meyerson, M., Aldape, K., Bigner, D., Mikkelsen, T., VandenBerg, S., Kahn, A., Penny, R., Ferguson, M.L., Gerhard, D.S., et al.: Comprehensive genomic characterization defines human glioblastoma genes and core pathways. Nature **455**(7216), 1061–1068 (2008)
9. Brennan, C.W., Verhaak, R.G.W., McKenna, A., Campos, B., Noushmehr, H., Salama, S.R., Zheng, S.Y., Chakravarty, D., Sanborn, J.Z., et al.: The somatic genomic landscape of glioblastoma. Cell **155**(2), 462–477 (2013)
10. Chung, I.F., Chen, C.Y., Su, S.C., Li, C.Y., Wu, K.J., Wang, H.W., Cheng, W.C.: DriverDBv2: a database for human cancer driver gene research. Nucleic Acids Res. **44**(D1), D975–D979 (2016)
11. Lu, M., Zhang, Q.P., Deng, M., Miao, J., Guo, Y.H., Gao, W., Cui, Q.H.: An analysis of human microRNA and disease associations. PLoS ONE **3**(10), e3420 (2008)
12. Li, Y., Qiu, C.X., Tu, J., Geng, B., Yang, J.C., Jiang, T.Z., Cui, Q.H.: HMDD v2.0: a database for experimentally supported human microRNA and disease associations. Nucleic Acids Res. **42**(D1), D1070–D1074 (2014)
13. Borgatti, S.P.: How to Explain Hierarchical Clustering (1994)
14. Zhang, Y., Wu, Z., Li, L., Xie, M.: miR-30a inhibits glioma progression and stem cell-like properties by repression of Wnt5a. Oncol. Rep. **38**(2), 1156–1162 (2017)
15. Wang, H., Sun, T., Hu, J., Zhang, R., Rao, Y., Wang, S., Chen, R., McLendon, R.E., Friedman, A.H., Keir, S.T., Bigner, D.D., Li, Q.-J., Wang, H., Wang, X.-F.: miR-33a promotes glioma-initiating cell self-renewal via PKA and NOTCH pathways. J. Clin. Invest. **124**(10), 4489–4502 (2014)
16. Ye, Z.-N., Liu, J.-P., Wu, L.-Y., Zhang, X.-S., Zhuang, Z., Chen, Q., Lu, Y., Liu, C.-G., Zhang, Z.-H., Zhang, H.-S., Hou, W.-Z., Hang, C.-H.: RETRACTED: downregulation of miR-204 expression correlates with poor clinical outcome of glioma patients. Hum. Pathol. **63**, 46–52 (2017)
17. He, Y., Zhao, C., Liu, Y., He, Z., Zhang, Z., Gao, Y., Jiang, J.: MiR-124 functions as a tumor suppressor via targeting hCLOCK1 in glioblastoma. Mol. Neurobiol. **54**(3), 2375 (2017). (Retraction of vol. 37, p. 6761, 2016)

18. Chao, T.-F., Zhang, Y., Yan, X.-Q., Yin, B., Gong, Y.-H., Yuan, J.-G., Qiang, B.-Q., Peng, X.-Z.: MiR-9 regulates the expression of CBX7 in human glioma. Zhongguo yi xue ke xue yuan xue bao. Acta Acad. Med. Sinicae **30**(3), 268–274 (2008)
19. Liu, Q., Tang, H., Liu, X., Liao, Y., Li, H., Zhao, Z., Yuan, X., Jiang, W.: miR-200b as a prognostic factor targets multiple members of RAB family in glioma. Med. Oncol. **31**(3), 859 (2014)
20. Cui, J.G., Zhao, Y., Sethi, P., Li, Y.Y., Mahta, A., Culicchia, F., Lukiw, W.J.: Micro-RNA-128 (miRNA-128) down-regulation in glioblastoma targets ARP5 (ANGPTL6), Bmi-1 and E2F-3a, key regulators of brain cell proliferation. J. Neurooncol. **98**(3), 297–304 (2010)
21. Hao, Y., Zhang, S., Sun, S., Zhu, J., Xiao, Y.: MiR-595 targeting regulation of SOX7 expression promoted cell proliferation of human glioblastoma. Biomed. Pharmacother. **80**, 121–126 (2016)
22. Wang, X.-F., Shi, Z.-M., Wang, X.-R., Cao, L., Wang, Y.-Y., Zhang, J.-X., Yin, Y., Luo, H., Kang, C.-S., Liu, N., Jiang, T., You, Y.-P.: MiR-181d acts as a tumor suppressor in glioma by targeting K-Ras and Bcl-2. J. Cancer Res. Clin. Oncol. **138**(4), 573–584 (2012)
23. Sun, X., Xu, M., Liu, H., Ming, K.: MicroRNA-219 is downregulated in non-small cell lung cancer and inhibits cell growth and metastasis by targeting HMGA2. Mol. Med. Rep. **16**(3), 3557–3564 (2017)
24. Xu, S., Wei, J., Wang, F., Kong, L.-Y., Ling, X.-Y., Nduom, E., Gabrusiewicz, K., Doucette, T., Yang, Y., Yaghi, N.K., Fajt, V., Levine, J.M., Qiao, W., Li, X.-G., Lang, F.F., Rao, G., Fuller, G.N., Calin, G.A., Heimberger, A.B.: Effect of miR-142-3p on the M2 macrophage and therapeutic efficacy against murine glioblastoma. JNCI-J. Nat. Cancer Inst. **106**(8) (2014)
25. Shang, C., Hong, Y., Guo, Y., Xue, Y.-X.: Mir-338-3p inhibits malignant biological behaviors of glioma cells by targeting MACC1 gene. Med. Sci. Monit. **22**, 710 (2016)
26. Smits, M., Nilsson, J., Mir, S.E., van der Stoop, P.M., Hulleman, E., Niers, J.M., Hamer, P. C.D.W., Marquez, V.E., Cloos, J., Krichevsky, A.M., Noske, D.P., Tannous, B.A., Wurdinger, T.: miR-101 is down-regulated in glioblastoma resulting in EZH2-induced proliferation, migration, and angiogenesis. Oncotarget **1**(8), 710–720 (2010)
27. Zhang, Q.-Q., Xu, H., Huang, M.-B., Ma, L.-M., Huang, Q.-J., Yao, Q., Zhou, H., Qu, L.-H.: MicroRNA-195 plays a tumor-suppressor role in human glioblastoma cells by targeting signaling pathways involved in cellular proliferation and invasion. Neuro-Oncol. **14**(3), 278–287 (2012)
28. Song, B., Long, Y., Liu, D., Zhang, W., Liu, C.: MicroRNA-582 promotes tumorigenesis by targeting phosphatase and tensin homologue in colorectal cancer. Int. J. Mol. Med. **40**(3), 867–874 (2017)
29. Hu, J.-Y., Yi, W., Wei, X., Zhang, M.-Y., Xu, R., Zeng, L.-S., Huang, Z.-J., Chen, J.-S.: miR-601 is a prognostic marker and suppresses cell growth and invasion by targeting PTP4A1 in breast cancer. Biomed. Pharmacother. **79**, 247–253 (2016)
30. Su, Y., He, Q., Deng, L., Wang, J., Liu, Q., Wang, D., Huang, Q., Li, G.: MiR-200a impairs glioma cell growth, migration, and invasion by targeting SIM2-S. NeuroReport **25**(1), 12–17 (2014)
31. Wang, Y., Wang, L.: miR-34a attenuates glioma cells progression and chemoresistance via targeting PD-L1. Biotech. Lett. **39**(10), 1485–1492 (2017)
32. Zhang, A., Hao, J., Wang, K., Huang, Q., Yu, K., Kang, C., Wang, G., Jia, Z., Han, L., Pu, P.: Down-regulation of miR-106b suppresses the growth of human glioma cells. J. Neurooncol. **112**(2), 179–189 (2013)
33. Zhao, S.G., Deng, Y.F., Liu, Y.H., Chen, X., Yang, G., Mu, Y.L., Zhang, D.M., Kang, J.H., Wu, Z.L.: MicroRNA-153 is tumor suppressive in glioblastoma stem cells. Mol. Biol. Rep. **40**(4), 2789–2798 (2013)

34. Tabas-Madrid, D., Nogales-Cadenas, R., Pascual-Montano, A.: GeneCodis3: a non-redundant and modular enrichment analysis tool for functional genomics. Nucleic Acids Res. **40**(w1), W478–W483 (2012)
35. Zhang, J.S., Gong, A., Gomero, W., Young, C.Y.: ZNF185, a LIM-domain protein, is a candidate tumor suppressor in prostate cancer and functions in focal adhesion pathway. Cancer Res. **64**(7), 619–620 (2004)
36. Lee, C., Fan, S., Sit, W., Jor, I.W., Wong, L.L., Man, K., Tan-Un, K., Wan, J.M.: Olive oil enriched diet suppresses hepatocellular carcinoma (HCC) tumor growth via focal adhesion pathway. Cancer Res. **67**(9 Suppl.), LB-60 (2007)
37. Ocak, S., Yamashita, H., Udyavar, A.R., Miller, A.N., Gonzalez, A.L., Zou, Y., Jiang, A., Yi, Y., Shyr, Y., Estrada, L.: DNA copy number aberrations in small-cell lung cancer reveal activation of the focal adhesion pathway. Oncogene **29**(48), 6331–6342 (2010)
38. Zhang, Q.: Role of Jak/Stat Pathway in the Pathogenesis of Breast Cancer (2010)

ENSEMBLE-CNN: Predicting DNA Binding Sites in Protein Sequences by an Ensemble Deep Learning Method

Yongqing Zhang[1,2], Shaojie Qiao[3(✉)], Shengjie Ji[1], and Jiliu Zhou[1]

[1] School of Computer Science, Chengdu University of Information Technology, Chengdu 610225, China

[2] School of Computer Science and Engineering, University of Electronic Science and Technology of China, Chengdu 610054, China

[3] School of Cybersecurity, Chengdu University of Information Technology, Chengdu 610225, China

sjqiao@cuit.edu.cn

Abstract. Detection of DNA binding sites in proteins plays an essential role in gene regulation processing. However, the difficult problem in developing machine learning predictors of DNA binding sites in protein is that: the number of DNA binding sites is significantly fewer than that of non-binding sites. Aiming to handle this issue, we propose a new predictor, named ENSEMBLE-CNN, which integrates instance selection and bootstrapping techniques for predicting imbalanced DNA-binding sites from protein primary sequences. ENSEMBLE-CNN uses a protein's evolutionary information and sequence feature as two basic features and employs sampling strategy to deal with the class imbalance problem. Multiple initial predictors with CNNs as classifiers are trained by applying SMOTE and a random under-sampling technique to the original negative dataset. The final ensemble predictor is obtained by majority voting strategy. The results demonstrate that the proposed ENSEMBLE-CNN achieves high prediction accuracy and outperforms the existing sequence-based protein-DNA binding sites predictors.

Keywords: Protein-DNA binding sites · Deep learning · Ensemble method Imbalance learning

1 Introduction

DNA-binding proteins are the proteins composed of DNA-binding domains. The interactions between these proteins and DNA play a crucial role in vital biological process [1, 2]. A number of high throughput experimental techniques have been developed to confirm the interactions between DNA and proteins, such as protein binding microarray(PBM) [3], ChIP-seq [4] and protein microarray assays [5]. However, the existing approaches are costly and time-consuming. Thus, there is urgent need to propose computational methods for predicting protein-DNA binding sites from sequences in an efficient and effective fashion.

D.-S. Huang et al. (Eds.): ICIC 2018, LNCS 10955, pp. 301–306, 2018.
https://doi.org/10.1007/978-3-319-95933-7_37

Currently, a series of computational methods have been proposed to predict DNA-binding sites in protein. Based on the discovered features for prediction, these methods can be partitioned into three groups: evolutionary features based methods, sequence features based methods, and the ones based on structure features. During the past decade, a number of machine learning algorithms have been used to predict DNA-binding sites from protein sequences, including BindN [6], BindN+ [7], ProteDNA [8], DP-Bind [9], MetaDBSite [10], DNABind [11] and TargetDNA [12]. These sequence-based predictors only utilize protein sequence information and recognize DNA-binding sites with one or more machine learning algorithms. Despite the promising results of these methods, there remains room for further improvements in accurately predicting DNA-binding sites form protein sequence.

Another important issue in machine learning predictors of protein-DNA binding sites is the severe intrinsic class imbalance problem: the number of DNA binding sites (minority class) is apparently fewer than that of non-binding sites (majority class). Re-sampling is the most straightforward strategy for dealing with the issue of class imbalance [13]. Based on the aforementioned problem, we proposed a sequence-based predictor, named ENSEMBLE-CNN, for the computational identification of DNA binding sites. First, we employ the protein evolutionary information and sequence features, which are determined solely from protein sequences. Next, SMOTE [14] is used to over-sample positive data. Then, we train multiple DNA binding site predictors with CNNs as a basic classifier by applying a bootstrap technique on the original imbalanced data. Lastly, we obtain the ensembled predictor by using the majority voting strategy.

2 Methods

2.1 Feature Descriptors

From the point of view of machine learning, prediction of DNA binding sites in proteins is actually a traditional binary classification problem. Various effective sequence-based feature, such as position specific scoring matrix (PSSM) [9], predicted secondary structure [15] and physicochemical properties [16], have been explored for predicting protein DNA binding residues. In this study, we employ PSSM feature and sequence feature for predicting DNA binding sites in proteins.

PSSM, being a very important type of evolutionary features, is obtain by running the PSI-BLAST [17] program to search the SwissProt database [18] via three iteration, with 10–3 as the E-value cutoff for multiple sequence alignment. In PSSM, there are 40 scores for each sequence position and each score implies the conservation degree of a specific residue type on that position. For each data instance, all the scaled scores in the PSSM are used as its evolution features. In this study, we use the window size w = 15.

Sequence features include local amino acid composition, predicted second structure and predicted solvent accessible area. Each probe sequence is converted into a 4 $\times$ L one-hot coded binary matrix (L is the probe length) and the intensity values are normalized.

2.2 SMOTE: Synthetic Minority Over-Sampling Technique

SMOTE [14] is an over-sampling approach in which the minority class is oversampled by creating "synthetic" examples rather than over-sampling with replacement. The minority class is over-sampled by taking each minority class sample and generating synthetic examples by the line segments joining approached on k nearest neighbors corresponding to each minority class. Depending upon the amount of over-sampling required, neighbors from the k nearest neighbors are randomly selected. In our model, we use 5 nearest neighbors.

2.3 Convolutional Neural Networks (CNN)

Most modern deep learning models are based on an artificial neural network [19–21]. Recently, deep learning techniques have demonstrated their capability of improving the discriminative power compared with other machine learning methods [22], and have been widely applied to the field of bioinformatics. The basic components of a CNN include convolutional, pooling and fully connected layers.

2.4 ENSEMBLE-CNN

We trained SMOTE and multiple different classifiers on balanced datasets obtained by applying random under-sampling method. Then, these trained classifiers are ensembled by the majority voting strategy. In this study, ENSEMBLE-CNN is trained by using the standard back-propagation algorithm and mini-batch gradient descent with the Adagrad variation. Dropout and early stopping strategies are used for regularization and model selection.

3 Experimental Results

3.1 The Datasets

In this study, three datasets, datasets PDNA-543 [12] and PDNA-TEST [12], were used to evaluate the performance of our method. PDNA-543 consists of 543 protein sequences; there are 9,549 DNA-binding residues as positive samples and 134,995 non-binding residues as negative samples. PDNA-TEST has 41 protein chains, which includes 734 positive samples and 14,021 negative samples.

Six famous evaluation measurements are used, which is the same as in [12]. Sensitivity (Sen), Specificity (Spe), Accuracy (Acc), Precision (Pre), the Mathew's Correlation Coefficient (MCC) and AUC are utilized to evaluate prediction performance.

3.2 Comparison with Different Features

This set of experiments examines the contributions of the three different kinds of features in ENSEMBLE-CNN for the DNA binding sites in proteins prediction on the training dataset. The detail results are given in Table 1. As mentioned above, Sen, Spe,

Acc, Pre, MCC and AUC are the main metrics. It can be observed that the PSSM2 + One-hot coding features outperforming the PSSM2 features by 5.04% for Sen, 15.79% for Spe, 14.49% for ACC, 29.81% for Pre, 0.276 for MCC and 0.114 for AUC. When the three kinds of features are combined, ENSEMBLECNN achieves 0.632 for MCC and 0.933 for AUC, which indicates that these features are complementary for each other and the one-hot coding feature is the important method for effectively predicting protein-DNA binding.

Table 1. The performance on PDNA-543 for various features by ten-fold cross-validation.

Feature	Sen (%)	Spe (%)	Acc (%)	Pre (%)	MCC	AUC
PSSM2	71.40	77.06	76.38	29.98	0.349	0.812
PSSM2 + One-hot coding	76.44	**92.85**	**90.87**	**59.79**	0.625	0.926
PSSM1 + PSMM2 + One-hot coding	**79.48**	92.23	90.69	58.70	**0.632**	**0.933**

3.3 Predicted Results on Independent Test

In this section, we demonstrate the effectiveness of the proposed method, ENSEMBLECNN, by comparing it with other commonly-used predictors of DNA binding sites in proteins, including BindN [6], BindN+ [7], ProteDNA [8], DP-Bind [9], MetaDBSite [10], DNABind [11] and TargetDNA [12], by performing independent validation tests on PDNA-TEST, the results of which are shown in Table 2.

Table 2. The predicting performance compared with other predictors on PDNATEST (the value of other predictors are from [12])

Predictor	Sen (%)	Sep (%)	Acc (%)	Pre (%)	MCC
Bind	45.64	80.90	79.15	11.12	0.143
ProteDNA	4.77	99.84	95.11	60.30	0.160
BindN + (FRP ≈ 5%)	24.11	95.11	91.58	20.51	0.178
BindN + (Sep ≈ 85%)	50.81	85.41	83.69	15.42	0.213
MetaDBSite	34.20	93.35	90.41	21.22	0.221
DP-Bind	61.72	82.43	81.40	15.53	0.241
DNABind	70.16	80.28	79.78	15.70	0.264
TargetDNA(Sen ≈ Spe)	60.22	85.79	84.52	18.16	0.269
TargetDNA(FPR ≈ 5%)	45.50	93.27	90.89	26.13	0.300
ENSEMBLE-CNN	48.10	91.20	89.08	21.99	0.274

Table 2 shows that ENSEMBLE-CNN achieves satisfactory results with the second-best MCC value of 0.274. When compared with TargetDNA(Sen ≈ Spe), ENSEMBLE-CNN achieves an high Sen value. As for BindN+, which is an improved version of BindN, ENSEMBLE-CNN achieves an improvement of 6.1% on MCC. By comparing with MetaDBSite, the proposed ENSEMBLE-CNN also achieves improvements of 13.9 and 5.3% on Sen and MCC, respectively. DNABind achieves the best performance on Sen (70.16%), but a much lower Spe value, implying too many false positive are incurred during prediction.

4 Conclusions

In this study, we proposed a new sequence-based predictor of protein-DNA binding sites, called ENSEMBLE-CNN. It is trained on the DNA-binding protein dataset collected from the most recently released PDB with a SMOTE, CNN, and the bootstrap classifier ensemble strategy. Experimental results with a training dataset and an independent validation dataset have demonstrated the effectiveness of the proposed ENSEMBLE-CNN. In terms of our future work, we will further investigate the applicability of our model to other types of molecules binding sites prediction problems.

Acknowledgement. This work was supported in part by the National Natural Science Foundation of China under Grants (No. 61702058, 61772091), the China Postdoctoral Science Foundation funded project (No. 2017M612948), the Scientific Research Foundation for Advanced Talents of Chengdu University of Information Technology under Grant (No. KYTZ201717, KYTZ201715, KYTZ201750), the Scientific Research Foundation for Young Academic Leaders of Chengdu University of Information Technology under Grant (No. J201701, J201706), the Planning Foundation for Humanities and Social Sciences of Ministry of Education of China under Grant (No. 15YJAZH058), and the Innovative Research Team Construction Plan in Universities of Sichuan Province under Grant (No. 18TD0027).

References

1. Si, J., Zhao, R., Wu, R.: An overview of the prediction of protein DNA-binding sites. Int. J. Mol. Sci. **16**(3), 5194–5215 (2015)
2. Wong, K.C., Li, Y., Peng, C., Wong, H.S.: A comparison study for DNA motif modeling on protein binding microarray. IEEE/ACM Trans. Comput. Biol. Bioinform. **13**(2), 261–271 (2016)
3. Berger, M.F., Philippakis, A.A., Qureshi, A.M., He, F.S., Estep, P.W., Bulyk, M.L.: Compact, universal DNA microarrays to comprehensively determine transcription-factor binding site specificities. Nat. Biotechnol. **24**(11), 1429–1435 (2006)
4. Valouev, A., Johnson, D.S., Sundquist, A., Medina, C., Anton, E., Batzoglou, S., Myers, R. M., Sidow, A.: Genomewide analysis of transcription factor binding sites based on chip-seq data. Nat. Methods **5**(9), 829–834 (2008)
5. Ho, S.W., Jona, G., Chen, C.T., Johnston, M., Snyder, M.: Linking DNA-binding proteins to their recognition sequences by using protein microarrays. Proc. Nat. Acad. Sci. U.S.A. **103** (26), 9940–9945 (2006)
6. Wang, L., Brown, S.J.: BindN: a web-based tool for efficient prediction of DNA and RNA binding sites in amino acid sequences. Nucleic Acids Res. **34**(Web Server issue), W243 (2006)
7. Wang, L., Huang, C., Yang, M.Q., Yang, J.Y.: BindN+ for accurate prediction of DNA and RNA-binding residues from protein sequence features. BMC Syst. Biol. **4**(S1), S3 (2010)

8. Chu, W.Y., Huang, Y.F., Huang, C.C., Cheng, Y.S., Huang, C.K., Oyang, Y.J.: ProteDNA: a sequence-based predictor of sequence-specific DNA-binding residues in transcription factors. Nucleic Acids Res. **37**(Web Server issue), W396 (2009)
9. Hwang, S., Gou, Z., Kuznetsov, I.B.: DP-bind: a web server for sequence-based prediction of DNA-binding residues in DNA-binding proteins. Bioinformatics **23**(5), 634–636 (2007)
10. Si, J., Zhang, Z., Lin, B., Schroeder, M., Huang, B.: MetaDBSite: a meta approach to improve protein DNA-binding sites prediction. BMC Syst. Biol. **5**(S1), S7 (2011)
11. Li, B.Q., Feng, K.Y., Ding, J., Cai, Y.D.: Predicting DNA-binding sites of proteins based on sequential and 3D structural information. Mol. Genet. Genomics **289**(3), 489–499 (2014)
12. Hu, J., Li, Y., Zhang, M., Yang, X., Shen, H.B., Yu, D.J.: Predicting protein-DNA binding residues by weightedly combining sequence-based features and boosting multiple SVMs. IEEE/ACM Trans. Comput. Biol. Bioinform. **PP**(99), 1389–1398 (2016)
13. Hu, J., Li, Y., Yan, W.X., Yang, J.Y., Shen, H.B., Yu, D.J.: KNN-based dynamic query-driven sample rescaling strategy for class imbalance learning. Neurocomputing **191**, 363–373 (2016)
14. Chawla, N.V., Bowyer, K.W., Hall, L.O., Kegelmeyer, W.P.: Smote: synthetic minority over-sampling technique. J. Artif. Intell. Res. **16**(1), 321–357 (2011)
15. Ahmad, S., Gromiha, M.M., Sarai, A.: Analysis and prediction of DNA-binding proteins and their binding residues based on composition, sequence and structural information. Bioinformatics **20**(4), 477–486 (2004)
16. Wong, K.C., Li, Y., Peng, C., Moses, A.M., Zhang, Z.: Computational learning on specificity-determining residue-nucleotide interactions. Nucleic Acids Res. **43**(21), 10180–10189 (2015)
17. Schffer, A.A., Aravind, L., Madden, T.L., Shavirin, S., Spouge, J.L., Wolf, Y.I., Koonin, E. V., Altschul, S.F.: Improving the accuracy of psi-blast protein database searches with composition-based statistics and other refinements. Nucleic Acids Res. **29**(14), 2994–3005 (2001)
18. Bairoch, A., Apweiler, R.: The SWISS-PROT protein sequence database and its supplement TrEMBL in 2000. Nucleic Acids Res. **28**(1), 45–48 (2000)
19. Huang, D.-S.: Radial basis probabilistic neural networks: model and application. Int. J. Pattern Recogn. Artif. Intell. **13**(07), 1083–1101 (1999)
20. Huang, D.S., Du, J.X.: A constructive hybrid structure optimization methodology for radial basis probabilistic neural networks. IEEE Trans. Neural Netw. **19**(12), 2099–2115 (2008)
21. Zhang, J.-R., Zhang, J., Lok, T.-M., Lyu, M.R.: A hybrid particle swarm optimization–back-propagation algorithm for feedforward neural network training. Appl. Math. Comput. **185** (2), 1026–1037 (2007)
22. Huang, D.-S.: A constructive approach for finding arbitrary roots of polynomials by neural networks. IEEE Trans. Neural Netw. **15**(2), 477–491 (2004)

Exploration and Exploitation of High Dimensional Biological Datasets Using a Wrapper Approach Based on Strawberry Plant Algorithm

Edmundo Bonilla-Huerta[1(✉)], Roberto Morales-Caporal[1], and M. Antonio Arjona-López[2]

[1] Tecnológico Nacional de México, Instituto Tecnológico deApizaco, Av. Instituto Tecnológico S/N, 90300 Apizaco, Tlaxcala, Mexico
edbonn@hotmail.fr

[2] Tecnológico Nacional de México, Instituto Tecnológico La Laguna, Av. Instituto Tecnológico de La Laguna S/N, Centro, 27000 Torreón, Coahuila, Mexico

Abstract. This paper presents a wrapper approach based on Strawberry Plant Algorithm (SPA) for gene selection in high dimension data classification problem by selecting the most relevant genes for each biological dataset. In order to perform an integrated exploration-exploitation approach to deal the near-optimal (small) gene subset problem obtained from high dimensional microarray data. First, a statistical filter is proposed for gene selection. After, a SPA is proposed to find the most informative genes from the previous gene selection, SPA is applied to explore and exploit new regions of this search and overall to overcome premature convergence. Empirical studies based in five public DNA-microarray datasets it is observed that our model gets the best performances using a smaller number of selected genes than other methods reported in the literature recently.

Keywords: Strawberry Plant Algorithm · Wrapper · Gene selection SVM

1 Introduction

In the last two decades many researchers have been tackle the high-dimensionality of microarray gene expression data by proposing various nature-inspired optimization algorithms for gene selection and classification of gene expression profiles. However, new hybrid algorithms based on this approach are reported in the literature to address the problem of select a suitable subset of informative genes from microarray data with high classification performance. In order to overcome this obstacle, we propose a wrapper approach. In the first step of our model, a preprocessing data is performed using a statistical filter to make an initial selection from biomedical databases. In the second step, the selection is effected by the Strawberry Plant Algorithm (SPA) proposed as local

D.-S. Huang et al. (Eds.): ICIC 2018, LNCS 10955, pp. 307–317, 2018.
https://doi.org/10.1007/978-3-319-95933-7_38

and global search method. Finally, performance is the selected genes is made with a support vector machine and validated using the 10 Fold Cross Validation method.

This model performs an intensive gene selection mechanism to find an optimal subset of genes with higher classification accuracy in five microarray datasets. The rest of the paper is organized as follows: Sect. 2, describe the related works on microarray gene expression data. Section 3, describes the statistical filter proposed to reduce the initial high dimensionality data. Our proposed wrapper approach is given in Sect. 4. In Sect. 5 provides an analysis of the experimental results. Conclusions are drawn in Sect. 6.

2 Related Works

There exists a vast body of literature available concerning new metaheuristics to solve many optimization problems. However, the new trend of research is totally focalised on improving the algorithms for many different problems of combinatorial optimization. However a new problem is still open the number parameters of each algorithm. To find the appropriate range for tuning these parameters is often based on a priori knowledge and expertise of the researchers, but in the most of cases according to the kind of problem, is often adjusted by trial and error. Thus this situation leads to obtain different solutions for the same problem by changing only a single value from an interval of parameter ranges.

In recent years many heuristics, metaheuristics and bioinspired algorithms are hybridized for the task of feature selection. Each method proposed is based on global and local search, however a single algorithm is not able to find a sufficiently good or near sufficiently good solution because several algorithms are trapped in a local minimum due to the diversity of the population decreases slightly after some generations. The first goal of the hybridization is to have a good balance between exploration and exploitation to avoid the premature convergence problem. In the first stage an algorithm will be able to explore the huge search space of microarray data in order discover the most promising regions, in other words, the best gene subsets for each biomedical dataset. In the second stage a new algorithm will be to exploit the information gathered by the first algorithm. Both methods influence the efficiency and the effectiveness of the search process. Nevertheless, conventional statistical methods have been used as filter to reduce the initial high dimensionality of microarray datasets. After a hybrid model coupled with a specific learning algorithm and validation method is incorporated into a framework to find a small subset of informative genes with the best prediction performance from the reduced search space.

Nowadays, in microarray high dimensional microarray datasets, the major challenge is to combine effectively and efficiently algorithms as an attempt to build a robust hybrid model in order to explore and exploit the huge search space and to discover the small gene subset with the higher performance. Recent articles reported in the literature, introduces new hybrid methods for gene selection and classification task with promising results. In [1] is proposed a new hybrid algorithm HICATS incorporating imperialist competition algorithm (ICA) which performs global search and tabu search (TS) that conducts fine-tuned search. Alshamlan et al. [2] proposed an algorithm that comprises ABC and mRMR. The new approach is based on an SVM algorithm to measure the classification accuracy for selected genes. Chuang et al. [3] proposed a hybrid method of binary particle swarm optimization (B4PSO) and a combat genetic algorithm (CGA) is to perform the microarray data selection. The K-nearest neighbor (KNN) method with leave-one-out cross-validation (LOOCV) served as a classifier which selects effective genes for better classification performance with low error rate. A hybrid approach composed of two-stages is reported by Elyasigomari et al. [4] as gene selection and classification process. In the first stage the minimum redundancy and maximum relevance (MRMR) filter is used to select a subset of relevant genes. In the second stage is hybridized the cuckoo optimization algorithm and harmony search using the support vector machine as a classifier. The performance was assessed by the leave one out cross-validation method. A hybrid gene selection approach is reported by Sharbaf et al. [5]. This paper proposes a filter to reduce the initial high-dimensional problem, after a wrapper approach which is based on cellular learning automata (CLA) optimized with ant colony method (ACO) is used to find the set of features which improve the classification accuracy. The results obtained shown that this process leads to find smallest set of genes with maximum accuracy. Lu et al. [6] introduced a novel hybrid gene selection for microarray data combining the mutual information maximization (MIM) and adaptive genetic algorithm (AGA) to reduce the dimension of the original gene expression dataset by removing the redundancies of the data. The genes selected by this approach show more accurate identification rates compared with existing feature selection approaches. Apolloni et al. [7] propose two hybrid methods used as efficient feature selection. Each hybrid approach is integrated in a two-stage algorithm: The first stage is a filter and the second stage is a wrapper, and both are based on a Binary Differential Evolution (BDE). The objective of this double hybridization is to identify the smallest subset of relevant features using the hybrid filter, and to obtain a classifier with the highest accuracy for classifying samples of microarray experiments using the hybrid models.

Hybrid approaches are very promising; however, we argue that to select a good of bio-inspired computation models is a very difficult task and challenge for researchers. Therefore, we focus on the application of a SPA as a single metaheuristic that can be used for an optimal balance of exploitation and exploration of high dimensional microarray datasets.

3 Reduction of High Dimension Datasets

Five high dimensional biological datasets from public domain have been analyzed in this study, and are available at http://sdmc.lit.org.sg/GEDdatasets/Datasets. The detailed description is provided in Table 1.

In order to reduce the initial high dimensional problem of microarray datasets we use mutual information (MI) filter. The goal of this filtering process is to identify relevant genes for each biomedical dataset. We select several thousands of top ranked genes to build the initial search space (see Fig. 1).

MI filter evaluates the features according to their information gain and considers a single feature at a time. The entropy measure is considered as a measure to rank

Table 1. High dimensional biological datasets

Dataset	Genes/samples	Class labels	Author
Leukemia	7129/72	25 AML/47ALL	Golub et al. [9]
Colon	2000/62	40 tumor/22 normal	Alon et al. [10]
Lung	12533/181	31 MPM/150 ADCA	Gordon et al. [11]
Prostate	12600/102	52 tumor/50 normal	Singh et al. [12]
CNS	7129/60	21 tumor/39 normal	Pomeroy et al. [13]

variables [8]. Consider A and B as two random variables with a different probability distribution and a joint probability distribution. The mutual information between the two variables I (A; B) is defined as the relative entropy between the joint probability and the product of the probabilities [8].

$$I(A;B) = \sum_{a_i}\sum_{b_j} P(a_i, b_j)\log\frac{P(a_i, b_j)}{P(a_i)P(b_j)} \tag{1}$$

Where $P(a_{i,}b_j)$, are the joint probabilities of the variables, $P(a_i)$ are the probabilities of the variable A and $P(b_j)$ is the probability of the variable B.

In the first stage, is applied a pre-processing technique for each high dimensional dataset. Four preprocessing steps was applied according to the authors [14, 15]. After a

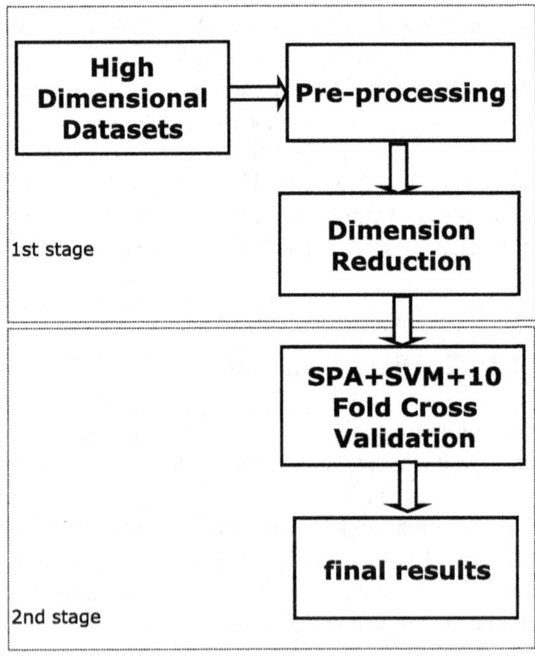

Fig. 1. Filter and wrapper approach based on Strawberry plant algorithm.

fixed number of genes for all datasets with the highest top ranking score are selected using MI filter. In a second stage, the exploration and exploitation process is proposed to discover the most relevant genes using a wrapper approach based on SPA.

4 Wrapper Approach

In a second stage a wrapper based on SPA is proposed to improve the exploration and exploitation process. This stage has as goal to discover potential biomarkers contained into different small gene subsets with high classification accuracy.

4.1 Strawberry Plant Algorithm

Recent scientific studies show that the plants exhibit intelligent and fascinating behaviors [16–19]. Strawberry Plant Algorithm (SPA) is a bio-inspired optimization method which imitates the propagation of the Strawberry Plant. From the optimization point of view strawberry plants simultaneously perform both the global and local search to find resources like water and minerals by developing runners and roots, respectively [17, 18]. We use SPA reported in [17] as our first gene selection process. Wrapper approach based on SPA for gene selection and classification of high dimensional microarray datasets is presented in Fig. 1. In this stage is defined the initial population that consist in a number of mother plants, the number of runners and roots and their lengths respectively. The roots guides the local search and the runners the global search. More precisely, the mother plant receives two column vectors. The duplicate plant is then evaluated according to a fitness function and then the elimination step begins, which according to the fitness value resulting there is a selection of the best vectors (plants) that will survive to form the new population of candidate solutions, and the least fit will be eliminated. The population then returns to its initial size. This procedure is repeated until the maximum number of iteration is reached.

5 Experimental Results

In order to evaluate our wrapper approach in terms of classification accuracy and size of gene subsets, specifically in situations where we select a high number of top ranked genes using the MI filter, five public microarray datasets were used. To assess the performance of the model, the SVM classifier was applied by using a 10 fold cross validation method. First, we examine a large number of top-ranked genes to explore more solutions and consequently obtain best performances for high dimensional microarray datasets. We noticed that our proposed model returns high predictive accuracy in 4 of 5 microarray datasets (see Table 2). For instance, 21 genes were sufficient to reach a classification performance of 100% for leukemia dataset, whereas the 100% accuracy was achieved for the colon tumor dataset using 35 genes. In the lung dataset, it is obtained 99.45% of accuracy with 20 genes. For prostate, our model can generate 96.32% classification accuracy by using 25 genes. However, 86.67% of classification was generated for CNS dataset with 31 genes.

Table 2. Results of wrapper approach using the first 60 top-ranked genes

Dataset	Accuracy	NG
Leukemia	100%	19
Colon	100%	35
Lung	99.45%	20
Prostate	96.35%	25
CNS	86.67	31

Table 3. Results of wrapper approach using the first 40 top-ranked genes

Dataset	Accuracy	NG
Leukemia	100%	14
Colon	98.39%	15
Lung	99.45	12
Prostate	95.59%	16
CNS	80%	15

In order to find a balance between exploration and exploitation for microarray dataset, we reduce slightly the large space of possible candidate solutions. We proposed the first 40 top ranked genes for finding a gene subset that maximizes the classification performance while minimizing the number of selected genes. Results are shown in Table 3. We observe that our method attained 100% of performance for Leukemia with 14 genes and 98.39% of classification accuracy for Colon tumor dataset with 8 relevant genes. In Prostate dataset, our model achieved a classification accuracy of 95.59, by using 16 genes. While in CNS dataset, we obtained 15 genes to generate 80% classification accuracy. We note that this dataset present the lower performance by using our model.

We compare our model with similar wrappers and hybrid algorithms (See Table 4). It can be seen than our approach may select a smaller gene subset with high classification performance than other methods published in the literature.

Table 4. Performance of related works

Method	Leukemia	Colon	Lung	Prostate	CNS
ACO/SVM	100% (8.6)	91.5%(**7.5**)	–	–	–
HICATS [1]	100% (3-5)	–	97.04%(7)	98.04%(5)	–
MRMR-ABC [2]	100%(14)	96.75%(15)	**100%(8)**	–	–
BPSO-CGA [3]					
PSO-GA [20]	97.2%(20.1)	91.9%(15.1)	–	–	–
MRMR-GA [46]	100%(15)	98.39%(15)	95.89%(15)		
Multiple Approaches [47]	100%(–)	75%(–)	100%(–)	93.85(–)	–
Our model	**100%**(19)	**100%**(35)	99.45%(20)	96.35%(25)	**86.65%**(31)
Our model	**100%**(14)	98.39%(15)	99.45%(12)	95.59%(16)	80%(15)

6 Biological Analysis of Genes

Tables 5 and 6, show the final gene subsets for all microarrays datasets by our wrapper approach and their corresponding indices. In Bold, the genes found out by other similar works. We present the indices and the authors who reported those genes in their works, for the two most common publicly datasets: Leukemia and Colon, respectively. For the rest of datasets we only list the indices.

For leukemia dataset using the first 60 top-ranked genes (see Table 5), 6 genes are considered relevant in the literature. Gene 760 (Cystatin A), gene 2121 (CTSD Cathepsin D), and gene 4366 (ARHG Ras homolog gene family) is reported by [20]. In other hand, Gene 4377 (ME491 gene extracted from H. Sapiens) is reported by the following authors [20–24, 36]. We observed that 5 genes are listed by [20] using a hybrid approach based on PSO/GA. In contrast, gene 5039 (LEPR Leptin Receptor) is has been found by other researchers [9, 25–29, 36]. In other hand, gene 4107 (PLECKSTRIN) is listed by the following works [20, 22, 30, 36]. Genes 3320 (LTC4 synthase) and 3847 (Homeo Box A9) are identified as relevant genes by [40], they use a tool for a deeper biological analysis, considering the leukemia dataset.

For Colon tumor dataset, 9 relevant genes obtained by our approach are also reported by Li and Tan [20]. The indices of those genes are: 14 (Myosin Light Chain Alkali), 67 (Cystatin C Precursor), 245 (Human cysteine—rich protein, exons 5 and 6), 249 (Humand desmin gene), 267 (Human cysteine-rich protein), 493 (Myosin Heavy Chain), 377 (H. Sapines mRNA for GCAP-II/uroguanylin precursor), 897 (Complement Factor D Precursor) and 1843 (Gelsolin Precursor). Genes reported by other authors are: gene 377 is also listed by [21, 24, 29, 31–33, 36]. Gene 918 (Human Mrna for transmembrane carcinoembyonic) is only listed by the authors [34, 36]. However, gene 267 (Human cysteine-rich protein) is reported in several works [20, 21, 23, 24, 29, 32, 34–36]. In other hand Gene 99 (Human mRNA for calmodulin, complete cds) have been recently discovery by [36].

The following genes have been listed in [45] with prior knowledge obtained from biological pathways for the lung dataset: 5727, 6571 and 8370. In this work the gene 11215 (Hs. 75842 Human mRNA for serine) for prostate cancer dataset is reported as relevant to define gene signatures for further clinical diagnostic and prognostic.

Table 6, show the final gene subsets by using the first 40 top-ranked genes obtained by our model. It can be seen that in Leukemia dataset 7 genes have been listed by other authors. Gene 1745 (LYN V-yes-1 Yamaguchi sarcoma viral related oncogene homolog), gene 1829 (PPGB Protective protein for beta-galactosidase), gene with index 1834 (CD 33CD33 antigen), gene 4377 (ME491 gene extracted from H. sapiens) and gene 6373 (PFC Properdin P factor, complement) are also listed by Li and Tang, [20]. Gene 4107 has been found in the following works [21, 22, 30, 36]. Gene 5039 is presented by the authors [9, 25–29, 36]. For the gene 6376 (PFC Properdin P factor, complement) is listed by [9, 10, 20–22, 36, 38]. Finally genes 1834 (CD33 Antigen), 3320 and 3847 are also reported in [37, 40] with biological relevance on the Leukemia dataset.

Table 5. Final gene subsets indices using the first 60 top-ranked genes

Dataset	Gene subset Indices
Leukemia	760, 1249, 2020, 2111, 2121, 3252, 3320, 3847, 4052, 4107, 4366, 4377, 5039, 6041, 6201, 6373, 6797, 6803, 6806
Colon	14, 67, 70, 99, 245, 249, 267, 354, 377, 389, 437, 493, 624, 740, 758, 822, 839, 897, 918, 1033, 1058, 1282, 1387, 1411, 1423, 1494, 1581, 1670, 1761, 1791, 1836, 1843, 1873, 1967, 1974
Lung	1192, 1431, 2244, 3361, 3490, 3513, 4424, 5727, 6571, 7000, 7765, 8172, 8370, 8564, 9131, 9228, 11620, 11958, 12153, 12429
Prostate	3417, 4243, 4365, 4726, 5757, 6185, 6865, 6866, 6930, 7315, 7405, 7623, 7905, 8168, 8759, 8965, 9354, 9735, 10749, 10753, 10996, 11202, 1121512148, 12454
CNS	110, 385, 538, 794, 812, 1352, 1507, 1641, 2149, 2412, 2511, 2803, 2929, 2930, 3063, 3587, 4010, 4116, 4130, 4202, 4695, 4750, 4829, 4951, 5061, 5302, 5600, 5704, 6058, 6179, 7052

Table 6. Final gene subsets indices using the first 40 top-ranked genes

Dataset	Gene subset indices
Leukemia	1249, 1745, 1829, 1834, 2020, 2111, 2186, 3320, 3847, 4107, 4377, 5039, 6345, 6376
Colon	14, 70, 201, 245, 286, 493, 918, 1058, 1210, 1247, 1280, 1494, 1581, 1750, 1974
Lung	869, 1192, 3334, 4336, 5854, 6164, 8370, 8393, 9707, 12152,12248,12308
Prostate	4243, 5757, 6185, 6930, 7428, 7718, 8123, 8168, 8498, 8850, 8986, 9442, 10833, 10996, 11202, 11215
CNS	385, 538, 544, 878, 1352, 2149, 2988, 3063, 3535, 4136, 4344, 5579, 5600, 6058, 6618

For colon tumor datasets genes with indices 14, 245 and 493 have been reported in [20]. Gene 918 is reported by the authors [34, 36]. Finally gene 1058 is presented in the works of [21, 32–34, 36, 39].

Most of genes obtained by our wrapper approach are reported in previous studies [40–45]. The genes found out with different similar approaches for each high dimensional biological dataset could be considered for diagnostic and prognostic of a specific disease: Leukemia, Colon, CNS, Prostate and Lung cancer.

7 Conclusions

This paper has presented a wrapper approach to explore and exploit the dimensionality of high-dimensional microarray datasets. In a first stage, the dimensionality is reduced by using a filtering process. In a second stage, a wrapper approach based on SPA is proposed in order to find a genes subset that maximizes the classification performance while minimizing the number of genes. We observed that the increase in the amount of top ranked genes in the selection process has made to get meaningful classification accuracies in 4 of 5 datasets. In other hand, a gradual decrease of the top-ranked genes offers small gene subsets with high classification accuracies. In our future work, we will study more effective and efficient heuristics to combine SPA and other recent local-search methods.

References

1. Wang, S., Aorigele, Kong, W., Zeng, W., Hong, X.: Hybrid binary imperialist competition algorithm and tabu search approach for feature selection using gene expression data. BioMed Res. Int. **2016**, 12 (2016)
2. Alshamlan, H., Badr, G., Alohalil, Y.: mRMR-ABC: a hybrid gene selection algorithm for cancer classification using microarray gene expression profiling. In: Hindawi Publishing Corporation BioMed Research International Volume (2015)
3. Chuang, L.-Y., Yang, C.-H., Li, J.-C., Yang, C.-H.: A hybrid BPSOCGA approach for gene selection and classification of microarray data. J. Comput. Biol. **19**(1), 68–82 (2012)
4. Elyasigomari, V., Lee, D.A., Screen, H.R.C., Shaheed, M.H.: Development of a two-stage gene selection method that incorporates a novel hybrid approach using the cuckoo optimization algorithm and harmony search for cancer classification. J. Biomed. Inf. **67**, 11–20 (2017)
5. Sharbaf, F.V., Mosafer, S., Moattar, M.H.: A hybrid gene selection approach for microarray data classification using cellular learning automata and ant colony optimization. Genomics **107**(6), 231–238 (2016)
6. Lu, H., Chen, J., Yan, K., Jin, Q., Xue, Y., Gao, Z.: A hybrid feature selection algorithm for gene expression data classification. Neurocomputing **256**, 56–62 (2017)
7. Apolloni, J., Leguizamón, G., Alba, E.: Two hybrid wrapper-filter feature selection algorithms applied to high-dimensional microarray experiments. Appl. Soft Comput. **38**, 922–932 (2016)
8. Bolón-Canedo, V., Sánchez-Maroño, N., Alonso-Betanzos, A.: Feature selection for high-dimensional data. Prog. Artif. Intell. **5**(2), 18 (2016)
9. Golub, T.R., Slonim, D.K., Tamayo, P., Huard, C., Gaasenbeek, M., Mesirov, J.P., Coller, H., Loh, M.L., Downing, J.R., Caligiuri, M.A., Bloomfield, C.D., Lander, E.S.: Molecular classification of cancer: class discovery and class prediction by gene expression monitoring. Science **286**, 531–537 (1999)
10. Alon, U., Barkai, N., Notterman, D.A., Gish, K., Ybarra, S., Mack, D., Levine, A.J.: Broad patterns of gene expression revealed by clustering analysis of tumor and normal colon tissues probed by oligonucleotide arrays. Proc. Natl. Acad. Sci. USA **96**, 6745–6750 (1999)
11. Gordon, G.J., Jensen, R.V., Ramaswamy, S., Richards, W.G., Sugarbaker, D.J., Bueno, R., et al.: Translation of microarray data into clinically relevant cancer diagnostic tests using gene expression ratios in lung cancer and mesothelioma. Can. Res. **17**(62), 4963–4967 (2002)
12. Pomeroy, S.L., Tamayo, P., et al.: Prediction of central nervous system embryonal tumour outcome based on gene expression. Nature **415**, 436–442 (2002)
13. Singh, D., Febbo, P., Ross, K., Jackson, D., Manola, J., Ladd, C., Tamayo, P., Renshaw, A., D'Amico, A., Richie, J.: Gene expression correlates of clinical prostate cancer behavior. Cancer Cell **1**, 203–209 (2002)
14. Dudoit, S., et al.: Comparison of discriminant methods for the classification of tumors using gene expression data. J. Am. Stat. Assoc. **9**, 77–87 (2002)

15. Tarek, S., Abd-Elwahab, R., Shoman, M.: Gene expression based cancer classification. Egypt. Inf. J. **18**(3), 151–159 (2017)
16. Yang, X.-S.: Nature-Inspired Metaheuristic Algorithms. Luniver Press, Bristol (2011)
17. Salhi, A., Fraga, E.: Nature-inspired optimisation approaches and the new plant propagation algorithm. In: Proceedings of 2011 International Conference on Numerical Analysis and Optimization (ICeMATH 2011), pp. K2-1–K2-8 (2011)
18. Merrikh-Bayat, F.: A Numerical Optimization Algorithm Inspired by the Strawberry Plant. arXiv preprint arXiv:1407.7399, pp. 10–36 (2014)
19. Akyol, S., Alatas, B.: Plant intelligence based metaheuristic optimization algorithms. Artif. Intell. Rev. **45**(4), 414–462 (2017)
20. Li, S., Tan, M.: Gene selection using hybrid particle swarm optimization and genetic algorithm. Soft. Comput. **12**, 1039–1048 (2008)
21. Ben-Dor, A., Bruhn, L., et al.: Tissue classification with gene expression profiles. J. Comput. Biol. **7**(3–4), 559–583 (2000)
22. Wang, Y., Makedon, F.S., Ford, J.C., Pearlman, J.: HykGene: a hybrid approach for selecting marker genes for phenotype classification using microarray gene expression data. Bioinformatics **21**(8), 1530–1537 (2005)
23. Wan, S.-L., Li, X., et al.: Tumor classification by combining PNN classifier ensemble with neighborhood rough set based gene reduction. Comput. Biol. Med. **40**, 179–189 (2010)
24. Wessels, L.F.A., Rain, J.T.M., et al.: Representation and classification for high-throughput data. In: Proceedings of the SPIE 4626, Biomedical Nanotechnology Architectures and Applications, vol. 4626, pp. 226–237 (2002)
25. Cho, S.-B., Won, H.-H.: Machine learning in DNA microarray analysis for cancer classification. In: Proceedings of the 1st Asia-Pacific bioinformatics conference on Bioinformatics, vol. 19, pp. 189–198 (2003)
26. Cho, S.-B., Won, H.-H.: Cancer classification using ensemble of neural networks with multiple significant gene subsets. Appl. Intell. **26**(3), 243–250 (2007)
27. Deb, K., Reddy, R.: Reliable classification of two-class cancer data using evolutionary algorithms. BioSystems **72**(1), 111–129 (2003)
28. Karimi, S., Farrokhnia, M.: Leukemia and small round blue cell tumor cancer detection using microarray gene expression data set: combining data dimension reduction and variable selection technique. Chemom. Intell. Lab. Syst. **139**, 6–14 (2014)
29. Tang, Y., Zhang, Y., Huang, Z.: Development of two-stage SVMRFE gene selection strategy for microarray expression data analysis. IEEE/ACM Trans. Comput. Biol. Bioinformat. **4**(3), 365–381 (2007)
30. Vinterbo, S.A., Kim, E.-Y., Ohno-Machao, L.: Small, fuzzy and interpretable gene expression based classifiers. Bioinformatics **21**(9), 1964–1970 (2005)
31. Chu, W., Ghahramani, Z., Falciani, F., Wild, D.L.: Biomarker discovery in microarray gene expression with Gaussian process. Bioinformatics **21**(16), 3385–3393 (2005)
32. Guan, Z., Zhao, H.: A semiparametric approach for marker gene selection based on gene expression data. Bioinformatics **24**(4), 529–536 (2005)
33. Hu, S., Rao, J.: Statistical redundancy testing for improved gene selection in cancer classification using microarray data. Cancer Informat. **2**, 29–41 (2007)
34. Arevalillo, J.-M., Navarro, H.: A new approach for detecting bivariate interactions in high-dimensional data using quadratic discriminant analysis. In: Proceedings of the 9th International Workshop Data Mining Bioinformatics, pp. 1–7 (2010)
35. Wan, X., Gotoh, O.: Microarray-based cancer prediction using soft computing approach. Cancer Informat. **7**, 123–139 (2009)

36. Bonilla-Huerta, E., et al.: Hybrid framework using multiple-filters and an embedded approach, for an efficient selection and classification of microarray data. IEEE/ACM Trans. Comput. Biol. Bioinf. **13**(1), 12–26 (2016)
37. Chen, D., et al.: Selecting genes by test statistics. J. Biomed. Biotechnol. **2**, 132–138 (2005)
38. Wang, S., et al.: Gene selection with rough sets for the molecular diagnosing of tumor based on support vector machines. In: Proceedings of the ICS, pp. 1368–1373 (2006)
39. Wang, S., Chen, H., Li, S.: Gene selection using neighborhood rough set from gene expression profiles. In: Proceedings of the International Conference on Computer Intelligent Security, pp. 959–963 (2007)
40. Luque-Baena, R.M., Urda, D., Subirats, J.L., Franco, L., Jerez, J.M.: Application of genetic algorithms and constructive neural networks for the analysis of microarray cancer data. Theoret. Biol. Med. Model. **11**(Suppl. 1), S7 (2014)
41. Vanitha, D.-A., Devarajb, D., Venkatesuluc, M.: Gene expression data classification using support vector machine and mutual information-based gene selection. Procedia Comput. Sci. **47**, 13–21 (2015)
42. Zhang, H., Wang, H., Dai, Z., et al.: Improving accuracy for cancer classification with a new algorithm for genes selection. BMC Bioinform. **13**, 298 (2012)
43. Gao, L., Ye, M., et al.: Hybrid method based on information gain and support vector machine for gene selection in cancer classification. Genomics Proteomics Bioinform. **15**, 389–395 (2017)
44. Mao, Z., Cai, W., Shao, X.: Hybrid method based on information gain and support vector machine for gene selection in cancer classification. J. Biomed. Inform. **46**, 594–601 (2013)
45. Luque-Baena, R.M., Urda, D., et al.: Robust signatures from microarray data using genetic algorithms enriched with biological pathway keywords. J. Biomed. Inform. **49**, 32–44 (2014)
46. Akadi, A.E., Amine, A., Ouardighi, A.E., Aboutajdine, D.: A two-stage gene selection scheme utilizing MRMR filter and GA wrapper. Knowl. Inf. Syst. **26**, 487–500 (2010)
47. Nanni, L., Brahnam, S., Lumini, A.: Combining multiple approaches for gene microarray classification. Bioinformatics **28**(8), 1151–1157 (2012)

Prediction of LncRNA by Using Muitiple Feature Information Fusion and Feature Selection Technique

Jun Meng[1], Dingling Jiang[1], Zheng Chang[1], and Yushi Luan[2(✉)]

[1] School of Computer Science and Technology, Dalian University of Technology, Dalian 116024, Liaoning, China
[2] School of Life Science and Biotechnology, Dalian University of Technology, Dalian 116024, Liaoning, China
luanyush@dlut.edu.cn

Abstract. Recent genomic studies suggest that long non-coding RNAs (lncRNAs) play an important role in regulation of plant growth. Therefore, it is important to find more plant lncRNAs and predict their functions. This paper presents an improved maximum correlation minimum redundancy method for lncRNAs recognition. Sequence feature, secondary structural feature and functional feature such as pseudo-nucleotides feature which is based on the physical and chemical properties between dimers dinucleotide of related RNA have been extracted. Then, using maximum correlation minimum redundancy method to integrate a variety of feature selection methods such as Pearson correlation coefficient, information gain, relief algorithm and random forest for feature selection. Based on the selected superior feature subset, the classification model is established by SVM. Experimental results on Arabidopsis sequence dataset show that pseudo-nucleotides feature reflects information of different RNA sequences and the classification model constructed according to the proposed method can be more accurate than other methods on identification of plant lncRNAs.

Keywords: Ensemble feature selection
Maximum correlation minimum redundancy · Pseudo nucleotides features
Classification · LncRNA

1 Introduction

The development of a new generation of deep sequencing technology enables people to discover tens of thousands of new transcripts [1]. Discovery of a large number of lncRNA in transcriptional process has attracted attention of researchers and has become one of the hotspots in genome research.

At first, lncRNA was thought to be a "dark matter", a noise without biological significance, and did not arouse scientists' attention [2, 3]. Now, researchers have found that lncRNA plays an important role in transcriptional, post transcriptional levels and epigenetic modifications. It is involved in genomic imprinting, chromatin modification, transcriptional interference, transcriptional activation, nuclear transport, and have a

D.-S. Huang et al. (Eds.): ICIC 2018, LNCS 10955, pp. 318–329, 2018.
https://doi.org/10.1007/978-3-319-95933-7_39

close connection with occurrence and development of many kinds of diseases [4, 5]. Until now, the prediction and research work of lncRNA has made some progress in human and animal, such as LncRNApred [6], CPAT [7] and COME [8]. But there are few studies on plant, and finding new plant lncRNAs through experimental method has the disadvantages of high cost and long time [9]. Therefore, it is necessary to develop a machine learning method to identify plant lncRNAs, and lay a foundation for further research on functional study, prevention and control of diseases and insect pests, the influence of environmental factors and so on.

Feature extraction is needed for lncRNAs identification by machine learning. Only with enough class information, can we classify samples correctly through classification algorithm. Few features were used for model construction in other methods. In this paper, we extracted sequence features, secondary structural features and functional features to describe lncRNAs more comprehensively and accurately from different aspects. Due to the imbalance problem of positive and negative samples, undersampling method can be used to improve minority classes performance. In this paper, we use improved K-means clustering algorithm to select representative points to reduce the impact of class imbalance [10].

Through multiple feature extraction, it can obtain sufficient sample information, but cause the increased dimension, calculation complexity and classifier over-fitting problems. Thus, feature selection is necessary. In order to select important and high efficiency features, we consider both the maximum correlation between features and categories and the minimum redundancy between features, integrating a variety of feature selection methods such as Pearson correlation coefficient, information gain, relief algorithm and random forest to evaluated comprehensively.

SVM is a classifier technology based on structural risk minimization theory and has a better generalization ability [11]. In this paper, SVM classification model is finally constructed to identify and predict plant lncRNAs.

The main contributions of this article are as follows:

(1) Several features of lncRNA are extracted from different aspects, especially the features of pseudo nucleotides based on physical and chemical properties between dinucleotide.
(2) Ensemble feature selection method is adopted. Features are comprehensively evaluated by the improved maximum correlation minimum redundancy method (p-mRMR) with multiple feature selection methods.

2 Feature Extraction

2.1 Sequence Feature

The sequence features of RNA are the simplest and most directly related features. k-mer sequence features are the most widely used. k represents k adjacent ribonucleotides, RNA are polymers of four ribonucleotides (A, C, G, U), so k adjacent ribonucleotides will have 4^k possible combinations. The corresponding frequency feature is:

$$D_k = [f_1^k, f_2^k, \ldots, f_{4^k-1}^k, f_{4^k}^k] \tag{1}$$

In this experiment, $k = 1$, 2 and 3. Meanwhile, C + G codon frequency has biological significance [12], CG codon frequency $f(C + G)$ can be used as an important sequence feature of RNA.

Open reading frame (ORF) is a class of sequence features. On mRNA chain, it has some longer ORFs while no ORFs on lncRNA chain. In this experiment, ORFs coverage length and coverage ratio features are extracted. The coverage length is the length of RNA covered by all ORFs, and the coverage ratio is the ratio of the coverage to the length of RNA sequence.

$$L_O = L_1 \cap L_2 \cap \ldots \cap L_i \tag{2}$$

$$L_C = \frac{L_O}{L} \tag{3}$$

L_i represents the length of ORFs in RNA i, L_O represents ORFs coverage length, L_C represents ORFs coverage ratio.

2.2 Secondary Structural Feature

RNA secondary structure can be calculated by ViennaRNA packet [13]. In point parenthesis representation, '(' or ')' indicates that the location has a matched ribonucleotide while '·' indicates that the location does not match a ribonucleotide. In this experiment, only considering the ribonucleotide whether has a matched ribonucleotide using '(' to represent, so, for any three adjacent nucleotides, there are 8 possible combinations of structures. The frequency probability of each structure state is calculated, and the 8 secondary structural feature vector is shown as:

$$D_S = [f("((("), f("((."), \ldots, f("..("), f("...")] \tag{4}$$

2.3 Functional Feature

The concept of PseAAC has been penetrating into almost all the fields of protein attribute predictions [14]. Encouraged by the successes of introducing the PseAAC approach into computational proteomics, pseudo nucleotide composition was proposed [15], and has been widely used in various field in DNA and RNA fields. The feature vector based on pseudo-nucleotide composition information is defined as:

$$D_k = [d_1, d_2, \ldots, d_{16}, \ldots, d_{16+\lambda}] \tag{5}$$

$$d_k = \begin{cases} \dfrac{f_k}{\sum_{i=1}^{16} f_i + w \sum_{j=1}^{\lambda} \theta_j} (1 \le k \le 16) \\ \dfrac{w\ \theta_{k-16}}{\sum_{i=1}^{16} f_i + w \sum_{j=1}^{\lambda} \theta_j} (17 \le k \le 16 + \lambda) \end{cases} \tag{6}$$

$f_k(k = 1, 2, \ldots, 16)$ represents the normalized dinucleotide frequency. w is a weighting factor in the range [0, 1]. The components from 16 + 1 to 16 + λ reflect effect of long-range sequence order. The definition of Θ_j is:

$$\theta_j = \frac{1}{L-j-1}\sum_{i=1}^{L-j-1} \theta(R_iR_{i+1}, R_{i+j}R_{i+j+1}) \tag{7}$$

$j = (1, 2, \ldots, \lambda; \lambda < L), \theta_j$ represents nucleotide relationship in the j-th layer sequence which can be calculated by physicochemical features between i and $i + j$ dinucleotides. The solution method of $\theta(R_iR_{i+1}, R_{i+j}R_{i+j+1})$ is shown as:

$$\theta(R_iR_{i+1}, R_{i+j}R_{i+j+1}) = \frac{1}{\mu}\sum_{v=1}^{\mu} \left[P_v(R_iR_{i+1}) - P_v(R_{i+j}R_{i+j+1})\right]^2 \tag{8}$$

Where μ represents the number of physicochemical properties between dinucleotides, in this paper, μ = 22 which means 22 physicochemical properties [16] are considered. $P_v(R_iR_{i+1})$ represents the value of dinucleotide located at the i-th position in the v-th physicochemical property.

In addition, the minimal free energy when RNA is folded to form a spatial structure is also an important functional feature of RNA [17].

3 Ensemble Feature Selection

3.1 Information Gain

Information gain is an evaluation method based on information entropy [18]. Assume that there are N samples in dataset D and m different class labels, then the expected information required for classifying N samples into m classes is:

$$Info(D) = -\sum_{i=1}^{m} p_i \log_2(p_i) \tag{9}$$

Where p_i represents the probability that a sample belongs to one class label.

Assume that A is a feature of dataset D, which has v different values $\{a_1, a_2, \ldots, a_v\}$. According to the value of feature A, D can be divided into v subsets $\{D_1, D_2, \ldots, D_v\}$, where each subset D_j has the same value on feature A. The expected information needed to classify N samples according to feature A is shown as:

$$Info_A(D) = \sum_{j=1}^{v} \frac{|D_j|}{|D|} \cdot Info(D_j) \tag{10}$$

Therefore, the information gain of feature A is defined as the difference between the information required to classify based on category and feature A, which is shown as:

$$Gain(D) = Info(D) - Info_A(D) \tag{11}$$

3.2 Pearson Correlation Coefficient

Pearson correlation coefficient [19] is a measure that reflects linear relationship and reflects degree of correlation between two random variables. Quotient of covariance between two random variables and its standard deviation product is defined as the Pearson correlation coefficient:

$$PCC = \frac{\mathrm{cov}(X, Y)}{\sigma_X \sigma_Y} = \frac{\sum_{i=1}^{n} (X_i - \overline{X})(Y_i - \overline{Y})}{\sqrt{\sum_{i=1}^{n} (X_i - \overline{X})^2}\sqrt{\sum_{i=1}^{n} (Y_i - \overline{Y})^2}} \tag{12}$$

When random variable X and Y have the same change, Pearson correlation coefficient between them does not change. Pearson correlation coefficient is in the range of $[-1, 1]$. It means that Y decreases with the increase of X when it is negative, Y increases with the increase of X when it is positive.

3.3 Relief Algorithm

Relief algorithm calculates correlation between features and categories based on discrimination between neighboring and heterogeneous sample, the distinguishing ability is reflected by distance.

Assume that there are N samples in dataset D, there are m features in each sample, let X_a^b represent the value of sample X_a on feature b, and the distance between sample X_i and X_j on feature t is shown as:

When the eigenvalue is discrete:

$$diff(X_i^t, X_j^t) = \begin{cases} 0 & X_i^t = X_j^t \\ 1, & otherwise \end{cases} \qquad (0 < i; j \leq N; 0 < t \leq m) \tag{13}$$

When the eigenvalue is continuous:

$$diff(X_i^t, X_j^t) = \left| \frac{X_i^t - X_j^t}{\max_t - \min_t} \right| \qquad (0 < i; j \leq N; 0 < t \leq m) \tag{14}$$

Where $\max_t$ represents the maximum of all samples on feature t, and $\min_t$ represents the minimum of all samples on feature t.

Randomly selecting a sample from the dataset, finding the nearest neighbor sample from the same category as X_i, denoted as *NearHit*, we find the nearest neighbor sample from different category as X_i, denoted as *NearMiss*. The weight δ^t of feature t is shown as:

$$\delta^t = \delta^t - \frac{diff(X_i^t, X_{NearHit}^t)}{r} + \frac{diff(X_i^t, X_{NearMiss}^t)}{r} \tag{15}$$

Where r is iterations number.

3.4 Random Forest

Random forest [20] is not only a classification method that fuse results of multiple decision trees, but also measures the importance of classification features.

The importance of feature I by random forest is calculated as follows:

(1) In the process of constructing K classification trees, K out-of-pocket data are composed by the samples that is not extracted each time, and error rate is calculated by the corresponding decision tree, which is denoted as $errOOB_1$.
(2) Adding interference to features I of all out-of-pocket data, randomly change eigenvalues. Then, error rate is calculated again by the corresponding decision tree, which is denoted as $errOOB_2$.
(3) The method for calculating the importance X_I of the feature is shown as:

$$X_I = \frac{\sum_{i=1}^{K} (errOOB_{2i} - errOOB_{1i})}{K} \tag{16}$$

We can see that when adding interference to a feature, if out-of-pocket data accuracy is significantly different, the feature has a greater impact on classification and is of higher importance.

3.5 Ensemble Feature Selection Based on Improved Maximum Relevance Minimum Redundancy

The maximum correlation minimum redundancy method proposed by Peng et al. [21] is a heuristic feature selection method. Its basic idea is to make the correlation larger between selected feature subset and categories, the correlation smaller between features at the same time.

In this paper, we use an improved maximum relevance minimum redundancy method. In the correlation calculation between features and categories, Pearson correlation coefficient, information gain, Relief algorithm and random forest are used.

$$maxD(S,C), D(S,C) = \frac{1}{N|S|}\sum_{m=1}^{N} \sum_{f_i \in S} I_m(f_i, C) \tag{17}$$

Where S is the feature set, C is the category information, and N is the number of feature selection methods. In this paper, N is 4, $I_m(f_i, C)$ represents the correlation between feature i and category calculated by *m-th* feature selection method.

Redundancy calculation between features is:

$$minR(S), R(S) = \frac{1}{|S|^2} \sum_{f_i, f_j \in S} |PCC(f_i, f_j)| \tag{18}$$

Where $PCC(f_i, f_j)$ represents the Pearson correlation coefficient between feature i and feature j.

Information difference is used as the evaluation function for feature subset selection.

$$max\phi_1(D,R), \phi_1 = D - R \tag{19}$$

Forward search method is used to find the feature subsets, all unselected features are evaluated, each time the best evaluated feature is selected from unselected feature subset. Assuming that the feature set S_{n-1} composed of the $n-1$ features has been selected, the *n-th* feature needs to be found from the remaining feature sets. The evaluation function is:

$$MID = \max_{f_j \in S - S_{n-1}} \left\{ \frac{1}{N} \sum\nolimits_{m=1}^{N} I_m(f_j, C) - \frac{1}{n-1} \sum\nolimits_{i=1}^{n-1} |PCC(f_j, f_i)| \right\} \tag{20}$$

The improved mRMR ensemble feature selection method is described as follows:

Input: extracted feature set S, category information set C and $N = 4$ feature selection methods.

Output: feature rating.

Step 1. Calculate the correlation between all features and categories based on correlation calculation formula (17).
Step 2. Select the most relevant feature as initial feature subset.
Step 3. Calculate the correlation between each unselected feature and categories, the redundancy among selected features.
Step 4. Unselected features is evaluated using formula (20), and each time feature with the highest evaluation is added to feature subset.
Step 5. Until all features have been evaluated, otherwise return to step3.

The overall experimental flow of this paper is shown in Fig. 1.

4 Experimental Results and Analysis

4.1 Dataset

In the paper, positive lncRNA dataset is downloaded from the Plant Non-conding RNA Database, the URL is http://structuralbiology.cau.edu.cn/PNRD. There are 2,545 validated Arabidopsis lncRNAs. The negative set is Arabidopsis encoded protein mRNA dataset, it is obtained from Arabidopsis Araport11 genome release in the Arabidopsis Information Resources (TAIR) database which contains 27,655 Protein coding gene, and the URL is http://www.arabidopsis.org/.

In order to solve imbalance between positive and negative samples, remove noise and redundant samples, avoid over-fitting of data at the same time, we use an improved K-means clustering method to cluster Arabidopsis mRNA dataset and select representation points of the same number as positive set.

In the selection of initial cluster center, Euclidean distance from each sample point to origin is calculated first according to one nucleotide frequency sequence feature, as

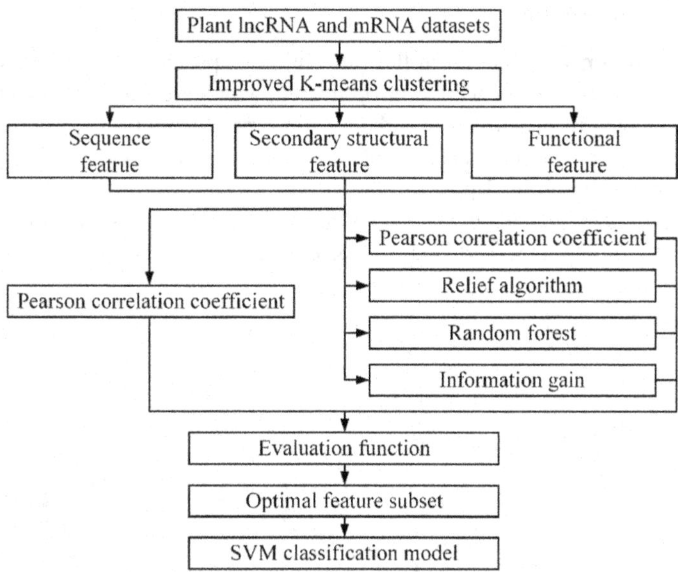

Fig. 1. Flow chart of prediction model. Mainly for feature extraction, feature selection and classifier construction three parts.

shown in Eq. (17), and then sort obtained Euclidean distances. Negative set is divided into 2545 clusters on average according to the sort. Taking the median of each cluster as the initial clustering center of K-means algorithm to cluster.

$$d_i = \sqrt{f_{ia}^2 + f_{ig}^2 + f_{ic}^2 + f_{iu}^2} \tag{21}$$

4.2 Data Preprocessing and Evaluation Index

In order to avoid classification error caused by inconsistent range of extracted feature, all extracted features are normalized to [−1, 1]. K-fold cross-validation, jackknife test and independent data test can be used to evaluate the performance of proposed models. However, among the three methods, the jackknife test is deemed the least arbitrary because it can always yield a unique outcome for a given benchmark dataset. Accordingly, the jackknife test has been increasingly used and widely recognized by investigators to examine the accuracy of various predictors. However, the jackknife test was time-consuming. For time saving, we used 5-fold cross-validation in this study. The classifier is constructed by SVM and RBF kernel function $K(x, y) = \exp(-\gamma\|x-y\|^2)$ was selected. Accuracy (ACC), Sensitivity (SN), Specificity (SP) and F1 value are used to evaluate the classification model.

4.3 Analysis of Results

Feature Extraction Results. From the calculation of pseudo-nucleotides, we can see that there are two parameters need to be optimized, which are λ and w. λ is a long-range information ranged from 1 to 50. w is the weight ranged from 0 to 1. In order to avoid over-fitting problem, besides, ensure accuracy of the prediction, we use grid search method to obtain the optimal parameters.
Using pseudo-nucleotide feature alone to construct a SVM classification model, the classification accuracy obtained is shown in Fig. 2.

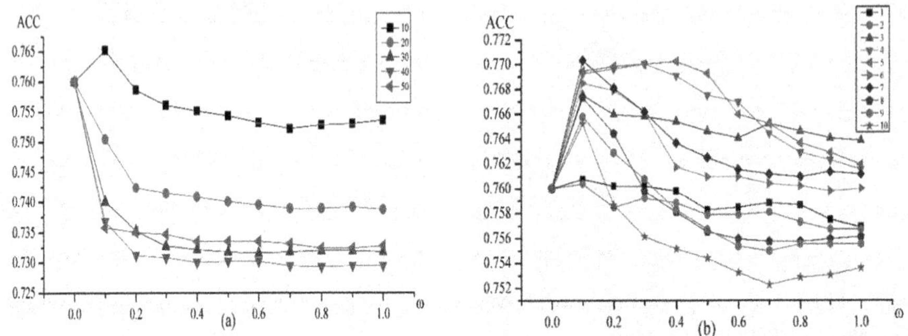

Fig. 2. Classification accuracy based on pseudo nucleotide features. (a) represents classification accuracy with the step length of λ is 10 from 10 to 50. (b) represents classification accuracy with the step length of λ is 1 from 1 to 10.

As we can see from Fig. 2(a), the accuracy of classifier decreases with the increase of λ from 10 to 40. In order to further find the optimal values of λ and w, the paper constructs a classification model of λ from 1 to 10. Classification results are shown in Fig. 2(b). When $\lambda = 7$ and $w = 0.1$, the classifier has the highest accuracy. Therefore, on the pseudo-nucleotide feature, taking $\lambda = 7$, $w = 0.1$, a total of 23 dimensional pseudo-nucleotide eigenvectors are obtained.

The 119-dimensional features extracted in this paper are shown in Table 1.

Ensemble Feature Selection Results. Figure 3 shows the accuracy of classifiers established by different feature selection algorithms to select different numbers of high ranked features. It can be seen from Fig. 3(a) that when the selected feature number is less than 60, classification accuracy constructed by p-mRMR feature selection method proposed in this paper is obviously higher than other methods. When the number of selected features is greater than 60, the classifier constructed by p-mRMR method also has the classification accuracy. As a whole, the method proposed in this paper has high classification accuracy and stability.

In order to find the optimal number of feature selection, we use p-mRMR method to select the feature numbers ranged from 25 to 40 and from 50 to 65. The experimental results are shown in Fig. 3(b) and (c). It can be seen from the comparison that when 34 features with higher rank are selected, the classification accuracy is the highest.

Table 1. LncRNA feature extraction results.

Feature category	Feature name	Number
Sequence feature	k-mer	84
	ORF	2
	CG content	1
Secondary structural feature	Three tuples	8
Functional feature	Pseudonucleotides	23
	Minimum free energy	1

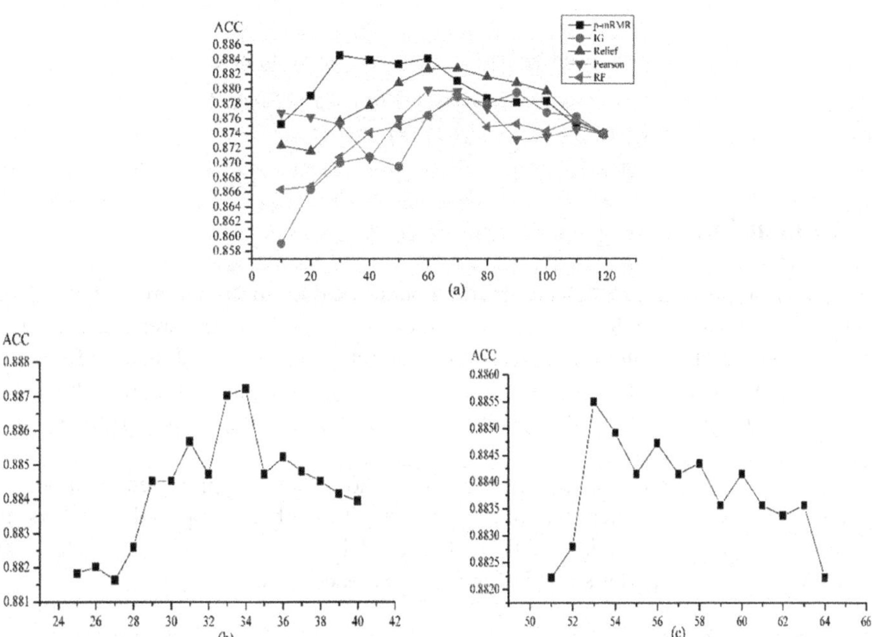

Fig. 3. Accuracy of different quantitative features for different feature selection methods. (a) represents classification accuracy with the step length of feature numbers is 10. (b) and (c) represent classification accuracy of feature numbers in the range from 25 to 40 and from 50 to 65.

Table 2 shows classification performance of classifiers constructed without using the RNA pseudonucleotide feature (NP), using all extracted features (AP), and using the 34 higher ranked features selected by p-mRMR method.

It can be seen from Table 2 that pseudo-nucleotide feature can reflect the information of different RNA sequences, and classification accuracy on positive and negative samples has also been improved with pseudo-nucleotide feature. Using p-mRMR method for feature selection, classification accuracy is improved about 1.5%, especially the classification performance for positive samples is improved in particular.

Table 2. Classification performance of different features.

Feature	ACC(%)	SN(%)	SP(%)	F1
NP	87.15	80.52	93.74	0.8713
AP	87.39	81.05	93.61	0.8732
p-mRMR	88.72	83.84	93.56	0.8871

Table 3 shows the results of comparative experiments with other methods. It can be seen from the table that in Arabidopsis dataset, the classification model constructed in this paper has a higher classification accuracy.

Table 3. Comparison of classification accuracy with other methods.

Method	p-mRMR	CPC	CPAT	LncRNApred
ACC(%)	88.72	69.47	81.29	83.64

5 Conclusion

This paper proposes a lncRNA prediction model based on multi-feature fusion. RNAs sequence features, secondary structure features, and functional features are extracted. Pearson correlation coefficient, information gain, relief algorithm and random forest are integrated to select features by using the maximum correlation minimum redundancy method. A 34-dimensional feature with a higher rank is selected, finally, using SVM to create a classification model.

Results on Arabidopsis sequence datasets show that the proposed method has certain accuracy of classification, and compared with methods proposed by others, its classification performance is the best. The model constructed in this paper is an lncRNA recognition model suitable for plant sequence data.

Acknowledgement. The current study was supported by the National Natural Science Foundation of China (Nos. 61472061 and 31471880), and the Graduate Educational Reform Fund of Dalian University of Technology (Jg2017015).

References

1. An, N., Palmer, C.M., Baker, R.L., et al.: Plant high-throughput phenotyping using photogrammetry and imaging techniques to measure leaf length and rosette area. Comput. Electron. Agric. **127**(C), 376–394 (2016)
2. Perron, U., Provero, P., Molineris, I.: In silico prediction of lncRNA function using tissue specific and evolutionary conserved expression. BMC Bioinform. **18**(5), 144 (2017)
3. Mercer, T.R., Mattick, J.S.: Structure and function of long noncoding RNAs in epigenetic regulation. Nat. Struct. Mol. Biol. **20**(3), 300 (2013)
4. Aryal, B., Rotllan, N., Fernández-hernando, C.: Noncoding RNAs and atherosclerosis. Current Atherosclerosis Rep. **16**(5), 1–11 (2014)

5. Lee, J.T., Bartolomei, M.S.: X-inactivation, imprinting, and long noncoding RNAs in health and disease. Cell **152**(6), 1308–1323 (2013)
6. Pian, C., Zhang, G., Chen, Z., et al.: LncRNApred: classification of long non-coding RNAs and protein-coding transcripts by the ensemble algorithm with a new hybrid feature. PLoS ONE **11**(5), e0154567 (2016)
7. Wang, L., Park, H.J., Dasari, S., Wang, S., Kocher, J.-P., Li, W.: CPAT: Coding-Potential Assessment Tool using an alignment-free logistic regression model. Nucleic Acids Res. **41** (6), e74 (2013)
8. Long, H., Xu, Z., Hu, B., et al.: COME: a robust coding potential calculation tool for lncRNA identification and characterization based on multiple features. Nucleic Acids Res. **45**(1), e2 (2017)
9. Schneider, H.W., Raiol, T., Brigido, M.M., et al.: A Support Vector Machine based method to distinguish long non-coding RNAs from protein coding transcripts. BMC Genom. **18**(1), 804 (2017)
10. Yen, S.J., Lee, Y.S.: Cluster-based under-sampling approaches for imbalanced data distributions. Expert Syst. Appl. **36**(3), 5718–5727 (2009)
11. Kumar, M., Gromiha, M.M., Raghava, G.P.: SVM based prediction of RNA-binding proteins using binding residues and evolutionary information. J. Mol. Recognit. **24**(2), 303–313 (2011)
12. Tatarinova, T., Brover, V., Troukhan, M., et al.: Skew in CG content near the transcription start site in, Arabidopsis thaliana. Bioinformatics **19**(Suppl. 1), i313 (2003)
13. Stadler, P.F., Hofacker, I.L., Lorenz, R., et al.: ViennaRNA Package 2.0. Algorithms Mol. Biol. **6**(1), 26 (2011)
14. Zhao, Y.W., Su, Z.D., Yang, W., et al.: IonchanPred 2.0: a tool to predict ion channels and their types. Int. J. Mol. Sci. **18**(9), 1838 (2017)
15. Chen, W., Feng, P.M., Lin, H., et al.: iRSpot-PseDNC: identify recombination spots with pseudo dinucleotide composition. Nucleic Acids Res. **41**(6), e68 (2013)
16. Liu, B., Liu, F., Fang, L., et al.: repRNA: a web server for generating various feature vectors of RNA sequences. Mol. Genet. Genomics **291**(1), 473–481 (2016)
17. Zuber, J., Sun, H., Zhang, X., et al.: A sensitivity analysis of RNA folding nearest neighbor parameters identifies a subset of free energy parameters with the greatest impact on RNA secondary structure prediction. Nucleic Acids Res. **45**(10), 6168–6176 (2017)
18. Dai, J., Xu, Q.: Attribute selection based on information gain ratio in fuzzy rough set theory with application to tumor classification. Appl. Soft Comput. J. **13**(1), 211–221 (2013)
19. Shin, J.H., Park, C.H., Yang, Y.J., et al.: Entropy-based analysis of the non-linear relationship between gene expression profiles of amplified and non-amplified RNA. Int. J. Mol. Med. **20**(6), 905 (2007)
20. Breiman, L.: Random forests. Mach. Learn. **45**(1), 5–32 (2001)
21. Peng, H., Long, F., Ding, C.: Feature selection based on mutual information: criteria of max-dependency, max-relevance, and min-redundancy. IEEE Trans. Pattern Anal. Mach. Intell. **27**(8), 1226–1238 (2005)

Tri-Clustering Analysis for Dissecting Epigenetic Patterns Across Multiple Cancer Types

Yanglan Gan[1], Zhiyuan Dong[1], Xia Zhang[2], and Guobing Zou[2(✉)]

[1] School of Computer Science and Technology, Donghua University, Shanghai, China
[2] School of Computer Engineering and Science, Shanghai University, Shanghai, China
gbzou@shu.edu.cn

Abstract. Tumor cells not only harbor genetic and epigenetic alterations, but also are regulated by various epigenetic modifications. Identification of tumor epigenetic similarities across different cancer types is useful for the discovery of treatments that can be extended to different cancers. Nowadays, abundant epigenetic modification profiles have provided good opportunity to achieve this goal. Here, we proposed a tri-clustering approach for integrative pan-cancer epigenomic analysis, named TriPCE. We applied TriPCE to uncover epigenetic mode among seven cancer types. This approach can identify significant cross-cancer epigenetic modification similarities. The associated gene analysis demonstrates strong relevance with cancer development and reveals consistent tendency among cancer types.

Keywords: Tri-clustering · Epigenetic pattern · Pan-cancer

1 Introduction

Aberrant epigenetic modification is a critical factor involving human diseases [1]. Tumor cells usually exhibit epigenetic abnormalities and further routinely use epigenetic processes to ensure their escape from various treatments [2]. Epigenetic modification patterns that lead to the corresponding dysregulation in cancers have become a critical research issue of cancer studies [3, 4].

BLUEPRINT, TCGA and the International Cancer Genome Consortium have integrated many epigenetic maps in normal and cancerous tissues [5–7]. It is urgent to decipher cancer common epigenetic patterns. Because DNA methylation in cancer is addressed elsewhere [8, 9], we focus on covalent histone modifications in cancers. Previous works mainly focus on identifying combinatorial epigenetic states. CoSBI captures epigenetic patterns based on correlations of histone signals [10]. ChromHMM and HiHMM apply a HMM model to annotate genomic sequences by co-occurrence of multiple epigenetic marks [11, 12]. RFECS is developed based on random forests [13]. IDEAS jointly characterizes epigenetic landscapes in many cell types and detects differential regulatory regions [14]. These methods successfully identify combinatorial

D.-S. Huang et al. (Eds.): ICIC 2018, LNCS 10955, pp. 330–336, 2018.
https://doi.org/10.1007/978-3-319-95933-7_40

epigenetic patterns among different cell types. However, the correlations among different regions are still need to be investigated.

Here, we proposed a tri-clustering approach TriPCE for integrative pan-cancer epigenomic analysis. We applied TriPCE to various epigenomic maps of seven cancer types and identified significant cross-cancer epigenetic modification similarities. Furthermore, the associated gene function analysis demonstrates strong relevance with cancer development and reveals consistent tendency among cancer types.

2 Materials and Methods

We analyzed the epigenomic maps of seven cancer types, including A549, K562, HepG2, HCT116, Hela-S3, multiple myeloma-Cell Line, sporadic Burkitt lymphoma-Cell Line. Totally, we obtained 42 datasets of six epigenetic modifications, including H3K4me1, H3K4me3, H3K9me3, H3K27ac, H3K27me3 and H3K36me3. RNA expression profiles of the seven cancer types were also collected. These dataset were downloaded from the website of NIH Roadmap Epigenome Project.

As shown in Fig. 1, the TriPCE model has three key components.

Step1. Preprocess the epigenetic modification data of different cancer types. Firstly, the genome was represented as consecutive genomic segments with size 200 bps. For each epigenetic mark, we computed the summary tag count of every segment. To remove noise, raw read counts were normalized by the total number of reads followed by arcsine transformation [15]. Further, the epigenetic profiles in the promoter regions were extracted. Then, for each epigenetic mark, the epigenetic profiles of different cancer types were represented as a matrix $E_{k,}$ where k is the index of the epigenetic mark ranging from 1 to K.

Step2. Identify BiClusters based on FP-growth algorithm for each epigenetic mark. We computed correlation coefficients of any two cancer types at every region and obtained a coefficient matrix. If the coefficient is higher than a given threshold, the epigenetic modifications of these two cancer types are regarded as coherent. Then we added the cancer type to the itemset. Based on the resulted itemset, we identified coherent epigenetic patterns using FP-growth algorithm. FP-growth is a data mining method that was originally developed for frequent itemset mining. Further, we inversely identified the corresponding gene set and determined the BiClusters.

Step3. Mine TriClusters with coherent epigenetic patterns across different cancers. Based on the BiClusters of each epigenetic mark, we enumerated the subsets of these epigenetic marks to obtain TriClusters. Each TriCluster represents as a gene set with similar epigenetic changes in different cancer types, which indicate conserved epigenetic signatures that shared by multiple cancer types.

Step1. preprocessing the epigenetic modification data of different tumors

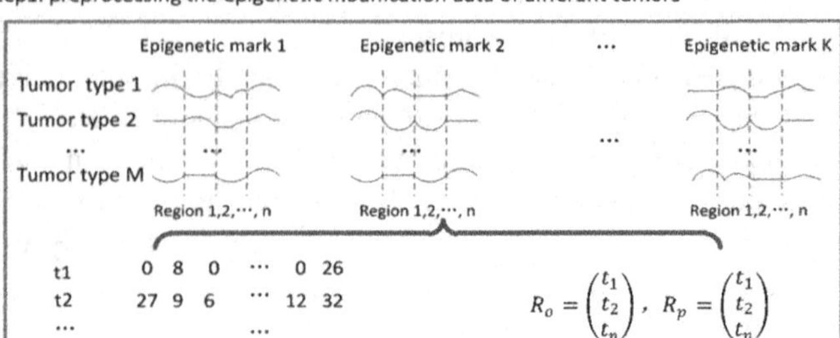

Step2. mining biclusters for each epigenetic mark

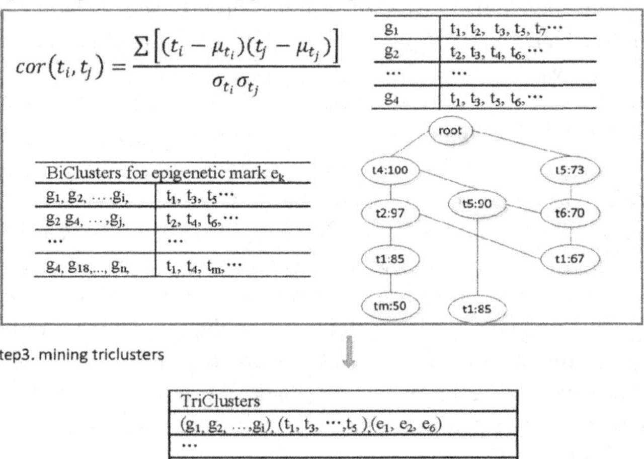

Step3. mining triclusters

TriClusters
$(g_1, g_2, \cdots, g_i)$, $(t_1, t_3, \cdots, t_5)$, (e_1, e_2, e_6)
···
$(g_1, g_2, \cdots g_n)$, $(t_2, t_4, \cdots t_m)$, (e_3, e_5, e_k)

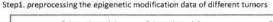

Fig. 1. The flowchart of the proposed TriPCE approach.

3 Results

3.1 Identifying Similar Epigenetic Patterns Across Different Cancer Types

We developed a tri-clustering approach, TriPCE, to capture similar epigenetic patterns among different cancer types. For each epigenetic mark, TriPCE first groups the regions based on the epigenetic modification profiles among different cancer types. Figure 2 shows a typical BiCluster of epigenetic mark H3K4me1, a gene set with similar modification pattern in cancer type Hela-S3, HepG2, K562 and A549. From this figure, we notice that the epigenetic profiles of these genes are similar in these cell types. Meanwhile, different cancers share similar epigenetic patterns. For examples, cancers (HepG2 and HCT116) are clustered together and share larger number of epigenetic marks, implying that they share more similar epigenetic regulation mechanisms. To get significant modification patterns, we set the minima support as 10% of the investigated genes. With diverse correlation coefficient thresholds, we respectively gained different numbers of BiClusters for the epigenetic marks. Among these epigenetic marks, H3K4me3 and H3K9me3 vary most. On the contrary, there are more similar epigenetic patterns of H3K4me1 and H3k27me3. This result is consistent with previous finding that H3K9me3/me2 and H3K36me3/me2 frequently observed in breast cancer [16], esophageal cancer [17] and MALT lymphoma [18]. As the threshold slightly affects the trend among different epigenetic marks, we chose the BiClusters with threshold 0.7 for further analysis.

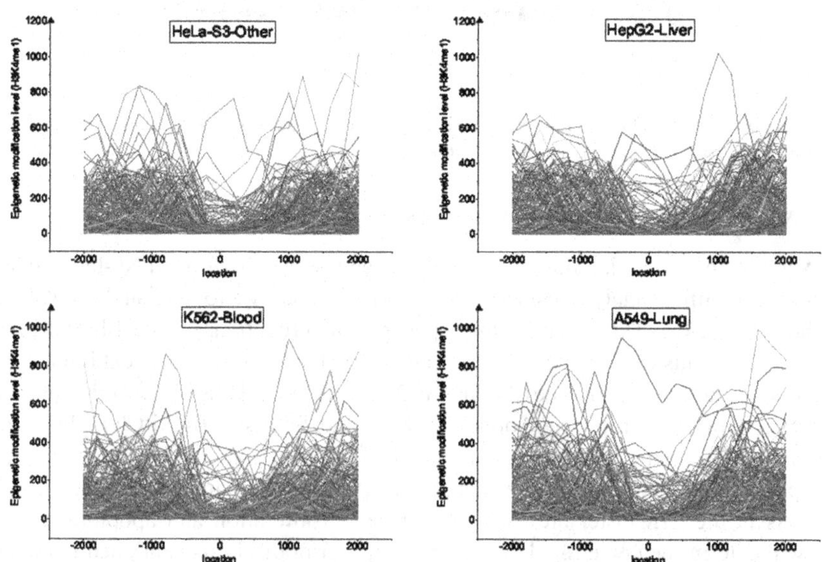

Fig. 2. Profiles of epigenetic modification H3K4me3 in a typical BiCluster display a similar pattern in four cancer types, including Hela-S3, HepG2, K562 and A549.

3.2 Identifying Coherent Patterns Among Different Epigenetic Marks

To identify conserved epigenetic states, we further clustered epigenetic marks based on the identified BiClusters. The TriClusters are represented as triples ('genomic regions', 'tumor types', 'epigenetic marks'). Initially, we obtained 175 TriClusters. Figure 3 shows the epigenetic marks, cancer types and supports of 15 typical clusters. There exist coherent epigenetic states across different cancers types. For example, the variation pattern of H3K4me1, H3K9me3, H3kK27me3 and H3K36me3 is shared in A549, HepG2 and K562. On the contrary, there are some epigenetic patterns are only coherent in certain cell types. We observed similar patterns of H3K36me3, H3K27ac and H3K27me3 among HepG2 and sporadic Burkitt lymphoma-Cell Line.

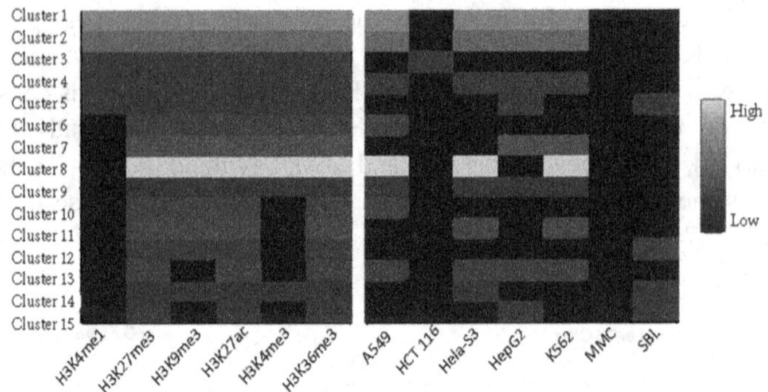

Fig. 3. Typical epigenetic TriClusters. (A) The epigenetic marks (column) in each cluster (row). (B) The cancer types (column) in each cluster (row).

3.3 Analyzing the Potential Roles of Associated Genes

To examine the potential functions of these genes, we performed systematic gene ontology enrichment analysis using DAVID tools. Overall, we found that the TriClusters enriched genes exhibited enrichment for cancer-related functions. Table 1 lists the result of a typical TriCluster (P value < 0.05). In this TriCluster, these genes exhibit coherent epigenetic pattern of H3K4me1, H3K4me3, H3K9me3, H3K27ac, H3K27me3 for HeLa-S3, HepG2, Multiple myeloma-Cell Line and Sporadic Burkitt lymphoma-Cell Line. In the table, term 'positive regulation of cell proliferation' and 'negative regulation of apoptotic process' are enriched in these gene sets. This result implies that the identified gene sets in the TriCluster are essential for cell proliferation and apoptotic process. Meanwhile, term 'negative regulation of gene transcription' is also enriched in the gene set, indicating these genes perform important regulation role in these cancers.

Table 1. Functional enrichment of genes in the identified TriClusters.

Term type	Term name	P-value	Term type	Term name	P-value
BP	positive regulation of cell proliferation	2.84E − 06	MF	glutathione binding	7.85E − 04
BP	protein targeting to Golgi	8.87E − 05	MF	glutathione transferase activity	8.00E − 03
BP	nitrobenzene metabolic process	1.14E − 04	MF	histone binding	1.16E − 02
BP	xenobiotic catabolic process	1.00E − 03	MF	peptidyl-prolyl cis-trans isomerase activity	1.35E − 02
BP	negative regulation of gene expression, epigenetic	1.39E − 03	MF	protein heterodimerization activity	3.32E − 02
BP	negative regulation of apoptotic process	1.88E − 03	CC	extracellular exosome	1.13E − 02

4 Discussion

Identifying epigenetic pattern is important to understand epigenetic mechanisms in various cancers. Our knowledge about the patterns of epigenetic modification and the cause and consequence of them are still limited. Computational approach that exploits the complex epigenomic landscapes and discovers significant signatures out of them are required. In this paper, we developed a tri-clustering approach for integrative pan-cancer epigenomic analysis, named TriPCE. We applied TriPCE to uncover epigenetic patterns of six epigenetic marks among seven cancer types. This approach identifies significant cross-cancer epigenetic modification similarities. The associated gene analysis demonstrates strong relevance with cancer development and reveals consistent tendency among cancer types. Different from existing methods, our approach enable researchers to explore the epigenetic patterns among different cancer types as well as the combinational mode of multiple epigenetic marks.

Acknowledgment. This work was supported in part by the Fundamental Research Funds for the Central Universities (2232016A3-05), the National Natural Science Foundation of China (61772128), and Shanghai Natural Science Foundation (17ZR1400200).

References

1. Jones, P.A., Issa, J.P.J., Baylin, S.: Targeting the cancer epigenome for therapy. Nat. Rev. Genet. **17**(10), 630–641 (2016)
2. You, J.S., Jones, P.A.: Cancer genetics and epigenetics: two sides of the same coin? Cancer Cell **22**(1), 9 (2012)
3. Dawson, M.A.: The cancer epigenome: concepts, challenges, and therapeutic opportunities. Science **355**(6330), 1147–1152 (2017)
4. Kelly, A.D., Issa, J.P.J.: The promise of epigenetic therapy: reprogramming the cancer epigenome. Curr. Opin. Genet. Dev. **42**, 68–77 (2017)
5. Kundaje, A., et al.: Integrative analysis of 111 reference human epigenomes. Nature **518**(7539), 317–330 (2015)
6. Weinstein, J.N., et al.: The cancer genome atlas pan-cancer analysis project. Nat. Genet. **45**(10), 1113–1120 (2015)
7. Beck, S., et al.: A blueprint for an international cancer epigenome consortium. a report from the AACR cancer epigenome task force. Can. Res. **72**(24), 6319–6324 (2012)
8. Kretzmer, H., et al.: Dna-methylome analysis in burkitt and follicular lymphomas identifies differentially methylated regions linked to somatic mutation and transcriptional control. Nat. Genet. **47**(11), 1316–1325 (2015)
9. Yang, X., et al.: Comparative pan-cancer dna methylation analysis reveals cancer common and specific patterns. Brief. Bioinform. **18**(5), 761 (2016)
10. Ucar, D., Hu, Q., Tan, K.: Combinatorial chromatin modification patterns in the human genome revealed by subspace clustering. Nucleic Acids Res. **39**(10), 4063–4075 (2011)
11. Ernst, J., et al., Coyne, M., et al.: Mapping and analysis of chromatin state dynamics in nine human cell types. Nature **473**(7345), 43 (2011)
12. Sohn, K.A., et al.: hiHMM: Bayesian non-parametric joint inference of chromatin state maps. Bioinformatics **31**(13), 2066–2074 (2015)

13. Rajagopal, N., Xie, W., Li, Y., Wagner, U., Wang, W., Stamatoyannopoulos, J., Ernst, J., Kellis, M., Ren, B.: RFECS: a random-forest based algorithm for enhancer identification from chromatin state. PLoS Comput. Biol. **9**(3), e1002968 (2013)
14. Zhang, Y., et al.: Jointly characterizing epigenetic dynamics across multiple human cell types. Nucleic Acids Res. **44**(14), 6721–6731 (2016)
15. Pinello, L., et al.: Analysis of chromatin-state plasticity identifies cell-type-specific regulators of H3K27me3 patterns. PNAS **111**(3), E344 (2014)
16. Liu, G., et al.: Genomic amplification and oncogenic properties of the GASC1 histone demethylase gene in breast cancer. Oncogene **28**(50), 4491 (2009)
17. Yang, Z.Q., et al.: Identification of a novel gene, GASC1, within an amplicon at frequently detected in esophageal cancer cell lines. Can. Res. **60**(17), 4735–4739 (2000)
18. Vinatzer, U., et al.: Mucosa-associated lymphoid tissue lymphoma: novel translocations including rearrangements of ODZ2, JMJD2C, and CNN3. Clin. Cancer Res. **14**(20), 6426–6431 (2008)

Efficient Framework for Predicting ncRNA-Protein Interactions Based on Sequence Information by Deep Learning

Zhao-Hui Zhan[1], Zhu-Hong You[2(✉)], Yong Zhou[1], Li-Ping Li[2], and Zheng-Wei Li[1]

[1] School of Computer Science and Technology, China University of Mining and Technology, Xuzhou 21116, China
[2] The Xinjiang Technical Institute of Physics and Chemistry, Chinese Academy of Science, Urumqi 830011, China
zhuhongyou@ms.xjb.ac.cn

Abstract. The interactions between proteins and RNA (RPIs) play a crucial role in most cellular processes such as RNA stability and translation. Although there have been many high-throughput experiments recently to detect RPIs, these experiments are largely time-consuming and labor-intensive. Therefore, it is imminent to propose an efficient computational method to predict RPIs. In this study, we put forward a novel approach for predicting protein and ncRNA interactions based on sequences information only. By employing the bi-gram probability feature extraction method and k-mer algorithm, the represent features from protein and ncRNA were extracted. To evaluate the performance of the proposed model, two widely used datasets named RPI1807 and RPI2241 were trained with the adoption of random forest classifier by using five-fold cross-validation. The experimental results with the AUC of 0.992 and 0.947 on dataset RPI1807 and RPI2241 respectively indicated the effectiveness of our experimental approach for predicting RPIs, which provided the guidance for reference for future research in the biological field.

Keywords: Protein-ncRNA interaction · Bi-gram · Deep learning Stacked autoencoder · PSSM

1 Introduction

In recent studies in the field of biological knowledge, more and more experiments have shown that ncRNA plays a vital role in the complex cell processes such as cellular proliferation and differentiation [1], chromatin modification [2], cellular apoptosis and so on [3]. At the meantime, a large number of ncRNA have been discovered with the development of modern advanced science and technology while their functions are not yet exactly known [4]. Therefore, it is imminent to make clear the functions of these ncRNAs. To learn about functions of these ncRNAs, researchers are required to identify whether these ncRNAs were able to interact with other proteins in some process of biological reactions [5–11]. However, there still are some shortcomings and improved space in the current prediction methods. Therefore, extracting feature

D.-S. Huang et al. (Eds.): ICIC 2018, LNCS 10955, pp. 337–344, 2018.
https://doi.org/10.1007/978-3-319-95933-7_41

information from sequences is a necessary method which can well identify the interactions proved by large number of research between ncRNA and protein [12–18].

In this study, we put forward a sequence-based method using deep learning model Stacked-autoencoder network combined with Random Forests (RF) classifier. We used K-mers sparse matrices to represent RNA sequences, and then extracted feature vector from matrix by Singular Value Decomposition (SVD). Position Specific Scoring Matrix (PSSM) was used to obtain evolutionary information from each sequence while Bi-gram was further used to get feature vector from PSSM. Then data and label was fed into RF classifier to classify whether a pair of protein and ncRNA interact or not. Furthermore, to evaluate the performance of our approach, five-fold cross validated and two widely used dataset RPI1807 and RPI2241 was used. The experimental results show that our method achieved high accuracy and robustness of the protein-ncRNA interaction prediction tasks.

2 Materials and Methods

2.1 Datasets

We executed experiments on two widely used public datasets including RPI1807 and RPI2241. The dataset RPI1807 consists of 1807 positive ncRNA-protein interaction pairs and 1436 negative ncRNA-protein pairs, including 1078 RNA chains, 1807 protein chains, 493 RNA chains and 1436 protein chains, respectively [19]. It is established by parsing the Nucleic Acid Database (NAD) which provides the RNA-protein complex data and protein-RNA interface database. The RPI2241 dataset is constructed in a similar way, and contains 2241 interacting RNA-protein pairs.

2.2 Features Extraction

To extracted features from ncRNA sequences, k-mer sparse matrix approach was used. A two-dimensional matrix deformation memory to store the features of ncRNA which can express much more useful and significant information such as frequency and location information [20]. An input ncRNA sequence is converted into a $4^k \times (L-k+1)$ matrix M can be defined as follow.

$$M = \left(a_{ij}\right)_{4^k} \times (L-k+1) \tag{1}$$

$$a_{ij} = \begin{cases} 1, & if\ m_j m_{j+1} m_{j+2} m_{j+3} = k - mer(i) \\ 0, & else \end{cases} \tag{2}$$

After obtaining the corresponding two-dimensional matrix from the original sequence of ncRNA, we transform this matrix with large amounts of data by way of singular value decomposition (SVD) [21].

And as well, we extracted protein features from the PSSM matrix calculated from the original protein sequence instead using it directly, since the combinations of amino acid cannot all be found in the original protein sequence [22]. To extract the features recognized from the protein fold, we proposed a bi-gram feature extraction technique computed through the representing information mainly contained from PSSM [23].

The bi-gram occurrence matrix B can be calculated as follows and $b_{m,n}$ be the element in the matrix B:

$$B = \{b_{m,n}, 1 \leq m \leq 20, 1 \leq n \leq 20\} \tag{3}$$

$$b_{m,n} = \sum\nolimits_{i=1}^{r-1} p_{i,m}p_{i+1,n}, i \leq m \leq 20, 1 \leq n \leq 20 \tag{4}$$

where $b_{m,n}$ can be interpreted as the occurrence probability of the transition from m_{th} amino acid to n_{th} amino acid which is able to calculated from the element $p_{i,j}$ in its PSSM matrix [24]. Let F be the bi-gram feature vector of the protein fold recognition which is as follows:

$$F = \{b_{1,1}, b_{1,2}, \cdots, b_{1,20}, b_{2,1}, \cdots, b_{2,20}, \cdots, b_{20,1}, \cdots, b_{20,20}\}^T \tag{5}$$

where the symbol T can be regarded as the transpose of the feature vector [25]. Then, the random forest classifiers were used to predict the interaction between ncRNA and protein.

2.3 Deep Learning Framework Based on Stacked Autoencoder

In order to improve the accuracy of the predicting performance, there had been many recent research which concentrated their attentions on automatic encoders and deep-learning networks [26–32]. In this study, we used the stacked auto-encoder network for deep learning and classification of training datasets to obtain an efficient deep learning network [33]. A complete stacked auto-encoder network consists of a sparse multilayer neural network auto-encoder which layer inputs can be obtained from the outputs of the previous layers [34]. With the hyper parameter optimization, we were able to get the best parameters of the stacked auto-encoder neural network suitable for our machine learning model [35]. The sparse auto-encoder network which was used to learn the feature changes is a single-layer automatic encoder as follows:

$$p_{(\alpha,\beta)}(x) = f(\alpha^T x) = f\left(\sum\nolimits_{i=1}^{n} \alpha_i x_i + \beta_i\right) \tag{6}$$

where the input x can be interpreted as the d-dimension dataset and $f(x)$ is an activation function. And the auto-encoder network maps X into the output $p(X)$. And Sigmoid was selected as activation function as follows:

$$f(y) = \frac{1}{1 + e^{-y}} \tag{7}$$

And consequently, the loss function is as follows:

$$H(X,\alpha) = \|\alpha p - X\|^2 + \omega \sum_j |p(j)| \tag{8}$$

The stacked neural network architecture is composed of multiple neural network layers which outputs of the previous layers are the inputs of next layers [36]. At the meantime, the keras library from Internet was used to implement stacked auto-encoder and the parameters *batch_size* and *nb_epoch* both set to be 100 [37]. The details about keras can be found in website http://github.com/fchollet/keras.

2.4 Stacked Ensemble

In order to find out the solution of assembling mechanism implementing to integrate every individual output from classifiers to implement multi-classifier assembling and obtain an approximately optimal objective function [8, 38–40], we regarded the outputs of all level 0 classifiers as predicted probability scores while the successive level 1 classifiers as logistic regression classifiers. The experimental results shown that stacked assembling was equal to the average individual model results strategy when score weights of logistic regression of all individual level 0 classifiers were same.

$$P_w(y = \pm 1|s) = \frac{1}{1 + e^{-yw^T s}} \tag{9}$$

where s is predicted probability scores of all level 0 classifiers vector outputs and w is the weight vector of corresponding classifiers [41].

3 Experimental Results

The five-fold cross-validation method is used to evaluate the performance of our study, which randomly divides all the data set into five equal parts [42–45]. We followed the widely used evaluation measures to evaluate our method, including accuracy, sensitivity, specificity, precision and AUC [46–50]. The experimental results in dataset RPI1807 and RPI2241 were shown in Table 1.

Table 1. The experimental results in RPI1807 and RPI2241.

	Accuracy	Sensitivity	Specificity	Precision	AUC
RPI1807	0.9600	0.9344	0.9989	0.9117	0.9920
RPI2241	0.9130	0.8772	0.9660	0.8590	0.9470

According to the Table 1, our method achieved a decent performance with an accuracy of 0.9600, sensitivity of 0.9344, specificity of 0.9989, precision of 0.9117 and AUC of 0.9920 in testing dataset RPI1807 and an accuracy of 0.9130, sensitivity of 0.8772, specificity of 0.9660, precision of 0.8590 and AUC of 0.9470 in testing dataset RPI2241.

4 Conclusions

In this study, we proposed a sequence-based method using deep learning model Stacked-autoencoder network combined with RF classifier. By employing the k-mers sparse matrix and bi-gram algorithm, the represent ncRNA and protein features were extracted from the corresponding sequence information. In the process of experiments, our method has shown a satisfying performance for predicting RPIs on each reference dataset which thanks to the contribution of the Stacked ensemble autoencoder framework using deep learning. In general, our method tried to extract protein features and automatic learn the advanced features with the use of random forests classifiers, but still do not had a very good breakthrough achievement from the perspective of biology. In future research, we expect to design a better network architecture for extracting hidden advanced features from the perspective of biology.

References

1. Wapinski, O., Chang, H.Y.: Long noncoding RNAs and human disease. Trends Cell Biol. **21**(6), 354–361 (2011)
2. Guttman, M., Amit, I., Garber, M., French, C., Lin, M.F., Feldser, D., Huarte, M., Zuk, O., Carey, B.W., Cassady, J.P.: Chromatin signature reveals over a thousand highly conserved large non-coding RNAs in mammals. Nature **458**(7235), 223 (2009)
3. Yu, F., Zheng, J., Mao, Y., Dong, P., Li, G., Lu, Z., Guo, C., Liu, Z., Fan, X.: Long non-coding RNA APTR promotes the activation of hepatic stellate cells and the progression of liver fibrosis. Biochem. Biophys. Res. Commun. **463**(4), 679–685 (2015)
4. Harrow, J., Frankish, A., Gonzalez, J.M., Tapanari, E., Diekhans, M., Kokocinski, F., Aken, B.L., Barrell, D., Zadissa, A., Searle, S.: GENCODE: the reference human genome annotation for The ENCODE Project. Genome Res. **22**(9), 1760–1774 (2012)
5. Chen, X., You, Z.H., Yan, G.Y., Gong, D.W.: IRWRLDA: improved random walk with restart for lncRNA-disease association prediction. Oncotarget **7**(36), 57919–57931 (2016)
6. Chen, X., Yan, C.C., Zhang, X., You, Z.H.: Long non-coding RNAs and complex diseases: from experimental results to computational models. Brief. Bioinform. **18**(4), 558 (2016)
7. Wang, Y.B., You, Z.H., Li, X., Jiang, T.H., Chen, X., Zhou, X., Wang, L.: Predicting protein-protein interactions from protein sequences by a stacked sparse autoencoder deep neural network. Mol. BioSyst. **13**(7), 1336–1344 (2017)
8. Li, S., You, Z.H., Guo, H., Luo, X., Zhao, Z.Q.: Inverse-free extreme learning machine with optimal information updating. IEEE Trans. Cybern. **46**(5), 1229 (2016)
9. Lei, W., You, Z.H., Xing, C., Li, J.Q., Xin, Y., Wei, Z., Yuan, H.: An ensemble approach for large-scale identification of protein-protein interactions using the alignments of multiple sequences. Oncotarget **8**(3), 5149–5159 (2016)
10. Huang, Q., You, Z., Zhang, X., Zhou, Y.: Prediction of protein-protein interactions with clustered amino acids and weighted sparse representation. Int. J. Mol. Sci. **16**(5), 10855–10869 (2015)
11. Huang, Y.A., You, Z.H., Chen, X.: A systematic prediction of drug-target interactions using molecular fingerprints and protein sequences. Curr. Protein Pept. Sci. **5**(19), 468–478 (2017)
12. You, Z.H., Huang, Z.A., Zhu, Z., Yan, G.Y., Li, Z.W., Wen, Z., Chen, X.: PBMDA: a novel and effective path-based computational model for miRNA-disease association prediction. PLoS Comput. Biol. **13**(3), e1005455 (2017)

13. Li, Z.W., You, Z.H., Chen, X., Li, L.P., Huang, D.S., Yan, G.Y., Nie, R., Huang, Y.A.: Accurate prediction of protein-protein interactions by integrating potential evolutionary information embedded in PSSM profile and discriminative vector machine classifier. Oncotarget **8**(14), 23638 (2017)
14. An, J.Y., You, Z.H., Chen, X., Huang, D.S., Yan, G., Wang, D.F.: Robust and accurate prediction of protein self-interactions from amino acids sequence using evolutionary information. Mol. BioSyst. **12**(12), 3702 (2016)
15. An, J.Y., You, Z.H., Chen, X., Huang, D.S., Li, Z.W., Liu, G., Wang, Y.: Identification of self-interacting proteins by exploring evolutionary information embedded in PSI-BLAST-constructed position specific scoring matrix. Oncotarget **7**(50), 82440–82449 (2016)
16. Lei, Y.K., You, Z.H., Ji, Z., Zhu, L., Huang, D.S.: Assessing and predicting protein interactions by combining manifold embedding with multiple information integration. BMC Bioinform. **13**(S7), S3 (2012)
17. You, Z.H., Lei, Y.K., Gui, J., Huang, D.S., Zhou, X.: Using manifold embedding for assessing and predicting protein interactions from high-throughput experimental data. Bioinformatics **26**(21), 2744 (2010)
18. You, Z.H., Zhu, L., Zheng, C.H., Yu, H.J., Deng, S.P., Ji, Z.: Prediction of protein-protein interactions from amino acid sequences using a novel multi-scale continuous and discontinuous feature set. BMC Bioinform. **15**(S15), S9 (2014)
19. Alipanahi, B., Delong, A., Weirauch, M.T., Frey, B.J.: Predicting the sequence specificities of DNA- and RNA-binding proteins by deep learning. Nat. Biotechnol. **33**(8), 831–838 (2015)
20. Pan, X., Fan, Y.X., Yan, J., Shen, H.B.: IPMiner: hidden ncRNA-protein interaction sequential pattern mining with stacked autoencoder for accurate computational prediction. BMC Genom. **17**(1), 582 (2016)
21. Chen, H., Huang, Z.: Medical image feature extraction and fusion algorithm based on K-SVD. In: Ninth International Conference on P2P, Parallel, Grid, Cloud and Internet Computing, 3PGCIC 2015, GuangDong, pp. 333–337 (2015)
22. Salwinski, L., Miller, C.S., Smith, A.J., Pettit, F.K., Bowie, J.U., Eisenberg, D.: The database of interacting proteins: 2004 update. Nucleic Acids Res. **32**, D449–D451 (2004)
23. Chatraryamontri, A., Breitkreutz, B.J., Oughtred, R., Boucher, L., Heinicke, S., Chen, D., Stark, C., Breitkreutz, A., Kolas, N., O'Donnell, L.: The BioGRID interaction database: 2015 update. Nucleic Acids Res. **43**, D470 (2015)
24. Suresh, V., Liu, L., Adjeroh, D., Zhou, X.: Revealing protein–lncRNA interaction. Brief. Bioinform. **17**, 106 (2015)
25. Paliwal, K.K., Sharma, A., Lyons, J., Dehzangi, A.: A tri-gram based feature extraction technique using linear probabilities of position specific scoring matrix for protein fold recognition. IEEE Trans. Nanobiosci. **13**(1), 44–50 (2014)
26. You, Z.H., Zhou, M.C., Xin, L., Shuai, L.: Highly efficient framework for predicting interactions between proteins. IEEE Trans. Cybern. **PP**(99), 1–13 (2016)
27. Huang, Y.A., Chen, X., You, Z.H., Huang, D.S., Chan, K.C.C.: ILNCSIM: improved lncRNA functional similarity calculation model. Oncotarget **7**(18), 25902–25914 (2016)
28. Zhu, L., You, Z.H., Huang, D.S., Wang, B.: t-LSE: a novel robust geometric approach for modeling protein-protein interaction networks. PLoS ONE **8**(4), e58368 (2013)
29. Zhu, L., You, Z.H., Huang, D.S.: Increasing the reliability of protein–protein interaction networks via non-convex semantic embedding. Neurocomputing **121**(18), 99–107 (2013)
30. You, Z.H., Yin, Z., Han, K., Huang, D.S., Zhou, X.: A semi-supervised learning approach to predict synthetic genetic interactions by combining functional and topological properties of functional gene network. BMC Bioinform. **11**(1), 1–13 (2010)

31. Xia, J.F., You, Z.H., Wu, M., Wang, S.L., Zhao, X.M.: Improved method for predicting phi-turns in proteins using a two-stage classifier. Protein Pept. Lett. **17**(9), 1117 (2010)
32. You, Z.H., Li, X., Chan, K.C.: An improved sequence-based prediction protocol for protein-protein interactions using amino acids substitution matrix and rotation forest ensemble classifiers. Neurocomputing **228**, 277–282 (2017)
33. Li, J.Q., Rong, Z.H., Chen, X., Yan, G.Y., You, Z.H.: MCMDA: matrix completion for MiRNA-disease association prediction. Oncotarget **8**(13), 21187 (2017)
34. Mchugh, C.A., Russell, P., Guttman, M.: Methods for comprehensive experimental identification of RNA-protein interactions. Genome Biol. **15**(1), 203 (2014)
35. Yi, H.-C., You, Z.-H., Huang, D.-S., Li, X., Jiang, T.-H., Li, L.-P.: A deep learning framework for robust and accurate prediction of ncRNA-protein interactions using evolutionary information. Mol. Ther. Nucleic Acids **11**, 337–344 (2018)
36. Vincent, P., Larochelle, H., Lajoie, I., Bengio, Y., Manzagol, P.A.: Stacked denoising autoencoders: learning useful representations in a deep network with a local denoising criterion. J. Mach. Learn. Res. **11**(12), 3371–3408 (2010)
37. Dahl, G.E., Sainath, T.N., Hinton, G.E.: Improving deep neural networks for LVCSR using rectified linear units and dropout. In: IEEE International Conference on Acoustics, Speech and Signal Processing, ICASSP 2013, Vancouver, pp. 8609–8613 (2013)
38. You, Z.H., Li, J., Gao, X., He, Z., Zhu, L., Lei, Y.K., Ji, Z.: Detecting protein-protein interactions with a novel matrix-based protein sequence representation and support vector machines. Biomed. Res. Int. **2015**(2), 1–9 (2015)
39. You, Z.H., Chan, K.C.C., Hu, P.: Predicting protein-protein interactions from primary protein sequences using a novel multi-scale local feature representation scheme and the random forest. PLoS ONE **10**(5), e0125811 (2015)
40. You, Z.H., Li, S., Gao, X., Luo, X., Ji, Z.: Large-scale protein-protein interactions detection by integrating big biosensing data with computational model. Biomed. Res. Int. (2) (2014). https://doi.org/10.1155/2014/598129
41. Pedregosa, F., Gramfort, A., Michel, V., Thirion, B., Grisel, O., Blondel, M., Prettenhofer, P., Weiss, R., Dubourg, V., Vanderplas, J.: Scikit-learn: machine learning in Python. J. Mach. Learn. Res. **12**(10), 2825–2830 (2012)
42. Yuan, H., You, Z.H., Xing, C., Chan, K., Xin, L.: Sequence-based prediction of protein-protein interactions using weighted sparse representation model combined with global encoding. BMC Bioinform. **17**(1), 184 (2016)
43. An, J.Y., You, Z.H., Meng, F.R., Xu, S.J., Wang, Y.: RVMAB: using the relevance vector machine model combined with average blocks to predict the interactions of proteins from protein sequences. Int. J. Mol. Sci. **17**(5), 757 (2016)
44. An, J.Y., Meng, F.R., You, Z.H., Fang, Y.H., Zhao, Y.J., Ming, Z.: Using the relevance vector machine model combined with local phase quantization to predict protein-protein interactions from protein sequences. Biomed. Res. Int. **2016**, 1–9 (2016)
45. Wong, L., You, Z.H., Ming, Z., Li, J., Chen, X., Huang, Y.A.: Detection of interactions between proteins through rotation forest and local phase quantization descriptors. Int. J. Mol. Sci. **17**(1), 21 (2015)
46. Wang, L., You, Z.H., Xia, S.X., Chen, X., Yan, X., Zhou, Y., Liu, F.: An improved efficient rotation forest algorithm to predict the interactions among proteins. Soft. Comput. **17**, 1–9 (2017)
47. Wang, L., You, Z.H., Chen, X., Yan, X., Liu, G., Zhang, W.: RFDT: a rotation forest-based predictor for predicting drug-target interactions using drug structure and protein sequence information. Curr. Protein Pept. Sci. **5**(19), 445–454 (2016)

48. Chen, X., Huang, Y.A., Wang, X.S., You, Z.H., Chan, K.C.: FMLNCSIM: fuzzy measure-based lncRNA functional similarity calculation model. Oncotarget **7**(29), 45948 (2016)
49. Luo, X., You, Z., Zhou, M., Li, S., Leung, H., Xia, Y., Zhu, Q.: A highly efficient approach to protein interactome mapping based on collaborative filtering framework. Sci. Rep. **5** (7702), 7702 (2015)
50. Lei, Y.K., You, Z.H., Dong, T., Jiang, Y.X., Yang, J.A.: Increasing reliability of protein interactome by fast manifold embedding. Pattern Recognit. Lett. **34**(4), 372–379 (2013)

BIOESOnet: A Tool for the Generation of Personalized Human Metabolic Pathways from 23andMe Exome Data

Marzio Pennisi[1], Gabriele Forzano[2], Giulia Russo[3], Barbara Tomasello[4], Marco Favetta[2], Marcella Renis[4], and Francesco Pappalardo[4](✉)

[1] Department of Mathematics and Computer Science, University of Catania, 95125 Catania, Italy
[2] BiONuMeRi ONLUS, 95127 Catania, Italy
[3] Department of Biomedicine and Biotechnological Science, University of Catania, 95123 Catania, Italy
[4] Department of Drug Sciences, University of Catania, 95125 Catania, Italy
francesco.pappalardo@unict.it

Abstract. The lowering of costs of whole exome sequencing (WES) services registered in the last two years has greatly increased the demand for managing different metabolic diseases, including autism spectrum disorders (ASD). WES allows the detection of a large part of exome single nucleotide polymorphisms (SNPs), whose expression can be in some cases modulated by epigenetics, life style and microbioma changes. However, such raw data usually needs to be manipulated in order to allow useful interpretation and analysis. We present BIOESOnet, a tool for the filtering and visualization of exome 23andMe raw data into a customized methylation pathway. The tool, available at: http://www.bionumeri.org/joomla/restricted-area/onecarbon-tool, enables a fast and extensive overview of possible mutations inside an extended metabolic pathway.

Keywords: Exome · Scalable Vector Graphics (SVG) · Pathway Whole Exome Sequencing (WES) · Autism Spectrum Disorders (ASD) Single Nucleotide Polymorphism (SNP)

1 Introduction

Whole Genome Sequencing (WGS) (about 3.2 billion nucleotides and about 23,500 genes) and WES (about 1% of the human genome, 180,000 exons, about 30 million nucleotides that can easily be sequenced in a single test) are clearly considered as a valid approach to extract biologically or clinically relevant information useful to effective personalized diagnosis and/or treatment [1].

The authors wish it to be known that, in their opinion, the first two authors should be regarded as Joint First Authors.

D.-S. Huang et al. (Eds.): ICIC 2018, LNCS 10955, pp. 345–352, 2018.
https://doi.org/10.1007/978-3-319-95933-7_42

There is substantial discussion in utilizing Next Generation Sequencing (NGS) in consideration of the tradeoff between the costs and health benefits, compared to standard care (i.e., gene panel test). In the last two years, the NGS, and in particular WES service costs, were significantly lowered. Consequently, the low cost and fast gene panels approach is considered underperforming first tier test, while WGS or WES are considered a more efficient and complete second tier test. In fact, they are able to provide complete coverage of the entire coding region of the genome and can be applied in different settings, from diagnosis, to therapy and research [2–4]. Nevertheless, WES approach has some limitations as it only detects individual genetic variations, hence not detecting large Copy Number Variations (CNV) or genetic variants of regulatory sequences located in intergenic regions [5].

Even if WES analysis has a relatively low cost, WES output raw data files are of little use if not further processed. Still, there is a lack of user-friendly and easily accessible IT tools to analyze and present raw data on genetic variants in an understandable way. Such tools would be useful to help day-to-day predictive/preventive personalized clinical practice.

There are a number of specialized tools developed for pathway enrichment analysis based on exome Single Nucleotide Polymorphisms (SNPs) calls along with methods that look at SNPs in the context of networks. For example, iCTNet [6] is a Cytoscape plugin that provides an interface to analyze and integrate genome-scale biological networks for human complex traits. NETVIEW P [7] is a visualization tool able to combine data quality control with the construction of population networks through mutual k nearest neighbors thresholds applied to genome wide SNPs.

Another recent tool that integrates genomic variation data with gene expression tissue information along with Boolean models is PATHiVar [8]. The usage of a WES service like 23andMe (https://www.23andme.com) allows to identify risk genes, unreported mutations as well as rare variants involved in ASD susceptibility, and to compare disease and health genetic pathways [9]. ASD is a multifactorial neurodevelopmental disorder in which familial clustering, genomic, phenotypic and clinical heterogeneity are present [10]. Recently, Codina-Solà et al. [11] made use of WES and blood transcriptome by RNAseq in a selected group of males with idiopathic ASD to detect putative causal genetic variants.

Here, we present BIOESOnet, a tool that automatically extracts the SNPs information relevant for ASD from 23andMe exome raw data files. When compared to the available tools, it adds the possibility to represent individualized SNPs information inside a specific and enriched human gene pathway, in a user-friendly environment that shows mutations with a "semaphore" metaphor along with comments that help the understanding of their damage meaning and relative phenotype changes.

2 Description of BIOESOnet

2.1 Selection of ASD Related SNPs

The initial goal of the project was to develop a graphical tool able to give a global overview about the status of the SNPs that may play a crucial role in ASD development and disease course. As such analysis had to be at everyone's reach even by an economical point of view, we focused on the exome data coming from the low cost 23andMe WES data service.

23andMe raw data files usually contain the alteration status of more than 60.000 SNPs, many of whom are not relevant for ASD. To this end, a first preliminary analysis was carried out to select, from 23andMe raw data, the subset of SNPs that are potentially relevant for ASD. Such initial subset included all the SNP that are represented inside the Yasko methylation cycle [12]. The choice of this pathway as starting point was mainly motivated by the fact that it already contains many SNPs that seem to be relevant for ASD spectrum disorders. Furthermore, the gene expression of many of them can be modulated by adequate dietary interventions, without the need of specific pharmaceutical treatments.

This pool of genes has been then enriched through an exhaustive literature research and the use of data mining algorithms with other SNPs that are considered relevant for ASD people management. These SNPs have been included into the Yasko pathway, leading to a customized "One Carbon Cycle" pathway. Such pathway, once personalized with patient data, represents the final outcome of the tool. The complete list of the used SNPs is presented in Table 1.

2.2 Implementation of the BIOESOnet Tool

BIOESOnet has been realized using PHP and D3.js, a javascript library developed to improve the visualization of organized numerical data combining HTML5, Scalable vector Graphics (SVG), and Cascading Style Sheet (CSS) standards.

The final result is represented by a web application that can be executed inside any browser, without the need of tedious software installation and configuration. The workflow implemented inside the tool is presented in Fig. 1.

The first step of the workflow relates to data filtering, and extracts from a 23andMe raw data file the status (no alteration, homozygous or heterozygous alteration) of the SNPs included in Table 1. The second step (data processing) personalizes an SVG template of the customized "One Carbon Cycle" pathway with the SNP alteration status coming from the first step. The final outcome is represented by an SVG image of the "One Carbon Cycle", in which the SNP alteration status is showed according to the 23andMe patient input file. This file can be both visualized by any web browser or downloaded.

2.3 BIOESOnet Usage and Output

In order to use the BIOESOnet processing workflow available at http://www.bionumeri.org/joomla/restricted-area/onecarbon-tool, the user has to upload a 23andMe raw data file and then click on the "Upload Exome" button.

After few seconds the SVG output will be presented inside the browser. The presented "One Carbon Cycle" SVG pathway will include all the information provided in the 23andMe input data file (see Fig. 2 for an example).

A semaphore light coloring system is used inside the pathway to represent gene alterations: green for no alteration; yellow for heterozygous alterations; red for

Table 1. ASD selected SNPs present in the "Yasko methylation cycle" metabolic pathway.

GENE/SNP	SNPEDIA link
MTHFR A1298	https://www.snpedia.com/index.php/rs1801131
MTHFR C677T	https://www.snpedia.com/index.php/rs1801133
BHMT 1	https://www.snpedia.com/index.php/rs585800
BHMT 2	https://www.snpedia.com/index.php/rs567754
BHMT 4	https://www.snpedia.com/index.php/rs617219
BHMT 8	https://www.snpedia.com/index.php/rs651852
CBS C699T	https://www.snpedia.com/index.php/rs234706
CBS A360A	https://www.snpedia.com/index.php/rs1801181
CBS N212 N	https://www.snpedia.com/index.php/rs2298758
SUOX S370S	https://www.snpedia.com/index.php/rs773115
MAO A	https://www.snpedia.com/index.php/rs6323
MAO B	https://www.snpedia.com/index.php/rs1799836
COMT V158 M	https://www.snpedia.com/index.php/rs4680
COMT H62H	https://www.snpedia.com/index.php/rs4633
COMT P199P	https://www.snpedia.com/index.php/rs769224
MTR A2756G	https://www.snpedia.com/index.php/rs1805087
MTRR A66G	https://www.snpedia.com/index.php/rs1801394
MTRR H595Y	https://www.snpedia.com/index.php/rs10380
MTRR K350A	https://www.snpedia.com/index.php/rs162036
MTRR R415T	https://www.snpedia.com/index.php/rs2287780
MTRR S257T	https://www.snpedia.com/index.php/rs2303080
MTRR 11	https://www.snpedia.com/index.php/rs1802059
AHCY 1	https://www.snpedia.com/index.php/rs819147
AHCY 2	https://www.snpedia.com/index.php/rs819134
AHCY 19	https://www.snpedia.com/index.php/rs819171
SHMT C1420T	https://www.snpedia.com/index.php/rs1979277
NOS3 D298E	https://www.snpedia.com/index.php/rs1799983
SOD2 A16 V	https://www.snpedia.com/index.php/rs4880
SOD3 C760G	https://www.snpedia.com/index.php/rs1799895
GSTP1 I105 V	https://www.snpedia.com/index.php/rs1695
GSTP1 A114 V	https://www.snpedia.com/index.php/rs1138272
GSTM1 -	https://www.snpedia.com/index.php/rs366631
VDR Taq	https://www.snpedia.com/index.php/rs731236
VDR Fok	https://www.snpedia.com/index.php/rs10735810
VDR Bsm	https://www.snpedia.com/index.php/rs1544410
CYP2R1 -	https://www.snpedia.com/index.php/rs2060793
APOA5 -	https://www.snpedia.com/index.php/rs12272004
FUT2 -	https://www.snpedia.com/index.php/rs602662
NBPF3 -	https://www.snpedia.com/index.php/rs4654748
GCH1 -	https://www.snpedia.com/index.php/rs10483639

(*continued*)

Table 1. *(continued)*

GENE/SNP	SNPEDIA link
GCH1 -	https://www.snpedia.com/index.php/rs3783641
GCH1 -	https://www.snpedia.com/index.php/rs8007267
PNP -	https://www.snpedia.com/index.php/rs104894460
ACAT -	https://www.snpedia.com/index.php/rs3741049
OTC -	https://www.snpedia.com/index.php/rs5963409
GAD1 -	https://www.snpedia.com/index.php/rs3749034
GAD1 -	https://www.snpedia.com/index.php/rs3828275
GAD1 -	https://www.snpedia.com/index.php/rs1978340
GAD1 -	https://www.snpedia.com/index.php/rs3791878
OTC -	https://www.snpedia.com/index.php/rs72554331
OTC -	https://www.snpedia.com/index.php/rs67960011

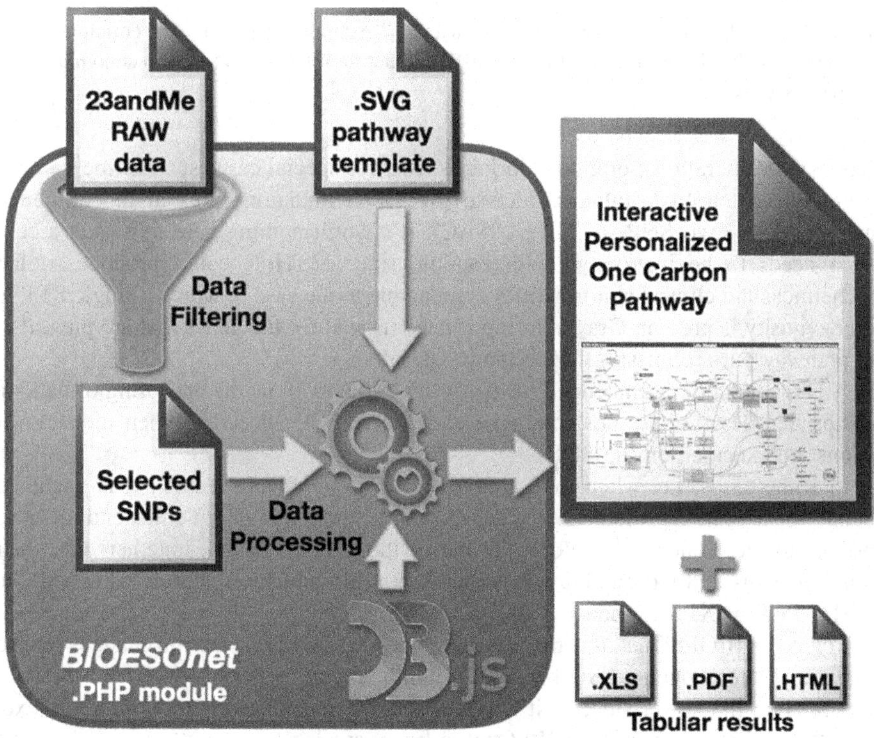

Fig. 1. The BIOESOnet processing workflow. Data coming from patient 23andMe raw datafile is filtered and only the mutation status of the SNPs in Table 1 is selected. Then, the D3.js script proceeds in personalizing a "One Carbon Cycle" SVG template with the SNPs mutation status. The outcome will be represented by an interactive personalized pathway, coupled with the results in different tabular formats.

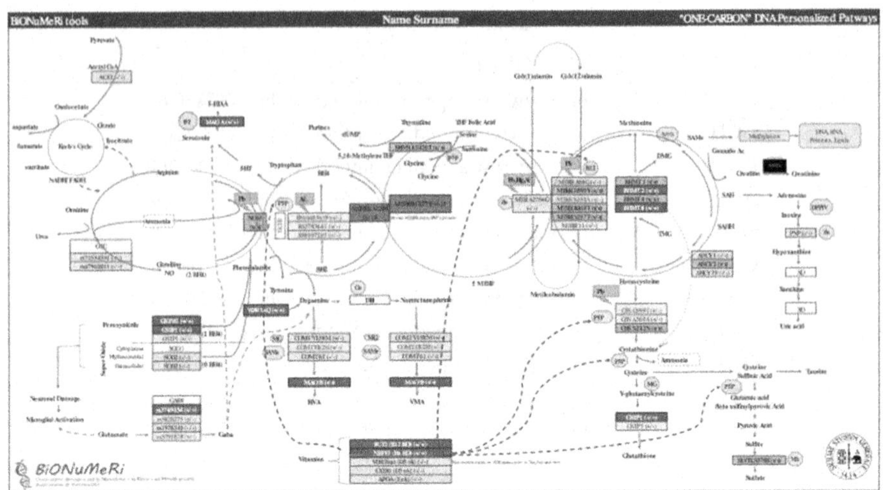

Fig. 2. BIOESOnet final output "one carbon cycle" example. The pathway enrichment and personalization has been obtained from the BIOESOnet tool using SNPs raw data coming from 23andMe service

homozygous alterations; orange coloring is used for special cases such as the variants of methylene tetrahydrofolate reductase (MTHFR) mutations present in compound heterozygosity (two SNPs 677CT/1298AC). It's worth nothing here that such a condition needs to be interpreted with caution, since MTHFR could produce similar biochemical and clinical abnormalities comparable to the case in which a single SNP in homozygosity is present. Gray coloring is used instead for the genes that are present in the pathway but are missing in the input data file.

A D3.js function is embedded inside the SVG file to allow the appearing of balloon tooltips that describe the possible gene correlations with diseases, when mouse-over actions are executed on the SNPs.

In some cases, i.e. when one wants to print the software results, it is useful to couple a schematic representation of the data to the graphical one. To this end it is also possible to view and/or download the list of the selected SNPs, together with their mutation status and the relative information about possible correlations with ASD, in HTML, PDF or XLS formats.

Thanks to BIOESOnet, the analysis and understanding of metabolic changes related to genomic alterations in ASD becomes within everyone's reach. Moreover, the software runs on the server-side, so it is platform independent and can be remotely executed through a web portal. BIOESOnet is freely available.

3 Conclusions

We presented BIOESOnet, a valuable tool that can facilitate the identification of genomic alterations in a disease-related metabolic pathway. BIOESOnet allows a reliable, fast, web-based approach for WES data management with the potential to facilitate personalized medicine and diagnosis.

BIOESOnet allows to extrapolate the genomic variations (e.g. SNPs) of genes involved in specific disease associated biochemical network, such as the "One carbon cycle" in ADS, by visualizing as output a personalized pathway from individual WES data. Furthermore, it offers an easy-to use graphical interface for data submission and an annotated visualization that uses a semaphore coloring to identify gene mutations. In addition, balloon tooltips are used to show disease-related variants information.

This tool assists non-IT mastered users to quickly draw and arrange specific genes on the selected pathway through minimal time and without any specialized bioinformatics staff. In particular, the implemented pipeline leads clinicians to obtain relevant information that may be useful to diagnosis, through simple steps.

Further improvements will be released in future versions of the tool, such as the evaluation of metabolites corresponding to the identified altered genes for a more integrative and global view of biochemical pathway. Also, BIOESOnet will allow to analyze raw data file formats from different available WES platforms.

Finally, BIOESOnet actually refers to the "One Carbon Cycle", but it will be soon tailored to other biochemical pathways and networks relevant for other diseases.

Acknowledgements. The raw results of the sample used in this work, related to one child with ASD diagnosis, were donated to BiONuMeRi by parents who had spontaneously acquired 23andMe kit for exome analysis. Moreover, the authors wish to thank Dr. Guanglan Zhang for her helpful contribution.

Authors' Contribution. MP: designed the tool, analysed data and wrote the manuscript. GF: designed the tool, analysed and provided data. GR: gave biological knowledge and wrote the manuscript. BT: gave biological knowledge and wrote the manuscript. MF: gave useful insights and wrote the manuscript. MR: supervised the whole project and drafted the manuscript. FP: supervised the whole project and drafted the manuscript.

References

1. Seidelmann, S.B, Smith, E., Subrahmanyan, L., Dykas, D., Ziki, M.D.A., Azari, B., et al.: Application of whole exome sequencing in the clinical diagnosis and management of inherited cardiovascular diseases in adults. Circ. Cardiovasc. Genet. **10**(1), pii, e001573 (2017)
2. Saudi Mendeliome Group: Comprehensive gene panels provide advantages over clinical exome sequencing for Mendelian diseases. Genome Biol. **16**, 134 (2015)
3. Meienberg, J., Bruggmann, R., Oexle, K., Matyas, G.: Clinical sequencing: is WGS the better WES? Hum. Genet. **135**, 359–362 (2016)

4. van El, C.G., Cornel, M.C., Borry, P., Hastings, R.J., Fellmann, F., Hodgson, S.V., et al.: Whole-genome sequencing in health care: recommendations of the European society of human genetics. Eur. J. Hum. Genet. **21**(6), 580–584 (2013)
5. Sener, E.F., Canatan, H., Ozkul, Y.: Recent advances in autism spectrum disorders: applications of whole exome sequencing technology. Psychiatr. Investig. **13**(3), 255–264 (2016)
6. Wang, L., Khankhanian, P., Baranzini, S.E., Mousavi, P.: iCTNet: a Cytoscape plugin to produce and analyze integrative complex traits networks. BMC Bioinform. **12**, 380 (2011)
7. Steinig, E.J., Neuditschko, M., Khatkar, M.S., Raadsma, H.W., Zenger, K.R.: NETVIEW P: a network visualization tool to unravel complex population structure using genome-wide SNPs. Mol. Biol. Resour. **16**(1), 216–227 (2015)
8. Hernansaiz-Ballesteros, R.D., Salavert, F., Sebastián-León, P., Alemán, A., Medina, I., Dopazo, J.: Assessing the impact of mutations found in next generation sequencing data over human signaling pathways. Nucleic Acids Res. **43**(W1), 270–275 (2015)
9. Scherer, S.W., Dawson, G.: Risk factors for autism: translating genomic discoveries into diagnostics. Hum. Genet. **130**, 123–148 (2011)
10. An, J.Y., Claudianos, C.: Genetic heterogeneity in autism: from single gene to a pathway perspective. Neurosci. Biobehav. Rev. **68**, 442–453 (2016)
11. Codina-Solà, M., Rodríguez-Santiago, B., Homs, A., Santoyo, J., Rigau, M., Aznar-Laín, G., et al.: Integrated analysis of whole-exome sequencing and transcriptome profiling in males with autism spectrum disorders. Mol. Autism **6**, 21 (2015)
12. Yasko, A.: Pathways to recovery, 3rd edn. Neurological Research Institute, Bethel (2009)

Accurate Prediction of Hot Spots with Greedy Gradient Boosting Decision Tree

Haomin Gan[1,2(✉)], Jing Hu[1,2], Xiaolong Zhang[1,2], Qianqian Huang[1,2], and Jiafu Zhao[1,2]

[1] School of Computer Science and Technology, Wuhan University of Science and Technology, Wuhan 430065, Hubei, China
1320095086@qq.com
[2] Hubei Laboratory of Intelligent Information Processing and Real-Time Industrial System, Wuhan, China

Abstract. Hot spot residues play a crucial role in protein-protein interactions, which are conducive to drug discovery and rational drug design. Only several amino acid residues provide most of the binding free energy for protein interface. These amino acids are called hot spots. This work is to predict hot spot residues by an ensemble machine learning method called Gradient Boosting Decision Tree in Alanine Scanning Energetics Database (ASEdb) and Structural Kinetic and Energetic database of Mutant Protein Interactions (SKEMPI). According to properties of amino acid and protein complex chain where the amino acid is, we design the a program that will not stop until the last most unimportant feature calculated in GBDT method is discarded in every iteration. Consequently, the greedy GBDT method can get a better prediction on hot spot residues after comparing the result, one of evaluation criteria F-score reach at 0.808 in the ASEdb dataset.

Keywords: Hot spot residues · Greedy Gradient Boosting Decision Tree
Amino acid properties · Protein complex chain

1 Introduction

Protein-protein interactions (PPIs) provide fundamental cellar functions as signal transduction, cellular motion and hormone-receptor interactions [1, 2]. The research has shown that the most of binding free energy is contributed by a small fraction of interface residues, which are clustered and packed tightly in the center of protein interface. These amino acid residues are regarded as hot spots. They maintain the stability of protein interactions effectively. Hot spots are enriched in tryptophan, tyrosine and arginine, with a rate of 21, 13.3 and 12.3%, and are surrounded by energetically less important residues that most likely serve to occlude bulk solvent from the hot spot [3]. The research on hot spots contributes to the comprehension of protein function and the development of drug design [4].

There are two kinds of methods to classify whether one amino acid is hot spot or not based on the existing databases. The first kind of method is based on energy. One energy-based method is through site-directed mutagenesis like alanine scanning,

D.-S. Huang et al. (Eds.): ICIC 2018, LNCS 10955, pp. 353–364, 2018.
https://doi.org/10.1007/978-3-319-95933-7_43

Ala-nine Scanning Energetics Database records data about the binding free energy of the mutated site [5]. Another energy-based method is by molecular dynamics simulations. The physical model which consists of an energy function and a measure of side-chain conformation entropy changes is used to predict hot spot residues [6]. The second kind of method is based on features. The feature-based method needs to collect amino acids' physicochemical attributes, sequence attributes, structural attributes, evolutional attributes and other attributes, then machine learning based method classify hot spot residues and non-hot spot residues. However, the performance of prediction on hot spot residues depends on machine learning algorithms and good features. A mass of features were used in machine learning, and a variety of machine learning methods were invoked to solve the problem. Tuncbag proposed an intuitive efficient method to predict hot spot residues on conservation, solvent accessibility and statistical pairwise residue potentials [7]. Then, Tuncbag built a web server HotPoint to predict hot spot residues using an empirical model incorporating occlusion from solvent and total knowledge-based pair potentials of residues [8]. Agrawal proposed a method called spatial interaction diagram (SIM) to predict hot spot residues from the structure [9]. Chen presented a method to predict hot spot residues through physicochemical properties of amino acid sequence [10]. Xia built a prediction method by support vector machine and extracted features through maximum relevance and minimum redundancy (mRMR) [11]. Recently, Huang proposed a method to predict hot spots by an ensemble leaning method adaboost, used mRMR to select features and dealt with the imbalance of hot spot residues and non-hot spot residues by SMOTE [12]. Hu presented a prediction method from sequence only by a new ensemble learning method [13].

It has been proved that an ensemble learning method was effective to predict hot spot residues. This paper proposes an ensemble learning method GBDT consists of gradient boosting and decision tree to predict hot spot residues [14, 15]. Amino acids features like structural properties, hydrophobicity and conservation from position specific scoring matrix are extracted. According to the contribution of every feature to predict hot spot residues, the greed algorithm is designed to discard the most unimportant feature and continue to calculate until the last feature is left. A best prediction result was selected as the last result. Experimental result shows that the machine learning method Greedy GBDT has a better performance on predicting the hot spot residues.

2 Method

2.1 Dataset

When one amino acid is mutated to alanine in the protein-protein interface, if the binding free energy changes are more than or equal to 2.0 kcal/mol, the amino acid is defined as hot spot residue, and if the binding free energy changes is less than 0.4 kcal/mol, the amino acid is defined as non-hot spot residue. The residues with binding free energy changes between 0.4 kcal/mol and 2.0 kcal/mol are discarded. In this study, the dataset that we used is from the database Alanine Scanning Energetics Database (ASEdb) and Energetic database of Mutant Protein Interactions (SKEMPI) [16]. ASEdb includes a vast of single alanine mutations whose binding affairs have been experimentally determined. The dataset from ASEdb contains 265 alanine mutation interface residues in 17 protein complexes. If protein sequence similarity is less than 35%, and SSAP (Secondary Structure alignment Program) is less than 80%, these proteins are considered as non-homologous proteins. The 17 experimental complexes are all non-homologous. In the ASEdb database, 65 hot spot residues and 90 non-hot spot residues are selected. SKEMPI contains 3047 experimental binding free energy changes. For the repetitive protein-protein interaction residues from the SKEMPI dataset, we just choose one of them. Then we get plenty of interface residues, only save those mutated to alanine and discard those amino acids whose binding free energy changes are between 0.4 kcal/mol and 2.0 kcal/mol. Finally, only 112 hot spot residues and 338 non-hot spot residues are selected. Table 1 shows the specific data.

Table 1. Databases **for hot spots prediction**

Databases	No. of hot spots	No. of non-hot spots	Total	Ratio[a]
SKEMPI	112	338	450	0.331
ASEdb	65	90	155	0.657

[a]The last column shows the ratio of the number of hot spots and that of non-hot spots

2.2 Feature Collection

Because protein chain exists two states: bound and unbound, we select structural properties like relative accessible surface area (RASA), accessible surface area (ASA), residues protrusion index (PI), residues depth index (DI) beyond two different states. We obtain these properties from Protein Structure and Interaction Analyzer (PSAIA) [17]. However, protrusion index is abbreviated to CX and depth index is abbreviated to DPX in PSAIA. It has been proved that these features are very effective in identifying hot spot residues and non-hot spot residues.

ASA has five derived attributes including total ASA (the sum of all atom values), backbone ASA (the sum of all backbone atom values), side-chain ASA (the sum of all side-chain atom values), polar ASA (the sum of all oxygen, nitrogen atom values), no polar ASA (the sum of all carbon atom values). RASA and protein complex chain ASA where the amino acid is have the same derived attributes as ASA.

DI has six derived attributes including average DI (the mean value of all atom values), average deviation DI (the mean deviation value of all atom values), average side-chain DI (the mean value of side-chain atom values), average deviation side-chain DI (the mean deviation of side-chain atom values), maximum DI (the maximum of all atom values), minimum DI (the minimum of all atom values). PI has the same attributes as DI. Otherwise, some features are derived from obtained features due to the different states. The calculating formulas (1)–(4) is showed below.

$$\text{RctASA} = \frac{[\text{unbound total ASA}] - [\text{bound total ASA}]}{[\text{unbound total ASA}]} \tag{1}$$

$$RcsASA = \frac{[\text{unbound side} - \text{chain ASA}] - [\text{bound side} - \text{chain ASA}]}{[\text{unbound side} - \text{chain ASA}]} \tag{2}$$

$$RctmPI = \frac{[\textit{unbound total mean PI}] - [\textit{bound total mean PI}]}{[\text{unbound total mean PI}]} \tag{3}$$

$$RctmDI = \frac{[\text{unbound total mean DI}] - [\text{bound total mean DI}]}{[\text{unbound total mean DI}]} \tag{4}$$

Every amino acid has different hydrophobicity due to different side-chain. Hydrophobicity amino acids generally tend not to contact with water molecules, while hydrophilic (polar and charged) amino acids prefer a high affinity for water. In the analysis of ASEdb, these hot spot residues are enriched in tryptophan, tyrosine and arginine, with a rate of 21, 13.3 and 12.3%. It is said that hot spots are related to the type of amino acid. The hydrophobicity can be obtained by PSAIA, too. Moreover, hot spot residues play a vital role in protein-protein interactions. Therefore, hot spot residue is hard to be changed due to mutation. If hot spot residue is mutated to other amino acid, the changed amino acid can not provide enough binding free energy to maintain protein complexes structure. That is very dangerous for creature. Many works have proved that hot spot residue is conserved [18–20]. We get every amino acid's conservation from position specific scoring matrices (PSSM). PSSM is generated from PSI-blast after 3 iterations [21]. We only select the conservation of amino acid in protein sequence. Finally, we collect 60 features.

2.3 GBDT and Feature Selection

Gradient Boosting Decision Tree is an ensemble machine learning which consists of gradient boosting and decision tree. Boosting is an algorithm that combines several weak learner to produce a strong learner. The weak learner is decision tree in GBDT. Normally, decision tree is substituted by cart tree. At each iteration, the base classifier is trained on a fraction subsample of the available training data with a slow learning rate.

$\{(x_1, y_1), \ldots, (x_i, y_i)\}$ are a set of random distributed incident data which consists of input x_i and output y_i. Given a historical training sample, our goal is to get a function F (x) that minimizes the expected value of loss function L(y, F(x)). In Gradient boosting Decision tree, decision tress is regarded as weak learner $h_m(x; w)$, Υ_m is the weight of the tree, w is the parameter of decision tree, the function is:

$$F(x) = \sum\nolimits_{m=1}^{M} \Upsilon_m h_m(x; w) \tag{5}$$

In every iteration, the function is to fit the tree, y_{im} is the negative steepest descent:

$$w = argmin \sum_{i=1}^{n} (y_{im} - h_m(x_i; w))^2 \tag{6}$$

In the additive mode, the decision tree $h_m(x; w)$ was chosen to minimize the loss function given the current model $F_{m-1}(x)$ and its fit $F_m(x)$:

$$F_m(x) = F_{m-1}(x) + \Upsilon_m h_m(x; w) \tag{7}$$

At each iteration m, a tree partitions the x-space into L-disjoint regions and predicts a separate constant value in each one.

$$\Upsilon_{im} = argmin_{\Upsilon} \sum_{x_i \epsilon R_{im}} L(y_i, F_{m-1}(x_i) + \Upsilon h_m(x_i; w)) \tag{8}$$

Gradient boosting tries to reduce the last residue to minimize the difficult problem via steepest descent. The steepest descent direction is the negative gradient of the loss function evaluated at the current model F_{m-1}, which can be calculated for any differentiable loss function. A shrinkage parameter was used to control the learning rate of the procedure. Through applying shrinkage strategy, reducing and shrinkage the impact of each additional tree avoid the over-fitting problem. Smaller shrinkage values should be better to minimize loss function, but the program will be slow if shrinkage is too small. The sample is randomly selected in the gradient boosting to avoid overfitting problem. We select 80% of samples to train the model. The algorithm is showed below.

---

algorithm GBDT

Specific steps of GBDT algorithm:
Input: A training dataset
Output: A multiple classifier
Step 1: training data (x_i, y_i) for i , … , n, loss function L(y, F(x)), for total iteration of M
Step 2: initialization:

$$F_0(x) = argmin_{\Upsilon} \sum_{i=1}^{n} L(y_i, \Upsilon)$$

Step3: for m = 1 to M do,

Step 3.1:calculate the response:

$$y_{im} = -\left[\frac{\partial L(y_i, F(x_i))}{\partial F(x_i)}\right]_{F(x)=F_{m-1}(x)} \quad \text{for } i = 1, \ldots, n$$

Step 3.2: train $h_m(x)$ using training data (x_i, y_i) for i = 1, … , n

$$w = \text{argmin} \sum_{i=1}^{n} (y_{im} - h_m(x_i; w))^2$$

Step 3.3: compute optimal solution of :

$$\Upsilon_m = argmin_{\Upsilon} \sum_{i=1}^{n} L(y_i, F_{m-1}(x_i) + \Upsilon h_m(x_i; w))$$

Step 3.4: update the model:

$$F_m = F_{m-1}(x_i) + \Upsilon h_m(x_i; w)$$

Step 4: end For
Step 5: output F_m

GBDT contains interpretable additive predictors. The influences of the predictor feature are different. The partial effect of predictors is used to estimate the importance of each feature. The values of the feature were calculated on this perturbed data set to measure the importance of every feature after training. The importance score for the feature was calculated by averaging the difference in out-of-bag error before and after the permutation over all trees. The score was normalized by the standard deviation of these differences.

Feature selection is a very important step to get good performance, reduce the program runtime and generate better model. We propose a greedy algorithm to select better features. According to GBDT method, we can calculate the importance of every feature. Then, the algorithm discard the most unimportant feature in every iteration, and continue to calculate in the remaining features. The feature importance calculation is below. Firstly, calculate every feature importance in a decision tree, like formula (9), the summation is over the non-terminal nodes t of the J-terminal node tree T, v_t is the splitting variable associated with node t, and $\hat{I}_t^2$ is the corresponding empirical improvement in squared error as a result of the split. The right part s associated with squared influence. For a collection of decision trees $\{T_m\}_1^M$, obtained through gradient boosting approach, feature importance can be generalized by its average over all of the trees in the sequence, like formula (10) shows.

$$\hat{I}_j^2(T) = \sum_{t=1}^{L-1} \hat{I}_t^2 l(v_t = j) \tag{9}$$

$$\hat{I}_j^2 = \frac{1}{M} \sum_{m=1}^{M} \hat{I}_j^2(T_m) \tag{10}$$

The feature selection is a NP problem. It is impossible to test a model in all combinations of features. The normal way is to find the correlation between features and samples. However, mRMR calculate the redundancy between features and the correlation between features and samples. Once the correlation of one feature reached at a threshold, the feature is identified to help to construct a better model. It has been proved that the importance of the feature calculated by the cart tree can select valid features for a better model in Xu et al. [22]. Because of the quality of GBDT algorithm to measure the importance of every feature, we attempt to discard the feature whose contribution to the model is least in current combination of features, and combine the remaining features for next iteration. To reduce the influence of discarded feature, we design the greedy algorithm. That is my way to select the valid features according to the performance of the model. Because it is noteworthy that GBDT adopt to subsample randomly for avoiding overfitting. It means that the most unimportant features may be not accurate because of the incomplete samples. According to the performance of model, the greedy algorithm is effective to select features based on GBDT model.

2.4 Evaluation Criteria

To evaluate the classification performance of GBDT method, several widely used measures are adopted including precision, recall or sensitivity, specificity, accuracy and F-score. In this study, F-score is the foremost evaluation criteria that gauges the balance between precision and recall rates.

$$Recall = \frac{\mathrm{TP}}{\mathrm{TP} + \mathrm{FN}} \tag{11}$$

$$\mathrm{Precision} = \frac{TP}{TP + FP} \tag{12}$$

$$\mathrm{Specificity} = \frac{TN}{TN + FP} \tag{13}$$

$$\mathrm{Accuracy} = \frac{TP + TN}{TP + FP + TN + FN} \tag{14}$$

$$\text{F-score} = \frac{2 \times Precision \times Recall}{Precision + Recall} \tag{15}$$

Where TP, FP, TN and FN denote the numbers of true positives (correctly predicted hot spot residues), false positives (non-hot spot incorrectly predicted as hot spot), true negatives (correctly predicted non-hot spot), false negatives (hot spot incorrectly predicted as non-hot spot). The confusion matrices is in Table 2. Roc curve usually is used to describe the relationship between the FP and the TP, and the area under the curve (AUC) is another important evaluation criteria to measure prediction performance.

Table 2. Confusion matrices

Actual	Predicted	
	NH	H
NH	True Negative (TN)	False Positive (FP)
H	False Negative (FN)	True Positive (TP)

3 Experiments

3.1 Performance on SKEMPI Dataset and ASEdb Dataset

In this experiment, we trained two models based on SKEMPI dataset and the ASEDB dataset. Every dataset was divided into training dataset and testing dataset, the ratio of two datasets is 07:0.3. Then, the training datasets was divided into ten sub datasets.

Nine datasets was taken out as training set, and the remaining one dataset was selected as a validation set. Then we used ten-fold cross validation, and repeated the process 10 times. The result of computing is Fig. 1 in different databases and different datasets. It is different for every feature to contribute to the prediction for hot spot

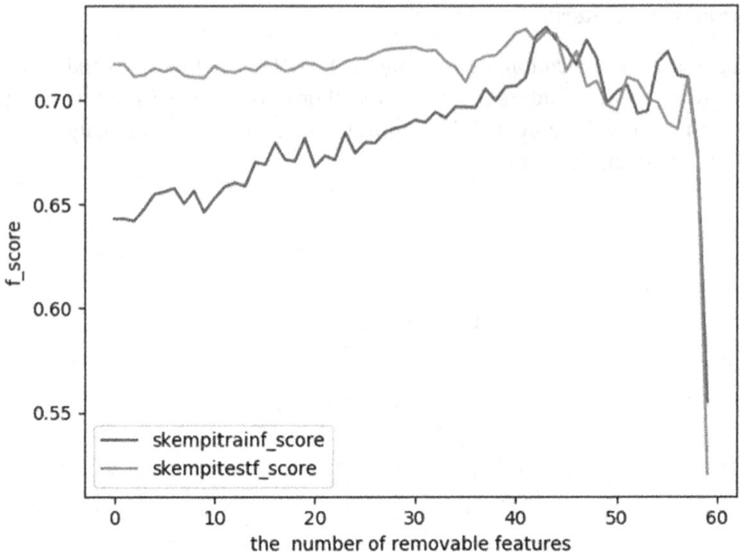

Fig. 1. After discard the most unimportant feature in every iteration, the prediction performance on ASEdb.

residues, so we discard the most unimportant feature and then go on predicting hot spot residues through formula (9) and (10). The broken lines show the variant tendency by the remaining feature. We can see that GBDT get the best prediction performance in the remaining 14 features from Fig. 1.

In the same 14 remaining features, we train a model in the ASEdb dataset to validate if these features are effective in other dataset. Table 3 shows the result in training dataset and testing dataset. Based on the features selected in the SKEMPI dataset, the model has a good performance in the ASEdb dataset. Due to the imbalance of hot spot residues and non-hot spot residues, the precision is low in the training dataset. Because there are about four hot spot residues to test in the ASEdb training datasets, precision will be lower due to the imbalanced distribution in the ten-fold cross validation. Respectively, there are about eight hot spot residues to test in the SKEMPI training datasets, precision is closed between training datasets and testing datasets.

Table 3. **Prediction** result on SKEMPI and ASEdb dataset

Dataset	Recall	Precision	F-score
SKEMPI_train	0.681	0.784	0.715
SKEMPI_test	0.667	0.815	0.732
ASEdb_train	0.73	0.762	0.746
ASEdb_test	0.746	0.886	0.808

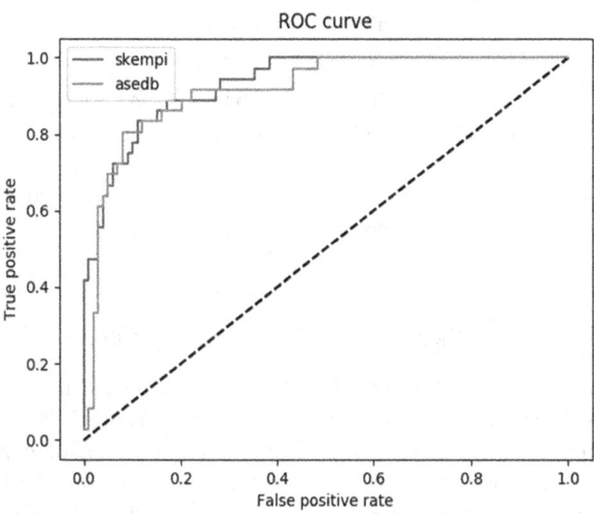

Fig. 2. The ROC curve of one random prediction on SKEMPI dataset and ASEdb dataset

Figure 2 is the first roc curve of one prediction selected under the remaining 14 features. The average AUC value of SKEMPI testing dataset is 0.92, and the average AUC value of ASEdb testing dataset is 0.91. Our method has a relative high AUC, and high AUC corresponds to good classification performance. Limited to the number of data, it is not smooth like other ROC curve with a big number of data.

3.2 Feature Analysis

Table 4 shows the importance of left features calculated by GBDT for the best accuracy. The information of protein complex chain is very important to predict hot spot except those effective proved features. We select 10 features of protein complex chain and three features are left. Our method select other features about protein complex chain except the features of amino acid. The experiment shows that the information of protein complex chain where the amino acid is helps to distinguish hot spot residues and non-hot spot residues. It has been proved that hot spot residues are conserved. So according to PSSM, we get conservation of every amino acid. The experiment proved that the amino acid's conservation is helpful to predict hot spot residues effectively.

Table 4. The importance of every feature

Feature	Importance	Feature	Importance
RctASA	0.072	Mean PI-bound	0.070
RcsASA	0.097	Mean deviation PI-bound	0.070
RctmPI	0.100	Mean side-chain PI-bound	0.058

(continued)

Table 4. (*continued*)

Feature	Importance	Feature	Importance
RctmDI	0.062	**Chain** Polar ASA-unbound	0.075
Chain back-bone ASA-bound	0.067	**Chain** non-polar ASA-unbound	0.079
Polar RASA-bound	0.046	Non-polar ASA-unbound	0.062
Mean deviation side-chain DI-bound	0.068	Conservation	0.071

3.3 Comparison with Other Methods in Different Databases

The best F-score (0.732) in the SKEMPI dataset is selected, we can see that our method can have a better performance in recall and F-score from Table 5. Our method has the highest recall (0.667) and F-score (0.732). It means that our method can find more hot spot residues. But precision of our method is lower than that of Hu. Our method is more possible to regard the non-hot spot residue as hot spot residue. For other methods, our method has a higher accuracy in every evaluation criteria. Table 4 shows that GBDT has a higher accuracy in predicting hot spot and non-hot spot with SKEMPI dataset.

Table 5. Prediction result on SKEMPI dataset

Method	Recall	Precision	F-score
Chen [10]	0.55	0.31	0.40
Huang [12]	0.583	0.772	0.665
Hu [13]	0.49	**0.98**	0.65
GBDT	**0.667**	0.815	**0.732**

Table 6 lists the results of prediction hot spots based on ASEdb dataset by different methods. From Table 6, Compared with Tuncbag and Hu's results, our precision is lower. But the recall (0.746) of our method is higher than that of Tuncbag and Hu. Meanwhile, F-score of our method is slightly higher than that of Hu. In conclusion, our method can find more hot spot residues, but regard non-hot spot residues as hot spot residues easily, too. Hence, according to evaluation criteria of F-score, our method has a better performance.

Table 6. Prediction result on ASEdb dataset

Method	Recall	Precision	F-score
Tuncbag [7]	0.122	**1.0**	0.40
Nan [23]	0.424	0.285	0.341
DICFC [24]	0.651	0.571	0.608
Hu [13]	0.63	**1.0**	**0.77**
GBDT	**0.746**	0.886	0.808

4 Conclusion

This paper proposed a greedy GBDT method to predict the hot spot residues according to the features of amino acid. We designed a greedy algorithm to select features according to the contribution of every features in the GBDT method. The recall of our method is highest. Our method can find more hot spot residues, but respectively the precision is lower than other methods. Due to build many models, the algorithm will be very slow if the data has many features. Because the valid features were less than all features that we collect, we chose to discard the most unimportant features according to formula (10) to calculate the contribution to the prediction. From the greedy algorithm, we can see the contribution of every feature to distinguish that a feature is valid or not. As Table 2 shows, protein complex chain ASA where the amino acid is and the conservation of amino acid are effective for predicting hot spot. Our greedy GBDT method can preserve exceptional accuracy on SKEMPI dataset. In ASEdb dataset, the performance of our method is slightly better than that of others' method. The performance in the ASEdb dataset is better than that in the SKEMPI dataset. The validation process is the same ten-fold cross validation. Our model shows the better overall prediction performance in SKEMPI and ASEdb datasets. In our future work, we will select more valid features to train models for predicting hot spot residues. Additionally, hot spot residue is conserved, we will detect the conservation of hot region which is consist of hot spot residues through PSSM.

Acknowledgment. The authors thank the members of Machine Learning and Artificial Intelligence Laboratory, School of Computer Science and Technology, Wuhan University of Science and Technology, for their helpful discussion within seminars. This work is supported by the National Natural Science Foundation of China (No. 61702385).

References

1. Chothia, C., Janin, J.: Principles of protein-protein recognition. Nature **256**(5520), 705 (1975)
2. Clackson, T., Wells, J.A.: A hot spot of binding energy in a hormone-receptor interface. Science **267**(5196), 383–386 (1995)
3. Bogan, A.A., Thorn, K.S.: Anatomy of hot spots in protein interfaces. J. Mol. Biol. **280**, 1–9 (1998)
4. Gul, S., Hadian, K.: Protein-protein interaction modulator drug discovery: past efforts and future opportunities using a rich source of low- and high-throughput screening assays. Expert Opin. Drug Discov. **9**(12), 1393–1404 (2014)
5. Thorn, K.S., Bogan, A.A.: ASEdb: a database of alanine mutations and their effects on the free energy of binding in protein interactions. Bioinformatics **17**(3), 284–285 (2001)
6. Kortemme, T., Baker, D.: A simple physical model for binding energy hot spots in protein-protein complexes. Proc. Natl. Acad. Sci. U. S. A. **99**(22), 14116–14121 (2002)
7. Tuncbag, N., Gursoy, A., Keskin, O.: Identification of Computational Hot Spots in Protein Interfaces: Combining Solvent Accessibility and Inter-residue Potentials Improves the Accuracy. Oxford University Press, Oxford (2009)

8. Tuncbag, N., Keskin, O., Gursoy, A.: Hotpoint: hot spot prediction server for protein interfaces. Nucleic Acids Research **38**(Web Server issue), 402–406 (2010)
9. Agrawal, N.J., Bernhard, H., Trout, B.L.: A computational tool to predict the evolutionarily conserved protein-protein interaction hot-spot residues from the structure of the unbound protein. FEBS Lett. **588**(2), 326–333 (2014)
10. Chen, P., Li, J., Wong, L., Kuwahara, H., Huang, J., Gao, X.: Accurate prediction of hot spot residues through physicochemical characteristics of amino acid sequences. Proteins Struct. Funct. Bioinform. **81**(8), 1351–1362 (2013)
11. Xia, J.F., Zhao, X.M., Song, J., Huang, D.S.: APIs: accurate prediction of hot spots in protein interfaces by combining protrusion index with solvent accessibility. BMC Bioinform. **11**(1), 174 (2010)
12. Huang, Q.Q., Zhang, X.L.: An improved ensemble learning method with SMOTE for protein interaction hot spots prediction. In: IEEE International Conference on Bioinformatics and Biomedicine, pp. 1584–1589 (2017)
13. Hu, S.S., Peng, C., Bing, W., Li, J.: Protein binding hot spots prediction from sequence only by a new ensemble learning method. Amino Acids **49**(1), 1–13 (2017)
14. Friedman, J.H.: Greedy function approximation: a gradient boosting machine. Ann. Stat. **29** (5), 1189–1232 (2001)
15. Ma, X., Ding, C., Luan, S., Wang, Y., Wang, Y.: Prioritizing influential factors for freeway incident clearance time prediction using the gradient boosting decision trees method. IEEE Trans. Intell. Transp. Syst. **18**(9), 2303–2310 (2017)
16. Moal, I.H., Fernándezrecio, J.: SKEMPI: a structural kinetic and energetic database of mutant protein interactions and its use in empirical models. Bioinformatics **28**(20), 2600–2607 (2012)
17. Mihel, J., Sikić, M., Tomić, S., Jeren, B., Vlahovicek, K.: PSAIA - protein structure and interaction analyzer. BMC Struct. Biol. **8**(1), 21 (2008)
18. Li, X., Keskin, O., Ma, B., Nussinov, R., Liang, J.: Protein-protein interactions: hot spots and structurally conserved residues often locate in complemented pockets that pre-organized in the unbound states: implications for docking. J. Mol. Biol. **344**(3), 781–795 (2004)
19. Jing, H., Li, J., Chen, N., Zhang, X.: Conservation of hot regions in protein-protein interaction in evolution. Methods **110**, 73–80 (2016)
20. Collins, J.C., Bedford, J.T., Greene, L.H.: Elucidating the key determinants of structure, folding, and stability for the, conformation of the b1 domain of protein g using bioinformatics approaches. IEEE Trans. Nanobiosci. **15**(2), 140–147 (2016)
21. Altschul, S.F., Madden, T.L., Schäffer, A.A., Zhang, J., Zhang, Z., Miller, W., et al.: Gapped blast and psi-blast: a new generation of protein database search programs. Nucleic Acids Res. **25**(17), 3389 (1997)
22. Xu, Z., Huang, G., Weinberger, K.Q., Zheng, A.X.: Gradient boosted feature selection, pp. 522–531. ACM (2014)
23. Nan, D., Zhang, X.: Prediction of hot regions in protein-protein interactions based on complex network and community detection. In: IEEE International Conference on Bioinformatics and Biomedicine, pp. 17–23. IEEE (2014)
24. Hu, J., Zhang, X., Liu, X., Tang, J.: Prediction of hot regions in protein-protein interaction by combining density-based incremental clustering with feature-based classification. Comput. Biol. Med. **61**(C), 127–137 (2015)

A Novel Efficient Simulated Annealing Algorithm for the RNA Secondary Structure Predicting with Pseudoknots

Zhang Kai[1,2(✉)] and Lv Yulin[1]

[1] School of Computer Science,
Wuhan University of Science and Technology, Wuhan 430081, China
zhangkai@wust.edu.cn
[2] Hubei Province Key Laboratory of Intelligent Information Processing
and Real-Time Industrial System, Wuhan 430081, China

Abstract. The pseudoknot structure of RNA molecular plays an important role in cell function. However, existing algorithms cannot predict pseudoknots structure efficiently. In this paper, we propose a novel simulated annealing algorithm to predict nucleic acid secondary structure with pseudoknots. Firstly, all possible maximum successive complementary base pairs would be identified and maintained. Secondary, the new neighboring state could be generated by choosing one of these successive base pairs randomly. Thirdly, the annealing schedule is selected to systematically decrease the temperature as the algorithm proceeds, the final solution is the structure with minimum free energy. Furthermore, the performance of our algorithm is evaluated by the instances from PseudoBase database, and compared with state-of-the-art algorithms. The comparison results show that our algorithm is more accurate and competitive with higher sensitivity and specificity indicators.

Keywords: RNA secondary structure · Pseudoknot
Simulated annealing algorithm · Minimum free energy

1 Introduction

RNA is a long chain of nucleotides acid molecule which consists of A (Adenine), U (Uracil), G (Guanine) and C (Cytosine). The four-base arrangement allows RNA to have a variety of functions that can play a role in genetic coding, translation, regulation, and gene expression. The search for the secondary structure of RNA sequence has been widely used as the first step in understanding biological functions [1]. The RNA secondary structure folds itself by forming hydrogen bonds between G-C, A-U, and G-U. Therefore, the prediction of RNA secondary structure is returned to predict all hydrogen connections from the primary structure of the sequence. Many components can be identified in the secondary structure, such as stacked pairs or stacks, hairpin loop, multi-branched loop or Multi-loops, bulge loop, and internal loop. The component structures can be represented by a schematic representation or arc representation, as shown in Fig. 1.

D.-S. Huang et al. (Eds.): ICIC 2018, LNCS 10955, pp. 365–370, 2018.
https://doi.org/10.1007/978-3-319-95933-7_44

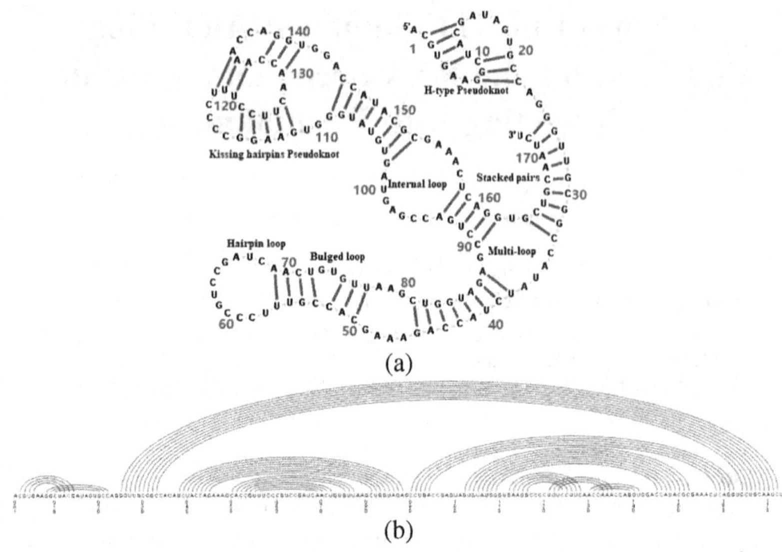

Fig. 1. (a) Typical RNA Secondary Structure. (b) Arc representation of a typical RNA secondary structure. This image was created using jViz.Rna [2].

Pseudoknots usually contains not well-nested base pairs, as shown in Fig. 1(b). These non-nested base pairs make the presence of pseudoknots in RNA sequences more difficult to predict by dynamic programming, which use a recursive scoring system to identify paired stems. The general problem of predicting minimum free energy structures with pseudoknots has been shown to be NP-complete [3].

The dynamic programming (DP) is the first computational approach used to predict RNA structure [4–8]. It can be seen that the temporal and spatial complexity of the prediction algorithm for dynamic programming is high, which is not good for the algorithm to make predictions for long sequence because it will take more time and resources. The other prediction approaches are based on heuristic methods and thermodynamics models [9–13].

In this paper, we propose a novel efficient simulated annealing algorithm to predict nucleic acid secondary structure with pseudoknots. The performance of our algorithm compared with RNA structure method using PseudoBase [14] benchmark instances. The comparison result shows that our algorithm is more accurate and competitive with higher sensitivity and specificity values.

2 Problem Defines

For a given RNA sequence X = 5'$-x_1x_2\ldots x_n-$3' of length n, $M(X)$ is the mapping string of complementary base-pairs of X, $M(X) = (m_1, m_2, \ldots, m_i, \ldots, m_n)$. Each m_i corresponds to the form of (i, j, k), which is called k successive base pairs, where i and j are the base position, where k is the number of successive base pair, and two constraints must be satisfied:

Base Pairs Constraint: If $(i, j, k) \in M$, then $\{(x_i, x_j), (x_{i+1}, x_{j-1}), \ldots, (x_{i+k-1}, x_{j-k+1})\} \in \{(A, U), (G, C), (G, U)\}$ in RNA.

K Successive Base Pairs Constraint: If $(i, j, k) \in M$, then $j - i > 2 * MinStem + Minloop$, $MinStem \geq 2$, $Minloop \geq 3$ and $k \geq MinStem$, where *MinStem* is the minimum number of stack and *MinLoop* is the minimum number of loop (Fig. 2). Such as there must be at least *Minloop* unpaired bases in a hairpin loop.

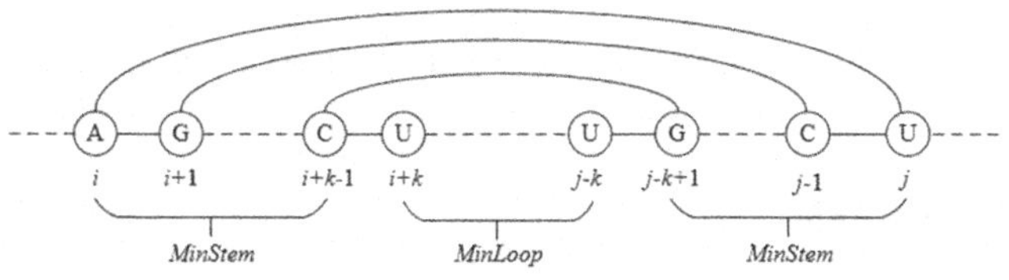

Fig. 2. A graphical illustration of a *MinStem* and *MinLoop*

3 The Proposed Approach

3.1 Set of K Successive Base Pairs

In computer-simulated base pairing, we do not pair individual bases but use successive base pairs. We reduced the range of all possible base pairs by setting *MinStem* and *MinLoop* parameters. Assume that there are three variables i, j, k, which *i* and *j* are the base position, where k is the number of successive base pair. According to the above Fig. 2, we can be seen that i, j, k need to satisfy the following three constraints:

$$1 \leq i \leq RNA.Length - 2 * MinStem - MinLoop + 1 \quad (2)$$

$$i + 2 * MinStem + MinLoop \leq j \leq n \quad (3)$$

$$MinStem \leq k \leq (j - i - MinLoop)/2 \quad (4)$$

3.2 Evaluation Function

For most MFE based RNA secondary structure prediction algorithm, the complex thermodynamic model is often used to evaluate candidate solutions [15]. There are no useful information to guide the candidate solution to find lower neighbor energy state. Consequently, the convergence of these MFE based prediction algorithms is very slow. However, among all of the secondary structure, only the successive base pairs stack structure ΔG_S provide negative free energy which contributes to the reduction of free energy. The stability of RNA sequence can also be approximately evaluated by successive base pairs stacks.

Let *M*(*X*) is the mapping string of complementary base pairs of *X*, $M(X) = (m_1, m_2, \ldots, m_i, \ldots, m_n)$. Each m_i corresponds to the form of (i, j, k), where $m_i.k$ equals *k*, *group* is the number of stems, then the following formula:

$$F(M(X)) = \begin{cases} TotalBP \times AverageBP^2, \textbf{\textit{if}}\ PesudoGroup < MaxPesudoGroup \\ TotalBP \times AverageBP^2 \times \frac{(TotalGroup - PesudoGroup)}{TotalGroup}, \textbf{\textit{else}} \end{cases} \quad (5)$$

Where *PseudoknotGroup* is the predicted number of pseudoknot by the algorithm, and *MaxPesudoKnot* is the expected number of pseudoknot.

$$TotalBasePair = \sum_{i=1}^{n} m_i.k \quad (6)$$

$$AverageBasePair = TotalBasepair/group \quad (7)$$

3.3 Overall Algorithm

The process of natural RNA folding to its minimal free energy state is very similar to the annealing process. In addition, compared with other heuristic prediction algorithms, such as genetic algorithms, the SA algorithm has faster convergence. Therefore, the paper proposes a new method to predict the RNA secondary structure with pseudoknots based on SA framework. This algorithm framework is as follows:

Algorithm:

```
-Initial Max_T, Min_T, CurrentPairs, MaxPairs.
-While(Temperature>Final_ Temperature) do: //T is current temperature;
    //The upper limit of i is the maximum value of MinLoop
    For (i=0 to RNA.Length-2*MinStem) do
        The new Pair is randomly generated from random set of K successive pairs.
        Remove the conflict match from the CurrentPairs.
        CurrentPairs.Add(Pair);
        ΔE = EnergyDelta(CurrentPairs, MaxPairs, maxPesudoKnot);.
        If(ΔE >=0 OR (Exp(ΔE/T)>Random(0,1)))
            MaxPairs = CurrentPairs;
        End If
    End For
    Decrease Temperature.
-End While.
-Return best solutions: MaxPairs. // MaxPairs is final solutions based on SA.
```

4 Experiments Result

The computational result of our algorithm is compared with IPknot [16], TT2NE [17], CyloFold [18] on 10 benchmark instances in PseudoBase RNA database. The evaluate indicators are *sensitivity* (SN) and *specificity* (SP) [19], as shown in Eq. (8).

$$SN = TP \div TP + FN, SP = TP \div TP + FP \tag{8}$$

Where TP represents the number of correctly predicted base pairs; FP represents the number of incorrectly predicted base pairs; FN represents the number of unpredicted base pairs compared to the known structure. When the prediction results are accurate, both SN and SP should be close to 100%.

The comparisons of the proposed method with the other methods are shown in Table 1. In terms of sensitivity, the proposed method provides the best results in six sequences, yields not the worst result in remaining sequences. In terms of specificity, the proposed method yields the best results in three sequences, similar result in five sequences, and inferior results in two sequences. On average, from all sequences, the proposed method outperforms the other methods in all measure. It has average sensitivity and specificity of 92.6% and 84.3% respectively.

Table 1. Comparison results with sensitivity and specificity indicator

Sequences	[18]	IPknot	TT2NE	OPA	[18]	IPknot	TT2NE	OPA
	Sensitivity				Specificity			
Ec_PK3	85.7	71.4	**100.0**	92.9	**100.0**	76.9	**100.0**	92.9
BEV	93.8	81.3	87.5	**100.0**	**100**	81.3	66.7	76.2
BaEV	86.7	0.0	**100.0**	93.3	**81.3**	0.0	65.2	70.0
VMV	100.0	50.0	92.9	**100**	**73.7**	38.9	65.0	70.0
ALFV	100.0	64.7	**100.0**	**100**	**73.9**	45.8	70.8	70.8
SARS-CoV	69.2	69.2	51.7	**84.6**	72.0	78.3	46.9	**100**
BCRV1	96.7	76.7	**100.0**	**100.0**	85.3	82.1	**96.8**	**96.8**
AMV3	71.8	74.4	74.4	**89.7**	80.0	96.7	72.5	**100**
RSV	97.4	71.8	**97.4**	92.3	88.4	90.3	**90.5**	90.0
CCMV3	66.7	**84.4**	71.1	73.3	66.7	**88.4**	71.1	76.7
Average	86.8	71.5	87.5	**92.6**	82.1	75.4	74.6	**84.3**

5 Conclusion

This paper proposes efficient SA algorithm for the RNA secondary structure predicting with pseudoknots, combined with the evaluation function to compensate for the high time complexity of the free energy calculation model. The algorithm sets the *MinStem* and *MinLoop* parameters to determine the pseudoknot structure formed by the base pair cross-combination, and optimizes the pool of candidate solutions, thereby reducing the time cost of the algorithm. We use the evaluation function to further reduce the time consumption of RNA secondary structure prediction algorithms. Moreover, the performance of our algorithm is compared with state of art algorithms using ten PseudoBase benchmark instances, and the comparison result shows that our algorithm is more accurate and competitive with higher sensitivity and specificity values.

Acknowledgement. This work was supported by the National Natural Science Foundation of China (Grant No. 61472293). Research Project of Hubei Provincial Department of Education (Grant No. 2016238).

References

1. Jr, T.I., Bustamante, C.: How RNA folds. J. Seq. Biol. **293**(2), 271–281 (1999)
2. Wiese, K.C., Glen, E.: jViz.Rna - an interactive graphical tool for visualizing RNA secondary structure including pseudoknots. In: IEEE Symposium on Computer-Based Medical Systacks, vol. 2006, pp. 659–664. IEEE Computer Society (2006)
3. Wang, C., Schröder, M.S., Hammel, S., Butler, G.: Using RNA-seq for analysis of differential gene expression in fungal species. Methods Mol. Biol. **1361**, 1–40 (2016)
4. Ray, S.S., Pal, S.K.: RNA secondary structure prediction using soft computing. IEEE/ACM Trans. Comput. Biol. Bioinform. **10**(1), 2–17 (2013)
5. Jiwan, A., Singh, S.: A review on RNA pseudoknot structure prediction techniques. In: International Conference on Computing, Electronics and Electrical Technologies, pp. 975–978. IEEE (2012)
6. Rivas, E., Eddy, S.R.: A dynamic programming algorithm for RNA structure prediction including pseudoknots. J. Seq. Biol. **285**(5), 2053–2068 (1999)
7. Reeder, J., Giegerich, R.: Design, implementation and evaluation of a practical pseudoknot folding algorithm based on thermodynamics. BMC Bioinform. **5**(1), 104 (2004)
8. Dirks, R.M., Pierce, N.A.: A partition function algorithm for nucleic acid secondary structure including pseudoknots. J. Comput. Chem. **24**(13), 1664–1677 (2003)
9. Ren, J., Rastegari, B., Condon, A., Hoos, H.H.: Hotknots: heuristic prediction of RNA secondary structures including pseudoknots. RNA **11**(10), 1494–1504 (2005)
10. Tsang, H.H., Wiese, K.C.: SARNA-Predict-pk: predicting RNA secondary structures including pseudoknots, pp. 1–8. IEEE (2008)
11. Wiese, K.C., Deschenes, A.A., Hendriks, A.G.: Rnapredict—an evolutionary algorithm for RNA secondary structure prediction. IEEE/ACM Trans. Comput. Biol. Bioinform. **5**(1), 25–41 (2008)
12. Tsang, H.H., Wiese, K.C.: Sarna-predict: accuracy improvement of RNA secondary structure prediction using permutation-based SA. IEEE/ACM Trans. Comput. Biol. Bioinform. **7**(4), 727 (2010)
13. Rastegari, B., Condon, A.: Linear Time Algorithm for Parsing RNA Secondary Structure. In: Casadio, R., Myers, G. (eds.) WABI 2005. LNCS, vol. 3692, pp. 341–352. Springer, Heidelberg (2005). https://doi.org/10.1007/11557067_28
14. PseudoBase. http://www.ekevanbatenburg.nl/PKBASE/PKB.HTML. Accessed 11 Mar 2018
15. Andronescu, M., Aguirrehernández, R., Condon, A., Hoos, H.H.: RNAsoft: a suite of RNA secondary structure prediction and design software tools. Nucleic Acids Res. **31**(13), 3416 (2003)
16. Sato, K., Kato, Y., Hamada, M., Akutsu, T., Asai, K.: IPknot: fast and accurate prediction of RNA secondary structures with pseudoknots using integer programming. Bioinformatics **27** (13), i85–i93 (2011)
17. Bon, M., Orland, H.: TT2NE: a novel algorithm to predict RNA secondary structures with pseudoknots. Nucleic Acids Res. **39**(14), e93–e93 (2011)
18. Bindewald, E., Kluth, T., Shapiro, B.A.: Cylofold: secondary structure prediction including pseudoknots. Nucleic Acids Res. **38**(Web Server issue), 368–372 (2010)
19. Baldi, P., Brunak, S.Y., Andersen, C., Nielsen, H.: Assessing the accuracy of prediction algorithms for classification: an overview. Bioinformatics **16**(5), 412 (2000)

TNSim: A Tumor Sequencing Data Simulator for Incorporating Clonality Information

Yu Geng[1,3,4], Zhongmeng Zhao[1,3], Mingzhe Xu[1,3], Xuanping Zhang[1,3], Xiao Xiao[2,3], and Jiayin Wang[1,3](✉)

[1] Department of Computer Science and Technology, School of Electronic and Information Engineering, Xi'an Jiaotong University, Xi'an 710049, China
wangjiayin@mail.xjtu.edu.cn

[2] School of Public Policy and Administration, Xi'an Jiaotong University, Xi'an 710049, China

[3] Shaanxi Engineering Research Center of Medical and Health Big Data, Xi'an Jiaotong University, Xi'an 710049, China

[4] Jinzhou Medical University, Jinzhou 121001, China

Abstract. In recent years, the next generation sequencing enables us to obtain high resolution landscapes of the genetic changes at single-nucleotide level. More and more novel methods are proposed for efficient and effective analyses on cancer sequencing data. To facilitate such development, data simulator is a crucial tool, which not only tests and evaluates proposed approaches, but provides the feedbacks for further improvements as well. Several simulators are released to generate the next generation sequencing data. However, based on our best knowledge, none of them considers clonality information. It is suggested that clonal heterogeneity does widely exist in tumor samples. The patterns of somatic mutational events usually expose a wide spectrum of variant allelic frequencies, while some of them are only detectable in one or multiple clonal lineages. In this article, we introduce a Tumor-Normal sequencing Simulator, TNSim, to generate the next generation sequencing data by involving clonality information. The simulator is able to mimic a tumor sample and the paired normal sample, where the germline variants and somatic mutations can be settled respectively. Tumor purity is adjustable. Clonal architecture is preassigned as one or more clonal lineages, where each lineage consists of a set of somatic mutations whose variant allelic frequencies are similar. A group of experiments are conducted to evaluate its performance. The statistical features of the artificial sequencing reads are comparable to the real tumor sequencing data whose sample consists of multiple sub-clones. The source codes are available at http://github.com/lnmxgy/TNSim and for academic use only.

Keywords: Cancer genomics · Cancer sequencing data · Data simulator Clonal structure

1 Introduction

Benefiting from the next generation sequencing technologies, we nowadays are able to interrogate the entire genomes, exomes and transcriptomes of tumor samples. One of the most important questions in cancer genomics is to differentiate the patterns of the

D.-S. Huang et al. (Eds.): ICIC 2018, LNCS 10955, pp. 371–382, 2018.
https://doi.org/10.1007/978-3-319-95933-7_45

somatic mutational events [1–3]. In some large cancer genome projects, such as The Cancer Genome Atlas (TCGA) and International Cancer Genome Consortium (ICGC), the sample collection is designed as following: each patient is collected at least two separate samples for sequencing and further data analyses. One sample is from tumor tissue, while the other sample is from adjacent normal tissue or blood. Such design facilitates the comparison of the variant calls at single-nucleotide resolution and the identification of both the germline variations and somatic mutational events.

Somatic mutations, especially the somatic driver events, are considered to govern the dynamics of clone birth, evolution and proliferation [1, 3]. Recent researches have revealed that the clonal evolution in cancer more frequently involves multiple simultaneous processes among co-existing sub-clones and bilaterally couples with the germline variations and other host environmental factors, rather than a single sequential process [1–4]. Although it is suggested that any two cancer cells should be different at the nucleotide level, these are such low-frequent events that may only be directly detectable when amplified in a clonal lineage. Recent studies based on cancer sequencing data, across a diversity of solid and hematological disorders, observe that tumor samples are usually both spatially and temporally heterogeneous and are frequently comprised of one or multiple founding clone(s) and a couple of sub-clones [1, 4–6]. Moreover, the sampled tumor tissue includes both cancerous and non-cancerous cells, which is known as the tumor purity problem, while the normal sample can also be contaminated by tumor cells due to permeation or metastasis for example [3, 7, 8].

Ding and others summarized four primary open problems in cancer clonality research [4], which include (1) the identification of clonality-related genetic alterations, (2) discerning clonal architecture, (3) understanding their phylogenetic relationships and (4) modeling the mathematical and physical mechanisms. Strictly speaking, none of these issues is completely solved, and these issues remain in the active areas of research, where powerful and efficient bioinformatics tools are urgently demanded for better analyses on rapidly accumulating data.

It is always important to improve and develop computational methods and algorithms under a simulated-based supervised learning framework. Existing next generation sequencing data simulators, such as pIRS [9], ART [10], GemSIM [11], mainly focus on modeling the sequencing errors. For example, ART is among the first ones introducing a read error model, while pIRS proposes an empirical base-calling profile and a GC%-depth profile with coverage bias. Without considering the structure of clonality, these approaches expose less fitness on simulating cancer sequencing data, especially when the tumor sample consists of multiple clonal lineages. Motivated by this, in this article, we propose TNSim, a Tumor-Normal sample-pair sequencing Simulator. This simulator simultaneously generates a tumor sample and the paired normal sample, where an adjustable tumor purity is considered. Clonal architecture is preassigned as one or more clonal lineages, where each lineage consists of a set of somatic mutations whose variant allelic frequencies are similar.

2 Methods and Features

TNSim consists of three components: NorSim, TumSim and ReadGen. The basic idea of TNSim is as follows: The function of NorSim is to generate a set of variants on a given reference genome to obtain the so-called normal sample. Those variants are considered as the germline variations. Subsequently, by given the generated variants, TumSim generates somatic mutations and spikes them into the genome, which already harbors the germline variations planted by NorSim, to obtain the tumor genome of the founding clone. By recurrently using TumSim, the simulator enables the user to obtain multiple tumor genomes corresponding to different clonal lineages. Once all of the tumor genomes are preassigned, ReadGen samples the paired-end reads with a predefined error model on the admixture of the normal and tumor genomes.

The planted mutations can be single base substitutions, common genomic structural variations, including insertions, deletions, inversions, repetitive elements, and their combinations, e.g. germline allelic imbalance, loss of heterogeneous, etc. [2, 12] In addition, the mutation rates of somatic mutational events often vary among different genomic regions. For example, the variants within the regions harboring repeat elements, such as mini-/micro-satellites and medium-to-large tandem repeats, have been found to occur at much higher rates (as high as 1% per locus) than those within single-copy regions [1, 4]. Recent integrated studies on germline and somatic variations across dozens of major cancer types summarizes that the spectrum of mutation rates also exposes diversity among different cancer types [1, 3]. To satisfy various features, TNSim enables the user to preset specific somatic mutation rate on each particular genomic region, which is also implemented in TumSim.

The somatic mutations which occur in the founding clone usually present in the tumor following a uniform distribution by propagating to the progeny of that one during clonal expansion [13], while the sub-clonal mutations that evolve in an existing neoplastic cell are passed on specific to the sub-population cells derived from it. To model the clonal architecture, TNSim keeps tracking on the evolutionary history and employs the conditional probability to artificially generate sub-clonal mutations. For each mutational descendant (sub-clone) from an existing clone, the mutations newly occur should be distinct from the existing ones in the ancestors or other clonal lineages, by either presenting on a different loci or haplotype or another variant type, behind which an implicit assumption is that it is rare to "repair" an existing somatic mutation by the same type of somatic mutational event.

D_{31}~D_{28}	D_{27}~D_8	D_7~D_6	D_5~D_4	D_3~D_0
length of an indel (inl_i)	inserted fragment (ins_i)	type (v_i)	genotype (g_i)	alternation (c_i)

Fig. 1. Encoding the attributes of variant i: c_i is the alternative allelic value at site i, g_i records the genotype, v_i denotes the type of this variant (such as SNV, insertion, deletion, etc.). If this is a structural variant, inl_i represents the length of the variant, while the context is shown in ins_i. For large structural variations, a pointer is placed at ins_i field which leads to the real context.

More detailed, TNSim first creates a .sim file complying with the variant call format (VCF file) to record the artificial germline variations, of which each record consists of the following attributes: location (chromosome and position), allelic contexts on the reference genome and the alternation, variant type, genotype and length (for structural variations). Similarly, for each clonal evolution process, TNSim generates a new .sim file to record the newly generated somatic mutations. Thus, considering a phylogenetic tree describes the step-wise clonal evolutionary, the root of the tree is the genome harboring the germline variations only, while each leaf represents a tumor genome and each .sim file is affiliated to the corresponding edge. The mutations belong to a leaf node (a sub-clone) are composed by merging the .sim files along with the path tracing back to the root. To facilitate the computation, TNSim encodes the attributes into two 32-bit variables in the implementation, shown as Fig. 1, where the attributes of variant i include: c_i is the alternative allelic value at site i, g_i records the genotype, v_i denotes the type of this variant (such as SNP, insertion, deletion, etc.). If this is a structural variant, inl_i represents the length of the variant, while the context is shown in ins_i. For large structural variations, a pointer is placed at ins_i field which leads to the real context. The framework of TNSim is illustrated in Fig. 2.

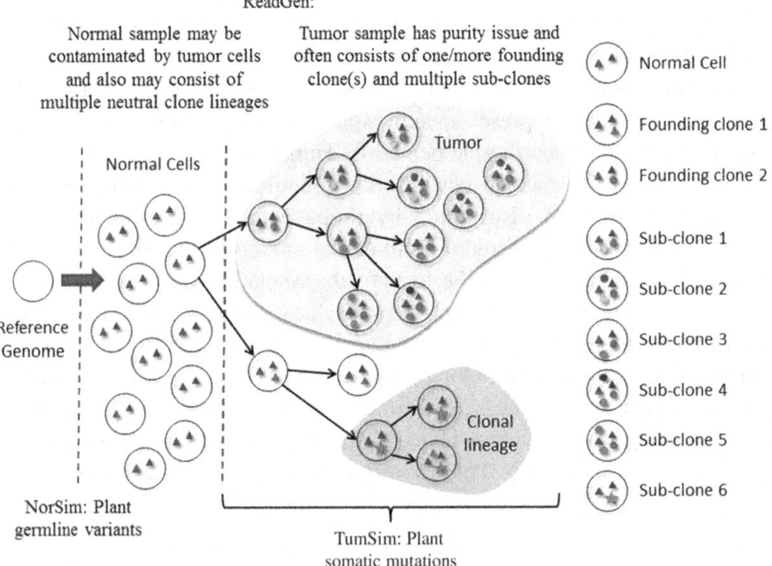

Fig. 2. The framework of TNSim: NorSim generates the germline variants on the given reference genome. TumSim evolves sub-clones based on one or more founding clones. ReadGen samples from a mixture of multiple clones and the normal genome.

Implemented in ReadGen component, TNSim allows the user to set the purity and the percentages of the content of each sub-clone. The paired-end reads are then sampled from the mixture and stored in a left.fq and a right.fq files separately. Similar to existing next-generation sequencing simulator, several parameters such as coverage, read length, the configuration of the insert size distribution, are all pre-set by the user.

3 Experiments

To examine the performance of TNSim, we examine the statistical features of the simulated data under different configurations. In the following experiments, the reference genome is the human genome GRCh37-lite chromosome 9. The alignment tool is *bwa aln* (version 0.7.5a-r705) under the default settings.

TNSim generates a set of paired tumor-normal sequencing data, in which the background mutation rate of the germline variants is 0.1%, while the somatic mutation rates are 0.6% in a preset interval and 0.001% in the other regions. In total four clones are evolved, where s_1 is the founding clone and sub-clones s_2 and s_3 are derived from s_1. In addition, sub-clone s_4 further develops from sub-clone s_2. A complex tandem repeat is planted, where a mutation occurs three out of total five repeats.

We first test the performance of ReadGen. We vary the preset coverages among 30×, 100× and 180×. 30× is the basic coverage for cancer sequencing. 100× is the popular coverage for the TCGA exome sequencing data. 180× is a widely used coverage setting for some recent exome sequencing project. We randomly exam the read depth at selected sites: The first site is randomly sampled at the beginning region of chromosome 9, and then sample one site per 10,000 bps. The results are shown in Fig. 3. Means of the read depths across the sampled sites are 29.8834, 100.0500 and 180.1250, respectively, while the standard deviations are 0.7239, 1.1871 and 1.6325, respectively.

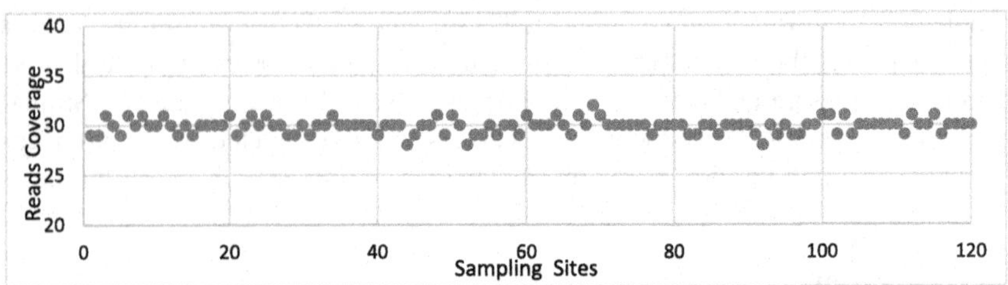

Read depths at sampled sites from the normal sample. x axis denotes the sites, while y axis denotes read depth at the corresponding site. (a) Preset coverage is 30×. Mean of the read depths across the sampled sites is 29.8834, while the standard deviation is 0.7239.

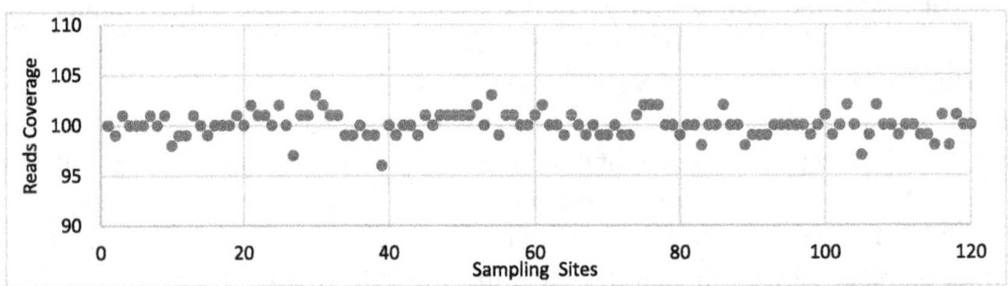

Read depths at sampled sites from the tumor sample. x axis denotes the sites, while y axis denotes read depth at the corresponding site. (b) Preset coverage is 100×. Mean of the read depths across the sampled sites is 100.0500, while the standard deviation is 1.1871.

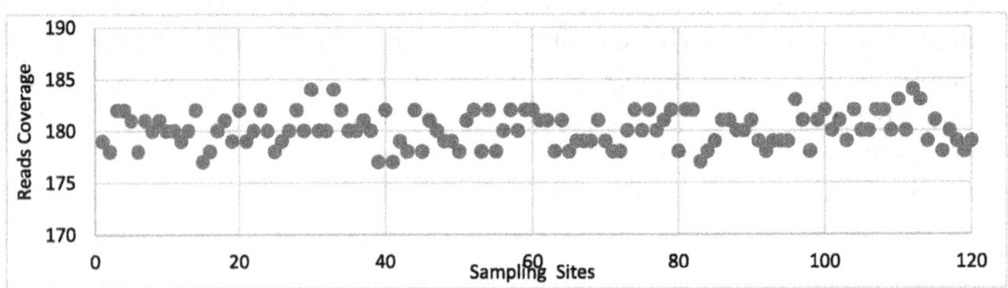

Fig. 3. Read depths at sampled sites from the tumor sample. x axis denotes the sites, while y axis denotes read depth at the corresponding site. (c) Preset coverage is 180×. Mean of the read depths across the sampled sites is 180.1250, while the standard deviation is 1.6325.

We then exam the statistics of the germline variants. We collect the distributions of variant allelic frequencies (VAFs) across all generated germline variants under different coverages. For each coverage configuration, two distributions are collected, one of which is the distribution of VAFs based on original .fq file, while the other is the distribution based on bwa aligned .sam file. We set the background mutation rate of the germline variants to be 0.1%. The heterozygous variants are preset as two thirds among all the generated variants. Here we still vary the preset coverages among 30×, 100× and 180×. The results are shown in Figs. 4, 5 and 6, respectively.

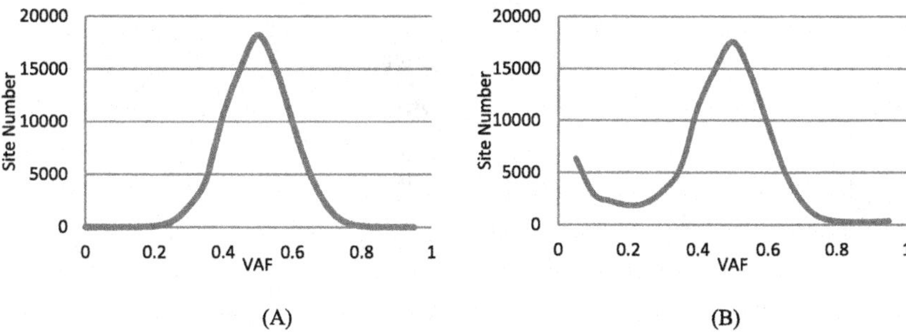

Fig. 4. The distribution of variant allelic frequencies (VAFs) across all generated germline variants. (**A**): the distribution of VAFs based on original .fq files, (**B**): the distribution based on bwa aligned .sam file. x axis denotes the VAF, while y axis denotes the number of variants at the corresponding VAF. Let the preset mutation rate of the germline variants be 0.1%, where the heterozygous variants are preset as two thirds among all the planted variants. The alignment tool is bwa aln (version 0.7.5a-r701) under the default setting. Coverage is 30×.

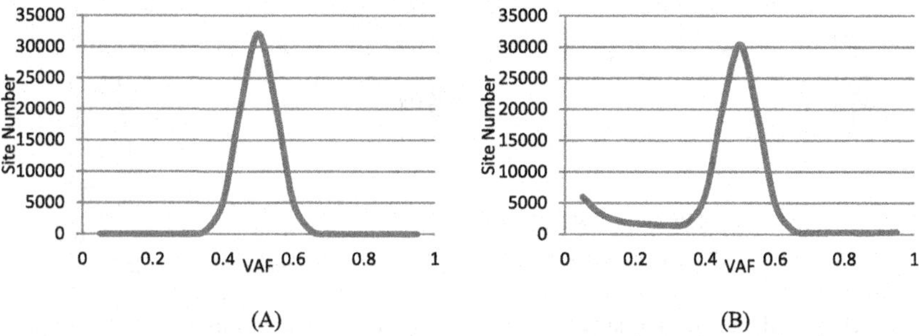

Fig. 5. The distribution of variant allelic frequencies (VAFs) across all generated germline variants. (**A**): the distribution of VAFs based on original .fq files, (**B**): the distribution based on bwa aligned .sam file. x axis denotes the VAF, while y axis denotes the number of variants at the corresponding VAF. Let the pre-set mutation rate of the germline variants be 0.1%, where the heterozygous variants is pre-set as two thirds among all the planted variants. The alignment tool is bwa aln (version 0.7.5a-r701) under the default setting. Coverage is 100×.

We then exam the statistics of the somatic mutations. We first generate a tumor sample which is set as a mixture of two clones s_1 (marked in dark red dots) and s_2 (marked in blue dots), whose percentages of the content are 30% and 70%,

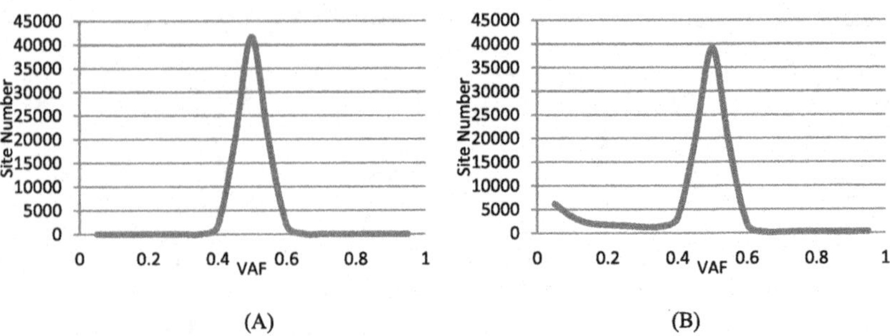

Fig. 6. The distribution of variant allelic frequencies (VAFs) across all generated germline variants. (**A**): the distribution of VAFs based on original .fq files, (**B**): the distribution based on bwa aligned .sam file. x axis denotes the VAF, while y axis denotes the number of variants at the corresponding VAF. Let the pre-set mutation rate of the germline variants be 0.1%, where the heterozygous variants is pre-set as two thirds among all the planted variants. The alignment tool is bwa aln (version 0.7.5a-r701) under the default setting. Coverage is 180×.

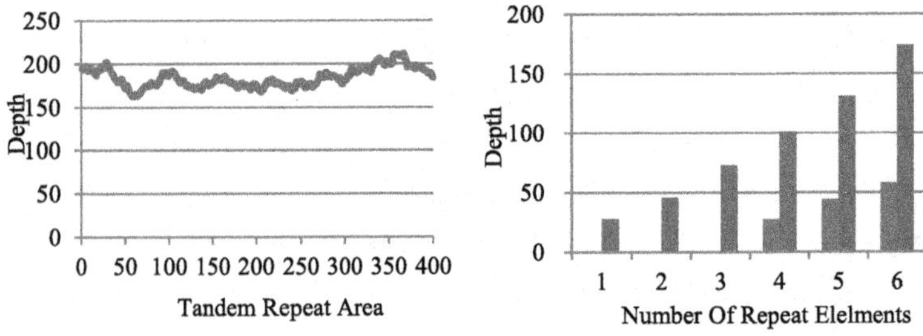

Fig. 7. The read depths on the tandem repeat region (belongs to s_1) in the tumor sample. The tandem repeat consists of 5 repeated elements, each of which copies the interval, Chr9: 22375100–22375499. A somatic occurs three out of five repeats. The percentage of s_1 is 30%. Tumor purity is set to 100%. Coverage is 100×.

respectively. s_1 is the founding clone, s_2 is a sub-clone derived from s_1. The somatic mutations are 0.6% in a preset interval and 0.001% in other regions. Tumor purity is set to 100%. The coverage is set to 100×. Figure 7 shows the read depths on the preset tandem repeat regions (belong to s_1) in the tumor sample. Figure 8 shows the comparison of the clonal structures before (a) and after (b) the read mapping and variant calling steps.

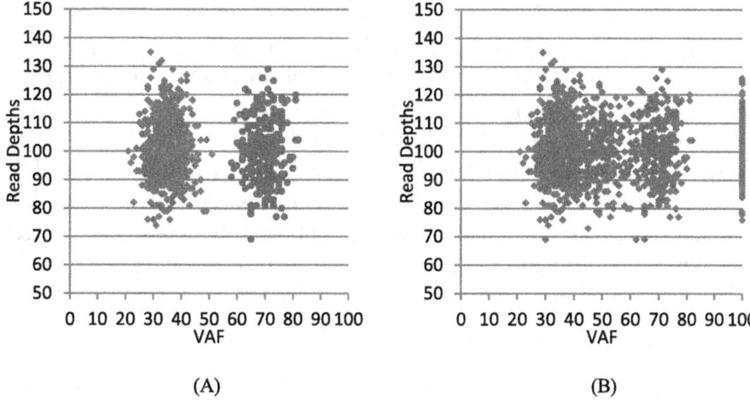

Fig. 8. Clonal architecture in a tumor sample. x axis denotes the VAF of the variants, while y axis denotes the corresponding read depths of the variants. (**A**): on original .fq files, (**B**): on bwa aligned .sam file. The tumor sample is set as a mixture of two clones s_1 (marked in red dots) and s_2 (marked in blue dots), whose percentages of the content are 30% and 70%, respectively. s_1 is the founding clone, s_2 is a sub-clone derived from s_1. Tumor purity is set to 100%. Coverage is 100×. (Color figure online)

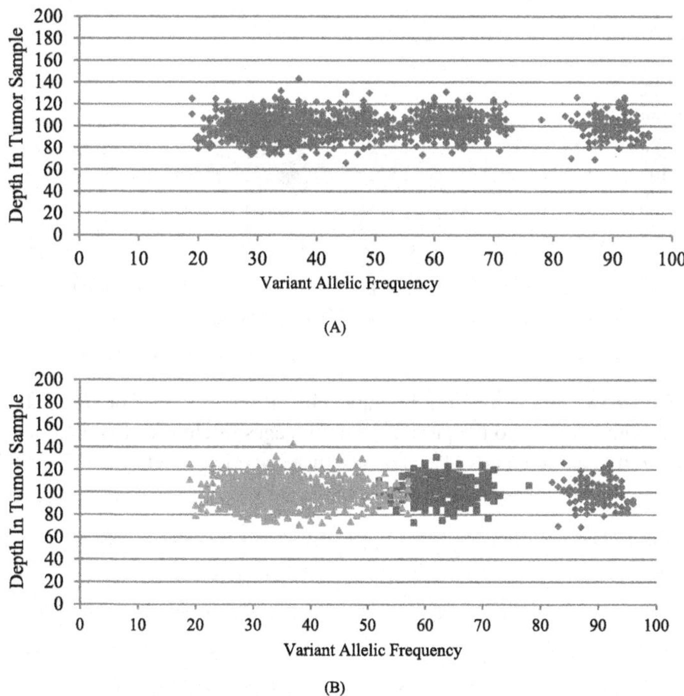

Fig. 9. Clonal architecture in the tumor sample with 90% purity. (**A**): on original .fq files, (**B**): on bwa aligned .sam file. Due to the purity, the percentages of clone s_1 (marked in red dots) and clone s_2 (marked in green dots) are 27% and 63%, respectively. Coverage is 100×. (Color figure online)

We also exam the tumor purity feature of the tumor sample. We preset the tumor purities to 90% and 80%, respectively. In the first case, the percentages of clone s_1 (marked in red dots) and clone s_2 (marked in green dots) are 27% and 63%, respectively. In the second case, the percentages of clone s_1 (marked in red dots) and clone s_2 (marked in green dots) are 24% and 56%, respectively. The results are shown in Figs. 9 and 10, respectively. In addition, Fig. 11 shows a complex tumor sample. The tumor sample is set as a mixture of four clones s_1 (marked in purple dots), s_2 (marked in red dots), s_3 (marked in green dots), and s_4 (marked in blue dots) whose percentages of the content are 10%, 20%, 40% and 30%, respectively. s_1 is the founding clone, sub-clones s_2 and s_3 are derived from s_1, while sub-clone s_4 further develops from s_2.

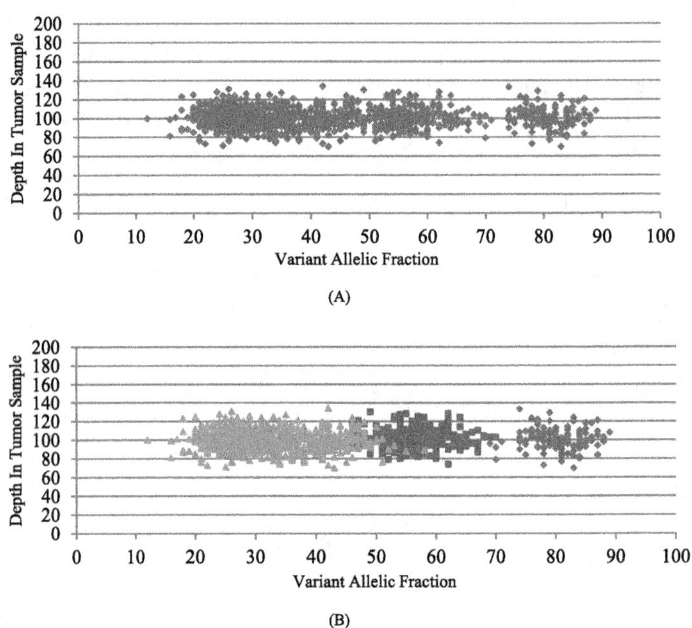

Fig. 10. Clonal architecture in the tumor sample with 80% purity. (**A**): on original .fq files, (**B**): on bwa aligned .sam file. Due to the purity, the percentages of clone s_1 (marked in red dots) and clone s_2 (marked in green dots) are 24% and 56%, respectively. Coverage is 100×. (Color figure online)

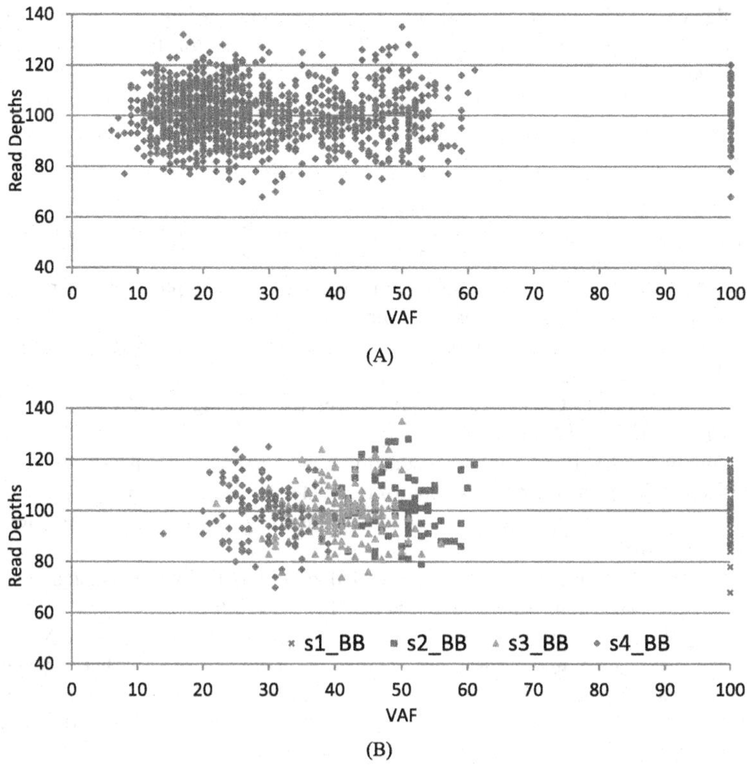

Fig. 11. Clonal architecture in a tumor sample. x axis denotes the VAF of the variants, while y axis denotes the corresponding read depths of the variants. (**A**): on original .fq files, (**B**): on bwa aligned .sam file. The tumor sample is set as a mixture of four clones s_1 (marked in purple dots), s_2 (marked in red dots), s_3 (marked in green dots), and s_4 (marked in blue dots) whose percentages of the content are 10%, 20%, 40% and 30%, respectively. s_1 is the founding clone, sub-clones s_2 and s_3 are derived from s_1, while sub-clone s_4 further develops from s_2. Tumor purity is set to 100%. Coverage is 100×. (Color figure online)

4 Conclusion

In this article, we design and implement an effective sequencing read simulator, TNSim, which is simple to use and can reflect both the tumor purity and the clonal architectures in tumor samples. TNSim consists of three components: NorSim, TumSim and ReadGen. NorSim generates the germline variations. TumSim recurrently generates somatic mutations corresponding to different clonal lineages. Then, ReadGen samples the paired-end reads with a predefined error model on the admixture of the normal and tumor genomes. We conduct a series of experiments to evaluate the performance of TNSim. According to the results, TNSim is not only suitable for generating the cancer sequencing data incorporating clonality information, but also can involve the tumor purity and other micro-environment features.

Acknowledgement. This work is supported by the National Science Foundation of China (Grant No: 31701150) and the Fundamental Research Funds for the Central Universities (CXTD2017003).

References

1. Kandoth, C., McLellan, M., Vandin, F., et al.: Mutational landscape and significance across 12 major cancer types. Nature **502**(7471), 333–339 (2013)
2. Lu, C., Xie, M., Wendl, M., et al.: Patterns and functional implications of rare germline variants across 12 cancer types. Nature Commun. **6**, 10086 (2015)
3. Huang, K., Mashl, R., Wu, Y., et al.: Pathogenic germline variants in 10,389 adult cancers. Cell **173**(2), 355–370 (2018)
4. Ding, L., Raphael, B., Chen, F., et al.: Advances for studying clonal evolution in cancer. Cancer Lett. **340**(2), 212–219 (2013)
5. The Computational Pan-Genomics Consortium: Computational pan-genomics: status, promises and challenges. Briefings Bioinform. **19**(1), 118–135 (2018)
6. Vijg, J.: Somatic mutations, genome mosaicism, cancer and aging. Curr. Opin. Genet. Dev. **26**(26C), 141–149 (2014)
7. Xie, M., Lu, C., Wang, J., et al.: Age-related cancer mutations associated with clonal hematopoietic expansion. Nature Med. **20**(12), 1472–1478 (2014)
8. Geng, Yu., Zhao, Z., Liu, R., Zheng, T., Xu, J., Huang, Y., Zhang, X., Xiao, X., Wang, J.: Accurately estimating tumor purity of samples with high degree of heterogeneity from cancer sequencing data. In: Huang, D.-S., Jo, K.-H., Figueroa-García, J.C. (eds.) ICIC 2017. LNCS, vol. 10362, pp. 273–285. Springer, Cham (2017). https://doi.org/10.1007/978-3-319-63312-1_25
9. Hu, X., Yuan, J., Shi, Y., et al.: pIRS: Profile-based Illumina pair-end reads simulator. Bioinformatics **28**(11), 1533–1535 (2012)
10. Huang, W., Li, L., Myers, J., et al.: ART: a next-generation sequencing read simulator. Bioinformatics **28**(4), 593–594 (2012)
11. McElroy, K., Luciani, F., Thomas, T.: GemSIM: general, error-model based simulator of next-generation sequencing data. BMC Genom. **13**(74), 1–9 (2012)
12. Geng, Y., Zhao, Z., Xu, J., et al.: Identifying heterogeneity patterns of allelic imbalance on germline variants to infer clonal architecture. In: Huang, D., Jo, K., Figueroa-García, J. (eds.) ICIC 2017. LNCS, vol. 10362, pp. 286–297. Springer, Cham (2017). https://doi.org/10.1007/978-3-319-63312-1_26
13. Miller, C., White, B., Dees, N., et al.: SciClone: Inferring clonal architecture and tracking the spatial and temporal patterns of tumor evolution. PLoS Comput. Biol. **10**(8), e1003665 (2014)

Optimizing HP Model Using Reinforcement Learning

Ru Yang[1], Hongjie Wu[1(✉)], Qiming Fu[2(✉)], Tao Ding[1], and Cheng Chen[1]

[1] School of Electronic and Information Engineering, Suzhou University of Science and Technology, Suzhou 215009, China
Hongjie.wu@qq.com
[2] Jiangsu Province Key Laboratory of Intelligent Building Energy Efficiency, Suzhou University of Science and Technology, Suzhou 215009, China
fqm_1@126.com

Abstract. Protein structure prediction has always been an important issue in bioinformatics field. This paper proposes an HP model optimization method based on reinforcement learning, which is a new attempt in the area of protein structure prediction. It does not require external supervision as the agent can find the optimal solution from the reward function in the training process. And the method also decreases computational complexity through making the time complexity of the algorithm has a linear relationship with the length of protein sequence.

Keywords: Reinforcement learning · HP model · Structure prediction

1 Introduction

The biological functions of proteins are determined by their spatial folding structure [1]. At present, to reduce the high complexity of prediction three-dimensional structure, many researches have proposed many computable theoretical models, and the HP model is a typical protein folding model that has been extensively studied. HP model can improve the accuracy of the three-dimensional model. And it has high reliability in predicting the protein helix structure.

Some methods based on machine learning for protein HP model optimization have poor universality, which makes the solution ineffective. Therefore, this paper presents an HP model optimization method based on reinforcement learning. The proposed method is universal and easy to calculate, which can predict the optimal structure for short sequences.

D.-S. Huang et al. (Eds.): ICIC 2018, LNCS 10955, pp. 383–388, 2018.
https://doi.org/10.1007/978-3-319-95933-7_46

2 Method

2.1 Reinforcement Learning

Reinforcement Learning (RL) refers to a class of online learning methods that map from the state of the environment to the action and obtain the maximum expected cumulative reward. The goal of reinforcement learning is to let the agent in the cognitive environment learns the optimal strategy to solve the problem by constantly interacting with the environment.

Markov decision process can be expressed as a four-dimensional array (*S*, *A*, *T*, *R*). *S* is the state set, *A* is the action set, *T* is the state transfer function, *R* is the immediate reward function obtained by taking action a under the state s [2]. Markovian refers to that the immediate reward which the agent gets by taking the current action a in the current state s to the next state s' is only related to the current state s and the current action a and has nothing to do with the previous states and actions.

Q-learning algorithm, also known as the off-policy TD control algorithm, its basic form can be expressed as:

$$Q(s_t, a_t) = (1 - \alpha)Q(s_t, a_t) + \alpha \left[r_{t+1} + \gamma \max_{a} Q(s_{t+1}, a) \right] \tag{1}$$

In this algorithm, regardless of which policy the agent is following during the training process, Q-learning algorithm adopts the optimal policy. That is, its behavioral policy and evaluation policy are not the same, so it is called an off-policy method.

2.2 The Framework Based on Reinforcement Learning

In recent years, the most typical simplified model for the protein folding problem is the two-dimensional HP lattice model proposed by Dill et al. [3]. According to different hydrophilic and hydrophobic properties, each amino acid can be divided into two categories: hydrophobic amino acid (H) and hydrophilic amino acid (P), so an arbitrary protein chain can be represented as a finite length string of H and P.

The use of reinforcement learning method to solve the HP two-dimensional sequence model optimization problem can be understood as Markov decision process. The framework of this method is shown in Fig. 1.

2.3 HP Model State Set S

The initial state of the agent in the environment is s_1. For a two-dimensional sequence of length n, its state space S consists of $\frac{4^n - 1}{3}$ states. When the state of the first amino acid is fixed, all possible states of the successor of each amino acid are the collection of four states (up, down, left, right) of the previous amino acid, that is, the number of all possible states of subsequent amino acid is four times the number of previous amino acids. The total number of the state set is shown in Eq. (2):

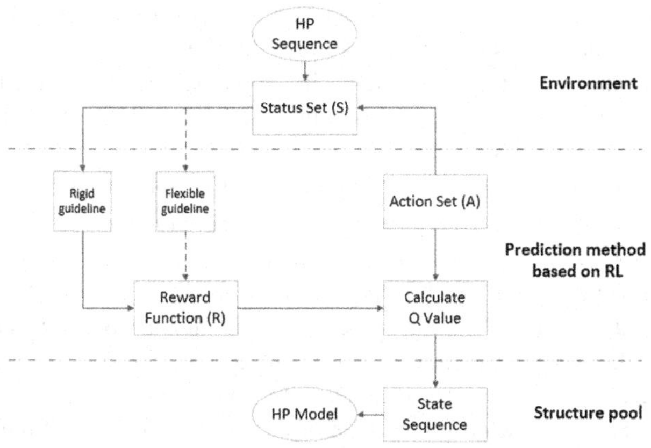

Fig. 1. A framework for HP model optimization based on reinforcement learning

$$S_n = \frac{1 * (1 - 4^n)}{1 - 4} = \frac{4^n - 1}{3} \tag{2}$$

So $S = \left\{s_1, s_2, \ldots, s_{\frac{4^n-1}{3}}\right\}$. At the same time, we need to define the state transfer function $T : s \rightarrow s'$ of the HP model, that is, $T(s, a) = s'$. The process that the agent takes the action a in the state s to the subsequent state s' can be written as the concrete expression as shown in formula (3):

$$T\left(s_{\frac{4k-1}{3}+i}, a_j\right) = s_{\frac{4k+1-1}{3}+4\cdot(i-1)+j} \tag{3}$$

where, $k \in [0, n-1]$, $i \in \left[1, 4^k\right]$, $j \in [1, 4]$.

This means that, the agent can move to one of four possible successor states from the state $s \in S$ by performing one of four possible actions. It should be noted that, each state $s' \in S$ can be accessed from the state s.

2.4 Reward Function Based on HP Model Energy Function

The energy value is only determined by the hydrophobic force [4]. Each pair of H that are not adjacent in sequence but are adjacent in two-dimensional space produces an energy of −1, and in other cases, the energy is calculated as 0. The energy value of the entire structure is calculated as follows:

$$E = \sum\nolimits_{i}^{n-1} \sum\nolimits_{j=i+1}^{n} W_{ij} \tag{4}$$

where, $W_{ij} = \begin{cases} -1, & i = j = H \, and \, i, \, j \, are \, adjacent \, in \, sequence \\ 0, & other \, cases \end{cases}$

On the definition of reward function, we divide it into two kinds.

The first, flexible criterions: (1) Before reaching the terminal state, the next state of the amino acid is placed in an invalid position with the reward set to −10. (2) Before reaching the terminal state, the next state of the amino acid is placed in a blank position with the reward set to 0. (3) When the terminal state is reached, the absolute value of the energy of the resulting structure will be rewarded.

$$\text{That is } R = \begin{cases} -10, & i \in (1 \sim n-1) \textit{ is at the invalid location} \\ 0, & i \in (1 \sim n-1) \textit{ is at the void location} \\ |E|, & i = n \end{cases}$$

The second, rigid criterions: Differing from the flexible criterion, the check matrix '*Check*' is introduced here. The lattice position where the amino acid has been placed is marked (also called invalid position), then in this episode, this position can no longer be placed. (1) Before reaching the terminal state, the reward is set to 0. (2) When the terminal state is reached, the absolute value of the resulting energy of the structure will be rewarded.

$$\text{That is } R = \begin{cases} 0, & i \in (1 \sim n-1) \\ |E|, & i = n \end{cases}$$

2.5 HP Model Training Algorithm Based on Reinforcement Learning

The training algorithm is shown minutely in Table 1.

Table 1. Training algorithm based on HP model

Algorithm: HP two-dimensional sequence training algorithm（Take rigid criterion for example）

C : check repeatability matrix R : reward function

s : current state a : current action s' : next state   TRML: terminal state

For one Episode:
  Initialize s
  While (s != TRML){
  $a \leftarrow$ Q(ε-greedy)
  $s' \leftarrow T(s, a)$
  If ($C[s']$==1)
    $a \leftarrow$ Q(ε-greedy)
    $s' \leftarrow T(s, a)$
  $r \leftarrow R(s, a)$
  $Q(s_t, a_t) = (1-\alpha)Q(s_t, a_t) + \alpha[r_{t+1} + \gamma \max_a Q(s_{t+1}, a)]$
  $s \leftarrow s'$}

3 Experiments

We use rigid criterion and flexible criterion for training respectively. In this paper, two amino acid sequences are selected as experimental subjects: sequence 1: HPPHHPH [5] and sequence 2: HPHHHPHHPH [6]. The parameters were set as follows: $\alpha = 0.01$, $\varepsilon = 0.5$, $\gamma = 0.9$.

Here, the number of iterations of all sequences is 3 million, and the training effect is tested every 10,000 times. As we can see from the Table 2, both reward settings can be trained to obtain the lowest energy, but the number of trainings required to train to convergence under the rigid criterion is much less than that of the flexible criterion. And as the length of the sequence increases, the advantage of the rigid criterion becomes more apparent. The comprehensive comparison shows that the results of the reward function using the rigid criterion are always better than the flexible criterion.

Table 2. Comparison of convergence required number of sequences (unit:/million)

Sequence	The number of convergence	1	2	3	4	5
1	Rigid	2	1	4	7	2
	Flexible	6	2	7	2	7
2	Rigid	3	14	4	7	11
	Flexible	47	109	30	57	23

4 Conclusion

This paper presents a model based on reinforcement learning to solve the HP model prediction problem. In this paper, two short-length sequences are selected as the experimental subjects, and it can be found that the reinforcement learning method using the rigid criterion can all converge stably, and different optimal structures of the HP model can be obtained. This article can be said to be a new attempt in the area of protein structure prediction by reinforcement learning, which will serve as a demonstration for subsequent further study in the field of protein three-dimensional structure prediction and other fields of biological information.

Acknowledgement. This paper is supported by the National Natural Science Foundation of China (61772357, 61502329, 61672371), Jiangsu 333 talent project and top six talent peak project (DZXX-010), Suzhou Foresight Research Project (SYG201704, SNG201610) and the Postgraduate Research & Practice Innovation Program of Jiangsu Province (SJCX17_0680).

References

1. Yan, S.M., Wu, G.: Detailed folding structures of M-lycotoxin-Hc1a and its mutageneses using 2D HP model. Mol. Simul. **38**(10), 809–822 (2012)
2. Pan, J., Wang, X., Cheng, Y., et al.: Multi-source transfer ELM-based Q learning. Neurocomputing **137**(11), 57–64 (2014)

3. Mann, M., Backofen, R.: Exact methods for lattice protein models. Bio-Algorithms Med-Syst. **10**(4), 213–225 (2014)
4. Tang, X., Wang, J., Zhong, J., et al.: Predicting essential proteins based on weighted degree centrality. IEEE/ACM Trans. Comput. Biol. Bioinform. **11**(2), 407–418 (2014)
5. Qin, Y.F., Zheng, X.Q., Wang, J., et al.: Prediction of protein structural class based on linear predictive coding of PSI-BLAST profiles. Open Life Sci. **1**, 11–15 (2015)
6. Huang, J.T., Wang, T., Huang, S.R., et al.: Prediction of protein folding rates from simplified secondary structure alphabet. J. Theor. Biol. **383**, 1–6 (2015)

Predicting Microbe-Disease Association by Kernelized Bayesian Matrix Factorization

Sisi Chen[1], Dan Liu[1], Jia Zheng[1], Pingtao Chen[2], Xiaohua Hu[1,3], and Xingpeng Jiang[1](✉)

[1] School of Computer, Central China Normal University, Wuhan 430079, Hubei, China
xpjiang@mail.ccnu.edu.cn
[2] School of Physical Sciences, University of Science and Technology of China, Hefei 230026, China
[3] College of Computing and Informatics, Drexel University, Philadelphia, PA 19104, USA

Abstract. The study of microbe-disease associations can be utilized as a valuable material for understanding disease pathogenesis. Developing a highly accurate algorithm model for predicting disease-related microbes will provide a basis for targeted treatment of the disease. In this paper, we propose an approach based on Kernelized Bayesian Matrix Factorization (KBMF) to predict microbe-disease association, based on the Gaussian interaction profile kernel similarity for microbes and diseases. The prediction performance of the method was evaluated by five-fold cross validation. KBMF achieved reliable results which is better than several state-of-the-art methods with around 8% improvement of AUC. Furthermore, case studies have demonstrated the reliability of the method.

Keywords: Microbe · Matrix factorization · Bayesian · Biological network

1 Introduction

Research has shown that a large number of microbes exist on different body surfaces or sites [1]. The relationship between human beings and microbial community can be symbiotic and mutually beneficial. For example, the metabolic capacity and the immune system are impacted by probiotic bacteria [2]. Simultaneously, some studies show that the interactions of the microbiome and cells can affect the health of the human body and cause diseases. Evidences have been found on the associations between microbes and human diseases, such as cancer, diabetes, obesity, kidney stones and cardiovascular disease [3]. The identification of microbe-disease associations will not only help to identify the pathogenesis of the disease, but also strengthen the diagnosis and treatment of the disease. By collecting the human microbe-disease associations of previous literatures, Ma et al. has established human microbial disease

S. Chen and D. Liu—Equal contribution

D.-S. Huang et al. (Eds.): ICIC 2018, LNCS 10955, pp. 389–394, 2018.
https://doi.org/10.1007/978-3-319-95933-7_47

association database (HMDAD) [4]. However, the number of known microbe-disease associations is less, leave lots of unknown microbe-disease pairs.

Developing the computational models can help us to identify novel disease-related microbes. Based on microbe-disease associations, some computational methods have been proposed to predict the microbe-disease associations. Huang et al. proposed PBHMDA and NGRHMDA to predict potential disease-related microbes [5, 6]. PBHMDA is a path-based method that utilizes a special depth-first search algorithm in the heterogeneous interlinked network [5]. NGRHMDA is neighbor-based collaborative filtering and a graph-based scoring method to compute association possibility of microbe-disease pairs [6]. The results of these two methods are evaluated using AUC (area under ROC curve) on five-fold cross validation, both around 0.90. Shen et al. proposed a method based on random walk to predict microbe-disease associations on the heterogeneous network [7]. Zou et al. proposed an extending approach based on bi-random walk on the heterogeneous network [8].

In this paper, we propose an approach based on Kernelized Bayesian Matrix Factorization (KBMF) to predict microbe-disease association. KBMF achieves better performances than other previous methods based on cross validation. Case studies of the asthma, IBD are implemented for further evaluation.

2 Materials

In the previous microbe-disease association studies [5, 6, 8], the dataset used was downloaded from the Human Microbe-Disease Association Database (HMDAD, http://www.cuilab.cn/hmdad), which has collected 450 verified human microbe-disease associations, including 292 microbes and 39 diseases. The microbe-microbe similarity and disease-disease similarity were calculated based on the Gaussian interaction profile kernel similarity [9].

3 Methods

3.1 Problem Formalization

In this paper, the set of microbes is denoted by $X_m = \{m_1, m_2, \ldots, m_{N_m}\}$, where N_m represents the number of microbes; and the set of diseases is denoted by $X_d = \{d_1, d_2, \ldots, d_{N_d}\}$, where N_d represents the number of diseases, respectively. The known associations between microbes and diseases are defined as an adjacency matrix $Y \in R^{N_m \times N_d}$, where each element $y_{ij} \in \{0, 1\}$. If a microbe m_i has been found to associate with a disease d_j in HMDA database, y_{ij} is set to 1; otherwise, y_{ij} is set to 0. The problem was described as a binary classification task, using only the similarity between microbes and the similarity between diseases to predict the association between microbes and diseases.

3.2 Kernelized Bayesian Matrix Factorization

The kernelized Bayesian matrix factorization technique has been successfully applied for drug-target interaction prediction in previous studies [10]. Now, we performed this method on disease and microbe association prediction. Both the drug-target interaction prediction and microbe-disease association prediction are in-matrix predictions. By formulating a fully conjugate probabilistic model and developing an inferred deterministic variational approximation mechanism, the Bayesian algorithm is efficient. The main idea is to project microbes and diseases kernel matrices (S_m and S_d) into a unified, low-dimensional subspace using two projection matrices A_m and A_d, and then estimate their associations based on these two low-dimensional spaces named U and V. Finally, the given association matrix $\mathbf{Y}$ is generated from the interaction fraction matrix $\mathbf{F}$ [10].

We will give the parameter descriptions and distribution assumptions of our proposed model. In addition to the notations we mentioned earlier, $\boldsymbol{r}$ is defined as the dimensionality of the projected subspace. The priors of corresponding projection parameters $\mathbf{A}_{\mathrm{m}}$ and $\mathbf{A}_{\mathrm{d}}$ are represented as $\boldsymbol{\Lambda}_{\mathrm{m}}$ and $\boldsymbol{\Lambda}_{\mathrm{d}}$, which the element $\lambda_k \sim \mathcal{G}(\cdot; \alpha_\lambda, \beta_\lambda)$. The matrix element of projected instances for microbes u_i^k and diseases v_j^k follow the normal distribution. The variances of U and V is denoted by σ_{g}^2. For simplicity, we denote all the priors in the model as $\boldsymbol{\Xi} = \{\boldsymbol{\Lambda}_{\mathrm{m}}, \boldsymbol{\Lambda}_{\mathrm{d}}\}$, where the rest variables by $\boldsymbol{\Theta} = \{\mathbf{A}_{\mathrm{m}}, \mathbf{A}_{\mathrm{d}}, \mathbf{U}, \mathbf{V}, \mathbf{F}\}$, and the hyper-parameters by $\zeta = \{\alpha_\lambda, \beta_\lambda\}$.

Based on these specific distributions, when we consider the random variables as deterministic values, the interaction score matrix can be decomposed as follows [10]:

$$\mathbf{F} = \mathbf{U}^{\mathrm{T}}\mathbf{V} = \mathbf{S}_{\mathrm{m}}^{\mathrm{T}}\mathbf{A}_{\mathrm{m}}\mathbf{A}_{\mathrm{d}}^{\mathrm{T}}\mathbf{S}_{\mathrm{d}} \tag{1}$$

The method represents the matrix of predicted instances in terms of kernel matrices $\mathbf{K}_{\mathrm{m}}$ and $\mathbf{K}_{\mathrm{d}}$, enabling the prediction for out-of-sample points by kernel functions. We utilize a deterministic variational approximation instead of Gibbs sampling, which use a lower bound on the marginal likelihood based on an ensemble of factored posteriors to find the joint parameter distribution [11]. Then we got the factorable ensemble approximation of the required posterior [10] and thereafter the approximate posterior distribution of a specific factor $\boldsymbol{\tau}$ can be established by maximizing with respect to each factor separately until convergence as follow [10]:

$$\mathrm{q}(\tau) \propto \exp\left(\mathbf{E}_{q(\{\boldsymbol{\Theta},\boldsymbol{\Xi}\}\setminus\tau)}[\log p(\mathbf{Y}, \boldsymbol{\Theta}, \boldsymbol{\Xi}|\mathbf{S}_{\mathrm{m}}, \mathbf{S}_{\mathrm{d}})]\right) \tag{2}$$

Since the truncated normal distribution has a closed form formula for its expectation, we can update the approximate posterior distributions of the projected instances for microbes and diseases to find their posterior expectations [10]. The procedure of the model algorithm can be summarized as constantly updating the approximate posterior distribution of model parameters and potential variables until convergence [10].

4 Result

4.1 Evaluation Measure and Parameter Setting

In previous studies [5, 6, 8], the prediction ability of microbe-disease association was evaluated using five-fold cross validation and leave-one-out method, with AUC as the evaluation measure. In this paper, we used five-fold cross validation approach. For full observation, all the microbe-disease pairs were randomly divided into 5 equal-sized parts. In each round, 4 parts were used for training data and the remaining one as test data. Moreover, we performed 100 times random divisions to reduce potential sample division bias. The hyper-parameters $(\alpha_\lambda, \beta_\lambda)$ are set to (1,1) and σ_g is set to 0.1, respectively.

4.2 Performance of KBMF

As a result, KBMF model achieved AUC values of 0.9895 ± 0.0004 in the 5-fold cross validation frameworks. In addition, we try to replace the microbe Gaussian kernel similarity matrix with microbial OUT (Operational Taxonomic Units) abundance data, however we get the AUC value of 0.7615 ± 0.0112 (see Fig. 1) which indicate that including the OTU abundance data may involve additional noise to the prediction.

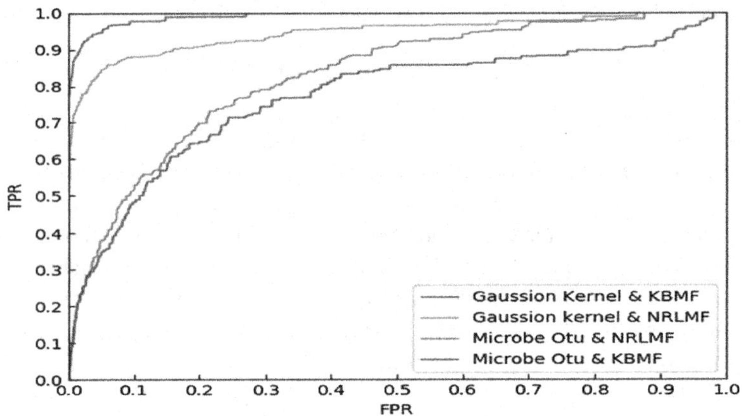

Fig. 1. Comparison results of KBMF and NRLMF in different similarity matrices in terms of ROC curve.

Note that there is an observed clear increasing trend in AUC with increasing the dimensionality of the latent space $\boldsymbol{r}$ in $[10, 100]$. Actually, the latent factors $\boldsymbol{r}$ has been set to 100 while implementing simple matrix factorization approach in microbe-disease association prediction [6]. Thus, we got the best AUC with $\boldsymbol{r} = 100$.

We compare several effective approaches to predict associations of diseases and microbes. NRLMF is another matrix factorization method integrating neighborhood regularized [12], which used as a comparison. The experimental result show that

KBMF achieved around 8% improvement in five-fold cross validation comparing to PBHMDA and NGRHMDA, and 4.8% to NRLMF (Table 1).

Table 1. Performance comparison among four different prediction models based on five-fold cross validation.

Method	Five-fold CV result
PBHMDA	0.9082 ± 0.0061
NGRHMDA	0.9023 ± 0.0031
BiRWHMDAD	0.8808 ± 0.0029
NRLMF	0.9412 ± 0.0016
KBMF	**0.9895 ± 0.0004**

4.3 Case Analysis

In order to further evaluate the proposed model, we obtained the top ranked unknown microbe-disease association pairs and analysis the top 30 novel associations, refer to the supporting for details. That 83.3% of the predictions (25 out of 30) are currently confirmed in experimental literatures, while 63.3% of the top 30 prediction by NRLMF are confirmed. Particularly, we implemented the case studies of Asthma by observing how many predicted microbes in the top 10 were verified in some previous experimental literatures. Predicting potentially relevant microbes is helpful to understand the pathology of the disease. In the previous studies [5, 8], asthma was always studied as a case. In KBMF, each of the microbes in the top 10 has been validated. For example, *Firmicutes*, *Actinobacteria*, and *Lachnospiraceae* are present in lower proportions in asthmatic patients [13, 14]. *Lactobacillus rhamnosus* is associated with asthma prevention [15]. The administration of killed *Propionibacterium acnes* suspensions has effect on Type I allergic asthma, enhances macrophage phagocytic and tumoricidal activities, acts as antibody-responsive auxiliary role; therefore, *Propionibacterium* is also considered to be associated with asthma [16].

5 Conclusion

With the development of sequencing technology and biology, there have been many experimental research literatures on exploring the relationships between diseases and microbes. However, the number of known microbe-disease pairs is less and many disease-related microbes are unknown. The proposed computational method–KBMF achieves better performance than these methods. This helps to select the most potential microbe-disease associations to further guide experimental validation.

Acknowledgement. This research is supported by the National Natural Science Foundation of China (No. 61532008), the Excellent Doctoral Breeding Project of CCNU, the Self-determined Research Funds of CCNU from the Colleges' Basic Research and Operation of MOE (No. CCN U16KFY04).

References

1. Holmes, E., Wijeyesekera, A., Taylor-Robinson, S.D., Nicholson, J.K.: The promise of metabolic phenotyping in gastroenterology and hepatology. Nat. Rev. Gastroenterol. Hepatol. **12**(8), 458–471 (2015)
2. Ventura, M., O'Flaherty, S., Claesson, M.J., Turroni, F., Klaenhammer, T.R., van Sinderen, D., O'Toole, P.W.: Genome-scale analyses of health-promoting bacteria: probiogenomics. Nat. Rev. Microbiol. **7**(1), 61–71 (2009)
3. Ettinger, G., MacDonald, K., Reid, G., Burton, J.P.: The influence of the human microbiome and probiotics on cardiovascular health. Gut Microbes **5**(6), 719–728 (2014)
4. Ma, W., Zhang, L., Zeng, P., Huang, C., Li, J., Geng, B., Yang, J., Kong, W., Zhou, X., Cui, Q.: An analysis of human microbe–disease associations. Briefings Bioinf. **18**(1), 85–97 (2016)
5. Huang, Z.A., Chen, X., Zhu, Z., Liu, H., Yan, G.Y., You, Z.H., Wen, Z.: PBHMDA: path-based human microbe-disease association prediction. Front. Microbiol. **8**, 233 (2017)
6. Huang, Y.-A., You, Z.-H., Chen, X., Huang, Z.-A., Zhang, S., Yan, G.-Y.: Prediction of microbe–disease association from the integration of neighbor and graph with collaborative recommendation model. J. Transl. Med. **15**(1), 209 (2017)
7. Shen, X., Chen, Y., Jiang, X., Hu, X., He, T., Yang, J.: Predicting disease-microbe association by random walking on the heterogeneous network. In: 2016 IEEE International Conference on Bioinformatics and Biomedicine (BIBM), pp. 771–774. IEEE, December 2016
8. Zou, S., Zhang, J., Zhang, Z.: A novel approach for predicting microbe-disease associations by bi-random walk on the heterogeneous network. PLoS ONE **12**(9), e0184394 (2017)
9. Chen, X., Yan, G.Y.: Novel human lncRNA-disease association inference based on lncRNA expression profiles. Bioinformatics **29**(20), 2617–2624 (2013)
10. Gonen, M.: Predicting drug-target interactions from chemical and genomic kernels using Bayesian matrix factorization. Bioinformatics **28**(18), 2304–2310 (2012)
11. Beal, M.J.: Variational Algorithms for Approximate Bayesian Inference. Ph.D. thesis, The Gatsby Computational Neuroscience Unit, University College London (2003)
12. Liu, Y., Wu, M., Miao, C., Zhao, P., Li, X.L.: Neighborhood regularized logistic matrix factorization for drug-target interaction prediction. PLoS Comput. Biol. **12**(2), e1004760 (2016)
13. Marri, P.R., Stern, D.A., Wright, A.L., Billheimer, D., Martinez, F.D.: Asthma-associated differences in microbial composition of induced sputum. J. Allergy Clin. Immunol. **131**(2), pp. 346–352, e341–e343 (2013)
14. Ciaccio, C.E., Kennedy, K., Barnes, C.S., Portnoy, J.M., Rosenwasser, L.J.: The home microbiome and childhood asthma. J. Allergy Clin. Immun. **133**(2) AB70
15. Yu, J., Jang, S.O., Kim, B.J., Song, Y.H., Kwon, J.W., Kang, M.J., Choi, W.A., Jung, H.D., Hong, S.J.: The effects of lactobacillus rhamnosus on the prevention of asthma in a murine model. Allergy Asthma Immunol. Res. **2**(3), 199–205 (2010)
16. Braga, E.G., Ananias, R.Z., Mussalem, J.S., Squaiella, C.C., Longhini, A.L.F., Mariano, M., Travassos, L.R., Longo-Maugéri, I.M.: Treatment with propionibacterium acnes modulates the late phase reaction of immediate hypersensitivity in mice. Immunol. Lett. **88**(2), 163–169 (2003)

Similarity-Based Integrated Method for Predicting Drug-Disease Interactions

Yan-Zhe Di[1], Peng Chen[2], and Chun-Hou Zheng[1,3(✉)]

[1] College of Computer Science and Technology, Anhui University, Hefei, Anhui, China
zhengch99@126.com

[2] Co-Innovation Center for Information Supply & Assurance Technology, Anhui University, Hefei, China

[3] Institute of Material Science and Information Technology, Anhui University, Hefei, Anhui, China

Abstract. The in silico prediction of potential interactions between drugs and disease is of core importance for effective drug development. Previous studies indicated that computational approaches for discovering novel indications of drugs by integrating information from multiple types have the potential to provide great insights to the complex relationships between drugs and diseases at a system level. However, each single data source is important in its own way and integrating data from different sources remains a challenging problem. In this article, we have proposed a new similarity combination method to integrate drug-disease association, drug chemical information, drug target domain information and target annotation information for drug repositioning. Specifically, we introduce interaction profiles of drugs (and of diseases) in a network, which are treated as label information and is used for model learning of new candidates. We compute multiple drugs and diseases similarity on these features, and use an integrated classifier for predicting drug-disease interactions. Comprehensive experimental results show that the proposed approach can serve as a useful tool in drug discovery to efficiently identify novel. Case studies show that our model has good performance.

Keywords: Drug-disease interaction · Drug repositioning · Similarity matrix

1 Introduction

At present, new drug research and development is an important issue in the world [1]. Drug repositioning is conducive to drug discovery and development. Previous studies have indicated that the integrative analysis methods can more efficiently explore multiple levels of drug information and usually promote the accuracy of drug indication prediction. Wang et al. constructed a triple-layer heterogeneous network with drug chemical similarity, target similarity and disease similarity for predicting drug indications [2]. Wu et al. unify the drug repositioning and drug combination to reveal new roles of drug interactions from a network perspective by treating drug combination as another form of drug repositioning [3]. By the construction of a joint model with Laplacian regularization terms and L1-norm constraint, LRSSL formulated a computational

D.-S. Huang et al. (Eds.): ICIC 2018, LNCS 10955, pp. 395–400, 2018.
https://doi.org/10.1007/978-3-319-95933-7_48

framework, a novel computational approach which integrates drug chemical information, drug target domain information and target annotation information for drug indications [4]. These methods have good performance in predicting performance. However, they still have problems in integrating datasets from multiple sources, and they rarely use the known drug-disease interaction information of the dataset to define similarity measures.

In this article, we propose a new similarity combination method to integrate drug-disease association, drug chemical information, drug target domain information and target annotation information. Then use machine learning method for drug-disease association prediction. Firstly, drug and disease are linked through drug-disease interactions. Secondly, calculating the similarity of the drug according these features (drug-disease association, drug chemical information, drug target domain information and target annotation information), and calculating the similarity of the disease according to drug-disease association. Thirdly, projecting drug and disease similarities into the same space and calculate drug-disease pairs similarity. Finally, an integrated classifier model is used to infer potential drug-disease interactions. We evaluated the drug-disease data set used in previous studies [4] and compared it with other drug indication predictions methods. The results show that the proposed method outperforms other methods.

2 Methods

We obtained the datasets from a published work [4]. There are 763 drugs and 681 diseases in the dataset with the total 3051 interactions. Each drug is described by three different kinds of feature profiles. Each disease is described by a phenotype profile based disease-disease similarity dataset which was downloaded from MimMiner [5].

2.1 Problem Formalization

We consider the problem of predicting new association using a drug-disease interaction network. Formally, we have a dataset $D = \{d_1, d_2, d_3, \ldots, d_n\}$ of drugs and a dataset $P = \{p_1, p_2, p_3, \ldots, p_m\}$ of diseases. There is also a set of known interactions between drugs and diseases. If we consider these associations as edges, then they form a bipartite network. We can characterize this network by the $n \times m$ adjacency matrix $Y = \{y_{ij}\}_{n \times m}$. That is, $y_{ij} = 1$ if drug d_i interacts with disease p_j and $y_{ij} = 0$ otherwise. Our task is now to rank all drug-disease pairs (d_i, p_j) such that highest ranked pairs are the most likely to interact.

2.2 Algorithm

The method we proposed is based on a bipartite network [6], which transforms the prediction problem into a binary classification problem. First, we construct a bipartite network according to drug-disease interactions. The *D-P* network of interactions can be described as a bipartite graph *G*(D, P, E), where $E = \{e_{ij} : d_i \in D, p_j \in P\}$. A link between d_i and p_j is drawn in the graph when the drug d_i is associated with the disease p_j.

Now, how to turn the network problem into a classification problem? If drug d_i interacts with disease p_j, give drug-disease pair (d_i,p_j) a positive label, for example (d_1,p_3), otherwise give it a negative label, for example (d_1,p_1). These labels are used for subsequent classification.

Our method is based on the assumption that the more the two diseases are treated by the same drug, the more related the two drugs, and vice versa. So we need to compute the similarity of drug and disease. In our study, each drug is represented by three different feature profiles. Each profile gives a feature matrix of drug in this work, denoted as chemical substructure matrix $X_1 \in R^{t_1 \times n}$, target domain matrix $X_2 \in R^{t_2 \times n}$, and target Gene ontology matrix $X_3 \in R^{t_3 \times n}$, where $t.$ is the number of features, n is the number of drugs. For each drug feature matrix X_p, a similarity matrix S_p is constructed. Its element $S_p(i,j)$ is:

$$S_p(i,j) = \frac{x_p(i) \cap x_p(j)}{x_p(i) \cup x_p(j)} \tag{1}$$

where $x_p(i)$ and $x_p(j)$ are vectors in the pth drug feature matrix.

We can treat drug-drug interactions as characteristics of drugs to calculate drug similarity matrix S_Y. Its element $S_Y(i,j)$ is:

$$S_Y(i,j) = \frac{Y_r(i) \cap Y_r(j)}{Y_r(i) \cup Y_r(j)} \tag{2}$$

where $R_r(i)$ and $Y_r(j)$ are row vectors in the drug-disease association matrix Y.

Each disease is described by a phenotype profile based disease-disease similarity dataset. Disease-disease similarity dataset is represented by a matrix $Z \in R^{m \times m}$, where m is the number of diseases. As above, we can treat disease-drug interactions as characteristics of diseases to calculate disease similarity matrix Z_Y. Its element $Z_Y(i,j)$ is:

$$Z_Y(i,j) = \frac{Y_c(i) \cap Y_c(j)}{Y_c(i) \cup Y_c(j)} \tag{3}$$

where $Y_c(i)$ and $Y_c(j)$ are column vectors in the drug-disease association matrix Y.

In the first step, we get the drug-disease pairs label, but we also need the drug-disease pairs feature when classifying. We use drug-disease pairs similarity as drug-disease pairs feature. Then, we map the similarity matrix S of drug and the similarity matrix Z of disease into the same space which constitutes the similarity matrix W of drug-disease pairs. We can use the following formula to calculate matrix $W \in R^{(m \times n) \times (m + n)}$.

$$W(k) = rbind(S(i), Z(j)), S \in \{S_1, S_2, S_3, S_Y\}, Z \in \{Z_1, Z_Y\} \tag{4}$$

where $W(k)$, $S(i)$ and $Z(j)$ are one row vector in the matrix W, S and Z, respectively. $rbind(A,B)$ indicates that vector A and vector B are combined by rows.

From the above we can obtain eight drug-disease pairs similarity matrices. Next, we learn a classifier by putting the labels and each set of features into a random forest [7].

Finally, the results of multiple classifiers were integrated to construct a model. The integrated method uses the simple averaging.

$$H(x) = \frac{1}{T}\sum\nolimits_{i=1}^{T} h_i(x) \tag{5}$$

where $h_i(x)$ is the result of a single classifier, T is the number of classifiers, $H(x)$ is the final result.

3 Results

We evaluate the performance of our algorithm in two kinds of settings: repositioning approved drugs to novel indications and therapeutic effect prediction of new drugs. In the first setting, we pick out drugs with more than one indication records and the disease associations of these drugs are divided into ten subsets randomly, then 10-fold cross validation is carried out. Under the second setting, the drugs in the dataset are randomly split into five subsets, then 5-fold cross validation is conducted while all disease associations of the test drugs are removed. We compare our method with two state-of-the-art methods: TL-HGBI [2] and LRSSL [4]. We use the area under the ROC curve (AUC) as evaluation metric for comparison.

First, we compared our method with TL-HGBI and LRSSL for the task of drug repositioning. We performed 5-fold cross-validation 20 times with different random seeds and calculated AUC value. Table 1 shows the average values of the evaluation metric for these methods under above setting.

Table 1. Performance of different methods for drug repositioning

Method	AUC
TL-HGBI	0.9011
LRSSL	0.9178
Our method	0.9708

TL-HGBI and LRSSL are developed for drug repositioning. Therefore, we also compared our approach with these two methods for the task of indication prediction of novel drugs. The average AUC was valued by 10 replicates of 10-fold cross-validation. As shown in Table 2, our method still has the best performance.

Table 2. Performance of different methods for predicting indication of novel drugs

Method	AUC
TL-HGBI	0.8111
LRSSL	0.9074
Our method	0.9253

In order to further test the performance of our proposed method, we used it to predict indications of the small molecule drugs which didn't have any therapeutic indication records in DrugBank. The drug-disease associations were predicted by our method, while use all drugs in the original dataset. We find that some of the higher-ranked forecasts match the records in the CTD [8] database. Table 3 shows the top 5 predictions for some novel drugs. The drug-disease associations which have been curated in CTD are indicated by boldface. The drug-disease association recorded in the CTD is indicated by a heavy font.

Table 3. Top 5 predictions for novel drugs with records in CTD

Drug	Disease (MeSH ID)
BIA3-335 (CID:4369285)	D006973 (Hypertension); **D001943 (Breast Neoplasms);** D001991 (Bronchitis); D006333 (Heart Failure); **D011471 (Prostatic Neoplasms)**
Aprindine (CID:2218)	**D010146 (pain); D003866 (Psychotic Disorders);** D011618 (Psychotic Disorders); **D012559 (Schizophrenia);** D000787 (Angina Pectoris)
Nomifensine (CID:4528)	**D003866 (Depressive Disorder); D012559 (Schizophrenia);** D016584 (Panic Disorder); D011618 (Psychotic Disorders); D010146 (pain)
Etorphine (CID:26721)	**D010146 (pain);** D001249 (Asthma); D012221(Rhinitis, Allergic, Perennial); D011537 (Pruritus); D003866 (Depressive Disorder)
Quercetin (CID:5280343)	**D006973 (Hypertension);** D006333 (Heart Failure); D001943 (Breast Neoplasms); **D009203 (Myocardial Infarction); D000787 (Angina Pectoris)**

4 Conclusions and Discussion

In this article, we propose a new similarity combination method for predicting drug-disease interactions. The most prominent innovation of this method is the construction of a model that integrates multiple classifiers based on multiple combination of the heterogeneous drug and disease similarity profiles.

The predictive performance of our method was measured by AUC on the task of predicting interactions between drugs and diseases. We obtained AUC scores of 97.08 on repositioning approved drugs and 92.53 on prediction of new drugs, which represents improvement of 5 and 2 points respectively over LRSSL [4]. Moreover, our approach can reduce time complexity and does not require high computational hardware requirements compared to many previous approaches. The AUC is a particularly relevant measure for this problem, because it is slightly affected by the unbalanced positive and negative samples. The large improvement in AUC suggests that the top ranked putative drug-disease interactions found by our method are more reliable than those found in previous methods.

Acknowledgment. This study was supported by the National Natural Science Foundation of China (No. 61672037), the Key Project of Anhui Provincial Education Department (No. KJ2017 ZD01), and the Key Project of Academic Funding for Top-notch Talents in University (No. gxbj ZD2016007).

References

1. Lipinski, C., Lombardo, F., Dominy, B., Feeney, P.: Experimental and computational approaches to estimate solubility and permeability in drug discovery and development settings. Adv. Drug Deliv. Rev. **64**, 4–17 (2012)
2. Wang, W., Yang, S., Zhang, X., Li, J.: Drug repositioning by integrating target information through a heterogeneous network model. Bioinformatics **30**, 2923–2930 (2014)
3. Wu, Z., Wang, Y., Chen, L.: Network-based drug repositioning. Mol. BioSyst. **9**, 1268–1281 (2013)
4. Liang, X., Zhang, P., Yan, L., Fu, Y., Peng, F.: LRSSL: predict and interpret drug-disease associations based on data integration using sparse subspace learning. Bioinformatics **33**, 1187–1196 (2017)
5. van Driel, M., et al.: A text-mining analysis of the human phenome. Eur. J. Hum. Genet. **14**, 535–542 (2006)
6. Yamanishi, Y., Araki, M., Gutteridge, A.: Prediction of drug-target interaction networks from the integration of chemical and genomic spaces. Bioinformatics **24**(2), 32–40 (2008)
7. Liaw, A., Wiener, M.: Classification and regression by randomforest. R News. **2**, 18–22 (2002)
8. Davis, A., et al.: The comparative toxicogenomics database's 10th year anniversary: update 2015. Nucleic Acids Res. **43**, D914–D920 (2014)

Nucleotide-Based Significance of Somatic Synonymous Mutations for Pan-Cancer

Yannan Bin, Xiaojuan Wang, Qizhi Zhu, Pengbo Wen, and Junfeng Xia(✉)

Institute of Physical Science and Information Technology, School of Computer Science and Technology, Anhui University, Hefei 230601, Anhui, China
jfxia@ahu.edu.cn

Abstract. Synonymous mutations have been identified to play important roles in cancer development. We investigated the characters of pathogenic and neutral somatic synonymous mutations identified by FATHMM across 15 cancer types from COSMIC. The comparisons of pathogenic synonymous mutations with neutral ones were performed with DNA-based characters to explore their functional mutations. Differences among pathogenic and neutral synonymous mutations are significant, for instance, pathogenic mutations were more conserved and with larger effect on splicing and translation. The function annotations of synonymous mutation were important mechanistic clues for downstream effects on gene and laid the groundwork for understanding the somatic synonymous mutations.

Keywords: Pan-cancer · Somatic · Synonymous mutation
DNA-based character

1 Introduction

Synonymous mutations, which occur in the gene-coding regions without changing the encoded amino acids, have been identified to affect biological processes in several ways, including by changing mRNA splicing and stability, protein expression and enzymatic activity [1]. Moreover, recent studies reported that synonymous mutations could act as drivers contributed to human cancers [2, 3].

As complex genetic diseases, cancers might be affected by a large number of variants in different genes. But to date, the targets of drugs and treatments associated with cancers are limited on a few variants or genes, so it is difficult to achieve the effective treatment for cancers. High throughput sequencing technology has enabled the systemic analyses of huge variants in large cohorts of cancer cases, e.g., The Cancer Genome Atlas (TCGA) [4] and International Cancer Genome Consortium [5]. COSMIC, using the latest sequencing and analysis methods to identify somatic variants across thousands of tumors, is found to meet the data needs [6]. Cancer-associated variants not only contain important drivers for cancer development, but also have many passengers with neutral function. It is a key step to identify pathogenic synonymous mutations for understanding cancer biology and evolving targeted treatments. There were several methods focused on predicting driver genes or mutations, such as,

D.-S. Huang et al. (Eds.): ICIC 2018, LNCS 10955, pp. 401–406, 2018.
https://doi.org/10.1007/978-3-319-95933-7_49

E-Driver [7], MuSiC [8], OncodriveCLUST [9], etc. Nevertheless, those studies mainly focused on the missense variants and ignored the potential functions of synonymous mutations.

In this study, we documented essentially the full repertoire of cancer associated synonymous mutations to investigate the mutational signatures of pathogenic synonymous mutations datasets. To acquire insight into the characters of harmful and neutral synonymous mutations for pan-cancer at DNA-based levels. The observation could add perspective to understand cancer-associated synonymous mutations.

2 Data and Methods

The cancer associated synonymous somatic mutations in TCGA were downloaded from COSMIC v83 GRCh37 (Catalogue of Somatic Mutations in Cancer, http://www.sangerac.uk/cosmic) [10]. We exclude mutations with improper information and got 90,904 pathogenic and 171,579 neutral synonymous mutations identified by FATHMM [11] from 6,012 samples across 15 cancer types: breast cancer (BRCA), central nervous system tumor (CNST), cervical adenocarcinoma (CEAD), endometrial adenocarcinoma (ENAD), haematopoietic and lymphoid tumor (HLTU), kidney carcinoma (KICA), large intestine adenocarcinoma (INAD), liver carcinoma (LICA), lung adenocarcinoma (LUAD), ovarian carcinoma (OVCA), prostate adenocarcinoma (PRAD), skin cancer (SKCA), stomach adenocarcinoma (STAD), thyroid carcinoma (THCA) and urinary tract carcinoma (UTCA). To further investigate the mutational signatures of pathogenic and neutral synonymous mutations, a dataset (total 88,326 mutations) with 1:1 ratio of pathogenic and neutral mutations was used to the analysis of mutation specific signatures. The detailed description of "close-by" datasets was shown in Ref. [12].

Then we assessed seven signatures based on DNA-based characters: distance to splice (DSP) was annotated by SeattleSeq [13]; conservation (GERP++), Splice-site sequence motif strength (MES), the mutation-causing changes in the secondary structure folding energy of pre-mRNA ($\Delta\Delta G$pre) and mature mRNA ($\Delta\Delta G$post), and the change in Relative synonymous codon usage ($|\Delta$RSCU$|$) were implemented by utilizing SilVA [14]; the different in translational efficiency caused by mutation ($|$Log(TAI)$|$) was calculated by using TAI [15].

3 Results and Discussion

3.1 Mutational Spectrum Among Different Cancer Types

As mentioned in Method, after filtering out mutations with imperfect information, 320,330 synonymous mutations were extracted from the TCGA data in COSMIC, and 90,904 pathogenic and 171,579 neutral mutations were applied in this work. For pathogenic mutations, Fig. 1a shows that HLTU and THCA have the lowest median number of mutations per sample (one mutation), and SKCA has the highest (45 mutations). Notable there are outliers in SKCA, LUAD, UTCA and STAD, which are twice the median number of synonymous mutations per sample for pan-cancer (5 mutations). Comparison of mutation patterns spectrum across 15 cancer types (Fig. 1b) reveals that the greatest frequently occurring mutation type is NoCpG_GCts with the percent range from 40.16% to 89.74% across the cancer types. which is possible to associate with the aberrant DNA methylation. SKCA contain increased NoCpG_GCts (G:C →A:T), a signature of ultraviolet light and deamination processes [16]. And LUAD with increased NoCpG_GCtv (G:C→T:A, G↔C) a signature of cigarette smoke exposure. At CpG dinucleotides, single-base substitutions are associated with the most common epigenetic modifications of DNA.

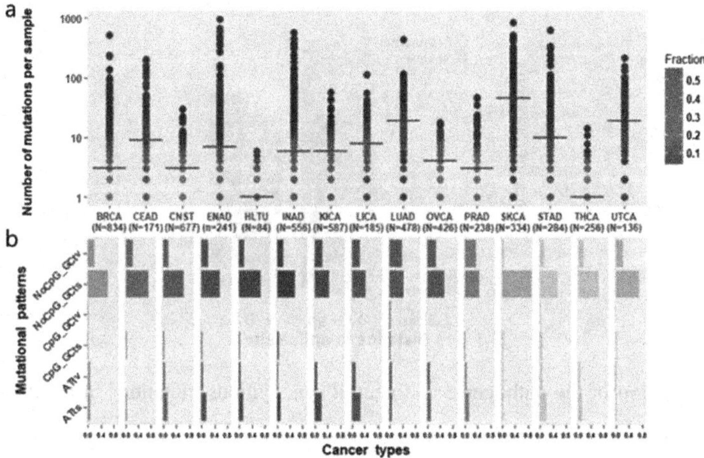

Fig. 1. Statistics of mutation frequencies and patterns of pathogenic synonymous mutations across 15 cancer types. (**a**) Distribution of mutations per sample in each cancer type. (**b**) Mutation spectrum of six possible synonymous mutation patterns of nucleotide changes across different cancer types. 15 cancer types: breast cancer (BRCA), central nervous system tumor (CNST), cervical adenocarcinoma (CEAD), endometrial adenocarcinoma (ENAD), haematopoietic and lymphoid tumor (HLTU), kidney carcinoma (KICA), large intestine adenocarcinoma (INAD), liver carcinoma (LICA), lung adenocarcinoma (LUAD), ovarian carcinoma (OVCA), prostate adenocarcinoma (PRAD), skin cancer (SKCA), stomach adenocarcinoma (STAD), thyroid carcinoma (THCA) and urinary tract carcinoma (UTCA).

3.2 Comparisons of Close-By Datasets at DNA-Based Level

In order to ensure the balance of pathogenic mutations dataset (90,904 synonymous mutations) and the neutral datasets (171,579 synonymous mutations) and correct analyzing of mutation-specific characters at nucleotide and protein levels, we selected the neutral synonymous mutations were located as close as possible to corresponding pathogenic mutations as “close-by” datasets (total 88,326 mutations, 1:1 ratio of pathogenic and neutral mutations) were used for the following analysis at DNA-based levels.

The disease-causing synonymous mutations may affect biological and disease processes by changing mRNA splicing and stability [3, 4], protein expression and enzymatic activity [5–7], therefor a broad array of analysis related to genomic characters were evaluated (detailed descriptions of the characters can be found in Addition file 3: Table S4). Figure 2 shows the distributions of pathogenic and neutral mutations in the immediate proximity (1–3 bp), the larger region to be implicated in splicing (4–69 bp) and the exon core (≥ 70 bp), defined based on the distance to splice site. Comparing with neutral mutations, pathogenic mutations are more closed to splice sites, and probably induce the influence on splicing regulation.

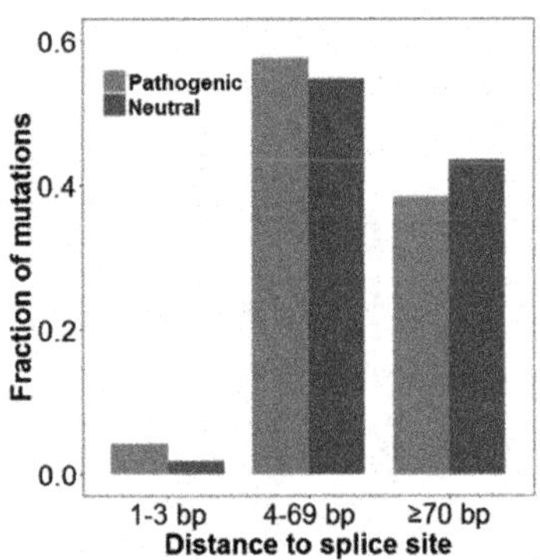

Fig. 2. Distribution of the pathogenic and neutral synonymous mutations in the three regions.

In addition, the pathogenic mutations tend to locate in the chromosomal sites with higher GERP++ score (t-test $p < 2.0e\text{-}16$, Fig. 3a) [17], and it is customary for mutations with important functional and evolutionary implications located in highly conserves regions. Based on the comparison of MES [18], pathogenic mutations are more likely to be with the higher probability and locate in the sequence with a true splice site, which is important for the splicing regulation (t-test $p < 2.0e\text{-}16$, Fig. 3b). $|\Delta RSCU|$ and $\Delta\Delta G$ (includes $\Delta\Delta G\_pre$ and $\Delta\Delta G\_post$) are all associated with translation effect, pathogenic mutations have higher scores in the three characters, indicating that they are prone to alter the translation efficiency greatly (t-test $p < 2.0e\text{-}16$ for the three characters, Fig. 3c, d and e). Consistent with these observations, pathogenic mutations tend to change the translation efficiency compared with neutral mutations (t-test $p < 2.0e\text{-}16$, Fig. 3f). The analyses of these genomic characters for pathogenic and neutral synonymous mutations suggest the potential pathogenicity of the mutations on splice regulation and translation efficiency.

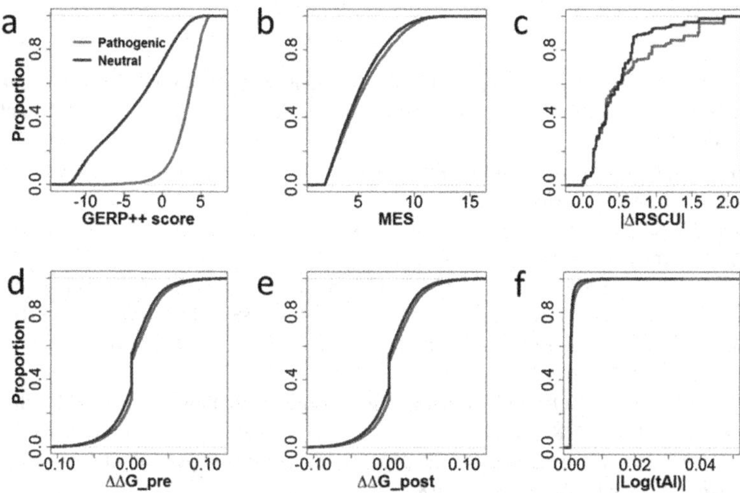

Fig. 3. Cumulative distribution function (CDF) curves of various genomic characters for pathogenic (red) and neutral (blue) synonymous mutations. (**a**) Conservation (GERP++ score). (**b**) Splice-site sequence motif strength (MES). (**c**) Change in Relative synonymous codon usage (|ΔRSCU|). (**d**), (**e**) Changes in the secondary structure folding energy of pre-mRNA (ΔΔG_pre) and mature mRNA (ΔΔG_post). (**f**) Different in translational efficiency (|Log(TAI)|). (Color figure online)

4 Conclusions

Synonymous variants have been identified to play important roles in cancer development, and landscape of somatic mutations associated with cancer is extraordinarily complex. We investigated the features of pathogenic synonymous somatic mutations identified by FATHMM across 15 cancer types from The Cancer Genome Atlas. The comparisons of pathogenic synonymous mutations with neutral ones were performed at DNA-based levels to explore their functional effect. Difference among pathogenic and neutral synonymous mutations are wide in the nucleotide characters, for instance, pathogenic mutations were significantly more conserved, with lower disorder scores and so on. The function predictions of synonymous mutation were important mechanistic clues about downstream effects on gene. Taken together, we illustrated the preferences of cancer associated synonymous mutations and laid the groundwork for understanding the synonymous mutations act as drivers in cancers.

Acknowledgment. The authors thank the members of our laboratory for their valuable discussions. This work has been supported by the grants from the National Natural Science Foundation of China (61672037 and 21601001), the Anhui Provincial Outstanding Young Talent Support Plan (No. gxyqZD2017005), the Young Wanjiang Scholar Program of Anhui Province and the Initial Foundation of Doctoral Scientific Research in Anhui University (J10113190035).

References

1. Soussi, T., Taschner, P.E., Samuels, Y.: Synonymous somatic variants in human can cer are not infamous: a plea for full disclosure in databases and publications. Hum. Mutat. **38**(4), 339–342 (2017)
2. Supek, F., et al.: Synonymous mutations frequently act as driver mutations in human cancers. Cell **156**(6), 1324–1335 (2014)
3. Gartner, J.J., et al.: Whole-genome sequencing identifies a recurrent functional synonymous mutation in melanoma. Proc. Natl. Acad. Sci. U.S.A. **110**(33), 13481–13486 (2013)
4. Cancer Genome Atlas Research Network: Comprehensive genomic characterization defines human glioblastoma genes and core pathways. Nature **455**(7216), 1061–1068 (2008)
5. Genomes Project Consortium: A map of human genome variation from population scale sequencing. Nature **467**(7319), 1061–1073 (2010)
6. Forbes, S.A., et al.: COSMIC: somatic cancer genetics at high-resolution. Nucleic Acids Res. **45**(D1), D777–D783 (2017)
7. Porta-Pardo, E., et al.: A pan-cancer catalogue of cancer driver protein interaction interfaces. PLoS Comput. Biol. **11**(10), e1004518 (2015)
8. Dees, N.D., et al.: MuSiC: identifying mutational significance in cancer genomes. Genome Res. **22**(8), 1589–1598 (2012)
9. Gonzalez-Perez, A., et al.: IntOGen-mutations identifies cancer drivers across tumor types. Nat. Meth. **10**(11), 1081–1082 (2013)
10. Forbes, S.A., et al.: COSMIC: exploring the world's knowledge of somatic mutations in human cancer. Nucleic Acids Res. **43**(D1), D805–D811 (2014)
11. Shihab, H.A., et al.: Predicting the functional consequences of cancer-associated amino acid substitutions. Bioinformatics **29**(12), 1504–1510 (2013)
12. Livingstone, M., et al.: Investigating DNA, RNA and protein-based features as a means to discriminate pathogenic synonymous variants. Hum. Mutat. **38**, 1336–1347 (2017)
13. http://snp.gs.washington.edu/SeattleSeqAnnotation138/index.jsp
14. Buske, O.J., et al.: Identification of deleterious synonymous variants in human genomes. Bioinformatics **29**(15), 1843–1850 (2013)
15. dos Reis, M., Savva, R., Wernisch, L.: Solving the riddle of codon usage preferences: a test for translational selection. Nucleic Acids Res. **32**(17), 5036–5044 (2004)
16. Sanchez, M.I., Grichnik, J.M.: Melanoma's high C > T mutation rate: is deamination playing a role? Exp. Dermatol. **23**(8), 551–552 (2014)
17. Davydov, E.V., et al.: Identifying a high fraction of the human genome to be under selective constraint using GERP ++. PLoS Comput. Biol. **6**(12), e1001025 (2010)
18. Yeo, G., Burge, C.B.: Maximum entropy modeling of short sequence motifs with applications to RNA splicing signals. J. Comput. Biol. **11**(2), 377–394 (2004)

Performance Analysis of Non-negative Matrix Factorization Methods on TCGA Data

Mi-Xiao Hou[1], Jin-Xing Liu[1(✉)], Junliang Shang[1(✉)], Ying-Lian Gao[2], Xiang-Zhen Kong[1], and Ling-Yun Dai[1]

[1] School of Information Science and Engineering, Qufu Normal University, Rizhao 276826, China
mixiaohou@163.com, sdcavell@126.com, shangjunliang110@163.com, kongxzhen@163.com, dailingyun_1@163.com

[2] Library of Qufu Normal University, Qufu Normal University, Rizhao 276826, China
yinliangao@126.com

Abstract. Non-negative Matrix Factorization (NMF) is recognized as one of fundamentally important and highly popular methods for clustering and feature selection, and many related methods have been proposed so far. Nevertheless, their performances, especially on real data, are still unclear due to few studies focusing on their comparison. This study aims at a assessment study of several representative methods from clustering and feature selection, including NMF, GNMF, MD-NMF, $L_{2,1}$NMF, LNMF, Convex-NMF and Semi-NMF, on the data of the Cancer Genome Atlas (TCGA), which is one of current research hotspot of bioinformatics. Specifically, three data types of four cancers are either separately or integratedly decomposed as the coefficient matrices and the basis matrices by these NMF methods. The coefficient matrices are evaluated by accuracies of clustered samples and the basis matrices are assessed by p-values of selected genes. Experiment results not only show merits and limitations of compared NMF methods, which may provide guidelines for applying them and proposing novel NMF methods, but also reveal several clues for the exploration of related cancers.

Keywords: Non-negative Matrix Factorization · Clustering · Genomic data Dimensionality reduction

1 Introduction

Nowadays, as the development of next generation sequencing, a steady flow of biological data is emerging. Since most of the data with "high dimension, small sample size", and many incurrent noises will cause confusion and interference on analysis, it is important for data to make dimensionality reduction and sparsity. Non-negative Matrix Factorization (NMF) [1, 2] is a mature method of data dimensionality reduction which can be used to reduce complexity of analysis about biological data. NMF can obtain two non-negative matrices to approximate the original data matrix, which reflects the concept of part-based representation in human thought. NMF can decompose original

D.-S. Huang et al. (Eds.): ICIC 2018, LNCS 10955, pp. 407–418, 2018.
https://doi.org/10.1007/978-3-319-95933-7_50

data into two matrices: the basis matrix contains all the genes that are commonly used in the analysis of genes (the selection of characteristic genes); the coefficient matrix contains all the samples, which are often applied in the analysis of samples (sample classification or clustering). NMF methods for analyzing genetic data have been studied for many aspects. The application of NMF for clustering on genomic data, feature gene selection [3, 4], protein function analysis [5], predicting drug side effects [6], which has been more and more impressive in the field of bioinformatics.

The Cancer Genome Atlas (TCGA, https://cancergenome.nih.gov/) dataset is the result of large-scale sequencing of human genes and also studied by NMF methods in recent years [7]. TCGA also provides integrated data with "high dimension, small sample size" characteristics [8, 9]. This paper focus on the analysis of NMF on TCGA data: one reason is that there are few synthesis literatures on TCGA data; for another, the consideration in the experimental point of view on NMF methods are not much. This paper is a comparative analysis of NMF algorithm and analysis of TCGA data mining information, which can be used as a reference for the improvement of NMF algorithm and the effective mining of TCGA cancer data.

In view of the fact that some experimental research have done about a variety of NMF methods before (with 9 different NMF models on leukaemia and colorectal cancer datasets for clustering samples) [10]. In this paper, several classic NMF methods are selected to make some relatively detailed analysis on the TCGA: NMF, Graph Regularized Non-negative Matrix Factorization (GNMF) [11], Manifold Regularized Discriminative Non-negative Matrix Factorization (MD-NMF) [12], Robust Non-negative Matrix Factorization using $L_{2,1}$-norm ($L_{2,1}$NMF) [13], Local Non-negative Matrix Factorization (LNMF) [14], Convex-NMF and Semi-NMF [15]. Four cancer datasets in the paper which all contain three types of data: Gene Expression (GE) data, Methylation (ME) data and Copy number variant (CNV) data. In order to verify the performances of these NMF methods, we will cluster samples of data based the coefficient matrix decomposed by NMF methods on integrative data. And experiments for feature selection are also given based on the basis matrix on GE.

This paper aims to do the related experiment analysis by utilizing the integrated data of several cancers from TCGA, and hope to provide novel ideas and values to the next work.

2 Methods

2.1 Basic NMF

NMF can obtain two non-negative matrices by factorization from one non-negative data matrix. It is can be defined as follows: given a non-negative matrix $\mathbf{X} \in R^{m \times n}$, there are two non-negative matrices $\mathbf{W} \in R^{m \times k}$ and $\mathbf{H} \in R^{n \times k}(k < \min\{m, n\})$ to approximate $\mathbf{X}$:

$$\mathbf{X} \approx \mathbf{W}\mathbf{H}^{\mathrm{T}}. \tag{1}$$

m represents the number of genes and n is the number of samples in our papers. Lee and Seung provided two models of NMF [1, 2]. One is to measure Euclidean distance of two matrices:

$$O_1 = \left\|\mathbf{X} - \mathbf{W}\mathbf{H}^{\mathbf{T}}\right\|^2 = \sum\nolimits_{i,j} \left(x_{ij} - \sum\nolimits_{k=1}^{K} w_{ik}h_{jk}\right)^2. \tag{2}$$

Another is based on Kullback-Leibler (KL) divergence of two matrices:

$$O_2 = D\left(\mathbf{X} \,\|\, \mathbf{W}\mathbf{H}^{\mathbf{T}}\right) = \sum\nolimits_{i,j} \left(x_{ij} \log \frac{x_{ij}}{y_{ij}} - x_{ij} + y_{ij}\right), \tag{3}$$

where $\left[y_{ij}\right] = \mathbf{Y} = \mathbf{W}\mathbf{H}^{\mathbf{T}}$. The purpose of NMF method is minimizing the results of above formulas. NMF is an NP problem which can be classified as optimization problem and solved by iterative method.

2.2 Graph Regularized Non-negative Matrix Factorization

GNMF [11] is a sophisticated method, which was firstly introduced by Cai et al. After that, many NMF methods with manifold learning have been generated in recent years [12, 16]. Manifold learning focus on internal spatial structure of the data which strengthen the relationship among data. Data can be mapped from the high dimension to the low dimension. The objective formula can be written as follows:

$$O = \left\|\mathbf{X} - \mathbf{W}\mathbf{H}^{\mathbf{T}}\right\|^2 + \lambda Tr\left(\mathbf{H}^{\mathbf{T}}\mathbf{L}\mathbf{H}\right). \tag{4}$$

$\mathbf{L}$ is graph Laplacian [17]. And the purpose of GNMF is minimizing the results of the formula.

2.3 Manifold Regularized Discriminative Non-negative Matrix Factorization

MD-NMF [12] is the model with manifold learning, the objective formula based on KL divergence is defined as below:

$$\begin{aligned} O = {} & D\left(\mathbf{X} \,\|\, \mathbf{W}\mathbf{H}^{\mathbf{T}}\right) + \frac{\alpha}{2} Tr\left(\mathbf{W}\mathbf{e}\mathbf{W}^{\mathbf{T}}\right) + \frac{\beta}{2} Tr\left(\mathbf{H}^{\mathbf{T}}\mathbf{H}\right) \\ & + \frac{\gamma}{2} Tr\left(\mathbf{H}^{\mathbf{T}}\left(\mathbf{L}_{\mathbf{c}}^{-\frac{1}{2}}\right)^{\mathbf{T}}\mathbf{L}_{\mathbf{g}}\mathbf{L}_{\mathbf{c}}^{-\frac{1}{2}}\mathbf{H}\right), \end{aligned} \tag{5}$$

inside $\alpha \geq 0$, $\beta \geq 0$ and $\gamma \geq 0$. And $\mathbf{e} = \overline{\mathbf{1}} - \mathbf{I}$, where $\overline{\mathbf{1}}$ is the matrix whose elements are all one and $\mathbf{I}$ is identity matrix. $\mathbf{L_g}$ is the graph Laplacian matrix of samples which with same labels as h_i. And $\mathbf{L_c}$ is the graph Laplacian matrix of samples which with different labels from h_i. And MD-NMF provides two Laplacian matrices about sample labels.

2.4 Robust Nonnegative Matrix Factorization Using $L_{2,1}$-norm

$L_{2,1}$NMF Changes the constrain of norm for NMF, which applies $L_{2,1}$-norm. $L_{2,1}$-norm [13] as a novel spare method which was raised in recent years, which will generate sparsity of rows. It combined with L_1-norm and L_2-norm, and $L_{2,1}$-norm of one matrix **F** is defined as follows:

$$\|\mathbf{F}\|_{2,1} = \sum_{i=1}^{n} \sqrt{\sum_{j=1}^{m} f_{ij}^2} = \sum_{i=1}^{n} \|\mathbf{f}^{\mathbf{i}}\|_2. \tag{6}$$

The equation is defined as follows:

$$O = \left\|\mathbf{X} - \mathbf{W}\mathbf{H}^{\mathbf{T}}\right\|_{2,1} \quad s.t. \quad \mathbf{W} \geq 0,\ \mathbf{H} \geq 0. \tag{7}$$

where $\left\|\mathbf{f}^{\mathbf{i}}\right\|_2$ is the L_2-norm of i-th row. The optimal solution of the formula is the minimum value. And $L_{2,1}$-norm presented strong sparsity compared with other sparse methods.

2.5 Local Non-negative Matrix Factorization

LNMF [14] is introduced based on KL divergence. What crux improvement of LNMF is by introducing the column orthogonality [18] for the basis matrix **W**. It can make the **W** get sparser but **H** become quite not sparse [19, 20]. The objective formula is as follows:

$$O = D\left(\mathbf{X} \,\|\, \mathbf{W}\mathbf{H}^{\mathbf{T}}\right) + \alpha \sum_{i,j} b_{ij} - \beta \sum_{i} u_{ii}, \tag{8}$$

inside $\mathbf{B} = [b_{ij}] = \mathbf{W}^{\mathbf{T}}\mathbf{W}$, $\mathbf{U} = [u_{ij}] = \mathbf{H}\mathbf{H}^{\mathbf{T}}$ and $\alpha,\ \beta > 0$. The purpose of the formula also is to get the minimum. The model makes the bases as orthogonal as they can. $\sum_{i,j} b_{ij}$ should take the minimum and $\sum_i u_{ii}$ should be maximum. It will make **W** contain more zero elements and **H** store more information [18].

2.6 Convex-NMF

Convex-NMF can be written as follows:

$$O = \left\|\mathbf{X}_{\pm} - \mathbf{W}_{\pm}\mathbf{H}_{+}^{\mathbf{T}}\right\|^2 = \left\|\mathbf{X}_{\pm} - \mathbf{X}_{\pm}\mathbf{A}_{+}\mathbf{H}_{+}^{\mathbf{T}}\right\|^2, \tag{9}$$

where **X** indicates the matrix contains mixed signs. **W** = **X**A is introduced as a set of convex combinations of the columns of **X**. Convex-NMF can apply to mixed sign and non-negative data matrices [15, 21].

2.7 Semi-NMF

Semi-NMF is a model which from the perspective of clustering. Clustering on **X** by using k-means algorithm and we can get a cluster centroids $\mathbf{W} = (\mathrm{w}_1, \ldots, \mathrm{w}_k)$. And let **H** denote the cluster indicators: if $h_{ik} = 1$ and it belongs to cluster c_k; otherwise $h_{ik} = 0$. The objection formula as:

$$O_{k-means} = \sum_{i=1}^{m} \sum_{k=1}^{K} h_{ik} \left\| x_i - w_k^{\mathbf{T}} \right\|^2 = \left\| \mathbf{X}_{\pm} - \mathbf{W}_{\pm} \mathbf{H}_{+}^{\mathbf{T}} \right\|^2. \quad (10)$$

The model regarded NMF as k-means. And values of h_{ik} are in $(0, \infty)$. Semi-NMF has no constraints on **X** and **W** but keeps **H** must be positive [15, 21].

3 Results

Clustering on the datasets decomposed by the NMF methods, which is a common ways to reflect the performances of these methods in most cases. In order to contrast performances of these NMF methods, clustering experiments are carried out from two aspects: samples clustering and integrated GE clustering. K-means method is used to cluster samples on above data. Feature selection is carried out based on GE datasets from the cancers. About parameters, convergence condition of all NMF methods is set to 10e-5 and the reduction dimension is 2 for all methods. Since NMF methods have a unified advantage: low dimension can restore the original data well, and can't affect the information mining.

3.1 Materials

The samples of the four cancers all cover two labels: Normal (negative) and Tumour (positive). There are four datasets from TCGA: Cholangiocarcinoma (CHOL) dataset, Pancreatic Adenocarcinoma (PAAD) dataset, HNSC (Head and Neck squamous cell carcinoma) and ESCA (Esophageal carcinoma). CHOL contains 65159 features divided into three types (GE, ME and CNV) in 45 samples (36 positive samples). Similarly, PAAD contains 65160 features in 180 samples (176 positive samples), HNSC contains 65160 features in 418 samples (398 positive samples) and ESCA contains 65160 features in 192 samples (183 positive samples). These datasets collect GE, ME, CNV, and all variables of these datasets are Level 3 processed. The experiments in this paper are performed on above datasets based on NMF methods.

3.2 Sample Clustering

In this part, clustering experiments are separately performed on three types of data: GE, ME and CNV. Each dataset contains more than 20 thousand genes, and samples are clustered into two categories (Normal and Tumour) on above data. Since CNV contains negative numbers, which are processed with absolute values. We utilized the accuracy (ACC) and F-measure to measure the clustering effect. The results are shown in Fig. 1.

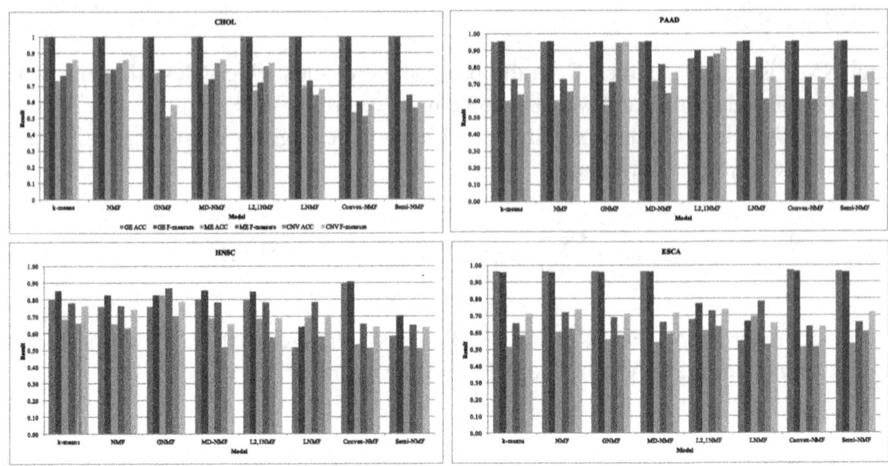

Fig. 1. The clustering results of NMF methods on three types of datasets of four cancers (k = 2)

On these single data sets, the performance of NMF, GNMF, and MD-NMF is stable, and it can be said that the manifold learning is suitable for clustering. For sparsely constrained NMF methods, sparse constraints are likely to reduce data integrity in theory. But the performance of $L_{2,1}$NMF on CHOL and PAAD is very considerable. It implies that the introduction of $L_{2,1}$-norm can enhance robustness.

Obviously, the clustering performances of NMF methods on GE data are better than the clustering accuracies on ME data and CNV data (Fig. 1). We assume that, the ability of GE data to distinguish diseases is powerful on these datasets from TCGA. And then, we integrate the GE datasets of four cancers and cluster the different subtypes.

Next, cluster all positive samples of these cancers (since the amount of the normal is very few): the four integrated GE datasets contains all cancer samples from TCGA, which exhibits the effects of NMF methods on integrated data (Fig. 2). In the integrated data, the difference of NMF effect is still quite obvious. The GNMF is still stable, but the convex-NMF looks weaker. In the case of complex data, excessive decomposition of convex-NMF can be regarded over-learning.

Overall, NMF, GNMF and MD-NMF are relatively stable methods for clustering and insensitive to the type of the data. $L_{2,1}$NMF and LNMF perform pretty well in most cases. But the effects of Convex-NMF and Semi-NMF fluctuated greatly and the performances of them have become polarized on the different datasets. In principle, GNMF makes points of samples to graph regularization in order to make related points easier to cluster. Thus GNMF is more applicable to clustering. MD-NMF introduces two Laplacian matrices about sample labels, which reflect advantages for clustering. And sparse degree of LNMF is larger, which is easy to lose information; but it retains local orthogonality, and it may ignore useful information which will affect clustering results. $L_{2,1}$NMF can generate sparse rows while maintaining the validity of the data from one side to reflect the robustness of $L_{2,1}$-norm. $L_{2,1}$NMF and LNMF are stressed with retaining the principal components while GNMF is biased on classification and

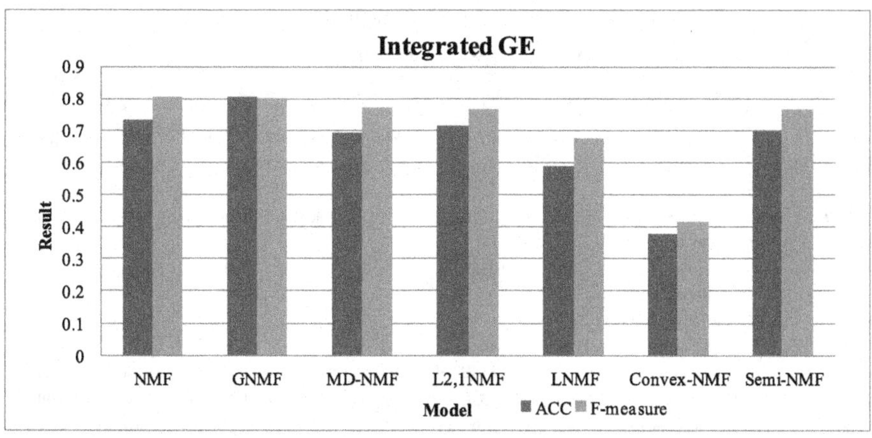

Fig. 2. The clustering results of NMF methods on integrated GE of four cancers (k = 4)

clustering of data, which explain the differences of their performances. Although Convex-NMF imposes the convex combination constraint on **X**, it has not obvious advantages for coefficient matrix to cluster in these experiments because of excessive decomposition. While Semi-NMF combined with k-means method and there will be better results in clustering experiments, the truth is not ideal (different datasets have different results of clustering).

3.3 The Feature Selection

About feature selection, we also compare the p-values of genes selected by different NMF methods on integrated GE datasets of four cancers. Our principle of selecting feature genes is adding the each element values of a column vector about the matrix and selecting bigger values to analyze. And then, order from big to small. Firstly, we put the top 1000 genes selected of each method into ToppFun (https://toppgene.cchmc.org/enrichment.jsp) to test. And then we set p-value to 0.01 and further analyze the performances of a variety of NMF methods. The genes separately selected whose p-values at top 10 from three aspects: Molecular Function, Biological Process and Cellular Component (Table 1). The p-values of genes selected by $L_{2,1}$NMF is smaller than other methods, indicating that $L_{2,1}$NMF has better performance about feature recognition (the smaller p-value means the higher amount of enrichment for genes). In this part, $L_{2,1}$NMF really highlights the advantages of the sparse approach due to sparse NMF can reduce a large unrelated elements of the data. But the p-values cannot strictly manifest the problem, which still need to reference some reasonable analysis.

We use two cancers as examples to conduct pathway mining studies. With genes related disease extracted by NMF methods, some pathways are found. On CHOL, the common genes extracted by these NMF are 579. Putting the overlapped genes on Kyoto Encyclopedia of Genes and Genomes (KEGG: http://www.genome.jp/kegg/) and obtaining the pathway which has the most overlap genes: Pathway in Cancer (Fig. 3). The genes marked with red that are extracted by the NMF methods.

Table 1. Gene selection of NMF methods on four cancers

ID	Name	NMF	GNMF	MD-NMF	$L_{2,1}$NMF	LNMF	Convex-NMF	Semi-NMF
GO:0044822	poly(A) RNA binding	1.258E-83	1.258E-83	2.484E-61	**1.051E-88**	6.026E-66	5.026E-52	2.837E-57
GO:0003723	RNA binding	2.366E-72	1.097E-71	6.254E-51	**2.497E-77**	2.295E-55	7.589E-34	1.62E-47
GO:00Z05198	Structural molecule activity	3.486E-41	7.416E-42	4.991E-33	**2.548E-42**	2.224E-35	2.609E-24	4.991E-33
GO:0006613	Cotranslational protein targeting to membrane	2.412E-78	2.412E-78	1.067E-76	1.863E-76	1.468E-76	9.535E-70	3.148E-73
GO:0006614	SRP-dependent cotranslational protein targeting to membrane	2.338E-75	2.338E-75	6.975E-72	1.912E-73	9.388E-72	2.412E-68	3.781E-70
GO:0045047	Protein targeting to ER	1.067E-73	1.067E-73	1.948E-70	6.853E-72	5.44E-72	1.835E-65	8.407E-69
GO:0006413	Translational initiation	1.818E-73	1.818E-73	1.869E-68	**5.592E-76**	1.437E-69	6.417E-61	5.859E-66
GO:0005925	Focal adhesion	2.69E-85	2.152E-86	1.768E-62	**7.989E-92**	1.971E-69	1.786E-42	1.389E-59
GO:0030055	Cell-substrate junction	1.263E-84	1.039E-85	4.706E-62	**4.453E-91**	6.226E-69	1.997E-41	3.445E-59

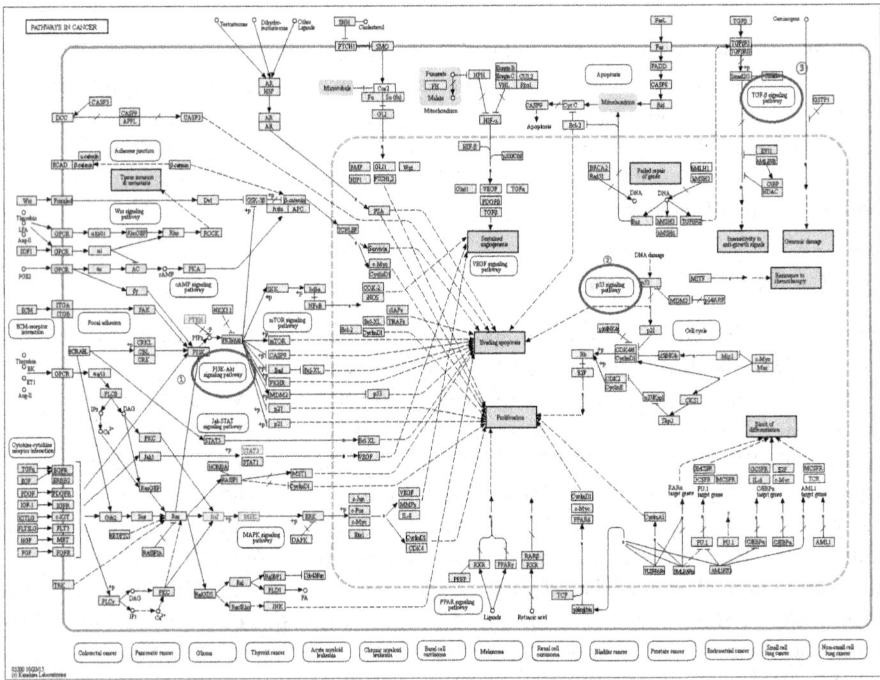

Fig. 3. The pathway in cancer found from CHOL with NMF methods (Color figure online)

It is generally known that many cancers are influenced by multiple genetic pathways. CHOL is a common malignant tumor in the bile duct, which has a low rate about radical resection and a poor prognosis. CHOL has been found to be due to mutations and allelic losses of genes and over-expression in multiple signal transduction pathways, accompanying methylation of several anti-oncogenes. There are three major pathways for CHOL: PI3K-Akt signaling pathway, p53 signaling pathway, TGF-signaling pathway. And we map common feature genes selected by NMF methods at them (Fig. 3). PTEN, which appears repeatedly in the figures, is one of the 8 potential drivers according to related research [22, 23]. PTEN is on human chromosome 10 and it is tumor anti-oncogene with phosphatase activity (protein consisting of 403 amino acid residues). And it can induce apoptosis and enhance sensitivity to apoptosis. The positive expression rate of PTEN protein in CHOL is significantly decreased, which suggests that the inactivation of PTEN may play an important role in the occurrence and development of CHOL. For example, PTEN is limited to the dephosphorylation, which causes decline about the positive expression rate of PTEN in extrahepatic carcinoma tissue (Fig. 4).

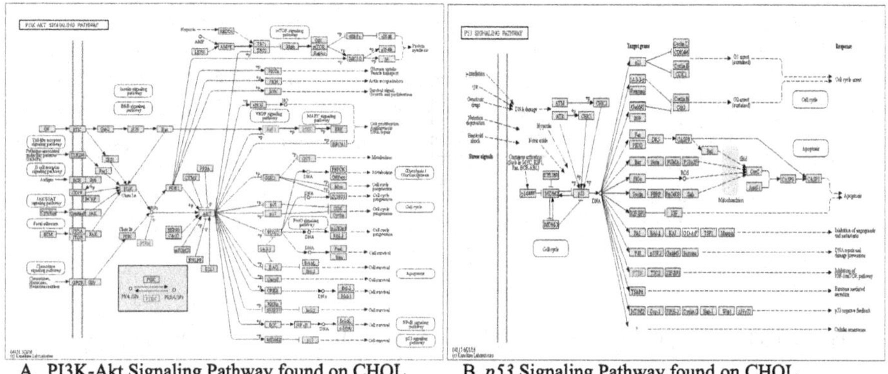

A. PI3K-Akt Signaling Pathway found on CHOL B. *p53* Signaling Pathway found on CHOL

Fig. 4. Signaling pathway found from CHOL with NMF methods

On the PAAD dataset, put the overlapped genes that selected from the NMF methods (925) on the KEGG to analyze pathways. And we find the pancreatic cancer pathway in the cancer pathway (Fig. 5). The red marker genes are the characteristic genes that extracted by NMF methods. There are four closely pathways related with PAAD: PI3K-Akt signaling pathway, MARK signaling pathway, JAK-STAY signaling pathway and TGF-β signaling pathway. In the marker genes, K-Ras (KRAS) is the most studied oncogene in development of PAAD. The research shows that 75%–90% cases of PAAD with this gene mutation [24].

In order to further prove the reliability of NMF models, it is necessary to check genes selected by these methods whether correlate related diseases. And the genes selected are specifically compared with genes from GeneCards (http://www.genecards.org/). Top 200 genes obtained by NMF methods are contrasted with genes of related disease records from GeneCards (Table 2). The integers in the table indicate the number of genes that are selected from the top 200 genes associated with disease. The values of the bracket are the highest values of the scores for the genes associated with the disease from the website. In Table 2, unfortunately, these NMF methods have not found the highest score of the gene at top 200 genes on these datasets. The NMF methods haven't showed competitive advantages in feature selection, perhaps they need to be improved about the ability of identifying the characteristic genes.

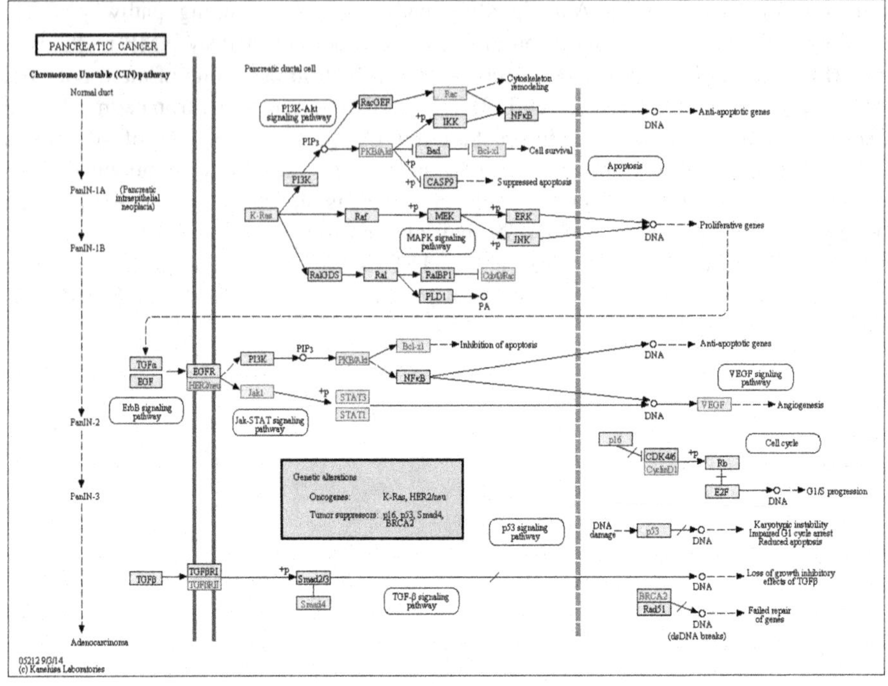

Fig. 5. Pancreatic cancer found from PAAD with NMF methods (Color figure online)

Table 2. Gene selection on four cancers

Data Method	CHOL(17.6)	PAAD(63.58)	HNSC(160.17)	ESCN(106.74)
NMF	44(12.62)	103(51.31)	111(65.68)	124(62)
GNMF	42(12.62)	103(51.31)	111(65.68)	124(62)
MD-NMF	42(8.17)	98(51.31)	131(65.68)	**126**(62)
$L_{2,1}$NMF	46(12.73)	**110**(47.77)	**149**(65.68)	122(62)
LNMF	44(10.58)	105(47.77)	119(48.38)	121(62)
Convex-NMF	31(7.17)	100(47.77)	111(48.38)	123(62)
Semi-NMF	**63**(8.17)	98(47.77)	107(48.38)	123(58.32)

4 Discussion

NMF methods decompose the data into two matrices about genes and samples, which is better to find differentially expressed genes and classified samples based on dimensionality reduction. Through the research in this paper, we find that NMF is still usable and effective method in terms of clustering; the effects of GNMF and MD-NMF are considerable for sample clustering; $L_{2,1}$NMF and LNMF perform pretty well in most instances. Convex-NMF and Semi-NMF are sensitive to the datasets; the effects of them are up and down on different datasets. But for large integrated data clustering, fundamental NMF seems to have advantages more than other improved methods. In the aspect of feature gene selection, most p-values of genes selected by $L_{2,1}$NMF are smaller and disease-related genes obtained by $L_{2,1}$NMF are also good, indicating that NMF with introduction of sparse constraints significantly highlights the differential expression values. GE dataset from TCGA is more conducive to identify differentially expressed genes compared with ME and CNV. We also find some pathways and differentially expressed genes reported previously on this dataset. Now, NMF is transforming from single data models to multi-view models (integrative models). The classic single data models involved in this paper can be integrated into integrative models, and there are already mature integrated models for NMF, such as iNMF [7] and jNMF [25]. The experimental investigation in this paper also provides experimental reference and basis for these classical single models can be integrated into an integrative model from specific functions and characteristics.

5 Conclusion

In this paper, we verified the effects of NMF methods from the two aspects of genes and samples, and found the pathways associated with the disease of cancers by gene expression analysis, which confirms the feasibility and effectiveness of application for NMF on TCGA once again. Of course, the results of the experiments will also be influenced by the data types, parameters setting and other related factors. There may be some poorly conceived with our research but we will continue to optimize ideas and strive to improve the NMF methods.

Acknowledgement. This work was supported in part by the NSFC under grant Nos. 61572284, 61502272 and 61702299.

References

1. Lee, D.D., Seung, H.S.: Learning the parts of objects by non-negative matrix factorization. Nature **401**(6755), 788–791 (1999)
2. Lee, D.D., Seung, H.S.: Algorithms for non-negative matrix factorization. In: Advances in Neural Information Processing systems, pp. 556–562 (2001)
3. Wang, D., Liu, J.X., Gao, Y.L. et al.: Characteristic gene selection based on robust graph regularized non-negative matrix factorization. IEEE/ACM Trans. Comput. Biol. Bioinform. **13**(6), 1059–1067 (2016)
4. Wang, S., Tang, J., Liu, H.: Embedded unsupervised feature selection (2015)
5. Sumanta, R., Sanghamitra, B.: A NMF based approach for integrating multiple data sources to predict HIV-1–human PPIs. BMC Bioinf. **17**(1), 1–13 (2016)
6. Zhang, W., Liu, X., Chen, Y., Wu, W., Wang, W., Li, X.: Feature-derived graph regularized matrix factorization for predicting drug side effects. Neurocomputing **287**, 154–162 (2018)
7. Yang, Z., Michailidis, G.: A non-negative matrix factorization method for detecting modules in heterogeneous omics multi-modal data. Bioinformatics **32**(1), 325–342 (2015)

8. Gao, J., Aksoy, B.A., Dogrusoz, U., Dresdner, G., Gross, B., Sumer, S.O., Sun, Y., Jacobsen, A., Sinha, R., Larsson, E.: Integrative analysis of complex cancer genomics and clinical profiles using the cBioPortal. Sci. Sign. **6**(269), 2383 (2013)
9. Zhu, Y., Qiu, P., Ji, Y.: TCGA-assembler: open-source software for retrieving and processing TCGA data. Nat. Meth. **11**(6), 599–600 (2014)
10. Hou, M.-X., Gao, Y.-L., Liu, J.-X., Shang, J.-L., Zheng, C.-H.: Comparison of Non-negative Matrix Factorization Methods for Clustering Genomic Data. In: International Conference on Intelligent Computing, pp. 290–299 (2016)
11. Cai, D., He, X., Han, J., Huang, T.S.: Graph regularized nonnegative matrix factorization for data representation. IEEE Trans. Pattern Anal. Mach. Intell. **33**(8), 1548–1560 (2011)
12. Guan, N., Tao, D., Luo, Z., Yuan, B.: Manifold regularized discriminative nonnegative matrix factorization with fast gradient descent. IEEE Trans. Image Process. **20**(7), 2030–2048 (2011)
13. Kong, D., Ding, C., Huang, H.: Robust nonnegative matrix factorization using l21-norm. In: Proceedings of the 20th ACM International Conference on Information and Knowledge Management, pp. 673–682 (2011)
14. Li, S.Z., Hou, X.W., Zhang, H., Cheng, Q.: Learning spatially localized, parts-based representation. In: Proceedings of the 2001 IEEE Computer Society Conference on Computer Vision and Pattern Recognition, CVPR 2001, vol. 1, I-207-I-212 (2001). vol. 201
15. Ding, C., Li, T., Jordan, M.I.: Convex and semi-nonnegative matrix factorizations. IEEE Trans. Softw. Eng. **32**(1), 45–55 (2010)
16. Long, X., Lu, H., Peng, Y., Li, W.: Graph regularized discriminative non-negative matrix factorization for face recognition. Multimed. Tools Appl. **72**(3), 2679–2699 (2014)
17. Chung, F.R.: Spectral Graph Theory. Volume 92 of CBMS Regional Conference Series in Mathematics. American Mathematical Society, Providence (1997)
18. Le, L., Yu-Jin, Z.: A survey on algorithms of non-negative matrix factorization. J. Acta Electronica Sinica. **36**(4), 737–743 (2008)
19. Chen, X., Gu, L., Li, S.Z., Zhang, H.-J.: Learning representative local features for face detection. In: 2001 Proceedings of the IEEE Computer Society Conference on Computer Vision and Pattern Recognition, CVPR 2001, vol. 1, I-1126-I-1131 (2001). vol. 1121
20. Pascual-Montano, A., Carazo, J.M., Kochi, K., Lehmann, D., Pascual-Marqui, R.D.: Nonsmooth nonnegative matrix factorization (nsNMF). IEEE Trans. Pattern Anal. Mach. Intell. **28**(3), 403–415 (2006)
21. Li, Y., Alioune, N.: The non-negative matrix factorization toolbox for biological data mining. Source Code Biol. Med. **8**(1), 10 (2013)
22. Wang, Q., Liu, X.D.: Genes and Cholangiocarcinoma Genesis and Development. Medical Recapitulate (2012)
23. Zou, S., Li, J., Zhou, H., Frech, C., Jiang, X., Chu, J.S., Zhao, X., Li, Y., Li, Q., Wang, H.: Mutational landscape of intrahepatic cholangiocarcinoma. Nat. Commun. **5**, 5696 (2014)
24. Biankin, A.V., Waddell, N., Kassahn, K.S., Gingras, M.C., Muthuswamy, L.B., Johns, A.L., Miller, D.K., Wilson, P.J., Patch, A.M., Wu, J.: Pancreatic cancer genomes reveal aberrations in axon guidance pathway genes. Nature **491**(7424), 399–405 (2012)
25. Zhang, S., Liu, C.C., Li, W., Shen, H., Laird, P.W., Zhou, X.J.: Discovery of multi-dimensional modules by integrative analysis of cancer genomic data. Nucleic Acids Res. **40**(19), 9379–9391 (2012)

Identifying Characteristic Genes and Clustering via an L_p-Norm Robust Feature Selection Method for Integrated Data

Sha-Sha Wu, Mi-Xiao Hou, Jin-Xing Liu(✉), Juan Wang, and Sha-Sha Yuan

School of Information Science and Engineering, Qufu Normal University, Rizhao 276826, China
wushashayx@126.com, sdcavell@126.com, mixiaohou@163.com, wangjuansdu@163.com, ssyuan@mail.qfnu.edu.cn

Abstract. In bioinformatics, feature selection is a good method for dimensionality reduction and has been widely used. However, the model of traditional feature selection method: Joint Embedding Learning and Sparse Regression (JELSR), whose the error term is in the form of a square term, which leads to the algorithm becoming extremely sensitive to noise and outliers and degrading the performance of the algorithm. Considering the above problem, we propose a new robust feature selection model by adding an L_p-norm constraint on error term, and name it as RJELSR, which improves the robustness of the algorithm. And we give an efficacious optimization strategy based on the augmented Lagrange multiplier method to get the optimal results. In the experimental section, we first preprocess different cancer data to obtain the integrated data, and then apply it to our algorithm for feature selection and sample clustering. Experiments on integrated data demonstrate that the performance of our method is superior to other compared methods and the selected characteristic genes are more biologically meaningful.

Keywords: L_p-norm constraint · Integrated Data · Feature selection Clustering

1 Introduction

With the continuous improvement of people's living standards, people are increasingly concerned about their own health. Cancer has become one of the major diseases affecting human health. According to the report of China's cancer statistics, there are up to 4.29 million new cases and 2.81 million deaths of cancer, and the annual morbidity and mortality increase at an alarming rate, but his pathogenic mechanism, treatment methods and other aspects of the study is not perfect [1]. Thus, it is vital and urgent to research on cancer.

With the rapid growth of massively parallel sequencing technologies, massive genomic data have been generated [2]. However, the main information and rules of cancer are hidden in these high-dimensional genomic data. These high-dimensional

D.-S. Huang et al. (Eds.): ICIC 2018, LNCS 10955, pp. 419–431, 2018.
https://doi.org/10.1007/978-3-319-95933-7_51

data not only bring the spatial complexity but also the complexity of time, so it must be reduced the dimension to mine the key information related to cancer. Feature selection algorithm is widely concerned as a dimensionality reduction method.

Feature selection is to select some of the most effective features from the original features, to effectively remove redundant and irrelevant features and reduce the dimensionality of the data and computational complexity [3]. In this paper, the key genes are selected by this method as features. And the selected key genes have clear biological significance, which can help biologists explore the potential expression mechanism of disease-related genes and be used in the prediction, diagnosis and prevention of diseases [4]. Therefore, it is necessary to find a suitable effective feature selection method for dealing with such high-dimensional data and searching for more meaningful characteristic genes. In addition, based on the similarity of the data, the clustering analysis of samples or genes for cancer data provides the basis for the early diagnosis and accurate subtypes of cancer [5].

The method of feature selection is generally divided into supervised and unsupervised feature selection methods based on whether the data has label information or not. Since there will be more and more data without category tags in the future and most current feature selection methods are supervised learning, it is very meaningful to research and apply unsupervised feature selection methods. The first unsupervised feature selection method is Pca Score (PcaScor) [6] proposed by Krzanowski based on principal component analysis. However, the PcaScor only uses the statistical properties of each feature and does not fully consider the manifold structure of the original data. Manifold can discover the internal geometric and differential structure of the data space [7]. Considering the above problem, He et al. proposed an unsupervised feature selection method based on Laplacian Eigenmaps (LE) [8] and local preserving mappings – Laplacian Score (LapScor) [9]. Zhao et al. proposed Spectral Feature Selection (SPEC) [10], which is similar to LapScor. The difference between the two is that the method of evaluating the weight of the feature is different. Although both LapScor and SPEC use the manifold structure in data space, they lack of a learning mechanism, resulting in insufficient accuracy of the selected features. The Multi-Cluster Feature Selection (MCFS) proposed by Cai et al. [11] and the Minimum Redundancy Spectral Feature Selection (MRSF) proposed by Zhao et al. [12] make up for the shortcoming of the above methods. Both of these methods perform the feature selection in two steps: the first step is to use LE to embed high-dimensional spatial data into low-dimensional space, and the second step is to use the regression coefficient to obtain the weight of each feature and accordingly sort all features for selection. The difference between MCFS and MRSF is that the sparse constraint condition of the regular term is different. If the above two independently executed steps can be performed jointly, it is possible to obtain better results. Hou et al. proposed Joint Embedding Learning and Sparse Regression (JELSR) [13] based on the above idea. Although JELSR is better than the previous feature selection algorithm, its error term is expressed as a square term, which makes the algorithm extremely sensitive to noise and outliers, and reduces the robustness of the algorithm. Therefore, we need to propose a more effective new method to enhance the performance of the algorithm, and accurately select the characteristic genes that are closely related to the diseases.

Recently, the L_p*-norm* has been widely used, and it has been proved to be more robust than other constraints ($0<p<1$) [14, 15]. Therefore, to improve the performance of the algorithm, we propose a new robust feature selection model by adding an L_p*-norm* constraint on the error term, and call it as RJELSR. In addition, changes in a single gene may affect multiple cancers, based on this idea, so different cancer data are first integrated to provide a new dataset that accurately facilitates analysis of oncogenes, and then the improved algorithm is applied to the integrated data, so that the final results are more biological significance.

The following is the main contributions of this article:

1. The L_p*-norm* constraint is introduced into the error term instead of the original norm constraint to reduce the effect of outliers and noise on the algorithm, thereby perfecting the efficiency of the algorithm.
2. A new optimization strategy is given based on the method of Augmented Lagrange Multiplier (ALM) [16] to get the optimal solutions.
3. Different cancer data are pretreated to get integrated data, and applying it to the improved algorithm to make the selected characteristic genes more biologically meaningful.

The following parts are arranged as follows: Sect. 2 introduces the related work. Section 3 describes and optimizes the improved algorithm. Section 4 gives the experimental results and analysis on the integrated data. Finally, this paper concludes with a summary.

2 Related Work

2.1 Related Notations

In the paper, the matrix $\mathbf{X} \in \mathbb{R}^{m \times n}$ is denoted as input data matrix, where each row of $\mathbf{X}$ represents one gene in n samples, and each column of $\mathbf{X}$ represents a sample in m genes. For the matrix $\mathbf{M} \in \mathbb{R}^{m \times n}$, its *i-th* column vector and *j-th* row vector are denoted as $\mathbf{m}_i$ and $\mathbf{m}^j$, respectively. Furthermore, the expression of different norm is different, and the general formula of the norm is denoted as follows.

$$\|\mathbf{M}\|_{r,q} = \left(\sum_{i=1}^{n}\left(\sum_{j=1}^{m}|m_{ij}|^r\right)^{\frac{q}{r}}\right)^{\frac{1}{q}} = \left(\sum_{i=1}^{n}\|m^i\|_r^q\right)^{\frac{1}{q}}. \quad (1)$$

For example, when $r = 2, q = 1$, Eq (1) is the expression of the $L_{2,1}$*-norm*, i.e.,

$$\|\mathbf{M}\|_{2,1} = \sum_{i=1}^{m}\sqrt{\sum_{j=1}^{n} m_{ij}^2}. \quad (2)$$

In the following paper, the norm is abbreviated as $\|\cdot\|_{r,q}$, for convenience.

2.2 Joint Embedding Learning and Sparse Regression

JELSR is an unsupervised feature selection method, and it makes up for the shortcoming of other traditional feature selection methods, which heightens the performance of the algorithm. To perfect the efficiency of the algorithm, embedding learning and sparse regression are combined in the process of the feature selection, where the regular term is constrained by the $L_{2,1}$-*norm*, and the error term is represented by the form of least-square, and its model is expressed as follows.

$$\arg \min_{\mathbf{W},\mathbf{Y}\mathbf{Y}=\mathbf{I}_{d\times d}} tr\left(\mathbf{Y}\mathbf{L}\mathbf{Y}^T\right) + \beta\left(\|\mathbf{W}^T\mathbf{X} - \mathbf{Y}\|_2^2 + \alpha\|\mathbf{W}\|_{2,1}\right), \tag{3}$$

where $\mathbf{W}$ is the sparse regression matrix, $\mathbf{Y}$ is the low dimensional embedding matrix, and α and β are two balance parameters.

3 Methodology

3.1 The Proposed Method

In reality, it is inevitably to contain some noise and outliers in the original data, which can affect the performance of the algorithm, so it is necessarily to find the way to reduce their influence. But the loss function of traditional feature selection method is usually expressed as a square term, which causes the algorithm to become extremely sensitive to noise and outliers. However, the introduction of L_p-*norm* can better solve the above problem [17]. Based on the above idea and JELSR proposed by Hou et al., we propose a new unsupervised robust feature selection method (RJELSR) that the original norm is replaced by L_p-*norm* on the error term, which perfects the performance of the algorithm and makes the final results more biologically meaningful. According to the above concept, we get the following objective function.

$$\begin{aligned} &\min_{\mathbf{W},\mathbf{Y}} \|\mathbf{W}^T\mathbf{X} - \mathbf{Y}\|_p + \alpha\|\mathbf{W}\|_{2,1} + \beta\,\mathrm{tr}\left(\mathbf{Y}\mathbf{L}\mathbf{Y}^T\right) \\ &s.t.\ \mathbf{Y}\mathbf{Y}^T = \mathbf{I}_{d\times d}. \end{aligned} \tag{4}$$

Enlightened by the process of Non-Negative Spectral Learning and Sparse Regression-Based Dual-Graph Regularized Feature Selection (NSSRD) proposed by Shang et al. [18], the constraint condition $\mathbf{Y}\mathbf{Y}^T = \mathbf{I}_{d\times d}$ is added to the target function to reduce the introduction of the parameters and convenient optimization, which can enhance the stability and capability of the algorithm and get a new objective function as below.

$$\min_{\mathbf{W},\mathbf{Y}} \|\mathbf{W}^T\mathbf{X} - \mathbf{Y}\|_p + \alpha\|\mathbf{W}\|_{2,1} + \beta\mathrm{tr}\left(\mathbf{Y}\mathbf{L}\mathbf{Y}^T\right) + \frac{\lambda}{2}\|\mathbf{Y}\mathbf{Y}^T - \mathbf{I}_d\|_2^2, \tag{5}$$

where $\mathbf{Y} \in \mathbb{R}^{d \times n}$ is the low-dimensional embedding matrix, $\mathbf{W} \in \mathbb{R}^{m \times d}$ is the sparse transformation matrix, α, β and λ are three reconcile parameters, $\mathbf{L} \in \mathbb{R}^{d \times d}$ is the graph Laplacian matrix.

3.2 Optimization

It is a trouble thing to straight update the function (5), so another strategy should be adopted to solve this problem. According to Rich Chartrand's method [19] of disposing the problem of the L_p*-norm*, we first introduce a supplementary variable $\mathbf{S} = \mathbf{W}^T\mathbf{X} - \mathbf{Y}$, and then we get:

$$\begin{aligned} &\min_{\mathbf{S},\mathbf{W},\mathbf{Y}} \|\mathbf{S}\|_p + \alpha\|\mathbf{W}\|_{2,1} + \beta \mathrm{tr}\left(\mathbf{Y}\mathbf{L}\mathbf{Y}^T\right) + \frac{\lambda}{2}\left\|\mathbf{Y}\mathbf{Y}^T - \mathbf{I}_d\right\|_2^2 \\ &s.t.\, \mathbf{S} = \mathbf{W}^T\mathbf{X} - \mathbf{Y}. \end{aligned} \tag{6}$$

The method of ALM [16] is a very effective way to tackle sub-problems. The optimization problem of the above objective function is addressed by ALM, i.e.

$$\begin{aligned} \mathrm{L}(\mathbf{S},\mathbf{W},\mathbf{Y},\boldsymbol{\Psi}) = &\|\mathbf{S}\|_p + \alpha\|\mathbf{W}\|_{2,1} + \beta \mathrm{tr}\left(\mathbf{Y}\mathbf{L}\mathbf{Y}^T\right) + \frac{\lambda}{2}\left\|\mathbf{Y}\mathbf{Y}^T - \mathbf{I}_d\right\|_2^2 \\ &+ \mathrm{tr}\left(\boldsymbol{\Psi}^T\left(\mathbf{S} - \mathbf{W}^T\mathbf{X} + \mathbf{Y}\right)\right) + \frac{\mu}{2}\left\|\mathbf{S} - \mathbf{W}^T\mathbf{X} + \mathbf{Y}\right\|_F^2, \end{aligned} \tag{7}$$

where μ is a balance parameter, $\boldsymbol{\Psi}$ is a augmented Lagrange multiplier. By further algebraic operation, the Eq (8) is obtained.

$$\begin{aligned} \mathrm{L}(\mathbf{S},\mathbf{W},\mathbf{Y},\boldsymbol{\Psi}) = &\|\mathbf{S}\|_p + \alpha\|\mathbf{W}\|_{2,1} + \beta \mathrm{tr}\left(\mathbf{Y}\mathbf{L}\mathbf{Y}^T\right) \\ &+ \frac{\lambda}{2}\left\|\mathbf{Y}\mathbf{Y}^T - \mathbf{I}_d\right\|_2^2 + \frac{\mu}{2}\left\|\mathbf{S} - \mathbf{W}^T\mathbf{X} + \mathbf{Y} + \frac{\boldsymbol{\Psi}}{\mu}\right\|_F^2. \end{aligned} \tag{8}$$

Next, we optimize these variables one by one and get their respective iteration formulas.

Optimizing **S**:

First, the variable **S** is updated when the rest of the variables are fixed. We define:

$$\begin{aligned} \mathrm{L}_\mathbf{S} &= \|\mathbf{S}\|_p + \frac{\mu}{2}\left\|\mathbf{S} - \mathbf{W}^T\mathbf{X} + \mathbf{Y} + \frac{\boldsymbol{\Psi}}{\mu}\right\|_F^2 \\ &= \|\mathbf{S}\|_p + \frac{\mu}{2}\left\|\mathbf{S} - \left(\mathbf{W}^T\mathbf{X} - \mathbf{Y} - \frac{\boldsymbol{\Psi}}{\mu}\right)\right\|_F^2. \end{aligned} \tag{9}$$

The upper formula is a function with regard to $\mathbf{S}$, and it can be resolved by the following $p - shrinkage$ operation [19]:

$$shrink_p(t, \gamma) := \max\left\{0, |t| - \gamma|t|^{p-1}\right\}\frac{t}{|t|}. \tag{10}$$

Let $\mathbf{T} = \mathbf{W}^T\mathbf{X} - \mathbf{Y} - \mathbf{\Psi}/\mu$ and $\gamma = 1/\mu$, the solution of Eq. (9) is obtained by the following shrinkage operation (also called as soft thresholding):

$$\mathbf{S}_{k+1} = shrink_p\left((\mathbf{W}_k)^T\mathbf{X} - \mathbf{Y}_k - \frac{\mathbf{\Psi}_k}{\mu_k}, \frac{1}{\mu_k}\right). \tag{11}$$

Optimizing $\mathbf{W}$:
From Eq. (8), we can get the following function on $\mathbf{W}$.

$$\mathrm{L_W} = \alpha\|\mathbf{W}\|_{2,1} + \frac{\mu}{2}\left\|\mathbf{S} - \mathbf{W}^T\mathbf{X} + \mathbf{Y} + \frac{\mathbf{\Psi}}{\mu}\right\|_F^2. \tag{12}$$

Because $\|\mathbf{W}\|_{2,1}$ is a convex problem, it is impossible to get the optimal result directly in Eq. (12). Nei et al. [20] propose a solution to solve the above problem:

$$\frac{\partial \mathrm{L_W}}{\partial \mathbf{W}} = \left(2\alpha\mathbf{U} + \mu\mathbf{X}\mathbf{X}^T\right)\mathbf{W} - \left(\mu\mathbf{X}\mathbf{S}^T + \mu\mathbf{X}\mathbf{Y}^T + \mathbf{X}\mathbf{\Psi}^T\right) = 0, \tag{13}$$

where $\mathbf{U} = diag(u_1, u_2, \cdots, u_m)$ is a diagonal matrix, $u_r = 1/2\|\mathbf{w}^i\|_2 (r = 1, 2, \cdots, m)$. Further operation on the Eq. (13) yields:

$$\mathbf{W}_{k+1} = \left(2\alpha\mathbf{U} + \mu_k\mathbf{X}\mathbf{X}^T\right)^{-1}\left(\mu_k\mathbf{X}(\mathbf{S}_k)^T + \mu_k\mathbf{X}(\mathbf{Y}_k)^T + \mathbf{X}(\mathbf{\Psi}_k)^T\right). \tag{14}$$

Optimizing $\mathbf{Y}$:
From Eq. (8), we obtain the following function with respect to $\mathbf{Y}$:

$$\mathrm{L_Y} = \beta\mathrm{tr}\left(\mathbf{Y}\mathbf{L}\mathbf{Y}^T\right) + \frac{\lambda}{2}\|\mathbf{Y}\mathbf{Y}^T - \mathbf{I}_d\|_2^2 + \frac{\mu}{2}\left\|\mathbf{S} - \mathbf{W}^T\mathbf{X} + \mathbf{Y} + \frac{\mathbf{\Psi}}{\mu}\right\|_F^2. \tag{15}$$

Next, $\mathbf{Y}$ is iteratively updated based on the above function. The Eq. (15) is executed the partial derivative on $\mathbf{Y}$ and let it equals 0 when other variables are fixed:

$$\frac{\partial \mathrm{L_Y}}{\partial \mathbf{Y}} = \mathbf{Y}\left(2\beta\mathbf{L} + 2\lambda\mathbf{Y}^T\mathbf{Y} - 2\lambda\mathbf{I}_n + \mu\mathbf{I}_n\right) - \left(\mu\mathbf{W}^T\mathbf{X} - \mu\mathbf{S} - \mathbf{\Psi}\right) = 0. \tag{16}$$

Thence, the following update formula of **Y** is attained by the above equation:

$$\mathbf{Y}_{k+1}=\left(\mu_k(\mathbf{W}_k)^T\mathbf{X}-\mu_k\mathbf{S}_k-\mathbf{\Psi}_k\right)\left(2\beta\mathbf{L}+2\lambda(\mathbf{Y}_k)^T\mathbf{Y}_k-2\lambda\mathbf{I}_n+\mu_k\mathbf{I}_n\right)^{-1}. \tag{17}$$

Optimizing $\mathbf{\Psi}$ and μ:

According to the algorithm of ALM [16], we can obtain the following optimization solutions with respect to $\mathbf{\Psi}$ and μ, respectively.

$$\mathbf{\Psi}_{k+1}=\mathbf{\Psi}_k+\mu_k\left(\mathbf{S}_{k+1}-(\mathbf{W}_k)^T\mathbf{X}+\mathbf{Y}_k\right), \tag{18}$$

$$\mu_{k+1}=\rho\mu_k, \tag{19}$$

where ρ ranges from 1 to 2. In a word, the process of RJELSR is summarized as follows.

Algorithm 1. Process of RJELSR

Input: $\mathbf{X},\alpha,\beta,\lambda,p$

Output: $\mathbf{S},\mathbf{W},\mathbf{Y},\mathbf{\Psi},\mu$

Initialize: $\mathbf{U}=\mathbf{I}_{m\times m}$

Repeat:

Loop

Update $\mathbf{S}_{k+1}$ by $\mathbf{S}_{k+1}=shrink_p\left((\mathbf{W}_k)^T\mathbf{X}-\mathbf{Y}_k-\frac{\mathbf{\Psi}_k}{\mu_k},\frac{1}{\mu_k}\right)$,

Update $\mathbf{W}_{k+1}$ by

$$\mathbf{W}_{k+1}=\left(2\alpha\mathbf{U}+\mu_k\mathbf{X}\mathbf{X}^T\right)^{-1}\left(\mu_k\mathbf{X}(\mathbf{S}_k)^T+\mu_k\mathbf{X}(\mathbf{Y}_k)^T+\mathbf{X}(\mathbf{\Psi}_k)^T\right),$$

Update $\mathbf{Y}_{k+1}$ by

$$\mathbf{Y}_{k+1}=\left(\mu_k(\mathbf{W}_k)^T\mathbf{X}-\mu_k\mathbf{S}_k-\mathbf{\Psi}_k\right)\left(2\beta\mathbf{L}+2\lambda(\mathbf{Y}_k)^T\mathbf{Y}_k-2\lambda\mathbf{I}_n+\mu_k\mathbf{I}_n\right)^{-1},$$

Update $\mathbf{\Psi}_{k+1}$ and μ_{k+1} by $\mathbf{\Psi}_{k+1}=\mathbf{\Psi}_k+\mu_k\left(\mathbf{S}_{k+1}-(\mathbf{W}_k)^T\mathbf{X}+\mathbf{Y}_k\right)$ and

$\mu_{k+1}=\rho\mu_k$.

Until Convergence

4 Experiments

In this section, we run RJELSR, JELSR, MCFS and LapScor to select characteristic genes and cluster sample on the integrated multiple cancer data. Firstly, the dataset used is given a brief description. Secondly, the selection of the main parameters is given. Next, we provide the results and analysis of the feature selection. Finally, the results and analysis of sample clustering are provided.

4.1 Data Description

In the experimental section, we use the dataset obtained from The Cancer Genome Atlas (TCGA). TCGA is a public and comprehensive database that contains a rich variety of cancer datasets, for instance, pancreatic adenocarcinoma (PAAD), esophageal carcinoma (ESCA), and cholangiocarinoma (CHOL) and so on, where it is available at https://cancergenome.nih.gov/. And the diseased and normal samples of PAAD, ESCA, and CHOL are 176 and 4,183 and 9, 36 and 9 respectively. Since a gene may be associated with a variety of cancers, multiple cancer data should be integrated to analyze the key genes. Consequently, we preprocessed the gene expression data of the above three data to obtain an integrated data (INDA), and then performed feature selection and clustering on this new dataset, so that the selected characteristic genes are more biological value. During the preprocessing of three cancer data, we deleted the normal sample that accounted for a small total sample and retained the diseased samples of the three data. And the simple sketch of INDA as Fig. 1.

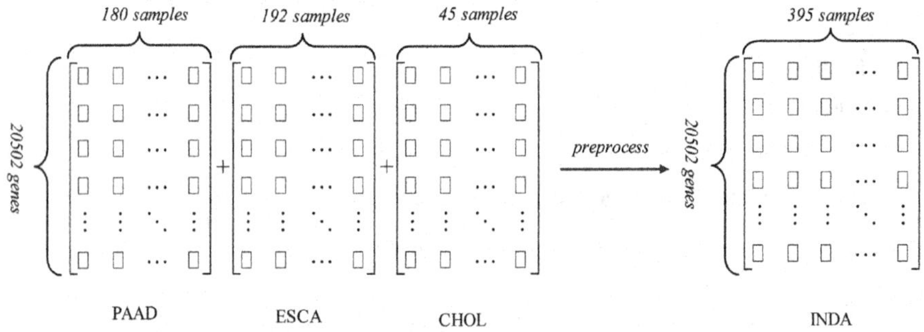

Fig. 1. A simple sketch of INDA.

4.2 Parameters Selection

In the experiment, we need to select the optimal parameters of each method. In RJELSR, the parameter p controls the error function and the effect of algorithm on noise and outliers, and $p \in [0.1, 0.2, 0.3, 0.4, 0.5, 0.6, 0.7, 0.8, 0.9]$ and the change of the accuracy of different p is shown in Fig. 2; α, β and λ are three reconcile parameters in the range of $10^t (t = 0, 1, 2, \cdots, 10)$. In JELSR, the main two parameters are α and β, and are adjusted within the range of $10^t (t = 0, 1, 2, \cdots, 10)$. The optimal values of the above parameters are obtained by the 5-fold cross-validation for fairness. Accordingly, throughout the experiment, the optimal values for the above two methods are $\alpha = 10^7$, $\beta = 10^8$, $\lambda = 10^3$ and $\alpha = 10^4$, $\beta = 10^{10}$, respectively. For MCFS, we use the default parameters obtained by [11]. For LapScor, it does not involve the parameter and is calculated by Laplacian score to feature selection.

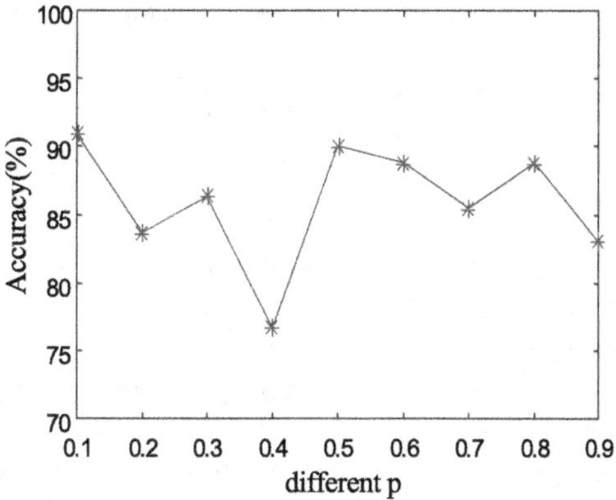

Fig. 2. The accuracy of different p.

4.3 Feature Selection Results

The feature selection method can not only reduce the dimension, but also select the key genes related to the diseases from the original data and analyze them. Moreover, it can help the treatment of the later disease and prove whether the performance of the method is valid.

To confirm the validity of our algorithm, RJELSR, JELSR, MCFS and LapScor are used to perform feature selection on INDA. The top 500 characteristic genes are selected by four methods, and put into ToppFun for analyzing, respectively. ToppFun is a public Gene Ontology (GO) tool, where it is available at https://toppgene.cchmc.org/enrichment.jsp. The results are arranged in ascending order of the p-values. Then we select the top ten GO items to show in Table 1 and the best results are bold. Under normal circumstances, the smaller the p-value corresponding to the same GO item, the better the performance of the algorithm. From Table 1, the p-values of the corresponding GO term gained by RJELSR are less than other three methods, so our method is demonstrated by this experiment that the performance outplay other compared methods in feature selection.

Furthermore, the first ten characteristic genes obtained by RJELSR are put into GeneCards to analyze the relationship between genes and diseases, and the results are shown in Table 2. GeneCards is an omnibus and open human genetic data pool that provides information about all predicted and known human genes, where it is available at http://www.genecards.org/. The related diseases, related GO annotations and paralog of genes are detailed in the Table 2. The official name of CEACAM5 is Carcinoembryonic Antigen Related Cell Adhesion Molecule 5, and it is a key gene that has a close relationship with multiple diseases. And a momentous paralog of this gene is CEACAM1. CEACAM5 is first found by [21], its abnormal expression can affect many diseases, such as breast and pancreatic cancer [22, 23]. MUC4 corresponds to the official name is Mucin 4, Cell Surface Associated, and the paralog of this gene is SUSD2. In normal cells, MUC4 is not expressed; if the expression of MUC4 is unregulated, the expression level of pancreatic cancer tumors will increase, leading to the occurrence of pancreatic cancer [24]. The relationship between the expression of

Table 1. The top ten p-values obtained by different method on INDA.

ID	Name	RJELSR	JELSR	MCFS	LapScor
GO:0005198	Structural molecule activity	**3.471E–29**	5.682E–20	1.823E–17	3.205E–13
GO:0006614	SRP-dependent cotranslational protein targeting to membrane	**2.136E–27**	5.191E–11	2.599E–12	2.270E–10
GO:0000184	Nuclear-transcribed mRNA catabolic process, nonsense-mediated decay	**9.945E–27**	3.039E–11	4.274E–11	1.261E–08
GO:0006613	Cotranslational protein targeting to membrane	**1.028E–26**	1.071E–10	5.804E–12	7.309E–10
GO:0045047	Protein targeting to ER	**1.274E–26**	1.182E–10	6.478E–12	1.021E–10
GO:0072599	Establishment of protein localization to endoplasmic reticulum	**2.933E–26**	1.739E–10	9.934E–12	2.311E–11
GO:0022626	Cytosolic ribosome	**2.23E–25**	4.403E–10	2.787E–11	5.366E–08
GO:0070972	Protein localization to endoplasmic reticulum	**1.002E–24**	8.935E–10	6.100E–11	4.456E–10
GO:0044391	Ribosomal subunit	**2.106E–23**	1.279E–09	1.135E–10	9.949E–06
GO:0090150	Establishment of protein localization to membrane	**2.199E–22**	8.793E–10	8.067E–15	4.471E–08

MUC4 and squamous dysplastic transformation [25], and its expression changes will cause the occurrence of Intrahepatic Cholangiocarcinoma and Cholangiocarcinoma. The table only gives partial functional explanations of some genes selected by RJELSR on INDA.

4.4 Clustering Results

The L_p*-norm* is introduced into our method to reduce the impact of outliers and noise on the algorithm, and ameliorate the performance of the algorithm. The clustering method is used to further prove the effectiveness of the proposed algorithm. Clustering is an unsupervised learning method and generally divided into two categories: clustering samples and clustering genes. There are many ways about clustering. In our experiments, we choose a classical clustering method – K-means and implement the sample clustering. The basic idea of K-means algorithm is to cluster with k points in the space as the center, classify the objects nearest to them, and iteratively update the value of each cluster center in sequence until the best clustering result is obtained. In our experiments, the samples of INDA are grouped into three categories, which are the diseased samples associated with PAAD, ESCA and CHOL, respectively. The clustering accuracy (ACC) [26] is used as a measure to evaluate the performance of all algorithms and is denoted as the following formula. And the value of ACC indicates the degree of the performance of the algorithm: the greater the value, the better the performance, and vice versa.

Table 2. The top ten genes selected by RJELSR on INDA.

Gene symbol	Related diseases	Related GO annotations	Gene paralog
CEACAM5	Rectal Neoplasm and Colorectal Cancer	Protein homodimerization activity and GPI anchor binding	CEACAM1
MUC2	Pseudomyxoma Peritonei and Signet Ring Cell Adenocarcinoma	A Protein Coding gene	DEFAULT
MUC4	Intrahepatic Cholangiocarcinoma and Cholangiocarcinoma	ErbB-2 class receptor binding and extracellular matrix constituent, lubricant activity	SUSD2
KRT19	Lung Cancer and Thyroid Cancer	Structural molecule activity and structural constituent of cytoskeleton	KRT15
PRSS1	Pancreatitis, Hereditary and Trypsinogen Deficiency	Serine-type endopeptidase activity	PRSS2
SST	Islet Cell Tumor and Somatostatinoma	Hormone activity	DEFAULT
KRT8	Cryptogenic Cirrhosis and Liver Cirrhosis	Structural molecule activity and scaffold protein binding	KRT5
MMP7	Brain Glioblastoma Multiforme and Light Chain Deposition Disease	Peptidase activity and metallopeptidase activity	MMP3
ALB	Analbuminemia and Hyperthyroxinemia, Familial Dysalbuminemic	Enzyme binding and chaperone binding	AFP
CRP	Acute Pancreatitis and Appendicitis	Calcium ion binding and cholesterol binding	APCS

$$ACC = \frac{\sum_{i=1}^{n} \delta(map(cl_i), tl_i)}{n}, \tag{20}$$

where n is the total number of samples, tl_i is the true label, cl_i is the cluster label, $map(cl_i)$ is the permutation mapping function that maps cl_i to the label of INDA, and $\delta(x, y)$ is the delta function that equals 1 if $x = y$ and equals 0 otherwise.

We execute RJELSR, JELSR, MCFS and LapScor to cluster sample by K-means. And the samples are aggregated into three categories. Table 3 shows that the values of ACC obtained by each algorithm on INDA. The best result is in bold. From Table 3 that we get the following conclusion: the values of ACC obtained by RJELSR are greater than other four methods. Therefore, this conclusion further validates that the effectiveness of our method outperforms the comparison method.

Table 3. ACC of different methods on INDA.

Methods	RJELSR	JELSR	MCFS	LapScor	K-means
ACC(%)	**82.53**	77.22	69.11	53.92	51.32

5 Conclusions

In this paper, we propose a new robust JELSR method by adding an L_p*-norm* on error function and apply it to integrated data. Firstly, the L_p*-norm* is introduced to reduce the sensitivity of the algorithm to outliers and improve the robust of the algorithm to a certain extent. Secondly, an iterative update strategy is given to solve the optimization problem of the objective function by using ALM. Thirdly, to better identify the relationship between genes and diseases, we have integrated the gene expression data of different cancers to get INDA. Finally, feature selection and sample clustering are performed on INDA, and the results show that the proposed method has better performance than other compared methods.

In the future, we hope to further reduce the introduction of parameters and the time complexity to better improve the stability and performance of the algorithm, and this is also an inadequacy of the proposed algorithm.

Acknowledgement. This work was supported in part by the NSFC under grant Nos. 61572284, 61502272, and 61701279.

References

1. Chen, W., Zheng, R., Baade, P.D., Zhang, S., Zeng, H., Bray, F., Jemal, A., Yu, X.Q., He, J.: Cancer statistics in China, 2015. CA Cancer J. Clin. **66**(2), 115 (2016)
2. Reis-Filho, J.S.: Next-generation sequencing. J. Biomed. Biotechnol. **11**(S3), S12 (2009)
3. D'Addabbo, A., et al.: SVD based feature selection and sample classification of proteomic data. In: Lovrek, I., Howlett, R.J., Jain, L.C. (eds.) KES 2008. LNCS (LNAI), vol. 5179, pp. 556–563. Springer, Heidelberg (2008). https://doi.org/10.1007/978-3-540-85567-5_69
4. Zheng, C.H., Yang, W., Chong, Y.W., Xia, J.F.: Identification of mutated driver pathways in cancer using a multi-objective optimization model. Comput. Biol. Med. **72**, 22–29 (2016)
5. Liu, J.X., Xu, Y., Zheng, C.H., Kong, H., Lai, Z.H.: RPCA-based tumor classification using gene expression data. IEEE/ACM Trans. Comput. Biol. Bioinform. **12**(4), 964–970 (2015)
6. Krzanowski, W.J.: Selection of variables to preserve multivariate data structure, using principal components. J. R. Stat. Soc. **36**(1), 22–33 (1987)
7. Cai, D., He, X., Han, J., Huang, T.S.: Graph regularized nonnegative matrix factorization for data representation. IEEE Trans. Pattern Anal. Mach. Intell. **33**(8), 1548 (2011)
8. Belkin, M., Niyogi, P.: Laplacian Eigenmaps for dimensionality reduction and data representation. Neural Comput. **15**(6), 1373–1396 (2006)
9. He, X., Cai, D., Niyogi, P.: Laplacian score for feature selection. In: International Conference on Neural Information Processing Systems, pp. 507–514 (2006)
10. Zhao, Z., Liu, H.: Spectral feature selection for supervised and unsupervised learning. In: Proceedings of the Twenty-Fourth International Conference on Machine Learning, pp. 1151–1157 (2007)

11. Cai, D., Zhang, C., He, X.: Unsupervised feature selection for multi-cluster data. In: ACM SIGKDD International Conference on Knowledge Discovery and Data Mining, pp. 333–342 (2010)
12. Zhao, Z., Wang, L., Liu, H.: Efficient spectral feature selection with minimum redundancy. In: Twenty-Fourth AAAI Conference on Artificial Intelligence, AAAI 2010, Atlanta, Georgia, USA, pp. 11–15, July 2011
13. Hou, C., Nie, F., Li, X., Yi, D., Wu, Y.: Joint embedding learning and sparse regression: a framework for unsupervised feature selection. IEEE Trans. Cybern. **44**(6), 793 (2014)
14. Nie, F., Huang, H., Ding, C.: Low-rank matrix recovery via efficient schatten p-norm minimization. In: Twenty-Sixth AAAI Conference on Artificial Intelligence, pp. 655–661 (2012)
15. Nie, F., Wang, H., Huang, H., Ding, C.: Joint schatten p-norm and ℓp-norm robust matrix completion for missing value recovery. Knowl. Inf. Syst. **42**(3), 525–544 (2015)
16. Chen, M., Lin, Z., Ma, Y., Wu, L.: The Augmented Lagrange Multiplier Method for Exact Recovery of Corrupted Low-Rank Matrices. Eprint Arxiv, vol. 9 (2010)
17. Liu, J., Liu, J.X., Gao, Y.L., Kong, X.Z., Wang, X.S., Wang, D.: A p-norm robust feature extraction method for identifying differentially expressed genes. PLoS ONE **10**(7), e0133124 (2015)
18. Shang, R., Wang, W., Stolkin, R., Jiao, L.: Non-negative spectral learning and sparse regression-based dual-graph regularized feature selection. IEEE Trans. Cybern. **PP**(99), 1–14 (2017)
19. Chartrand, R.: Nonconvex splitting for regularized low-rank + sparse decomposition. IEEE Trans. Signal Process. **60**(11), 5810–5819 (2012)
20. Nie, F., Huang, H., Cai, X., Ding, C.H.: Efficient and robust feature selection via joint ℓ2,1-norms minimization. In: Advances in Neural Information Processing Systems, pp. 1813–1821 (2010)
21. Gold, P., Freedman, S.O.: Specific carcinoembryonic antigens of the human digestive system. J. Exp. Med. **122**(3), 467–481 (1965)
22. Gebauer, F., Wicklein, D., Horst, J., Sundermann, P., Maar, H., Streichert, T., Tachezy, M., Izbicki, J.R., Bockhorn, M., Schumacher, U.: Carcinoembryonic antigen-related cell adhesion molecules (CEACAM) 1, 5 and 6 as biomarkers in pancreatic cancer. PLoS ONE **9**(11), e113023 (2014)
23. Blumenthal, R.D., Leon, E., Hansen, H.J., Goldenberg, D.M.: Expression patterns of CEACAM5 and CEACAM6 in primary and metastatic cancers. BMC Cancer **7**(1), 2 (2007)
24. Choudhury, A., Moniaux, N., Winpenny, J.P., Hollingsworth, M.A., Aubert, J.P., Batra, S. K.: Human MUC4 mucin cDNA and its variants in pancreatic carcinoma. J. Biochem. **128** (2), 233–243 (2000)
25. Lópezferrer, A., Alameda, F., Barranco, C., Garrido, M., De, B.C.: MUC4 expression is increased in dysplastic cervical disorders. Hum. Pathol. **32**(11), 1197–1202 (2001)
26. Huang, J., Nie, F., Huang, H.: A new simplex sparse learning model to measure data similarity for clustering. In: International Conference on Artificial Intelligence, pp. 3569–3575 (2015)

Using Novel Convolutional Neural Networks Architecture to Predict Drug-Target Interactions

ShanShan Hu[1], DeNan Xia[1], Peng Chen[2(✉)], and Bing Wang[3]

[1] School of Computer Science and Technology, Anhui University, Hefei 230601, Anhui, China
[2] Institute of Physical Science and Information Technology, Anhui University, Hefei 230601, Anhui, China
pchen.ustc10@yahoo.com
[3] School of Electrical and Information Engineering, Anhui University of Technology, 243032 Ma'anshan, Anhui, China

Abstract. Identifying potential drug-target interactions (DTIs) are crucial task for drug discovery and effective drug development. In order to address the issue, various computational methods have been widely used in drug-target interaction prediction. In this paper, we proposed a novel deep learning-based method to predict DTIs, which involved the convolutional neural networks (CNNs) to train a model and yielded robust and reliable predictions. The method achieved the accuracies of 92.0%, 90.0%, 92.0% and 90.7% on enzymes, ion channels, GPCRs and nuclear receptors in our curated dataset, respectively. The experimental results indicated that our methods improved the DTIs predictions in comparison with the state-of-the-art computational methods on the common benchmark dataset.

Keywords: Drug-target interactions (DTIs) · CNNs · Ensemble method

1 Introduction

Discriminating potential interactions between small molecules and proteins is a crucial step in the whole drug discovery field, which can provide insights into the mechanisms of drug actions and reduce drug development expense [1]. In addition, accurately identify drug-target interactions also fascinate the understanding of protein-protein interaction mechanism and protein secondary structure [2, 3]. As experimental approaches for the verification of drug-target interactions (DTIs) remain many challenges, a growing number of researchers have developed appropriate and powerful computational machine learning methods to infer drug-target associations effectively on a large scale. Yamanishi et al. [4] developed a supervised bipartite graph technique to predict the probability of interacting pairs. A semi-supervised learning approach, Laplacian regularized least square (NetLapRLS) [5], exploited small amount of labeled data and sufficient unlabeled data to reach the maximum generalization ability. More recently, with the accumulation of large amounts of biomedical data, deep learning techniques have been taken advantages over traditional state-of-the-art machine

D.-S. Huang et al. (Eds.): ICIC 2018, LNCS 10955, pp. 432–437, 2018.
https://doi.org/10.1007/978-3-319-95933-7_52

learning methods. A framework based on deep-belief network (DBN) was first applied to extract representations from input vectors and constructed a prediction model using known compound-protein interactions [6].

In this paper, we built an effective and robust predictive model based on deep learning approach to infer potential drug-target associations. Each one of 34 low-correlated properties extracted from AAindex1 database was used to calculate target descriptors; while a 1444-dimensional vector calculated by PaDEL-Descriptor software was used to represent drug candidate. Afterwards, the concatenating vectors of random selected descriptors from drug descriptors and 442 descriptors from target were all reshaped into 28 × 28 matrices, which were input to the model of convolutional neural networks. An ensemble of the prediction results was collected to discriminate DTIs by majority voting technique. As a result, our methods yields stable and highlighted performance compared with other machine learning methods.

2 Methods

2.1 Datasets

We used two drug-target datasets that were derived from KEGG BRITE [7] database (http://www.kegg.jp/kegg/brite.html), in which drugs without structural information or target proteins without primary sequence were discarded. One was provided from reference [8], called as *Dataset1*. Another one called *Dataset2* was manually collected, in which DTIs that are redundant and overlapping with *Dataset1* were omitted. The corresponding negative samples of the two datasets were generated in the following steps: (i) All drugs and targets in the benchmark dataset were integrated into pairs together without the known drug-target interactions; (ii) Randomly selecting pairs until the number of negative samples reached exactly two times as many as that of positive samples. Figure 1 illustrates the number of drugs, target proteins as well as drug-target pairs on both *Dataset1* and *Dataset2*.

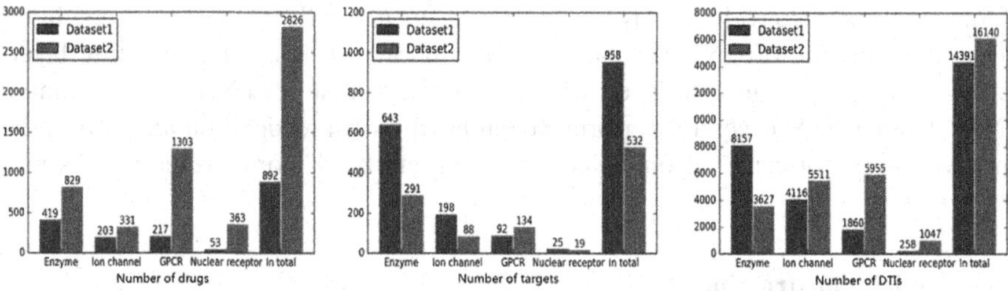

Fig. 1. The distribution of the numbers of drugs, targets and drug-target pairs on both *Dataset1* and *Dataset2*.

2.2 Drug Descriptors

PaDEL-Descriptor is a free and open-source software to calculate 1D, 2D and 3D descriptors and 10 types of fingerprints of chemical small molecules [9]. In this study, only 1D and 2D descriptors are calculated. Finally, 1444-dimensional vectors formulated as D [D_1, D_2, D_3, …, D_{1444}] is the final result to characterize a drug candidate.

2.3 Target Descriptors

We utilized 544 amino acids physicochemical properties in AAindex1 database to encode each sequence of target. In order to avoid predictive bias and obtain successful performance, properties with the correlation coefficient more than 0.5 were removed. The correlation coefficients between all two properties were computed and ranked in descend. From the top one property, the corresponding properties were consequently removed step by step until only the properties with correlation coefficient lower than 0.5 existed. In this way, 34 properties were retained, of which each one was used to encode protein sequences through the following formula [10, 11],

$$T(d) = \frac{\frac{1}{N-d}\sum_{i=1}^{N-d}(P_i - \bar{P})(P_{i+d} - \bar{P})}{\frac{1}{N}\sum_{i=1}^{N-d}(P_i - \bar{P})^2}, \tag{1}$$

where P_i and P_{i+d} are property values in one of 34 amino acid properties at sequence positions i and $i + d$, respectively; d is the distance between the two neighborings residue (in this work is set to be 13); N is the length of a protein sequence; $\bar{P}$ is the average value of P_i, i.e. $\bar{P} = \left(\sum_{i=1}^{N} P\right)/N$. Then 34 vectors are concatenated so that target descriptors are characterized by vector T [T_1, T_2, T_3, …, T_{442}].

2.4 Convolutional Neural Network

Convolutional Neural Networks (CNNs) are a type of deep learning architecture, which are prevailing in image and video recognition [12]. Generally, CNNs' topology consists of three key parts: convolution layers, pooling layers and fully connected layers. In this study, the structure of CNNs contains 3 convolutional layers, 2 max-pooling layers, and only one fully connected layer, which is similar to that of LeNet-5 [13], a kind of CNNs that reached successful performance in handwritten recognition area. To further improve the performance of our model, dropout and batch normalization tricks were also adopted.

2.5 Model Construction

First, the 342-dimesional vectors randomly picked up from 1444 drug descriptors were jointed to the 442-dimensional descriptors of targets. Thus, each drug-target pair is characterized by 784-dimensional features [D, T] = D[D_1, D_2, D_3, …, D_{342}, T_1, T_2, T_3, …,T_{442}]. In this way, the random selection process was repeated n times until the number of all drug-target pairs is around 40000–50000. Afterwards, the feature vector of each drug-target pair was reshaped into a 28 × 28 matrix to train a model of CNNs algorithm. The ensemble of the n-times prediction values yields the final predictions. In the ensemble, one drug-target pair is predicted to be interacting if at least half of the n-pairs were predicted as a positive sample, otherwise it is a non-interaction pair.

2.6 Measurement of Prediction Quality

Four widely used metrics were employed to assess the performance of our proposed model objectively: accuracy (Acc), precision (Pre), sensitivity (Sen) and F1 score (F1). The details of the calculations are briefly described as below:

$$\mathrm{Acc} = \frac{\mathrm{TP} + \mathrm{TN}}{\mathrm{TP} + \mathrm{FP} + \mathrm{TN} + \mathrm{FN}} \tag{2}$$

$$\mathrm{Pre} = \frac{\mathrm{TP}}{\mathrm{TP} + \mathrm{FP}} \tag{3}$$

$$\mathrm{Sen} = \frac{\mathrm{TP}}{\mathrm{TP} + \mathrm{FN}} \tag{4}$$

$$\mathrm{F1} = \frac{2 \times \mathrm{Sen} \times \mathrm{Pre}}{\mathrm{Sen} + \mathrm{Pre}}, \tag{5}$$

where TP (True Positive) and TN (True Negative) respectively represent the number of the correct predicted DTIs and non-interaction pairs. FP (False Positive) and FN (False Negative) represent the number of incorrectly predicted DTIs and non-interaction pairs.

3 Results

Four different predictors using the same parameters were constructed by 10-fold cross-validation to evaluate the performance of our models. That is to say, our dataset was randomly partitioned into 10 disjoint subsets, where one subset was considered as test set while the remaining subsets were regarded as training set. This progress was repeated 10 times until all instances were tested.

3.1 Performance Evaluation on Drug-Target Interactions

First, *Dataset2* was adopted to train the model to distinguish DTIs. As shown in Table 1, the model for Enzymes yields the highest performance among the four DTIs classes, with an accuracy of 0.920, a sensitivity of 0.881, a precision of 0.880, an F1 of 0.881. It is noticed that both enzyme and GPCR classes achieve the highest accuracy value (Acc = 0.920). It is well known that GPCRs are the most difficult cases in the identification of DTIs due to a few of known 3D structure information for GPCRs. That is to say, some unknown or noise features are existed to characterize GPCR targets. The results indicated that our model has a strong ability to discriminate DTIs on GPCR.

Subsequently, *Dataset1* was used to further evaluate the generalization ability of our model by the same parameters and neural network topology on different protein

Table 1. The detailed performance for the four protein families on both *Dataset1* and *Dataset2* by 10-fold cross validation.

Type		Acc	Sen	Pre	F1
Dataset1	Enzymes	0.943	0.927	0.903	0.915
	Ion channels	0.919	0.894	0.867	0.881
	GPCRs	0.884	0.818	0.831	0.824
	Nuclear receptors	0.884	0.872	0.798	0.833
Dataset2	Enzymes	0.920	0.881	0.880	0.881
	Ion channels	0.900	0.948	0.792	0.863
	GPCRs	0.920	0.899	0.866	0.882
	Nuclear receptors	0.907	0.891	0.841	0.865

families. The performances on the nuclear receptors and GPCRs of *Dataset1* are worse than those on *Dataset2*, whose F1 are respectively falling into 0.833 and 0.824. The possible reason might be that the number of DTIs in nuclear receptors and GPCRs classes is smaller than others, especially for nuclear receptors.

3.2 Prediction Comparison with Other Methods

Our method was compared with the work in reference [8] on the *Dataset1*. The comparative results showed that our model outperformed the methods in reference [8], which achieves higher accuracy values of 8.8%, 11.1%, 9.9% and 2.7% on enzymes, ion channels, GPCRs and nuclear receptors, respectively (Table 2). These results illustrated that more-refined features were extracted by CNNs model to obtain better predictive performance.

Table 2. Performance comparison in accuracy of our method with two methods on *Dataset1*

Methods	Enzymes	Ion channels	GPCRs	Nuclear receptors
Our method	0.943	0.919	0.884	0.884
Ref.[8]	0.855	0.808	0.785	0.857

4 Conclusion

In this work, we proposed a novel CNNs structure to discriminate possible drug-target associations, which contributes to drug reposition and drug discovery. The descriptors from drugs and target proteins are all reshaped into an image-like matrices as inputs to train the CNNs model. To our knowledge, it is the first time to adopt matrix characterized interaction pairs through CNNs to predict DTIs. Two different benchmark datasets have been carried out on our model and the outcomes proves the strength of our model in the predictions of DTIs. Compared with traditional machine learning methods, our model achieved highlighted performance, which demonstrated that informative and useful features can be extracted from massive features by deep learning technique. So it is promising to apply deep learning method to many biological areas and obtain powerful and reliable predictions.

Acknowledgement. This work was supported by the National Natural Science Foundation of China (Nos. 61672035, 61300058 and 61472282).

References

1. Dai, Y.F., Zhao, X.M.: A survey on the computational approaches to identify drug targets in the postgenomic era. Memórias Do Instituto Oswaldo Cruz **2015**, 1–9 (2015)
2. Huang, D.S., Zhang, L., Han, K.: Prediction of protein-protein interactions based on protein-protein correlation using least squares regression. Curr. Protein Pept. Sci. **15**(6), 553–560 (2014)
3. Huang, D.S., Huang, X.: Improved performance in protein secondary structure prediction by combining multiple predictions. Protein Pept. Lett. **13**(10), 985–991 (2006)
4. Yamanishi, Y., Araki, M., Gutteridge, A., Honda, W., Kanehisa, M.: Prediction of drug-target interaction networks from the integration of chemical and genomic spaces. Bioinformatics **24**, i232–i240 (2008)
5. Xia, Z., Wu, L.Y., Zhou, X., Wong, S.T.: Semi-supervised drug-protein interaction prediction from heterogeneous biological spaces. BMC Syst. Biol. **4**(Suppl 2), S6 (2010)
6. Wen, M., Zhang, Z., Niu, S., Sha, H., Yang, R., Yun, Y., Lu, H.: Deep-learning-based drug-target interaction prediction. J. Proteome Res. **16**, 1401–1409 (2017)
7. Kanehisa, M., Goto, S., Hattori, M., Aoki-Kinoshita, K.F., Itoh, M., Kawashima, S., Katayama, T., Araki, M., Hirakawa, M.: From genomics to chemical genomics: new developments in KEGG. Nucleic Acids Res. **34**, D354–D357 (2006)
8. He, Z., Zhang, J., Shi, X.H., Hu, L.L., Kong, X., Cai, Y.D., Chou, K.C.: Predicting drug-target interaction networks based on functional groups and biological features. PLoS ONE **5**, e9603 (2010)
9. Yap, C.W.: PaDEL-descriptor: an open source software to calculate molecular descriptors and fingerprints. J. Comput. Chem. **32**, 1466–1474 (2011)
10. Moran, P.A.P.: notes on continuous stochastic phenomena. Biometrika **37**, 17 (1950)
11. Li, Z.R., Lin, H.H., Han, L.Y., Jiang, L., Chen, X., Chen, Y.Z.: PROFEAT: a web server for computing structural and physicochemical features of proteins and peptides from amino acid sequence. Nucleic Acids Res. **34**, W32–W37 (2006)
12. Grinblat, G.L., Uzal, L.C., Larese, M.G., Granitto, P.M.: Deep learning for plant identification ssing vein morphological patterns. Comput. Electron. Agric. **127**, 418–424 (2016)
13. LeCun, Y., Bottou, L., Bengio, Y., Haffner, P.: Gradient-based learning applied to document recognition. Proc. IEEE **86**, 2278–2324 (1998)

Computational Prediction of Driver Missense Mutations in Melanoma

Haiyang Sun[1], Zhenyu Yue[2], Le Zhao[2], Junfeng Xia[1], Yannan Bin[1], and Di Zhang[2(✉)]

[1] Institute of Physical Science and Information Technology, Anhui University, Hefei, Anhui, China
[2] School of Computer Science and Technology, Anhui University, Hefei, Anhui, China
zdbbyy@163.com

Abstract. Discovering driver mutations used as the diagnostic and prognostic biomarkers is important for the treatment of cancer, including melanoma. Although during the last decade several computational methods have been developed to predict the effect of missense mutations in cancer, only a few have been specifically designed for identifying driver mutations in a specific disease context. To take into consideration of disease-specific factor, here we made efforts to prioritize missense mutations presented in melanoma. We collected 385 pathogenic mutations from the database of curated mutations (DoCM), and 392 benign mutations filtered from a benchmark neutral database (VariSnp), respectively. To evaluation of the model effect, we also selected 45 mutations from other databases. Then a random forest classifier was constructed to prioritize melanoma pathogenic mutations based on conservation, functional region annotation, protein secondary structure, protein domain, physicochemical features, and splicing information. The proposed method achieved an AUC of 0.94 on both training and test sets. When compared with previous developed algorithms, our method obtained a higher accuracy in identifying driver missense mutations in melanoma, along with a more balanced sensitivity and specificity than the other prediction methods.

Keywords: Melanoma · Missense mutation · Pathogenicity prediction

1 Introduction

Melanoma is a malignant neoplasm of melanocytes, known as its sufferings brought to patients and its poor prognosis. It is estimated that 74,680 new diagnoses of melanoma in 2017 in the United States [1]. As a complex genetic disease, melanoma contains numerous accumulated mutations. These mutations are comprised of a small number of driver mutations that have a switching performance to carcinogenesis, whereas the remaining bulky ones were regarded as passengers that have no direct effect on cancer [2]. So far, several strategies have been applied to distinguish driver mutations from passenger ones. Firstly, functional experiments have shed light on many disease-causing mutations, such as the V600E mutation on *BRAF* and the L576P mutation on

D.-S. Huang et al. (Eds.): ICIC 2018, LNCS 10955, pp. 438–447, 2018.
https://doi.org/10.1007/978-3-319-95933-7_53

KIT in melanoma [3, 4]. However, it was limited by the low throughput when confronting with massive mutation data produced by the sequencing technology. Secondly, statistical approaches characterized driver mutation by its prevalence [5], but a moderate and relevant mutation data is essential. Under these circumstances, developing the computational method is necessary for prioritizing mutations in melanoma.

Up to now, multiple in silico methods have been proposed for identifying the biological impact of missense mutations. Most of them designed for generic disease, such as CADD [6] and SIFT [7]. Notably, they have no regards for the fact that the driver mutations usually occurred in a specific genetic or disease context. For example, the A617T alteration on *CDH1* presents as pathogenicity in endometrial carcinoma while presents as benign in gastric cancer [8, 9], so that the prediction result maybe not accurate without taking into account disease-specific factors. Among the published algorithms, CHASM [10] and CanDrA [11] were the only two that specially considered the information of cancer type. CHASM was trained utilizing the mutation data mainly from the catalogue of somatic mutations in cancer (COSMIC), and each mutation was characterized by a set of 49 predictive features. Compared with CHASM, CanDrA added some new features. However, CanDrA has some hide flaws. First, CanDrA collected deleterious mutations located upon proximity, meaning that the positive samples may abundant in some common genes. At the same time, for passenger mutations, the authors set criteria that they should not be located in any cancer genes (cancer gene census, CGC) [12]. Obviously, CanDrA should be confounded by type 2 of circularity ("excellent accuracies can be achieved by predicting the status of a mutation based on the other mutations in the same protein") [13]. Second, CanDrA defined the driver and the passenger mutations based on the frequency of the mutation in tumor samples. However, mutation rate of melanoma is so high that it may be inaccurate to define drivers and passengers based only on prevalence [11]. Considering the two above-mentioned situations, we carried out this study to improve the reliability of driver mutations prediction and assess whether a better performance can be achieved.

In this work, we firstly derived a set of mutations from DoCM [14] and VariSNP [15], and then assigned to training and test sets. In addition, we collected an independent dataset for validating our model and comparing with the other existing tools. We then computed 20 attributes, which were informative for mutations pathogenicity prediction, for all the mutations employed in this study. Our model got an AUC of 0.94 by 10-fold cross validation on the training set and showed the same result on the test set. In comparison, our method showed a higher specificity along with a balanced sensitivity and achieved a more competitive accuracy than other tools. In all, our method obtained a higher accurate and more reliable results than the existing tools.

2 Methods and Materials

2.1 Datasets

The datasets of mutations were derived from those reported in Database of Curated Mutations (DoCM) (version 3.2) and VariSNP (20170216 release): (a) DoCM represents a highly curated set of disease causing mutations, including 1364 total mutations.

We marked a driver mutation if it presents as missense mutation and related with melanoma. (b) VariSNP is a benchmark database containing neutral mutations from dbSNP after filtering out deleterious mutations. We collected the neutral missense mutations only located in genes appeared in dataset that filtered from DoCM. Therefore, 777 mutations (385 pathogenic and 392 neutral mutations) were collected. We randomly chose seven-tenths of mutations from the pathogenic and neutral dataset as the training set, and the remaining three-tenths were used as the test set.

To validate our method and compare with other related predictors, we also obtained an independent test set from several authoritative databases (Landrum et al. 2016; Lovly et al. 2016; Griffith et al. 2017; Yue et al.). The mutations were gathered by the rule listed as follows. (a) melanoma related. (b) pathogenic missense mutation. (c) no overlapping of the test and training set. So that, the positive subset contains 23 mutations, where 12 of them come from ClinVar [16], 10 from My Cancer Genome [17], and 1 from CIViC [18]. Similarly, the negative subset contains 22 mutations, where 15 of them from dbCPM (Yue et al.) and 7 from ClinVar. All these datasets used in our study were showed in Table 1.

Table 1. Summary of mutation datasets used in this study

Dataset	Sample	Mutation	Source
Training set	Positive	271	DoCM
	Negative	275	VariSNP
Test set	Positive	114	DoCM
	Negative	117	VariSNP
Independent test set	Positive	23	ClinVar; My Cancer Genome; CIViC
	Negative	22	ClinVar; dbCPM

2.2 Feature Quantification

As some features is transcript specific, we adopted two approaches to obtain the mutations' transcript information: if the transcript information for each mutation was available in the original database, we maintained the information, otherwise we curated the longest transcript by means of using SnpEff (Version 4.3) [19]. In this study, we encoded 7 features on the DNA level, and 13 features on the protein level as showed in Table 2. Two kinds of conservation scores (GerpS and mamPhyloP) for each mutation were scored by the MyVariant tool [20]. Four attributes of functional region annotation, were encoded based on the BED files from ENCODE [21]. For the splicing feature, the genome distance to the nearest splice site for the mutation was from SettleSeqAnnotation 138 [22]. The mutations of Pfam domain were annotated using Ensembl GRCh37 [23], if the mutation locates in a specific domain it will be encoded as 1, otherwise 0. Protein secondary structure features, which consist of 7 attributes, were obtained from RaptorX-Property [24]. Physicochemical changes of amino acid (amino acid factors, AAFactors) were measured with five properties: polarity, molecular volume, secondary structure, electrostatic charge, and codon diversity [25]. Finally, the

Table 2. Summary of the attributes annotated for each mutation

Level	Feature class	Type	Dimension
DNA	Conservation	Numeric	2
	Functional region annotation	Discrete	4
	Distance to nearest splice site	Discrete	1
Protein	Pfam domain	Discrete	1
	Protein secondary structure	Numeric	7
	AAFactor	Numeric	5

raw values for each feature were scaled to Z-scores, which were used in subsequent classifier construction and performance evaluation.

2.3 Model Construction

The classification model for predicting driver mutations in melanoma was based on random forest [26], which is an effective and widely used method. In this study, package "Random forest" (version 1.0.10) in Weka (Version 3.8) [27] was employed and implemented with 10-fold cross validation. To obtain good prediction results, we optimized two major parameters, the number of trees of the forest (numIterations), and the maximum depth of the tree (maxDepth), by using grid search based on the results of 10-fold cross validation on the training set.

2.4 Statistics To Evaluate Prediction Performance

We measured the performance of prediction methods by the Sensitivity, Specificity, and Accuracy:

$$\text{Sensitivity} = \frac{\text{TP}}{\text{TP} + \text{FN}}$$

$$\text{Specificity} = \frac{\text{TN}}{\text{TN} + \text{FP}}$$

$$\text{Accuracy} = \frac{\text{TP} + \text{TN}}{\text{TP} + \text{TN} + \text{FP} + \text{FN}}$$

where:

TP: True positives, correctly predicted pathogenic mutations.
FP: False positives, neutral mutations predicted as disease one.
TN: True negatives, correctly predicted neutral mutations.
FN: False negatives, pathogenic mutations predicted as neutral

In addition, the AUC (the area under Receiver Operating Characteristic (ROC) curve) was also used as a measure of prediction results.

3 Results and Discussions

3.1 The Interrogation for the Training Dataset

As Grimm et al. cautioned in the previous research, there are two types circularity which may brought an unrealistic result of the prediction tools [13]. Type 1 circularity means the existence of the overlap between the training set and the test set samples, and we have avoided this problem. As for type 2 circularity, it happens when mutations presented on the same protein are together labeled as pathogenic or neutral, which may mislead predictors over rely on protein-level information to classify new mutations. Findings about type 2 circularity opened the masquerade that some prediction tools brought a statistically successful but unreasonable ultimately. Beware of this problem, the mutations used for model constructed in this study are from proteins mixed with neutral and pathogenic mutations. As showed in Fig. 1, the much larger fraction proteins of our training set are comprised of both pathogenicity and the neutral mutations: 64.8% of all mutations (354 out of 546) can be found in the category of mixed proteins.

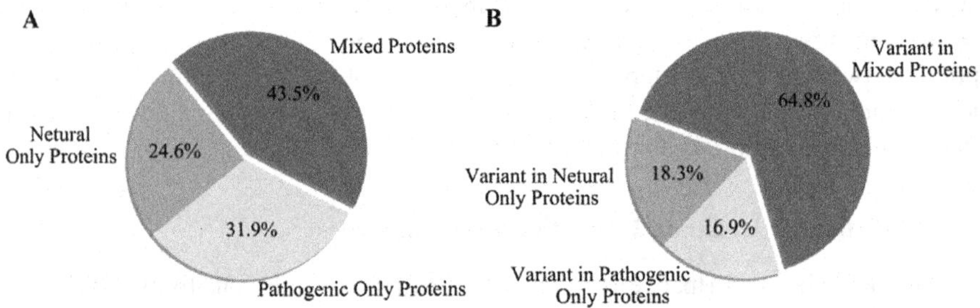

Fig. 1. In our training set, most samples are in proteins mixed with neutral and pathogenic mutations. A: Protein perspective: proportion of proteins containing only neutral mutations ("neutral-only"), only pathogenic mutations ("pathogenic-only"), and both types of mutations ("mixed"). Nearly half of the proteins are mixed. B: Mutation perspective: proportions of mutations in each of the three categories of proteins. Most of mutations are in mixed proteins.

3.2 Explorations for the Optimal Model

In this study, in order to identify the best machine learning technique suitable for predicting driver mutations in melanoma, we comprehensively evaluated the performances of random forest, SVM, neural network, linear discriminant analysis, and Naive bayes, which are commonly used on binary classification [28]. These classifiers are listed as: (a) Random forest (Forest), using the RandomForest package [26]. (b) Support vector machine (SVM), using the LibSVM package, version 1.0.10 [29]. (c) Neural network (NNet), using the MultilayerPerceptron package, version 1.0.10. (d) Linear discriminant analysis (LDA), using the discriminant Analysis package, version 1.0.2. (e) Naive bayes (Bayes), using the NaiveBayes package. All these five algorithms were implemented with the parameters optimized by using grid search.

Finally, we made a comparison of their AUC results on the training set with 10-fold cross validation and the test set respectively. As showed in Fig. 2, we can see that the five classifiers (Forest, SVM, NNet, LDA, Bayes) achieved an AUC of 10-fold cross validation of 0.94, 0.90, 0.84, 0.81, and 0.78 on training set respectively, and achieved an AUC of 0.94, 0.86, 0.82, 0.81, and 0.76 on test set respectively. It is obviously seen that random forest achieved the best result. Note that the two optimized parameters, numIterations = 500 and maxDepth = 18, were applied to Forest. In consideration of the data distribution was done in a random way, we cannot rule out that the best model may acquire a fortuitous result. Therefore, we repeated data distribution nine times, and then constructed the nine models with random forest. The result based on these models got the proximate result (the variance is less than 1×10^{-4}) as the previous random forest classifier performed (AUC = 0.94).

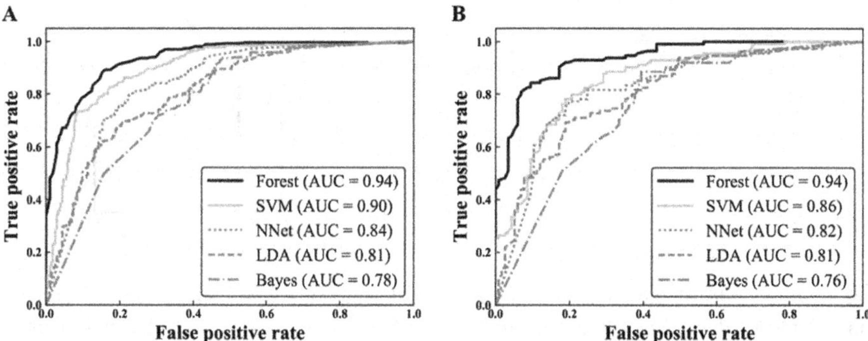

Fig. 2. ROC curves comparing the performance of several machine learning methods on training set (A) and test set (B). (Forest: random forest; SVM: support vector machine; NNet: neutral network; LDA: linear discriminant analysis; Bayes: naive bayes).

3.3 Analysis of Quantification Result

To a certain extent, an outperformance model may be explained by the contribution of the features adopted. As for the encoded 20 features, we interpreted it from a statistical perspective and performed the Wilcoxon signed-rank test, which is a non-parametric statistical hypothesis test. The statistical results of these features and their corresponding underlying biological implications are summarized as follows.

There are 14 features with a statistically significant classify ability for distinguish disease-causing mutations from neutral one. Note that all conservation and protein structures, counted into 10 features, performed a P-value less than 0.05. Maybe we can draw a conclusion that disease-causing mutations tended to have a high conservation and or locate in an important position of protein structure. Firstly, as we known, different locations in the genome are under different selection pressures during evolution. That is, important regions are not easy to accept alterations and non-important regions are easy to accept alterations. Although conservative scores may vary, depending on species and method of sequence alignment, but generally, deleterious mutation sites tend to have higher conservative scores [30]. Next, changing on protein structure is critical for the normal functioning of cellular machinery. These structural changes directly affect the functional protein biosynthesis, and or affect the binding to targeted biological macromolecules. Here, we also selected two features with the smallest P-value from the DNA and protein level respectively. The Z-score distribution of all the pathogenic and neutral mutations in the training set for the selected features was depicted in Fig. 3.

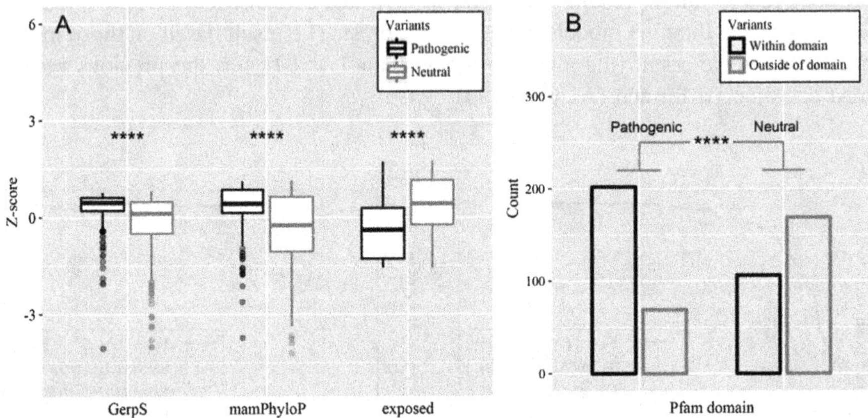

Fig. 3. Z-score distribution of all the pathogenic and neutral mutations on the training set for the four attributes which selected based on *P*-value. The symbol of '****' indicating statistical significance: $p \leq 0.0001$.

3.4 Performance Evaluation of Our Method and Comparison with Other Tools

To validate our method on unseen examples, we have been collected an independent test set. Here, for comparison, we adopted several representative and widely used predictors that meet the following criteria to make comparison. (a) the tools were designed for cancer. (b) the tools can predict the missense mutations. (c) the training data or its code sources was open access. Finally, four tools were collected according to these selecting criteria: CScape [31], Mutation Assessor [32], CanDrA(general) and CanDrA(melanoma) [11]. As for CanDrA, there were three samples predicted into the class of 'no-call' which mean equivocal, we regarded it as the wrong prediction. Finally, these classifiers prediction results on the independent test set were displayed in Table 3. Our method got the highest specificity and acquired more accurate results than the other tools at the expense of lowering sensitivity. We also note that the two pathogenic mutations of the independent test set: the R183C on *GNA11* and the R183Q on *GNAQ*, were predicted to the neutral mutations by our method, which are in part consistent with previous study [33].

Table 3. Performance comparison of different tools on the independent test set

Predictor	Our method	CScape	CanDrA[#1]	CanDrA[#2]	MutationAssessor
Sensitivity	0.61	0.91	1	1	0.86
Specificity	0.82	0.27	0	0.32	0.05
Accuracy	0.71	0.60	0.50	0.66	0.45

Note: #1 means CanDrA predict for general; #2 means CanDrA predict for melanoma.

4 Conclusion

The missense mutations identified in cancer genomes have been greatly increased with the development of sequencing technology. The immediate priority is to understand the functional impacts on these mutations, and depict which alterations are detrimental and which are less effect on cancer promoting. Although many computational methods were designed to distinguish pathogenic mutations from neutral ones, most of them do not consider the specific genetic or disease context in which a mutation occurs. In this article, we present a method for the prioritization of driver missense mutations in melanoma. The training set we used that are more reliable, because we have considered two types of circularity [13]. In our analysis, pathogenic mutations of melanoma tended to high conservative and played a key role for the protein structure. As compared with several other tools, our method obtained a higher accuracy, and a balanced sensitivity and specificity results. In the future work, we will collect more reliable mutations of melanoma and add more informative features to improve the prediction performance.

Acknowledgements. The authors thank the members of our laboratory for their valuable discussions. This work has been supported by the grants from the National Natural Science Foundation of China (61672037 and 21601001) and the Anhui Provincial Outstanding Young Talent Support Plan (gxyqZD2017005), and the Young Wanjiang Scholar Program of Anhui Province, China.

References

1. Siegel, R.L., Miller, K.D., Jemal, A.: Cancer statistics. CA Cancer J. Clin. **67**(1), 7–30 (2017)
2. Greenman, C., et al.: Patterns of somatic mutation in human cancer genomes. Nature **446** (7132), 153–158 (2007)
3. Shtivelman, E., et al.: Pathways and therapeutic targets in melanoma. Oncotarget **5**(7), 1701 (2014)
4. Lovly, C.M., et al.: Routine multiplex mutational profiling of melanomas enables enrollment in genotype-driven therapeutic trials. PLoS ONE **7**(4), e35309 (2012)
5. Xia, J., et al.: A meta-analysis of somatic mutations from next generation sequencing of 241 melanomas: a road map for the study of genes with potential clinical relevance. Mol. Cancer Ther. **13**(7), 1918–1928 (2014)
6. Kircher, M., et al.: A general framework for estimating the relative pathogenicity of human genetic variants. Nat. Genet. **46**(3), 310–315 (2014)

7. Kumar, P., Henikoff, S., Ng, P.C.: Predicting the effects of coding non-synonymous variants on protein function using the SIFT algorithm. Nat. Protoc. **4**(7), 1073–1081 (2009)
8. Suriano, G., et al.: Identification of CDH1 germline missense mutations associated with functional inactivation of the E-cadherin protein in young gastric cancer probands. Hum. Mol. Genet. **12**(5), 575–582 (2003)
9. Suriano, G., et al.: E-cadherin germline missense mutations and cell phenotype: evidence for the independence of cell invasion on the motile capabilities of the cells. Hum. Mol. Genet. **12**(22), 3007–3016 (2003)
10. Carter, H., et al.: Cancer-specific high-throughput annotation of somatic mutations: computational prediction of driver missense mutations. Cancer Res. **69**(16), 6660–6667 (2009)
11. Mao, Y., et al.: CanDrA: cancer-specific driver missense mutation annotation with optimized features. PLoS ONE **8**(10), e77945 (2013)
12. Futreal, P.A., et al.: A census of human cancer genes. Nat. Rev. Cancer **4**(3), 177–183 (2004)
13. Grimm, D.G., et al.: The evaluation of tools used to predict the impact of missense variants is hindered by two types of circularity. Hum. Mutat. **36**(5), 513–523 (2015)
14. Ainscough, B.J., et al.: DoCM: a database of curated mutations in cancer. Nat. Methods **13**(10), 806–807 (2016)
15. Schaafsma, G.C., Vihinen, M.: VariSNP, a benchmark database for variants from dbSNP. Hum. Mutat. **36**(2), 161–166 (2015)
16. Landrum, M.J., et al.: ClinVar: public archive of interpretations of clinically relevant variants. Nucleic Acids Res. **44**(D1), D862–D868 (2016)
17. My Cancer Genome Homepage. https://www.mycancergenome.org/content/disease/melanoma/. Accessed 21 Nov 2017
18. Griffith, M., et al.: CIViC is a community knowledgebase for expert crowdsourcing the clinical interpretation of variants in cancer. Nat. Genet. **49**(2), 170–174 (2017)
19. Cingolani, P., et al.: A program for annotating and predicting the effects of single nucleotide polymorphisms, SnpEff: SNPs in the genome of Drosophila melanogaster strain w1118; iso-2; iso-3. Fly (Austin) **6**(2), 80–92 (2012)
20. Xin, J., et al.: High-performance web services for querying gene and variant annotation. Genome Biol. **17**(1), 91 (2016)
21. ENCODE Project Consortium: An integrated encyclopedia of DNA elements in the human genome. Nature **489**(7414), 57–74 (2012)
22. Ng, S.B., et al.: Targeted capture and massively parallel sequencing of 12 human exomes. Nature **461**(7261), 272–276 (2009)
23. Flicek, P., et al.: Ensembl. Nucleic Acids Res. **42**(Database issue), D749–D755 (2014)
24. Wang, S., et al.: RaptorX-Property: a web server for protein structure property prediction. Nucleic Acids Res. **44**(W1), W430–W435 (2016)
25. Atchley, W.R., et al.: Solving the protein sequence metric problem. Proc. Natl. Acad. Sci. U.S.A. **102**(18), 6395–6400 (2005)
26. Breiman, L.: Machine Learning. Kluwer Academic Publishers, The Netherlands (2001)
27. Frank, E., Hall, M.A., Witten, I.H.: The WEKA Workbench. Fourth edn. Burlington (2016)
28. Buske, O.J., et al.: Identification of deleterious synonymous variants in human genomes. Bioinformatics **29**(15), 1843–1850 (2013)
29. Chang, C.-C., Lin, C.-J.: LIBSVM: a library for support vector machines. ACM Trans. Intell. Syst. Technol. **2**(3), 1–27 (2011)
30. Fraser, H.B., et al.: Evolutionary rate in the protein interaction network. Science **296**(5568), 750–752 (2002)

31. Rogers, M.F., et al.: CScape: a tool for predicting oncogenic single-point mutations in the cancer genome. Sci. Rep. **7**(1), 11597 (2017)
32. Reva, B., Antipin, Y., Sander, C.: Predicting the functional impact of protein mutations: application to cancer genomics. Nucleic Acids Res. **39**(17), e118 (2011)
33. Van Raamsdonk, C.D., et al.: Mutations in GNA11 in uveal melanoma. New Engl. J. Med. **363**(23), 2191–2199 (2010)

Further Evidence for Role of Promoter Polymorphisms in *TNF* Gene in Alzheimer's Disease

Yannan Bin[1], Ling Shu[2], Qizhi Zhu[1], Huanhuan Zhu[2], and Junfeng Xia[1(✉)]

[1] Institute of Physical Science and Information Technology, School of Computer Science and Technology, Anhui University, Hefei 230601, Anhui, China
jfxia@ahu.edu.cn

[2] School of Life Sciences, Anhui University, Hefei 230601, China

Abstract. Tumor necrosis factor (*TNF*) expression level is associated with regulating effects on Alzheimer's disease (AD) development. And several *TNF* SNPs have been reported to associate with AD, however, it is unclear whether *TNF* SNPs could affect *TNF* signaling. In this study, the effects of AD related *TNF* promoter SNPs (rs361525, rs1800629, rs1799724, rs1800630 and rs1799964) on the gene expression were explored by multiple large-scale expression quantitative trait loci (eQTL) datasets. We found that the five SNPs with minor allele could significantly regulate reduced *TNF* expression on different brain regions or whole blood sample in European population. Consistent with the result of eQTL analysis, we found rs1800630 A allele and rs1799964 C allele were significantly associated with increasing the volumes of brain amygdala based on the Enhancing Neuroimaging Genetics through Meta-Analysis database. These findings suggest that the promoter SNPs in *TNF* may play protective roles on AD risk. In addition, there was an interesting discovery that rs4248161 C allele significantly reduced the volumes of five brain regions, suggesting its potential risk for AD or other neuropathogenic diseases. Our studies could advance the understanding of the impact of *TNF* promoter SNPs in AD.

Keywords: Tumor necrosis factor · Alzheimer's disease
Expression quantitative trait loci · Brain region

1 Introduction

Alzheimer's disease (AD) is the most common form of dementia in the elderly with neuropathological characteristics of abnormal amyloid β (Aβ) accumulation and neurofibrillary tangles [1]. With years of AD pathogenesis research, a component of AD genetic susceptibility has been established by case-control studies, and the genes of *APOE*, *PSEN1*, *APP*, *PICALM* and *CLU* are firmly associated with AD risk [2, 3]. As a genetically complex and highly heritable disease, drugs striking one or a few gene targets may not bring AD cure [4], and therefore more information related to AD is needed to reveal new paradigms for the disease as well as for treatment.

D.-S. Huang et al. (Eds.): ICIC 2018, LNCS 10955, pp. 448–459, 2018.
https://doi.org/10.1007/978-3-319-95933-7_54

Recent studies have concentrated on neuroinflammation as a major brain pathology contributes to AD pathogenesis [5]. As the most extensively investigated proinflammatory cytokine, tumor necrosis factor (TNF) protein encoded by *TNF* gene is associated with modulating effects on memory and synaptic functions in AD patients and animal models [4, 6]. It was found that the levels of *TNF* in cerebrospinal fluid and serum of AD patients were significantly higher than the levels of healthy adults [7, 8]. The clinical and animal studies have shown excess *TNF* level in AD patients brain can enhance Aβ production and decrease Aβ clearance, leading to neuronal loss and cell death. It follows that *TNF* driving processes is involved in multiple stages of AD pathophysiology and progression [6].

TNF gene (ENSG00000232810), located on chromosome 6p21.231, consists of four exons. Several SNPs in the promoter region of *TNF* have been identified to be associated with AD and play dissimilar roles in different population. For example, a significant increased risk of AD was observed with rs1800629 A allele (OR = 12.36, 95% CI 7.47-20.45, $p < 0.0001$) in Iran [9], while Gnjec et al. revealed the potential protective role of A allele carried at rs1800629 against AD in Australian (OR = 0.75, 95% CI 0.57-0.99, $p = 0.048$) [10]. Gnjec et al. also investigated the relationship between rs1799724 C allele and AD risk (OR = 1.90, 95% CI 1.39-2.59, $p < 0.001$) [10], however, Ma et al. reported rs1799724 C allele plays a neuroprotective role against AD in Chinese (OR = 0.46, 95% CI 0.23-0.90, $p = 0.030$) [11]. *TNF* expression dysregulation could cause chronic inflammation and tissue damage, different functional polymorphisms within genes modulating expression has been demonstrated to alter the disease risk [12]. However, it is still unclear whether there SNPs could influence *TNF* signaling and further change Aβ produce and metabolism.

In this study, we investigated the effects of five AD related *TNF* SNPs on regulating gene expressions in multiple human brain regions and whole blood based on four large-scale eQTL resources, including Brain eQTL Almanac (Braineac) [13], Genotype-Tissue Expression project (GTEx) [14], Blood eQTL data [15] and Brain Expression Genome-Wide Association Study (Brain eGWAS) [16]. In addition, the associations between SNPs and the volumes of different brain regions were investigated by using Enhancing Neuroimaging Genetics through Meta-Analysis (ENIGMA) datasets. Understanding the relationship between SNPs and AD should be possible to develop anti-*TNF* therapeutics that likely help delay or slow AD progression.

2 Material and Methods

2.1 Identifying *TNF* SNPs Associated with AD

We searched all possible studies with keywords including "Alzheimer's disease" or "AD", "tumor necrosis factor" or "*TNF*", and "polymorphism" in PubMed (https://www.ncbi.nlm.nih.gov/pubmed) and AlzGene database (http://www.alzgene.org/) [2]. The literature search was updated on May 15, 2017, and then we selected the studies provides an odd ratio with 95% CI as well as p-value to identify the *TNF* SNPs associated with AD.

2.2 eQTLs Analysis

The *TNF* SNPs identified to be associated with AD were taken as potential candidates for eQTL analysis. We performed *cis*-eQTL analysis on three sources used human brain tissues (Braineac [13], GTEx [14] and eGWAS [16]) and one used blood sample [15], and a brief description of these gene expression datasets is provided below, and more detailed information can be found in the original papers.

Braineac (http://caprica.genetics.kcl.ac.uk/BRAINEAC/) includes genome-wide genotype data and whole transcriptome expression data of 134 neuropathological normal human individuals with European descent. The effects of SNPs on *TNF* expression were determined by eQTL analysis on the expression levels of different genotypes. In the database, the mRNA expression of *TNF* was exhibited at exon-specific and transcript levels in ten different brain regions [13]. The eQTL data search was conducted using the R package Matrix EQTL [17]. All p-values were corrected for multiple comparisons using the FDR with Benjamini-Hochberg procedure, and the associations with FDR < 5% were considered as significant.

GTEx (v6 release) database (https://gtexportal.org/home/) provided *cis*-eQTL analysis for 7,051 samples from 44 tissues of 449 donors which combine genotype data from whole exome and genome sequencing as well as expression data from microarray and RNA sequencing [14, 18]. In this study, we selected ten human brain regions (Table 1), and the eQTL analysis was performed using the R package Matrix EQTL [23]. FDR < 5% was used as significant.

Brain eGWAS database, consisted six brain-expression GWAS datasets, provided expression levels on cerebellum and temporal cortex of autopsied subjects with AD (ncerebellar = 197 and ntemporal cortex = 202) and with other brain pathologies such as Lewy body disease and multiple system atrophy (as non-AD or Control, ncerebellar = 177 and ntemporal cortex = 197) [16]. The combined cerebellar/temporal cortex AD and non-AD subjects could provide complementary information in discovery of variants with functional implications [8]. The comparison among the most relevant regions in different datasets was shown in Table 1.

Blood eQTL dataset reported by Westra et al. was conducted by a large-scale meta-analysis of whole-genome eQTL data from 5,311 peripheral blood samples of 7 cohorts in European population [15]. Blood eQTL dataset is a large-scale meta-analysis and also a supplement for the eQTL analysis on brain regions. We obtained the *cis*-eQTL signals of SNPs by entering gene name (*TNF*) in Query eQTL Results at https://molgenis58.target.rug.nl/bloodeqtlbrowser/.

2.3 Genetic Influences on the Brain Structures

To further identify the function of the *TNF* promoter SNPs on the brain, we investigated the effect of these SNPs on the volumes of brain structures based on the data extracted from Enhancing Neuroimaging Genetics through Meta-Analysis (ENIGMA, http://enigma.ini.usc.edu/) consortium [19]. The ENIGMA2 GWAS (Subcortical), one

Table 1. Brain regions in Braineac, GTEx, eGWAS and ENIGMA databases

Braineac (10 brain regions)	GTEx (10 brain regions)	eGWAS (2 brain regions)	ENIGMA (8 brain regions)
NA	NA	NA	Accumbens (ACCU)
NA	Anterior cingulate cortex (ACC)	NA	NA
NA	NA	NA	Amygdala (AMYG)
NA	Caudate basal ganglia (CAUD)	NA	Caudate
Cerebellar cortex (CRBL)	NA	NA	NA
NA	Cerebellar hemisphere (CEHE)	NA	NA
NA	Cerebellum (CERE)	Cerebellar	NA
NA	Cortex (CORT)	NA	NA
Frontal cortex (FCTX)	Frontal cortex	NA	NA
Hippocampus (HIPP)	Hippocampus	NA	Hippocampus
NA	Hypothalamus	NA	NA
NA	NA	NA	Intracranial volume (ICV)
NA	Nucleus accumbens basal ganglia (NABG)	NA	NA
Medulla (MEDU)	NA	NA	NA
Occipital cortex (OCTX)	NA	NA	NA
NA	NA	NA	Pallidum (PALL)
Putamen (PUTM)	Putamen	NA	Putamen
Substantia nigra (SNIG)	NA	NA	NA
Temporal cortex (TCTX)	NA	Temporal cortex	NA
Thalamus (THAL)	NA	NA	Thalamus
White matte (WHMT)	NA	NA	NA

of the ENIGMA working groups, included 50 cohorts with 30,717 samples [20]. There are seven genome-wide association studies' data sets about the volumes of seven subcortical regions (thalamus, caudate nucleus, putamen, pallidum, hippocampus, amygdala, nucleus accumbens) and intracranial volume. The associations with FDR < 5% were considered as significant.

3 Results and Discussion

3.1 *TNF* Promoter Polymorphisms Associated with AD

There are nine SNPs (rs1799964, rs1800630, rs1799724, rs4248158, rs4248161, rs1800750, rs1800629, rs361525 and rs4645838) within *TNF* promoter region. In human genetic studies, evidence shows that genetic variants could modify gene expression and cause disease risk [23]. We obtained 35 articles concerning the correlation of *TNF* polymorphisms with AD based on the criteria mentioned on Methods Section. For the further analysis, we selected the most common five *TNF* promoter SNPs associated with AD in European population. The information of these SNPs was shown in Table 2 and Fig. 1.

Table 2. Basic information of five AD associated promoter SNPs in *TNF*

SNP	Chr:Pos[a]	A_{12}^{b}	MAF[c]	OR[d] [95% CI]	p-value	Ref.
rs361525	chr6:31543101	G/A	0.13	1.64 [0.79, 3.39]	2.30E−01	[21]
rs1800629	chr6:31543031	G/A	0.14	0.75 [0.57, 0.99]	4.80E−02	[10]
			0.11	1.23 [1.01, 1.49]	3.70E−02	[22]
rs1799724	chr6:31542482	C/T	0.08	0.46 [0.23, 0.90]	3.00E−02	[11]
rs1800630	chr6:31542476	C/A	0.15	0.24 [0.12, 0.50]	3.10E−02	[11]
rs1799964	chr6: 31542308	T/C	0.22	4.62 [2.22, 9.57]	< 1.00E−03	[11]

[a] based on GRCh37; [b] Reference and alternate alleles; [c] Minor allele frequency; [d] Odd ratios.

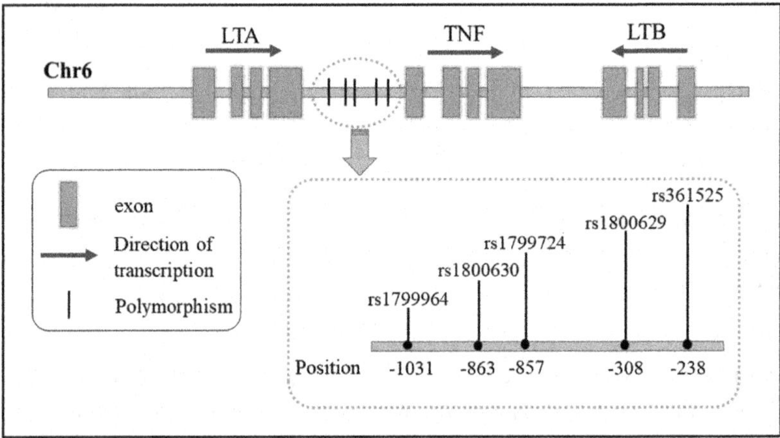

Fig. 1. Representation of *TNF* and its AD associated promoter SNPs with their positions. *TNF* gene, flanked by lymphotoxin α (*LTA*) and β (*LTB*) genes, consists of four exons.

3.2 eQTL Analysis of the Effect on *TNF* Expression

As a neurodegenerative disease, the pathological changes of AD patients mainly locate on brain regions. Therefore, it is imperative to use disease-relevant tissues in different brain regions to investigate the effects of SNPs on gene expression. From the expression levels in human brain (Fig. 2), we noted that *TNF* was uniquely expressed in difference brain regions. For example, at the transcript level (transcript ID: t2902416), the highest expression appears in MEDU and the lowest expression in CRBL. The fold change between MEDU and CRBL was 1.25 (p-value = 7.47E−12). At exon-specific levels for different probes, the *TNF* expression levels in each brain region were different for each other (data not shown).

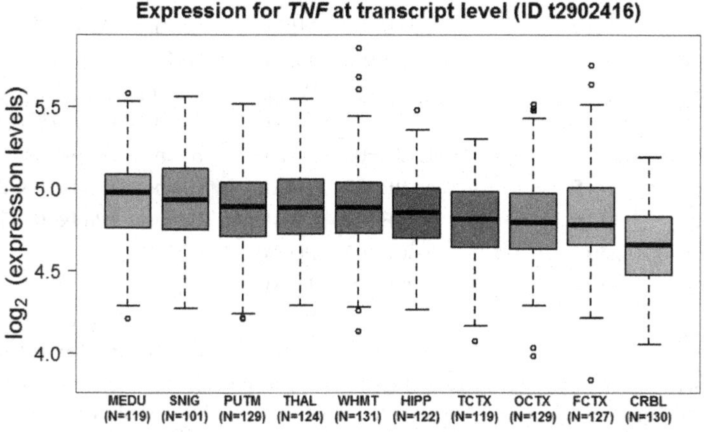

Fig. 2. Expression levels of *TNF* in different brain regions for transcript probe ID t2902416. CRBL, cerebellar cortex; FCTX, frontal cortex; HIPP, hippocampus; MEDU, medulla; OCTX, occipital cortex; PUTM, putamen; SNIG, substantia nigra; TCTX, temporal cortex; THAL, thalamus; WHMT, white matter. Number in parentheses indicates the sample number. The TNF expression data is from Braineac database.

In addition, the SNPs with eQTL signals in multiple human brain regions and whole blood were exhibited in Table 3 (unsignificant data with p-value > 0.05 not shown). It was worth to note that the *TNF* region-specific eQTL signals of the five promoter SNPs for difference probes from three data sources. From the data shown in Table 3, we identified that these five SNPs could significantly regulate reduced *TNF* expression in two tissues in the GTEx dataset, seven tissues in Braineac dataset and blood eQTL dataset. Among these eQTL signals, except rs1799964 C allele in PUTM from GTEx with $\beta > 0$, the other SNPs with alternative allele were all with $\beta < 0$. As regression coefficient based on the alternative allele, $\beta > 0$ and $\beta < 0$ mean that the alternative allele regulates increased and reduced gene expression, respectively [23].

rs1799964 C allele significantly regulated reduced *TNF* expression in different brain regions (FCTX, THAL and CRBL) and whole blood sample. As an exception, we found that the rs1799964 C allele could significantly regulate increased *TNF* expression

Table 3. SNPs and *TNF* expression in eQTL datasets (p value < 0.05)

SNP	Alt[a]	β[b]	p-value[c]	Tissues[d]	Data source
rs361525	A	−0.13	3.39E−02	THAL	Braineac (Probe ID: t2902416)
		−0.26	2.00E−03	THAL	Braineac (Probe ID: 2902418)
		−0.22	4.77E−02	THAL	Braineac (Probe ID: 2902421)
rs1800629	A	−0.19	4.64E−02	TCTX	Braineac (Probe ID: 2902423)
		<0 (z-score[e] = −5.28)	1.28E−07	Whole blood	Westra et al. 2013 [15] (Probe ID: 26403016)
rs1799724	T	−0.26	4.20E−03	FCTX	Braineac (Probe ID: 2902417)
		−0.17	4.78E−02	PUTM	Braineac (Probe ID: 2902417)
		−0.19	1.12E−02	WHMT	Braineac (Probe ID: 2902418)
rs1800630	A	−0.20	8.91E−03	CRBL	Braineac (Probe ID: 2902420)
		−0.20	1.37E−02	MEDU	Braineac (Probe ID: 2902420)
		−0.53	1.00E−02	THAL	GTEx
		<0 (z-score = −9.44)	3.71E−21	Whole blood	Westra et al. 2013 [15] (Probe ID: 26403016)
rs1799964	C	−0.12	1.77E−02	FCTX	Braineac (Probe ID: 2902418)
		−0.15	4.19E−03	THAL	Braineac (Probe ID: 2902418)
		−0.14	3.13E−02	CRBL	Braineac (Probe ID: 2902420)
		0.39	3.40E−02	PUTM	GTEx
		−0.41	2.90E−02	THAL	GTEx
		<0 (z-score = −9.59)	9.02E−22	Whole blood	Westra et al. 2013 [15] (Probe ID: 26403016)

[a] Alt, alternative allele; [b] β, regression coefficient based on overall estimated effect for alternative allele; [c]Significance level is 0.05; [d]Tissues: the abbreviations were listed in legend of Fig. 1; [e]z-score = β/SE, SE is the standard error for β estimate.

only in PUTM in GTEx datasets. One possible reason is the small sample size (81 samples) inducing this error. A choice of small sample sizes, though sometimes necessary, can result in wide confidence intervals or risks of errors in statistical hypothesis testing. And increasing sample size would lead to different result. However, as a whole, rs1799964 C allele might be protective against AD risk.

The SNPs in the 5'-flanking region of the *TNF* (such as rs361525 and rs1800629) have been reported to affect the *TNF* expression. rs361525 has been approved to associated with decreased transcriptional activity [24]. In this work, based on eQTL analysis, we discovered that rs361525 with A allele was significantly associated with reducing the mRNA expression in THAL at both transcript and exon-specific levels, and the result is consisted with previous studies [24]. Based on meta-analysis, Tengfei Wang observed a significant association between rs1800629 A allele and AD risk in Chinese, but a significantly protective effect in the occurrence of AD among European populations [24]. However, it is unclear whether rs1800629 could affect *TNF* signaling and further change its function in AD for European. In addition, by analyses using datasets from Braineac, blood eQTL and GTEx, we found in European population rs1800629 A allele significantly reduced *TNF* expression level in TCTX (Fig. 3A) and

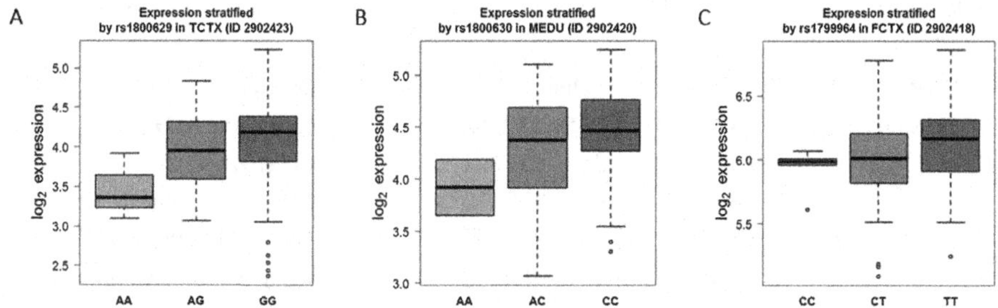

Fig. 3. The significant associations between SNPs and *TNF* expression in human brain regions at exon-specific level. (A) *TNF* expression affected by rs1800629 (probe ID 2902423) in TCTX (p-value = 4.64E−02). (B) *TNF* expression affected by rs1800630 (probe ID 2902420) in MEDU (p-value = 1.37E−02). (C) *TNF* expression affected by rs1799964 (probe ID 2902418) in FCTX (p-value = 1.77E−02). The expression data was extracted from Braineac.

whole blood. The outcome could be the explication for the protective effect on reducing the AD risk in European ethnicity. Reversely, the rs1800629 A allele might play a role on heightening *TNF* expression for Chinese raising the AD risk.

Other SNPs rs1799724, rs1800630 and rs1799964 were also proved to regulate the transcriptional activity of *TNF* and associated with different AD risk in different population [25]. Based on the eQTL analysis using Braineac, GTEx and blood eQTL data, we discovered that rs1799724 T allele, rs1800630 A allele and rs1799964 C allele regulated reduced the *TNF* expression in brain regions and whole blood (Fig. 3B and C). It is possible that these SNPs with alternative allele play roles on protecting in European population for AD risk, different from other population.

3.3 *TNF* Expression in GWAS Datasets

To further validate the effect of promoter SNPs on *TNF* expression, we analyzed the eQTL data on brain eGWAS source with six brain-expression GWAS datasets ('AD', 'Control' and 'All' for cerebellar and temporal cortex, respectively) [16]. Based on the eGWAS datasets, it was discovered that rs1800639 with A allele significantly regulated *TNF* expression for AD and combined samples (p-value < 0.01) (Table 4). For Cerebellar AD and Combined datasets, rs1800629 A allele significantly reduced *TNF* expression with $\beta = -4.88\text{E}{-02}$ (p-value = 6.21E−03) and $\beta = -3.59\text{E}{-02}$ (p-value = 4.25E−03), respectively. It was consistent with the result of eQTL analysis by using Braineac and blood datasets. In addition, among these brain eGWAS datasets, there was none significant association between SNPs rs1800630 A allele, rs1799964 C allele and *TNF* expression, and none expression data for rs361525 A allele and rs1799724 T allele.

Table 4. Five SNPs regulated *TNF* expression in brain expression GWAS datasets of AD and control

SNP	Alt	β	p value	Sample number	Dataset
rs1800629	**A**	***−4.88E−02***	***6.21E−03***	**188**	**Cerebellar AD**
		−3.59E−02	***4.25E−03***	**361**	**Cerebellar All[a]**
		−1.89E−02	2.94E−01	173	Cerebellar Control[b]
		−2.311E−02	5.80E−01	195	Temporal cortex AD
		−4.38E−02	2.00E−01	379	Temporal cortex All
		−3.79E−02	4.78E−01	184	Temporal cortex Control
rs1800630	A	−7.39E−03	7.00E−01	190	Cerebellar AD
		5.48E−03	6.85E−01	366	Cerebellar All
		2.40E−02	2.13E−01	176	Cerebellar Control
		−3.85E−02	3.93E−01	196	Temporal cortex AD
		1.21E−03	9.75E−01	383	Temporal cortex All
		2.26E−02	7.10E−01	187	Temporal cortex Control
rs1799964	C	1.18E−02	4.77E−01	189	Cerebellar AD
		1.16E−02	3.23E−01	364	Cerebellar All
		1.34E−02	4.37E−01	175	Cerebellar Control
		−3.63E−02	3.52E−01	195	Temporal cortex AD
		−2.49E−03	9.40E−01	382	Temporal cortex All
		1.32E−02	8.08E−01	187	Temporal cortex Control

[a] All the combined AD and Control samples; [b] Control subjects with non-Alzheimer's disease pathology.

3.4 Volumes Analysis of Subcortical Brain Structures

Beside the eQTL analysis, we also investigated the effect of SNPs variants on the structures of subcortical and intracranial brain regions. Using eight genome-wide association studies' datasets about the volumes of seven subcortical regions and intracranial volume (ICV) in ENIGMA, we found that rs1800630 with A allele and rs1799964 C allele significantly affected the structure of brain amygdala with $\beta = 7.69$ (p-value = 1.76E−02) and $\beta = 9.08$ (p-value = 1.86E−03), respectively (Fig. 4 and Table 5). The two SNPs with minor alleles elevating the volume of brain amygdala to inhibit AD development supported the protection effect of rs1800630 A allele and rs1799964 C allele on AD risk, and the result was consistent with the eQTL analysis. The other three SNPs did not significantly influence the brain structures.

Interestingly, we discovered that rs4248161 with C allele significantly reduced the volumes of five subcortical regions (accumberns, amygdala, hippocampus, putamen and thalamus) with $\beta < 0$ and p-value < 0.05 (Table 5). Based on literature search with keyword "rs4248161", there are only a few reports about the association between rs4248161 and pulmonary tuberculosis [26], postmenopausal osteoporosis [27], but no

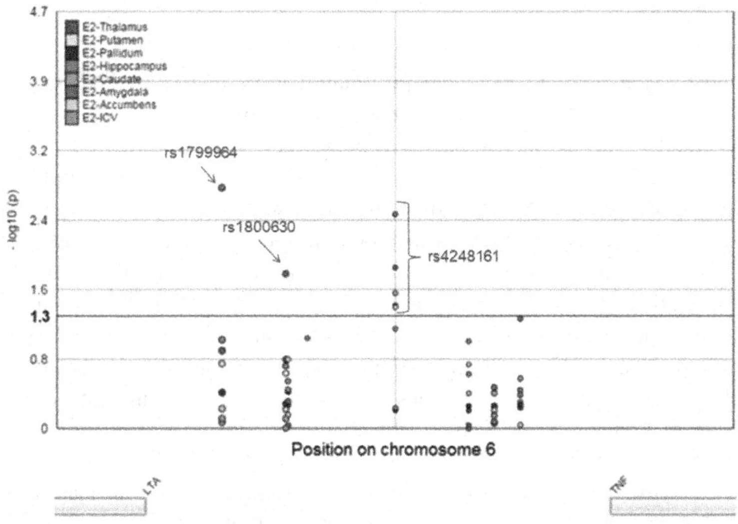

Fig. 4. Regional association plot of locus 6p21.33 associated with subcortical volumes and intracranial volumes. The circles above the line 1.3 (p values < 0.05) were considered statistically significant. ICV, intracranial volume. The circle represents SNP. The data were got from ENIGMA.

Table 5. The influences of *TNF* SNPs on the volumes of subcortical brain regions (p value < 0.05)

SNP	Alt[a]	Brain regions[b]	β[c]	SE[d]	p-value	Sample number
rs1800630	A	AMYG	7.69	3.24	1.76E−02	13,160
rs1799964	C	AMYG	9.08	2.92	1.86E−03	13,160
rs4248161	C	ACCU	−13.91	6.38	2.83E−02	13,112
		AMYG	−34.33	14.12	1.50E−02	13,160
		HIPP	−78.41	27.02	3.71E−03	13,163
		PUTA	−68.44	33.67	4.21E−02	13,145
		THAL	−74.22	36.20	4.03E−02	13,193

[a] Alt, alternative alleles; [b] Brain region: AMYG, amygdala; ACCU, accumberns; HIPP, hippocampus; PUTA, putamen; THAL, thalamus; [c] β, regression coefficient; [d] SE standard error.

reports for relationship between rs4248161 and neurogenic disease. We prognosticated that rs4248161 with C allele could be an important variant for AD or other neurogenic diseases on account of the significant association with affecting brain structures. Of course, it needs a lot of biological experiments and medical data to authenticate the prediction.

4 Conclusion

As the most extensively investigated proinflammatory cytokine, TNF expression is associated with modulating effects on memory and synaptic functions in AD. In this study, the effect of AD associated five *TNF* promoter SNPs on the gene expression was explored by integrating four different eQTL datasets in European population. We discovered these promoter SNPs with minor allele significantly reduced *TNF* mRNA expression on different brain regions or blood sample and could play protective roles for AD risk. Consistent with eQTL analysis, we found rs1800630 A allele and rs1799964 C allele were significantly associated with increasing the volumes of brain amygdala by using the ENIGMA database. In addition, rs424161 C allele significantly reduced the volumes of five brain regions and was potential risk for AD or other neuropathogenic diseases. Our studies could advance the understanding of the involvement of *TNF* promoter SNPs mechanisms in AD.

Acknowledgement. The authors thank the members of our laboratory for their valuable discussions. This work has been supported by the grants from the National Natural Science Foundation of China (61672037 and 21601001), the Initial Foundation of Doctoral Scientific Research in Anhui University (J01001319) and the Initial Foundation of Postdoctoral Scientific Research in Anhui University (J01002047).

References

1. Hardy, J., Selkoe, D.J.: The amyloid hypothesis of Alzheimer's disease: progress and problems on the road to therapeutics. Science **297**(5580), 353–356 (2002)
2. Bertram, L., et al.: Systematic meta-analyses of Alzheimer disease genetic association studies: the AlzGene database. Nat. Genet. **39**(1), 17–23 (2007)
3. Lambert, J.C., et al.: Genome-wide association study identifies variants at CLU and CR1 associated with Alzheimer's disease. Nat. Genet. **41**(10), 1088–1093 (2009)
4. Rezazadeh, M., et al.: Genetic factors affecting late-onset Alzheimer's disease susceptibility. NeuroMol. Med. **18**(1), 37–49 (2016)
5. Bagyinszky, E., et al.: Role of inflammatory molecules in the Alzheimer's disease progression and diagnosis. J. Neurol. Sci. **376**, 242–254 (2017)
6. Chang, R., Yee, K.L., Sumbria, R.K.: Tumor necrosis factor alpha inhibition for Alzheimer's disease. J. Cent. Nerv. Syst. Dis. **9**, 1–5 (2017)
7. Tarkowski, E., et al.: TNF gene polymorphism and its relation to intracerebral production of TNFalpha and TNFbeta in AD. Neurology **54**(11), 2077–2081 (2000)
8. Tarkowski, E., et al.: Intrathecal inflammation precedes development of Alzheimer's disease. J. Neurol Neurosur. Ps. **74**(9), 1200–1205 (2003)
9. Ardebili, S.M., et al.: Genetic association of TNF-alpha-308 G/A and -863 C/A polymorphisms with late onset Alzheimer's disease in Azeri Turk population of Iran. J. Res. Med. Sci. **16**(8), 1006–1013 (2011)
10. Gnjec, A., et al.: Association of alleles carried at TNFA -850 and BAT1 -22 with Alzheimer's disease. J. Neuroinflamm. **5**, 36 (2008)
11. Ma, S.L., et al.: Association between tumor necrosis factor-alpha promoter polymorphism and Alzheimer's disease. Neurology **62**(2), 307–309 (2004)

12. Kollias, G.: TNF pathophysiology in murine models of chronic inflammation and autoimmunity. Semin. Arthritis Rheum. **34**(5 Suppl1), 3–6 (2005)
13. Ramasamy, A., et al.: Genetic variability in the regulation of gene expression in ten regions of the human brain. Nat. Neurosci. **17**(10), 1418–1428 (2014)
14. Consortium, G.: The Genotype-Tissue Expression (GTEx) pilot analysis: multitissue gene regulation in humans. Science **348**(6235), 648–660 (2015)
15. Westra, H.J., et al.: Systematic identification of trans eQTLs as putative drivers of known disease associations. Nat. Genet. **45**(10), 1238–1243 (2013)
16. Zou, F., et al.: Brain expression genome-wide association study (eGWAS) identifies human disease-associated variants. PLoS Genet. **8**(6), e1002707 (2012)
17. Shabalin, A.A.: Matrix eQTL: ultra fast eQTL analysis via large matrix operations. Bioinformatics **28**(10), 1353–1358 (2012)
18. Murthy, M.N., et al.: Increased brain expression of GPNMB is associated with genome wide significant risk for Parkinson's disease on chromosome 7p15.3. Neurogenetics **18**(3), 121–133 (2017)
19. Thompson, P.M., et al.: ENIGMA and the individual: predicting factors that affect the brain in 35 countries worldwide. Neuroimage **145**(Pt B), 389–408 (2017)
20. Adams, H.H.H., et al.: Novel genetic loci underlying human intracranial volume identified through genome-wide association. Nat. Neurosci. **19**(12), 1569–1582 (2016)
21. Culpan, D., et al.: Tumour necrosis factor-alpha gene polymorphisms and Alzheimer's disease. Neurosci. Lett. **350**(1), 61–65 (2003)
22. Flex, A., et al.: Effect of proinflammatory gene polymorphisms on the risk of Alzheimer's disease. Neurodegener Dis. **13**(4), 230–236 (2014)
23. Liu, G., et al.: Genetic variant rs763361 regulates multiple sclerosis CD226 gene expression. Proc. Natl. Acad. Sci. U.S.A. **114**(6), E906–e907 (2017)
24. Wang, T.: TNF-alpha G308A polymorphism and the susceptibility to Alzheimer's disease: an updated meta-analysis. Arch. Med. Res. **46**(1), 24–30 (2015)
25. Bona, D.D., et al.: Systematic review by meta-analyses on the possible role of TNF-α polymorphisms in association with Alzheimer's disease. Brain Res. Rev. **61**(2), 60–68 (2009)
26. Qidwai, T., Jamal, F., Khan, M.Y.: DNA sequence variation and regulation of genes involved in pathogenesis of pulmonary tuberculosis. Scand. J. Immunol. **75**(6), 568–587 (2012)
27. Jin, X., Zhou, B., Zhang, D.: Replication study confirms the association of the common rs1800629 variant of the TNF gene with postmenopausal osteoporosis susceptibility in the han chinese population. Genet. Test Mol. Biomark. **22**(4), 246–251 (2018)

Verifying TCM Syndrome Hypothesis Based on Improved Latent Tree Model

Nian Zhou[1,2,3], Lingshan Zhou[4], Lili Peng[1,2,3], Bing Wang[1,2,3](✉), Peng Chen[5], and Jun Zhang[6]

[1] School of Electronics and Information Engineering, Tongji University, Shanghai, China
zhounian@tongji.edu.cn, wangbing@ustc.edu
[2] The Advanced Research Institute of Intelligent Sensing Network, Tongji University, Shanghai, China
[3] The Key Laboratory of Embedded System and Service Computing, Tongji University, Shanghai, China
[4] Neurology Department, Jinzhou Medical University, Shenyang, Liaoning, China
ls_zhou17@163.com
[5] Institute of Physical Science and Information Technology, Anhui University, Hefei 230601, China
[6] School of Electronic Engineering and Automation, Anhui University, Hefei 230601, China

Abstract. Traditional Chinese Medicine (TCM) is a significant channel for the prevention and treatment of Chinese diseases and is increasingly popular among non-Chinese people. However, it suffered serious credibility problems. The fundamental question is that TCM syndrome differentiation is it a totally subjective question or is it based on evidence? In recent years, a method called latent tree analysis (LTA) has been put forward. The main idea is, based on statistical principles for cluster analysis of the epidemiological survey symptoms data, to discover latent variables implicated in the data and compare them with TCM syndromes. However, LTA has its own limitations. It states that one manifest variable in the latent tree model (LTM) can only correspond to one latent variable. This is inconsistent with the theory of traditional Chinese medicine. Therefore, this paper proposed an improved LTA, based on the LTM obtained from the original LTA, adding arrows between symptoms and syndromes. The current analysis used the improved LTA to study a dataset of 37,624 patients with hepatopathy. The latent variables found here well match the latent factors of TCM, in addition, there are also some symptoms associated with multiple syndromes, it not only provides evidence for the validity of the relevant TCM hypothesis in the case of hepatopathy and helps to classify these patients into TCM syndromes, but also proved that the improved LTM has a higher degree fitting to the original data.

Keywords: Latent tree model · Syndrome differentiation · TCM

D.-S. Huang et al. (Eds.): ICIC 2018, LNCS 10955, pp. 460–469, 2018.
https://doi.org/10.1007/978-3-319-95933-7_55

1 Introduction

Traditional Chinese Medicine is a significant channel for the prevention and treatment of Chinese diseases and is increasingly popular among non-Chinese people. However, it suffered serious credibility problems, especially in the West. One of the reasons is the lack of rigorous randomized trials to support the efficacy of TCM treatment [1]. Another equally important reason for this article is the lack of validation of TCM theory.

The diagnosis of TCM begins with a comprehensive observation of symptoms (including signs) using four diagnostic methods: examination, hearing, questioning and palpation. Based on the collected information, patients are classified as various types of TCM vocabulary collectively referred to as "Zheng" [2]. "Zheng" is usually translated as "Traditional Chinese Medicine Syndrome." The process of classifying patients into various syndromes is called TCM syndrome differentiation.

TCM syndrome such as Yang deficiency, Yin deficiency is the hypothesis of traditional Chinese medicine to explain the occurrence and co-occurrence of signs and symptoms. For example, TCM believes that Yang and Yin are essential substances for the human body. They have the function of warming and nourishing the body [3]. Deficiency of yang can lead to cold performance, such as chilling. Therefore, patients with these symptoms are often classified as Yang class. Similarly, deficiency of Yin may lead to dry mouth, throat fever, fever of hands and feet, and other symptoms [4]. Therefore, patients with these symptoms are often divided into Yin deficiency and Yang deficiency class.

Western medicine divides patients into different categories according to disease types or subtypes and treats them accordingly. Instead, TCM classifies patients into different categories based on the type of symptoms and treats them accordingly. The difference between Chinese and Western medicine is that western medicine is treated by the type of disease, and TCM is treated by co-occurrence of symptoms [5–7].

Two fundamental questions about TCM syndromes are usually raised: (1) Do they correspond to real-world entities, or are they purely subjective? (2) TCM Syndrome Differentiation is it a totally subjective question or is it based on evidence? For more than half a century, researchers have been seeking answers to these questions through laboratory experiments [5, 6] However, the question remains unanswered [7–9].

Recently, a different method has been proposed [10, 11]. It distinguishes between two variables of traditional Chinese medicine. Clinically, symptoms such as "cold", "dry mouth" and "throat dryness" can be directly observed and are therefore referred to as observation variables. On the other hand, the syndrome factors such as Yang deficiency and Yin deficiency cannot be directly observed and must be determined indirectly according to the symptoms. Therefore, they are latent factors.

Zhang et al. speculated that certain conceptions of syndromes such as Yang deficiency and Yin deficiency originated from the observed regularity of symptoms and co-occurrence in ancient times. They proposed a new TCM syndrome research method, in which the researchers collected data on the patient's symptoms and ruled out the doctor's diagnosis, and re-extracted, from the unlabeled data collected, the latent factors postulated in TCM. Since the purpose of this method is to provide objective evidence for the diagnosis of TCM, the diagnostic results will not be collected in the first step. The second step is accomplished by using a new type of probability model called the Latent Tree Model, which was developed specifically for the study of TCM syndromes [10]. This method is called latent tree analysis (LTA).

They have tested the LTA method on the kidney deficiency data set. The latent variables they found were in good agreement with the relevant latent factors of TCM. (Note that in this article, the term "latent factor" refers to a factor that is not observed in TCM, and "latent variable" refers to a variable that is not observed in the statistical model.) This provides statistical validation of the relevant TCM hypothesis. However, this latent tree analysis has its own limitations. It states that one observed variable in the latent tree model can only correspond to one latent variable. This is inconsistent with the theory of traditional Chinese medicine. According to traditional Chinese medicine theory, a symptom may be caused by multiple Syndromes, that is, one symptom can be connected to multiple syndromes. Therefore, based on the model of LTM, by increasing the connection between symptoms and syndromes, the arrow-adding (LTM-AA) operation is performed to improve the fitting degree between the model and the data. Experiments have proved that the improved latent tree model has a higher degree fitting to the original data.

The current analysis used the improved LTA to study a dataset of 37,624 patients with hepatopathy. The latent variables found here well match the latent factors of TCM. This provides evidence for the validity of the relevant Chinese medicine hypothesis in the case of hepatopathy and helps to classify these patients into TCM syndromes.

2 Improved Latent Tree Model

2.1 Review Latent Tree Model

The latent tree model is the simplest latent structural model. They were previously called "hierarchical latent class models." [12] The "latent tree model" is a tree-structured Bayesian network [13–15] in which the leaf node variables are observed and formed as " manifest variables," whereas variables at internal nodes are latent and hence are called "latent variables." All variables are assumed to be discrete. Arrows indicate direct probability dependence. In the model shown on the left side of Fig. 1, there is an arrow from variable Y1 to variable Y2. This means that Y2 is directly dependent on Y1. The dependence is characterized by a conditional distribution P (Y2| Y1) which gives the Y2 distribution of each value of Y1.

As an example, consider high school students who need several subjects (such as math, science, literal, and history). Grades are influenced by latent factor analysis skills and literal skills, which in turn are affected by general intelligence. These relationships form a latent tree structure, as shown on the right side of Fig. 1.

The interest in this article is how to induce a latent tree model from the data. This can be divided into two sub-problems. First, which one is the best among all possible models? This is a matter of model selection. Zhang [12] empirically examined several criteria, namely, Bayesian Information Criterion (BIC) score [16], Akaike Information Criterion (AIC) score [17], Cheeseman-Stutz score [18], and holdout-likelihood [19].

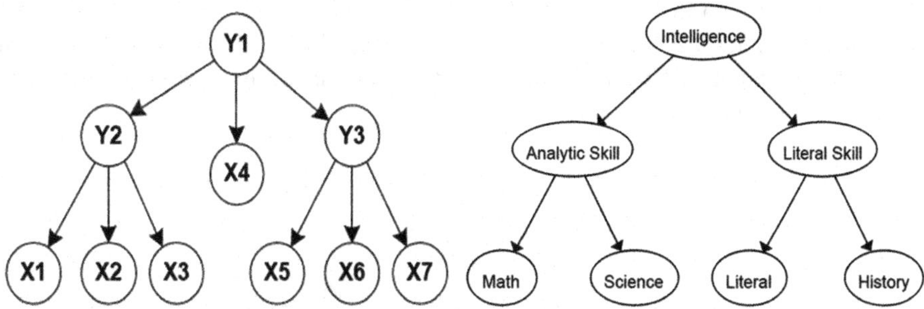

Fig. 1. Left: A latent tree model. The Xi's are manifest variables and the Yi's are latent variables. Right: An example of latent tree model.

Bayesian Information Criteria (BIC) score turns out to be the most appropriate for the task. The BIC score for the latent tree model is given by formula (1):

$$\mathrm{BIC(m|D)} = \log(D|m, \theta^*) - \frac{d(m)}{2} logN \tag{1}$$

Where D is the data set, m is the structure of the model, θ^* is the maximum likelihood estimate of the model parameters, d(m) is the number of free parameters, and N is the sample size. Please note that this definition of BIC score is used in the machine learning community rather than social science researchers who often use negatives.

The second sub-question is how to find the model with the highest BIC score in all possible model spaces. The first algorithm for this task was DHC [13]. It can handle data sets with only about six manifest variables. As for it, it is a concept test algorithm. The second algorithm is called SHC [20]. It can handle about a dozen manifest variables. It is a springboard for a more efficient EAST (Expansion Adjustment Simplification until Termination) [21] algorithm that can handle about 100 manifest variables. EAST has been tested on synthetic data and real-world datasets through market research [22] and from a social survey. It finds interesting latent structures in all situations.

2.2 The Arrow-Adding Algorithm

Based on the latent tree model obtained by the EAST algorithm, considering that one symptom is often associated with multiple syndromes in the TCM theory, this paper uses the arrow-adding (LTM-AA) to improve the original latent tree model, Improvement is made to enable one manifest variable to correspond to multiple latent variables, that is, one symptom corresponds to multiple syndromes. However, considering the complexity of the improved algorithm, this paper considers one syndrome can only add arrow once.

The LTM-AA algorithm is another search process under the guidance of the BIC score, adding one edge at each step. For example, for the latent tree model shown in left of Fig. 1, taking the latent variable Y1 as an example, the AA algorithm is connected to X1, X2, X3, X5, X6, X7 via Y1, respectively. Then there are six candidate models, comparing the BIC scores of the six candidate models and the initial model. and then select the model with the largest BIC score as the initial model for the next step, the other latent variables Y2, Y3 search process is the same. In the end, the model shown in Fig. 2 may be obtained. This model is no longer a tree and is a more complex Bayesian network than the latent tree model.

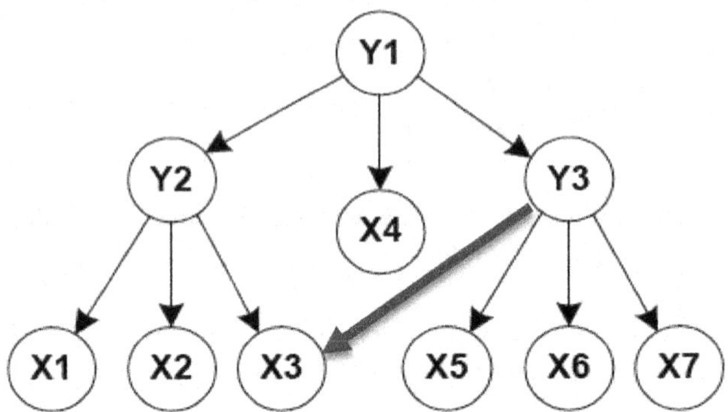

Fig. 2. The improved latent tree model which obtain by adding arrow algorithm, and the red arrow is the added edge. (Color figure online)

The LTM-AA algorithm is given as follows:

LTM-AA (m, θ^*, D):
Repeat all latent variables:
  Repeat all manifest variables:
    (m_1, θ_1^*) ← Add arrow from latent variable to manifest variable.
    If BIC (m, θ^*) ≤ BIC (m_1, θ_1^*),
      (m, θ^*) ← (m_1, θ_1^*).
  Return (m, θ^*).

3 Analysis of TCM Data Set

3.1 Data Collection

The dataset was collected in 2010–2014. It contains 37624 patient cases and 150 symptoms. These symptom variables were selected according to the Chinese National TCM Clinical Terminology Criteria [23] and some TCM diagnostic textbooks [24]. These variables are the most important factors for Chinese medicine doctors in determining whether a patient has hepatopathy. Therefore, we call this dataset hepatopathy data. Each symptom variable has four possible values, namely "no", "light", "medium" and "serious". Operational criteria [25] are defined when collecting data to determine the consistency of the severity of the symptoms.

The data was from a Shanghai hospital and all subjects were inpatients or outpatients who had suffered from hepatopathy. Therefore, all the conclusions drawn are about this group.

3.2 Data Processing

Since the data sample is too large and the data itself is sparse, when the symptoms' frequency is less than 5%, the symptoms were removed from the data set, and there are remaining 135 symptoms. However, there are still too many symptoms, and the improved latent tree analysis can't handle such a big data, so we also removed some other symptoms which can be considered irrelevant symptoms. We removed them by measuring the similarity through Euclidean distance between the symptoms. After that, there are 55 symptoms contained.

In order to decrease the algorithm runtime, there are only 11854 patient cases chosen and the removed cases are which contains too few symptoms. Finally, 11854 patient cases and 55 symptoms included in further analysis.

3.3 The Result of LTA

We used the improved EAST algorithm to analyze hepatopathy data. The result of the analysis is a latent tree model that will be referred to as a hepatopathy model. The best model's BIC Score is -248106 and the structure of the model is shown in Fig. 3.

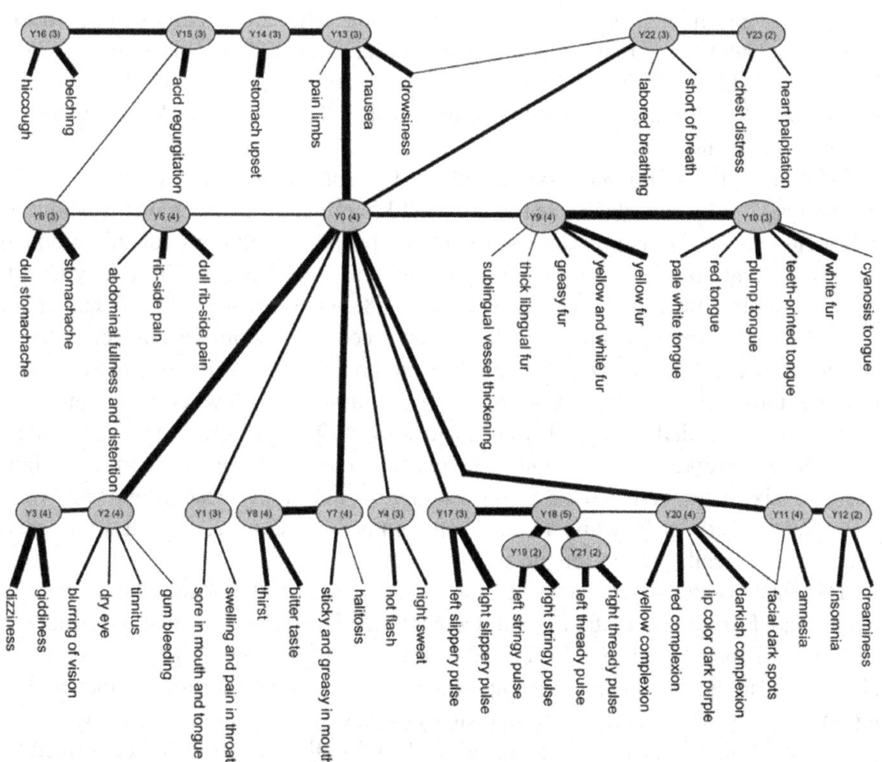

Fig. 3. The structure of the best model M found for hepatopathy data.

In the model, nodes marked with English phrases represent symptom variables. Each of them has four possible values that represent the degree of symptoms. The symptom variable comes from the data set. Nodes marked with a capital letter 'Y' and an integer index are latent variables. They are not from data sets. Instead, they are introduced in the data analysis process to explain the patterns in the data. Each latent variable has an integer next to it. This is the number of possible values for the latent variable.

The edges in the model represent probability dependencies. Each edge is characterized by a conditional probability distribution. The width of the edge indicates the strength of the correlation between the variables. For example, Y5 is strongly related to "rib-side pain", related to "dull rib-side pain", and weakly related to "abdominal fullness and distention." In this article, we will focus on the links between variables and the advantages of these links. The conditional probability distribution contains quantitative information that can be used as a syndrome differentiation. We will discuss them in future work.

4 Discussions

4.1 Latent Variables as Evidence for TCM Hypothesis

In the latent tree model, the set of manifest variables that are directly connected to a particularly latent variable is called a sibling cluster. The siblings cluster together with the latent variables constitute a family. For example, two symptom variables under Y3, "dizziness" and "giddiness" form a sibling cluster. Together with Y3, they form a family and are said to be headed by Y3.

Why are some symptom variables grouped to form sibling clusters during latent tree analysis? Why introduce latent variables? Examination of the model (qualitative and quantitative information) shows that there are three conditions for this problem. First, some symptom variables are grouped into a sibling cluster because they tend to occur at the same time. In the family headed by Y8, one example is "thirst" and " bitter taste." In this case, the latent variable is introduced to explain the co-occurrence of symptoms. Second, some symptom variables are grouped into a sibling cluster because they are mutually exclusive. One of the examples is "yellow complexion", "red complexion" and "darkish complexion" headed by Y20. In this case, the latent variable is introduced to represent a partition of the patients based on those symptoms. The third case is a mix of the first two cases. An example is the family headed by Y10, where "red tongue" and "pale white tongue" are mutually exclusive, and they all occur together with "teeth-printed tongue".

The latent variable in the first case is evidence of the validity of the TCM hypothesis. For example, traditional Chinese medicine believes that syndrome of deficiency of liver blood may lead to "dizziness" and "giddiness." The implication of this hypothesis is that these two symptoms tend to co-occur in clinical practice. The introduction of Y3 in the latent tree analysis has confirmed that "dizziness" and "giddiness" really do tend to co-occur in the data. In other words, it has confirmed the meaning of this hypothesis. In this sense, it provides support for the assumption of traditional Chinese medicine.

4.2 Evidence for the Validity of TCM Hypothesis

Starting from the lower center of the model M, we noticed that 'hot flash' and 'night sweat' are grouped under Y4. This and related quantitative information suggest that these two symptoms tend to occur together in the data. On the other hand, the co-occurrence of these two symptoms is the hypothesis that fire excess from yin deficiency will lead to "hot flash" and "night sweat". So, Y4 variable has confirmed this meaning and therefore proves the validity of this hypothesis.

The family headed by Y13 showed three symptoms of "pain limbs", "nausea" and "drowsiness" tended to occur in the data at the same time. The evidence to support this hypothesis is that phlegm-heat blocking orifices may cause "pain limbs", "nausea" and "drowsiness".

The family headed by Y23 indicated that there was "chest distress" and "heart palpitation" co-occurrence in the data. The evidence supporting this hypothesis is that stagnant blockade of heart blood may lead to "chest distress" and "heart palpitation". At the same time, family headed by Y22 show that in the data, together with "chest distress" and "heart palpitations", there are also symptoms of " labored breathing ", "short of breath" and "drowsiness" occurred together. This is evidence supporting the hypothesis that chest disorder of qi movement may lead to "chest distress", "heart palpitation", "labored breathing", "short of breath" and "drowsiness".

In addition, we found that the symptoms of "drowsiness" were linked to Y13 and Y22 at the same time, and that the interpretation of Y13 and Y22 could both correspond to the syndromes in TCM theory. Y22 and Y13 were related to each other, and it was consistent with the hypothesis that phlegm-heat blocking orifices is relevant to chest disorder of qi movement. This not only verifies the correctness of the latent tree model, but also shows that the improved latent tree model has higher degree of fitting to the original data.

Family headed by Y12 showed two symptoms: "insomnia" and "dreaminess" tend to occur in the data at the same time. The evidence supporting this hypothesis is that yin deficiency can lead to "insomnia" and "dreaminess". The family headed by Y5 indicated that there is "stomachache", "dull stomachache", "abdominal fullness and distention", "rib-side pain" and "dull rib-side pain" co-occurrence in the data. The theory of syndrome differentiation in traditional Chinese medicine generally considered that: (1) "stomachache" and "dull stomachache" suggest that the lesion is located on the stomach, which is a manifestation of stomach qi stagnation. There are many causes of stomach qi stagnation, such as phlegm, drinking, hydrosphere, moisture, retained food, cold, heat and other evil stagnation in the stomach, or stomach yang deficiency, can lead to stomach qi stagnation and painful. (2) abdominal fullness and distention prompted lesions in the spleen, but also in the large intestine, liver, etc. it often due to a variety of deficiency, such as yang, qi deficiency and real qi stagnation, coagulation cold, congestion and other factors cause local blood block. (3) "rib-side pain" and "dull rib-side pain" suggest that the lesions are mostly in the liver and gallbladder, are often caused by pathogenic factors such as qi stagnation, hot and humid condition, heat pathogen, etc. which hinder the liver and gallbladder vent. The evidence supporting this hypothesis is that spleen and stomach qi stagnation and liver qi stagnation may lead to the above five symptoms.

The other sibling clusters in the model are also clearly meaningful. The variables under Y9 are mostly about the color of fur; the variables under Y10 are mostly about the situation of the tongue; And the variables of Y20 represents a partition of “color of complexion”.

5 Conclusion

Based on the latent tree model obtained by EAST algorithm, an arrow- adding (LTM-AA) algorithm is used to achieve a single manifest variable that can correspond to multiple latent variables, that is, one symptom corresponds to multiple syndromes in TCM theory.

An improved LTM was performed on the symptom data of 11854 patients with hepatopathy. This article introduces the model obtained and explains how to understand and appreciate the qualitative aspects of the model. In particularly, this report discusses how and in what sense data analysis provides evidence for confirming the TCM syndrome hypothesis. All about this evidence is determined through systematic inspection of the model.

This analysis shows that according to the data, a symptom corresponds to multiple syndromes with their realistic basis. For example, for symptoms of drowsiness, before the arrow-adding algorithm is performed, the drowsiness is only associated with the latent variable Y13, but by the arrow-adding algorithm, it can find that it is also related to the latent variable Y22, in addition, for the latent variables Y13 and Y22, we can find the corresponding syndrome in the TCM. This not only verifies the correctness of the traditional Chinese medicine theory for chest disorder of qi movement, phlegm-heat blocking orifices, but also reflects the improvement of the latent tree model is more fit to the original data. In the same way, this work has provided statistical validation to TCM hypothesis about yin deficiency, fire excess from yin deficiency, chest disorder of qi movement, phlegm-heat blocking orifices, stagnant blockade of heart blood, stomach qi stagnation and liver qi stagnation.

References

1. Normile, D.: The new face of traditional Chinese medicine. Science **299**(5604), 188–190 (2003)
2. World Health Organization: WHO International Standard Terminologies on Traditional Medicine in the Western Pacific Region. WHO Regional Office for the Western Pacific, Manila (2007)
3. Wu, D.X., Li, D.X., Yan, S.Y.: Fundamental Theories of Traditional Chinese Medicine. Science and Technology Press, Shanghai (1994)
4. Zhu, B., Wang, H.: Diagnostics of Traditional Chinese Medicine. Singing Dragon, London (2011)
5. Wang, H., Xu, Y.: The Current State and Future of Basic Theoretical Research on Traditional Chinese Medicine. Military Medical Sciences Press, Beijing (1999)
6. Feng, Y., Wu, Z., Zhou, X., Zhou, Z., Fan, W.: Knowledge discovery in traditional Chinese medicine: state of the art and perspectives. Artif. Intell. Med. **38**(3), 219–236 (2006)

7. Liang, M., Liu, J., Hong, Z., Xu, Y.: Perplexity of TCM Syndrome Research and Countermeasures. People's Health Press, Beijing (1998)
8. Wang, B., Shen, H., Fang, A., D.-s., H., Jiang, C., Zhang, J., et al.: A regression model for calculating the second dimension retention index in comprehensive two-dimensional gas chromatography time-of-flight mass spectrometry. J. Chromatogr. A **1451**, 127–134 (2016)
9. Wang, B., Chen, P., Wang, P., Zhao, G., Zhang, X.: Radial basis function neural network ensemble for predicting protein-protein interaction sites in heterocomplexes. Protein Pept. Lett. **17**(9), 1111–1116 (2010)
10. Zhang, N.L., Yuan, S., Chen, T., Wang, Y.: Latent tree models and diagnosis in traditional Chinese medicine. Artif. Intell. Med. **42**(3), 229–245 (2008)
11. Zhang, N.L., Yuan, S., Chen, T., Wang, Y.: Statistical validation of traditional Chinese medicine theories. J. Altern. Complement. Med. **14**(5), 583–587 (2008)
12. Zhang, N.L.: Hierarchical latent class models for cluster analysis. J. Mach. Learn. Res. **5**(6), 697–723 (2004)
13. Pearl, J.: Probabilistic Reasoning in Intelligent Systems: Networks of Plausible Inference. Elsevier, New York (2014)
14. Chen, P., Hu, S., Zhang, J., Gao, X., Li, J., Xia, J., et al.: A sequence-based dynamic ensemble learning system for protein ligand-binding site prediction. IEEE/ACM Trans. Comput. Biol. Bioinform. (TCBB) **13**(5), 901–912 (2016)
15. Xia, S., Chen, P., Zhang, J., Li, X., Wang, B.: Utilization of rotation-invariant uniform LBP histogram distribution and statistics of connected regions in automatic image annotation based on multi-label learning. Neurocomputing **228**, 11–18 (2017)
16. Schwarz, G.: Estimating the dimension of a model. Ann. Stat. **6**(2), 461–464 (1978)
17. Akaike, H.: A new look at the statistical model identification. IEEE Trans. Autom. Control **19**(6), 716–723 (1974)
18. ACSR: Bayesian classification (autoclass): theory and results (1996)
19. Cowell, R.G., Dawid, P., Lauritzen, S.L., Spiegelhalter, D.J.: Probabilistic Networks and Expert Systems: Exact Computational Methods for Bayesian Networks. Springer, New York (2006). https://doi.org/10.1007/b97670
20. Zhang, N.L., Kocka, T., (eds.): Efficient learning of hierarchical latent class models. In: 16th IEEE International Conference on Tools with Artificial Intelligence, ICTAI 2004. IEEE (2004)
21. Chen, T., Zhang, N.L., Liu, T., Poon, K.M., Wang, Y.: Model-based multidimensional clustering of categorical data. Artif. Intell. **176**(1), 2246–2269 (2012)
22. Zhang, N.L., Yi, W., Tao, C.: Discovery of latent structures: experience with the CoIL challenge 2000 data set. J. Syst. Sci. Complex. **21**(2), 172–183 (2008)
23. Supervision CSBoT: National Standards on Clinic Terminology of Traditional Chinese Medicinal Diagnosis and Treatment—Syndromes. China Standards Press, Beijing (1997)
24. Yang, W.M.F., Jiang, Y.: Diagnostics of Traditional Chinese Medicine. Academy Press, Beijing (1998)
25. Yan, S.L., Zhang, L.W., Wang, M.H., Yuan, S.H.: Operational standards for determining the severity levels of kidney deficiency symptoms. J. Chengdu Univ. Chin. Med. **24**(1), 56–59 (2001)

acsFSDPC: A Density-Based Automatic Clustering Algorithm with an Adaptive Cuckoo Search

Chang Liu[1], Junliang Shang[1,2](✉), Xuhui Zhu[1], Yan Sun[1], Jin-Xing Liu[1], Chun-Hou Zheng[3], and Junying Zhang[4]

[1] School of Information Science and Engineering, Qufu Normal University, Rizhao 276826, China
lllcccgreat@163.com, shangjunliang110@163.com, bxyxzhuxuhui@163.com, sunyan225@126.com, sdcavell@126.com

[2] School of Statistics, Qufu Normal University, Qufu 273165, China

[3] School of Computer Science and Technology, Anhui University, Hefei 230601, China
zhengch99@126.com

[4] School of Computer Science and Technology, Xidian University, Xi'an 710071, China
jyzhang@mail.xidian.edu.cn

Abstract. Clustering has gained increasing attention in the data mining field since it plays an important role in the unsupervised classification of samples. While numerous clustering methods have been proposed, these suffer from various limitations including sensible parameter dependence and difficult identification of the number of clusters. In this paper, an automatic density-based clustering method based on an adaptive cuckoo search (acsFSDPC) was proposed. Data points with higher density and larger distance from other data points are assumed as clustering centers. Firstly, an adaptive cuckoo search algorithm is employed using clustering evaluation index as fitness function to determine the optimal cutoff distance for each cluster number. Then, the cluster number with the minimal fitness function value is chosen as the best number of clusters and the corresponding clustering result is the optimal clustering results. The benefits of acsFSDPC are automatic estimation of cluster number and optimal cutoff distance. Experiments of acsFSDPC and its comparison with other recent methods STClu, ACND, CH-CCFDAC, and LR-CFDP are performed on five simulation data sets and a real data set. Results show that the acsFSDPC is promising in estimating the appropriate number of clusters automatically and effectively.

Keywords: Adaptive cuckoo search · Automatic clustering · Cutoff distance Fitness function

D.-S. Huang et al. (Eds.): ICIC 2018, LNCS 10955, pp. 470–482, 2018.
https://doi.org/10.1007/978-3-319-95933-7_56

1 Background

Following the rapid expansion of data scale and deficiency of priori knowledge (such as class tags), there has been a rapid increase in the availability of clustering. Clustering divides numerous different samples (cases or patterns) into multiple clusters (or categories) according to their similarity and thus, enables samples in the same cluster to achieve high similarity. However, samples of different clusters are not similar. Clustering is widely used in different subjects and fields, including bioinformatics, pattern recognition, and image segmentation, etc.

In general, automatic detection of clustering results is a significant challenge. The first challenge is the difficulty of cluster center determination. For example, the clustering centers are artificially selected in the clustering algorithm proposed by Rodriguez and Laio in 2004 [1]. This may lead to identification errors in outliers and clustering centers. The second challenge is that the clustering result is sensitive to a preassigned parameter. Existing methods typically require the user to set appropriate threshold parameters and those parameters are typically not adaptive. The parameters set by the users are different and the obtained clustering results are largely different.

Although many methods have been performed to obtain clustering results, most of these remain limited to cluster center determination difficulty and sensible parameter dependence. In 2014, Rodriguez and Laio introduced the new density-based clustering algorithm FSDPC [1], which utilized the Euclidean distance to evaluate the similarity between samples. The algorithm is based on the assumption that the center of a cluster is surrounded by neighbors with lower local density and that it has a relatively large distance from any point with a higher density. After the center of a cluster has been found, each of the remaining points is then distributed to a cluster of the nearest neighbors with higher density. FSDPC can find density peaks fast and determine clustering centers. Moreover, this algorithm can still be improved. Firstly, FSDPC used a proportion t to confirm the cutoff distance (d_c) and the author suggested the t to range between 1% and 2%. This is definitely based on the experience value of several data sets, but it may not be adaptable to other application problems. Secondly, for clustering center identification, FSDPC still needs to select the clustering center artificially through the decision map, which is entirely subjective.

To solve these deficiencies of FSDPC, several methods are presented to optimize the solution of d_c and to automatically cluster the data. These appear promising in detecting clustering results since they are based on a well-developed theory. Chen *et al.* [2] considered singular points outside the confidence interval of normal distribution curves as the clustering centers and used a mountain climbing algorithm to realize a self-adaptive density radius. Although the obtained clustering results are convincing, the preassigned parameter confidence interval needs to be determined subjectively. Guo *et al.* [3] proposed an algorithm improving location stability of clustering centers by measuring the point density through the nearest neighbors information of a sample. This method determined clustering centers fast and automatically by applying linear regression and residual analysis. However, it still requires the artificial selection of the clustering center via the residual figures. Shi *et al.* [4] presented an adaptive clustering algorithm to determine the cluster number and automatically remove noise based on

KNN and density. Wang *et al.* [5] designed further automatic clustering via outward statistical testing on density metrics; however, the performance of the proposed algorithm is related to the number of the nearest neighbors. Zhou *et al.* [6] introduced an alternative definition of the indicators and the threshold of cluster centers is decided by an improved canopy algorithm. Yan *et al.* [7] designed a statistic-based method to identify the cluster centers from the decision graph.

Based on these observations, we develop the adaptive cuckoo search clustering method acsFSDPC, which is based on the density peak. First of all, the preassigned parameter t is set as the bird's nest position in acsFSDPC. Consequently, acsFSDPC is not sensitive to the parameter t. Then, the adaptive cuckoo search algorithm is utilized to improve the precision of optimization and speed of convergence. The iterative solution minimizes the fitness function to find the optimal t (the optimal bird's nest position) and the optimal d_c is also obtained accordingly. Finally, for the different clustering number, the number of clusters corresponding to the minimum value of the optimal fitness function is the optimal number of clusters, at the same time, the clustering results are automatically determined. In summary, the algorithm acsFSDPC proposed here was easy to gather samples in the low-dimension space (especially for the two-dimensional space). The experimental results indicated that the effectiveness and robustness of this algorithm were superior to STClu [5], ACND [4], CH-CCFDAC [2], and LR-CFDP [3].

2 Methods

2.1 Introduction of the FSDPC Clustering Algorithm

FSDPC is the density-based clustering algorithm based on the assumption that the clustering center generally has a high local density and that there is a large difference between samples with a high local density [1]. This algorithm includes three steps: matrix calculation, clustering center identification, and sample clustering.

For each sample, FSDPC defines two indexes ρ and δ, which stand for the local density and minimal distance of each sample, respectively. The definition of local density in FSDPC is shown as follows:

$$\rho_i = \sum_{i=1}^{n} \chi(d_{ij} - d_c) \tag{1}$$

here, the function $\chi(x) = \begin{cases} 1, x<0 \\ 0, x\geq 0 \end{cases}$, d_{ij} represents the distance between sample i and sample j. The parameter d_c greater than zero represents the cutoff distance, which leads to the mean neighboring number in each data point ranging between 1% and 2% of the total data points.

For the data set to include N data points, the distance $d_{ij}(i<j)$ between two arbitrary data points is sorted in a descending order, thus obtaining the sequence as $d_1 \leq d_2 \leq d_3 \ldots\ldots \leq d_M (where, M = N(N-1)/2)$. Assuming that $d_k(k \in \{1,2,3\ldots M\})$ as the d_c, the proportion of distance between two data points is less than d_c, which is about k/M, and then, the number is about $kN(N-1)/M$. Then, each data point is $k(N-1)/M \approx kN/M$, let

$$t = k/M \tag{2}$$

thus $k = Mt$. Then, the d_k is the d_c.

Based on the local density of samples, both the minimal distance between points and points with high density are defined as follows:

$$\delta_i = \begin{cases} \min\limits_{j \in I_S^i}\{d_{ij}\}, I_S^i \neq \varnothing \\ \max\limits_{j \in I_S}\{d_{ij}\}, I_S^i = \varnothing \end{cases} \tag{3}$$

where, $I_S = \{1, 2, \ldots, N\}, I_S^i = \{j \in I_S : \rho_j > \rho_i\}$.

FSDPC constructs each sample into a two-dimensional data point (ρ_i, δ_i) and forms the decision map. In the decision map, the data points with the large ρ and δ are selected as clustering centers. After confirming the clustering center, residual samples are then distributed to the nearest cluster with high density.

A data set may contain outliers. In general, points with small ρ and large δ are selected as outliers.

3 Adaptive Step Length Cuckoo Search Algorithm

Before introducing the adaptive step length cuckoo search algorithm, the general cuckoo search algorithm (CS) is described. The name of the CS algorithm is inspired by a bird species known as the cuckoo due to their particular lifestyle and reproduction strategy. For simplicity in describing the CS, the following three ideal rules are assumed [8]: (1) Each cuckoo lays one egg at a time, and introduces this egg into randomly chosen nest; (2) The best nests with high quality of eggs will carry over to the next generations; (3) The number of available host nests is fixed, and the egg laid by a cuckoo is discovered by the host bird with the probability $p_a \in [0, 1]$. In this case, the host bird can either remove the egg, or abandon the nest and build a completely new nest. For simplicity, the last assumption can be approximated by assuming that the fractions p_a of all n nests are replaced by new nests.

For simplicity, the following representations are used: each egg in a nest represents a solution and each cuckoo egg represents a new solution. In the simplest form, each nest contains one egg. If the cuckoo egg is very similar to that of the host, the chance that the cuckoo egg is discovered is small; thus, the fitness function should be related to the difference in solutions. Through iteration, new and potentially better solutions replace poor solutions in the nests.

When generating the new solution x_i^{t+1} for a cuckoo i, a Lévy flight is performed using the following equation [9]:

$$x_i^{(t+1)} = x_i^{(t)} + a \cdot S \tag{4}$$

where, $\alpha > 0$ represents the step size that should be related to the scale of the problem of interest. The parameter S described in Eq. (4) represents the length of random walk with Lévy flights according to Mantegna's algorithm:

$$S = \frac{u}{|v|^{1/\beta}} \tag{5}$$

where, β represents a parameter between the [1, 2] interval, which is considered to be 1.5; u and v are drawn from a normal distribution:

$$u \sim N(0, \sigma_u^2), v \sim N(0, \sigma_v^2) \tag{6}$$

where, $\sigma_u = \left\{ \frac{\Gamma(1+\beta)\sin(\pi\beta/2)}{\beta\Gamma((1+\beta)/2)2^{(\beta-1)/2}} \right\}^{1/\beta}, \sigma_v = 1.$

In addition, the p_a fraction of the worst nests is defined and is replaced by new ones according to Eq. (7). Through continuous iteration, the global optimal value and global optimal solution can be solved.

$$x_i^{(t+1)} = x_i^{(t)} + \alpha \cdot S \otimes H(p_a - \varepsilon) \otimes (x_j^t - x_k^t) \tag{7}$$

where, x_j^t and x_k^t are two different solutions that are chosen by random substitution; $H(u)$ is a Heaviside function; ε is the random number drawn from the uniform distribution; $\otimes$ is the dot product of two vectors.

Yang *et al.* have proved that CS is more efficient in finding the global optima with higher success rates, which outperforms both genetic algorithms and particle swarm algorithms [8]. Therefore, if given sufficient computation time, it is guaranteed to converge to an optimal solution [10]. However, the search process may be time consuming due to associated random walk behavior. In the search process, the larger the step length is, the easier the global search optimization will be. However, the search precision is reduced. Sometimes, the phenomenon of oscillation interferes. The smaller the step length is, the slower the search speed will be. Nevertheless, the precision of the solution is improved. As a result, the step length generated by a Lévy flight has randomness, but lacks adaptability.

The key difference between CS and ACS (Adaptive Cuckoo Search) lies in the way of adjusting the step length. To improve the performance of the CS algorithm, the adaptive step length strategy [11] is introduced:

$$stepsize = (1/iter)^{(abs((fit(j) - best\_fit)/(best\_fit - worst\_fit)))} \tag{8}$$

in the formula, *iter* represents the iterations. $fit(j)$ represents the fitness function of the jth bird's nest. *best_fit* represents the best state of the bird's nest position. *worst_fit* represents the worst state of the bird's nest position. When the bird's position j is close to the best position, the step length will be smaller. With increasing iterations, the step length will become smaller. In this way, the step length of this iteration is dynamically updated according to the result of the previous iteration. The search speed and optimization accuracy of the algorithm are greatly improved.

The bird's nest position is updated by Eq. (9):

$$s = s + stepsize * randn(size(s)) \tag{9}$$

Based on the above-mentioned rules, the basic steps of ACS can be summarized with pseudo code in Fig. 1.

```
Adaptive cuckoo search without Lévy Flights
begin
The target function f(x), x = (x_1,...,x_d)^T
Generate initial population of n hosts nests x_i (i = 1,2,...,n)
Evaluate quality of the solution or target function value f_i
while (t < MaxGeneration) or (stop criterion) do
Generate a new bird's nest position x_i' by (8)
Evaluate its quality/fitness f_i' of the new solution x_i'
if f_i' <= f_i then
Replace x_i by x_i'
end
A fraction (p_a) of worse nests are abandoned
Generate the new bird's nest
Keep the best solution (or nest with high-quality solution)
Rank the solution and find out the best one
Update t→t+1
end while
Postprocess results and visualization
end
```

Fig. 1. The pseudo code of ACS

3.1 acsFSDPC: Density-Based Clustering Algorithm Based on an Adaptive Step Length CS

In this section, we propose a novel clustering algorithm (named acsFSDPC) to overcome two deficiencies of the FSDPC mentioned above. Firstly, acsFSDPC attempts to find the optimal d_c and the corresponding fitness function value. Then, acsFSDPC tries to obtain the optimal cluster number based on the optimal fitness function value for each cluster number. The entire procedure of acsFSDPC is illustrated in Fig. 2. The following two subsections are innovation points of the acsFSDPC.

The Bird's Nest Position - t**.** In FSDPC, the average percentage of neighbours per sample t is established to calculate the cutoff distance d_c. Changing t will directly affect d_c, thus affecting the final clustering results. The different clustering results corresponding to different t are shown in Fig. 3.

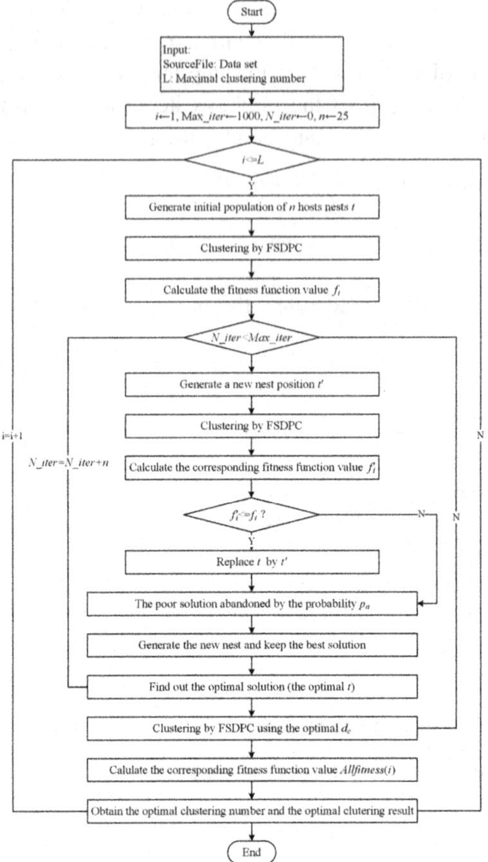

Fig. 2. Procedure overview of acsFSDPC

In Fig. 3, the clustering result is presented in Fig. 3a for $t = 0.2$. The result is optimal. The clustering result is presented in Fig. 3b for $t = 0.02$. The clustering result is poor.

In the FSDPC algorithm, the author locked t to range from 1%–2% based on the experience of several data sets; however, this may not be suitable for other application problems. In the above, when the parameter t is determined appropriately, the parameter d_c is obtained by Eq. (2) and clustering centers can be effectively extracted via FSDPC. Therefore, the selection of optimal t is very important to the determination of the optimal d_c.

To select the optimal t, we treat t as the bird's nest position. Through iteration, the optimal solution of fitness function is solved, and the corresponding bird's nest position is the optimal solution (the optimal t). Then the optimal d_c can be determined. acsFSDPC can automatically solve the optimal d_c and has no need for artificial parameter setting; thus, the clustering results are independent of preassigned parameters, increasing the precision and practicality of clustering.

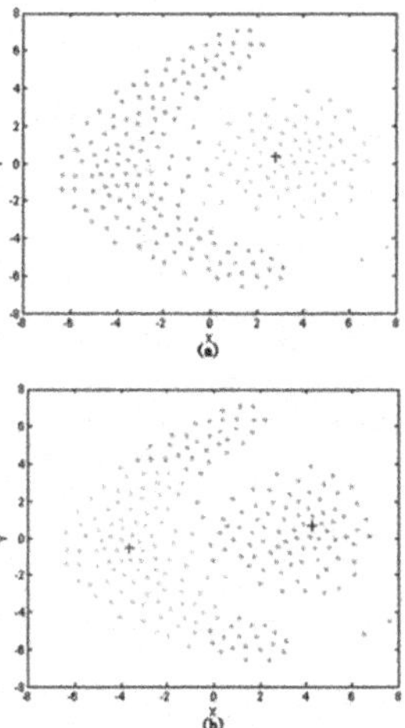

Fig. 3. Influences of different t.

Fitness Function – The Clustering Evaluation Index. When the adaptive step length is used to update the position of each bird's nest t, d_c is changed accordingly. In this paper, to measure the clustering effect of different d_c, the clustering evaluation index is designed as a fitness function.

This fitness function is defined as follows:

$$Fitness = \max_{i \in [1,k]} \frac{1}{k-1} \sum_{j=1, j \neq i}^{k} \frac{\overline{C_i} + \overline{C_j}}{\left\| w_i - w_j \right\|_2} \tag{10}$$

where, $\overline{C_i}$ represents the mean distance in the cluster i, $\left\| w_i - w_j \right\|_2$ represents the distance between two cluster centers i, j, and k represents the clusters number.

The fitness function value (the clustering evaluation index) is smaller, implying a smaller distance in the cluster. Moreover, the distance between clusters is larger, therefore, the clustering effect is better.

Determining the Optimal Clusters Number Based on the Fitness Function. Via iteration of the adaptive cuckoo search algorithm, the optimal d_c corresponding to the minimum fitness function value is obtained. For each clustering number, the optimal d_c is leveraged in FSDPC and the corresponding fitness function value is obtained. Then, the clustering number that corresponds to the minimum value of the fitness function is the optimal number of clusters, at the same time, the clustering results are automatically determined.

4 Results and Discussion

Evaluation Measures: *Accuracy*
In this algorithm, we use *Accuracy* to analyze the pros and cons of clustering experimental results. *Accuracy* is defined as follows:

$$Accuracy = \frac{\sum_{i=1}^{k} TruePositive(i)}{SampleNum} \tag{11}$$

where, *i* represents the label of a cluster. $TruePositive(i)$ represents the correct sample size in the cluster *i*. *SampleNum* represents the total sample number. *k* represents the clustering number.

4.1 Experimental Results on Synthetic Data Sets

The experiment of acsFSDPC is performed on five data sets: Flame, Spiral, Aggregation, R15, and Square1. Square1 is available on http://en.pudn.com/Download/item/id/1030717.html and others are available on http://cs.uef.fi/sipu/datasets/. The clustering effect is evaluated via comparative studies with several existing methods, using above simulation data. These are: STClu [5], ACND [4], CH-CCFDAC [2], and LR-CFDP [3]. For acsFSDPC, the number of nests *n* is 25, the probability p_a is set to 0. 25, the iteration number *Max_iter* is set to 1000 [11], and *t* ranges from 0.0001 to 1.

The clustering results of the compared methods on five data sets are shown in Figs. 4, 5, 6, 7, and 8. In the cluster graphs of five artificial data sets, each cluster with different color and shape marks a cluster center except for the ACND algorithm.

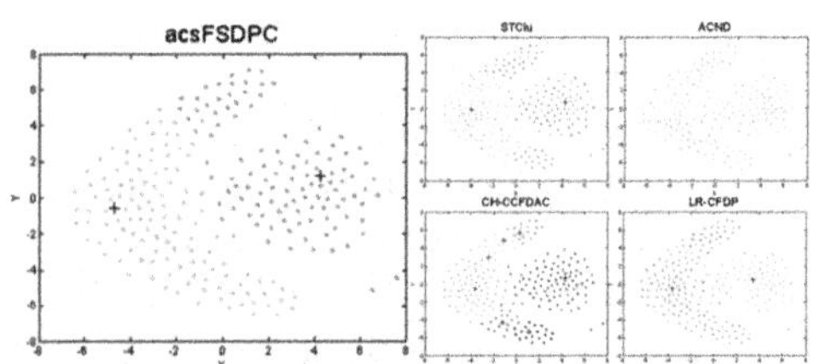

Fig. 4. Clustering results of Flame (Color figure online)

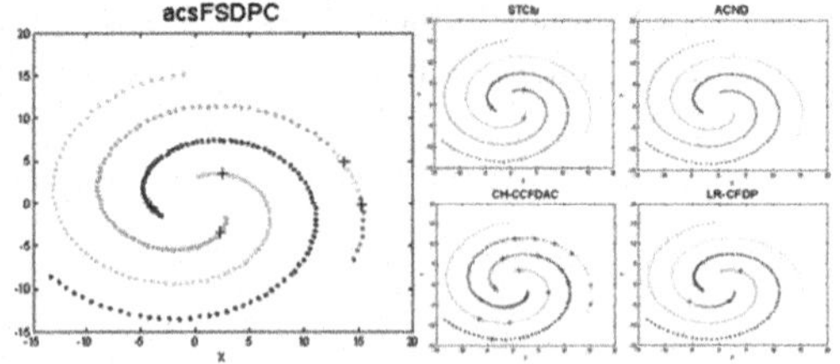

Fig. 5. Clustering results of Spiral (Color figure online)

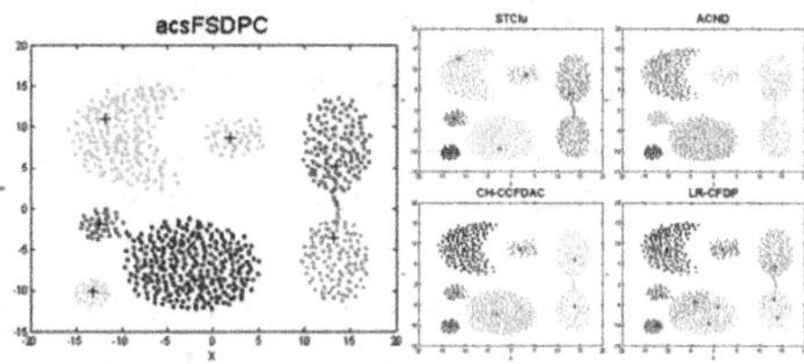

Fig. 6. Clustering results of Aggregation (Color figure online)

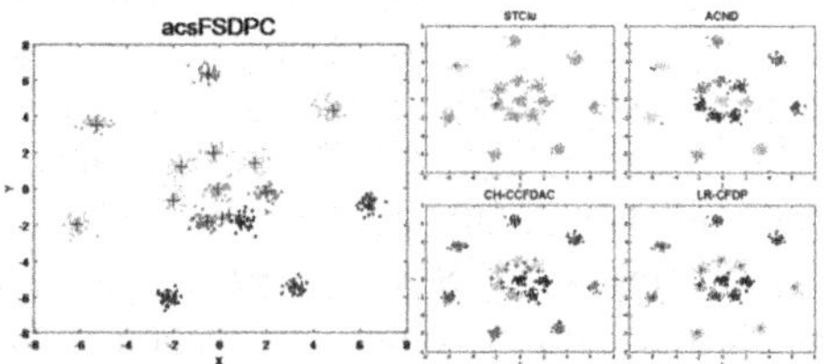

Fig. 7. Clustering results of R15 (Color figure online)

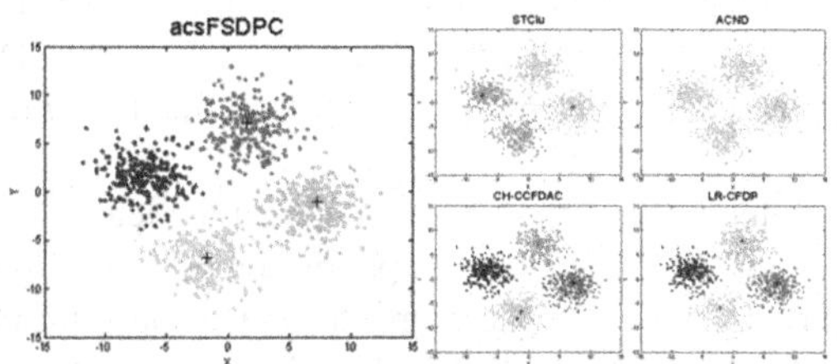

Fig. 8. Clustering results of Square1 (Color figure online)

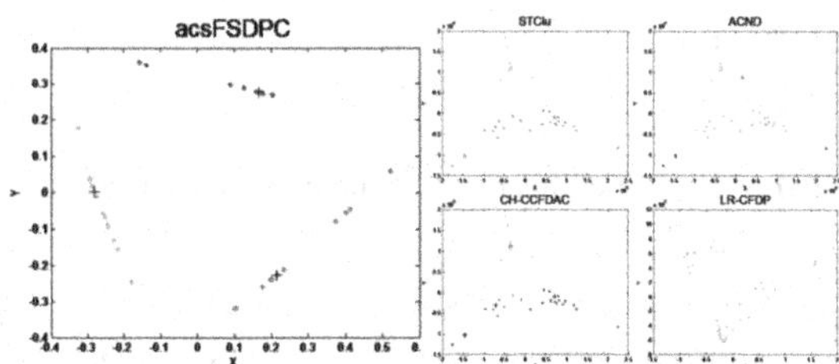

Fig. 9. Clustering results of the leukemia data set

Figure 4 shows that in the cluster graph of the Flame data set, acsFSDPC recognizes two clusters. STClu obtains two clusters and ACND obtains only one cluster. Seven points are identified as clustering centers by CH-CCFDAC. The two clusters obtained in LR-CFDP are completely consistent with the clustering results of Flame. Figure 5 shows the results for the data set Spiral; acsFSDPC recognizes five clusters. STClu and ACND can detect all correct clustering centers. CH-CCFDAC, which divides the data set into 18 clusters, performs poorly. For the LR-CFDP algorithm, four clusters are obtained. Figure 6 shows results of Aggregation, showing that acsFSDPC recognizes seven clusters, which are slightly different to the clustering results of Aggregation. STClu obtains six clusters and ACND obtains five clusters. Seven clusters obtained in CH-CCFDAC are completely consistent with the clustering results of Aggregation. For the LR-CFDP algorithm, 11 clusters are obtained. Figure 7 shows the results for the data set R15, showing that acsFSDPC recognizes 17 clusters and LR-CFDP recognizes 15 clusters, both of which are slightly different than the clustering results of R15. STClu obtains only two clusters, which is slightly consistent with the true clustering results. ACND divides samples into 13 clusters and CH-CCFDAC divides samples into 28 clusters. Figure 8 shows the results for data set Square1, showing that acsFSDPC and LR-CFDP recognize four clusters, which are slightly different than the clustering results of Square1. STClu and ACND, both of which divide samples into two clusters, perform poorly. Five points are identified as clustering centers by CH-CCFDAC.

4.2 Experimental Results on the Leukemia Data

The leukemia data set contains 5000 genes in 38 samples, consisting of 19 cases of B cell ALL, 8 cases of T cell ALL, and 11 cases of AML. For acsFSDPC, the correlation coefficient is used as the distance formula in the fitness function, t ranges from 0.001 to 1 and the other parameters remain unchanged.

The clustering results of the compared methods on the leukemia data set are shown in Fig. 9. It shows that in the cluster graph of the leukemia data set, acsFSDPC recognizes three clusters which is the same as the actual clusters number. STClu obtains two clusters. Four points are identified as clustering centers by both ACND and CH-CCFDAC. Only one cluster is obtained in LR-CFDP.

To quantitatively visualize the advantage of acsFSDPC, the experimental results of acsFSDPC and compared methods are shown in the Table 1, from which, we obtain the following observations.

Table 1. The comparison of the experimental results

Methods	Flame	Spiral	Aggregation	R15	Square1	Leukemia
acsFSDPC	0.883	0.949	0.999	0.997	0.992	0.921
STClu	0.788	1.000	0.871	0.133	0.499	0.684
ACND	0.630	1.000	0.973	0.780	0.254	0.763
CH-CCFDAC	0.663	0.356	1.000	0.918	0.922	0.842
LR-CFDP	1.000	0.894	0.631	0.997	0.992	0.500

acsFSDPC achieves a higher accuracy than STClu on Flame, Aggregation, R15, Square1, and the leukemia data sets. It is seen that acsFSDPC performs better than CH-CCFDAC on Flame, Spiral, R15, Square1 and the leukemia data sets. The accuracy of LR-CFDP on Flame data set and the accuracy of ACND on Spiral data set are higher than that of acsFSDPC. The results show that the accuracy of acsFSDPC is higher than that of methods on most of the data sets.

5 Conclusions

In this paper, an adaptive CS clustering method based on the density peak is proposed, namely acsFSDPC. This method does not require preassigned parameter and can automatically determine clustering number and optimal d_c. Firstly, the optimal d_c corresponding to the minimum fitness function leveraging the clustering evaluation index is obtained by an adaptive step length CS algorithm. Based on the optimal d_c, the corresponding fitness function values of all clustering numbers are calculated by acsFSDPC. Then, for different cluster numbers, the minimum value of the fitness function is selected as the optimal function value and the corresponding clustering number indicates the optimal clustering number. Finally, the clustering result obtained by the FSDPC algorithm is the optimal clustering result. acsFSDPC produces very good clustering results on low-dimensional data, especially in a data set that is concentrated in each cluster. The experimental results show both the effectiveness and robustness of the proposed algorithm acsFSDPC.

Although the results demonstrate that acsFSDPC performs well on the above data sets, several limitations remain. Firstly, the optimal number of clusters is obtained by acsFSDPC via iteration and the distance between any two samples is calculated at each clustering, therefore the time complexity of acsFSDPC is relatively high. Secondly, acsFSDPC cannot effectively analyze or deal with the noise points of the data set. These limitations have inspired us to continue this work in future.

Acknowledgments. This work was in part supported by the National Natural Science Foundation of China (61502272, 61572284), the Science and Technology Planning Project of Qufu Normal University (xkj201410), the Scientific Research Foundation of Qufu Normal University (BSQD20130119).

References

1. Rodriguez, A., Laio, A.: Clustering by fast search and find of density peaks. Science **344** (6191), 1492 (2014)
2. Chen, J., Lin, X., Zheng, H., Bao, X.: A novel cluster center fast determination clustering algorithm. Appl. Soft Comput. **57**, 539–555 (2017)
3. Guo, P., Wang, X., Wang, Y., Cheng, Y., Zhang, Y.: Research on automatic determining clustering centers algorithm based on linear regression analysis. In: International Conference on Image, Vision and Computing, pp. 1016–1023 (2017)
4. Shi, B., Han, L., Yan, H.: Adaptive clustering algorithm based on kNN and density. Pattern Recognit. Lett. **104**, 37–44 (2018)

5. Wang, G., Song, Q.: Automatic clustering via outward statistical testing on density metrics. IEEE Trans. Knowl. Data Eng. **28**(8), 1971–1985 (2016)
6. Zhou, R., Zhang, S., Chen, C., Ning, L., Zhang, Y., Feng, S., Liu, Y., Luktarhan, N.: A Distance and density-based clustering algorithm using automatic peak detection. In: IEEE International Conference on Smart Cloud, pp. 176–183 (2016)
7. Yan, H., Lu, Y., Ma, H.: Density-based clustering using automatic density peak detection. In: International Conference on Pattern Recognition Applications and Methods, pp. 95–102 (2018)
8. Yang, X.S., Deb, S.: Cuckoo search via Lévy flights. In: World Congress on Nature & Biologically Inspired Computing, NaBIC 2009, pp. 210–214 (2010)
9. Yildiz, A.R.: Cuckoo search algorithm for the selection of optimal machining parameters in milling operations. Int. J. Adv. Manuf. Technol. **64**(1–4), 55–61 (2013)
10. Ong, P.: Adaptive cuckoo search algorithm for unconstrained optimization. Sci. World J. (9) (2014). https://doi.org/10.1155/2014/943403
11. Naik, M., Nath, M.R., Wunnava, A., Sahany, S., Panda, R.: A new adaptive Cuckoo search algorithm. In: IEEE International Conference on Recent Trends in Information Systems, pp. 1–5 (2015)

Epileptic Seizure Detection Based on Time Domain Features and Weighted Complex Network

Hanyong Zhang[1,2], Qingfang Meng[1,2](✉), Bo Meng[3], Mingmin Liu[1,2], and Yang Li[1,2]

[1] School of Information Science and Engineering, University of Jinan, Jinan 250022, China
ise_mengqf@ujn.edu.cn

[2] Shandong Provincial Key Laboratory of Network Based Intelligent Computing, Jinan 250022, China

[3] Institute of Jinan Semiconductor Elements Experimentation, Jinan 250014, China

Abstract. Epileptic seizure detection is one of the important steps in diagnosis of epilepsy. Excellent automatic detection algorithm of epilepsy will help healthcare workers to better treat epilepsy patients, which has important study significance. In this paper, we proposed a new epileptic seizure detection method based on time domain features and weighted complex network of electroencephalogram (EEG) signals. Firstly, each EEG segment is divided into four sub-segments and each sub-segment is divided into thirty-two clusters. A set of time domain features is extracted from each cluster. Then, each set of this features is used as a node of complex network. Features sets are converted into weighted horizontal visibility graph. Thirdly, average weighted degree of complex network is extracted as the classification feature. Finally, average weighted degree is inputted into a linear classifier to classify epileptic EEG signals. The experimental result shows that the classification accuracy is up to 96.5%. The obtained result indicates that the proposed method is effective in epileptic seizure detection.

Keywords: Epileptic EEG signals · Time domain features
Weighted complex network · Averaged weighted degree

1 Introduction

Epilepsy is a common brain disease and millions people in the world are suffering from epilepsy. The exact cause of epilepsy is still unknown. However, a large number of epileptic seizure detection methods have been proposed so far. Clinically, electroencephalogram (EEG) is the most common and important tool in the diagnosis of different brain diseases [1]. The professional doctors detected epilepsy by recognizing EEG with their eyes in the beginning. It requires a lot of manpower and time. Hence, the automatic detection method of epilepsy EEG signals has important significance for clinical research.

D.-S. Huang et al. (Eds.): ICIC 2018, LNCS 10955, pp. 483–492, 2018.
https://doi.org/10.1007/978-3-319-95933-7_57

Time domain methods was first used for epilepsy detection. Altunay et al. [2] utilized linear energy prediction method to detect epilepsy. Acharya et al. [3] utilized PCA for the epileptic detection and found that different EEG signals have obvious differences. Kaplan [4] studied the spectral characteristics of epileptic EEG and used LDA to analysis it. Frequency domain and time-frequency domain methods were also introduced into detecting epilepsy. Polat et al. [5] used the Welch for two-class epilepsy detection. Kumar et al. [6] used DWT in their study and found that different signals have different wavelet coefficients. Li et al. [7] proposed feature extraction method of ictal EEG using EMD.

Nonlinear methods became more and more popular in epileptic seizure detection. Acharya et al. [8] proposed a detection method utilizing the approximate entropy and confirmed that different EEG signals have different entropies. Song and Lake used sample entropy [9, 10], which improves the robustness of the algorithm and achieves better classification results. Many other nonlinear methods were also used for EEG classification [11–13]. Performances of conventional approximate entropy, sample entropy and some other nonlinear methods are unsatisfactory, which can't meet the actual clinical diagnosis needs. In recent years, complex network theories provide a new method for time series analysis. Lacasa et al. [14] proposed the visibility graph (VG) algorithm, which could convert arbitrary time series into a complex network. Then, Lacasa et al. [15] proposed the horizontal visibility graph (HVG) algorithm, which has a simpler construction regulation.

Tang et al. [16] applied visibility graph to recognize the high frequency EEG signals and extracted sequence degree as the feature to classify EEG signals. Zhu et al. [17] proposed weighted HVG approach to identify seizures and obtained better results. Supriya et al. [18] proposed weighted VG approach to detect epilepsy and modularity and average weighted degree were extracted in epileptic seizures recognition.

In this paper, we propose a new method that convert EEG time series into WHVG. For the first time, we combine time domain information with weighted horizontal visibility graph. Firstly, each single channel EEG signals is divided into four sub-segments, and each sub-segment is divided into 32 clusters. Then, 12 time domain statistical features of each cluster are extracted. Finally, 1536 time domain features are used as nodes of complex network and converted into WHVG. Average weighted degree of the WHVG is extracted as the last classification feature.

The rest of this paper is organized as follows: experimental data is described in the next section. Our proposed method and materials are described in the third section. Experiment results and analysis are in the fourth part. The last section includes the conclusions of our work.

2 Experimental Data Description

In this study, we use a clinical epileptic EEG data set from the University of Bonn, Germany. There are five datasets in the data. Datasets A and B were acquired from five healthy people. Datasets C, D, E were acquired from five epileptic patients. Dataset C and dataset D were from patients during seizure free. Dataset E was obtained from patients exhibiting ictal activity. Only two sets were used in this study: interictal (set D) and ictal (set E). This two sets are the key point to classify epilepsy seizure. Each dataset contains 100 segments of single channel EEG with 23.6 s duration. Every segment is sampled at a rate of 173.61 Hz, and has 4096 points.

3 Methodology

Figure 1 shows a block diagram of the method proposed in this paper, which includes dimensionality reduction, construction of WHVG, extraction of the feature, and the classification based on this feature.

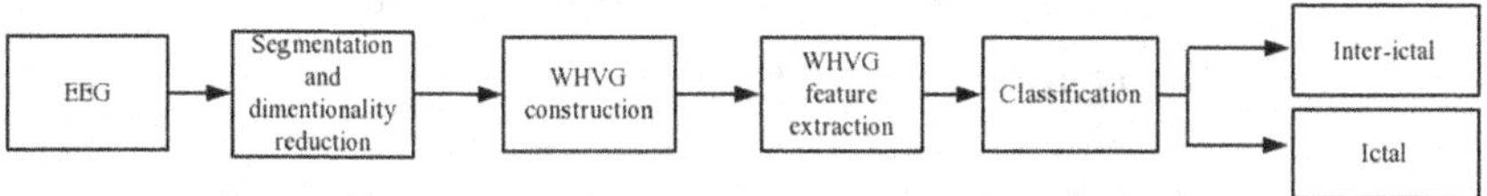

Fig. 1. The block diagram of the proposed method

3.1 EEG Dimensionality Reduction

In order to reduce the dimensionality of EEG signals and make the important information retained, we improve the dimensionality reduction method proposed by Siuly [19]. In the experimental data, each single EEG segment contains 4097 data points. This 4097 points are divided into four sub-segments of 1024, 1024, 1024 and 1025, respectively. To extract time domain statistical features, each sub-segment is partitioned into 32 clusters. The division of the interval is based on experience. The feature set is {median, maximum, minimum, mean, mode, range, first quartile, second quartile, standard deviation, variation, skewness, kurtosis}. Finally, 12 features are extracted from each cluster. All the features from one segment is regarded as a vector to represent the segment. The dimensionality of EEG segment is reduced from 4096 to 1536. Figure 2 shows the process of segmentation and dimensionality reduction of EEG signals.

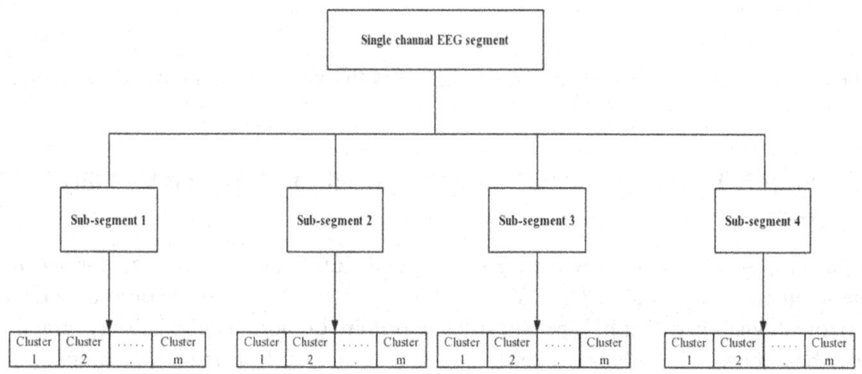

Fig. 2. The segmentation and dimensionality reduction

Table 1 provides a short explanation of the 12 features. In the time domain statistical characteristics of time series, maximum, minimum, variation, skewness and kurtosis are used for basic mathematical statistics. The mean and the standard deviation are used for measures of symmetric distribution of time series. While median, range and quartile are used to measure the center and the spread of time series. In addition, the mode is used to describe the most frequent number of a dataset. According to the previous references, these 12 features are the most important statistical characteristics of time series and can contain all the important information of time series. Besides, the 12 features are simple and easy to understand.

Table 1. The short explanation of the features

No.	Feature	Formula
1	Max	$X_{Max} = \max[x_n]$
2	Mean	$X_{Mean} = \frac{1}{n}\sum_{1}^{N} X_n$
3	Median	$X_{Me} = \left(\frac{N+1}{2}\right)^{th}$
4	First quartile	$X_{Q1} = \frac{1}{4(N+1)}$
5	Variation	$X_{Var} = \sum_{n=1}^{N}(Xn - AM)\frac{2}{N-1}$
6	Kurtosis	$X_{Ku} = \sum_{n=1}^{N}(Xn - AM)\frac{4}{(N-1)SD^4}$
7	Min	$X_{Min} = \min[x_n]$
8	Mode	*
9	Range	$X_{Ra} = X_{Max} - X_{Min}$
10	Standard deviation	$X_{SD} = \sqrt{\sum_{n=1}^{N} Xn - AM\frac{2}{N-1}}$
11	Skewness	$X_{Ske} = \sum_{n=1}^{N}(X_n - AM)\frac{3}{(N-1)SD^3}$
12	Second quartile	$X_{Q1} = \frac{4}{4(N+1)}$

where $X_n = 1, 2, 3, \ldots, n$ is a sample series, N is the number of sample points, AM is the mean of the sample.

3.2 Feature Extraction Method Based on Weighted Horizontal Visibility Graph

In our proposed methodology, firstly the original EEG time series is transformed into time-domain feature vector. Then, we convert time domain feature vector into weighted horizontal visibility graph. The specific algorithm of horizontal visibility graph is described below. Considering a series $\{x_1, x_2, \ldots, x_i, \ldots x_n\}$, x_i is the *ith* sample point and is considered as a node n_i of graph $G(V, E)$, where V is the node set and

$V = \{n_i\}, i = 1, 2, \ldots, N$. E is the edge set and E_{ij} is 1 if edges of two nodes exist. For mapping the time series data into HVG, each data point of the vector is considered as a node of the graph. The edge between two nodes are determined according to the following equation:

$$n_{\mathrm{k}} < \min(n_{\mathrm{i}}, n_{\mathrm{j}})\, \forall k,\; i<k<j \tag{1}$$

where n_i is the ith node in HVG. It means that the edge between n_i and n_j exists if one can draw a line from the vertex of the smallest of the two points but cutting no intermediate data n_k. Figure 3 presents an example of HVG constructed from 20 points. In Fig. 3, the upper part is a time series of 20 data and the bottom part is the complex network constructed through the HVG algorithm. Each point in the time series corresponds to a node in the complex network and the edge between two nodes are determined according to the criterion equation.

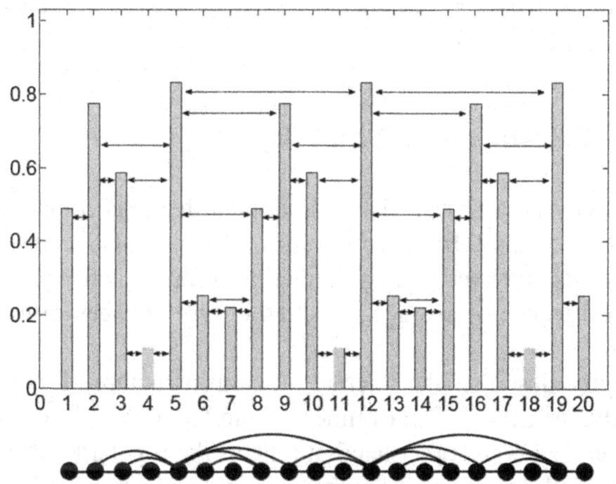

Fig. 3. Sample points of 20 data and its associated horizontal visibility graph

HVG has its disadvantages such as it does not contain weight information between two nodes of complex network. The weight between the nodes of complex network reflects the dynamic information of the original data series and it is an important topological structure of complex network. Weighted horizontal visibility graph was first used for epileptic seizure detection in [17] and obtained satisfactory results. In this paper, we improve the WHVG by using a new edge weight which has explicit mathematical significance. The edge weight is calculated by the following equation:

$$w_{ij} = n_j - n_i \tag{2}$$

where, w_{ij} is the difference between node i and j.

In the weighted complex network, weighted degree is the most analyzed and significant feature. Weight degree reflects the structure of the edges of the weighted complex network. The weighted degree of the node i is the total weights of all the edges attached to node i which is represented by:

$$W(i) = \sum_{j \in B(i)} w_{ij} \tag{3}$$

where, $B(i)$ is the total neighborhood of node i. The average weighted degree (AWD) is defined as:

$$AWD = \frac{1}{N} \sum_{i=1}^{N} W(i) \tag{4}$$

where, N is total number of nodes in the graph. It should be explained that there are sudden changes in the waveform of epilepsy EEG signals, the edge weight of different signals will have obvious differences.

4 Experiment Results and Analysis

After the above analysis, we know that time-domain features contain all the important information of the original EEG time series. WHVG also contains the essential information of these feature vectors. By extracting complex network feature of WHVG, we find that the difference of average weighted degree between different EEG signals are very obvious.

Weighted degree describes the internal topology structure of WHVG. On the one hand, it reflects the internal contact of time-domain features and dynamics information of original time series. On the other hand, it reflects the regularity of complex network. According to the Eq. 4, the improved average weighted degree of WHVG is calculated. Then the average value of each sample is extracted as a feature. Finally, the 200 features extracted from 200 samples are inputted into a linear classifier. The classification accuracy is 96.5%. In order to get the more intuitive classification result visually, Fig. 4 plots the classification result with average weighted degree. In Fig. 4, each '+' presents the ictal signal and each '*' presents the interictal signal. The black line below the '*' is used as the classification threshold. It means that the two kinds of signals are separated to the maximum extent at this time.

From Fig. 4, we can find that the feature values of interictal signals are all near 0 and the values of ictal signals are scattered and far from 0. Using the classification threshold given by the linear classifier, that is, the dividing line in the Fig. 4, more sample points are classified correctly except for a few points. It means that values of different interictal EEG signals are more concentrated, so different samples have similar feature values. However, the feature values of different ictal EEG signals are very sparse and have big difference. It also means that there are essential differences between the two EEG signals. From EEG we can find that there is greater fluctuation in the brain waves when the seizures occur. Dramatic fluctuations in EEG signals lead to large

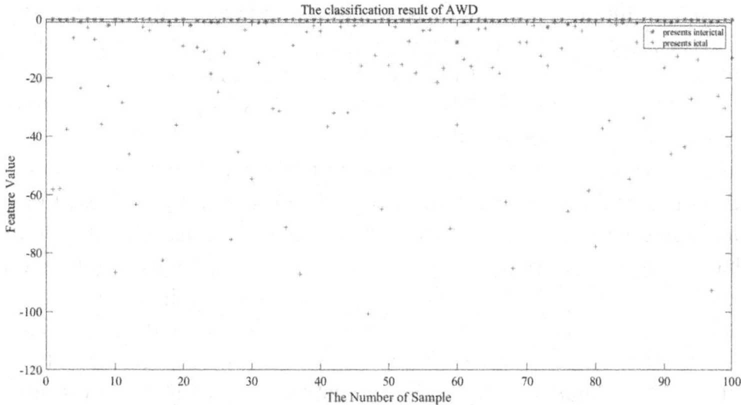

Fig. 4. The classification result using AWD

changes in time-domain statistical characteristics. These changes in turn affect the internal structure of complex network.

In order to more vividly describe the differences of feature values corresponding to EEG signals in different states, Fig. 5 plot the distribution of AWD of complex network in the boxplot graph.

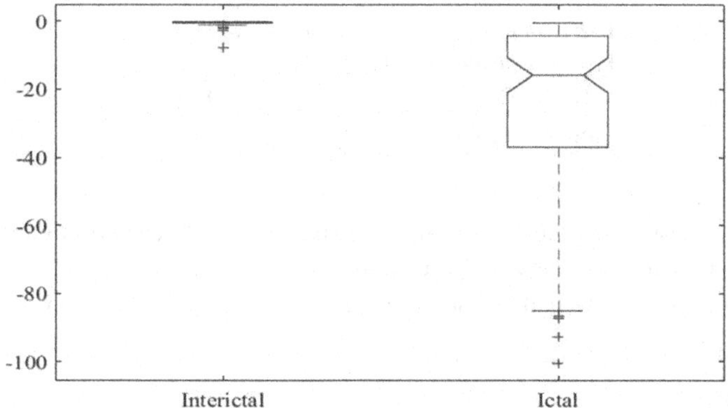

Fig. 5. The boxplot of the distribution of AWD

From Fig. 5, we can clearly find that the values of the interictal are higher than the ictal generally. It means that this feature can be used for epileptic seizure detection. It also means that the single feature proposed in this paper can become an excellent classification feature, so we can classify the EEG signals by studying the different values of AWD of complex network.

The reason that the characteristics of different EEGs are significantly different lies in two points. One reason is that the waveforms of different EEG signals have different shapes, which leads to a big difference in the 12 time domain statistical characteristics.

Then, we convert the time domain features into WHVG and extract complex network feature. This step makes the difference between the two EEG signals more obvious. By extracting statistical features of EEG signals in time domain and complex network domain, we obtain the dynamic characteristics of EEG signals. The feature reveals essential differences of different EEG signals.

Table 2 shows the classification accuracy of epileptic EEG signals based on the proposed method and other existing methods. It shows that our proposed method has better classification accuracy than other methods obtains a better classification result. It means that the method proposed in the paper has a stronger universality and applicability. Besides, it increases computational efficiency through dimensionality reduction while including more data. The combination of time-domain statistical features and WHVG provides a new direction for the study of signal processing. The 12 time-domain statistical features have been proven that they can contain a lot of information of time series. So when using them for EEG signals, useful statistics of EEG signals in the time domain can be obtained.

Table 2. The classification results of the proposed method and other existing methods

Method	Data length	ACC (%)
DWT [6]	4096	95
WVG [18]	1024	93.25
Clustering [19]	4096	93.6
Sample entropy [10]	2048	91.0
RQA [20]	1024	94
Proximity network [21]	2048	94.5
WHVG [17]	2048	93.5
Proposed method	4096	96.5

By converting the extracted time-domain statistical features into complex network, the essential difference between the statistical characteristics of EEG signals is revealed. So that we can better distinguish the interictal and ictal EEG signals.

5 Conclusions

A lot of epilepsy detection algorithms have been proposed. In this paper, we propose a new epileptic seizure detection method to classify ictal EEG signals from interictal EEG signals. In the proposed method, each single EEG segment containing 4097 points is divided into four sub-segments of 1024, 1024, 1024 and 1025, respectively. To extract time domain statistical features, each sub-segment is partitioned into 32 clusters. Then, 12 features are extracted from each cluster. All the features from one segment is regarded as a vector to represent the segment. The feature vectors are transformed into complex network by using the weighted horizontal visibility graph algorithm. The edge weight of WHVG is the difference between two connected nodes, which reflects the numerical difference between nodes. Finally, average weighted degree is extracted from WHVG, which is used as a single classification feature of interictal and ictal EEG signals and last classification accuracy is 96.5%. The experimental results showed that a combination of time domain characteristics and complex network provide more accurate classification results. Based on experimental results, we can reach the conclusion that the method proposed in this study can distinguish ictal EEG from interictal EEG with high accuracy. With the increasing incidence of epilepsy, the proposed method will help to better diagnosis and treatment of epilepsy.

Acknowledgments. This work was supported by the National Natural Science Foundation of China (Grant No. 61671220, 61701192, 61201428), the National Key Research and Development Program of China (No. 2016YFC0106000), the Project of Shandong Province Higher Educational Science and Technology Program, China (Grant No. J16LN07), the Shandong Province Key Research and Development Program, China (Grant No. 2016GGX101022), the Natural Science Foundation of Shandong Province, China, (Grant No. ZR2017QF004).

References

1. Tzallas, A.T., Tsipouras, M.G., Fotiadis, D.I.: Epileptic seizure detection in EEGs using time–frequency analysis. IEEE Trans. Inf. Technol. Biomed. **13**(5), 703–710 (2009)
2. Altunay, S., Telatar, Z., Erogul, O.: Epileptic EEG detection using the linear prediction error energy. Expert Syst. Appl. **37**(8), 5661–5665 (2010)
3. Acharya, U.R., Sree, S.V., Alvin, A.P.C., et al.: Use of principal component analysis for automatic classification of epileptic EEG activities in wavelet framework. Expert Syst. Appl. **39**(10), 9072–9078 (2012)
4. Kaplan, A.Y.: Segmental structure of EEG more likely reveals the dynamic multistability of the brain tissue than the continual plasticity one. In: Conference on Neural Information Processing, vol. 2, pp. 633–638. IEEE (1999)
5. Polat, K., Güneş, S.: Classification of epileptiform EEG using a hybrid system based on decision tree classifier and fast Fourier transform. Appl. Math. Comput. **187**(2), 1017–1026 (2007)
6. Kumar, Y., Dewal, M.L., Anand, R.S.: Epileptic seizure detection using DWT based fuzzy approximate entropy and support vector machine. Neurocomputing **133**, 271–279 (2014)
7. Li, S., Zhou, W., Yuan, Q., et al.: Feature extraction and recognition of ictal EEG using EMD and SVM. Comput. Biol. Med. **43**(7), 807–816 (2013)
8. Acharya, U.R., Molinari, F., Sree, S.V., et al.: Automated diagnosis of epileptic EEG using entropies. Biomed. Signal Process. Control **7**(4), 401–408 (2012)
9. Song, Y., Crowcroft, J., Zhang, J.: Automatic epileptic seizure detection in EEGs based on optimized sample entropy and extreme learning machine. J. Neurosci. Methods **210**, 132–146 (2012)
10. Lake, D.E., Richman, J.S., Griffin, M.P., Moorman, J.R.: Sample entropy analysis of neonatal heart rate variability. Am. J. of Physiol. Regul. Integr. Comp. Physiol. **283**, 789–797 (2002)
11. Acharya, U.R., Sree, S.V., Suri, J.S.: Automatic detection of epileptic EEG signals using higher order cumulant features. Int. J. Neural Syst. **21**(05), 403–414 (2011)
12. Du, X., Dua, S., Acharya, R.U., Chua, C.K.: Classification of epilepsy using high-order spectra features and principle component analysis. J. Med. Syst. **36**(3), 1731–1743 (2012)

13. Nurujjaman, M., Narayanan, R., Iyengar, A.S.: Comparative study of nonlinear properties of EEG signals of normal persons and epileptic patients. Nonlinear Biomed. Phys. **3**(1), 6 (2009)
14. Lacasa, L., Luque, B., Ballesteros, F., Luque, J., Nuno, J.C.: From time series to complex networks: the visibility graph. Proc. Natl. Acad. Sci. U.S.A. **105**(13), 4972–4975 (2008)
15. Luque, B., Lacasa, L., Ballesteros, F., Liuque, J.: Horizontal visibility graphs: exact results for random time series. Phys. Rev. E **80** (2009). Article ID 046103
16. Tang, X., Xia, L., Liao, Y., Liu, W., Peng, Y., Gao, T., Zeng, Y.: New approach to epileptic diagnosis using visibility graph of high-frequency signal. Clin. EEG Neurosci. **44**(2), 150–156 (2013)
17. Zhu, G., Li, Y., Wen, P.P.: Epileptic seizure detection in EEGs signals using a fast weighted horizontal visibility algorithm. Comput. Methods Programs Biomed. **115**(2), 64–75 (2014)
18. Supriya, S., Siuly, S., Wang, H., et al.: Weighted visibility graph with complex network features in the detection of epilepsy. IEEE Access **4**, 6554–6566 (2016)
19. Li, Y., Wen, P.P.: Clustering technique-based least square support vector machine for EEG signal classification. Comput. Methods Programs Biomed. **104**(3), 358–372 (2011)
20. Meng, Q.F., Chen, S.S., Chen, Y.H., et al.: Automatic detection of epileptic EEG based on recurrence quantification analysis and SVM. Acta Physica Sinica **63**(5) (2014). https://doi.org/10.7498/aps.63.050506
21. Wang, F., Meng, Q., Chen, Y., et al.: Feature extraction method for epileptic seizure detection based on cluster coefficient distribution of complex network. WSEAS Trans. Comput. **13**, 351–360 (2014)

Multi-path 3D Convolution Neural Network for Automated Geographic Atrophy Segmentation in SD-OCT Images

Rongbin Xu, Sijie Niu(✉), Kun Gao, and Yuehui Chen

Shandong Provincial Key Laboratory of Network Based Intelligent Computing, School of Information Science and Engineering, University of Jinan, Jinan, China
ise_niusj@ujn.edu.cn

Abstract. To automatically segment the geographic atrophy (GA) in spectral-domain optical coherence tomography (SD-OCT) images, we propose a novel segmentation method by designing a multi-path 3D convolution neural network (CNN) model in this paper. Firstly, the 3D patch was fed into the multi-path 3D CNN model as sample to preserve spatial features and overcome the excessive dependence of layer segmentation. Then, an improved classifier was trained by the optimization of network structure and the combination of softmax loss and center loss. The proposed method has been evaluated in two data sets, including fifty-five and fifty-six cubes respectively. For the two data sets, our method obtained the mean overlap ratio (OR) 87.24% ± 7.95% and 75.89% ± 15.11%. Compared with the state-of-the-art-algorithms on these two data sets, the mean OR of our results have been improved 5.38% and 5.89% respectively, indicating that our method can get higher segmentation accuracy.

Keywords: Image segmentation · SD-OCT · Multi-path 3D CNN Geographic atrophy

1 Introduction

Age related macular degeneration (AMD) is a chronic progressive disease in the elderly. According to its characteristics, it is mainly divided into two kinds: choroid neovascular disease (CNV) and geographic atrophy (GA). However, more than 85% to 90% of the AMD patients are GA. The characteristic appearance of GA results from the loss of the photoreceptor layer, retinal pigment epithelium (RPE), and choriocapillaris [1–3]. Currently, the treatment of retinal diseases have made great development, however there are still have no specific methods for the vision loss caused by geographic atrophy. So it's important to segment the disease region accuracy in order to analyze the influences of GA and predict its future development. However manual segmentation by experts is a time-consuming and burdensome task. It's easy to cause the problem of inaccurate boundary segmentation because of the different subjectivity of different experts. So it's very meaningful to realize automatic segmentation of GA lesion.

D.-S. Huang et al. (Eds.): ICIC 2018, LNCS 10955, pp. 493–503, 2018.
https://doi.org/10.1007/978-3-319-95933-7_58

For the analysis of the retinal diseases, there are several fundus imaging modalities, such as color fundus photographs [4], fundus auto fluorescence (FAF) [5]. However these methods are just applied to quantify atrophic area and are unable to identify retinal structure axially in fundus imaging modalities. Compared with the traditional fundus imaging modalities, the spectral-domain coherence tomography (SD-OCT) has become a key diagnostic technology in retinal structures. The volumetric OCT images can accurately identify imaging characteristics of GA and provide detailed anatomic assessments. Besides, the SD-OCT allows the axial differentiation of retinal structures, generating volumetric image data and visualizing GA region is SD-OCT by considering an axial projection of the volumetric data (*en face* OCT fundus image) [6, 7]. The pervious methods use graph theory and dynamic programming to segment the intra-retinal layers in images with GA and the segmentation results are used to measure the thickness and volume of RPE which can be as the evaluation of GA [8, 9]. Although the thickness and volume can be measured in SD-OCT images, it's hard to get the boundary of GA lesion. To directly identify the GA lesion, there are many state-of-the-art methods segment the GA lesion based on the projection images generated from SD-OCT volumetric data. Tsechpenakis et al. [10] propose the method that using geometric deformable model driven by dynamically updated probability fields to segment the GA in dry AMD of human eyes. A semi-automated GA segmentation algorithm with the geometric active contour model was proposed by Chen et al. [11] to segment the GA region in the projection images of SD-OCT. For better segmentation result, Niu et al. [12] proposed an automatic method combines a region-based C-V model with a local similarity factor in projection images of a choroid sub-volume. However all these methods identify GA regions based on the projection fundus images, which are sensitive to the layer segmentation result.

With the development of neural network, the deep learning algorithm has been widely used in image recognition [13], classification [14] and segmentation [15]. The deep learning has been successfully applied to medical image analysis, for example, liver segmentation, pulmonary nodules segmentation, and brain tissue, breast segmentation. Ji et al. [16] proposed that using voting strategy with deep AutoEncoder for GA lesion segmentation. They directly processed the image axial data as a processing sample without using the layer segmentation. However, this method ignores the spatial information.

In this paper, we proposed the multi-path 3D convolution neural network for the automatic segmentation of SD-OCT retina images. First, we construct the 3D data set without using layer segmentation and taking the spatial features into account, providing reliable input for the neural network. Then the multi-path 3D convolution neural network (CNN) models which was optimized by softmax loss and center loss was proposed to segment the GA region. Finally, we validated models on two data sets and obtain higher accuracy than the state-of-the-art-algorithms.

The structure of this paper is organized as follows. We introduce the multi-path 3D CNN model in Sect. 2. In Sect. 3, the detail of experiment and results will be explained.

2 Methodology

As Fig. 1 shown, an automatic GA segmentation model for SD-OCT images based on multi-path 3D CNN was proposed. First, in order to provide the appropriate data for the network, image denoising and constructing restricted regions are applied to the pre-processing stage. Then the multi-path 3D CNN supervised by the softmax loss and center loss is constructed. Finally, the final segmentation results are obtained by feeding the testing data into segmentation network.

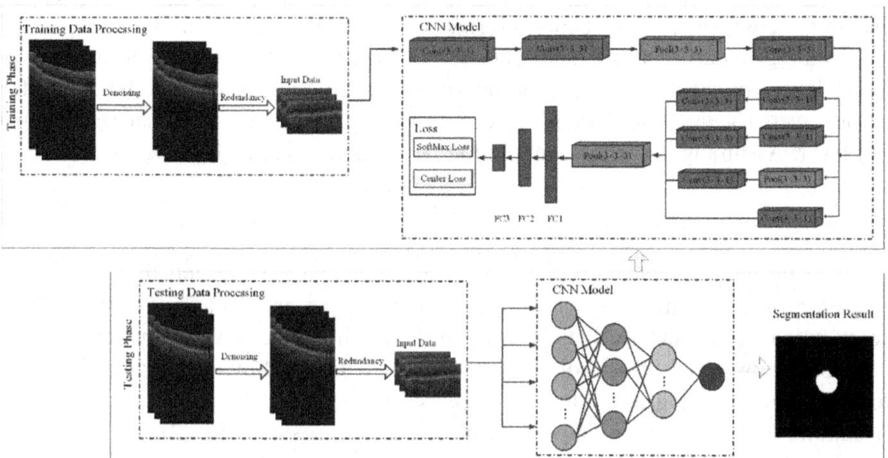

Fig. 1. Flow chart of the segmentation model where the red cubes represent the convolution operation, the green cubes represent the pooling operation and blue cubes represent the full connection operation (Color figure online)

2.1 Preprocessing

In the process of imaging, there are many image noise produced because of the imaging devices and other causes [17]. In order to reduce the influence of the noise in SD-OCT images, we used the bilateral filter for volumetric data denoising.

After the image denosing, reducing the redundant operation of the original images was proposed. In our experiment, we use the 3D volume data as samples, however there are so many redundant data in every column which not only cause the increase of computation, but also influence the segmentation result. So we restrict the sub-volume beneath the segmented internal limiting membrane (ILM) layer. Besides, in order to ensure that the edge data was not omitted in the process of window sliding, we were filled around the data set with 0 after eliminating redundancy.

2.2 Multi-path 3D Convolution Neural Network

In recent years, CNN has become one of the research hotspots in many fields of science. Especially in the field of pattern classification, because the network avoids the complex preprocessing of images, it can directly input original images, so it has been widely applied [18]. Similar to the ordinary neural network, CNN is composed of neurons with learning weights and bias constants. Each neuron receives input data and does dot operating, and the output is the score of each classification [19].

As Fig. 1 shown, our model consists of one input layer, multi convolution layers, pooling layers, full connection layers and one output layer. In the sixth layer of the network, we try to use multiple branches to replace one layer of classic network. The multi-path network is used to extract different features by using different convolution kernel to achieve the goal of enriching features. In addition, in this process, we try to use convolution to avoid the destruction of spatial features caused by pooling and to ensure the completeness of spatial information as much as possible.

As the general CNN, the multi-path 3D CNN includes the input layer, the convolution layer, the down sampling (the pooling) layer, the full connection layer and the output layer. The input of CNN is image *X*, and H_i represent the feature map ($H_0 = X$) of *i*-th layer. Assuming that H_i is convolution layer, the H_i can be described as:

$$H_i = f(H_i - 1 \otimes W_i + b_i) \tag{1}$$

W_i is the weight vector of *i*-th convolution kernel and $\otimes$ represents the convolution operation of convolution kernel with (*i* – *1*)-th images or feature maps, the sum of convolution output and the bias b_i through the nonlinear activation function f(x) can obtain the value of H_i. The pooling layer is usually followed by the convolution layer, for the down sampling to the feature maps according to certain down sampling rules. Assuming that H_i is pooling layer, the H_i can be described as Eq. (2):

$$H_i = subsampling(H_{i-1}) \tag{2}$$

After multiple convolution and pooling alternating transmission, CNN is classified by the features extracted by the full connection layer. And the probability distribution *Y* (l_i represent the *i*-th tag category) based on input can be obtained.

$$Y(i) = P(L = l_i | H_0 : (W, b)) \tag{3}$$

The purpose of training CNN is to minimize the loss function *L*(*W*, *b*) of the network. The input H_0 after the forward conduction and passed by the loss function can get the loss value, then the residual error can be calculated by the difference between the expected value and loss value. In this paper, besides the softmax loss, we also use the center loss [20] to calculate the distance between feature and feature center to reduce the intra class distance and update the trainable parameters (*W* and *b*) layer by layer with gradient descent as Eq. (4)

$$L = L_s + \lambda L_c \tag{4}$$

$$L_s = -\sum\nolimits_{i=1}^{m} log \frac{e^{W_{y_i}^T x_i + b_{y_i}}}{\sum\nolimits_{j=1}^{n} e^{W_j^T x_i + b_i}} \tag{5}$$

$$L_c = \frac{\lambda}{2}\sum_{i=1}^{m} \left|\left|x_i - c_{y_i}\right|\right|_2^2 \quad (6)$$

where L_s is the softmax loss function and L_c is the center loss function. In Eq. (6), the c_{yi} represents the feature center of y_i-th class, x_i represents the feature after full connection and m is the size of batch. In this paper, the λ was 0.001, setting manually by experiment.

3 Experiment and Results

3.1 Data Sets and Evaluation Criterions

In this paper, two different datasets were used to evaluate the performance of our proposed method where all these data contain GA lesion. Each SD-OCT image set was acquired over a 6 × 6 mm macular area and a 2 mm axial depth (corresponding to 1024 pixels) with the Cirrus (Carl Zeiss Meditec, Inc., Dublin, CA) device. The first data set consists of fifty-five cubes belonged to eight patients. The second data set consist of fifty-six cubes belonged to fifty-six patients. For the first data set, two independent experts manually drew the outlines of GA based on the B-scan images in two repeated separate session, which used to generate the ground truth. For the second data set, the expert drew the outline based on the FAF images and then mapped to the projection images to get the ground truth.

We constructed two training data for these two data sets. Each training set includes 200 thousand training samples, including 100 thousand positives samples and 100 thousand negative samples, and the size of each sample is 300 × 3 × 3. And all the cubes of two data sets are used as testing data.

We used three indicators to evaluate the performance of each method: overlap ratio (OR), absolute area difference (AAD) and correlation coefficient (cc). The detailed introduction of the evaluation indicator can be referred to [16]

3.2 Testing I: Segmentation Results on the Dataset with a Size of 1024 × 512 × 128

In this part, we validate our model on first data set which consists of fifty-five cubes from eight patients. In Fig. 2, we select eight samples to show the segmentation performance of our model. The outline of our segmentation results and ground truths are mapped to the projection images of SD-OCT images cubes. The red line is the boundary of ground truth and the green line represents the boundary of our segmentation result. Though the SD-OCT images are low contrast and there are so many similar features between normal region and lesion region, our method can achieve the better results. For the first data set, the training time of this model is 3.27 h and the testing time of 55 cubes is 0.75 h.

Figure 3 shows the comparison of our segmentation results and other two methods [11, 12]. The red line, green line, blue line and white line represent ground truth, our results, Niu's results and Chen's results respectively. For the first, second, and forth

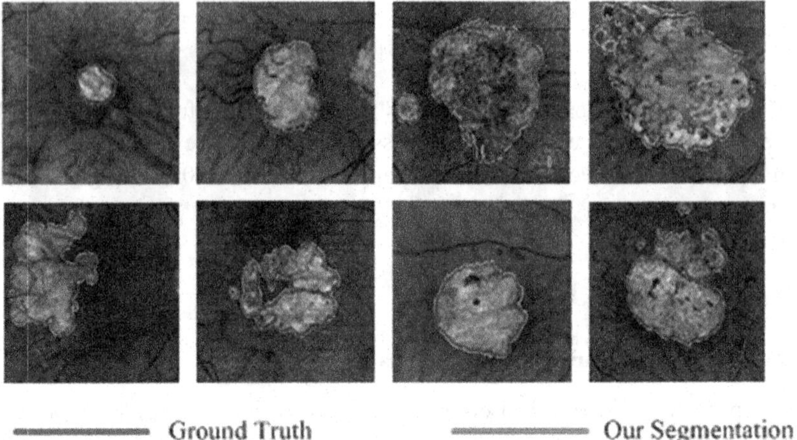

Fig. 2. The segmentation results and ground truth of eight patients overlaid on first data set's projection images (Color figure online)

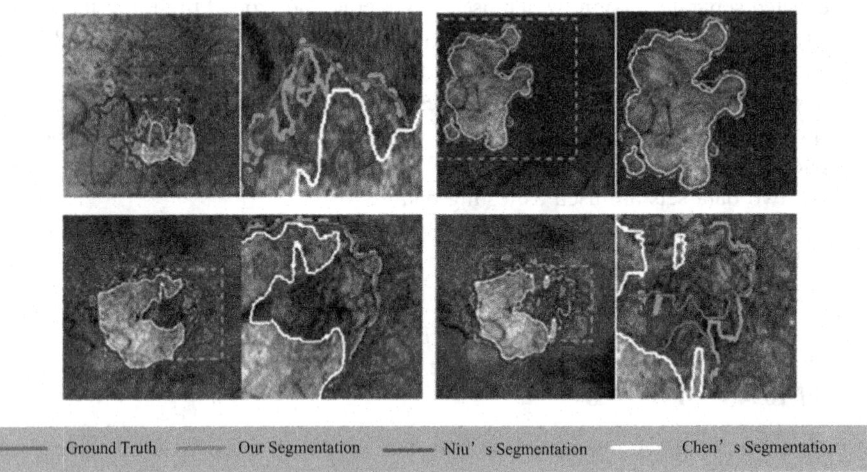

Fig. 3. Comparison of different method's segmentation results overlaid on first data set's projection images (Color figure online)

cubes, all three methods can obtain a better performance. However for other cubes, influenced by image quality and resolution, some segmentation errors are produced. In the third and fifth cubes, all of Niu's and Chen's methods misclassify the lesion region into normal region. And in the sixth cube, Chen's method misclassifies the lesion region into normal region while Niu's method misclassifies normal region into lesion region. Compared with the two methods, our segmentation model can have a better performance.

Table 1 summarized the quantitative comparison of different methods' segmentation results and ground truth [11, 12]. Compared with ground truth, our segmentation model has higher OR, cc and lower AAD than Chen's and Niu's methods, it indicates that our results are more similar to the ground truth.

Table 1. The quantitative summarize (mean ± standard deviation) of segmentation results and ground truth on first data set

Methods		Chen's method	Niu's method	Our method
Avg. expert	Cc	0.970	0.979	**0.9971**
	AAD [mm^2]	1.438 ± 1.26	0.811 ± 0.94	**0.83 ± 0.49**
	AAD [%]	27.17 ± 22.06	12.95 ± 11.83	**13.03 ± 9.67**
	OR [%]	72.60 ± 12.01	81.86 ± 12.01	**87.24 ± 7.95**
Expert A_1	Cc	0.967	0.975	**0.9931**
	AAD [mm^2]	1.308 ± 1.28	0.758 ± 0.99	**0.45 ± 0.51**
	AAD [%]	25.23 ± 22.71	12.62 ± 12.86	**6.68 ± 7.56**
	OR [%]	73.26 ± 15.61	81.42 ± 12.12	**89.52 ± 7.69**
Expert A_2	Cc	0.964	0.976	**0.9928**
	AAD [mm^2]	1.404 ± 1.31	0.853 ± 1.04	**0.38 ± 0.55**
	AAD [%]	26.14 ± 21.48	13.32 ± 12.74	**5.97 ± 8.77**
	OR [%]	73.12 ± 15.15	81.61 ± 12.29	**89.85 ± 7.95**
Expert B_1	Cc	0.968	0.976	**0.9939**
	AAD [mm^2]	1.597 ± 1.33	0.984 ± 1.08	**0.34 ± 0.48**
	AAD [%]	29.21 ± 22.17	14.91 ± 12.65	**5.15 ± 6.78**
	OR [%]	71.16 ± 15.42	80.05 ± 13.05	**89.01 ± 7.94**
Expert B_1	Cc	0.977	0.975	**0.9968**
	AAD [mm^2]	1.465 ± 1.14	0.897 ± 1.05	**0.32 ± 0.28**
	AAD [%]	27.62 ± 20.57	14.07 ± 11.78	**5.04 ± 5.50**
	OR [%]	72.09 ± 14.82	80.65 ± 12.51	**89.59 ± 7.22**

3.3 Testing II: Segmentation Results on the Dataset with a Size of 1024 × 200 × 200

In this part, we validate our segmentation model on the second data set which consists of fifty-six cubes from fifty-six patients. In Fig. 4, eight segmentation results were selected to show the performance of our segmentation model. The red line and green line are the boundary of ground truth and our segmentation results. Through observation, our model has achieved good segmentation results. For the second data set, the training time of this model is 3.21 h and the testing time of 56 cubes is 0.30 h.

Figure 5 shows the comparison of different segmentation models [11, 12], the red line, green line, blue line and white line are the boundary of ground truth, our results, Niu's results and Chen's results respectively. For the third cube, all these three methods have better performance, however for the first, second and forth cubes, some misclassifications were occurred, Chen's method misclassifies the normal region into lesion region in first cube, Niu's method misclassifies the lesion region into normal region in forth cube and both Chen's and Niu's methods misclassify the lesion region

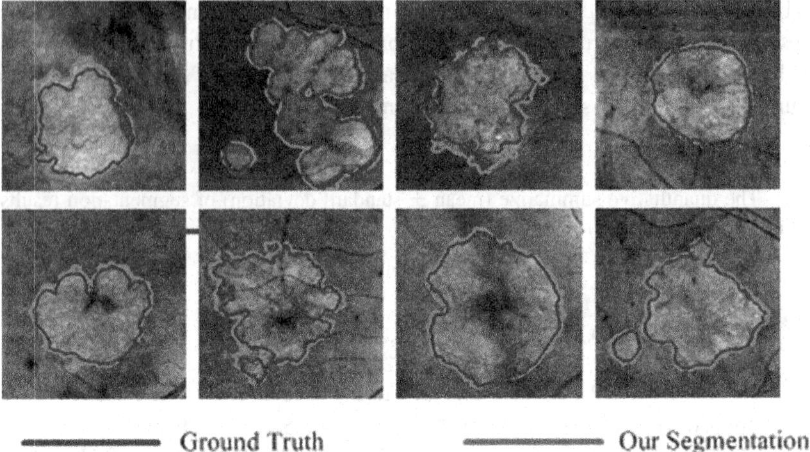

Fig. 4. The segmentation results and ground truth of eight patients overlaid on second data set's projection images (Color figure online)

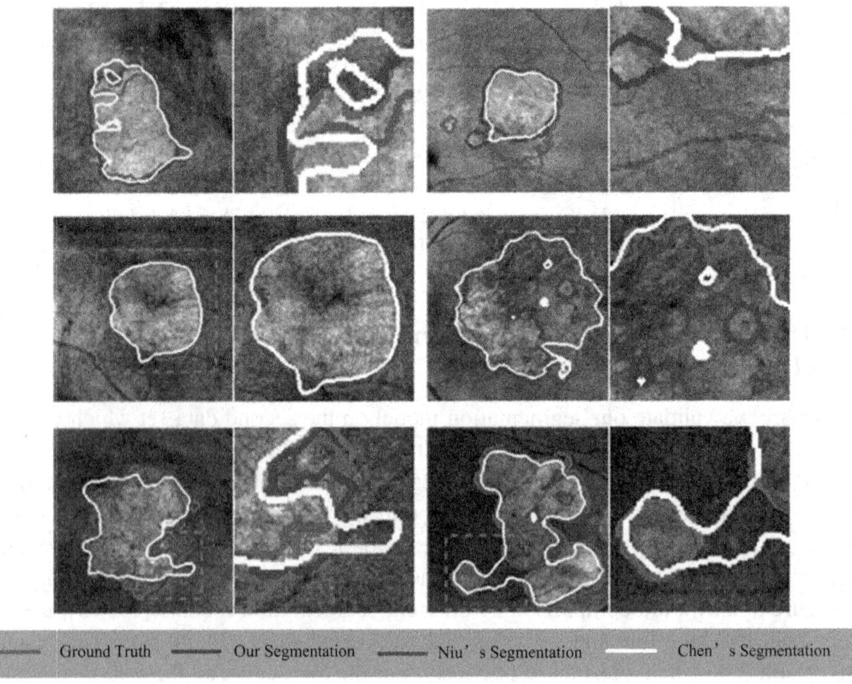

Fig. 5. Comparison of different method's segmentation results overlaid on second data set's projection images (Color figure online)

into normal region in second and fifth cubes. What's more, for the sixth cube, Chen's method misclassifies the normal region into lesion region while Niu's method misclassifies lesion region into normal region.

Table 2 summarized the quantitative comparison of different methods' segmentation results and ground truth [11, 12]. Compared with ground truth, our segmentation model has higher OR, cc than Chen's and Niu's methods, it indicates that our results are more similar to the ground truth.

Table 2. The quantitative summarize (mean ± standard deviation) of segmentation results and ground truth on second data set

Methods	Chen's method	Niu's method	Our method
cc	0.970	0.979	**0.9942**
AAD [mm^2]	0.951 ± 1.28	1.215 ± 1.58	**1.39 ± 1.12**
AAD [%]	19.68 ± 22.75	22.96 ± 21.74	**22.85 ± 22.71**
OR [%]	65.88 ± 18.38	70.00 ± 15.63	**75.89 ± 15.11**

4 Conclusion

This paper presents a novel automated segmentation algorithm for GA in SD-OCT images. The 3D patch samples constructed from the sub-volume of the retina beneath the IML to reduce the interference of redundant data. And the multi-path 3D CNN was proposed to extract the spatial features and overcome the problem of overdependence layer segmentation. The combination of softmax loss and center loss is more effective for the optimization of the model.

As the above tables shown, we evaluated our model on two data sets. From the Tables 1 and 2, we can observe that the comparison of different methods in OR, ADD and cc. For the first data set, we improved 5.38% and for the second data set, we improved 5.89%. On the other hand, our method has limitations as well. The combination of softmax loss and center loss can reduce the intra class distance, however for the normal region which is similar to the lesion region, the model can't distinguish well. The future work will focus on how to increase inter class distance and reduce the false positives.

Acknowledgement. The work is supported by the National Natural Science Foundation of China under Grant No. 61701192, the Natural Science Foundation of Shandong Province, China, under Grant No. ZR2017QF004, China Postdoctoral Science Foundation under Grants No. 2017M612178, the Shandong Provincial Key R&D Program (2016ZDJS01A12), the National Key Research and Development Program of China (No. 2016YFC0106000), Shandong Province Natural Science Foundation (No. ZR2018LF005).

References

1. Bressler, N.M.: Age-related macular degeneration is the leading cause of blindness. JAMA **291**(15), 1900–1901 (2004)
2. Bhutto, I., Lutty, G.: Understanding age-related macular degeneration (AMD): relationships between the photoreceptor/retinal pigment epithelium/Bruch's membrane/choriocapillaris complex. Mol. Aspects Med. **33**(4), 295–317 (2012)
3. Nunes, R.P., Gregori, G., Yehoshua, Z., et al.: Predicting the progression of geographic atrophy in age-related macular degeneration with SD-OCT en face imaging of the outer retina. Ophthalmic Surg. Lasers Imaging Retina **44**(4), 344–359 (2013)
4. Niemeijer, M., Van Ginneken, B., Staal, J., et al.: Automatic detection of red lesions in digital color fundus photographs. IEEE Trans. Med. Imaging **24**(5), 584–592 (2005)
5. Spaide, R.F.: Fundus autofluorescence and age-related macular degeneration. Ophthalmology **110**(2), 392–399 (2003)
6. Garvin, M.K., Abramoff, M.D., Wu, X., et al.: Automated 3-D intraretinal layer segmentation of macular spectral-domain optical coherence tomography images. IEEE Trans. Med. Imaging **28**(9), 1436–1447 (2009)
7. Quellec, G., Lee, K., Dolejsi, M., et al.: Three-dimensional analysis of retinal layer texture: identification of fluid-filled regions in SD-OCT of the macula. IEEE Trans. Med. Imaging **29** (6), 1321–1330 (2010)
8. Chiu, S.J., Izatt, J.A., O'Connell, R.V., et al.: Validated automatic segmentation of AMD pathology including drusen and geographic atrophy in SD-OCT images. Invest. Ophthalmol. Vis. Sci. **53**(1), 53–61 (2012)
9. Yamashita, T., Yamashita, T., Shirasawa, M., et al.: Repeatability and reproducibility of subfoveal choroidal thickness in normal eyes of Japanese using different SD-OCT devices. Invest. Ophthalmol. Vis. Sci. **53**(3), 1102–1107 (2012)
10. Tsechpenakis, G., Lujan, B., Martinez, O., et al.: Geometric deformable model driven by CoCRFs: application to optical coherence tomography. In: International Conference on Medical Image Computing and Computer-Assisted Intervention, pp. 883–891 (2008)
11. Chen, Q., de Sisternes, L., Leng, T., et al.: Semi-automatic geographic atrophy segmentation for SD-OCT images. Biomed. Opt. Express **4**(12), 2729–2750 (2013)
12. Niu, S., de Sisternes, L., Chen, Q., et al.: Automated geographic atrophy segmentation for SD-OCT images using region-based CV model via local similarity factor. Biomed. Opt. Express **7**(2), 581–600 (2016)
13. Wang, S., Chen, L., Xu, L., et al.: Deep knowledge training and heterogeneous CNN for handwritten Chinese text recognition. In: 15th International Conference on Frontiers in Handwriting Recognition (ICFHR), pp. 84–89. IEEE (2016)
14. Qi, C.R., Su, H., Nießner, M., et al.: Volumetric and multi-view CNNs for object classification on 3D data. In: Proceedings of the IEEE Conference on Computer Vision and Pattern Recognition, pp. 5648–5656 (2016)
15. Kamnitsas, K., Ledig, C., et al.: Efficient multi-scale 3D CNN with fully connected CRF for accurate brain lesion segmentation. Med. Image Anal. **36**, 61–78 (2017)
16. Ji, Z., Chen, Q., Niu, S., et al.: Beyond retinal layers: a deep voting model for automated geographic atrophy segmentation in SD-OCT images. Transl. Vis. Sci. Technol. **7**(1), 1 (2018)
17. Cameron, A., Lui, D., Boroomand, A., et al.: Stochastic speckle noise compensation in optical coherence tomography using non-stationary spline-based speckle noise modelling. Biomed. Opt. Express **4**(9), 1769–1785 (2013)

18. Garcia-Garcia, A., Orts-Escolano, S., Oprea, S., et al.: A review on deep learning techniques applied to semantic segmentation. arXiv preprint arXiv:1704.06857 (2017)
19. Yu, L., Chen, H., Dou, Q., et al.: Integrating online and offline three-dimensional deep learning for automated polyp detection in colonoscopy videos. IEEE J. Biomed. Health Inform. **21**(1), 65–75 (2017)
20. Wen, Y., Zhang, K., Li, Z., Qiao, Yu.: A discriminative feature learning approach for deep face recognition. In: Leibe, B., Matas, J., Sebe, N., Welling, M. (eds.) ECCV 2016. LNCS, vol. 9911, pp. 499–515. Springer, Cham (2016). https://doi.org/10.1007/978-3-319-46478-7_31

Multi-information Fusion Based Mobile Attendance Scheme with Face Recognition

Likai Dong[1], Qinlin Li[1], Tao Xu[1,2], Xuesong Sun[1(✉)], Dong Wang[1,2], and Qingqing Yin[1]

[1] School of Information Science and Engineering, University of Jinan, Jinan 250022, Shandong, China
ise_sunxs@ujn.edu.cn
[2] Shandong Provincial Key Laboratory of Network Based Intelligent Computing, Jinan 250022, Shandong, China

Abstract. There are some problems in traditional classroom attendances, such as complex interaction, masquerading, waste of time and information out of sync. A novel class attendance scheme is proposed based on face recognition and the service of location for these problems. The proposed scheme has the functions of face detection, altitude detection and position location, which is better for saving the time of attendance, reducing student impostor, convenient for teachers to check on work attendance and improve the effective feedback of information. Experimental results show the convenience and effectiveness of the proposed scheme.

Keywords: Mobile attendance scheme · Face recognition
Multi-information fusion

1 Introduction

Class attendance is an important content of the construction of study style in university, which can guarantee the teaching order effectively and the quality of personnel training. So far most of the colleges and universities in our country still adopt the traditional way like check-in and roll call to check on attendance, however people often use pseudonym in this way and always time-consuming, and classroom attendance equipment tend to have obviously deficiencies in high implementation cost, difficulties of installation, news out of sync and so on. With the popularity of mobile terminal, the method of check on work attendance research papers about solutions [1–4] are mostly based on the mobile end face recognition, this way only guarantee the reliability of the identity, and does not have effective solutions about some vulnerability such as a portable mobile terminal, different check-in, across the floor sign and this system uses the face recognition, latitude and longitude positioning, altitude positioning to solve the vulnerability. Attendance data about the feature of report form improves the information feedback effectively, the system is an open source, low cost in implement of a portable Android system, convenient to install, easy to use, does not depend on the special equipment.

D.-S. Huang et al. (Eds.): ICIC 2018, LNCS 10955, pp. 504–514, 2018.
https://doi.org/10.1007/978-3-319-95933-7_59

2 The Mobile Attendance Scheme

The most important function of the system's design is the interaction of the process of the roll call. The interaction based on the teacher end and the student end, in which the teacher call the roll and then the current system will get the current location and the information about height for using json to transfer data to the server later, the server send information message to the student end. The student end will get the current location and height information when signing in to certify. By measuring the height differences between the geographical position of the teacher end and the student end and then calculate the height differences. It will return the verification results to the server and update data by teachers if the current distance and height is beyond the scope of threshold value, if the current distance and height in the range of the threshold, it will call the face verification module to verify and the cloud will return the verification information to the student end and in turn back and finally the teacher end will update the data. The proposed scheme is shown in Fig. 1.

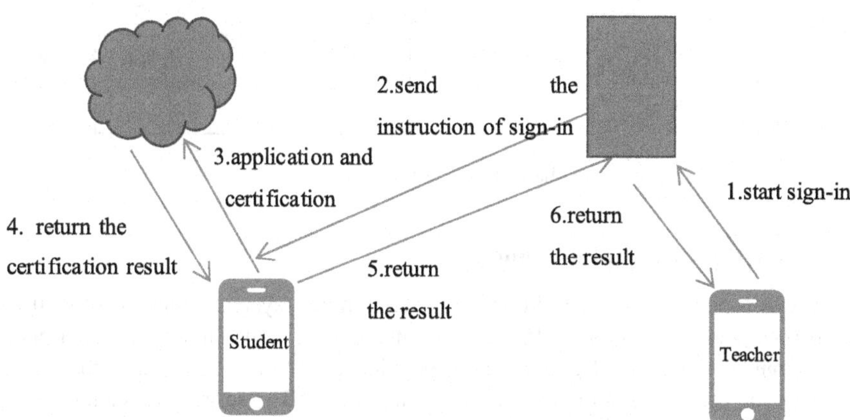

Fig. 1. The diagram of mobile attendance scheme.

The whole system according to the recognition module is divided into face recognition module and position recognition module and high recognition module, the following is the specific process description about the face recognition module.

2.1 Face Recognition

Face recognition module is divided into two parts: face verification and face identification [5]. The program makes face login interface as the entrance, when a user face swiping, the client will uploaded image of face to the cloud, the cloud firstly registers module for mobile user and then upload user face image and cell phone number to the cloud images and generate a unique identity-userInfo. When a user logs in by face swiping, it will upload the image to the cloud database and does face alignment operation for face image in the cloud and marks the position of feature points of images shot from different angles of users. Then the cloud face information database will identify face, if get the information will back to the userInfo, if not, then start the pop up window of registration for face registration. When the class call the roll and sign, face image carries on the userInfo information which currently logged on and upload the information to the cloud for face verification. The face recognition process is shown in Fig. 2.

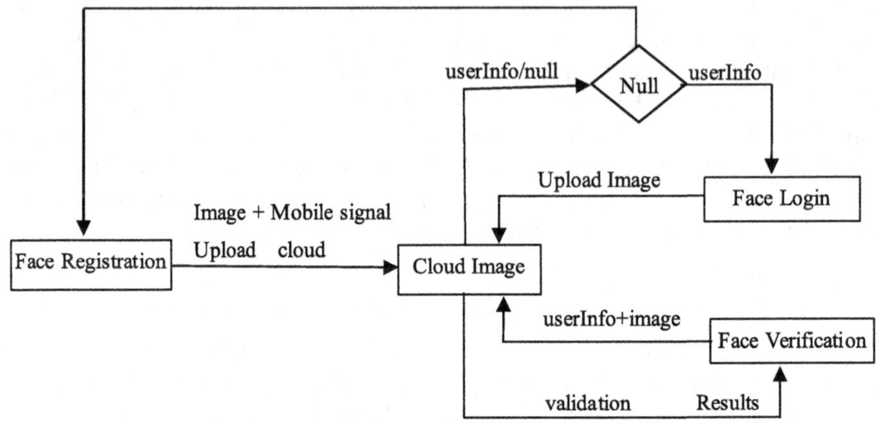

Fig. 2. The flow chart of face recognition and face verification.

2.2 Three-Dimensional Positioning

Geographical position and height measurement in the system constitute the three-dimensional positioning module [6, 7]. when the module signs on at the teacher end, it acquires the information of geographical position and height and then will sent the information to the server in the form of encoding of json. The server sends a message to the student end and notify the student end to sign in and then compares the deviation between geographical position and height information acquired by the student end and relevant information request server to obtain during students signing in, if the deviation in the scope which meets the requirement and then accept it within the prescribed scope, if there is any discontent will be regard as beyond the scope of regulation, and remind the user of error messages, whether successful or not, data of check-in upload to the server and notify the teacher end to update. This procedure is illustrated in Fig. 3.

2.3 Face Recognition

Comparison of Face Recognition Methods. Authentication technology is an effective solution in the computer network in the process to determine the identity of the operator, and the basic method of identity authentication can be divided into authentication based on secret information, authentication based on trust object, authentication based on biological characteristics.

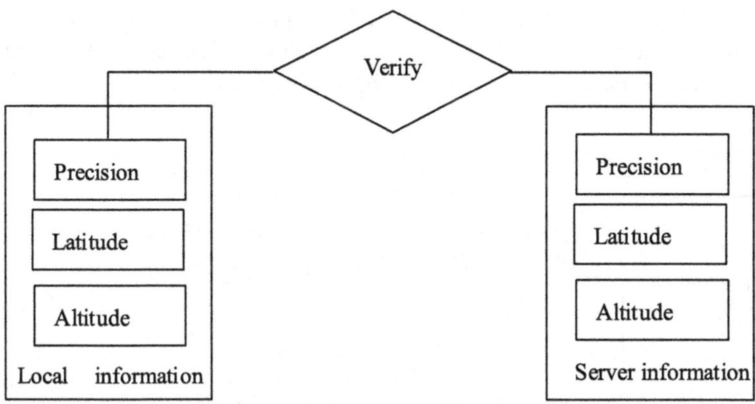

Fig. 3. Three-dimensional positioning module.

With the development of science and technology and the improvement of information protection consciousness, different verification methods show certain defects during authentication (Table 1).

Table 1. Comparison of authentication methods.

	Advantages	Disadvantages	Identity risk
Based on secret information	Practical, convenient	Easy to be forgot	Easy to be embezzled
Based on trust object	Practical, convenient	Need to be carried, easy to be lost	Easy to be embezzled
Based on biology feature	Unique, accurate	Many factors causing wrong recognition	Hard to be embezzled

As an emerging authentication method based on biology feature, compared with the fingerprint recognition needing to integrate the fingerprint recognition hardware, the face recognition is better on the mobile phones with Android (Table 2).

Table 2. Comparison of authentication methods based on biology features.

	Hardware dependence	Contact or not	Realization cost	Time cost	Recognition accuracy
Fingerprint recognition	Dedicated equipment	Contact	Medium	Medium	Medium
Face recognition	Camera	Not contact	Low	Low	High

Principle of Face Recognition. The generalized face recognition process includes face image acquisition and prepossessing, face detection and feature extraction and face comparison and recognition. The principle is shown in Fig. 4.

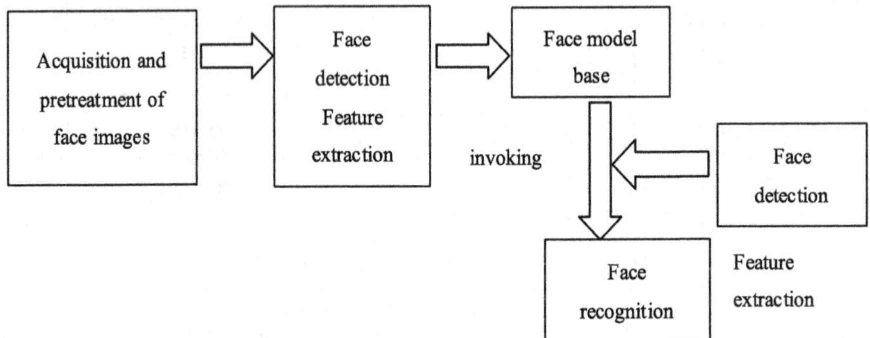

Fig. 4. The principle diagram of face recognition.

Face Detection and Feature Extraction. Principal Component Analysis(PAC), which is a common algebraic feature extraction method and widely used because it is efficient in the face information extraction and dimension reduction. The main idea of the PAC is: to achieve the purpose of the reduction of image dimension with relatively small data features to describe samples, mainly by using maximum direction of variance and the distribution of its location in space of the sample points to calculate the difference vector, so as to complete the feature extraction. The face principal component extraction is mainly obtained by using L - L transformation.

Face Recognition. We project registered face sample vectors towards feature vector space, then get the projection coefficient of samples, when the user login next time, we will get the sample and project it towards feature vector space, to get the projection coefficient of login sample. We calculate the norm of, the projection coefficient in feature of the face under test and all the registered face. Select the minimum norm, then the identity of login sample can be determined (see Fig. 5).

Location. The system hopes to use LBS as a basis for obtaining location, because of the scene using is mainly indoor environment and single based on GPS positioning error does not meet this requirement, owing to the system needs to calculate whether the two coordinate values within the prescribed scope, so we use Gaode map's more precise positioning way which based on the network and GPS and measuring the distance with the function "calculateLineDistance ()" which provided by Gaode map [8, 9].

Height Measurement. There are usually two methods to measure elevation, one is using Global Position System, and the other is to calculate the elevation according to the air pressure through the measured atmospheric pressure [6].

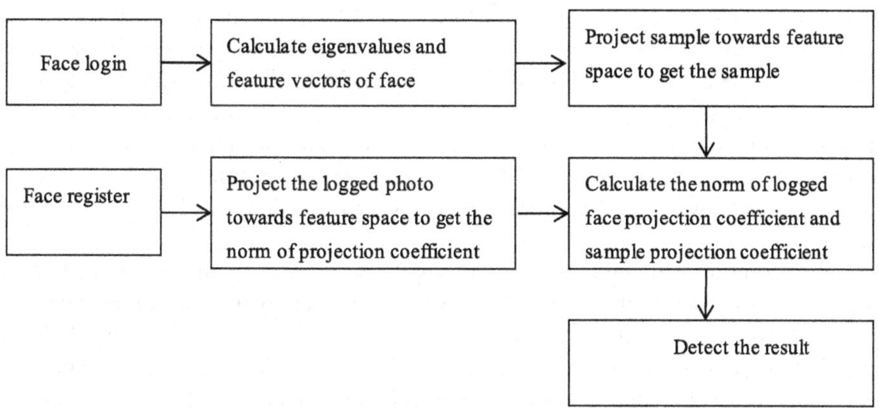

Fig. 5. The flow chart of face recognition.

Due to the limitation of technology and other reasons, there are about ten meter's error by using the GPS to calculate elevation, in order to meet the indoor precision, firstly using pressure sensor as the basis of height. Currently most of the mobile devices have air pressure sensor and the mobile phone which have this components can use the interface provided by Android system to get the atmospheric pressure, with the following formula:

$$P = P_0\left(1 - \frac{A}{44300}\right)^{5255} \quad (1)$$

In which P is the current atmospheric pressure, P_0 is the standard atmospheric pressure, A is elevation. Formulas are obtained by formula conversion:

$$A = 44300 * \left[1 - \left(\frac{p}{p_0}\right)^{\frac{1}{5255}}\right] \quad (2)$$

When the current equipment has no built-in air pressure sensor, it can obtain the elevation through the global positioning system (GPS) which adopts Gaode Map's api to get the height information in order to obtain high accuracy elevation.

Image Compression. In order to reduce the delay caused by server interacting and the server load of large files, and to accelerate the checking speed of attendance system, the image compression technology are adopted to reduce the size of image and accelerate the average response speed of attendance system.

The compression of images becomes an important issue as the resolution of the camera's camera increases. Image processing in Android uses an open source image processing engine called Skia, which is located in the Android source/external/ Skia. Usually, the image processing program we used in Java is actually using some related functions in it. In the applications of Android, we often use quality compression and reduced sampling compression to compress. One of them, the quality compression, is using the engine Skia to process. It often uses a dynamic library to compress named libjpeg.so. In the early Android ages, considering the efficiency, they used fixed length code to encode Android jpeg compression. Thus, we transplant libjpeg, and compile, and we get a dynamic library from it. We use JNI to achieve using Hoffman algorithm to do quality compression. Through the measured, we find that there are 5 to 10 times size difference between used and unused Hoffman to encode files which has the same quality.

Considering in reducing sampling rate ways, because we use mobile phone to take photos, the proportion of the picture(the short edge divided by the long edge) usually is 16:9 or 16:10, which is suit to mobile phone used in daily life. We use the following algorithm to decide its length, width, size, and other values. We define the short edge of the picture as width, and the long edge as height, then calculate the scale via width/height.

According to the interval of the scale, we compare the height with boundaries values in forms. If the height is bigger than one of the boundaries values, then assigns the compressing edge thumbW, thumbH to width/pow(2,n–1), height/pow(2,n–1). Then get the real size of the picture based on relation formula. Judge if the size is too small, and assign the real size to true size based on the relation. Trans the thumbW, thumbH and true size into the compression process to compress, until the picture is suitable.

3 Experimental Results and Analysis

3.1 System Operation Results

The experimental terminal using the mobile phone called Nexus 6p, which is produced by Huawei LTD. Nexus 6p has a Qualcomm Valon 810(MSM8994) CPU, which include 8 cores, and 3 GB of memory. It also has front-camera with 8 million pixels. And the operating system is Android 6.0. The testing environment is the classroom of college about 70 m^2. It has 82 seats in total, and the height is 3.8 m. The light of this environment is normal. The expression of observed is normal, too. There is also no obvious rotation of their head.

This test chooses ten users with no twins or similar look to register and test authentication. The people who have registered will test three times, first text is the legal test, and the following two times are repeating tests, authentication for 5 times each, record the number of initial registration, the number of repeat registration, the rate of certification pass and the specific time. The specific test results are shown in Table 3.

According to the experimental results, the feedback of recognition result can keep within 2 s, the deviation of recognition is in the acceptable range. In order to ensure the recognition rate of secondary recognition, the system will recognize the process again after rejection, and the whole process can basically achieve the actual use process [10].

Table 3. Performance test results.

Test project	Test result	The average time/ms
Initial registration pass rate.	1	1760
Repeat registration pass rate.	0	1200
Certification rejection rate	0.06	1256
Certification rate	0.96	1680
Certification error rate	0.04	1489

3.2 Image Compression Data Analysis

In order to test the actual compression quality about the use of mobile phones, different sizes, pictures of different sizes. This article compares and texts the size of picture before you send it by WeChat's Android, and the test data is shown in Table 4.

Table 4. Image compression quality.

Image	Awork	Luban	Wechat
Screen shot 720P	720*1280 390k	720*1280 87k	720*1080 56k
Screen shot 1080p	1080*1920 2.21M	1080*1920 104K	1080*1920 112K
Photograph 13M (4:3)	3096*4128 3.12M	1548*2064 141K	1548*2064 147K
Photograph 9.6M (16:9)	4128*2322 4.64M	1032*581 97k	1032*581 74k

The test results show that the compression algorithm can effectively compress the picture and avoid the occurrence of the situation of OOM, simply observation from the test result, the algorithm's compression rate is higher for large image and has the similar effect compared with WeChat, the level of compressing picture can effectively reduce bandwidth consumption and communication consumption time.

3.3 Multi-platform Data Analysis

Considering that the user's degree of freedom will be higher in actual use, the recognition rate of the distance will be affected by the user's environment, the user's expression and so on.

This paper has carried on more demanding text for the three platforms which have a good reputation about face recognition in the clouds currently (SuperID, Tencent picture, Baidu AI) using the face library of ideal test. This article selects the actual situations that often appear of the light is dark, whether wear glasses, facial expressions, and head movements and those situation about 13special cases and 2 normal, the specific test results are as follows.

Normal Text. In the beginning, this article tested 2 cases in the most of the time of the actual use, one is the recognition rate which the same person with different photos and the other is that the different people misjudged to the same person, the specific test results are shown in Table 5.

Table 5. Recognition performance test.

Text Project	SuperID		Tencent picture		Baidu AI	
	Recognition rate	Time-consuming/ms	Recognition rate	Time-consuming/ms	Recognition rate	Time-consuming/ms
Same people	1	2740	1	3650	1	526
Different people	0	2810	0	3540	0	498

From the testing results we found that for the normal situation the test accuracy of three platforms are very high and can meet requirements of the actual use, testing speed in the range that users can accept, It can be said that all three platforms can be selected in actual use.

Special Case Analysis. In order to test the processing ability of three platforms in some special cases, 13 kinds of special cases of life has been selected and tested separately. The specific test results are shown in Table 6.

Table 6. Platform specific processing performance test.

Test Project	SuperID		Tencent picture		Baidu AI	
	Recognition rate	Time-consuming/ms	Recognition rate	Time-consuming/ms	Recognition rate	Time-consuming/ms
Normal<->Glasses	0.423	1800	0.805	2740	0.943	530
Normal<->Laugh	0.504	1060	0.806	2260	1.000	240
Normal<->left 30	0.512	1150	0.902	2040	0.992	860
Normal<->left 45	0.568	1170	0.642	2210	0.976	870
Normal<->left 90 Side face	0.000	1160	0.089	2780	0.626	580
Normal<->left 90 Turned	0.650	1140	0.670	2680	0.959	810
Normal<->Rise	0.390	1130	0.504	1960	0.991	820
Normal<->Bow	0.268	1140	0.707	2550	0.967	800
Closeeyes<->Glasses	0.413	2030	0.578	2820	0.943	330
Left 30<->Right 30 Side face	0.626	2490	0.772	2100	0.991	240
Left 45<->Right 45 Side face	0.781	2490	0.228	1790	0.943	310
Left 45<->Right 45	0.439	1350	0.618	1780	0.959	300
Rise<->Bow	0.508	3070	0.114	1300	0.919	270

After testing we found that the recognition ability and speed of the three platforms have influence on the special condition test, and the three platforms are different in the direction of emphasis. Baidu AI is better simply from the text results, whether the identification rate or time consuming are more obvious priority than the other two platforms. Comparing Tencent picture and Baidu AI, they are similar in identification rate while the former is not so high-speed as the latter in recognition. So we can see that compared with the other two, SuperID lacks in recognizing these special cases.

However, SuperID has convenience in integration and according to the test results above, it is enough to deal with most of the situations in actual use.

4 Conclusions

The new method of class attendance put forward in this article depends on the image compression algorithm. It uploaded local photos of authentication to the cloud server with low footprint, low bandwidth and low load and it also reduces the time, memory, bandwidth consumption of bandwidth. Place the authentication server in the cloud can effectively identify students, increase the portability of the system, solve the problem of the performance of the mobile intelligent terminals, reduces the dependence on the client's hardware, dependent on LBS and barometer can effectively guarantee the user's sign-in place. Sign in a certain scope can effectively reduce the imposter and the situation of sign-in in different locations. The development of system based on mobile intelligent terminal is convenient to the roll call process, increases the speed, saves the time, saves the teacher's time effectively and is convenient for teachers to manage the student's information, solves the disadvantages of the traditional ways of attendance.

Acknowledgement. This research was supported by Shandong Provincial Natural Science Foundation (No. ZR2018LF005), the Scientific Research Fund of Jinan University (No. XKY1711, No. XKY1622, No. XBS1653), Industry-University Cooperative Education Project of Ministry of Education (No. 201601023018) and Teaching Research Project of Jinan University (No. J1638).

References

1. Jing, W.: The design and realization of image attendance system based on face recognition. Wirel. Internet Technol. **10**, 52–53 (2015)
2. Dong, L., Cui, X., Zhang D., Zhang, H.: A student attendance system based on face recognition technology. J. Daqing Norm. Univ. (03), 15–18 (2014)
3. Yan, H., Li, C.: Design and implementation of face recognition attendance system. J. Tonghua Norm. Univ. **37**(12), 1–3 (2016)
4. Li, G.: A study based on face recognition enterprise attendance system. Comput. Age (04), 53–55 (2017)
5. Yong, Z., Jielin, Z., Guizhen, W., Shengnan, Z.: Research and implementation of face recognition system in Android platform. J. Nanjing Inst. Eng. (Nat. Sci.) **01**, 53–57 (2013)
6. Hong, H., Wu, G., Chen F., Chen, Y.: Indoor pedestrian 3D localization algorithm based on smart phone sensor. Sci. Surv. Mapp. (07), 47–52 (2016)

7. Hua, S., Ci, S., Tao, P.: Research of indoor positioning data analysis and application. Prog. Geogr. **35**(05), 580–588 (2016)
8. Xiaorui, W., Yilin, L.: Design and implementation of mobile campus information push system based on indoor positioning technology. Fujian Comput. **31**(03), 5–6 (2015)
9. Yi, J., Lu, A.: Research on indoor 3D navigation system based on Android platform. Jiangxi Sci. (3), 446–450 (2017)
10. Yongping, L., Fatu, Z.: Research on the data collection and analysis of college students' attendance data based on the Android platform. J. Ningde Norm. Univ. (Nat. Sci. Edit.) **28** (03), 255–259 (2016)

The Wide and Deep Flexible Neural Tree and Its Ensemble in Predicting Long Non-coding RNA Subcellular Localization

Jing Xu[1,2], Peng Wu[1,2(✉)], Yuehui Chen[1,2(✉)], Hussain Dawood[3], and Dong Wang[1,2]

[1] School of Information Science and Engineering, University of Jinan, Jinan, China
{ise_wup, yhchen}@ujn.edu.cn

[2] Shandong Provincial Key Laboratory of Network Based Intelligent Computing, Jinan, China

[3] Faculty of Computing and Information Technology, University of Jeddah, Jeddah, Saudi Arabia

Abstract. The long non-coding RNA (lncRNA) is a hot research topic among researchers in the field of biology. Recent studies have illustrated that the subcellular localizations carry salient information to understand the complex biological functions. However, the experimental setup cost and the computational cost to identify the subcellular localization of lncRNA is too high. Therefore, there is a need of some efficient and effective methods to predict the lncRNA subcellular locations. In this paper, a wide and deep flexible neural tree (FNT) is proposed to predict the subcellular localization of lncRNA. The wide component has ability to memorize the original input features, while the deep component has ability to automatically extract hidden features. To fully exploit lncRNA sequence information, we have extracted seven features which are further fed to four wide and deep FNT classifiers respectively. By ensemble four classifiers, it can predict 5 subcellular localizations of lncRNA, including cytoplasm, nucleus, cytosol, ribosome and exosome.

Keywords: Flexible neural tree · Wide and deep learning · Ensemble learning LncRNA subcellular localization

1 Introduction

The genomic plan showed that there are 3 billion base pairs, which are used to make up the human genome. Only 1.5% of the nucleic acid sequence are used for protein coding, and the remaining 98.5% of the genome are a non-protein coding sequence. These sequences are known as a "garbage sequence" and not considered. However, in the enlisted ENCODE research program, it was found that 75% of the genomic sequences could be transcribed into RNA, and nearly 74% of the transcripts were non-coding RNAs (ncRNAs). In non-coding RNAs, most transcripts are more than 200 bases long. These "long non-coding RNAs (lncRNAs)" regulate the expression of protein-coding genes at the transcriptional and post-transcriptional levels [1, 2], so that

D.-S. Huang et al. (Eds.): ICIC 2018, LNCS 10955, pp. 515–525, 2018.
https://doi.org/10.1007/978-3-319-95933-7_60

they participate extensively in important life processes including cell differentiation and individual development [3, 4]. In addition, the abnormal expressions are also closely related to the occurrence of various major human diseases [5–7].

Long RNAs can occur in different subcellular structures, where the long non-coding RNAs located in the nucleus account for the largest proportion. For example, the long non-coding RNA MEN ε/β is mainly located in the nucleus and is an important component of the nuclear substructure paraspeckles; MALAT-1 (Metastasis-associated lung adenocarcinoma transcript 1) and Neat1 (Nuclear enriched abundant transcript 1) are mainly located in the nuclear speckle of the nucleus and are involved in the cleavage of the precursor mRNA [8]. In addition, Cesana et al. [9] found that linc-MD1 is mainly expressed in the cytoplasm of differentiated muscle cells and acts as a competitive RNA (ceRNA) to regulate skeletal muscle differentiation. In 2011, Rackham et al. [10] first identified three long-chain non-coding RNAs encoded by the mitochondrial genome, lncND5, lncND6, and lncCytb, when analyzing high-throughput sequencing data. It can be seen that the long non-coding RNAs may exist in many subcellular structures of cells, and the special subcellular localization has important significance for the biological function of long non-coding RNAs.

The long non-coding RNA (lncRNA) have been hot topics in the area of RNA biology [11, 12]. Recent studies have illustrated that their subcellular localizations carry important information for understanding their complex biological functions [13, 14]. Considering the costly and time-consuming experiments for identifying subcellular localization of lncRNA, computational methods are urgently desired. However, there are no computational tools for predicting the lncRNA subcellular locations.

Chen [15, 16] has proposed an architecture called flexible neural tree (FNT). The flexible neural tree (FNT) is a special type of neural network that allows over-layer connections and input variables selection. FNT has successfully solved the problems of designing the architecture of neural networks [17]. The structure of FNT is automatically selected by the evolutionary algorithm.

In this paper, a wide and deep flexible neural tree (FNT) is proposed to predict the subcellular localization of lncRNA. The wide component has the function to memorize the original input features and it can be regarded as a linear model. While the deep component has ability to extract hidden features, which shows the generalization of a model. To combine the advantages of memorization and generalization of a model [18–20], the wide and deep FNT is proposed. To attain the compact lncRNA sequence information, we have extracted seven features which are used as the input to four wide and deep FNT classifiers, respectively. By ensemble four classifiers, we got better results to predict 5 subcellular localizations of lncRNA.

The rest of paper is organized as: Sect. 2 provides a detail description of wide and deep flexible neural tree. The ensemble learning is presented in Sect. 3. The detailed experimental results and analysis is provided in Sect. 4. At last, in Sect. 5 the conclusion is included.

2 Wide and Deep Flexible Neural Tree

2.1 Wide and Deep Flexible Neural Tree Model

The wide and deep flexible neural tree is a special type of neural network. The function set F and terminal instruction set T are used to generate a FNT model, which is defined as follows:

$$S = F \cup T = \{+_2, +_3, \ldots, +_N\} \cup \{x_1, \ldots, x_n\} \tag{1}$$

where $+_i$ (i = 2, 3, . . ., N) represent instructions of non-leaf nodes with i parameters. $x_1, x_2 \ldots, x_n$ are instructions of leaf nodes without parameters. In the generation of flexible neural tree, if a nonterminal instruction, $+_i$ (i = 2, 3, 4, . . ., N) is considered, in which i values are randomly generated for non-leaf node and its connection weights between children [17]. In addition, two parameters a_i and b_i are randomly created as flexible activation function parameters. The following flexible activation function is considered.

$$f(a_i, b_i, x) = e^{-\left(\frac{x-a_i}{b_i}\right)^2} \tag{2}$$

The output of a flexible neuron $+_n$ can be produced as follows.

$$net_n = \sum\nolimits_{j=1}^{n} w_j * x_j \tag{3}$$

Where x_j (j = 1, 2,..., n) are the inputs. The output of the node $+_n$ is computed as

$$out_n = f(a_n, b_n, net_n) = e^{-(net_n - a_n/b_n)^2} \tag{4}$$

The wide component acts as a linear model of the form $y = w^T x + b$, which includes raw input features and transformed features. If a model only has the deep component, the original input features can be lost during the process of training, but sometimes it has a significant effect on the final output.

The deep component of our model is a FNT, which automatically extracts features. The parameters are changed to minimize the loss function during training process. The wide component and deep component are combined using a weighted sum of their output, which is then used as the input to one loss function for joint training [20]. The common structure of wide and deep FNT is illustrated in Fig. 1.

Grammar guided genetic programming is used to choose the best architecture of wide and deep FNT, automatically. The parameters of FNT are specified by particle swarm optimization algorithm.

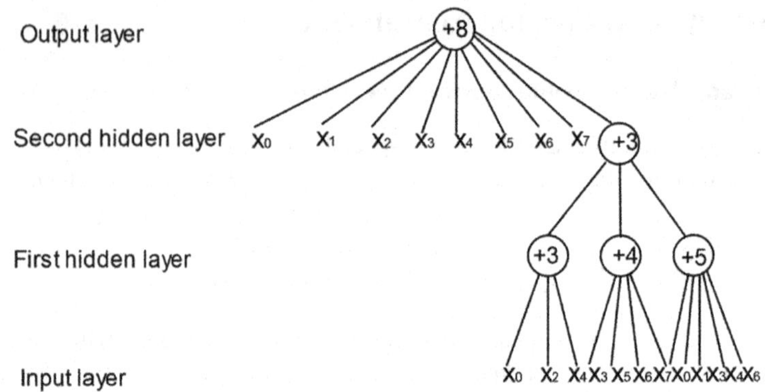

Fig. 1. A typical representation of wide and deep FNT

2.2 Architecture Optimization by Grammar Guided Genetic Programming

Genetic programming has made a breakthrough based on genetic algorithm(GA). It operates on groups of independent computer programs rather than fixed-length binary strings and overcomes the limitations of traditional genetic algorithms.

The individuals in the initial population are evaluated by fitness function. Individuals are randomly selected from the initial population for a tournament in which the program with the best fitness score is copied into the subsequent generation, following application of a genetic operator:

With reproduction, the program is copied without modification.
With mutation, a portion of the program is randomly modified.
With crossover, two programs are selected to contribute a portion of their own code to a new program.

Each mutation and crossover might result in a reduction or improvement in the fitness score for each individual program. For the initial population, all programs in each new generation are evaluated and randomly selected for tournaments, which are evolved for subsequent generation. These programs are selected to exhibit the highest fitness score, their contribution to the next generation enables an overall improvement. This process continues until the user-defined termination criteria is met.

GP requires the input of the same type for contents of the function sets and the terminator character sets. Otherwise, an invalid representation tree may be generated during the crossover and mutation operations easily, that can limit the GP application to a non-limited type. This limitation triggers a grammar guided genetic programming (GGGP). Using grammars to set syntactical constraints was first introduced by Whigham: where context-free grammars were used.

A context-free grammar G is defined by the 4-tuple: $\mathrm{G} = \{\mathrm{N}, \mathrm{T}, \mathrm{P}, \Sigma\}$, where N is called nonterminal characters, and T is a set of terminals characters. The members of P are called rules of the grammar. Σ is the start symbol and an element of N. The rules of the grammar are expressed as $\mathrm{x} \rightarrow \mathrm{y}$, where x belongs to N and y belongs to $\mathrm{N} \cup \mathrm{T}$.

There are four basic steps to generate grammar guided genetic programming: (1) Generate an initial population. In this process, individual trees are randomly produced. The production of trees is based on the grammar model. (2) Evaluate each tree in the current generation. Each tree has a specific value, these values are the fitness of individuals. (3) Apply one of three genetic operators to produce next generation: reproduction, mutation and crossover. Then evaluate all trees in the new generation. (4) Repeat until the best tree is found or termination criteria is met.

2.3 Parameter Optimization by Particle Swarm Optimization

The particle swarm optimization(PSO) algorithm searches the best solution using particles. The initial particles are randomly generated. Each particle represents a potential solution and has a position represented by a position vector x_i. A population of particles move through the problem space, with the moving velocity represented by a vector v_i. At each step, a function f_i represents a fitness value. Each particle keeps track of its own best position, and the best fitness of particle is in a vector p_i. Moreover, the best position among all the particles is kept track of as p_g. At each time step t, a new velocity for particle i is calculated by

$$v_i(t+1) = v_i(t) + c_1\varphi_1(p_i(t) - x_i(t)) + c_2\varphi_2(p_g(t) - x_i(t)) \tag{5}$$

where φ_1 and φ_2 are random number in [0,1], c_1 and c_2 are limited factors of position. Based on the changed velocities, each particle changes its position according to the following equation:

$$x_i(t+1) = x_i(\mathrm{t}) + v_i(t+1) \tag{6}$$

In experiments, the parameters in wide and deep flexible neural tree are optimized by PSO.

3 Ensemble Learning

3.1 Base Classifier Construction

Ensemble learning is to train multiple base classifiers, then integrating several individual classifiers and determining the final classification result by combining the results of multiple classifiers [21, 22]. When there is divergence in the prediction results, and the error rate of each classifier is less than 0.5, then an ensemble classifier can achieve better performance than a single classifier.

The results of the base classifiers directly affect the result of the final integration. At present, the construction methods of several base classifiers that are commonly used are universal and suitable for many machine learning algorithms. Common methods for constructing a combined classifier are the following:

(1) By processing training data sets - bagging and boosting;
(2) By processing input features;

(3) By processing class labels - error correcting output coding;
(4) By processing learning algorithms;

In this paper, we use procession of class labels to improve prediction accuracy.

Dietterich and Bakiri described the use of error correcting output codes (ECOC) to generate multiple different base classifiers [22]. The error correction output coding method can be used to convert the multi-classification problem into multiple two classification problems. Moreover, the error correction output code itself has the characteristics of error correction capability, so the prediction accuracy of the supervised learning algorithm can be improved. At present, error correction output coding has become a research hotspot in supervised classification.

The basic principle of the error correction output encoding method is: For multi-class problems, a certain coding method is used to binary code the target classification, so that each category is coded with a binary string of length L bits, and it is necessary to ensure the hamming code distance between each binary code of the class is large enough. So the classification problem can be used to learn L different binary classifiers on the same training set, and finally put the results of each base classifier together, then the decision algorithm finds the target classification whose coding value is closest to this result as the output of the entire classifier.

The specific method of the error correction output encoding method is as follows: First, the categories are encoded, and each category corresponds to a bit string of length L, forming a m*L two-dimensional encoding matrix, m is the number of categories; then each column of the encoding matrix is learned and each gets a binary classifier. For the class with the ith ($1 \leq i \leq L$) bit of the encoded string being 0, the output produced by the ith column of the encoding matrix is 0, otherwise the output is 1; in the classification, each binary classifier classifies the unknown sample and obtains a binary string with a code length of L. The minimum hamming distance method is used to determine that the bit string is the closest to the code of that class, thus obtaining the category of sample.

Error correcting coding cannot only improve the classification accuracy of base classifiers effectively, also has the following advantages: Firstly, when the number of training sets is small, the error correcting output coding method is still effective. Secondly, the method does not depend on the specific coding value of each target classification.

3.2 Base Classifier Ensemble

Now we have obtained the classification results of several base classifiers, however, to combine the individual classifier results to form the final decision is still not easy. Different integration methods have been used in literature, the most common used are: voting method, stack integration method, cascade combination method and algorithm correlation method.

The voting method is the simplest integration method. In this paper, the voting method is considered. The basic idea behind the voting method is to consider the performance of multiple base classifiers classified prediction, and then counting the prediction results of each classifier, if most classifiers predict the sample X belong to

C_i, the ensemble classifier classifies the category of sample X as C_i. According to different voting principles, the voting method can be divided into one-vote veto, unanimous vote, minority majority, threshold voting, etc. [23, 24].

Another voting method is weighted voting, which also called linear combination method. The basic principle is that when a single classifier participates in a vote, it is assigned a weight to them, indicating that the result of the classifier accounts for the final decision. The key to weighted voting is to determine the weight of each classifier. A relatively simple method is based on the classification accuracy of the training sample set by each classifier, setting the weight of each classifier is proportional to its classification accuracy.

4 Experimental Results and Analysis

The considered dataset has 5 kinds of subcellular localizations of lncRNA, including cytoplasm, nucleus, cytosol, ribosome and exosome. These data are taken from the website (http://www.csbio.sjtu.edu.cn/bioinf/lncLocator/Data.htm). The dataset is an lncRNA sequence, so the k-mers method was chosen to extract 7 features according to frequency statistics. The data processing flow is shown in Fig. 2. Data is separated into two parts: 455 training data and 152 test data.

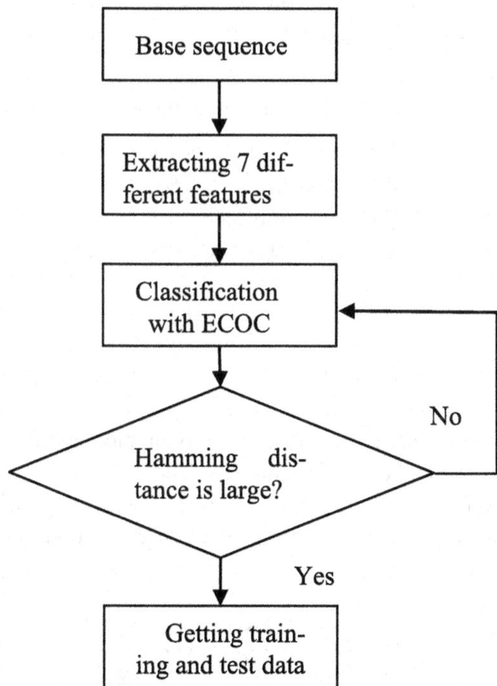

Fig. 2. Data processing flow chart

The design of the base classifier uses error correcting output coding methods that deal with class labels. The idea of error correcting output coding is to encode the output class, which transforms 5 classified problems into multiple 2 classified problems. The wide and deep FNT models is the base classifier in experiments, which is used for dealing with two classification problems. ECOC coding is defined according to the hamming code. Five classification problem must be represented by at least 3 binary digits, and at the same time, the Hamming code distance must be kept large enough, so 4 base classifiers are selected. Table 1 lists the coding of the 4 base classifiers.

Table 1. The coding of the 4 base classifiers.

Five categories	The coding of first classifier	The coding of second classifier	The coding of third classifier	The coding of fourth classifier
Cytoplasm	1	0	0	1
Cytosol	0	1	0	0
Exosome	1	1	0	0
Nucleus	0	0	1	0
Ribosome	1	0	1	0

In experiments, we used grammar guided genetic programming to optimize the architecture of individual trees. As for the parameters of trees, it can be specified by PSO algorithm. The corresponding parameters are listed in Table 2.

Table 2. parameter settings.

Parameter	Value
Population size	50
Crossover probability	0.4
Mutation probability	0.01
C1	2
C2	2
Vmax	2

Seven different features are extracted by k-mers method and the result is calculated by the voting of four wide and deep FNT models. The instruction sets of wide and deep FNT model is $S = F \cup T = \{+_2, +_3, \ldots, +_8\} \cup \{x_0, x_1, \ldots, x_6\}$. To verify the effectiveness of proposed model, a common evaluation method known as accuracy is used.

$$\text{accuracy} = \frac{1}{n}\sum\nolimits_{i=1}^{n} 1\left(Y_i' = Y_i\right) \tag{7}$$

Where Y_i is the actual value, Y_i' is the predicted value, 1(x) is the indicator function, and n is the total number of data samples.

Table 3 shows the results of the four base classifiers of wide and deep FNT and its ensemble in predicting the subcellular localization of lncRNA. The result of the ensemble classifier is closely related to the results of the four base classifiers and the accuracy of the five classification problems has reached 0.605. The reason why it gets better result is that the wide and deep flexible neural tree can intelligently handle discrete-valued and vector-valued samples such as the lncRNA dataset. The wide component has the function to memorize the original input features and it can be regarded as a linear model. While the deep component can automatically extract hidden features, which shows the generalization of a model. In addition, that the five classified problems are transformed into two classified problems is also an important reason for the improvement of the prediction accuracy.

Table 3. The accuracy of predicting the subcellular localization of lncRNA.

Classifier	Accuracy
The first base classifier	0.612
The second base classifier	0.783
The third base classifier	0.665
The fourth base classifier	0.599
Ensemble classifier	0.605

As is shown in Table 4, we can find that the wide and deep flexible neural tree combined with ECOC method can improve the accuracy. The reason why it gets better result is that ECOC itself has ability to correct mistakes. Therefore, the lncRNA subcellular localization can achieve better results using proposed method.

Table 4. The comparison of classifier with ECOC.

Classifier	Accuracy
classifier without ECOC	0.480
classifier with ECOC	0.605

5 Conclusion

In this paper, using the wide and deep flexible neural tree model and its ensemble to predict the subcellular localization of lncRNA. The wide component has the function to memorize the original input features, while the deep component can automatically extract hidden features, so the wide and deep FNT is proposed. The experiments show that the wide and deep FNT model and its ensemble can achieve better results in predicting lncRNA subcellular localization. The architecture of the wide and deep FNT is developed by grammar guided genetic programming (GGGP) and the parameters are optimized by particle swarm optimization algorithm (PSO). The experimental results indicate that the ensemble of FNT can get better accuracy in multiple classification problems.

In future, we intend to do further research on using other methods to optimize the architecture of the wide and deep FNT.

Acknowledgments. This work was supported by the National Natural Science Foundation of China(Grant No. 61671220, 61640218, 61201428), the National Key Research and Development Program of China (2016YFC106000) , the Shandong Distinguished Middle-aged and Young Scientist Encourage and Reward Foundation, China (Grant No. ZR2016FB14), the Project of Shandong Province Higher Educational Science and Technology Program, China (Grant No. J16LN07), the Shandong Province Key Research and Development Program, China (Grant No. 2016GGX101022), the Doctoral Foundation of University of Jinan, the Shandong Province Natural Science Foundation (No. ZR2018LF005).

References

1. Martens, J.A., Laprade, L., Winston, F.: Intergenic transcription is required to repress the Saccharomyces cerevisiae SER3 gene. Nature **429**(6991), 571–574 (2004)
2. Jiang, W., Liu, Y., Liu, R.: The lncRNA DEANR1 facilitates human endoderm differentiation by activating FOXA2 expression. Cell Rep. **11**(1), 137–148 (2015)
3. Yang, L.Q., Lin, C.R., Rosenfeld, M.G.: A lincRNA switch for embryonic stem cell fate. Cell Res. **21**(12), 1646–1648 (2011)
4. Wamstad, J.A., Alexander, J.M., Truty, R.M., Shrikumar, A., Li, F.G., Eilertson, K.E., Ding, H.M., Wylie, J.N., Pico, A.R., Capra, J.A., Erwin, G., Kattman, S.J., Keller, G.M., Srivastava, D., Levine, S.S., Pollard, K.S., Holloway, A.K., Boyer, L.A., Bruneau, B.G.: Dynamic and coordinated epigenetic regulation of developmental transitions in the cardiac lineage. Cell **151**(1), 206–220 (2012)
5. Faghihi, M.A., Modarresi, F., Khalil, A.M., Wood, D.E., Sahagan, B.G., Morgan, T.E., Finch, C.E., St Laurent 3rd, G., Kenny, P.J., Wahlestedt, C.: Expression of a noncoding RNA is elevated in Alzheimer's disease and drives rapid feed-forward regulation of β-secretase. Nat. Med. **14**(7), 723–730 (2008)
6. Dereure, O.: Role of non-coding RNA ANRIL in the genesis of plexiform neurofibromas in neurofibromatosis type 1. Ann. Dermatol. Vénéréol. **139**(5), 421–422 (2012)
7. Gibb, E.A., Brown, C.J., Lam, W.L.: The functional role of long non-coding RNA in human carcinomas. Mol. Cancer **10**(1), 38 (2011)
8. Tripathi, V., Ellis, J.D., Shen, Z., Song, D.Y., Pan, Q., Watt, A.T., Freier, S.M., Bennett, C. F., Sharma, A., Bubulya, P.A., Blencowe, B.J., Prasanth, S.G., Prasanth, K.V.: The nuclear-retained noncoding RNA MALAT1 regulates alternative splicing by modulating SR splicing factor phosphorylation. Mol. Cell **39**(6), 925–938 (2010)
9. Cesana, M., Cacchiarelli, D., Legnini, I., Santini, T., Sthandier, O., Chinappi, M., Tramontano, A., Bozzoni, I.: A long noncoding RNA controls muscle differentiation by functioning as a competing endogenous RNA. Cell **147**(2), 358–369 (2011)
10. Rackham, O., Shearwood, A.M., Mercer, T.R., Davies, S.M., Mattick, J.S., Filipovska, A.: Long noncoding RNAs are generated from the mitochondrial genome and regulated by nuclear-encoded proteins. RNA **17**(12), 2085–2093 (2011)
11. Zhang, R.K., Zhang, L., Yu, W.Q.: Genome-wide expression of non-coding RNA and global chromatin modification. Acta Biochim. Biophys. Sin. **44**(1), 40–47 (2012)
12. Muers, M.: RNA: genome-wide views of long non-coding RNAs. Nat. Rev. Genet. **12**(11), 742–743 (2011)

13. Derrien, T., Johnson, R., Bussotti, G.: The GENCODE v7 catalog of human long noncoding RNAs: analysis of their gene structure, evolution, and expression. Genome Res. **22**(9), 1775–1789 (2012)
14. Van Heesch, S., van Iterson, M., Jacobi, J.: Extensive localization of long noncoding RNAs to the cytosol and mono-and polyribosomal complexes. Genome Biol. **15**(1), 1–12 (2014)
15. Chen, Y., Yang, B., Dong, J.: Time-series forecasting using flexible neural tree model. Inf. Sci. **174**(3–4), 219–235 (2005)
16. Chen, Y., Yang, B., Abraham, A.: Flexible neural trees ensemble for stock index modeling. Neurocomputing **70**(4–6), 697–703 (2007)
17. Chen, Y., Yang, B., Meng, Q.: Small-time scale network traffic prediction based on flexible neural tree. Appl. Soft Comput. **12**(1), 274–279 (2012)
18. Chen, C.L.P., Liu, Z.: Broad learning system: a new learning paradigm and system without going deep. In: Automation, pp. 1271–1276. IEEE Press (2017)
19. Chen, C., Liu, Z.: Broad learning system: an effective and efficient incremental learning system without the need for deep architecture. IEEE Trans. Neural Netw. Learn. Syst. **29**, 1–15 (2017)
20. Cheng, H.T., Koc, L., Harmsen, J., et al.: Wide & deep learning for recommender systems. In: The Workshop on Deep Learning for Recommender Systems, pp. 7–10. ACM (2016)
21. Dietterich, T.G.: Ensemble methods in machine learning. In: Multiple Classier Systems, Cagliari, Italy (2000)
22. Valentini, G., Masulli, F.: Ensembles of learning machines. In: Marinaro, M., Tagliaferri, R. (eds.) WIRN 2002. LNCS, vol. 2486, pp. 3–20. Springer, Heidelberg (2002). https://doi.org/10.1007/3-540-45808-5_1
23. Xu, L., Krzyzak, A., Suen, C.Y.: Methods of combining multiple classifiers and their applications to handwriting recognition. IEEE Trans. Cybern. **22**(3), 418–435 (1992)
24. Bahler, D., Navarro, L.: Methods for combining heterogeneous sets of classiers. In: National Conference on Artificial Intelligence (2000)

Research on Auto-Generating Test-Paper Model Based on Spatial-Temporal Clustering Analysis

Yuling Fan[1], Likai Dong[1], Xuesong Sun[1](✉), Dong Wang[1,2], Wang Qin[1], and Cao Aizeng[1]

[1] School of Information Science and Engineering, University of Jinan, Jinan 250022, China
ise_sunxs@ujn.edu.cn

[2] Shandong Provincial Key Laboratory of Network Based Intelligent Computing, University of Jinan, Jinan 250022, China

Abstract. In the process of auto-generating test-paper, the category and the difficulty of the title plays a key role in the quality of generating test-paper. It will produce low quality questions and hard to popularize when used the methods of artificial generating test-paper and random generating test-paper, because considering less on the knowledge point classification and difficulty in the subject. To improve the quality of auto-generating test-paper, this paper takes the evaluation data of ACM Online Judge system as the research object. After normalization, (1) we can get three different results by the K-means clustering analysis based on the temporal and spatial characteristics of time variance and average time; (2) On the basis of clustering, the difficulty index of each topic of all the categories is calculated by using the number of submissions and the number of submissions to solve the problem. The ratio of the two is proportional to the difficulty of the problem. In this paper, the ratio of the two to determine the degree of difficulty index; (3) The Gaussian stochastic process is used to extract numbers of questions of each knowledge point, and calculated the difficulty index which were extracted to make sure they are within range to complete the auto-generating test-paper. In the experiment, we try to train and test the automatic test paper model by the number of professional problems (about 50000 data) in the C language test question of the university OJ system. The average difficulty index of the test paper was 0.4663, which meet the requirements, and the difficulty index of the title fit in with the normal distribution. Compared with the traditional generating test-paper method, the automatic test paper model is based on the difficulty and discrimination of the subject, and it can evaluate the level of tester scientifically. The experimental results show that the proposed automatic test model is simple and effective.

Keywords: Spatial-temporal feature · Clustering algorithm
Difficult coefficient · Auto-generating test paper

D.-S. Huang et al. (Eds.): ICIC 2018, LNCS 10955, pp. 526–535, 2018.
https://doi.org/10.1007/978-3-319-95933-7_61

1 Introduction

Online learning platform has three important links: learning, discussion and assessment, of which examination is an important means to examine the effect of the participants, and the test paper is the most important factor of the participants' learning level in a fair test. High quality papers should take into account the difficulty, distinction and coverage of the questions. At present, online platform testing is difficult to ensure the quality of test questions, and can't truly reflect the knowledge level of the tested people.

At present, the popularity rate of OJ (Online Judge) system is high in Colleges and universities. A large number of topics are preserved in the database, and each topic includes the evaluation data of all participants. This article takes the online evaluation data set as the research object, extracts the time variance as the spatial feature of the topic, and takes the time characteristic of the topic as the time characteristic of the topic, and based on the temporal and spatial characteristics to carry on the K-means clustering [1], and uses the Davies-Bouldin (DB) index to obtain the optimal classification. When auto generating test paper, the difficulty coefficient of all subjects is calculated by the total number of submissions and the number of times of submission, setting up an automatic test paper model through Gauss random process.

Many scholars at home and abroad make statistical and in-depth analysis of massive data generated by online platforms. Chen Chi et al. of Tsinghua University thinks that big data research in online education is very important for the efficiency of online education. The learning analysis technology uses the data produced by the learners and the analysis model to find information and social relations, so as to achieve the purpose of predicting and improving people's learning. The test questions, such as Zong Yang, Beijing Normal University, think that the test questions with appropriate difficulty can stimulate the potential and enthusiasm of the learners, and put forward the learning model of the difficulty coefficient of MOOCs formative test. Zhang Renlong and others of Jiangxi Normal University put forward that the difficulty level of test questions usually follows normal distribution and the normal distribution numerical method in discrete definition field is used to determine the number of test questions to be extracted from each difficulty level, so as to generate test papers.

This paper takes the online evaluation data as the analysis object, extracts the characteristics to cluster, selects the optimal clustering results, and then extracts the difficulty coefficient of each type of topic, according to the knowledge point distribution and the difficulty range of the test paper input, and finally forms the group model, The specific process is shown in Fig. 1.

(1) Data set: The sample data include subjects covered by all majors covered all knowledge points, and this data information is used as the original evaluation data [2]. In the original data, the detection and cleaning of incomplete data and error data, the isolation point [3] is removed, the influence of noise data is reduced to a minimum, and data prepossessing is completed to form a data set.
(2) Normalization of data [4]: The "completion time standard deviation" and "average completion time" were extracted from data as classification features. The features are normalized to improve data comparability between.

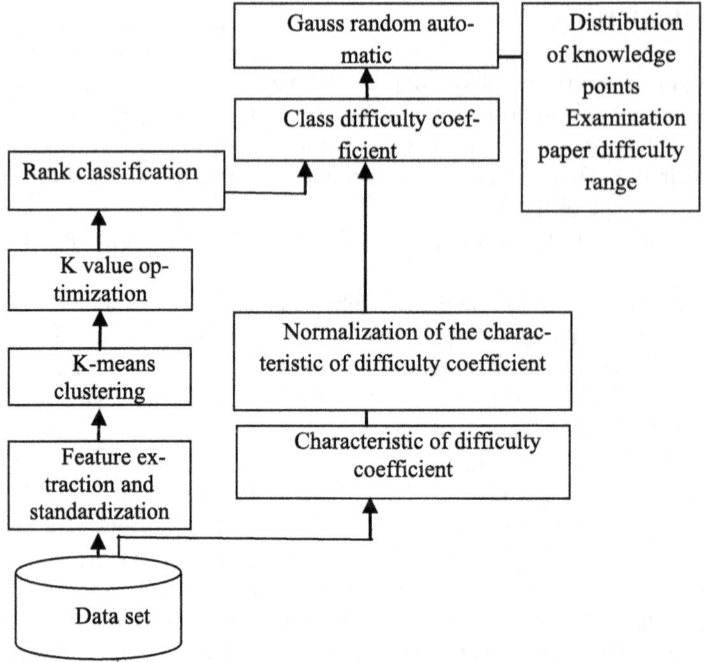

Fig. 1. Automatic test paper model.

(3) Cluster analysis: Select different parameters to carry out K-means clustering experiments, analyze and compare different clustering results, and determine the best classification.
(4) Determination of the degree of difficulty: After obtaining the best classification, the number of submission and the number of submitted solutions is extracted, and the difficulty coefficient is finally calculated.
(5) The test paper model: Based on the distribution of knowledge points and the coefficient of difficulty, Gauss's random process is used to select different difficult subjects to complete automatic test paper generation.

The first part of this paper is an overview of the paper model, the second part introduces the relevant algorithms and key technologies, the third part introduces the experimental setup and results analysis, and finally summarizes the work of this paper.

2 Algorithms and Key Technologies

2.1 Spatio-Temporal Characteristics

In the cluster analysis of the topic, we mainly refer to two indicators: the average time and the time variance. Among them, the average time of the subject is the average time of the subject that the subject is tested. The characteristic is aimed at the evaluation index of the difficulty of the test question. The time of the subject's completion of the subject is proportional to the difficulty of the subject, so the time characteristic of the choice of the subject is the average time. Two kinds of features need to be normalized respectively based on temporal and spatial clustering analysis.

2.2 Clustering Algorithm

K-means Clustering Algorithm
As a classical unsupervised clustering algorithm [5], K-means has the advantages of simplicity, small footprint and fast computation speed. But the disadvantages are also very obvious, such as the sensitivity to isolated points, and very few outliers will directly affect the location of the center. In addition, the selection of cluster number K needs to be determined before executing algorithm.

In view of the above shortcomings, we first clean up most of the outliers in the data pre-processing part, so as not to have a greater impact on the clustering results. The next thing to solve is to determine the best clustering number, that is, the K value. Firstly, the range of K is determined in [k1, k2], and the same clustering algorithm is performed using different K values for data sets. The results are evaluated using DB index, and the value of each index is compared, and the best clustering K value is obtained.

K value Optimization
The key to solve the best clustering number K is to construct appropriate clustering validity function [6]. There are two kinds of evaluation criteria for clustering validity: First, the external standard is used to evaluate the quality of clustering results by measuring the consistency of clustering results and reference standards. The other is the internal index, which is used to evaluate the good degree of the clustering results of the same clustering algorithm under different cluster numbers, and is usually used to determine the best clustering number of the data sets.
In this paper, we use DB index to describe the intra class divergence of samples and the distance between each cluster center. The definition of DB is shown in Formula (1).

$$\mathrm{W_1} = \frac{1}{\mathrm{K}}\sum_{i=1}^{k} \max_{j=1 \sim k, j \neq i}\left(\frac{W_i + W_{\mathrm{j}}}{C_{ij}}\right) \tag{1}$$

Among them, K is the number of clustering, representing the average distance of all samples from the class to its cluster center, representing the average distance of all the samples from the class to the center of the class, indicating the distance between the class and the center. It can be seen from formula (1) that the smaller the DB is, the lower the similarity between class and class is, the better the corresponding clustering result is.

2.3 Difficulty Coefficient

Feature Extraction of Difficulty Coefficient
In the OJ system, the difficulty coefficient is determined by the total number of submitted questions and the number of submitted solutions. Among them, the total number of submissions includes the number of erroneous answers submitted and the number of correct answers. The number of times to submit the answer number to indicate the correct answer is submitted. Taking the two as the feature of the difficulty coefficient, the ratio between the two is directly proportional to the difficulty of the topic. In this paper, the difficulty coefficient is calculated based on the ratio of the two.

The total number of submission of the title P_i is sub_i, the number of errors submitted is $unsol_i$, and the number of submission times is sol_i, that is, $sub_i = sol_i + unsol_i$. The difficulty coefficient of each item is dd_i, which is defined by formula (2).

$$dd_i = \frac{sub_i}{sol_i} = \frac{sol_i + unsol_i}{sol_i} (i = 1 \cdots n) \quad (2)$$

In which 'i' represents the title number, and 'n' represents the total number of topics.

Standardization of the Characteristic of Difficulty Coefficient
After determining the best difficulty classification, we calculate the mean coefficient of difficulty of all subjects in each category, and record it as $\overline{dd_j}(j = 1...K)$. It is normalized to map to [0 and 1], and the difficulty coefficient is $DD_j(j = 1...K)$. It is defined as:

$$DD_j = \frac{\overline{dd_j} - \overline{dd_{min}}}{\overline{dd_{max}} - \overline{dd_{min}}} \quad (3)$$

$$\overline{dd_j} = sumdd_i/m \quad (4)$$

Among them, $sumdd_i = \sum_{i=1}^{n} dd_i$, 'i' represents the title number, 'j' indicates the classification number, 'm' indicates the number of topics included in the j classification, $\overline{dd_{min}} = \min(\overline{dd_j})(j = 1 \cdots K)$, $\overline{dd_{max}} = \max(\overline{dd_j})(j = 1 \cdots K)$, and 'k' indicates the total number of categories.

3 Experimental Setup and Result Analysis

3.1 Evaluation Data Processing

Take more than 2000 subjects in the C language test database of a OJ system as an example., each topic has saved all the students' learning behavior, including the completion time, the number of submissions, the ranking and other data. In the experiment, 300 subjects (about 50000 data) of 160 professional students were selected as samples for model training.

Generally speaking, indicators for evaluating the quality of papers [2] include difficulty index, discrimination index and coverage rate. In the classical test theory [7], the determination of the difficulty of the test is usually expressed by the formula $q = 1 - \overline{X}/X_{full}$, in which 'q' indicates the difficulty of the test, $\overline{X}$ indicates the average value of the score of the test in the subject, and X_{full} indicates the full score of the question. Choosing the average completion time of each topic to indicate the difficulty of the topic is more difficult and the longer it takes.

Discriminant is used to evaluate students' knowledge level and distinguish different level of subjects. Because the time variance represents the distance between the original time and the average time, it is a standardized deviation based on the standard deviation, which is used to represent the index of area division. Raw data and normalized data, as shown in Table 1.

Table 1. Normalization table.

Subject	Original evaluation data		Normalized data	
	Average time	Time variance	Average time	Time variance
P1	60.8312	28.13344	0.17667	0.39251

3.2 Analysis of Clustering Results

Clustering

The samples are respectively clustered according to K = 5, K = 6 and K = 7 respectively. The clustering results are shown in Fig. 2, and the left map shows the cluster number of 5. The middle graph shows the result of the classification number of 6, and the right graph shows the clustering result of the classification number of 7.

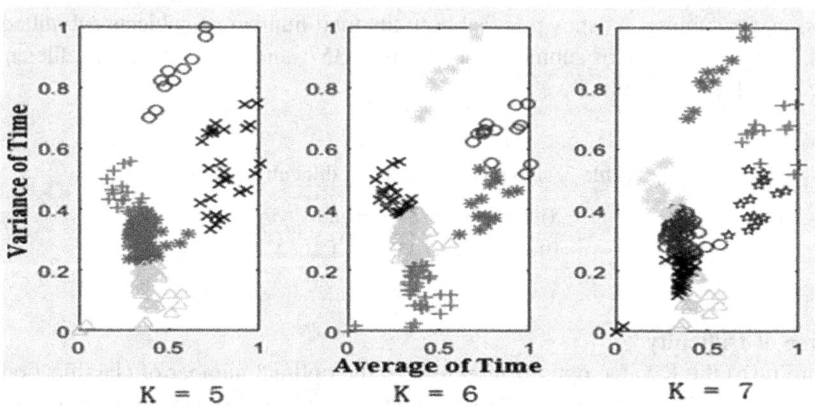

Fig. 2. Clustering results.

Experimental Results of K Value Selection

Three K value experiments were compared with DB [13] index to evaluate the experimental results. It is defined as:

$$\mathrm{DB}(k) = \frac{1}{k}\sum_{i=1}^{k} j = \max_{1 \sim k, j \neq i} \left(\frac{w_i + w_j}{c_{ij}}\right) \tag{5}$$

In which K is clustering number, class Wi said all the samples in the Ci to its average distance clustering center, Wj said class all the samples in the Ci to the average distance from the center of the class Cj, Cij said kind of distance between Ci and Cj center. It can be seen that the smaller DB is, the lower the similarity between classes and classes, and thus the better clustering results.

The evaluation result is shown in Table 2. When K = 6, CriterionValues [14] value is the smallest, so when the classification number is 6, the classification effect is the best.

Table 2. DB index evaluation

Classification number	CriterionValues
5	0.8049
6	0.6867
7	0.7093

3.3 Degree of Difficulty

Characteristic of Difficulty Coefficient

Each item in the evaluation data has two characteristics: the total number of submission (submit) and the solved number. The quotient of the two is dd of the difficulty coefficient of the subject. As shown in Table 6, the total number of subjects submitted by P1 is 395, the number of submitted solutions is 357, and the difficulty coefficient of subjects is 1.1064 (Table 3).

Table 3. Original features and difficulty values

Title	Submit	Solved	dd
P1	395	357	1.1064

Degree of Difficulty

According to the K value optimization results, the optimal number of classifications is 6. Using formula (3) difficulty coefficient DD calculated by category, as shown in Table 4. Classification 2 is the most difficult, the difficulty coefficient is 1, the classification 4 is the simplest, and the difficulty coefficient is 0.

Table 4. The various degree of difficulty coefficient

Classification	Class difficulty coefficientDD
1	0.8468
2	1
3	0.5242
4	0
5	0.2731
6	0.9486

3.4 Automatic Test Paper Model Algorithm

Based on all kinds of difficulty coefficients shown in Table 4, we use Gauss stochastic process to build automatic generating test paper model. Among them, there are 13 knowledge points in the evaluation data. Each knowledge point contains different topics.

The automatic test paper model could be obtained as follows:

1. extract the corresponding number of knowledge points based on Gauss stochastic process;
2. calculate the average difficulty coefficient by pumping problem;
3. Judge whether the average difficulty coefficient is in the difficulty range of examination paper. If it is, turn to 4; otherwise, adjust the problem of the highest or lowest difficulty coefficient to 1.
4. Confirm the number of all the subjects being selected in the test paper, and generate the test paper according to the number of the title.

The input of the model consists of the number of questions in the test paper (Table 5) and the difficulty range [0.4, 0.6] of the test paper.

Table 5. The input of the number of questions in the test paper

Knowledge point sequence number	The total number of questions	Number of questions
1	34	2
5	287	1
6	169	2
8	289	1
12	66	1

As shown in Table 6, the test paper is composed of 10 questions, and the distribution of knowledge points is determined by testers. The average difficulty coefficient is 0.4663, which is in line with the difficulty range of the examination paper. The difficulty coefficient of each topic conforms to the normal distribution.

Table 6. Automatic test paper results

Knowledge point	The title of the subject	Classification	Difficulty
1	P10	4	0
1	P30	5	0.2731
4	P189	3	0.5242
4	P202	5	0.2731
5	P287	4	0
6	P567	3	0.5242
6	P612	5	0.2731
8	P769	6	0.9486
10	P811	1	0.8468
12	P1067	2	1

4 Conclusions

This paper presents an automatic test paper model based on temporal and spatial clustering analysis, which uses time and time variance as temporal and spatial characteristics, and obtains the best classification results according to the DB index. The auto generating test paper model is generated by Gauss random process based on the difficulty coefficient. The experimental results verify the effectiveness of the proposed method, which is consistent with the normal distribution of the difficulty coefficient of subjects, and can scientifically evaluate the level of knowledge that a testers hold.

Acknowledgments. This research was supported by Shandong Provincial Natural Science Foundation (No. ZR2018LF005), Industry-University Cooperative Education Project of Ministry of Education (No. 201601023018), the Scientific Research Fund of Jinan University (No. XKY1711, No. XKY1622, No. XBS1653) and Teaching Research Project of Jinan University (No. J1638).

References

1. Guo, C., Tian, F., Jin, X.: Data Mining Tutorial, pp. 107–121. Tsinghua University Press (2005)
2. Duda, R.O., Hart, P.E., Stork, D.G.: Model Classification, 2nd edn., pp. 11–12. Machinery Industry Press, Beijing (2003). (Li hongdong, yao tianxiang)
3. Zhou, A., Chen, B., Wang, Y.: Research and improvement of k-means algorithm. J. Comput. Res. Dev. **22**(10), 101–104 (2012)
4. Sun, J., Liu, J., Zhao, L.: Research on clustering algorithm. J. Softw. **19**, 48–61 (2008)
5. Zhijie, Li, Yuanxiang, Li, Feng, Wang, Li, Kuang: Accelerated multi-task online learning algorithm for big data stream. J. Comput. Res. Dev. **52**(11), 2545–2554 (2015)
6. Yang, Y., Jin, F., Kamel, M.: Evaluation of clustering effectiveness. Appl. Res. Comput. **25** (6), 1630–1632 (2008)
7. Sun, J., Li, X.: The algorithm and design of the difficulty coefficient of the test in the classical test theory. China Sci. Technol. Inf. (19), 44–45 (2009)

8. Asifa, R., Merceronb, A., Abbas Alic, S., Haidera, N.G.: Analyzing undergraduate students' performance using educational data mining. Comput. Educ. **113**, 177–194 (2017)
9. Chen, C., Wang, Y., Li, C., Zhang, Y., Xing, C.: The research and application of big data in the field of online education. J. Comput. Res. Dev. **51**, 67–74 (2014)
10. Li, T., Wang, T.: Study and analysis technology research and application status review. China Educ. Technol. **8**, 129–133 (2012)
11. Zong, Y., Zheng, Q., Zhang, X., Chen, L.: The research about difficulty coefficient of MOOCs' formative tests in the perspective of learning analytics. J. Distance Educ. **3**, 96–103 (2016)
12. Zhang, R., Zhou, Q., Hu, B., Jiang, J., Wen, S.: Research and implemnetation of test paper intellectual difficulty adaptively generating. Comput. Mod. **3**, 8–10 (2012)
13. Tang, X., Qiu, G., Zhuang, L.: Clustering method for irregular and uncompact data a clustering method for the distribution of non-dense spatial distribution data. Comput. Sci. **36** (3), 167–169 (2009)
14. Korhonen, P., Salo, S., Steuer, R.E.: A heuristic for estimating nadir criterion values in multiple objective linear programming. Oper. Res. **45**(5), 751–757 (1997)

A Semantic Context Model for Automatic Image Annotation

Xin Fu[1], Dong Wang[2], Sijie Niu[2], and Hengcai Zhang[3](✉)

[1] School of Resources and Environment, University of Jinan, Jinan 250022, China

[2] School of Information Science and Engineering, University of Jinan, Jinan 250022, China

[3] State Key Lab of Resources and Environmental Information System, Institute of Geographic Sciences and Natural Resources Research, Chinese Academy of Sciences, Beijing 100101, China

zhanghc@lreis.ac.cn

Abstract. How to retrieve a required image from tens of thousands of images is challenging task. In this paper, we proposed a novel semantic context model for automatic annotation with context and spatial information. We reconstructed the image annotation as a multi-class classification problem and assign each object a label considering each object as an individual in both learning and annotation stage. And then, the class distribution of query region is estimated using Gaussian mixture model whose parameters are learned by expectation maximum algorithm. The posterior probabilities of all the concepts are obtained according to modified Bayesian rule. In the experiment, we conduct the performance evaluation on LabelMe image databases including 2651 images and 25653 regions. The experimental results illustrated that our proposed model effectively improve the performance of image annotation system, and the context information and height information could improve the precision of image annotation separately.

Keywords: Image annotation · Machine learning · Semantic context Color · Texture

1 Introduction

How to effectively retrieve our favorite contents from the tens of thousands of images has received an increasing amount of attention in the field of computer vision and geographical information system [1–3]. The automatic image annotation is an essential step toward retrieving the desired images. The aim is to assign images a set of labels automatically that can describe the images at semantic levels [4–6]. The existing approaches could be classified into two categories [7–10], probabilistic modeling method and classification method. The probabilistic modeling method is to determine the joint probabilities between labels and low-level features. The classification method is to associate the semantic gap between low-level features and semantic concept. For any query image, the association is applied to annotate the low-level features automatically.

D.-S. Huang et al. (Eds.): ICIC 2018, LNCS 10955, pp. 536–542, 2018.
https://doi.org/10.1007/978-3-319-95933-7_62

Although there has been much work on annotating the images. The existing approaches are still face many challenges, for example, the labeled samples are too small. In additional, manual annotation for huge archive is labor-intensive and time-consuming.

In this paper, we proposed a novel methodology of semantic image annotation with context and spatial information. We reconstructed the image annotation as a supervised multi-class classification problem. The class density is determined by Gaussian mixture model whose parameters are learned by expectation maximum algorithm, and the prior information of each label is obtained. We modified Bayesian rule using context and height information to optimize and improve the algorithm accuracy effectively. Finally, the concepts compete to annotate the given region where maximum a posterior (MAP) is used to choose the annotation.

2 Methodology

In this section, we proposed a new methodology of semantic image annotation, which include three steps. The first step is supervised multi-class learning stage to label a region feature set with semantic concept. The second step is to estimate the class distribution of a query corresponding to each category by using Gaussian Mixture Model. In the final step, the regional posterior probability was obtained based on Bayesian rule where the prior probability of each category is the percentage of each category in training dataset. Simultaneously, many optimization and improvements of method were proposed.

2.1 Supervised Multi-class Learning

Our proposed method reconstructed the semantic image annotation as a supervised multi-class classification problem. That is, we adopted Gaussian Mixture Model to estimate the class distribution of the query image, the posterior probability of each concept was obtained based on Bayesian rule which can compare with each other since they were from the same multi-class classifier. It is possible to directly compare all the semantic concepts and annotate the query image directly. The query image could be annotated by the semantic concept with the largest posterior probability w.

$$w = \arg\max_{i} p(w_i|I) \tag{1}$$

where w_i denotes the semantic concept, I denotes the query image.

2.2 Estimate Class Distribution

Each semantic concept corresponds to an object in an image, and the object generally occupies a region in the image. Given a dataset of images $T = \{I_1, I_2, \ldots, I_N\}$, where N denotes the size of the dataset, the image I_i was segmented into different object regions, and then we classify all the regional features according to their labels into a region feature set X_k which corresponds to the semantic concept w_k. We applied the Gaussian mixture model to estimate the class density of each query region $P(x|w_k)$.

$$
\begin{aligned}
&P(x|w_k) \\
&= GMM(x|\theta) \\
&= \sum_{i=1}^{K} \lambda_i G\left(x \middle| \mu_i, \sum_i\right) \\
&= \sum_{i=1}^{K} \frac{\lambda_i}{\sqrt{(2\pi)^d |\sum_i|}} \exp\left\{-\frac{(x-\mu_i)^T \sum_i^{-1} (x-\mu_i)}{2}\right\}
\end{aligned} \tag{2}
$$

where x denotes the query regional feature vector, λ_i, μ_i and $\sum_i$ denote the weight, average value and covariance matrices of ith Gaussian component respectively, d represent the dimension of the feature vector. The number of Gaussian components K is fixed according to the results of clustering algorithms. We use K-means to initialize the parameters K and the iteration can be terminated referring to a predefined threshold for the difference between parameters obtained from adjacent iterations. The parameters θ, are obtain from X_k using expectation maximum (EM) algorithm which consists of two steps (E-step and M-step).

2.3 Estimate Posterior Probability

For any region, B_i in a query image I, the posterior probability of w_k could be calculated based on Bayesian rule using class distribution obtained from GMM.

$$
P(w_k|B_i) = \frac{P(B_i|w_k)P(w_k)}{P(B_i)} = \frac{P(x|w_k)P(w_k)}{P(x)} \tag{3}
$$

where x denotes the feature vector extracted from region B_i. $P(x|w_k)$ denotes the class distribution of region B_i given semantic concept w_k, $P(x)$ is a constant, and $P(w_k)$ denotes the prior probability of each category in training set, which means the percentage of each category in training dataset.

2.4 Optimization Using Context and Spatial Information

2.4.1 Context Model

The main idea of context model is to make the most of semantic relations between regions in one image. For example, 'sea' and 'beach', 'road' and 'car', 'sky' and 'bird' are all pairs that occur together. The context model constructed the semantic similarity of the different regional feature vectors in the procedure of semantic image annotation. The region relevance context model is defined as follows:

Let $N_i = \{n_{ij}, j = 1, 2, .., J-1\}$ where J is the number of regions in a test image, denote the set of the other regions for the current region B_i in the image. The probability of N_i given w_k denotes the likelihood of the other regions under the condition that the current region is annotated with w_k.

$$\begin{aligned} P(N_i|w_k) &= \prod_j P(n_{ij}|w_k) \\ &= \prod_j \sum_m P(n_{ij}|w_m)P(w_m|w_k) \\ &= \prod_j \sum_m P(x_{ij}|w_m)P(w_m|w_k) \end{aligned} \tag{4}$$

where w_m denotes the possible label of region n_{ij}, x_{ij} dentoes the corresponding feature vector of region n_{ij}, and $P(w_m|w_k)$ represents the percentage of images labeled with w_m in the images labeled with w_k in training dataset.

$$P(w_m|w_k) = \frac{P(w_m, w_k)}{P(w_k)} = \frac{Q_{mk}}{Q_k} \tag{5}$$

where Q_{mk} is the number of images labeled with w_m in the images labeled with w_k, and Q_k is the number of images labeled with w_k in training dataset.

2.4.2 Spatial Model

We choose the average height of a region to represent the height information hidden in images. Gaussian distribution is used to learn the distribution of height information from training dataset. For each concept category w_k, the parameters μ_k and σ_k of Gaussian distribution are learned from each set of regions labeled with w_k in training dataset. Then for a test region B_i, hi denotes its average height, and the probability of this height given concept w_k can be expressed as,

$$p(h_i|w_k) = \frac{1}{\sqrt{2\pi}\sigma_k} \exp\{-\frac{(h_i - \mu_k)^2}{2\sigma_k^2}\} \tag{6}$$

2.4.3 Modified Bayesian Model

In Bayesian rule, based on context and spatial model detailed above, w_k is used instead of $P(w_k|B_i)$, i.e., $P(B_i, N_i, h_i|w_k)$ is used to substitute $P(B_i|w_k)$. Therefore, with the assumption that B_i, N_i, h_i are independent with each other, the modified Bayesian rule can be expressed as,

$$\begin{aligned} P(w_k|B_i, N_i, h_i) &= \frac{P(B_i, N_i, h_i|w_k)P(w_k)}{P(B_i, N_i, h_i)} \\ &= \frac{P(B_i|w_k)P(N_i|w_k)P(h_i|w_k)P(w_k)}{P(I)} \\ &\propto P(x|w_k)P(N_i|w_k)P(h_i|w_k)P(w_k) \end{aligned} \tag{7}$$

where $P(I)$ is assumed as a constant, x is the feature vector of the current region B_i.

2.5 Feature Extraction

We use color and texture features extracted from each segmented region to describe the characters in the region. Color feature vector is extracted for each region in HSV color space using the HSV histograms for the H, S and V channels separately. The number of histogram bins used should be $16 \times 4 \times 4 = 256$ bins. Hue channel has 16 levels so that each level is assigned a code in the range 0–15. Saturation channel has 4 levels so that each level is assigned a code in the range 0–3. Value channel has 4 levels so that each level is assigned a code in the range 0–3. And then, a combination of the 3 codes for H, S and V would be regarded as a separate bin, which represents a separate color. In each color channel color features are obtained including mean and standard deviation of the channel and the color value referring to the bin with the maximum size. A 9-dimension color vector is obtained to express the color character of the region. Texture feature vector is obtained using Gabor filter with 4 scales and 6 orientations. Mean and standard deviation of the coefficients in each filter results are calculated and obtain a 48-dimension texture vector to express the texture feature of the region.

3 Experiments

3.1 Experimental Settings

The experimental data were selected from a subset of LabelMe image databases for computer vision research from MIT, including 2651 images and 25653 regions. We choose regions whose label occurring frequency was more than 290 as the training database. There are 2686 images segmented into concept regions and labels of segmented regions are given. The top 14 categories including building, car, fence, ground, mountain, person, pole, rock, sign, sky, streetlight, tree, and water category are selected as real training regions with 25653 regions in total.

3.2 Experimental Results

Considering that the quality of clustering impact the results of GMM seriously, the first experiment was conducted on different clustering number K to analyze the performance. Figure 1 shows the F scores of each category using different number of Gaussian component K. As illustrated in Fig. 2, the more the clustering number K, we could achieve the better performance, and the characteristics of each category should be considered simultaneously.

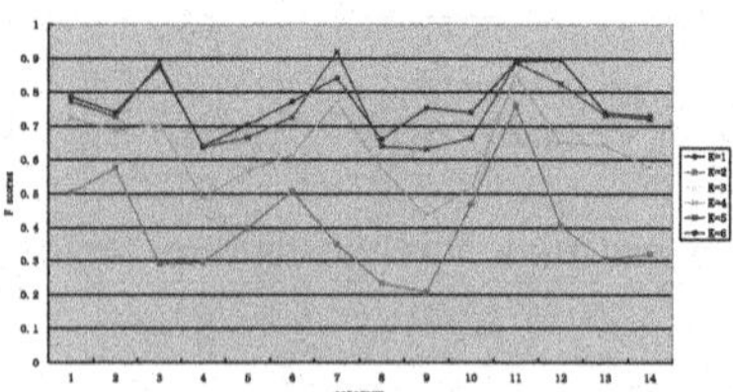

Fig. 1. The F scores using different number of Gaussian component K

Figure 2 shows that the F scores of annotating testing database based on Bayesian rule $p(w|x)$, on modified Bayesian rule using context information $p(w|x,n)$, on modified Bayesian rule using height information $p(w|x,h)$, and on modified Bayesian rule using both context and height information $p(w|x,n,h)$. As illustrated in Fig. 2, that context information and height information can improve the performance of semantic image annotation separately. If both the two information are taken into consideration, the performance could be improved more.

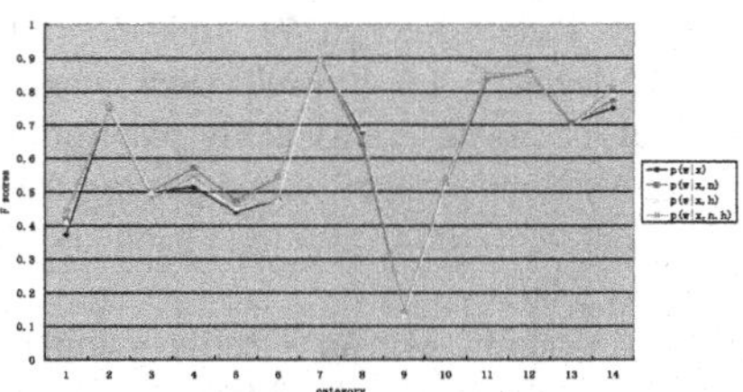

Fig. 2. The F scores based on Bayesian rule and modified Bayesian model.

4 Conclusion

In this paper, we proposed a novel semantic context model for automatic annotation with context and spatial information. Some machine learning techniques including supervised multi-class learning, EM algorithm, GMM model, Bayesian rule, were applied. We consider each concept as a category in which segmented regions from training database labeled with the concept are included. We reconstructed the image annotation as a multi-class classification problem and assign each object a label considering each object as an individual in both learning and annotation stage. To improve the performance of our semantic image annotation system, context information and special distribution are used to modify the result of Bayesian model. The experiment results on LabelMe database demonstrated that context information and special distribution are effective to improve the performance of Bayesian-based semantic image annotation system.

Several directions for future work are worthy of attention. The metadata of images belongs to the important semantic information for automatic image annotation and should be used to overcome the limitation of visual features in the future work.

Acknowledgments. This research was supported by the National Natural Science Foundation of China (Grant No. 41701521, 41771436), A Project of Shandong Province Higher Educational Science and Technology Program (Grant No. J15LH08) and Shandong Provincial Natural Science Foundation, China (Grant No. ZR2018LF005). We also thank the anonymous referees for their helpful comments and suggestions.

References

1. Zhang, D., Islam, M.M., Lu, G.: A review on automatic image annotation techniques. Pattern Recogn. **45**(1), 346–362 (2012)
2. Tao, D., Cheng, J., Gao, X., Li, X., Deng, C.: Robust sparse coding for mobile image labeling on the cloud. IEEE Trans. Circuits Syst. Video Technol. **27**(1), 62–72 (2017)
3. Carneiro, G., Chan, A., Moreno, P., Vasconcelos, N.: Supervised learning of semantic classes for image annotation and retrieval. IEEE Trans. Pattern Anal. Mach. Intell. **29**(3), 394–410 (2007)
4. Uricchio, T., Ballan, L., Seidenari, L., Del Bimbo, A.: Automatic image annotation via label transfer in the semantic space. Pattern Recogn. **71**, 144–157 (2017)
5. Zhang, J., Wu, Q., Shen, C., Zhang, J., Lu, J.: Multi-label image classification with regional latent semantic dependencies, pp. 1–11 (2016). arXiv preprint arXiv:1612.01082
6. Liu, C., Yuen, J., Torralba, A.: Nonparametric scene parsing: label transfer via dense scene alignment. In: 2009 IEEE Conference on Computer Vision and Pattern Recognition, CVPR 2009, pp. 1972–1979. IEEE (2009)
7. Vatani, A., Ahvanooey, M.T., Rahimi, M.: An effective automatic image annotation model via attention model and data equilibrium. order **9**(03) (2018)
8. Li, J., Wang, J.: Real-time computerized annotation of pictures. IEEE Trans. Pattern Anal. Mach. Intell. **30**(6), 985–1002 (2008)
9. Shi, F., Wang, J., Wang, Z.: Region-based supervised annotation for semantic image retrieval. AEU-Int. J. Electron. Commun. **65**(11), 929–936 (2011)
10. Torralba, A., Murphy, K., Freeman, W., Rubin, M.: Context-based vision system for place and object recognition. In: 2003 Proceedings of the Ninth IEEE International Conference on Computer Vision, pp. 273–280. IEEE (2003)

Identifying Transportation Modes Using Gradient Boosting Decision Tree

Xin Fu[1], Dong Wang[2], and Hengcai Zhang[3](✉)

[1] School of Resources and Environment, University of Jinan, Jinan 250022, China

[2] School of Information Science and Engineering, University of Jinan, Jinan 250022, China

[3] State Key Lab of Resources and Environmental Information System, Institute of Geographic Sciences and Natural Resources Research, Chinese Academy of Sciences, Beijing 100101, China

zhanghc@lreis.ac.cn

Abstract. Identifying the transportation modes could be applicable to many applications including personalized recommendation, transportation planning. The existing studies had not fully considered the impact of geographical information. In this paper, we propose a novel approach to detect transportation modes from massive trajectories using Gradient Boosting Decision Tree (GBDT), which adopted and estimated the impact of geographical information to achieve a better performance. In the experiments, we conduct the performance evaluation using the Geolife dataset which collected by 182 users over five years. The dataset contains 8347 trajectories with transportation mode such as driving, taking a bus, riding a bike and walking. 60% of trajectories are randomly chosen as training dataset, and then we tested on the remaining dataset. The experimental results showed that our proposed approach considering geographical information by using gradient boosting decision tree method achieve the precision of 84%, with the maximum increase of 6.83% to the traditional identifying transportation modes method. In addition, the geographical information contributed over 12% to improve the precision of recognition.

Keywords: Trajectories · Transportation mode · GBDT · Pattern recognition Geographical features

1 Introduction

Motivated by the explosion growth of massive trajectories due to the development of location technology and location-based services, the studies on human mobility has attracted more and more attention in the field of geographical information system and urban computing. As the hot topics of human mobility, identifying the transportation modes and understanding the user mobility brings new opportunities to uncover the movement and behavioral patterns of people. The information of transportation modes belongs to one of the important semantic and is applicable to many behavior research field, such as personalized recommendation, transportation planning, and intelligent traffic management.

D.-S. Huang et al. (Eds.): ICIC 2018, LNCS 10955, pp. 543–549, 2018.
https://doi.org/10.1007/978-3-319-95933-7_63

There has been a lot of related work on the transportation mode identification over the past few years, such as Decision Tree [1–4], Support vector machines [5, 6], random forest and hidden markov models [7, 8]. In addition, many deep neural networks by converting the raw trajectory data structure into image data are usually adopted into the identification processing [9]. Generally, the process of inferring the transportation modes includes two steps. The first step is to determe whether the user is moving or staying. And then, the trajectories are divided into trajectory segments. The second step is to identify the specifically transportation mode of each segment. Among them, extracting more discriminative feature from trajectories play an important role in improving the performance and precision. Many common statistical features such as mean velocity and more advanced features such as the stop rated and velocity change rate were widely adopted.

It appears that these studies on identifying transportation modes belongs to geo-related research. The geographical information such as road network, transit network, or subway network could effectively improve the performance of detection, in comparison with only trajectories. Although Stenneth et al. tried to introduce real time bus locations, spatial rail, and bus stop information to improve the classification effectiveness [4], it's not enough to detect transportation modes, because that some redundant and unimportant features are adopted, and more features could lead to the curse of dimensionality.

In this paper, we proposed a novel approach to detect transportation modes from massive trajectories using Gradient Boosting Decision Tree (GBDT), which adopted and estimated the impact of geographical information to achieve a better performance, in comparison with Random Forest, DT, GMM, and SVM. We adopted some dimensionality reduction to avoid the curse of dimensionality caused by the import of geographical information. Finally, we evaluate our proposed approach using a real-life trajectories dataset collection by Geolife project, which contains 8347 trajectories with transportation mode such as driving, taking a bus, riding a bike and walking.

2 Methodology

In this section, our identification approach is proposed to identify five different transportation modes containing driving a car, taking a bus, riding a bike, taking a taxi and walking. Figure 1 shows the architecture of our approach. As shown in figure, we employed some data processing techniques to remove the redundant data and outlier in the dataset.

And then, the trajectory was divided into segments according to the speed threshold. In our approach, the threshold was set as 9 m/s, if the average speed of trajectory segments exceeded the threshold, the segment was marked as walking. On the contrary, the segment was marked as no-walking. The second is to build the feature engineering. To achieve a better performance, the feature engineering was divided into geographical features and statistical features. Finally, we train the prediction model of identifying transportation modes using the Gradient Boosting Decision Tree.

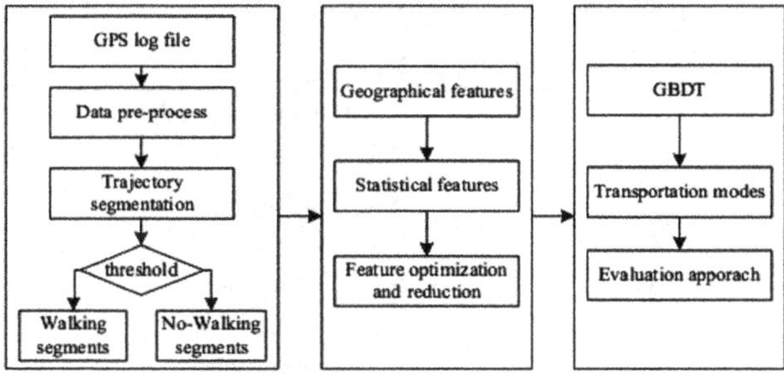

Fig. 1. The architecture of our approach

2.1 Feature Engineering

A summary of transportation features is shown in Fig. 2. The distribution of transportation features in trajectories is mentioned. Figure 2 shows the distribution of sub-trajectory length, sub-trajectory time, mean speed, mean acceleration of different transportation modes. The mean differences between distance, acceleration and speed suggest that these may be good discriminators as the separations are large.

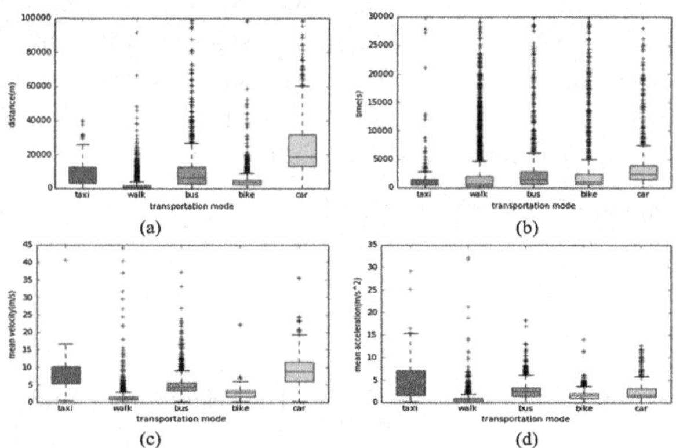

Fig. 2. The distribution of transportation modes features

Table 1 shows the statistical features including path length, activity time, mean speed, desired speed, speed variance, mean acceleration, and acceleration variance of sub-trajectory. Figure 2 illustrates the geographical features. The raw trajectories have no semantic information and don't reflect the user's activity information. In our approach, we extracted the O-D pairs of each segment. According to the O-D pairs, some geographical features were calculated, such as travel distance, travel time, the landmark architectures and the like in the process of travel.

Table 1. The statistical features

Traditional features	Meaning
D	Sub-trajectory length
T	Sub-trajectory time
v	Mean speed
E_v	Desired speed
S_v	Speed variance
a	Mean acceleration
S_a	Acceleration variance

2.2 Identifying Transportation Model

As an iterative decision tree algorithm, Gradient Boosting Decision Tree was proposed by Jerome Friedman [10], which was composed of Decision Tree and Gradient Boosting. Gradient boosting belonged to a machine learning technique for regression and classification problems, which produced a prediction model in the form of an ensemble of weak prediction models, typically decision trees. It build the model in a stage-wise fashion like other boosting methods, and it generalized them by allowing optimization of an arbitrary differentiable loss function. GBDT allows the combination of different features to have different discriminant other than SVM, which only had a unique global discriminant. So, it is particularly suited to different combinations of features that produce different results. This is more closed to the essence of identifying transportation modes. The input is defined as training set $\{(x_i, y_i)\}_{i=1}^n$, the loss function is $L(y, F(x))$. And the M is the number of iterations. It includes four steps.

(a) The initialization of prediction model $F_o(x)$.

$$F_o(x) = \arg\min \sum_{i=1}^{n} L(y_i, \gamma) \tag{1}$$

(b) The negative gradient of loss function L. The residual is estimated as:

$$r_{im} = -\left[\frac{\partial L(y_i, F(x_i))}{\partial F(x_i)}\right]_{F(x)-F_{m-1}(x)} \tag{2}$$

(c) Fit a base learner $h_m(x)$ to pseudo-residuals, i.e. train it using the training set $\{(x_i, r_{im})\}_{i=1}^n$. Compute multiplier γ_m by solving the following one-dimensional optimization problem:

(c) residual coefficients y_m is calculated according to the formulations:

$$y_m = \arg\min \sum_{i=1}^{n} L(y_i, F_{m-1}(x_i) + \gamma h_m(x_i)) \tag{3}$$

(d) Update the model.

$$F_m(x) = F_{m-1}(x) + \gamma_m h_m(x) \tag{4}$$

The $F_m(x)$ is the trained prediction model. It need to set parameters that include learning speed (μ), decision tree number (n), and tree depth (d). In our approach, the grid search algorithm is used to find the optimum parameter combination: $\mu = 0.2$, $n = 200$, $d = 5$.

3 Experiments

3.1 Experimental Settings

The proposed approach was implemented using Java and python as the main programming language and Eclipse Oxygen as the development environment. Some machine learning library including TensorFlow, Keras and Scikit-learn were adopted. The experiments were conducted on personal computer with Intel Xenon CPU E5-2620 v4 @2.10 GHz, 8G memory and 512G Solid State Drives, which was Linux-base installation of operation system. We conducted the performance evaluation using the Geolife dataset which collected by 182 users over five years. The dataset contains 8347 trajectories with transportation mode such as driving, taking a bus, riding a bike and walking. 60% of trajectories are randomly chosen as training dataset, and then we tested on the remaining dataset. The geographical information dataset includes the road network of Beijing with 26,220 segments and 18,856 intersections, transit network, metro network, and land use data.

3.2 Experimental Results

The contribution rates of traditional features are 88%, and the contribution rate of geographic information is about 12%, as shown in Fig. 3.

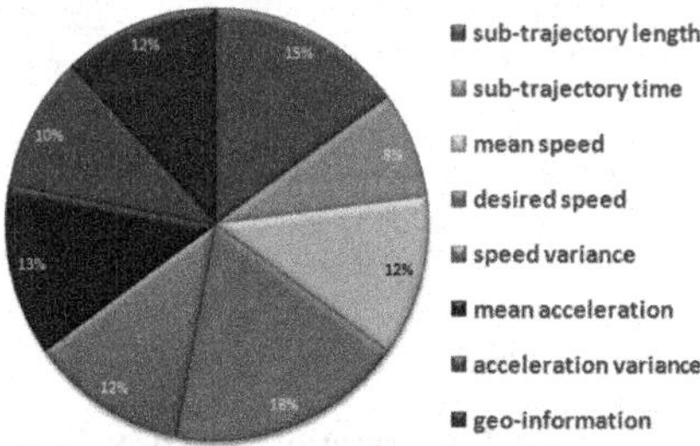

Fig. 3. Contribution rates of each feature

The whole identification accuracy of the transportation modes is 84%, and the identification accuracies of single transportation mode is showed in Fig. 4. The identification accuracies of walk and bike is obviously higher than that of other motor vehicle travel patterns, and the difference in accuracy of motor vehicles is not particularly obvious. Because the speed and acceleration of walking and cycling are obviously lower than those of motor vehicles.

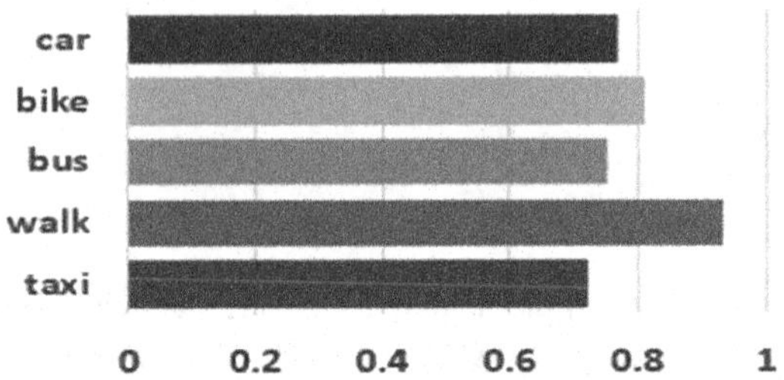

Fig. 4. Identification accuracies of single transportation mode

Four different kinds of classification models are compared with two criteria that is having geographic information and no geographic information. It can be seen from Fig. 5 that GBDT is superior to other models, and the identification accuracies of feature vectors having geographic information are better than those without geographical information.

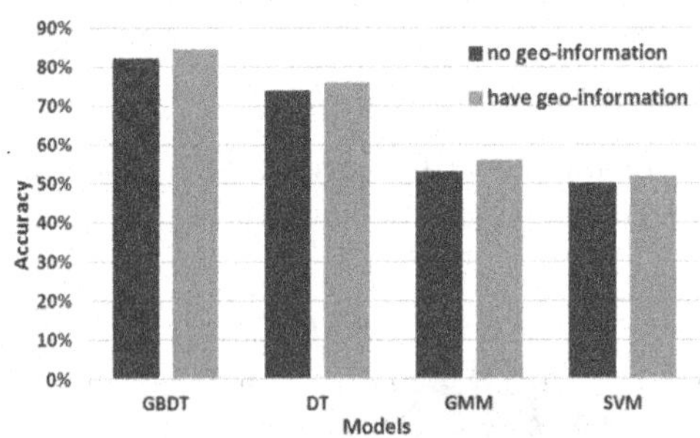

Fig. 5. Identification accuracies of different models

4 Conclusion

In this paper, by using geographical information, we proposed an approach to identify transportation modes from massive trajectories using Gradient Boosting Decision Tree. In addition to the traditional features such as distance, time, speed and acceleration, we build the feature engineering including geographical features and statistical features to improve the performance. Some dimensionality reduction was employed to avoid the curse of dimensionality. Finally, we carried out performance evaluation on a real-life trajectories dataset collection by Geolife project. The results show that adding geographic information to the features can effectively improve the identification accuracy of transportation modes, the identification accuracy is 84%, and the contribution rate of geographic information in this experiment is 12%. In the future, several directions are worthy of attention. Some sophisticated graphical model and deep learning algorithms should be developed to infer the transportation modes.

Acknowledgments. This research was supported by the National Natural Science Foundation of China (Grant No. 41701521, 41771436), A Project of Shandong Province Higher Educational Science and Technology Program (Grant No. J15LH08) and Shandong Provincial Natural Science Foundation, China (Grant No. ZR2018LF005). We also thank the anonymous referees for their helpful comments and suggestions.

References

1. Zheng, Y., Li, Q., Chen, Y., Xie, X., Ma, W.-Y.: Understanding mobility based on GPS data. In: Proceedings of the 10th International Conference on Ubiquitous Computing, Seoul, Korea, 21–24 September 2008, pp. 312–321 (2008)
2. Zheng, Y., Liu, L., Wang, L., Xie, X.: Learning transportation mode from raw GPS data for geographic applications on the web. In: Proceedings of the 17th International Conference on World Wide Web, Beijing, China, 21–25 April 2008, pp. 247–256 (2008)
3. Stenneth, L., Wolfson, O., Yu, P.S., Xu, B.: Transportation mode detection using mobile phones and GIS information. In: Proceedings of the 19th ACM SIGSPATIAL International Conference on Advances in Geographic Information Systems, Chicago, IL, USA, 1–4 November 2011, pp. 54–63 (2011)
4. Reddy, S., Mun, M., Burke, J., Estrin, D., Hansen, M., Srivastava, M.: Using mobile phones to determine transportation modes. ACM Trans. Sens. Netw. (TOSN) **6**, 13 (2010)
5. Dodge, S., Weibel, R., Forootan, E.: Revealing the physics of movement: comparing the similarity of movement characteristics of different types of moving objects. Comput. Environ. Urban Syst. **33**, 419–434 (2009)
6. Bolbol, A., Cheng, T., Tsapakis, I., Haworth, J.: Inferring hybrid transportation modes from sparse GPS data using a moving window SVM classification. Comput. Environ. Urban Syst. **36**, 526–537 (2012)
7. Bashir, F.I., Khokhar, A.A., Schonfeld, D.: Object trajectory-based activity classification and recognition using hidden markov models. IEEE Trans. Image Process. **16**, 1912–1919 (2007)
8. Widhalm, P., Nitsche, P., Brändie, N.: Transport mode detection with realistic smartphone sensor data. In: Proceedings of the 21st International Conference on Pattern Recognition (ICPR), Tsukuba, Japan, 11–15 November 2012, pp. 573–576 (2012)
9. Endo, Y., Toda, H., Nishida, K., Kawanobe, A.: Deep feature extraction from trajectories for transportation mode estimation. In: Bailey, J., Khan, L., Washio, T., Dobbie, G., Huang, J.Z., Wang, Ruili (eds.) PAKDD 2016. LNCS (LNAI), vol. 9652, pp. 54–66. Springer, Cham (2016). https://doi.org/10.1007/978-3-319-31750-2_5
10. Friedman, J.: Greedy function approximation: a gradient boosting machine. Ann. Stat. **29**(5), 1189–1232 (2001)

A Multi-object Optimization Cloud Workflow Scheduling Algorithm Based on Reinforcement Learning

Wu Jiahao[1(✉)], Peng Zhiping[2], Cui Delong[2], Li Qirui[2], and He Jieguang[2]

[1] Department of Computer, Guangdong University of Technology, Guangzhou 510006, China
469981325@qq.com
[2] Department of Computer and Electronic Information, Guangdong University of Petrochemical Technology, Maoming 525000, China

Abstract. In this paper, for the problem of long task scheduling time and unbalanced system load in the task scheduling of cloud workflow. To minimize the task scheduling time and optimize load balancing as the scheduling goal, a Markov decision process model conforming to the cloud workflow environment is established. Based on this, a multi-objective optimization cloud workflow scheduling algorithm based on reinforcement learning is proposed. The algorithm combines Q_Learning features, adding a function with a weighted fitness value function in the Q_Learning reward function so that it can apply multi-objective optimization. The set of scheduling schemes is a Pareto optimal solution set, which can select the optimal scheduling scheme according to the user's preference. Compared with other methods, this algorithm can reduce the execution time and optimize the system load. And this paper uses the real cloud workflow data to carry out the simulation experiment, and carries on the experiment through the simulation platform WorkflowSim. The result proves the effectiveness of this algorithm.

Keywords: Cloud workflow · Reinforcement learning · Task scheduling Load balancing

1 Introduction

1.1 Research Background and Significance of Cloud Workflow

Cloud computing was first proposed by Google at the search engine strategy conference in 2006. After that, the concept of cloud computing reached a preliminary consensus in the debate. After several years of development, cloud computing is being implemented. Google, Amazon and other companies have adopted cloud computing as their core strategy for the future. Cloud computing will subvert the existing IT product landscape in technologies, services, business models, etc., and will be the core of the next-generation information technology revolution and IT application innovation, and will bring fundamental changes in the working methods and business models. There are

D.-S. Huang et al. (Eds.): ICIC 2018, LNCS 10955, pp. 550–559, 2018.
https://doi.org/10.1007/978-3-319-95933-7_64

three levels of resource services provided by cloud computing: Infrastructure-as-a-service (IaaS), Platform-as-a-service (PaaS), and Software-as-a-service (SaaS) [1]. Although these three levels have different focuses on computing resources, the task scheduling mechanism is the same. Therefore, workflow task scheduling not only has theoretical research value, but also has higher application value in industry [2].

Workflow, as a complex application with complex business processes, large amounts of data, and computational time-consuming processes, has always been a research focus for researchers. The mathematical model of the relationship of workflow priorities is accustomed to Directed Acyclic Graphs (DAG). To express. With the continuous advancement of cloud computing technology, the implementation of large-scale workflows requires powerful computing power, complex data structures, and huge storage capacity. The cloud computing platform can take on such a role and open up a resource pool. The resource pool runs the operational specification of this workflow, which improves efficiency and reduces costs. This is the cloud workflow. The execution steps of the cloud workflow generally include two major stages of task allocation and resource supply. The workflow scheduling problem has been proved to be a NP-complete problem, and the operating environment is a distributed cloud computing platform, which adds to its computational complexity.

1.2 Related Work

At present, there are traditional scheduling algorithms, list scheduling algorithms and meta-heuristic scheduling algorithms for cloud workflow task scheduling. Traditional scheduling algorithms, such as round robin scheduling algorithm, MIN-MIN algorithm and MIN-MAX algorithm, have the advantages of simple implementation and low algorithm complexity, but can only be applied to specific scenarios. The most widely used algorithms in industrial scenes are mainly list scheduling algorithms [3, 4], such as: Heterogeneous earliest completion time algorithm, mapping heuristic algorithm, and so on. Another widely used algorithm is based on metaheuristic scheduling algorithms [5, 6], the main idea of which is based on random and local search algorithm, and this algorithm can give a feasible solution in a certain operation time and space cost.

This paper mainly studies the scheduling algorithm based on stochastic search technology, which gives a feasible solution to the optimization problem to be solved given the limited cost (calculation time, memory space), but the optimality of the solution cannot be given. Guarantee. The main algorithm can only be GA, PSO and RL algorithms. (1) Taking performance as the scheduling goal: Peng et al. [7] designed a task scheduling scheme based on reinforcement learning and queuing theory, optimized task scheduling under resource constraints, accelerated the learning process using state aggregation technology, and proved the algorithm through comparative experiments. Effectiveness. In terms of scheduling resource optimization: Tian et al. [8] first proposed the "relatively stringent degree" indicator for measuring the emergency level of DAG deadlines, which can reasonably handle the emergency level relationship among multiple DAGs, thus achieving the maximum dispatching of DAG throughput aims. In terms of scheduling costs: Guo et al. [9] solved the problem of workflow cost optimization under the constraint of communication overhead through the idea of hierarchical scheduling.

2 Model of Cloud Workflow Scheduling

2.1 Model of Cloud Workflow

Cloud workflow is a set of tasks defined by the user in the cloud computing environment with a certain timing correlation, because this task set has a certain timing relationship. Academically, the general abstraction is a Directed Acyclic Graph (DAG) [10]. As shown in Fig. 1, this DAG has seven cloud workflows with constraints. Any DAG can be represented as a triple DAG = {T,E,A}, where

(1) $T = \{t_1, t_2, \ldots, t_n\}$ represents a set of tasks consisting of n tasks, where t_i is the i-th task.
(2) $E = \{(t_i, t_j)\} i \geq 1, j \leq n$, represents task dependency set, the timing relationship between the tasks, mainly indicating that the task t_i is executed before the task t_j, where t_i is the precursor node of t_j. $t_i = \text{pre}(t_j)$, or $t_j = suc(t_i)$, means t_j is the successor of t_i.
(3) $A = \{a_{ij} | 1 \leq i \leq \text{m}, 1 \leq j \leq \text{n}\}$, where a_{ij} indicates the time when the processor u_i executes the task t_i.

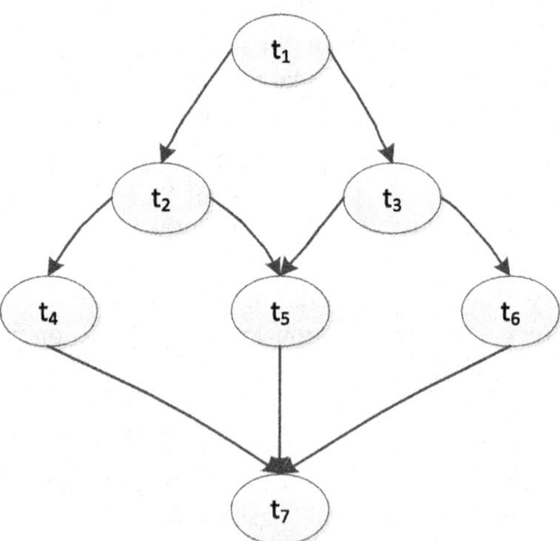

Fig. 1. DAG of cloud flow

2.2 Model of Reinforcement Learning (RL)

RL is a search algorithm based on interaction with the environment, as shown in Fig. 2. This algorithm learns which behaviors can be most rewarded in a particular environment when they are constantly trying. In many cases, the current actions will not only affect the current Reward, but also affect the subsequent status and a series of Rewards. The three most important RL features are: (1) Basically it is a closed-loop form. (2) It does not directly indicate which action to choose. (3) A series of actions and reward signals will affect the longer state.

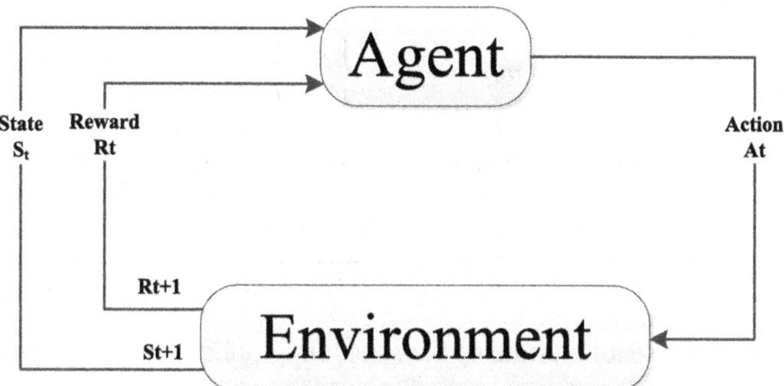

Fig. 2. Interaction between agent and environment

As shown in the Fig. 2, t indicates time, where t = 0,1,2,3…

$S_t \in S$, where S indicates environment set.
$A_t \in A(S_t)$, where $A(S_t)$ is the action set under action space.
$R_t \in R$, where R indicates data-type Reward.

3 Multi-objective Cloud Workflow Task Scheduling Algorithm Based on Reinforcement Learning

3.1 MDP Model of Cloud Workflow Task Scheduling

Cloud workflow task scheduling is based on the constraint optimization problem of DAG, and RL refers to the agent obtaining the optimal strategy through continuous interaction and trial and error with the environment. Before this, the Markov decision process (MDP) model of the cloud workflow task scheduling problem is needed, which is indicated as follow:

$$\{S, A, r(S_t, a_t), f(S_t, a_t); S_t, S_{t+1} \in S, a_t \in A\} \tag{1}$$

S is the state space, and A is the action space.
$r(S_t, a_t)$ is the reward function of the state S_t for Reward by selecting the a_t action.

$f(S_t, a_t)$ is a state transfer function.

According to the above definition, this paper gives a description of the MDP model of cloud workflow task scheduling:

(1) The state space S is represented by a task assignment matrix $Matrix_{m \times n}$, which corresponds to the feasible solution set of the task scheduling problem of the cloud workflow. The transition is the solution of the search process in this solution space set.
(2) Action space A is represented by a set of tasks. For any state, it corresponds to n tasks, that is, n actions. If the current state is $m_{ij} = 1$, then action i will transfer the job to the next processor unit, as shown below:

$$m_{kj} = 1, k = \begin{cases} i + 1 & i < m \\ 1 & i = m \end{cases} \tag{2}$$

(3) The reward function $r(S_t, a_t)$ can be expressed as follow:

$$r(S_t, a_t) = \frac{1}{t(S_{t+1})} \tag{3}$$

where $t(x_{t+1})$ is the execution time of the scheduling policy after the current state S_t is transferred to the state S_{t+1}.

(4) The state transition function $f(S_t, a_t))$ is calculated from the formula in Q_Learning.

3.2 Q_Learning Algorithm for Cloud Workflow Task Scheduling

The classic Q_Learning was proposed by Watkins et al. [11]. and proved its convergence, and was then widely applied in the field of reinforcement learning. Q_Learning is an offline reinforcement learning algorithm based on stochastic dynamic process. It is mainly composed of episodes, and each episode consists of several steps. The Q_Learning algorithm training process is to iterate the following steps:

(1) Observe the current state S_t and select the appropriate a_t.
(2) From state S_t to state S_{t+1}, instant reward $r(S_t, a_t)$ is calculated.
(3) The value function is represented as $f(S_t, a_t)$, and the adjustment value function formula is as follows:

$$f_{t+1}(S_t, a_t) = (1 - \theta)f_t(S_t, a_t) + \theta\left[r_t + \gamma \max_{a_{t+1}} f_t(S_{t+1}, a_{t+1})\right] \tag{4}$$

Among the formula above, θ is the learning rate, and the range is $0 \leq \theta \leq 1$. The larger the learning rate is, the faster convergence speed will be, but the oscillation is easy to occur, while the learning rate gets lower, the convergence speed will reduce. γ is a discount factor, and the range is $0 \leq \gamma \leq 1$. When the discount factor has a large value, we consider the long-term Reward, and when the value of the discount factor is small, we consider the short-term Reward.

3.3 Multi-objective Optimization of Cloud Workflow Task Scheduling

The two goals of this paper's optimization are the execution time and load balancing minimization. The execution time refers to the time required for all the processors to process the entire task. The load balancing of this document refers to preventing the processor from overloading the processor due to the processing sequence of tasks by effectively allocating tasks to the most suitable processor., unable to complete or affect the progress of completion. For solving multi-objective optimization problems, Q_Learning can't strictly compare which one is the best for a Pareto optimal solution. Because this solution set cannot be dominated by any other solution, this paper uses the method that the user can specify the weight value to optimize the multi-objective problem [12]. The following is the formulation of these two optimization goals:

(1) Time Objective Formula:

$$O_T(M) = T_{total}(M) \tag{5}$$

(2) Load balancing function:

$$O_C(M) = C_{total}(M) \tag{6}$$

In the cloud workflow scheduling problem, it is very difficult to find solutions for achieving $\min[O_T(M)]$ and $\min[O_C(M)]$. In order to solve this multi-objective optimization problem, this paper introduces a feasible adaptive-value function with weight value:

$$\mathrm{F}(X) = \omega O_T(M) + (1-\omega)O_C(M) \tag{7}$$

ω represents user's attention to time and load balancing. It determines the degree of contribution of each goal. You can get the best result by adjusting ω.

It is known from the function F(*X*) that the reward function $r(S_t, a_t)$ should be changed accordingly:

$$r(S_t, a_t) = \frac{\omega}{t(S_{t+1})} + \frac{(1-\omega)}{c(S_{t+1})} \tag{8}$$

Among them, the reason that $O_T(M)$ and $O_C(M)$ take the inverse is to control the two numerical values within the range of $(0, 1)$, which is convenient for uniform dimension and calculation.

3.4 Steps of our Algorithm

Step 1: Initialize all the $f(S_t, a_t)$;
Step 2: Randomly initialize the state space S;
Step 3: Randomly select an action a_r from the actions space A;
Step 4: Update the value function $f(S_t, a_t)$, and let $S_t = S_{t+1}$;
Step 5: If the expected goal is not reached, go back to Step 3, otherwise, output the result.

4 Simulation and Analysis

4.1 Experimental Data and Parameter Settings

In order to reflect the practicality of the algorithm, this paper uses the open source workflow simulator WorkflowSim [13] developed by the University of California, Chen and Deelman, as an experimental verification tool. WorkflowSim extends CloudSim and provides simulations at the level of the scientific workflow. The workflow selected in this paper is the workflow DAX file provided by WorkflowSim for simulation experiments. These scientific workflows are workflow data in the real environment. The goal of this article is to minimize execution time and load balancing. The degree of load balancing is expressed using the load balance difference, and the load balance difference formula is as follows.

$$\phi = \sqrt{\frac{\sum_{j=1}^{m} \left(LB_j - \bar{LB_j}\right)^2}{m-1}} \tag{9}$$

Among them, the load balancing factor LB_j is the number of tasks on a certain processor j, and $\bar{LB_j}$ is the average value of the load balancing factor LB_j on the processor j.

The comparison algorithm uses three kinds of algorithms with better performance in the limited processor environment to compare, namely HEFT algorithm, MIN-MIN algorithm and MAX-MIN algorithm. The HEFT algorithm selects high-priority tasks for scheduling according to the constraints of the tasks, and schedules the tasks on the processor that can perform the earliest time of completion. In the scheduling process, the algorithm searches for the earliest free time of the processor and puts the appropriate Tasks are inserted into it. The MIN-MIN algorithm belongs to a batch processing algorithm. The algorithm first calculates the running time of each task on the processor, selects the processor that corresponds to the minimum completion time of each task, forms pairwise mapping pairs, constitutes a mapping set, and then in the slave mapping. The task in the set looking for the minimum completion time is the processor mapping pair. According to the mapping pair, the task is allocated to the processor for execution. After the execution is completed, the selected pair is deleted in the task set, and the steps are repeated until all the mapping pairs are executed. The MAX-MIN algorithm also belongs to the batch processing algorithm, which is based on the MIN-MIN algorithm. The main difference lies in forming the mapping set of the task-processor's earliest completion time, and selecting the latest completion task from the mapping set. - Processor mapping is performed first.

The parameters of the experimental simulation platform for this article are shown in Table 1:

According to the definition of the formula (4),The algorithm parameter settings in this paper are shown in Table 2:

Table 1. Parameters of the experimental simulation platform

Platform parameters	Value
Number of VM	10
Number of core	1
Processing capacity	[500–1000]
RAM	1024
Bandwidth	1000

Table 2. Algorithm parameter settings

Parameters	Value
State space S	Random initialization
Learning rate θ	0.75
Discount factor γ	0.75
Threshold ω	0.5

4.2 Experimental Results and Analysis

This experiment uses workflow applications with different numbers of tasks as experimental simulation data. The experimental data is that the Inspiral workflow is a gravitational wave phenomenon proposed by the laser interferometer gravity wave observatory based on Einstein's general theory of relativity. This workflow is characterized by a large number of tasks, each task is relatively small, and the amount of calculation is large.

Inspiral workflow scheduling is shown in Table 3, Fig. 3, Table 4, Fig. 4.

Table 3. Different tasks for Inspiral workflow scheduling completion time

Number of tasks	RL	HEFT	MAX-MIN	MIN-MIN
	time (s)	time (s)	time (s)	time (s)
30	1427.25	1427.81	1573.33	1850.03
50	2034.93	2159.15	2771.57	2636.27
100	3250.84	3632.75	3395.89	3827.62
1000	30391.58	31536.41	31742.38	32494.26

Experimental results show that a multi-objective optimization cloud workflow scheduling algorithm based on reinforcement learning proposed in this paper is superior to the other three algorithms in the real cloud workflow environment, both in optimizing the completion time and minimizing load balancing. Only when the number of tasks is 1000, the load balance deviation is close to other algorithms. The main reason is that there are only 10 virtual machines in the experiment, and 1000 tasks can't be compensated by the algorithm's schedule for 10 virtual machine loads, which completely exceeds the virtual machine operating load. The algorithm in this paper has the ability to search multiple solutions at the same time.

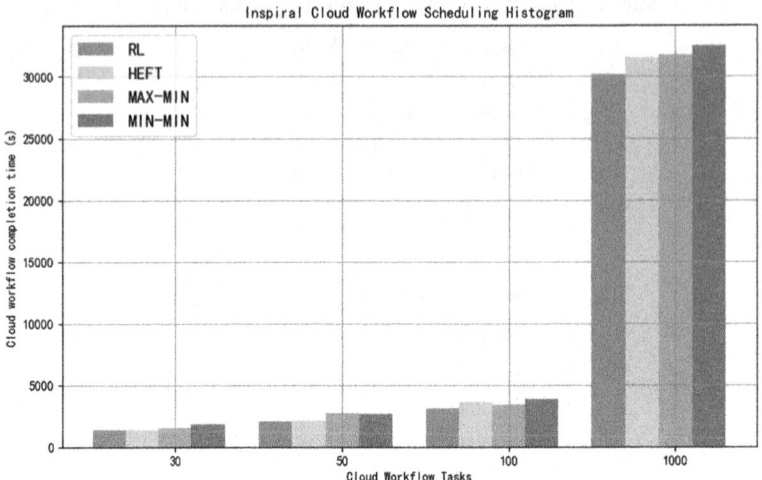

Fig. 3. Inspiral cloud workflow scheduling histogram

Table 4. Different tasks for Inspiral workflow scheduling load balance deviation

Number of tasks	RL	HEFT	MAX-MIN	MIN-MIN
	value	value	value	value
30	0.23	0.25	0.34	0.35
50	0.26	0.28	0.36	0.36
100	0.45	0.51	0.56	0.52
1000	0.98	0.98	0.99	0.99

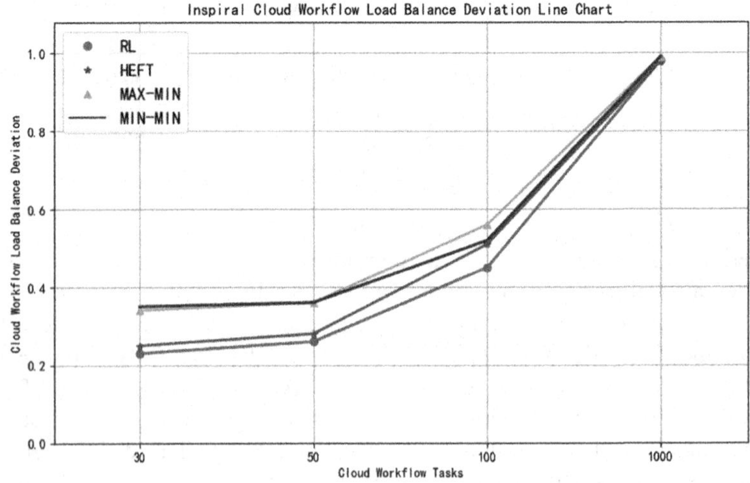

Fig. 4. Inspiral cloud workflow schedule line chart

5 Concluding Remarks

In this paper, aiming at the multi-objective optimization cloud workflow scheduling problem, an MDP model of cloud workflow scheduling is established. Based on this, a multi-objective optimization cloud workflow scheduling algorithm based on reinforcement learning is proposed. In combination with Q_Learning algorithm with weighted fitness value function, the algorithm can search for the two goals of completion time and minimization of load balancing under the real cloud workflow environment. Since the improved Q_Learning algorithm proposed in this paper is an offline scheduling algorithm, the next research direction is to focus on solving the multi-objective optimization of online scheduling algorithm.

Acknowledgement. Fund Project: National Natural Science Foundation of China (61772145, 61672174, 61272382), Guangdong Province Science and Technology Plan Project (2015B020233019, 2014A020208139)

References

1. Zuo, L.Y., Cao, Z.B.: An overview of research on scheduling problems in cloud computing. Appl. Res. Comput. **11**(29), 4023–4027 (2012)
2. Lin, W.W., Qi, D.X.: Survey of cloud computing resource scheduling. Comput. Sci. **10**(39), 1–6 (2012)
3. Topcuoglu, H., Hariri, S., Wu, M.Y., et al.: Performance-effective and low-complexity task scheduling for heterogeneous computing. IEEE Trans. Parallel Distrib. Syst. **13**(3), 260–274 (2002)
4. El-Rewini, H., Lewis, T.G.: Scheduling parallel program tasks onto arbitrary target machines. J. Parallel Distrib. Comput. **9**(2), 138–153 (1990)
5. Salza, P., Ferrucci, F., Sarro, F., et al.: Deploy and execute parallel genetic algorithms in the cloud. In: Proceedings of the 2016 on Genetic and Evolutionary Computation Conference Companion, pp. 121–122. ACM, Denver (2016)
6. Li, H.H., Chen, Z.G., Zhan, Z.H.: Renumber coevolutionary multiswarm particle swarm optimization for multi-objective workflow scheduling on cloud computing environment. In: Proceedings of the Companion Publication of the 2015 Annual Conference on Genetic and Evolutionary Computation, pp. 1419–1420. ACM, Madrid (2015)
7. Peng, Z.P., Cui, D.L., Zuo, J.L., et al.: Random task scheduling scheme based on reinforcement learning in cloud computing. Cluster Comput. **18**(4), 1595–1607 (2015)
8. Tian, G.Z., Xiao, C., Xie, J.Q.: Scheduled multi-DAG shared resource scheduling and fair cost optimization method. Chin. J. Comput. **37**(7), 1607–1619 (2014)
9. Guo, T., Chen, Z., Yu, Y.L.: Workflow cost optimization model and algorithm for DAG with communication cost. J. Comput. Res. Dev. **52**(6), 1400–1408 (2015)
10. Li, X.: Cloud computing task scheduling, computer measurement and control based on dependent task and saras(λ) algorithm. **23**(8), 2809–2813 (2015)
11. Watkins, C.J.C.H., Dayan, P.: Q-learning. Mach. Learn. **8**, 279–292 (1992)
12. Hong, L.: Research on Workflow Scheduling Algorithm Based on Multi-objective Particle Swarm Optimization in Cloud Environment. Beijing Jiaotong University, Beijing (2015)
13. Chen, W., Deelman, E.: WorkflowSim: a toolkit for simulating scientific workflows in distributed environments. In: Proceedings of the IEEE International Conference on E-Science, pp. 1–8. IEEE, Bangalore (2012)

A AprioriAll Sequence Mining Algorithm Based on Learner Behavior

Zhenghong Yu[1(✉)] and Dan Li[2]

[1] City College of Wuhan University of Science and Technology, Wuhan, China
yzh@wic.edu.cn
[2] The College of Post and Telecommunication of WIT, Wuhan, China

Abstract. Although massive learning resources in online education platform provide users with more learning opportunities, users are also faced with new challenges of information overload. At present, most of the personalized recommendation related research on educational resources is based on the campus application or the traditional online learning website design personalized recommendation algorithm for educational resources. It does not take into account the new characteristics of user behavior in online learning, and does not make full use of the collective wisdom embodied in the educational resources under the internet background.

In view of the shortcomings of personalized recommendation technology of educational resources, we put forward a learner model based on AprioriAll mining algorithm on the basis of analyzing the characteristics of user learning behavior in the Internet. It concretely attributes learners' attributes and understands learners' behaviors according to learner models. According to the established learner model, the learners' behavior is tracked, and the potential relationship between courses is found through the use of sequence mining algorithm based on the behavior of the learners, and the courses that are more in line with the learners' interest are recommended.

Keywords: AprioriAll sequence mining algorithm · Learner behavior
Autonomous learning

1 Introduction

With the rise of network education, many experts and scholars have begun to pay attention to the study of learner autonomy in the network environment, and introduce computer technology and network technology into the network learning support system. The learning effect of autonomous learning system is not as good as what one wishes. An important reason is whether the autonomous learning system can meet the individualized requirements of the learners, in addition to whether the content and arrangement are appropriate and the examples and questions and questions and answers are correct or not.

These problems lead to the low quality of learning, inefficient learning and low utilization of resources in autonomous learning system.

D.-S. Huang et al. (Eds.): ICIC 2018, LNCS 10955, pp. 560–569, 2018.
https://doi.org/10.1007/978-3-319-95933-7_65

User behavior data analysis is one of the main research directions at home and abroad. Such user behavior analysis provides a good condition for building a reasonable user interest model. Compared to e-commerce sites, the number of educational resources in the Internet is less and the subject classification of resources is easier to apply to the research and practice of the recommended algorithms. On the other hand, in the research field of recommendation algorithm, many scholars are applying the idea of recommendation algorithm to a wider range of scenarios. Many scholars have conducted exploratory experiments around the theme of Educational Resource Recommendation Technology. The recommendation technology of educational resources is the combination of related research and recommendation algorithm research of educational resource platform. User behavior data in education resource platform is an important foundation of recommendation algorithm. Reasonable recommendation algorithm design and application can help users to efficiently screen resources and learn independently in the Internet education platform.

But at present, some research results still have some shortcomings. First of all, the main purpose of educational resource related research on user behavior analysis is to design reasonable courses. The data analysis only pays attention to the characteristics and rules of the learning behavior of the universal users, and lacks the analysis of the user's interactive data. Therefore, it cannot establish a personalized interest model for the users.

Secondly, in the background of the Internet, the educational resources platform embodies the collective wisdom of the users, such as the custom tags added to the educational resources, but the current research lacks the semantic analysis of the user label, and does not make full use of the collective wisdom.

In addition, the personalized recommendation algorithms for educational resources are mostly limited to recommending a single kind of educational resources for users, without considering the theme characteristics of comprehensive educational resources projects, and cannot recommend educational resources for users from different levels.

In this paper, we propose a mining algorithm based on AprioriAll sequential pattern. It can concretely attribute learners' attributes and understand learners' behaviors according to learners' models. We track learners' behaviors based on the established learner model. At the same time, based on learner behavior, we use sequence mining algorithm to identify potential relationships between courses to recommend courses that are more in line with learners' interests. Through the depth learning algorithm, we excavate the student's record making, analyze the cause of the students' mistakes, the degree of knowledge system mastery, the evaluation of innovation ability and so on, so as to provide the decision-making basis for the teaching.

2 Learner Model Design Based on Sequence Mining

The learner model based on sequential mining algorithm should achieve this function: by defining the learner model, we can sequence mining the learner's behavior according to the relevant parameters of the model, find out the maximum frequent sequence, and propose a learning sequence suitable for learners.

(1) Learners

Learners are the main body of learning. Sequential mining algorithm aims at mining the same learner's behavior without paying attention to the relationship between the learner and other learners.

(2) Learning courses

The learning course, the constituent element of the learning sequence, is organized into a learning sequence in a different order.

(3) Learning time

Sequential mining algorithm emphasizes the order and causality of things, so learning time is an important parameter of sequential mining algorithm. We can determine the learning sequence that the learners have completed by comparing the time of comparison.

(4) Sequence duration

As time goes on, people's worldview will change. Knowledge will also appear alternately. Once used learning sequence may not be suitable for the new era, we must determine whether it satisfies the current teaching requirements for the sequence beyond the sequence duration.

(5) Event folding window

If the event-folding window is T, the learner learns the course A and B in same time, it is considered that the course A&B is simultaneous, that is, A&B should be recorded as a project.

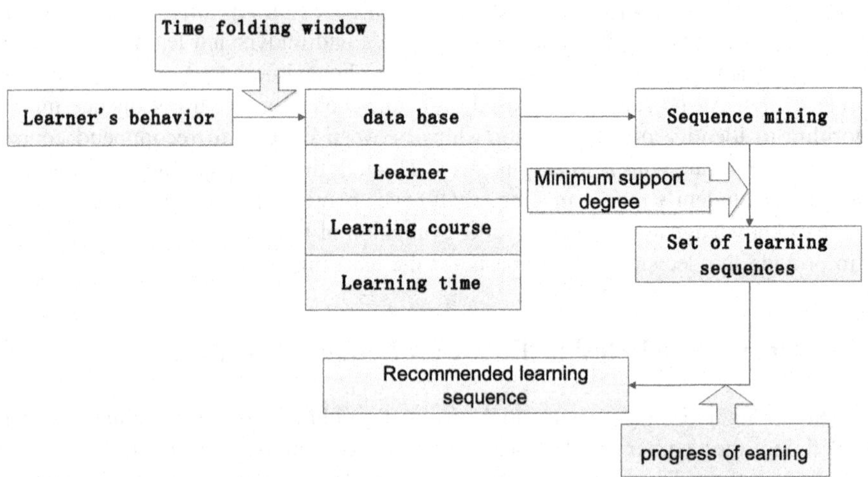

Fig. 1. Learner model based on sequential mining algorithm

(6) Minimum support degree

A threshold used to measure support is customer-defined, and only a set of items larger than the minimum support is adopted by the system.

Combining the above factors, the learner model based on the sequential mining algorithm is shown in Fig. 1.

3 AprioriAll Sequence Mining Algorithm Based on Learner Behavior

The sequence-mining algorithm based on learner behavior can describe the algorithm in the following form according to the learner model based on the sequential mining algorithm: Collecting the information of learners→Information processing→Sequence Mining→Find frequent item set→matching learning progress→Recommended course of study, as shown in Fig. 2.

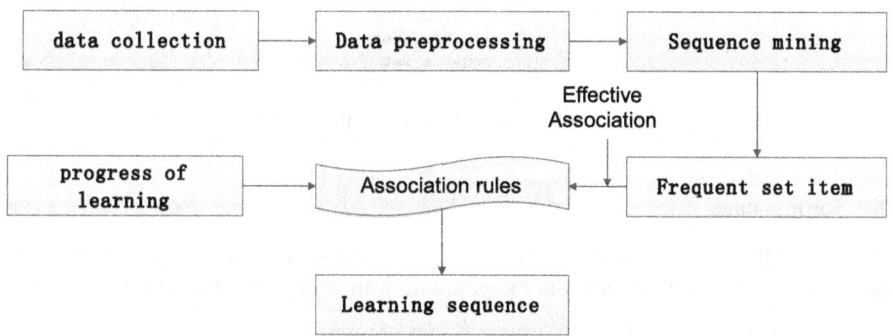

Fig. 2. The process of sequence mining based on learner behavior

The basic idea of the AprioriAll algorithm is to traverse the sequence database to generate candidate sequences and to prune the frequent sequences by using the properties of the Apriori. Each traversal is generated by connecting the frequent sequences obtained last time to generate candidate sequences with new length plus one, and then scanned each candidate sequence to verify whether it is a frequent sequence.

AprioriAll algorithm divides sequential pattern mining process into five stages: sorting stage, finding frequent item set stage, transforming stage, sequential pattern mining phase, and finding out the maximum frequent sequence set stage. The learning sequence mining process based on learner behavior is shown in Fig. 3.

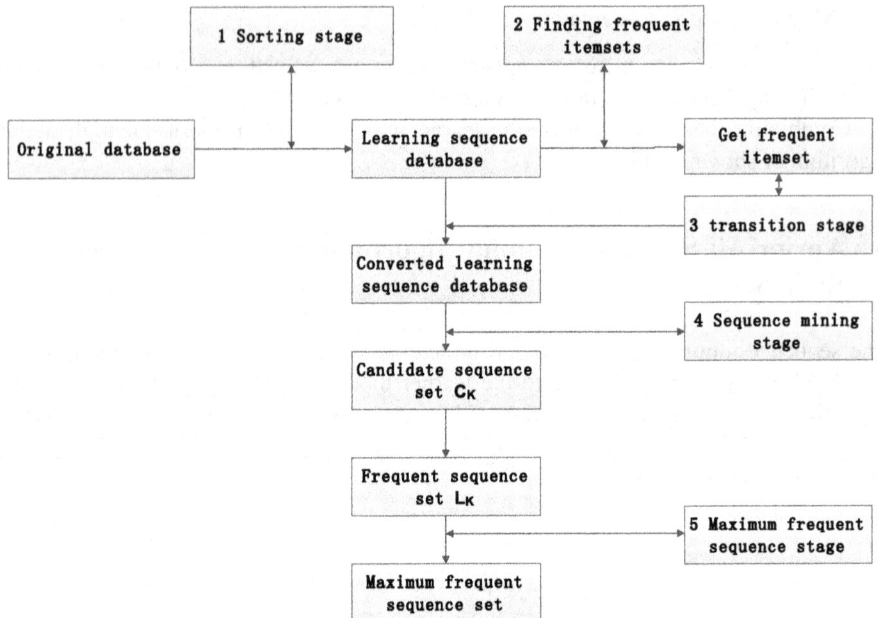

Fig. 3. Flow chart of AprioriAll algorithm

(1) Sorting stage

We sort the data in the database according to the primary key of the learner's ID and the learning time as a secondary key, and turn the original database into a database of things. As shown in Table 1.

Table 1. Learner's behavior database

Learner ID	Learning course	Learning time
2	A,B	2017-06-07
5	H	2017-06-09
2	C	2017-06-12
2	D,F,G	2017-06-17
4	C	2017-06-22
4	C,E,G	2017-06-22
1	C	2017-06-22
1	H	2017-06-27
4	D,G	2017-06-27
4	H	2017-07-22

After obtaining the learning record, we need to set up a database of things. This requires sorting items in a simple way, and ascending ranks of item set according to learners' ID and learning time, as shown in Table 2.

Table 2. Sorting the set of learners' behavior

Learner ID	Learning course	Learning time
1	C	2017-06-22
1	H	2017-06-27
2	A,B	2017-06-07
2	C	2017-06-12
2	D,F,G	2017-06-17
3	C,E,G	2017-06-22
4	C	2017-06-22
4	D,G	2017-06-27
4	H	2017-07-22
5	H	2017-06-09

We combine the sequence set of learners according to the sequence of ID learners, and get the sequence of things. The order of things is determined by learning time, because session access is different sequence. AB and BA are two different access sequences, as shown in Table 3.

Table 3. Transactional database

Learner ID	Learning content sequence
1	<(C) (H)>
2	<(A,B) (C) (D,F,G)>
3	<(C,E,G)>
4	<(C) (D,G) (H) >
5	<(H)>

(2) Finding frequent item sets

In the process of association rule mining, algorithm also needs to find frequent item sets. Sequential pattern mining is similar to the process of association rule mining, which differs from the definition of support. In the transaction database, the number of transactions that support the use of an item set is supported in the definition of association rules, and the number of customers is used in the definition of sequence pattern mining. In the learner's model, we use the course to replace it. The algorithm of sequence pattern mining based on learning behavior can be obtained by modifying the algorithm of finding frequent item sets in association rules mining. Then the frequently found frequent item set are mapped into continuous integers to reduce the cost of calculating the learning content of the learner. To scan the transaction database, find out the frequent item sets for (C), (DG) and (H), according to a subset of frequent item sets and frequent item sets, dismantling of frequent item sets, get (D), (G) for frequent item sets, and then each frequent item sets for mapping as shown in Table 4:

Table 4. Frequent item sets and their mappings

Frequent item sets	Mapping
(C)	1
(D)	2
(G)	3
(DG)	4
(H)	5

(3) Conversion stage

Learners' learning courses are replaced by frequent item sets. If no frequent item sets are included, delete this learning content, because it doesn't produce any association and does not constitute a sequence. If learning sequence includes frequent sequences, mapping is used instead. In the transformed learning sequence, we should use all frequent item sets set instead.

In the transformation stage, if an original sequence does not contain any frequent item sets, it should be deleted, and every learning sequence is replaced by a large item sets containing it.

(4) The stage of sequential pattern mining

Every time we scan the database, we use the frequent sequence generated from the database to generate candidate sequences. We calculate the support degree of candidate sequences and delete the incomplete sequences at the same time. At the same time, we choose the candidate sequences that are greater than the support as the next large sequence. In the first scan of the database, a series of frequent sequences with length of 1 is used as the initial large sequence.

The minimum support degree we set is 2, and its solution process is shown in the following list. In the first scan, the frequent 1 sequence is used as the L1 sequence and the support of the L1 sequence is obtained. The L1 sequence is connected to itself to get the candidate sequence set C2. The sequence that does not meet the minimum support sequence in the candidate sequence set C2 is deleted, and the satisfied sequence is used as L2 sequence. The L2 sequence is used as the large sequence, and the candidate sequence set C3 is generated, and the sequence of the candidate sequence set C3 satisfies the minimum support sequence is used as the L3 sequence. The generation process of the L4 sequence is identical.

(5) Find the maximum sequence stage

This step has been integrated into the fourth step, but if you want to leave all the large sequences, then the large sequences cannot be deleted in the fourth step. In the final stage, the hash tree is traversed. By deleting the subsequences in the large sequence, the real maximum sequence is obtained.

In the last step, we delete the subsequences in large sequences, such as <1 2 3><1 2 4><1 3 4><2 3 4>in L3 sequence, which are all subsequences of L4 sequences, and delete them to get the last maximal frequent subsequence. Finally, we get the learning sequence <1 23 4><1 3 5><4 5>, that is, to dig out three learning sequences suitable for different interest students to learn.

The implementation of pseudo code is as follows:

```
(1) L1={large 1-sequences}; // The result of the large item set phase
(2) FOR (k=2; Lk-1 ¹ Æ; k++) DO
BEGIN
(3)   Ck=aprioriALL_generate(Lk-1); // Ck is a new candidate from Lk-1
(4)   //Scan the database to get the support count for each Ck
      FOR eachcustomer-sequencec in DTDO
              Sum the count of allcandidates inCk that are contained inc;
(5)    Lk= Candidates inCk with minimum support
(6) END;
(7) Answer = Maximal Sequences in ∪kLk;
```

4 Algorithm Evaluations

The experiment uses the sequence-mining algorithm based on the learner's behavior to recommend the learning sequence for the learners. The experimental data come from the construction project of education information platform. According to the data collected from learners' behavior information data, we extract four sequences from frequent sequence sets to evaluate. The frequent sequence set must satisfy the minimum support and minimum confidence, and there is no wrong sequence. After finding the sequence, all the sequences satisfy the above requirements, so the algorithm is stable and reliable. By communicating with the members of the expert group, in view of the trend of the increase in the number of courses, we set the system support at 30%, and the confidence level is set at 30% (Table 5).

Table 5. Correct rate assessment

Evaluating indicator	Sequence 1	Sequence 2	Sequence 3	Sequence 4
Minimum support degree	44.6%	40.4%	33.1%	31.2%
Minimum confidence	60.2%	41.9%	39.6%	41.3%

Evaluation of sequence usage: for two classes, 80 students selected 4 of them to carry out the evaluation of the sequence usage. Assessment score = $(P_1 + P_2 + \ldots + P_n)/n$, where P_n represents the probability of students using learning sequence n, n represents the number of sequences to participate in the evaluation, and the higher the score is, the higher the probability of using the recommended sequence is.

The accuracy rate of the student's selection of the recommended course is shown in Table 6.

Table 6. Utilization evaluation

	Sequence 1 utilization ratio	Sequence 2 utilization ratio	Sequence 3 utilization ratio	Sequence 4 utilization ratio
Class 1	77.50%	85.00%	80.00%	72.50%
Class 2	82.50%	87.50%	85.00%	65.00%
Average probability	80.00%	86.25%	82.50%	68.75%

The average score of algorithm is = 79.375. 80 learners hold a relatively positive attitude towards the learning sequence recommended by the system, and reach a satisfactory level, which proves that the algorithm is acceptable. Through experimental verification, the algorithm can excavate the sequence pattern of the learning sequence in the database and discover the regularity of the sequence pattern. We got a satisfactory result.

5 Summary and Prospect

With the unique advantages of web-based autonomous learning, such as flexibility, diversity, openness and so on, under the network environment reflects the learner's subjectivity, we through the autonomous learning traces, based on data mining technology to obtain the learning sequence from the sequence of things, can provide the recommended courses for later learners. First of all, this paper investigates the advantages and disadvantages of the common autonomous learning platform at home and abroad, and puts forward the advantages of using the sequential mining algorithm for autonomous learning platform. Then, by analyzing the concepts and related principles of network teaching, combining the needs of personalized teaching and referring to the norms of national learner models, we set up a web-based learner model suitable for Chinese teaching. According to the established learner model, we conduct personality tracking for learners' behavior, grasp learner's behavior, and recommend a curriculum that is more suitable for learners' interest through sequential mining algorithm.

Based on the structure of the learner model set the database tables and fields, we are stored in the database through the platform module learning behavior of learners, and then convert the database into object sequence database, algorithm of data mining on the data collected through sequential pattern mining, to meet the personalized needs of learners recommended courses. At the end of the article, we experimentation and analysis of the algorithm, and put forward the evaluation model and the improvement scheme.

Acknowledgment. This work was supported in part by Hubei Province Natural Science Foundation of China (No. 2018CFB526), by National Natural Science Foundation of China (No. 61502356).

References

1. Wang, D., Yu, G., Bao, Y., Wang, G.: A classification of domain knowledge and representation for data mining preprocessing. School of Information Science and Engineering, Northeastern University (2003)
2. Zhang, Z., Ji, G.: Design and implementation of Apriori algorithm based on MapReduce (NATURAL SCIENCE EDITION). J. Huazhong Univ. Sci. Technol. (S1) (2012)
3. Zhang, T.: Research on the algorithm of personalized recommendation system. Harbin University of Science and Technology (2017)
4. Cui, W.: Personalized cloud service recommendation algorithm based on association rules and user interest model. Beijing University of Posts and Telecommunications (2017)
5. Zhou, J.F.: Design and implementation of a hybrid recommendation system for personalized learning resource sharing. Beijing University of Posts and Telecommunications (2015)
6. Miao, X.L.: Comparative study on pattern mining of gap constraint sequence pattern. Netw. Secur. Technol. Appl. (02) (2017)
7. Wu, H., Zhu, J., Gao, G., Cheng, Z.: Study on web sequential pattern mining based on improved AprioriAll algorithm. Comput. Eng. Des. **31**(5) (2010)

Emotion Recognition Based on Electroencephalogram Using a Multiple Instance Learning Framework

Xiaowei Zhang(✉), Yue Wang, Shengjie Zhao, Jinyong Liu, Jing Pan, Jian Shen, and Tingzhen Ding

School of Information Science and Engineering, LanZhou University, Lanzhou 730000, China
{zhangxw, wangyue15, zhaoshj15, liujy2016, panj17, shenj17, dingtzh16}@lzu.edu.cn

Abstract. Electroencephalogram (EEG)-based emotion recognition has been widely researched in the field of affective computing. Nevertheless, EEG signals which reflect brain activity are always unstable, it is inappropriate for traditional analysis methods to treat each sliding time window of signals as independent sample during classification. In this study, we employ a multi-instance learning (MIL) framework for EEG-based emotion recognition and regard sliding time windows from the same EEG signal as a whole by learning two MIL models based on Citation-kNN and mi-SVM algorithms. Experiment results show that our methods can achieve higher classification accuracy of 74.21% and 77.50% on two affective dimensions (valence and arousal) respectively when comparing with traditional single-instance classification algorithms. We believe that MIL framework can improve the generalization performance of EEG-based emotion recognition further, and provide new inspiration for affective computing.

Keywords: Affective computing · Emotion recognition
Electroencephalogram · Multi-instance Learning

1 Introduction

Affective computing is a key technology for advanced human-computer interaction (HCI). Picard first proposed the concept of "Affective Computing" and interpreted it as "emotional-related, emotionally or apparently emotional-affected computing" [1]. The purpose of affective computing is to give the computer similar ability in perceiving, understanding or expressing feelings of various emotions just like human beings [2], its implementation involves a series of related techniques, such as emotion signal acquisition, emotion recognition, feedback and expression. Thereinto, modeling emotion signals and recognizing people's emotion state have become the most crucial factors in the field of affective computing [3].

As emotion recognition becomes a multi-disciplinary research topic, there are still numerous problems to be solved. Early researches focused on facial expressions or voice signals to recognize different affective states [4, 5], although these two kind of signals are easier to obtain, they are more likely to be disguised especially when

D.-S. Huang et al. (Eds.): ICIC 2018, LNCS 10955, pp. 570–578, 2018.
https://doi.org/10.1007/978-3-319-95933-7_66

subjects don't want others to notice their true emotion. Therefore, some researchers have proposed the use of physiological signals to avoid disguise, such as electroencephalogram (EEG), electromyogram, galvanic skin response, etc. [6–8]. Among them, EEG has been widely used for emotion recognition in recent years because it reflects the activities of central nervous system and provides an important window for researchers to gain insight into the internal functions of brain [9]. It is also considered as the most effective source of information for emotion recognition [10].

Generally, modeling for EEG signals is based on the assumption that these signals are stationary, whereas brain activity is inherently unstable [11], especially with some external emotional stimulus. A common way for resolving this issue is to segregate EEG signals into short periods to satisfy piecewise stable conditions, such as fixed-size segmentation methods [12]. However, these homogeneous segments are not necessarily related to subject's emotional tag. On the one hand, they are likely to be affected by noise, different artifacts, or dominant area of physiological function associated with unrelated brain activities [13]. On the other hand, although each emotional stimulus lasts for a while, for example, one minute, it is impossible to ensure subjects experiencing same emotion during the whole process. Therefore, it is inappropriate to assign same emotional tag to all segments and treat these segments as independent samples during modeling.

In order to overcome above drawbacks, we need to investigate these segments from an integral perspective and multi-instance learning (MIL) [14] provides a solution due to its ability to learn from bags composed of independent samples. MIL is being widely researched in the analysis of videos, pictures, vocal signals, or even EEG signals. Kandemir et al. quantified several MIL methods by benchmarking on Barrett's cancer diagnosis and diabetic retinopathy screening [15]. Huo et al. proposed a novel method termed Multi-Instance Dictionary Learning (MIDL) for detecting abnormal events in crowded video scenes [16]. Fang et al. obtained a steady structure based on sparse representation by considering facial expression recognition as a MIL problem [17]. Lee et al. extracted features from vocal signals to predict the affective state of married couples with MIL classifiers, and reached a recognition accuracy of 53.93% [18]. Wu et al. proposed a novel Bayesian model using sentence-level music and lyrics features for affective recognition from a multi-instance perspective [19]. Jafari et al. decomposed EEG signals into independent components and form bags to identify the noise components for removal using a multi-instance learning algorithm, and they achieved 91.2% artifact identification accuracy [20]. Sadatnejad et al. treated each subject as a bag of covariance matrices of homogenous segments and applied multi-instance (MI) framework to relieve the non-stationary of the EEG signal in their EEG-based Computer Aided Diagnosis [13]. In this study, we employed multi-instance learning (MIL) framework for EEG-based emotion recognition and investigate sliding time windows from an integral perspective by learning two MIL models based on Citation-kNN and mi-SVM algorithms. Experiment on DEAP dataset showed that our methods can improve generalization performance of EEG-based emotion recognition further and achieve higher classification accuracy of 74.21% and 77.50% on two affective dimensions respectively.

The reminder of this paper is structured as follows. Section 2 introduces the mathematical preliminaries of MIL framework and two typical algorithms employed in this paper. Section 3 describes the EEG database used in experiment and data processing methods. The modeling and classification results are shown in Sect. 4 with the conclusion presented in Sect. 5.

2 Multi-instance Learning Framework

Multi-instance Learning (MIL) is a special learning framework which deals with uncertainty of instance labels. Specifically, training instances in MIL are arranged in sets called bags and the labels of the bags are provided, but the labels of instances in the bags are unknown. There is a standard hypothesis in the relationship between instances and labels of the bags: Assuming that each instance has an implicit tag, if at least one of the instances within the bag is positive, the label of the bag is positive; a bag is labeled as negativity if all instances in this bag are negative. Method of time window segmentation can solve the problem of instability for EEG signals better, while employing MIL framework is to from the aspect with instances rather than independent samples, and find instances which are consistent with emotion stimuli exactly via a corresponding relationship of 1: n between the object and instances, then it can predict emotion labels of new samples.

In MIL framework, the algorithms attempt to learn a classification function that can predict the labels of bags and/or instances in the testing data. Let $x = R^d$ be a D-dimensional feature space, $\Omega = \{-1, +1\}$ denotes the label space of binary classification. For a given dataset $\{(X_1, y_1), (X_2, y_2), \ldots, (X_n, y_n)\}$, the goal is to learn a multi-sample learning mapping $2^x \rightarrow \Omega$ [21], as formula (1):

$$f_{MIL}(X_i) = \begin{cases} +1, \exists f(x_{ij}) = 1 \\ -1, \forall f(x_{ij}) = -1 \end{cases} (1 \leq i \leq n, 1 \leq j \leq n_i) \tag{1}$$

where $\{X_{i1}, X_{i2}, \ldots, X_{in_i}\}$, $X_i \in X$ denotes a bag of instances set, and a instance is denoted as x_{ij}, where $x_{ij} \in x = R^d$ $(1 \leq i \leq n, 1 \leq j \leq n_i, 1 \leq n_i \leq \infty)$, n is the number of bags in the dataset,n_i represents the number of instances contained in bag X_i,d is feature dimension of the instance x_{ij}, the label of bag X_i is expressed as y_i, $y_i \in \Omega$.

2.1 Bag-Level Classification

The general multiple-instance assumption requires algorithms to solve multi-instance problem in the bag-level space [22]. In other words, the algorithms must treat bag as a whole to extract its global bag-level information. A typical bag-level classification algorithm termed Citation-kNN was proposed by J. Wang et al. based on extension of traditional k-Nearest Neighbor (kNN) algorithm [23]. The basic theory of this algorithm is borrowed from the concept of "citation" in scientific literature, and it applies kNN algorithm to bag-level and predicts an unknown bag using voting method. During classification, in order to increase the robustness, the labels of those bags that are closest to the test sample, and the labels of these bags which take the test sample as the nearest neighbor, are taken into account simultaneously, and then carry out the voting. In order to measure degree of similarity between any two bags, this algorithm introduces a modified hausdorff distance.

Suppose there are two bags $X = \{x_1, x_2, \ldots, x_n\}$ and $Y = \{y_1, y_2, \ldots, y_m\}$. $x_i, i = 1, 2, \ldots, n$ and $y_j, j = 1, 2, \ldots, m$ are instances of each bag respectively. The Hausdorff distance is defined as:

$$HD(X, Y) = \max\{h(X, Y), h(Y, X)\} \tag{2}$$

where

$$h(X, Y) = \max_{x \in X} \min_{y \in Y} \|x - y\|$$

The Hausdorff distance is more sensitive to noise, in order to increase the ability to resist noise, it has been modified as:

$$HD_{\min}(X, Y) = \max\{h1(X, Y), h1(Y, X)\} \tag{3}$$

where

$$h1(X, Y) = \min_{x \in X} \min_{y \in Y} \|x - y\|$$

Utilizing this minimum hausdorff distance as bag-level distance, we can employ traditional k-NN algorithm to predict untagged bags.

2.2 Instance-Level Classification

Instance-level classification is different from bag classification because while training is performed using data arranged in sets, the objective is to classify instance individually. For instance, mi-SVM algorithm explicitly treated the instance labels y_i as latent variables subject to constraints defined by their bag labels Y_I and added these latent variables to the following optimization problem [24]:

$$\begin{gathered} \min_{\{y_i\}} \min_{w,b,\xi} \frac{1}{2} \|w\|^2 + C \sum_i \xi_i \\ s.t.\ \forall i : y_i(\langle w, x_i \rangle + b) \geq 1 - \zeta_i, \zeta_i \geq 0, \\ \sum_{i \in I} \frac{y_i + 1}{2} \geq 1, \forall I\ s.t.\ Y_I = 1,\ and\ y_i = -1,\ \forall I\ s.t.\ Y_I = -1 \end{gathered} \tag{4}$$

During optimization iteration, mi-SVM algorithm trained an instance-level standard Support Vector Machine (SVM) based on the latent labels y_i, and then updated these labels by predicting with the learned SVM. Finally, a linear decision boundary was found which can separate at least one instance of each positive bag from all the instances belonging to the negative bags. It shows that mi-SVM algorithm can set the latent instance labels freely under the constraints of their bag labels to maximize the margin, and it is suitable for tasks in which users care about instance labels.

3 Experiment

3.1 Dataset

The EEG signals used in our experiment are from the Database for Emotion Analysis Using Physiological Signals (DEAP) that was published by Koelstra et al. [25]. For the expression of emotion, researchers use different emotional representation pointing at various research purposes. At present, there are two widely accepted methods in the field of psychology: one is the discrete emotional category (for example, happy, sad, disgust, etc.), the other is the continuous emotional dimension representation [26]. The investigators found that each of the discrete emotional states can be represented by a two-dimensional valence-arousal model [27] which was used in this study. Valence corresponds to the emotional state ranging from negative to positive, and the corresponding emotional state of arousal dimension is ranging from calm to excited. 32 healthy volunteers (16 males and 16 females) participated in the experiment which aged between 19 and 37 (mean age 26.9), the EEG (32 channels according to international 10–20 system) and physiological signals of them were recorded during watching 40 one-minute video clips, thus 40 trails per subject were obtained. After each video, self-assessment for arousal, valence, liking and dominance (rating from 1 to 9) were reported by themselves. In our experiment, we performed binary bag-level and instance-level classifications on the arousal and valence respectively and divided self-ratings on each dimension into high (rating 6–9) and low (rating 1–5) classes/levels.

3.2 Preprocessing

At first, all the EEG signals were downsampled from 512 Hz to 128 Hz. Then we used FIR band-pass filter to intercept the waveform of 4–47 Hz and divided it into four basic frequency bands: theta (4–8 Hz), alpha (8–13 Hz), beta (13–30 Hz), gamma (30–47 Hz). After that, we used a hybrid de-noising method combining Discrete Wavelet Transformation and Adaptive Predictor Filter to remove the eye movement [28].

3.3 Feature Extraction

In accordance with traditional analyzing methods, for each 60-s trail of 32 subjects (40 trails per subject), we computed nine features on signals recorded from 32 channels with 4-s sliding time windows. In addition to seven traditional power spectrum characteristics of different frequency bands, there were two global characteristics which extracted from the whole frequency band including spectral entropy and renyi entropy [29, 30]. The summary of extracted features is shown as Table 1.

Traditional analyzing methods treated each time window as an independent sample, it neglected relevance and integrity among samples which belonging to a same trail. In MIL framework, we treated each sliding time window as an instance, and instances derived from the same trail constituted a bag. From the above, there were 40 trails (bags) $\times$ 15 segments = 600 instances for each subject, and each instance consisted of 9 $\times$ 32 channels = 288 features. Before modeling, all the features were normalized into the range of [0, 1].

Table 1. Features extracted from each time window

Feature name	Feature description
Abs_powerθ\α\β, Max_powerθ\α\β	Absolute spectral power, maximum power of theta, alpha, beta respectively
Abs_powerβ/Abs_powerθ	Energy ratio of beta to theta bands
Spectral_entropy	The spectral entropy of full-wave band
Renyi_entropy	The renyi entropy of full-wave band

4 Modeling and Classification

Different from traditional classification algorithms which learned the models based on training samples with explicit labels, we constructed a MIL framework for EEG-based emotion recognition in which the labels of instances in the bags were neglected and the labels of the bags are provided. Then we learned classification models that can predict the labels of bags and instances in the testing data respectively using Citation-kNN and mi-SVM algorithms. In our MIL framework, leave-one-person-out cross-validation scheme was used to evaluate the classification performance. Each subject was selected as the testing set each time, the remaining 31 subjects as the training set, and this process was repeated 32 times to ensure that each subject's EEG signals have the opportunity to be tested. For different folds, two parameters (RefNum and CiterRank) of Citation-kNN were optimized in the range of [2, 10] and [2, 4] respectively with step size of 1. Linear kernel was used in mi-SVM and the penalty factor was confirmed by optimizing in the range of [0.1, 10] with step size of 0.8. In order to compare the classification performance of our MIL framework with traditional methods, we also trained models using standard kNN and SVM algorithms with the same cross-validation scheme.

Considering the influence of unbalanced sample size on binary classification, we introduced F1-score in performance evaluation except for classification accuracy. This measurement considers both the precision and the recall of the test to compute the score as:

$$F_1 = 2 \cdot \frac{precision \cdot recall}{precision + recall} \tag{5}$$

Simultaneously, it is worth noting that the standardized SVM and kNN algorithms must predict final accuracy and F-score values at the bag-level as comparative methods. We used the voting strategy and set the label of a bag in testing set as same as the majority of instance labels belong to this bag. Figure 1 showed the comparison of mean accuracy and F1-score on valence and arousal respectively. It can be seen that our MIL framework outperformed traditional methods significantly both on mean accuracy and F1-score. The Citation-kNN algorithm achieved higher classification accuracy (about 74.21%) on valence dimension, and the mi-SVM achieved higher classification accuracy (about 77.50%) on arousal dimension. For F1-scores, the highest values on the two emotional dimensions were all obtained by mi-SVM algorithm, which were 72.50% and 73.65% respectively.

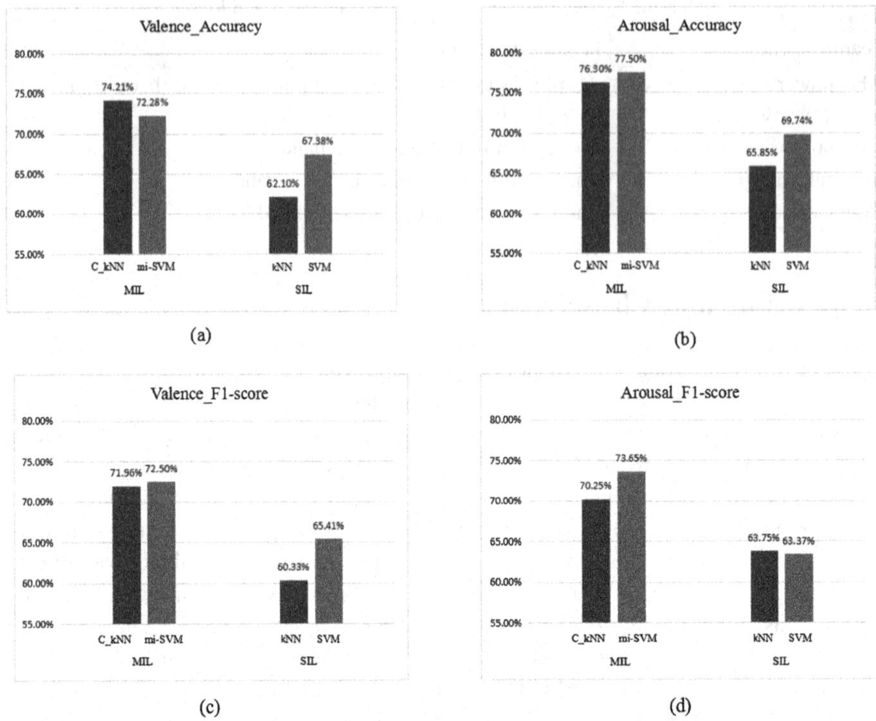

Fig. 1. Performance comparison for two-dimensional emotion recognition (a) accuracy on valence. (b) accuracy on arousal. (c) F1-score on valence. (d) F1-score on arousal.

In addition, we gained better recognition performance on arousal dimension than valance dimension, partly indicating that subjects experienced high or low pleasure much easier than positive or negative emotions. It also demonstrated the validity of EEG signals in discriminating high and low arousal states.

5 Conclusion

In this study, we employed a multi-instance learning (MIL) framework for EEG-based emotion recognition and overcome the drawbacks of traditional methods which treated each sliding time windows of signals as independent samples during classification and neglected correlation among samples belong to a same trail. In our MIL framework, each sliding time window was regarded as an instance and power spectrum characteristics and global features were extracted from it, then all the instances were grouped into a set of bags corresponding to trails, and we learned two typical MIL models using these bags and corresponding instances. The experiment results showed that our methods can improve generalization performance of EEG-based emotion recognition to some extent, and provide new inspiration for future research in the field of affective computing.

Acknowledgement. This work was supported by the National Basic Research Program of China (973 Program) (No.2014CB744600), the state key development program of China (No.2017YFE0111900), the National Natural Science Foundation of China (grant No.61402211, No.61210010) and the Fundamental Research Funds for the Central Universities (lzujbky-2017-196, lzujbky-2017-kb08). The authors acknowledge European Community's Seventh Framework Program (FP7/2007-2011) for their DEAP database.

References

1. Picard, R.W.: Affective Computing, vol. 1, 1st edn, pp. 71–73. IGI Global, Hershey (1997)
2. Calvo, R.A., D'Mello, S.: Affect detection: an interdisciplinary review of models, methods, and their applications. IEEE Trans. Affect. Comput. **1**(1), 18–37 (2010)
3. Tkalčič, M., Burnik, U., Košir, A.: Using affective parameters in a content-based recommender system for images. User Model. User-Adapt. Interact. **20**(4), 279–311 (2010)
4. Anderson, K., Mcowan, P.W.: A real-time automated system for the recognition of human facial expressions. IEEE Trans. Syst. Man Cybern. Part B Cybern. Publ. IEEE Syst. Man Cybern. Soc. **36**(1), 96–105 (2006)
5. van der Wal, C.N., Kowalczyk, W.: Detecting changing emotions in human speech by machine and humans. Appl. Intell. **39**(4), 675–691 (2013)
6. Wagner, J., Kim, N.J., Andre, E.: From physiological signals to emotions: implementing and comparing selected methods for feature extraction and classification. In: IEEE International Conference on Multimedia and Expo, pp. 940–943. IEEE (2005)
7. Mao, C., et al.: EEG-based biometric identification using local probability centers. In: International Joint Conference on Neural Networks, pp. 1–8. IEEE (2015)
8. Chen, J., et al.: Feature-level fusion of multimodal physiological signals for emotion recognition. In: IEEE International Conference on Bioinformatics and Biomedicine, pp. 395–399. IEEE (2015)
9. Paus, T., Sipila, P.K., Strafella, A.P.: Synchronization of neuronal activity in the human primary motor cortex by transcranial magnetic stimulation: an EEG study. J. Neurophysiol. **86**(4), 1983–1990 (2001)
10. Chanel, G., et al.: Short-term emotion assessment in a recall paradigm. Int. J. Hum Comput Stud. **67**(8), 607–627 (2009)
11. Kaplan, A.Y., et al.: Nonstationary nature of the brain activity as revealed by EEG/MEG: methodological, practical and conceptual challenges. Signal Process. **85**(11), 2190–2212 (2005)
12. Sanei, S., Chambers, J.A.: EEG signal processing. In: The Fernow Watershed Acidification Study, pp. 207–236. Springer, Netherlands (2013)
13. Sadatnejad, K., et al.: EEG Representation Using Multi-instance Framework on The Manifold of Symmetric Positive Definite Matrices for EEG-based Computer Aided Diagnosis (2017)
14. Dietterich, T.G., Lathrop, R.H., Lozano-Pérez, T.: Solving the multiple instance problem with axis-parallel rectangles. Artif. Intell. **89**(1–2), 31–71 (1997)
15. Kandemir, M., Hamprecht, F.A.: Computer-aided diagnosis from weak supervision: a benchmarking study. Comput. Med. Imaging Graph. **42**, 44–50 (2015)
16. Huo, J., Gao, Y., Yang, W., Yin, H.: Abnormal event detection via multi-instance dictionary learning. In: Yin, H., Costa, J.A.F., Barreto, G. (eds.) IDEAL 2012. LNCS, vol. 7435, pp. 76–83. Springer, Heidelberg (2012). https://doi.org/10.1007/978-3-642-32639-4_10

17. Fang, Y., Chang, L.: Multi-instance feature learning based on sparse representation for facial expression recognition. In: He, X., Luo, S., Tao, D., Xu, C., Yang, J., Hasan, M.A. (eds.) MMM 2015. LNCS, vol. 8935, pp. 224–233. Springer, Cham (2015). https://doi.org/10.1007/978-3-319-14445-0_20
18. Lee, C.-C., et al.: Affective state recognition in married couples' interactions using PCA-based vocal entrainment measures with multiple instance learning. In: D'Mello, S., Graesser, A., Schuller, B., Martin, J.-C. (eds.) ACII 2011. LNCS, vol. 6975, pp. 31–41. Springer, Heidelberg (2011). https://doi.org/10.1007/978-3-642-24571-8_4
19. Wu, B., et al.: Music emotion recognition by multi-label multi-layer multi-instance multi-view learning. In: ACM International Conference on Multimedia, pp. 117–126. ACM (2014)
20. Jafari, A., et al.: An EEG artifact identification embedded system using ICA and multi-instance learning. In: IEEE International Symposium on Circuits and Systems, pp. 1–4. IEEE (2017)
21. Maron, O., Lozano-Pérez, T.: A framework for multiple-instance learning. In: Advances in Neural Information Processing Systems, vol. 200, no. 2, pp. 570–576 (1998)
22. Weidmann, N., Frank, E., Pfahringer, B.: A two-level learning method for generalized multi-instance problems. In: Lavrač, N., Gamberger, D., Blockeel, H., Todorovski, L. (eds.) ECML 2003. LNCS (LNAI), vol. 2837, pp. 468–479. Springer, Heidelberg (2003). https://doi.org/10.1007/978-3-540-39857-8_42
23. Wang, J., Zucker, J.D.: Solving the multiple-instance problem: a lazy learning approach. In: Seventeenth International Conference on Machine Learning, pp. 1119–1126. Morgan Kaufmann Publishers Inc. (2000)
24. Andrews, S., Tsochantaridis, I., Hofmann, T.: Support vector machines for multiple-instance learning. In: Advances in Neural Information Processing Systems, vol. 15, no. 2, pp. 561–568 (2002)
25. Koelstra, S., et al.: DEAP: a database for emotion analysis; using physiological signals. IEEE Trans. Affect. Comput. **3**(1), 18–31 (2012)
26. Bos, D.O.: EEG-based Emotion Recognition (2008)
27. Lang, P.J.: The emotion probe. Studies of motivation and attention. Am. Psychol. **50**(5), 372 (1995)
28. Zhao, Q., et al.: Automatic identification and removal of ocular artifacts in EEG–improved adaptive predictor filtering for portable applications. IEEE Trans. Nanobiosci. **13**(2), 109–117 (2014)
29. Seitsonen, E.R., et al.: EEG spectral entropy, heart rate, photoplethysmography and motor responses to skin incision during sevoflurane anaesthesia. Acta Anaesthesiol. Scand. **49**(3), 284–292 (2005)
30. Inuso, G., et al.: Brain activity investigation by EEG processing: wavelet analysis, kurtosis and Renyi's entropy for artifact detection. In: International Conference on Information Acquisition, pp. 195–200. IEEE (2007)

Using Hybrid Similarity-Based Collaborative Filtering Method for Compound Activity Prediction

Jun Ma(✉), Ruisheng Zhang, Yongna Yuan, and Zhili Zhao

School of Information Science and Engineering, Lanzhou University, Lanzhou 730000, China
{junma, zhangrs}@lzu.edu.cn

Abstract. It is important for researchers to predict compound activity to the targets quickly and effectively in the field of drug design. In the paper, the problem of compound activity prediction is converted to the recommendations in the field of e-commerce, compounds are viewed as users, and protein targets are viewed as items. A rating matrix is extracted by IC50 of each compound to targets, there are four filtering recommendation algorithms could be used for predicting compound activity. In order to improve the accuracy of prediction, the hybrid similarity-based Collaborative Filtering (HybridSimCF) Method is proposed, the method will combine the similarity of the compound structure and the similarity based on the rating matrix to predict the activity. Through compared with other three collaborative filtering methods, HybridSimCF has better results. It not only improves the values of RMSE and MAE, but also effectively solves the cold start problem. The method can quickly and effectively solve the prediction of compound activity.

Keywords: Drug design · QSAR · Compound activity · Rating prediction
Machine learning · Collaborative filtering

1 Introduction

With the development of chemical informatics and computer-aided drug design, the prediction of the activity of small molecules can be demonstrated through computational models, and QSAR (Quantitative structure–activity relationship) model is often used. Currently, in these models there are three categories of machine learning algorithms to be used for predicting the compound activity, the single-target-oriented models, the chemogenomics-based models and the multi-targets-oriented models.

In the single-target-oriented QSAR model, the support vector machine (SVM) [1, 2], Partial-Least-Squares Regression [3, 4], neural networks [5–7], decision trees [8], recursive partitioning [9, 10], Bayesian model [11, 12], random forest [13, 14], and kernel-based SAR [15, 16] can be used to build QSAR models. However, the single target-based models tend to ignore other protein target resources. The prediction result is not ideal. The chemogenomics-based model compensates for the drawback of the single-target-oriented models. Dumitru et al. [17, 18] constructed collaborative filtering algorithms based on multi-task neural networks and multi-task cores for protein targets in a family.

D.-S. Huang et al. (Eds.): ICIC 2018, LNCS 10955, pp. 579–588, 2018.
https://doi.org/10.1007/978-3-319-95933-7_67

The multi-target model compensates for the deficiencies of the chemogenomics-based model. Ning et al. [19] obtained highly related target proteins by comparing the similarity of protein sequences and the similarity of the active compounds of these proteins, they used a multi-task model constructed by SVM to learn the activity relationship between the targets and the compounds. In order to prove the result, Zhang et al. [20] use the same data to predict SSR. In multi-target model, the multiple targets are belong one family, L Rosenbaoum et al. [21] used 112 human kinases as targets. Therefore, the utilization rate of the target is still not high.

In the paper, it is focused that how to use the recommendation method to predict the compound activity in target-unrelated data set. There are three innovated ideas in the paper. Firstly, the recommendation idea is imported in compound activity prediction. Users, items, and ratings in the recommendation system are corresponded to compound molecules, targets, and values of the activity in chemical informatics, respectively. Secondly, a sparse compounds-protein targets matrix based on activity rules is constructed from the ChemBL open database. Thirdly, the HybridSimCF recommendation algorithm is proposed and after compared with other three algorithms, the HybriSimCF algorithm can improve prediction accuracy in the task.

In the paper, the second section describes the two data sets used in the paper, one is the rating matrix, and the other is the structure data of compounds. The third section describes the HybrdSimCF algorithms and the other three comparison algorithms. The fourth section analyzes the calculation results of the four algorithms on the data set based on two evaluation parameters. The summary is given in the fifth section.

2 Data Set

In the paper, all targets data and compounds are downloaded from ChEMBL [22]. The ChemBL database contains compound bioactivity data against drug targets. The bioactivity values of compounds in ChemBL have Ki, Kd, IC50, and EC50. In the paper, IC50 value [23] is selected as the activity of a compound (i.e., the concentration of the compound that is required for 50% inhibition of the target under consideration, and lower IC50 values indicate higher activity). In drug design, the higher the activity value of a drug molecule, the drug is more likely to be effective in the body. In CHEMBL, 57 targets are selected randomly and these targets both are single protein. The data set is named the gold standard data set.

In the collaborative filtering recommendation algorithm, the user's similarity is calculated by the rating matrix. When this idea is converted to the prediction of the activity of the compound, the rating matrix is indispensable. In drug design, a compound is considered to be active for a given target if its IC50 value for that target is less than 0.1 um [24]. Therefore, the rating matrix is based on this rule. All compounds in the gold standard data set are divided into three categories according the value of IC50. When the IC50 value of the compound is less than 0.1 um, the activity value of the compound is a strong activity, and its score value is 3. When the IC50 value of the compound is between 0.1 and 1, the activity value of the compound is weak activity, and its score value is set to 2. When the IC50 value is greater than or equal to 1 um, the compound is inactive and its score value is set to 1. Finally, the gold data set has 57 protein targets, 871 compound molecules, 1906 relationships, and 3 scoring values. Some statistic information is below in Table 1.

Table 1. The statistic information of the gold standard data set

Number of compounds	871
Number of targets	57
Number of rating	1906
Number of relationships of 1	496
Number of relationships of 2	629
Number of relationships of 3	781
Sparsity	0.0384

In the Fig. 1, it describes the statistic distribution of compounds' activity values, and the statistic distribution of targets which is scored by compounds. Each target has a score of one compound at least. Each compound scores for two targets at least. From the Fig. 1, in the 871 compounds there are 242 compounds have strong activity, there are 402 compounds have weak activity, there are 227 compounds have no activity.

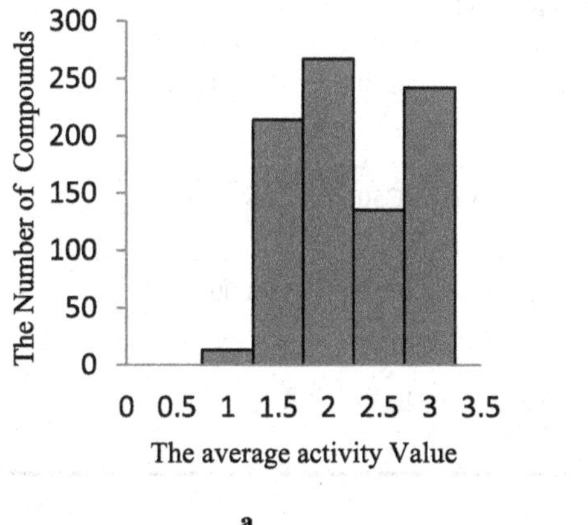

a

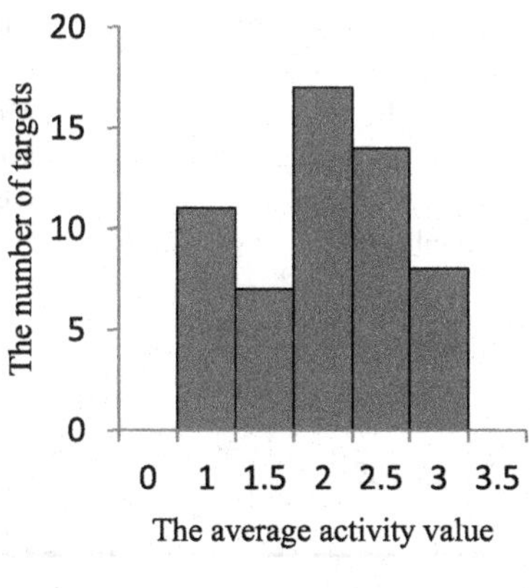

b

Fig. 1. The statistic distribution about compounds and targets (a. is the distribution of average activity values of compounds. b. is the distribution of average activity values of targets which has active compounds)

In the paper, each compound's structure is obtained by DRAGON [25]. DRAGON is an application for the calculation of molecular descriptors originally developed by the Milano Chemometrics and QSAR Research Group. These descriptors can be used to evaluate molecular structure-activity or structure-property relationships, as well as for similarity analysis and high throughput screening of molecule databases. In the paper, each compound's 2D structure with 726 dims is selected from DRAGON.

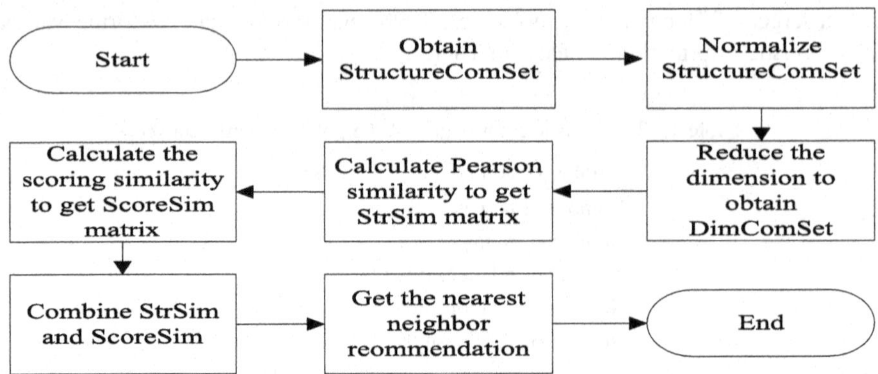

Fig. 2. The flow chart of HybriSimCF

Table 2. Five compounds in the gold standard data set

CHEMBLID	Compound structure	Target name	bioType	Type value	Rating
Chembl194837		Muscarinic acetyl-choline receptor M2	IC50	0.0045um	3
Chembl716		Muscarinic acetylcholine receptor M2	IC50	3.3447um	1
Chembl119634		Acetylcholinesterase	IC50	68.8um	1
Chembl131579		Acetylcholinesterase	IC50	0.0007um	3
Chembl300982		Cyclooxygenase-2	IC50	0.37um	2

In Table 2, there are 5 records selected from the gold standard data set. The table shows IC50 values and the structures of the five compounds.

3 Proposed Method

In the paper, four methods are used to compare the prediction results. There are common User-based collaborative filtering algorithm, Slope-One algorithm, PWSlope-One algorithm and hybrid similarity-based collaborative filtering (HybridSimCF) algorithm.

User-based collaborative filtering is a class of collaborative filtering algorithms [26, 27] and is the most widely used in recommendation system. The basic idea is that the prediction results are come from the analysis of the user's previous interests, preferences, and other historical behavior. In this paper, Pearson similarity [28] calculation method is used. The formula is as follows

$$Sim_{xy} = \frac{\sum_{i=1}^{n} (x_i - \bar{x})(y_i - \bar{y})}{\sqrt{\sum_{i=1}^{n} (x_i - \bar{x})^2} \sqrt{\sum_{i=1}^{n} (y_i - \bar{y})^2}} \tag{1}$$

where n is the sample size, x_i, y_i, are the single samples indexed with i, $\bar{x}$ is the sample mean. The equation is below

$$\bar{x} = \frac{1}{n} \sum_{i=1}^{n} x_i \tag{2}$$

and analogously for $\bar{y}$.

After calculating the similarity of each compound in the gold standard data set, the formula (3) is used for the recommendation.

$$\mathrm{P}(u, i) = \sum_{v \in S(u,K) \cap N(i)} w_{uv} r_{vi} \tag{3}$$

where $S(u,K)$ contains K compounds in which the activity values are most close to compound user u, $N(i)$ is a set of compounds which have activity values on item i, w is the similarity between compound u and compound v, r_{vi} is the activity value which compound v give the target i.

Because the Slope-One algorithm is fit for the data set in which the number of items is less than the number of users and it is based on items, the Slope-One algorithm is used to predict the task. In the paper an improved Slope-One [29] named PWSlope-One will also be used to the task. Slope-one is a logical analysis based on linear regression, the compound's activity value on the item is calculated from the average deviation. The formula is as follows.

$$\mathrm{R}(ij) = \frac{\sum_{u \in N(i) \cap N(j)} (r_{ui} - r_{uj})}{|N(i) \cap N(j)|} \tag{4}$$

where r_{ui} is the compound u's activity value of the target i, r_{uj} is the compound u's activity value of the item j. $N(i)$ records the compounds which scored target i, $N(i) \cap N(j)$ denotes the compounds which have activity value both target i and target j, $|N(i) \cap N(j)|$ denotes the number of the compounds which have activity value both item i and item j.

$$\mathrm{P}(u,j) = \frac{\sum_{i \in N(u)} |N(i) \cap N(j)| (r_{ui} + R(ij)) w_{ij}}{\sum_{i \in N(u)} |N(i) \cap N(j)|} \tag{5}$$

where $N(u)$ records the targets which is scored by the compound u.

The PWSlope-One combines the item's similarity to the Formula (5), and the item's similarity is computed from the rating matrix, and Pearson method is used to compute the item's similarity. The new formula is as follows.

$$\mathrm{P}(u,j) = \frac{\sum_{i \in N(u)} (r_{ui} + R(ij)) w_{ij}}{\sum_{i \in N(u)} w_{ij}} \tag{6}$$

where w_{ij} is the similarity of the item i and item j.

One of the main problems that influence the results of the collaborative filtering algorithms is the cold start problem [30]. In the paper, the cold start problem will also reduce the recommendation effect of this data set. In this experiment, in order to improve the accuracy, the structural information of each compound was extracted as a method to solve the cold start problem. In Sect. 2, the compound's structure data set came from DRAGON is name as StructureComSet. The HybridSimCF algorithm proposed in the paper will use the similarity of the compound's structure and the similarity of the compound in rating matrix to improve the recommended results. In Table 3, it describes the steps of the HybridSimCF algorithm.

Table 3. Some predicted activity value of compounds and targets

Compound	Target	Predicted activity value
Chembl1231	Chembl211	2.0565
Chembl184115	Chembl301	1.3645
Chembl1922211	Chembl308	1.6925
Chembl1922209	Chembl301	1.5370
Chembl1922123	Chembl308	1.6923

In Fig. 3, StructureComSet is the structure information of all compounds, DimComSet is the data set of reducing the dimension of StructureComSet, StrSim matrix is the Pearson similarity of compound's structure, ScoreSim matrix is the compound similarity based on the rating matrix. HybridSim is the sum of the two similarities which is based on formula (6).

$$HybridSim = \alpha ScoreSim + \alpha StrSim \tag{7}$$

where *ScoreSim* is calculated by formula (1), α is a threshold which is used to control *HybridSim* value not more than 1 and the two similarities are important in improving the accuracy of recommendation, so α is set to 0.5.

Secondly, in order to obtain the underlying features of the compound structure data set, a non-line model is used for the dimensionality reduction task. A three-layer automatic encoder reduces the dimensions of the 726-dimensional compound structure data. In the experiment of this paper, the data is divided into training set and testing set randomly. The training set is used to train the model, and the testing set is used to predict the compound activity and test the prediction accuracy.

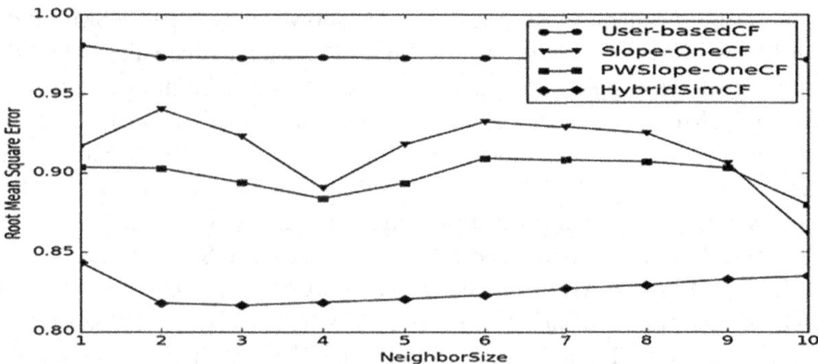

Fig. 3. The four methods of RMSE

4 Experimental Results and Discussion

In this section the details of the experimental results based on the gold standard data set are presented. 5-fold cross validation to evaluate the performance of three methods is used in the experiment, each fold includes the training data set and the test data set. In order to compare the accuracy of the four algorithms, the two indicators are used to evaluate the accuracy of the method, one is RMSE, and the other is MAE. The formulas are as follow.

$$RMSE = \frac{\sqrt{\sum_{u,i \in T} (r_{ui} - \hat{r}_{ui})^2}}{|T|} \tag{8}$$

$$MAE = \frac{\sum_{u,i \in T} |r_{ui} - \hat{r}_{ui}|}{|T|} \tag{9}$$

where r_{ui} is real score that compound u gave to target i, $\hat{r}_{ui}$ is prediction score that is computed by the method (Fig. 4).

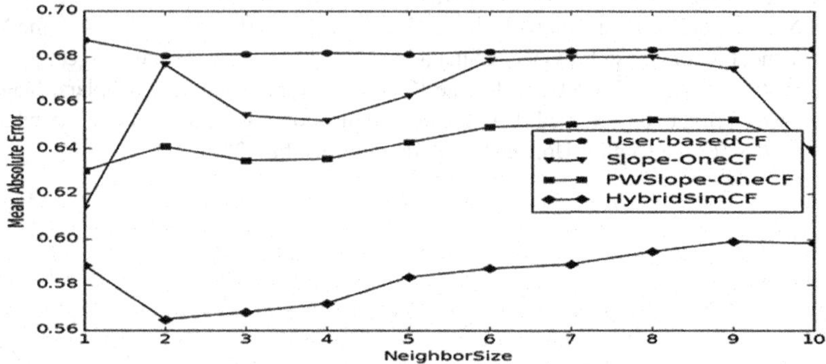

Fig. 4. The four methods of MAE

For the recommendation task, Figs. 2 and 3 show the averaged RSME and MAE values of 5 runs of the cross validation. Because of the smaller the values of RMSE and MAE, it means that the higher the accuracy of the algorithm. From Figs. 2 and 3, all of the values of RSME and MAE in the four algorithms are smaller than 1, that means the four methods are fit for the recommendation task. However, HybridSimCF has best performance in four methods. The curve of HybridSimCF always keeps the minimum in the nearest neighbor. When the nearest neighbor is 2, the RMSE and MAE of User-basedCF are 0.9729 and 0.6804, the RMSE and MAE of Slop-one are 0.9401 and 0.6765, the RMSE and MAE of PWSlope-one are 0.9028 and 0.6405, and the RMSE and MAE of HybridSimCF are 0.8178 and 0.5646, respectively. When the nearest neighbor is 10, the RMSE and MAE of HybridSimCF are the maximum while they are still smaller than other values of the three algorithms.

In Table 2, there are some recommendation results in the gold standard data set are listed in Table 2. Therefore HybridSimCF combined compounds structure into collaborative filtering algorithm can improve the accuracy effectively.

5 Conclusions

First of all, in the paper it is proposed that the problem of compound activity prediction can be converted to a recommendation task based on collaborative filtering. Through analyzing the characteristic of compound activity, a rating matrix based on IC50 value is obtained. Because of the rating matrix, the idea of recommendation can be carried out to predict the compound activity. However in drug design, the requirements for the accuracy of recommendations are very high, one of main problems impacted the accuracy of prediction is the cold start. In order to resolve the problem, the compound's structure is extracted, and the Pearson similarity based on the compound's structure is calculated. The algorithm named HybridSimCF based on the two similarities of compound is used for compound activity prediction. Through comparison of the four methods, HybridSimCF can effectively solve the problem of cold start, and improve the accuracy of prediction. Therefore the new algorithm, HybridSimCF, can be used for compound activity prediction.

Acknowledgement. This work was supported by the Fundamental Research Funds for the National Natural Science Foundation of China (Grant No. 21503101, No. 61702240), the Natural Science Foundation of Gansu Province, China (Grant No. 1506RJZA223), the Project Sponsored by the Scientific Research Foundation for the Returned Overseas Chinese Scholars, State Education Ministry (Grant No. External department of Education [2015] 311) and the Fundamental Research Funds for the Central Universities (Grant No. lzujbky-2017-191).

References

1. Darnag, R., Mazouz, E.L.M., Schmitzer, A., Villemin, D., Jarid, A., Cherqaoui, D.: Support vector machines: development of QSAR models for predicting anti-HIV-1 activity of TIBO derivatives. Eur. J. Med. Chem. **45**, 1590 (2010)
2. Afantitis, A., Melagraki, G., Sarimveis, H., Koutentis, P.A., Igglessi-Markopoulou, O., Kollias, G.: A combined LS-SVM & MLR QSAR workflow for predicting the inhibition of CXCR3 receptor by quinazolinone analogs. Mol. Divers. **14**, 225–235 (2010)
3. Mehmood, T., Liland, K.H., Snipen, L., Sæbø, S.: A review of variable selection methods in Partial Least Squares Regression. Chemom. Intell. Lab. Syst. **118**, 62–69 (2012)
4. Sharma, M.C., Sharma, S., Sahu, N.K., Kohli, D.V.: QSAR studies of some substituted imidazolinones angiotensin II receptor antagonists using Partial Least Squares Regression (PLSR) method based feature selection. J. Saudi Chem. Soc. **17**, 219–225 (2013)
5. Dahl, G.E., Jaitly, N., Salakhutdinov, R.: Multi-task Neural Networks for QSAR Predictions. Computer Science (2014)
6. Myint, K.Z., Wang, L., Tong, Q., Xie, X.Q.: Molecular fingerprint-based artificial neural networks QSAR for ligand biological activity predictions. Mol. Pharm. **9**, 2912–2923 (2012)
7. Dearden, J.C., Rowe, P.H.: Use of artificial neural networks in the QSAR prediction of physicochemical properties and toxicities for REACH legislation. In: Cartwright, H. (ed.) Artificial Neural Networks. MMB, vol. 1260, pp. 65–88. Springer, New York (2015). https://doi.org/10.1007/978-1-4939-2239-0_5
8. Gupta, S., Basant, N., Singh, K.P.: Estimating sensory irritation potency of volatile organic chemicals using QSARs based on decision tree methods for regulatory purpose. Ecotoxicology **24**, 873–886 (2015)
9. Burton, J., Danloy, E., Vercauteren, D.P.: Fragment-based prediction of cytochromes P450 2D6 and 1A2 inhibition by recursive partitioning. SAR QSAR Environ. Res. **20**, 185–205 (2009)
10. Choi, S.Y., Shin, J.H., Ryu, C.K., Nam, K.Y., No, K.T., Choo, H.Y.P.: The development of 3D-QSAR study and recursive partitioning of heterocyclic quinone derivatives with antifungal activity. Bioorgan. Med. Chem. **14**, 1608–1617 (2006)
11. Chandrasekaran, M., Sakkiah, S., Lee, K.W.: Combined chemical feature-based assessment and Bayesian model studies to identify potential inhibitors for Factor Xa. Med. Chem. Res. **21**, 4083–4099 (2012)
12. Yang, Y., Zhang, W., Cheng, J., Tang, Y., Peng, Y., Li, Z.: Pharmacophore, 3D-QSAR, and Bayesian model analysis for ligands binding at the benzodiazepine site of GABAA receptors: the key roles of amino group and hydrophobic sites. Chem. Biol. Drug Des. **81**, 583–590 (2013)
13. Kim, J.H., Chong, H.C., Kang, S.M., Lee, J.Y., Lee, G.N., Hwang, S.H., Kang, N.S.: The predictive QSAR model for hERG inhibitors using Bayesian and random forest classification method. Bull. Korean Chem. Soc. **32**, 1237–1240 (2011)
14. Singh, H., Singh, S., Singla, D., Agarwal, S.M., Raghava, G.P.S.: QSAR based model for discriminating EGFR inhibitors and non-inhibitors using Random forest. Biol. Direct **10**, 10 (2015)
15. Fechner, N., Hinselmann, G., Jahn, A., Zell, A.: Kernel-based estimation of the applicability domain of QSAR models. J. Cheminform. **2**, 1 (2010)
16. Tebby, C., Mombelli, E.: A kernel-based method for assessing uncertainty on individual QSAR predictions. QSAR Comb. Sci. **31**, 741–751 (2015)
17. Erhan, D., L'Heureux, P.J., Shi, Y.Y., Bengio, Y.: Collaborative filtering on a family of biological targets. J. Chem. Inf. Model. **46**, 626 (2006)

18. Erhan, D.: Collaborative filtering techniques for drug discovery (2006)
19. Ning, X., Rangwala, H., Karypis, G.: Multi-assay-based structure-activity relationship models: improving structure-activity relationship models by incorporating activity information from related targets. J. Chem. Inf. Model. **49**, 2444 (2009)
20. Zhang, R., Li, J., Lu, J., Hu, R., Yuan, Y., Zhao, Z.: Using deep learning for compound selectivity prediction. Curr. Comput. Aided Drug Des. **12**, 1 (2016)
21. Rosenbaum, L., Dörr, A., Bauer, M.R., Boeckler, F.M., Zell, A.: Inferring multi-target QSAR models with taxonomy-based multi-task learning. J. Cheminform. **5**, 33 (2013)
22. Gaulton, A., Bellis, L.J., Bento, A.P., Chambers, J., Davies, M., Hersey, A., Light, Y., Mcglinchey, S., Michalovich, D., Allazikani, B.: ChEMBL: a large-scale bioactivity database for drug discovery. Nucleic Acids Res. **40**, 1100 (2012)
23. Cheng, Y.C., Prusoff, W.H.: Relation between the inhibition constant (K1) and the concentration of inhibitor which causes fifty percent inhibition (I50) of an enzymic reaction. Biochem. Pharmacol. **22**, 3099–30108 (1973)
24. Xia, N.: Machine learning and data mining methods for recommender systems and chemical informatics. University of Minnesota (2012)
25. DRAGON Homepage: http://www.talete.mi.it/
26. Schafer, J.B., Konstan, J., Riedl, J.: Recommender systems in e-commerce. In: Proceedings of the 1st ACM Conference on Electronic Commerce, vol. 158 (1999)
27. Shi, J., Chen, J., Bao, Z.: An application study on collaborative filtering in e-commerce. In: International Conference on Service Systems and Service Management, vol. 1 (2011)
28. Stigler, S.M.: Francis Galton's account of the invention of correlation. Stat. Sci. **4**, 73 (1989)
29. You, H., Li, H., Wang, Y., Zhao, Q.: An improved collaborative filtering recommendation algorithm combining item clustering and slope one scheme. Lect. Notes Eng. Comput. Sci. **2215**, 313–316 (2015)
30. Sedhain S., Braziunas D., Braziunas D., Christensen J., Christensen J.: Social collaborative filtering for cold-start recommendations. In: ACM Conference on Recommender Systems, vol. 345 (2014)

Improving Initial Model Construction in Single Particle Cryo-EM by Filtering Out Low Quality Projection Images

Zhijuan Wang and Yonggang Lu(✉)

School of Information Science and Engineering, Lanzhou University, Lanzhou 730000, Gansu, China
ylu@lzu.edu.cn

Abstract. An important problem in the single-particle 3D reconstruction by cryo-electron microscopy (cryo-EM) is to construct the initial model of a macromolecule from its 2D noisy projection images at unknown random orientations. The methods for initial model construction are often based on "Angular Reconstruction" that computes the directions of the projection images by establishing a coordinate system. However, it is difficult to obtain the projection angles of the projection images which have low signal-to-noise ratio. In this paper we propose a method to improve the initial model construction by filtering out low quality projection images. The projection angles are usually represented by Euler angles α, β and γ. It is found that the quality of a projection image can be evaluated in the process of estimating its Euler angle γ. After the low quality projection images are removed, the rest of the projection images are used to construct the initial model in our method. Based on the synchronization method for initial model construction proposed by Yoel Shkolnisky, it is found that using the proposed filtering method can successfully improve the initial model construction. It is also found that filtering using Euler angle γ estimation is better than filtering using good common line estimation in the initial model construction.

Keywords: Initial model construction · 3D reconstruction
Cryo-Electron Microscopy · Common lines

1 Introduction

Structure determination is crucial for studying the molecular mechanism of macromolecules, which is necessary for a comprehensive understanding of biochemical and cellular processes. Effective methods for macromolecular structure determination include X-ray crystallography, Nuclear Magnetic Resonance (NMR) and electron cryo-Electron Microscopy (cryo-EM) [1–4]. X-ray crystallography is limited by the requirement for large amounts of sample and by the bottleneck of protein crystallization [5]. NMR may provide unique information about dynamics and interactions, but atomic structure determination is restricted to small complexes with molecular weights (MWs) below 40–50 kDa. Compared with X-ray and NMR, cryo-EM requires much less sample, poses fewer restrictions on sample purity, and does not require crystallization.

D.-S. Huang et al. (Eds.): ICIC 2018, LNCS 10955, pp. 589–600, 2018.
https://doi.org/10.1007/978-3-319-95933-7_68

In recent years, both the hardware and the software of single-particle cryo-EM have been greatly improved [6], such as the improvement of the electron microscope detectors and the image processing methods. This makes single-particle cryo-EM method rivals X-ray crystallography in the field of macromolecular structure determination [7]. The main process of the single-particle cryo-EM reconstruction is embedding the molecules in a thin layer of vitreous ice with a range of orientations, imaging these molecules using an electron beam to obtain the projection images, and then obtain the 3D models of the molecules through three-dimensional reconstruction using the projection images. Due to the high noise of the projection images, a high resolution model is usually obtained by iterative projection matching (PM) technique [8] based on an initial model. But a low quality initial model may cause the final model bias [9, 10], such as the Einstein from noise pitfall [11]. So, improving the quality of the initial model is of great significance in single-particle 3D reconstruction.

There are two well known methods to construct the initial model. The first method is based on collecting multiple exposures of the same object at different angles as the specimen is tilted in the microscope. The drawbacks of the method are that it is difficult to record good quality high-tilt data and that Fourier space coverage is generally incomplete due to experimental limitations [12]. So, the second method based on Angular Reconstruction [13] is also proposed, which calculates the orientation of the projection images by estimating the common lines among the images according to the Central Section Theorem [14]. Although the method is free of the limitations of the first method, the quality of the initial model constructed by using the Angle Reconstruction method may be affected by the low signal-to-noise ratio (SNR) and the presence of outliers of the projection images [15]. A method proposed in 2012 [16] has described a synchronization algorithm to determine the projection angles of the images. The method can work well with low SNR projections. However, there usually exist low quality projection images whose common lines and Euler angles are difficult to estimate. The including of these projections in the computation may bring errors in the construction of the initial model. So, in this paper, we propose a method to improve the initial model construction by filtering out low quality projection images. The proposed method is also based on Angular Reconstruction. The method first uses the Euler angles α, β, and γ of one projection image to calculate the angle γ' of another projection image, then uses the consistency of γ' to estimate the quality of the projection images. Finally, a part of low quality projection images are filtered out before constructing the initial model.

The rest of this paper is organized as follows: In Sect. 2, we introduce some related works including the conception of the common lines, the Angular Reconstruction method and an algorithm using a global self-correcting voting procedure to select common lines. In Sect. 3, the proposed method for filtering the projection images by computing the consistency of the projection angles is described. The experimental results are shown in Sect. 4.

2 Related Works

The proposed filtering method uses the consistency of the projection angles to estimate the quality of the projection images based on the common line estimation and the Angular Reconstruction method. In this section, some related works are introduced.

2.1 The Angular Reconstruction Method

The most important problem of single-particle cryo-EM reconstruction is to find the projection angles of two-dimensional images produced at unknown random orientations from the three-dimensional macromolecule structure. The angular reconstruction method can be used to solve the problem by estimating common lines between projections. The theoretical basis for common line estimation is the Central Section Theorem: the Fourier transform of a two-dimensional projection image is a section of the Fourier transform of the three-dimensional structure which produces the projection image [13, 14]. According to the theorem, in Fourier space, any two different projection images produced from the same structure share only one common line.

The method based on normalized cross correlation can be used to find the common line between two projection images [12]. Sampling the Fourier transform of images along L radial lines, at n equispaced points along each radial line, results in L vectors:

$$\Lambda_{kl} = (\hat{P}_k(\frac{B}{n}, \frac{2\pi l}{L}), \hat{P}_k(\frac{2B}{n}, \frac{2\pi l}{L}), \ldots, \hat{P}_k(B, \frac{2\pi l}{L}))$$

where $\hat{P}_k$ is the polar Fourier translation of the *k-th* projection image. The normalized cross correlation between each pair of L radial lines which come from two different images is computed. The pair of the radial lines which has maximum normalized cross correlation is considered as the common lines between the two images. The common line matrix C is an N-by-N matrix whose (i, j) and (j, i) entries store the indices of the pair of the radial lines, which are presented as l_1 and l_2. The common line matrix C [17] can be obtained by:

$$(C(i,j), C(j,i)) = \underset{0 \le l_1 < L/2, 0 \le l_2 < L}{\arg\max} \frac{< \Lambda_{i,l_1}, \Lambda_{j,l_2} >}{\|\Lambda_{i,l_1}\| \|\Lambda_{i,l_1}\|}, \text{for all } i \neq j \tag{1}$$

After the common lines are estimated, the projection angles represented by Euler angles can be estimated using the common lines. The definition of the Euler angle is shown in Fig. 1. The Euler angles of a projection image reflect how the image is rotated before it enters the three-dimensional space for 3D reconstruction. Because the relative position of a projection image can be determined by the common lines between the image and other images, the Euler angles α, β, and γ can be estimated using the relationship between the common lines [13].

2.2 The Voting Algorithm

Finding common lines between projection images is an important step for the Angular Reconstruction method. However, it is difficult to determine all the common lines due to the high noise of the projection images. The voting algorithm proposed in [17] uses a

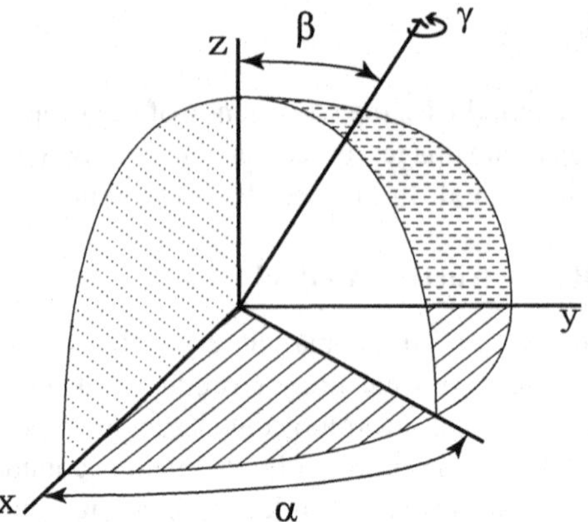

Fig. 1. The definition of the Euler angles α, β and γ.

Bayesian approach to establish a global voting procedure for identifying which common lines are detected correctly and which are spurious.

The angle α_{12} between a pair of projection images k_1 and k_2 can be estimated by a third image k_3 using the common lines between the three images. If there are N projection images, the angle between each pair of images can be estimated separately by the rest of the N–2 images. Using all the N–2 estimations, the algorithm produces a histogram using Gaussian smoothing to analyze the results of the estimations for each pair of images. The common lines between a pair of images which have large values of the histogram peak are considered as good common lines. The algorithm is described in the Algorithm 1.

Algorithm 1. Voting Algorithm

Input: The common line matrix C defined in (1).
Output: The matrix VotingScore.
Define T equally spaced angles between 0° and 180°: $\alpha_t \leftarrow 180t/T$, $t \leftarrow 0,\ldots,T\text{-}1$.
for $k_1 \leftarrow 1$ to N **do**
  for $k_2 \leftarrow k_1+1$ to N **do**
    Initialize the histogram vector h of length T to zero.
    for $k_3 \leftarrow 1$ to N such that $k_3 \neq k_1, k_2$ **do**
      if the angle α_{12} can be computed using $C[k_1, k_2]$, $C[k_2, k_1]$, $C[k_1, k_3]$, $C[k_3, k_1]$, $C[k_2, k_3]$, $C[k_3, k_2]$ **then**
        Update the histogram h using Gaussian smoothing:

$$h(t) \leftarrow h(t) + \frac{1}{\sqrt{2\pi\sigma^2}} e^{-(\alpha_t - \alpha_{12})^2/(2\sigma^2)}, t \leftarrow 0,\ldots,T-1, \sigma \leftarrow 180/T.$$

      end if
    end for
    Find and store the mode of the histogram: VotingScore$[k_1, k_2] \leftarrow \max_t h(t)$.
  end for
end for
Return VotingScore

---

3 Method

The value of the histogram peak computed using the voting algorithm introduced in Subsect. 2.2 is stored in a matrix called VotingScore, where VotingScore[*i*, *j*] is the histogram peak between image *i* and image *j* calculated by the voting algorithm. Then the VotingScore is used to select good common lines. The detail is shown in the Algorithm 2 which returns the matrix GoodCL that stores the flags of the good common lines.

3.1 Filtering Projection Images Using Good Common Lines

It is found that the common lines selected by the voting algorithm between the projection images can be used to estimate the quality of the images. The more the good common lines a projection image has with the other projections, the higher quality the projection image is considered to have. So the method for filtering the projection images using the good common lines is developed. We refer to the method as the common line filtering method (CLFiltering). The detail of the method is described in the Algorithm 3.

The Algorithm 3 returns the filtered projections which are identified as the high-quality projections. Half of the projection images with low quality are filtered out.

Algorithm 2. SelectGoodCL

```
Input: The matrix VotingScore.
Output: The matrix GoodCL.
m ← median of VotingScore.
for i ← 1 to N do
  for j ← 1 to N do
    if VotingScore[i, j] > m then
      GoodCL[i, j] ← 1.
    else
      GoodCL[i, j] ← 0.
    end if
  end for
end for
Return GoodCL
```

Algorithm 3. CLFiltering

Input: The matrix GoodCL;
    The projection images.
Output: Filtered projection images.
Clfilter[1...N] ← 0.
for i ← 1 to N **do**
  for j ← 1 to N **do**
    if GoodCL[i, j] == 1 and GoodCL[j, i] == 1 **then**
      CLfilter[i] ← CLfilter[i] + 1.
    end if
  end for
end for
m ← median of CLfilter.
for i ← 1 to N **do**
  if CLfilter[i] > m **then**
    Add image i to the filtered projection images.
  end if
end for

3.2 Filtering Projection Images by Calculating the Euler Angle γ

Although the common line filtering method can be used to select the good projection images, it is found that filtering using the Euler angle γ can produce better results. This may be because the computation of the Euler angle γ is more sensitive to the quality of the projection images. So, another projection image filtering method using Euler angle γ is developed. We refer to the method as the angle filtering method (AngleFiltering). The method is described as follows.

Considering a pair of projection images i and j, using the Euler angles α_i, β_i and γ_i of image i and the Euler angles α_j and β_j of image j, the Euler angle γ_j of image j can be computed.

For image i and image j, the common line between the two images in the local coordinate system of image i is:

$$\hat{C}_{ij} = \begin{pmatrix} \cos\varphi_i \\ \sin\varphi_i \\ 0 \end{pmatrix} \tag{2}$$

The vector lies in the X-Y plane. And φ_i equals the index of the common line between image i and image j on the plane of image i. Similarly, the common line between image i and image j in the local coordinate system of image j is:

$$\hat{C}_{ji} = \begin{pmatrix} \cos\varphi_j \\ \sin\varphi_j \\ 0 \end{pmatrix} \tag{3}$$

The rotation which brings the vector $\hat{C}_{ij}$ to the normal coordinate system is:

$$R_i = R_{\gamma_i} R_{\beta_i} R_{\alpha_i}$$

So, the common line between image i and image j in the normal coordinate system is at the direction:

$$\begin{aligned} C_{ij} &= R_i \hat{C}_{ij} \\ &= \begin{pmatrix} \cos\gamma_i & \sin\gamma_i & 0 \\ -\sin\gamma_i & \cos\gamma_i & 0 \\ 0 & 0 & 1 \end{pmatrix} \begin{pmatrix} \cos\beta_i & 0 & -\sin\beta_i \\ 0 & 1 & 0 \\ \sin\beta_i & 0 & \cos\beta_i \end{pmatrix} \begin{pmatrix} \cos\alpha_i & \sin\alpha_i & 0 \\ -\sin\alpha_i & \cos\alpha_i & 0 \\ 0 & 0 & 1 \end{pmatrix} \begin{pmatrix} \cos\varphi_i \\ \sin\varphi_i \\ 0 \end{pmatrix} \\ &= \begin{pmatrix} (\cos\gamma_i \cos\beta_i \cos\alpha_i - \sin\gamma_i \sin\alpha_i)\cos\varphi_i + (\cos\gamma_i \cos\beta_i \cos\alpha_i + \sin\gamma_i \cos\alpha_i)\sin\varphi_i \\ (-\sin\gamma_i \cos\beta_i \cos\alpha_i - \cos\gamma_i \sin\alpha_i)\cos\varphi_i + (-\sin\gamma_i \cos\beta_i \sin\alpha_i + \cos\gamma_i \cos\alpha_i)\sin\varphi_i \\ \sin\beta_i \cos\alpha_i \cos\varphi_i + \sin\beta_i \sin\alpha_i \sin\varphi_i \end{pmatrix} \end{aligned} \tag{4}$$

The rotation which brings the common line C_{ij} to the local coordinate system of the image j is:

$$R_j = R_{\beta_j} R_{\alpha_j}$$

Then, the rotated common line C_{ij} in the local coordinate system of image j is:

$$\begin{aligned} \hat{C}'_{ji} &= R_j C_{ij} \\ &= \begin{pmatrix} \cos\beta_j & 0 & -\sin\beta_j \\ 0 & 1 & 0 \\ \sin\beta_j & 0 & \cos\beta_j \end{pmatrix} \begin{pmatrix} \cos\alpha_j & \sin\alpha_j & 0 \\ -\sin\alpha_j & \cos\alpha_j & 0 \\ 0 & 0 & 1 \end{pmatrix} C_{ij} \\ &= \begin{pmatrix} \cos\theta \\ \sin\theta \\ 0 \end{pmatrix} \end{aligned} \tag{5}$$

The difference between $\hat{C}_{ji}$ and $\hat{C}'_{ji}$ is the Euler angle γ_j. More specifically, the angle θ is produced from formula (5), the angle φ_j in formula (3) can be produced from the index of the common line between image i and image j at the plane of image j, and then the Euler angle γ_j is computed by (φ_j–θ). The method for computing Euler angle γ_j is described in Algorithm 4.

Based on the above observation, the algorithm description for filtering the projection images using the Euler angle γ is outlined in Algorithm 5.

In the Algorithm 5, the parameter $\varepsilon = 6^{\circ}$ is used, which is shown to be a good setting in our experiments. Using different values of ε between 6° and 15° produces similar results.

Algorithm 4. ComputeGamma

Input: The angles α_i, β_i and γ_i of the projection i;
The angles α_j, β_j of the projection j;
The common line matrix C defined in (1);
The matrix GoodCL.
Output: The angle γ_j of the projection j.
if GoodCL[i, j] == 1 and GoodCL[j, i] == 1 **then**
  Compute the φ_i and φ_j using $C[i, j]$, $C[j, i]$.
  Compute the $\hat{C}_{ij}$ using (2).
  Rotate $\hat{C}_{ij}$ to the normal coordinate system using (4).
  Compute the angle θ using (5).
  $\gamma_j \leftarrow (\varphi_j - \theta)$.
else
  $\gamma_j \leftarrow$ INF.
end if

4 Experimental Results and Analysis

In the experiments, 1000 simulated projection images are produced with random orientations for 70S subunit of the E. coli ribosome, with SNR = 1, SNR = 1/2, SNR = 1/4, SNR = 1/8, and SNR = 1/16 respectively. Examples of the projection images produced with different SNR are shown in Fig. 2. Using the synchronization method [16], the projection angles can be estimated, and then the initial models can be constructed from the projection images. To evaluate the proposed method, the initial models are constructed with three different filtering settings: without filtering (Original), with the common line filtering (CLFiltering), and with angle filtering (AngleFiltering). Then the Fourier shell correlation (FSC) [18] of the initial model is computed against the reference density map. Figure 3 shows the FSC for the various initial models.

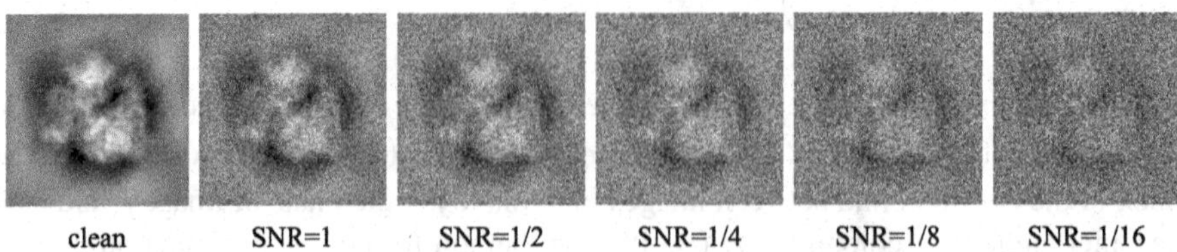

Fig. 2. Simulated projection images with different SNR.

From Fig. 3a to d, the results of FSC show that the common line filtering method and the angle filtering method both can enhance the accuracy of the initial model compared with the original method, for that they both produce FSC curves which have larger values than the one produced using the original method. The FSC curves produced by the angle filtering method all have larger values than the corresponding ones produced by the common line filtering method. So, the filtering of the projections by computing the Euler angel γ is more effective than the filtering by good common lines.

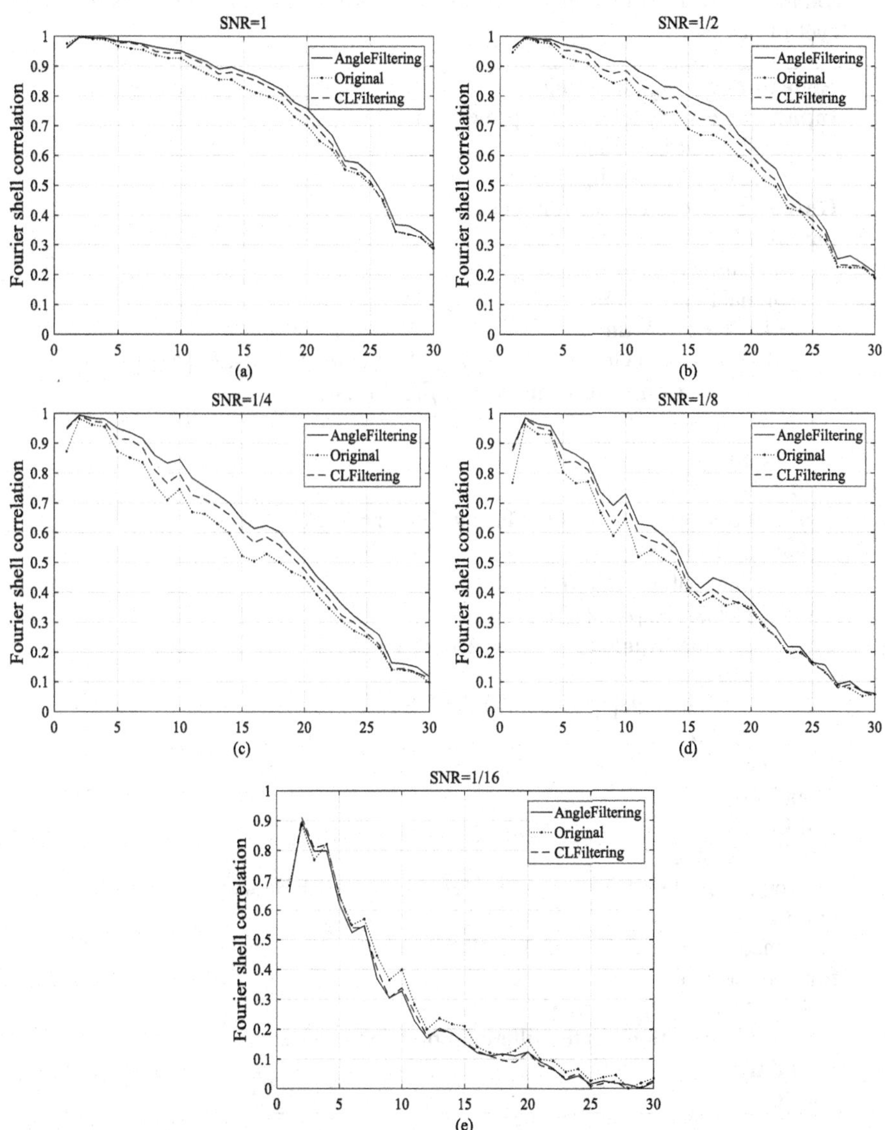

Fig. 3. Fourier shell correlations between various initial models and the reference model.

It can also be seen from the Fig. 3 that, with the increasing of the noise, the FSC values usually decrease, which shows that higher level of noise may cause more errors in the initial model construction. When the SNR is decreased to 1/16 (as shown in Fig. 3e), it is found that the FSC curves produced using the two filtering methods mostly have lower values than the corresponding ones produced without filtering. This is different from the results produced with SNR values greater that 1/16 (as shown in Fig. 3a–d). The reason may be that, with the high level of noise, both common line and projection angles are difficult to be estimated accurately, so the filtering depending on both common lines and Euler angles cannot work as effectively as in the cases with the low level of noises.

Algorithm 5. AngleFiltering

Input: The Euler angles of the projections;
  The matrix GoodCL;
  The projection images.
Output: Filtered projection images.
for $k_1 \leftarrow 1$ to N **do**
  ProQuality[1...N] $\leftarrow$ 0.
  Gamma[1...N] $\leftarrow$ INF.
  for $k_2 \leftarrow 1$ to N **do**
    if $k_2 \neq k_1$ and GoodCL[k_1, k_2] == 1 and GoodCL[k_2, k_1] == 1 **then**
      $\gamma_{k_1} \leftarrow$ **ComputeGamma**($\alpha_{k_2}, \beta_{k_2}, \gamma_{k_2}, \alpha_{k_1}, \beta_{k_1}$).
      Gamma[k_2] $\leftarrow \gamma_{k_1}$.
    end if
  end for
  $\bar{\gamma} \leftarrow$ mean of the non-INF values in Gamma[1...N].
  for $i \leftarrow 1$ to N **do**
    if Gamma[i] $\neq$ INF **then**
      if $|$Gamma[i]- $\bar{\gamma}| \leq \varepsilon$ **then**
        GoodAngle[i] $\leftarrow$ GoodAngle[i] + 1.
      else
        BadAngle[i] $\leftarrow$ BadAngle[i] + 1.
      end if
    end if
  end for
end for
for $i \leftarrow 1$ to N **do**
  ProQuality[i] $\leftarrow$ GoodAngle[i] / (GoodAngle[i] + BadAngle[i]).
end for
$m \leftarrow$ median of all the elements in ProQuality.
for $i \leftarrow 1$ to N **do**
  if ProQuality[i] > m **then**
    Add projection i to the filtered projection images.
  end if
end for

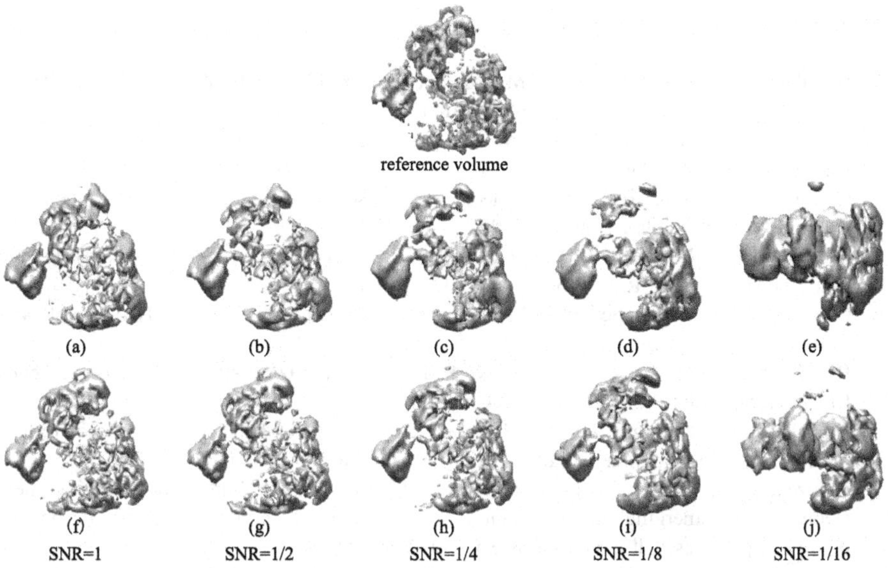

Fig. 4. Comparison of the reference model with the initial models constructed using the original method (a–e) and the angle filtering method (f–j) at different noise levels.

To further study the proposed method, the initial models constructed using the original method and the initial models constructed using the angle filtering method are shown in Fig. 4 for comparison. Figure 4(a–e) shows the initial models constructed using the original method with different SNR. Figure 4(f–j) shows the initial models constructed by the angle filtering method at different noise levels. For the initial models produced with the same SNR, the model produced using the angle filtering method contains more details compared with the one produced using the original method. It can also be seen that with the increasing of the noise, the quality of the initial models produced using the original method decreases more obviously compared to the initial models produced using the angle filtering method. Both the observations show that the proposed angle filtering method can enhance the initial model construction.

5 Conclusion

In this paper, we have proposed a new method to filter out a part of the projection images which bring large errors in the initial model construction. It is found that filtering out low quality projection images can enhance the initial model construction, and filtering the projections using the Euler angle γ is more effective than the filtering using good common lines. In the future work, we will improve the proposed method to deal with projection images at high level of noise by more accurately identifying the high quality projections.

References

1. Nogales, E., Scheres, S.H.: Cryo-EM: a unique tool for the visualization of macromolecular complexity. Mol. Cell **58**(4), 677–689 (2015)
2. Nogales, E.: The development of cryo-EM into a mainstream structural biology technique. Nat. Methods **13**(1), 24–27 (2016)
3. Carroni, M., Saibil, H.R.: Cryo electron microscopy to determine the structure of macromolecularcomplexes. Methods **95**, 78–85 (2016)
4. Vinothkumar, K.R., Richard, H.: Single particle electron cryomicroscopy: trends, issues and future perspective. Q. Rev. Biophys. **49**, e13 (2016)
5. Cheng, Y.: Single-particle cryo-EM at crystallographic resolution. Cell **161**(3), 450–457 (2015)
6. Bai, X., McMullan, G., et al.: How cryo-EM is revolutionizing structural biology. Trends Biochem. Sci. **40**(1), 49 (2015)
7. Liu, Z., Gutierrez-Vargas, C., Wei, J., et al.: Determination of the ribosome structure to a resolution of 2.5 Å by single-particle cryo-EM. Protein Sci. **26**(1), 82–92 (2017)
8. Penczek, P.A., Grassucci, R.A., Frank, J.: The ribosome at improved resolution: new techniques for merging and orientation refinement in 3D cryo-electron microscopy of biological particles. Ultramicroscopy **53**, 251–270 (1994)
9. Elmlund, H., Elmlund, D., Bengio, S.: PRIME: probabilistic initial 3D model generation for single-particle cryo-electron microscopy. Structure **21**(8), 1299–1306 (2013)
10. Rosenthal, P.B., Rubinstein, J.L.: Validating maps from single particle electron cryomicroscopy. Curr. Opin. Struct. Biol. **34**, 135–144 (2015)
11. Henderson, R.: Avoiding the pitfalls of single particle cryo-electron microscopy: Einstein from noise. Proc. Natl. Acad. Sci. U. S. A. **110**(45), 18037 (2013)
12. Lyumkis, D., Vinterbo, S., Potter, C.S., et al.: Optimod–an automated approach for constructing and optimizing initial models for single-particle electron microscopy. J. Struct. Biol. **184**(3), 417–426 (2013)
13. Van, H.M.: Angular reconstitution: a posteriori assignment of projection directions for 3D reconstruction. Ultramicroscopy **24**(1), 111–123 (1987)
14. Bracewell, R.N.: Strip integration in radio astronomy. Austr. J. Phys. **9**, 198–217 (1956)
15. Penczek, P.A., Asturias, F.J.: Ab initio cryo-EM structure determination as a validation problem. In: IEEE International Conference on Image Processing, pp. 2090–2094. IEEE (2015)
16. Shkolnisky, Y., Singer, A.: Viewing direction estimation in cryo-EM using synchronization. SIAM J. Imaging Sci. **5**(3), 1088–1110 (2012)
17. Singer, A., Coifman, R.R., Sigworth, F.J., et al.: Detecting consistent common lines in cryo-EM by voting. J. Struct. Biol. **169**(3), 312–322 (2010)
18. Van, M.H., Schatz, M.: Fourier shell correlation threshold criteria. J. Struct. Biol. **151**(3), 250 (2005)

I-MMST: A New Task Scheduling Algorithm in Cloud Computing

Belal Ali Al-Maytami[1], Pingzhi Fan[1(✉)], and Abir Hussain[2(✉)]

[1] Institute of Mobile Communication, Southwest Jiaotong University, Chengdu, China
belal@my.swjtu.edu.cn, pzfan@swjtu.edu.cn

[2] Department of Computer Science, Liverpool John Moores University, Liverpool L33AF, UK
A.Hussain@ljmu.ac.uk

Abstract. In grid network and heterogeneous computing systems, the scheduling algorithms are important for obtaining high performance through transferring the data. In this paper, we present a new scheduling algorithm for a bounded number of fully connected graph based on Improve Max-Min, Min-Min and MiM-MaM scheduling task, (I-MMST) to optimize a new task scheduling algorithm for a specific data over cloud computing. Also, we offer significant makespan improvements by introducing a look-ahead feature without increasing the time complexity associated with computation cost by using the principle of components analysis algorithm (PCA). The analysis and experiments based on randomly generated graphs with various characteristics, show that our scheduling algorithm significantly surpass previous approaches in term of makespan, speedup, and efficiency.

Keywords: Machine learning · Denial of service · IPV6 · DAG Task scheduling

1 Introduction

In grid computing, various problems are encountered in the task scheduling which tends to choose the best resources for a given task. In addition, the task scheduling problem for heterogeneous systems is more complex than that for Homogeneous Computing (HC) systems due to the different execution rates among processors and possibly different communication rates among various processors.

There are many scheduling heuristics (online and batch mode) and heterogeneous available, however they are not able to achieve the maximum objectives such as makespan, time of transferring and complexity [1–4, 7, 9, 16]. Quality of Service (QoS) based approach is required to achieve the maximum objectives. QoS based approach keeps into account the QoS based characteristics for both the tasks and the resources. In this paper, we study the different heuristics techniques for resources selection using Principle Component Analysis (PCA) in Grid Computing. Furthermore, we go through the need and significance of this QoS based (**I-MMST**) approach in Grid Computing.

D.-S. Huang et al. (Eds.): ICIC 2018, LNCS 10955, pp. 601–612, 2018.
https://doi.org/10.1007/978-3-319-95933-7_69

The key goal of grid computing is to design a system that can provide improved efficiency and a platform for proper utilization of all computing resources within an enterprise or extended enterprise to meet end-user demands [5]. But the heterogeneous and decentralized nature of Grid makes it very complex and selecting appropriate resources for jobs has become a critical issue due to the rapid increase of resources in the grid.

The task scheduling issue has been extensively studied, and various approaches have been proposed because of its key influence on performance [5, 11]. There are two types of scheduling algorithms, a heuristic algorithm, which is based on list scheduling strategies [5], and is one of the classical scheduling algorithms for cluster system environments. This algorithm has low time complexity but poor convergence performance and low universality, and the second are the heterogeneous algorithms. In general, a task scheduling problem in heterogeneous cluster systems is computationally intractable even under simplifying assumptions; it is NP-hard in the general case and in as some restricted cases [6, 8].

2 Related Work

Many scheduling algorithms have been designed for grid environment to solve the problem of mapping a set of tasks to a set of machines [14, 15].

Min-Min Heuristic

Min-Min algorithm finds the task which has minimum execution time and assigns the task to the resource that produces minimum completion time. This procedure is repeatedly executed until all tasks are scheduled. In this case, Min-Min scheduling algorithm chooses the smaller task first [14].

Max-Min Heuristic

Max-Min scheduling algorithm is similar to Min-Min scheduling algorithm, but it schedules the larger task first. The ready time of the resource is updated. This process is repeated until all unmapped tasks are assigned. Min-Min and Max-Min are used for small scale distributed system [13, 15].

This method is appropriate when most of the jobs arriving in the grid system are short ones. Thus, Max-Min would try to schedule at the same time all the short jobs and the longest ones while Min-Min would schedule first the shortest jobs and after that the longest ones, implying thus a larger make span.

MiM-MaM

MiM-MaM concerned with the merits and drawbacks of two well known traditional algorithms, Max-Min and Min-Min. In future, the deadline of each task, arriving rate of the tasks, cost of the task execution on each resource, cost of the communication are to be considered [17].

Table 1, summarize the difference between several heuristics algorithms in the task scheduling. In addition, the task scheduling problem is broadly classified into two major categories, namely Static Scheduling, and Dynamic Scheduling. In the Static category, all information about tasks such as execution and communication costs for each task and the

Table 1. Objectives of heuristics algorithms

Heuristics	Objectives		
	Makespan	Resource utilization	Load balancing
MET [3, 4]	Yes	No	No
MCT [3, 4]	Yes	No	Yes
Min-Min [14]	Yes	No	No
Max-Min [15]	Yes	Yes	No
QoS priority based scheduling Heuristic [10]	Yes	Yes	Yes
QoS Guided Min-Min Heuristic [14]	Yes	Yes	Yes
QoS priority grouping scheduling heuristic [9]	Yes	Yes	Yes
QoS guided weighted mean Heuristic	Yes	Yes	Yes
QoS Suffrage heuristic	Yes	Yes	Yes

relationship with other tasks are known beforehand; in the dynamic category, such information is not available, and decisions are made at runtime. Moreover, Static scheduling is an example of compile-time scheduling, whereas, Dynamic scheduling is representative of run-time scheduling. Static scheduling algorithms are universally classified into two major groups, namely Heuristic-based and Guided Random Search based algorithms. Heuristic-based algorithms allow approximate solutions, often good solutions, with polynomial time complexity [9].

As such, our proposed solution provide the following benefits:

- Reduce the time of task scheduling.
- Obtain the optimal scheduling algorithm for specific data on the data center.
- Reduce the computation cost, communication cost and tasks scheduling in the data center.
- Use PCA to reduce the required matrix.
- Makespan factor is utilised to analyze the performance of the algorithm.

The problem addressed in this paper is the static scheduling of a single application in a heterogeneous system with a set P of processors and set of tasks V. We consider that P processors are available for the tasks V and that they are not shared during the execution of the task. Therefore, with the system and tasks parameters are known at compile time, all the studies focus on how to reduce the makespan, the time of tasks execution in datacenter.

3 Computation of the Task Graph

Directed Acyclic Graph (DAG), G = (V, E), as illustrated in Fig. 1, is the application that represents the base for this paper, where V refers to the set of v nodes, and each node $v_i \in$ V represents an application task, which includes instructions that must be executed on the same machine. E is the set of e communication edges between tasks;

each represents the task-dependency constraint such that task v_i should complete its execution before task v_j startes. The DAG is complemented by a matrix W that is a $v \times p$ computation cost matrix, where v is the number of tasks and p is the number of processors in the system. In his case, $w_{i,j}$ gives the estimated time to execute task v_i on machine p_j. The mean execution time of task v_i is calculated using Eq. 1.

$$\overline{wi} = \left(\sum_{j \in P} w_{i,j}\right) / p \tag{1}$$

The simple model in Fig. 1 is a common in this scheduling problem [9–12], and we consider it to permit a fair comparison with state-of-the art algorithms as these simplifications correspond to real systems (Table 2).

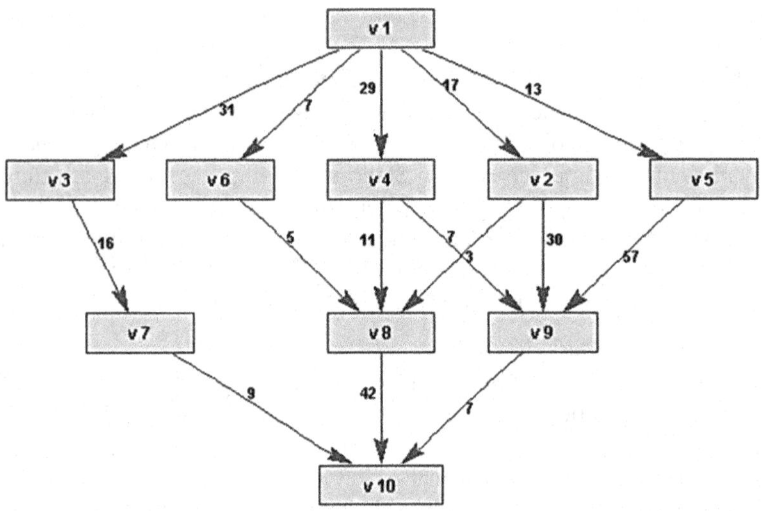

Fig. 1. A simple task graph with 10 tasks

4 The Proposed Algorithm I-MMST

The efficient scheduling of tasks is paramount to maximizing the benefits of executing an application in Heterogeneous Computing. Hence, the task scheduling problem has been extensively explored and consequently a variety of scheduling algorithms such as list scheduling and clustering, task duplication based scheduling and the algorithms based on genetic approaches have been proposed in the literature.

Interestingly, significant works have been proposed for the development of task scheduling algorithms for homogeneous processors [3–5, 13] and the same has not been fully explored for heterogeneous processors [14, 15]. List-scheduling algorithms generally perform well at a relatively low cost compared to other categories. Clustering

Table 2. Computation costs of the tasks in Fig. 1 on three processors

Edge(E)	Node(V)
(1,2)	12
(1,3)	22
(1,4)	11
(1,5)	4
(1,6)	17
(3,7)	5
(2,8)	13
(4,8)	12
(6,8)	22
(2,9)	20
(4,9)	5
(5,9)	33
(7,10)	44
(8,10)	18
(9,10)	21

Task	P1	P2	P3
v1	12	15	10
v2	23	43	16
v3	24	12	22
v4	16	27	33
v5	33	32	17
v6	12	17	12
v7	9	21	33
v8	13	18	43
v9	14	22	11
v10	20	44	14

algorithms are well suited for homogeneous processors. Task duplication-based scheduling algorithms are used when the program is communication intensive.

To define the objective of the task scheduling problem, two attributes namely, Earliest Start Time (EST) and Earliest Finish Time (EFT) are defined. The EST of task v, on processor p, is represented as EST(v, p). Likewise the EFT of task vi, on processor pj is represented as EFT(vi, p_j). Let EST(v_i) and EFT(v_i) represent the earliest start time upon any processor and the earliest finish time upon any processor, respectively. For the entry task ventry the EST(v_{entry}) = 0, for other tasks in the task graph the EST and EFT values are computed starting from the entry task to exit task by traversing the task graph from top to bottom. To compute the EST of a task v_i all immediate predecessor tasks of vi should be scheduled. The task scheduling problem is mathematically defined as follows:

$$EST(v_i, p_j) = \max\{p\_avail\,[v_i, p_j], \max(EFT(v_p, p_k) + C(v_p, v_i)\} \quad (2)$$

Where $v_p \in pred\ (\mathrm{v}_i), EFT(v_p, p_k) = EFT(v_p)$

$$C(v_p, v_i) = 0 \text{ when } k = j \quad (3)$$

$$EFT(v_i, p_j) = T(v_i, p_j) + EST(v_i, p_j) \quad (4)$$

P_avail [v_i, p_j] is defined as the earliest time that processor p_j will be available to begin executing task v_i. The inner *max* clause in the EST equation finds the latest time that a predecessor's data will arrive at processor p_j. If the predecessor finishes earliest on a processor other than p_j, communication cost must also be included at this time. EST(v_i, p_j) is the maximum of times at which processor p_j becomes available and the time at which the last message arrives from any of the predecessors of v_i.

$$ETC = \begin{bmatrix} ETC_{1,1} & ETC_{2,1} & . & . & ETC_{i,1} \\ ETC_{2,1} & ETC2,2 & . & . & ETC_{i,2} \\ . & . & . & . & . \\ . & . & . & . & . \\ ETC_{i,1} & . & . & . & ETC_{i,j} \end{bmatrix}$$

$$PCA(ETC) = \begin{bmatrix} ETC_{1,1} & . & . & . & . \\ . & ETC_{2,2} & . & . & . \\ . & . & . & . & . \\ . & . & . & . & . \\ . & . & . & . & ETC_{i,j} \end{bmatrix} \quad (5)$$

$$makespan = EFT(v_i, p_j),$$

Where v_i is the exit task. From the matrix PCA(ETC) we observe that reduce the matrix from m × n to vector m, which leads to reduce the accuracy as well.

The main goals of the proposed algorithm (**I-MMST**) are to minimize the scheduling rate (makespan) as shown in Algorithms 1 and 2 (Fig. 2).

Algorithm 1. Main algorithm to calculate the expected computing time of task i on host j

```
Input:
p = hosts;
v = task;
Output:
PCA(ETC)
While there tasks to schedule
    For all v_i to schedule
        For all p_j
          Compute CT_i,j=CT(v_i,p_j)
        End for
        Compute metric i= F(CT_i,1,CT_i,2, ........ )
    End for
    Select the best metric match m
    Compute minimum CT_m,n
    Schedule task m on n
End while.
```

Algorithm 2. Details of I-MMST

- Generate random data
- Group the data in S group randomly and each group can include high and low QoS
  - For each group of data, run six algorithms and output the makespan time For each algorithm P_i={p_1,p_2,…….,p_m), m is number of machines i =1..6 (for all algorithms)

- Use PCA
- $\psi = \frac{1}{m}\sum_{i=1}^{M} h_i$ mean value
- $\phi_i = |h_i - \psi|$
- Set $A = [\phi_1, \phi_2, \ldots\ldots.\phi_M]$
- Covariance C = A AT
- Find the Eigen vector (A AT) U

- Select the best vector (minimum makespan (P_k,F_k) and which denote to the best algorithm k.
- Save <U,k> in the data base
- B is evaluated data $B \in F^{N \times M}$
- Find the Eigen vector (B BT)=W
- Find min ||W – Ui|| (the closet Eigen vector in the data base) and choose its algorithm F_k that is suitable predicated algorithm.

G = group of data high QoS and low QoS
P = hosts
F = algorithms
k = index of algorithm

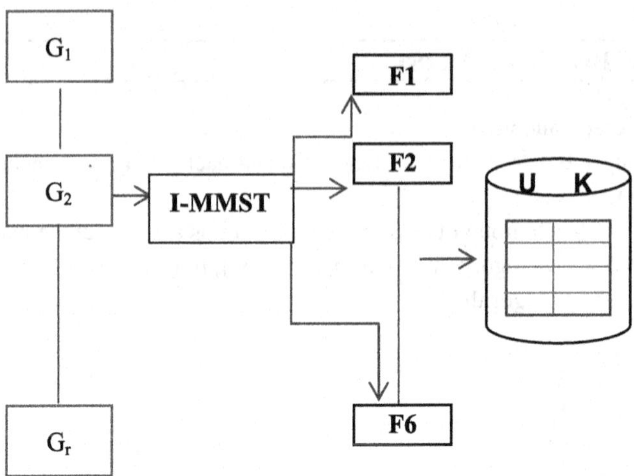

Fig. 2. I-MMST algorithm

5 Simulation Result and Discussions

This section presents the comparison of the proposed heuristics with three well known task scheduling algorithms Min-Min [14], Max-Min [15] and MiM-MaM [17], which are developed for heterogeneous system. The experiments were carried out using MATLAB R2013a on an Intel Core i3 processor, 2.40 GHz CPU and 4 GB RAM running on Microsoft Windows 7 platform. The main performance metric chosen for the comparison is the makespan. Since a large set of application graphs with different properties is used.

Makespan: Is the main performance metric which gives the overall completion time for all the tasks in a given graph, Eq. (5).

Speedup: The Speedup value of a schedule is defined as ratio of the sequential schedule length obtained by assigning all tasks to the fastest processor, to the parallel execution time of the task schedule (makespan).

$$Speedup = \frac{min_{p \in Q}\left(\sum_{n_i \in V} w_{i,j}\right)}{makespan} \tag{6}$$

Efficiency: It is the ratio of speedup with total number of used processors to schedule the entire DAG application and values of Table 3. It always tends to increasing in nature.

$$Eff = \frac{Speedup}{p} \tag{7}$$

Table 3. The results of Makespan for Four algorithms

Task	Min-Min	Max-Min	Mam-MiM	I-MMST
10	2.9213	2.811	2.666	**2.601**
20	6.2041	4.678	4.324	**4.003**
30	7.1933	5.6	5.455	**4.65**
40	9.0017	7.118	6.256	**5.304**
50	10.3585	9.84	8.622	**6.102**
60	12.4807	10.767	9.12	**7.325**
70	14.7421	11.239	10.505	**8.55**
80	15.896	12.884	11.805	**9.434**
90	16.0835	14.005	12.335	**10.566**

Here, we compare the performance of **I-MMST** algorithm with three well-known scheduling algorithms in grid systems: the Min-Min [14], Max-Min [15] and MiM-MaM [17] algorithms. To make the comparison efficient, we used different scenarios of scheduling problem. As per the scheduling objective, the proposed algorithm mainly looking for minimization of cloud makespan. Figures 3, 4 and 5 show the performance of the proposed algorithm (makespan, speedup and efficiency) compared with the three algorithms. The results of comparisons are intended not only to present quantitative results, but also to analyze the results qualitatively and to suggest explanations, for a better insight in the overall scheduling problem.

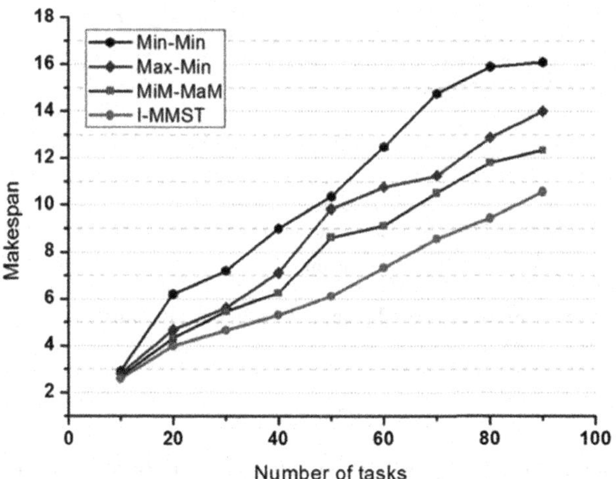

Fig. 3. The Makespan generated different algorithms

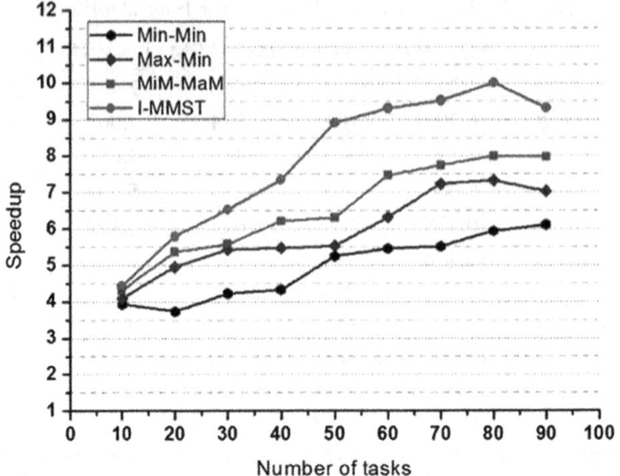

Fig. 4. The speedup of different algorithms

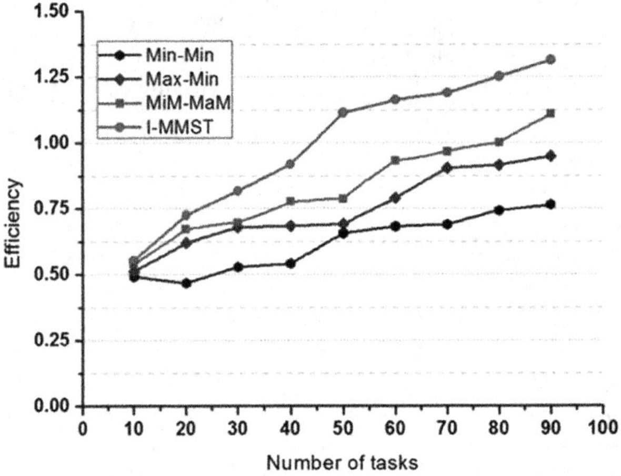

Fig. 5. The efficiency of different algorithms

6 Conclusions

Quality of service of Grid computing is a challenging but timely important problem. In addition, we improved the performance by using PCA approach to reduce the size of matrix.

In conclusion, I-MMST reduced the execution time of the tasks and overall makespan, which could be beneficial in a cloud computing context. For future work, we can improve the performance of I-MMST by using a more sophisticated approach to decide and predict in advance the optimal algorithms for specific data.

References

1. Masood, A., et al.: HETS: heterogeneous edge and task scheduling algorithm for heterogeneous computing systems. In: Proceeding of 2015 IEEE 17th International Conference on High Performance Computing and Communications, 2015 IEEE 7th International Symposium on Cyberspace Safety and Security, and 2015 IEEE 12th International Conference on Embedded Software and Systems (2015)
2. Hoffmann, R., Prell, A., Rauber, T.: Dynamic task scheduling and load balancing cell processors. In: 18th Euromicro International Conference on Parallel, Distributed and Network-Based Processing (PDP), pp. 205–212 (2010)
3. Munir, E.U., Li, J., Shi, S.: QoS sufferage heuristic for independent task scheduling in grid. Inf. Technol. J. **6**(8), 1166–1170 (2007)
4. Buyya, R., Yeo, C., Venugopal, S., Broberg, J., Brandic, I.: Cloud computing and emerging IT platforms: vision, hype, and reality for delivering computing as the 5th utility. Futur. Gener. Comput. Syst. **25**(6), 599–616 (2009)
5. Bawa, R.K., Sharma, G.: Modified min-min heuristic for job scheduling based on QoS in Grid environment. In: 2nd International Conference on Information Management in the Knowledge Economy (IMKE). IEEE (2013)
6. Napper, J., Bientinesi, P.: Can cloud computing reach the top500? In: Proceedings of the Combined Workshops on Unconventional High Performance Computing Workshop Plus Memory Access Workshop, Ischia, pp. 17–20 (2009)
7. Wang, E.D., Li, X.: QoS-oriented monitoring model of cloud computing resources availability. In: International Conference on Computational and Information Sciences (2013)
8. Zhang, C., Huang, R., Zhang, J.: Distributed adaptive consensus tracking of unknown heterogeneous linear systems via output feedback. In: Proceedings of the 35th Chinese Control Conference, 27–29 July 2016, Chengdu (2016)
9. Feng, C., Xu, H., Li, B.: An alternating direction method approach to cloud traffic management. arXiv preprint arXiv:1407.8309 (2014)
10. Begum, S., Prashanth, C.S.R.: Stochastic based load balancing mechanism for non-iterative optimization of traffic in cloud. In: International Conference on Wireless Communications, Signal Processing and Networking. IEEE (2016)
11. Smirnov, A.V., et al.: Network traffic processing module for infrastructure attacks detection in cloud computing platforms. In: XIX IEEE International Conference on Soft Computing and Measurements (SCM). IEEE (2016)
12. Kang, L., Ting, X.: Application of adaptive load balancing algorithm based on minimum traffic in cloud computing architecture. In: International Conference on Logistics, Informatics and Service Sciences (LISS). IEEE (2015)
13. Rajendra, S., Chaturvedi, A.K.: Many-objective comparison of twelve grid scheduling heuristics. Int. J. Comput. Appl. (0975–8887), **13**(6) (2011)
14. Amudha, T., Dhivyaprabha, T.T.: QoS priority based scheduling algorithm and proposed framework for task scheduling in a grid environment. In: IEEEs - International Conference on Recent Trends in Information Technology, ICRTIT 2011. Department of Computer Application, School of Computer Science & Engineering, Bharathiar University, Coimbatore – 46, MIT, Anna University, Chennai, 3–5 June 2011

15. Konjaang, J.K., Maipan-uku, J.Y., Kubuga, K.K.: An efficient max-min resource allocator and task scheduling algorithm in cloud computing environment. arXiv preprint arXiv:1611.08864 (2016)
16. He, X., Sun, X., Von Laszewski, G.: QoS guided min-min heuristic for grid task scheduling. J. Comput. Sci. Technol. **18**(4), 442–451 (2003)
17. Kfatheen, S.V., Banu, M.N.: MiM-MaM: a new task scheduling algorithm for grid environment. In: Computer Engineering and Applications, pp. 695–699. IEEE (2015)

Recognition and Severity Rating of Parkinson's Disease from Postural and Kinematic Features During Gait Analysis with Microsoft Kinect

Ilaria Bortone[1], Marco Giuseppe Quercia[2], Nicola Ieva[2], Giacomo Donato Cascarano[2], Gianpaolo Francesco Trotta[2,3], Sabina Ilaria Tatò[4], and Vitoantonio Bevilacqua[2(✉)]

[1] Institute of Clinical Physiology, National Research Council, Pisa, Italy
[2] Department of Electrical and Information Engineering, Polytechnic University of Bari, Bari, Italy
vitoantonio.bevilacqua@poliba.it
[3] Department of Mechanics, Mathematics and Management, Polytechnic University of Bari, Bari, Italy
[4] Medica Sud s.r.l, Viale della Resistenza n. 82, Bari, Italy

Abstract. When diagnosing Parkinson's disease (PD), medical specialists normally assess several clinical manifestations of the patient and rate a severity level according to established criteria. They refer to the Movement Disorder Society – sponsored revision of Unified Parkinson's Disease Rating Scale (MDS-UPDRS), the most widely adopted scale for rating PD. Since gait patterns differ between healthy elders and those with PD, we implement a simple, low-cost clinical tool that can extract kinematic and postural features through Microsoft Kinect v2 sensor to classify and rate PD. Thirty participants were enrolled for the purpose of the present study: sixteen PD patients, rated according to MDS-UPDRS Part IV for motor complications, and fourteen healthy paired subjects. Several gait cycles were extracted for each patient to improve the reliability of the methods and sixteen kinematic and postural features were considered. After preliminary feature selection, several classifier families were trained (both Support Vector Machine, SVM, and Artificial Neural Networks, ANN) and evaluated for the best solution. Results showed that the ANN classifier performed the best by reaching 89,40% of accuracy with only nine features in diagnosis PD and 95,02% of accuracy with only six features in rating PD severity.

Keywords: Classification · Artificial Neural Network · Parkinson's disease Gait · Posture · Microsoft Kinect

1 Introduction

Parkinson's disease (PD) is a degenerative brain disorder characterized by a loss of midbrain dopamine (DA) neurons [1]. The main clinical PD features related to body movement include tremor, rigidity, bradykinesia, and gait abnormalities. This loss of

D.-S. Huang et al. (Eds.): ICIC 2018, LNCS 10955, pp. 613–618, 2018.
https://doi.org/10.1007/978-3-319-95933-7_70

mobility includes symptoms, such as, flexed posture, lack of spinal flexibility and trunk rotation, and reduced joint range of movement during gait cycles [2].

When assessing the severity level of Parkinson disease, numeric scales are preferred. The Movement Disorder Society – sponsored revision of Unified Parkinson's Disease Rating Scale (MDS-UPDRS) [3] is widely adopted. Although several scientific results support the validity of the MDS-UPDRS for rating, subjectivity and low efficiency are inevitable as most of the diagnostic criteria use descriptive symptoms, which cannot provide a quantified diagnostic basis. Therefore, the development of computer-assisted diagnosis and computer-expert system is very important. Since the release of the Microsoft Kinect SDK, the Kinect sensor has been widely utilized for PD-related research. Several projects focused on rehabilitation and they proposed experimental ways of monitoring patients' activities [4–7].

In our work, we aim at providing clinicians with a mainstream sensor with a two-fold objective: detect and recognize typical PD motor issues and facilitate assessment of gait alterations both at the hospital and at home. In the paper, we detail the use of the MS Kinect system for movement-data acquisition, the detection of gait features (both in terms of spatiotemporal parameters and postural variables), and the analysis of gait disorders via selected processing methods.

The paper is organized as follows. Section 2 introduces the proposed approach and depicts the experiments, and Sect. 3 shows results. We conclude the paper and discuss possible future work in Sect. 4.

2 Materials and Methods

2.1 Population

We recruited thirty elderly participants for this study from a local clinical center (Medica Sud s.r.l., Bari, Italy): 14 healthy subjects (10 male and 4 female, 73.57 ± 6.47 years, range 65–82 years) and 16 idiopathic Parkinson patients (13 male and 3 female, 74.94 ± 7.68 years, range 63–87 years). Right and left sides of each patient were considered separately in the study. All participants provided written informed consent. The subject were asked to walk straight towards the device, with their normal walking rhythm (Fig. 1). Several trials were acquired for each patient, changing the starting foot, to have at least one gait cycle for each side without errors.

2.2 Microsoft Kinect Framework

The Kinect sensor samples at a frequency of approximately 30 Hz and video frames are captured both in color and depth. Using captured frames, the Kinect SDK segments and tracks human skeletons and gives the output of a human skeleton represented by 25 nodes or control points in the Kinect's own reference frame known as the skeleton space. Motion analysis data collection started with the subject standing in a T-pose for one second to facilitate the skeleton tracking. Subjects then walked toward the Kinect sensor, which was placed 3.5 m away from the subject's starting point at a height of 0.75 m. The 3.5-m distance was selected to guarantee that the recorded gait cycle,

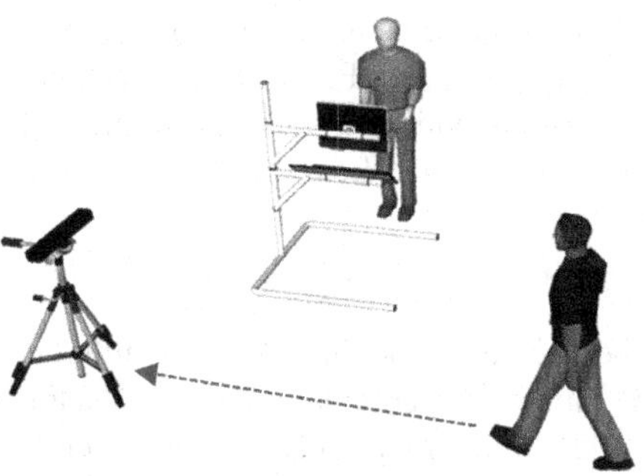

Fig. 1. Representation of the proposed set-up in the clinical center. The dotted red line indicates the walking direction (one-way walk). (Color figure online)

which began when the subject was about 2.5–3 m from the Kinect, did not include the acceleration/deceleration phases of walking that are anticipated during the initiation or completion of the gait task.

Three categories of features have been considered:

- Temporal: to assess the duration of gait phases in seconds and in percentage compared to the duration of gait cycle (Stance and Swing Phase/Time, Double Support Phase, Stride Time);
- Spatial: to estimate length, width and velocity of movements, normalized by the height or the lower limb length of the subject (Stride Cadence/Length/Velocity, Step Length/Width, Swing Velocity);
- Angular: to assess the degree of rotation for specific postures and movements, typical of Parkinsonian patients (Trunk/Neck Flexion, Pisa Syndrome, Arm Swing).

2.3 Feature Selection

The dataset resulting from the previous analyses was constituted by sixteen different features, grouping, for each subject, gait parameters. For each patient several gait cycles were extracted, to have more reliable values. According to the UPDRS-IV part scores obtained from the clinical assessment, the sample consisted of fourteen healthy, nine mild Parkinson's disease and seven moderate/severe Parkinson's disease subjects. We analyzed two different subgroups for the classification analysis as follow:

- **Case A: Healthy (14) versus Parkinson's Disease (16)**. Dataset consists of a total of 30 records, 16 PD patients and 14 older age normal subjects. Right and left sides of each patient were considered separately in the study. So, the final dataset is composed by 60 instances.
- **Case B: Mild (9) versus Moderate/Severe (7) Parkinson's Disease**. Dataset consists of a total of 16 records, 9 slight and 7 moderate PD patients. Right and left sides of each patient were considered separately in the study. So, the final dataset is composed by 32 instances.

The values included in each dataset record refer to average measures on all the gait cycles considered for the single subject. For each approach, a classification was carried out, before with all the features and then with several sets of selected features, identified using the Open Source Machine Learning Software Weka.

2.4 Classification Algorithms and Techniques

All the analyses have been conducted following two different classification strategies based on supervised learning. In particular, Support Vector Machines (SVMs) and Artificial Neural Networks (ANNs) have been used, which are state-of-the-art classifiers that have gained popularity within pattern recognition tasks [8–10].

Considering the easy tuning of training parameters, SVMs classifiers [11] have been considered to realize a preliminary inspection of the processed data. SVM is a classifier whose goal is to find the best decision hyper-plane that separates the training features space.

SVMs have high generalization capability because they can be extended to separate a space of non-linear input features [12]. Evolutionary approaches, and optimization strategies based on probabilistic graphical models, for the design of classification architectures are becoming very popular thanks to their ability to automatically optimize the classifiers topologies to improve the overall classification performance.

In this work, the implemented genetic algorithm is an improved version of the one reported in [10], able to search for ANN topologies for multi-class discrimination. The fitness function maximized by the GA consists in the mean value of Accuracy reached by each ANN classifier on a fixed number of iterations in which the ANN is trained, validated and tested using a random permutation of the dataset, maintaining constant the ratio among the number of samples of the different classes (Eq. 1). The Accuracy value is the ratio between the number of instances correctly classified for each class, and the total number of instances in the test set. Considering a binary classifier, with Positive and Negative classes, the Accuracy is computed following Eq. 2.

$$Fitness = \frac{\sum_{n=1}^{iterations} Accuracy_n}{iterations} \tag{1}$$

In this work, the GA configuration parameters were the same used in Bortone et al. [8], whereas the number of iteration was set to 250. In both the ANN and SVM strategies, the performance of the classifiers were evaluated in terms of Accuracy, Specificity and Sensitivity (Eq. 2), considering the confusion matrix as the one reported in Table 1 for a binary classifier example.

$$\boldsymbol{Accuracy} = \frac{TP+TN}{TP+TN+FP+FN};\ \boldsymbol{Specificity} = \frac{TN}{TN+FP};\ \boldsymbol{Sensitivity} = \frac{TN}{TP+FN} \tag{2}$$

Table 1. ANN and SVM performance comparison with only selected features.

		Accuracy	Sensitivity	Specificity
Case A.2 (9 Features)	SVM	0.785 ± 0.034	0.817 ± 0.049	0.748 ± 0.055
	ANN	**0.894 ± 0.082**	**0.870 ± 0.127**	**0.918 ± 0.110**
Case B.2 (6 Features)	SVM	0.887 ± 0.039	0.789 ± 0.060	0.963 ± 0.051
	ANN	**0.950 ± 0.071**	**0.900 ± 0.157**	**0.990 ± 0.043**

3 Results

All the participants were able to complete both clinical and instrumented evaluations. We then reported and compared the results obtained with both SVM and optimized ANN classifiers. In detail, the comparison has been evaluated analyzing the average values of Accuracy, Sensitivity and Specificity across 250 different training iterations.

The results showed that the ANN classifier performed the best in both cases and in both configurations (all features versus reduced features). In particular, when diagnosing PD, the ANN reached 89,4% (±8,6%) of Accuracy, 87,0% (±12,7%) of Sensitivity and 91,8% (±11,1%) of Specificity with only 9 selected features; while, the ANN reached 95,0% (±7,1%) of Accuracy, 90,0% (±15,7%) of Sensitivity and 99,0% (±4,3%) of Specificity with 6 selected features in classifying mild to moderate PD patients.

4 Conclusion

ANN classifier reached the best performance: 89,40% of accuracy with only nine features in diagnosis PD and 95,02% of accuracy with only six features in rating PD severity. Postural features were relevant in both cases and to our knowledge no previous studies have investigated tin depth he role of these components in classification and rating of PD.

Acknowledgment. This work has been funded from the Italian Project ROBOVIR (INAIL, BRIC-2017).

References

1. Twelves, D., Perkins, K.S.M., Uk, M., Counsell, C.: Systematic review of incidence studies of parkinson's disease. Mov. Disord. **18**, 19–31 (2003)
2. Horváth, K., Aschermann, Z., Ács, P., Deli, G., Janszky, J., Komoly, S., Balázs, É., Takács, K., Karádi, K., Kovács, N.: Minimal clinically important difference on the motor examination part of MDS-UPDRS. Parkinsonism Relat. Disord. **21**, 1421–1426 (2015)
3. Goetz, C.G., Tilley, B.C., Shaftman, S.R., Stebbins, G.T., Fahn, S., et al.: Movement disorder society-sponsored revision of the unified parkinson's disease rating scale (MDS-UPDRS): scale presentation and clinimetric testing results. Mov. Disord. **23**, 2129–2170 (2008)

4. Eltoukhy, M., Kuenze, C., Oh, J., Jacopetti, M., Wooten, S., Signorile, J.: Microsoft Kinect can distinguish differences in over-ground gait between older persons with and without Parkinson's disease. Med. Eng. Phys. **44**, 1–7 (2017)
5. Bevilacqua, V., Nuzzolese, N., Barone, D., Pantaleo, M., Suma, M., D'Ambruoso, D., Volpe, A., Loconsole, C., Stroppa, F.: Fall detection in indoor environment with kinect sensor. In: Proceedings of the 2014 IEEE International Symposium on Innovations in Intelligent Systems and Applications (INISTA), pp. 319–324 (2014)
6. Springer, S., Seligmann, G.Y.: Validity of the kinect for gait assessment: a focused review. Sensor **16**(2), 194 (2016)
7. Manghisi, V.M., Uva, A.E., Fiorentino, M., Bevilacqua, V., Trotta, G.F., Monno, G.: Real time RULA assessment using Kinect v2 sensor. Appl. Ergon. **65**, 481–491 (2017)
8. Bortone, I., Trotta, G.F., Brunetti, A., Cascarano, G.D., Loconsole, C., Agnello, N., Argentiero, A., Nicolardi, G., Frisoli, A., Bevilacqua, V.: A novel approach in combination of 3D gait analysis data for aiding clinical decision-making in patients with parkinson's disease. In: Huang, D.-S., Jo, K.-H., Figueroa-García, J.C. (eds.) ICIC 2017, Part II. LNCS, vol. 10362, pp. 504–514. Springer, Cham (2017). https://doi.org/10.1007/978-3-319-63312-1_44
9. Bevilacqua, V., Tattoli, G., Buongiorno, D., Loconsole, C., Leonardis, D., Barsotti, M., Frisoli, A., Bergamasco, M.: A novel BCI-SSVEP based approach for control of walking in virtual environment using a convolutional neural network. In: 2014 International Joint Conference on Neural Networks (IJCNN), pp. 4121–4128 (2014)
10. Bevilacqua, V., Brunetti, A., Triggiani, M., Magaletti, D., Telegrafo, M., Moschetta, M.: An optimized feed-forward artificial neural network topology to support radiologists in breast lesions classification. In: Proceedings of the 2016 on Genetic and Evolutionary Computation Conference Companion, pp. 1385–1392 (2016)
11. Cortes, C., Vapnik, V.: Support-vector networks. Mach. Learn. **20**, 273–297 (1995)
12. Bevilacqua, V., Pannarale, P., Abbrescia, M., Cava, C., Tommasi, S.: Comparison of data-merging methods with SVM attribute selection and classification in breast cancer gene expression. In: Huang, D.-S., Gan, Y., Premaratne, P., Han, K. (eds.) ICIC 2011. LNCS, vol. 6840, pp. 498–507. Springer, Heidelberg (2012). https://doi.org/10.1007/978-3-642-24553-4_66

Rhino-Cyt: A System for Supporting the Rhinologist in the Analysis of Nasal Cytology

Giovanni Dimauro[1(✉)], Francesco Girardi[2], Matteo Gelardi[3], Vitoantonio Bevilacqua[4], and Danilo Caivano[1]

[1] Dipartimento di Informatica, Università degli Studi di Bari 'Aldo Moro', Bari, Italy
giovanni.dimauro@uniba.it
[2] UVARP ASL Bari, Lungomare Starita, 4, 70123 Bari, Italy
[3] Department of Basic Medical Sciences, Neuroscience and Sense Organs, Università degli Studi di Bari 'Aldo Moro', Bari, Italy
[4] Dipartimento di Ingegneria Elettrica e dell'Informazione, Politecnico di Bari, Bari, Italy

Abstract. In recent years, cytological observations in the rhinological field are being increasingly utilized. This development has taken place over the last two decades and has proven to be fundamental in defining new nosological entities and in driving changes in the previous classification of rhinopathies. The simplicity of the technique and its low invasiveness make rhinocytology a practical diagnostic tool practical for all rhinoallergology services. Furthermore, since it allows the monitoring of responses to treatment, this method plays an important role in guiding a more effective and less expensive diagnostic program. Microscopic observation requires prolonged effort by a specialist, but the modern scanning systems for cytological preparations allow scanning of an entire preparation enlarged to 400x. By means of the system presented in this paper, it is possible to automatically identify and classify cells present on a rhinocytologic preparation based on a digital image of the preparation itself. Thus, pivotal diagnostic support has been made available to the rhinocytologist, who can quickly verify that the cells have been correctly classified by observation on a monitor. In the system presented herein, image processing and image segmentation techniques have been used to find images of cellular elements within the preparation. Cell classification is based on a convolutional neural network composed of three blocks of main layers. Cell identification (first step, image segmentation) exhibits accuracy greater than 90%, while cell classification (second step, seven cytotypes) attained a mean accuracy of approximately 98%. Finally, the classified cell images are shown to a specialist for rapid verification. This complete system supports clinicians in the preparation of a rhinocytogram report.

Keywords: Nasal cytology · Automatic cell recognition · Rhinologic Image analysis

D.-S. Huang et al. (Eds.): ICIC 2018, LNCS 10955, pp. 619–630, 2018.
https://doi.org/10.1007/978-3-319-95933-7_71

1 Introduction

1.1 Background

Scientific and technological progress over the last decades has radically innovated medical practice, both in terms of treatments and diagnostic methods as well as in the fundamental contributions of digital technologies. Alongside technologies well-known to the general public, such as MRI, PET or gene sequencing, other techniques, perhaps less known, have profoundly modified diagnostic procedures in some fields of medicine. Take, for instance, the advent of immunophenotyping in haematology for the characterization of different leukaemic forms. In the diagnostic field, the role assumed by cytopathology has become increasingly important. For example, consider the importance of screening cervical tumours by means of the Pap test, consisting of microscopic observation of cells taken from the cervix stained using the Papanicolau method. The development of imaging diagnostics and modern endoscopes allows cellular sampling from previously unattainable lesions. Rhinocytology, a study of the cells of the nasal mucosa, has been assuming an increasingly important role in the field of otolaryngology over the last twenty years thanks to the efforts of one of the authors of this paper, Matteo Gelardi, the founder of the Italian Academy of Rhinocytology (AICNA), the first rhinocytology academy in the world. Gelardi began to organize master courses in rhinocytology in 2003, and since then, he has held 60 basic master courses in Italy, five in Europe (Madrid, Paris, London, Berlin and Lugano) and five advanced master courses, attended by approximately 1500 physicians. Rhinocytology is a diagnostic technique characterized by its simplicity and low invasiveness, qualities that allow much greater diffusion [1]. This method enables evaluation of cellular behaviour in the most varied conditions taking into account that the nasal mucosa is in direct contact with the external environment and is subjected to various physical, chemical, and biological agents (viruses, bacteria, fungi). Recently, with the introduction of scanning systems for cytological preparations, such as Optika Optiscan 10, the Pathoscan scanner, the Metafer platform or the D-SIGHT platform, and the development of new image formats such as jpeg2000, it has become possible to analyse an entire preparation using automatic image analysis techniques.

1.2 Cells of the Nasal Mucosa

The nasal mucosa cells belonging to a cytotype possess some elements with high similarity; however, each cytotype appears very different from all the others: these features of the nasal mucosa cells allow their automatic classification. To sufficiently understand this work, a brief description of the appearance of cells found in the nasal mucosa smear is described below, and corresponding sample images are shown in Table 1.

- *Ciliated*: among the most common cytotypes of the nasal mucosa are ciliated cells. They possess a polygonal shape and a nucleus situated at various heights from the basement membrane. At 400x magnification, the apical region, the seat of the ciliary apparatus, is recognized as a well-represented body that includes a large part of the cytoplasm and the nucleus. Discrimination of the basal region is also possible.

- *Muciparous*: the muciparous cell is in the shape of a cup and is a unicellular gland. The nucleus is always situated in the basal position, while the vacuoles, containing mucinous granules, are located above the nucleus, giving the mature cell its characteristic chalice shape.
- *Basal*: the basal cell adheres to the "basal membrane", but unlike the other cells, it does not reach the superficial portion of the airways. It is smaller than other cells.
- *Striated*: a type of columnar cell having a nucleus positioned towards the lower pole. It contains microfilaments forming decrypts in the cytoplasm next to rough endoplasmic reticulum cisterns and abundant glycogen granules.
- *Neutrophil*: characterized by a polylobate nucleus, whose lobes are joined by very thin strands of nuclear material within the cytoplasm, which contains finely coloured granules
- *Eosinophil*: usually has a bilobed nucleus and acidophilous granules that intensely stain with eosin (hence the name) as an orange-red colour.
- *Mast cell*: a granulocyte with an oval nucleus covered in purple basophil granules. The metachromatic colouring of these granules almost completely masks the nucleus
- *Lymphocyte*: a nucleus that occupies most of the cell with a thin cytoplasmic rim of "light blue" colour surrounding the nucleus.
- *Metaplastic*: defined as a cellular element that does not have specific characteristics of the cytotypes that normally constitute the nasal mucosa. Usually, the expression of a chronic inflammatory process, both infectious and flogistic.

Table 1. Cells of the nasal mucosa.

Epithelial cells				
 Ciliated	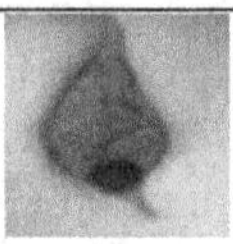 Muciparous	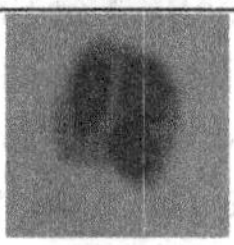 Basal	Striated	
Cells of the immunoflogosis				
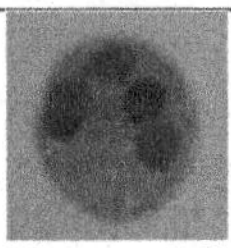 Neutrophil	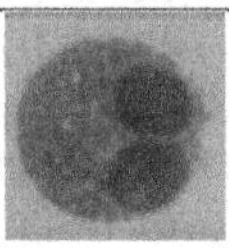 Eosinophil	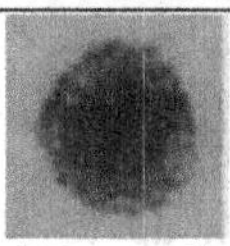 Mast cell	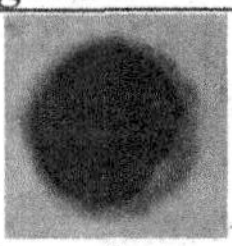 Lymphocyte	Metaplastic

1.3 Related Works

This is the first study with the aim of automatically classifying the cells present in the nasal mucosa. Almost all studies on automatic classification of cellular elements are in the field of haematology. In 2004, Piuri and Scotti used microscopic images to classify white blood cells [2]. In the literature, many papers treat specific aspects of the WBC automatic classification problem. In some studies [3, 4], for example, only segmentation aspects are discussed. Bevilacqua et al. [5] obtained over 95% accuracy in the classification of cytotypes present in a haematological smear of a healthy subject starting from digital scans of haematological preparations using neural network-based classifiers. Unlike in a haematological smear, where the white blood cells to be classified appear in almost all cases as isolated from each other, in the smears of nasal mucosa the cells appear in most cases as massively amassed. Blood cells circulate within the blood vessels, which can be considered a closed system. In contrast, the nasal mucosa is not only in direct contact with the external environment, but since a healthy adult person breathes approximately 8 litres of air per minute, the nasal epithelium comes into contact with powders, pollen, and bacterial/fungal spores normally present in the air. In fact, among the tasks of the nasal cavities is to prevent toxic and pathogenic agents from reaching the lungs, which is accomplished thanks to the mucociliary system of the nasal mucosa that holds and remove these elements from the respiratory tract.

2 Materials and Methods

2.1 Sample Collection and Preparation

The cytological technique includes the following steps: withdrawal (also called sampling), processing (which includes fixation and colouring), and microscopic observation. Cytological sampling consists of the collection of superficial cells from the nasal mucosa, which can be performed either with the aid of a sterile swab or by use of a small curette (scraping) in disposable plastic material e.g., nasal scraping. Sampling should be performed at the middle portion of the inferior turbinate, home to the correct ratio between ciliate and muciparous cells. For this study, ten samples were taken at the Rhinology Clinic of the Otolaryngology Department of the University of Bari. After sampling, cellular material was spread over a microscopic slide, fixed by drying in the air and coloured according to the method of May Grunwald-Giemsa (MGG), in which three dyes are used: eosin red-orange, methylene blue, blue and azur II of grey-blue colour. Using this method, all cellular components of the nasal mucosa, immunoflogosis cells, bacteria, fungal spores and fungal hyphae are stained. This colouring technique requires approximately 30 min. For each slide, the rhinocytogram was analysed as usual by observing 50 fields under an optical microscope provided with a lens capable of magnification up to 1000x to evaluate the number of cell populations present and identify cellular elements important for diagnosis.

2.2 Acquisition and Scanning

Each smear was digitally acquired using the microscope D-Sight 200 (www.menarinidiagnostics.it) available at the laboratory of Pathological Anatomy of the Institute of Hospitalization and Care at Scientific Character *Giovanni Paolo II* in Bari. This system allows selection of lenses with different magnifications, up to a maximum of 400x, focusing mode, fast or precise, and for each glass, the areas to scan.

Some attempts were made to evaluate scanning times and characteristics of scanned images with different magnifications and focus modes. We chose the maximum magnification allowed by the system, equal to 400x, with a fast focus mode, selecting the areas within each slide where an adequate number of not excessively packed cells were present in order to facilitate recognition of elements present over a reasonable period of scanning and subsequent processing. The image file size and the acquisition time depend on the amount of area selected with the D-Sight acquisition interface. The digital acquisition of the rhinologic smears may be affected by a non-uniform illumination due to the irregularity of the rhinologic smears in term of thickness. Since the acquisition was performed with the automatic focus in order to speed up the acquisition phase some areas are not perfectly focused.

2.3 Software

The system was implemented using *Python* [6] programming language because many libraries are available to facilitate matrix operations and others for the implementation of machine learning algorithms. Regarding the IDE, *Pycharm* [7] was used as an integrated development environment used specifically for Python. Among the libraries available in Python we used the following:

- The *os* library was used to choose images to process and for storing tiles.
- *GDAL* was primarily used to manage the images in JPEG 2000 format and to divide images into smaller regions. Each region was elaborated through the methods offered by the Pillow library in order to improve image colour.
- *OpenCV* is a cross-platform open-source software library used in the field of computer vision. It was developed in the C++ language but can interface with C, Python and Java. Resizing and conversion functions in the RGB model were used here.
- The *Keras* [8] library consists of a set of APIs to implement neural networks. It runs on other libraries such as TensorFlow, CNTK or Theano. *Keras* was developed to ensure faster experimentation with neural networks. Furthermore, it is focused on being user-friendly, modular and extensible as both convolutional and recurrent neural networks. In addition, it allows *data augmentation* operations. The data structure that represents the core of this library is the model, representing the way in which the various layers of a neural network are organized. In this study, we use the sequential model.
- *Hyperas* [9] is an open source library used for the optimization of hyperparameters in a model created with Keras. In particular, it uses the syntax provided by the hyperopt library and allows a choice of different options for hyperparameters. Therefore, the optimization algorithm evaluates the model realized with Keras through the different indicated hyperparameters, returning the best combination thereof to obtain the best results from the model.
- *Scipy* [10] is an open-source library for the Python programming language primarily dedicated to scientific computing. The packages used in this work are as follows:

- *NumPy*: the main package for scientific calculation in Python. In particular, it provides many functions for operations between matrices. It was useful for making comparisons in cell identification.
- *Scikits* [11]: scikit-image is a collection of algorithms used for image processing in Python. The scikit-learn sub-package, which provides machine learning support algorithms, was used in this work. Specifically, functions for calculating statistic metrics were used.

- *Matplotlib* is a library for creating charts and was used for creating comparison graphs of loss function and accuracy for the training and validation phase.
- *Time* and *datetime* were used to track the time to complete the entire first phase.

3 System Description

In this section image extraction from the jp2 files and their classification among nasal cytotypes will be described. We conclude this section showing how a specialist can verify the classification results.

3.1 File Processing and Cell Image Extraction

Jpeg files obtained from scanning each preparation were approximately 500 MB, and colours were not sufficiently contrasted, then it was necessary to divide the original file into smaller files and balance the original colours.

Subdivision in Tiles and Colour Processing. The whole digital image was divided into smaller regions, called tiles. In terms of pixels, all tiles have the same size. Many attempts were made to identify an optimal dimension to speed up creating and processing each tile. Ultimately, a partitioning in tiles of 1500 × 1500 pixels proved to be a fair compromise. For each scan from the original file in jpeg2000 format, approximately 500 files in png format were generated, each having sizes in the range of 0.7–3.5 Mb. In the original images, colours were not perfectly differentiated; therefore, it was necessary to elaborate them using the *PIL.ImageEnhance.Color* class of the Pillow library, which balances image colours. The improvement factor 3.0t was adopted.

Finding ROIs and Extracting Cell Images. The Otsu algorithm was used to separate coloured areas from the background, applying the thresholding algorithm on the red channel of the RGB, using the colour of cells from eosin, one of the components of the MGG stain used for slide preparation. Morphological opening operations were then applied to separate weakly connected regions, followed by labelling, marking the different 'objects' with different shades of colour to facilitate subsequent classification.

Among previously labelled regions having dimensions greater than or equal to 200 pixels, the pixels belonging to the edges were obtained and then all the operations previously carried out were repeated to prevent parts of the same cell from being identified with different labels (Fig. 1).

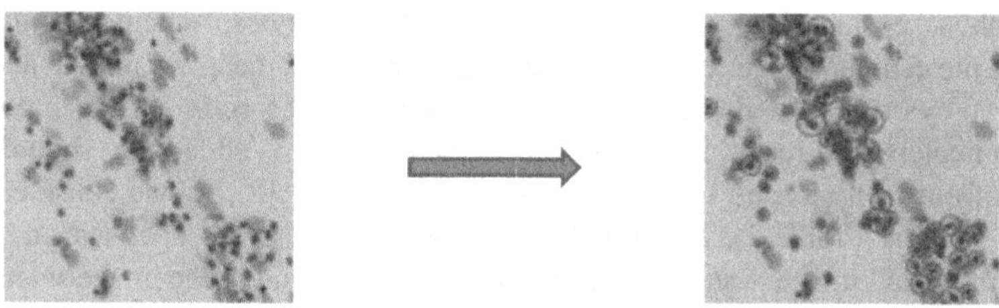

Fig. 1. Cell extraction. On the left, the original image; on the right the elements, recognized as cells, are surrounded by red circles (Color figure online)

3.2 Classification

The challenge of image classification includes assigning a label to an input image that corresponds to one of the categories of a pre-established set. Specifically, to determine the category of membership of an image, the probability that it belongs to each of the prefixed categories is calculated. The label of the most likely category will be assigned to the image. A convoluted neural network (CNN or ConvNet) consists of layers containing neurons that can learn weights. Weights are key parameters in this type of classification because they express how much a given "feature" can influence the classification of a given object.

CNN Optimization Techniques. To improve CNN accuracy, two different techniques were used. The first one consists of increasing the size of the dataset, starting from pre-existing images i.e., image augmentation that increases the dataset and reduces over-fitting. Image augmentation consists of geometric transformations such as reflection, rotation, translation or cutting parts of an image. The second method is called hyperparameter-optimization and evaluates the neural network on different hyperparameters to choose the model that yields the highest possible precision. A data driven approach was used to design the classifier. We selected images of different cell types from the scanned slides and labelled them to create a training set that was passed on to the classifier. The training phase was followed by the validation phase (on the validation set) and then the test was run on a new set of images (test set) by the classifier to verify the correct classification rate.

Implementation. The project was organized into three primary modules, independent of each other, and related to the following:

- implementation of the CNN, including the training and testing phases
- implementation of image augmentation operations
- implementation of the CNN hyper-parameters optimization algorithm.

Five of the ten scanned preparations exhibiting superior coloration, with better focusing and with greater numbers of cellular elements from the different cytotypes were considered. From these, the images to be assigned to the different classes were selected to create the training-set and the test-set and then to design the classification algorithm. Basal and striated cells, normally present in the mucosa, were excluded as being not useful for diagnostic purposes. The cytotypes considered are metaplastic, neutrophils, eosinophils, ciliated, mast cells, lymphocytes, muciparous (goblets). Among the cellular images extracted from a nasal mucosa smear, we find, in addition to the seven cytotypes listed above, elements not representing cells, such as colour spots or bare nuclei. As a result, a further class named *other* was added that should contain objects from the extracted images that do not represent cells. So, we considered eight classes in total (Table 2). Because the implemented classifier operates on 28 × 28 pixel images, all images were resized. The dataset, at the end of the image selection and augmentation steps, consisted of 8000 images, evenly distributed among the eight classes. Randomly, 80% were assigned to the training-set and the 20% the test-set for each class.

Table 2. Examples of elements from the different classes of the dataset

Metaplastic	*Mast cell*
Neutrophil	*Lymphocyte*
Eosinophil	*Muciparous*
Ciliated	*Others*

Description of the Classification Algorithm. Starting with the CNN model, consisting of two blocks of primary layers proposed by Agarap [12], we implemented a CNN based on three blocks. We defined the eight categories to be classified, giving a label to each category. Table 3 shows the classification model, consisting of 3 blocks of main layers, each one containing the following sequence: convolution, activation, max-pooling, and dropout.

In each block, a convolution operation was performed using 5 × 5 pixel filters in order to extract low-level, middle-level and high-level features, followed by use of the 'ReLu' activation function for the introduction of non-linearity. Subsequently, max-pooling operations were applied using a 2 × 2 filter to reduce computational effort during the training phase and dropout with a probability equal to 0.25 to reduce overfitting. The number of filters used in the convolution operation was 32 in the first block, 64 in the second and 128 in the third. This resulted in a 3-D activation map that was converted to a single dimension because the fully connected layer needs single feature vectors to be used for the final classification. A fully connected layer of 512 neurons was added, followed by the application of batch normalization to speed up training time. Then, we applied a dropout with a probability of 0.7 and a further fully

Table 3. On the left is the CNN model, consisting of two blocks of main layers proposed by Agarap. On the right is the three block model used in this study.

Agarap model	3-block model
INPUT: 32 × 32 × 1	INPUT: 28 × 28 × 1
CONV: 5 × 5 size, 32 filters	**CONV: 5 × 5 size, 32 filters** + ReLU
ReLU: max(0, hθ(x))	Pool: 2 × 2 size
Pool: 2 × 2 size, 1 stride	**CONV: 5 × 5 size, 64 filters** + ReLU
CONV: 5 × 5 size, 64 filters	Pool: 2 × 2 size
ReLU: max(0, hθ(x))	**CONV: 5 × 5 size, 128 filters** + ReLU
Pool: 2 × 2 size	Pool: 2 × 2 size
Fully connected	Fully connected
Dropout	Dropout
Softmax classifier	Softmax classifier

connected layer consisting of a number of neurons equal to the number of categories, in this case eight. Finally, the classifier Softmax was added, which returns a probability index for each category. The classification model was configured by setting the loss function, the optimization algorithm and the type of metric used to evaluate the classifier. To calculate the performance of the neural network, we compared the labels that were predicted by the model with those associated with the images of the training set used, using as a loss function for the classification of several classes the categorical cross entropy. Optimization was performed by applying the *Adam* algorithm, considered state of the art. After configuring the model to perform the training, the following parameters were set: batch size (64), which is the number of images presented to the classifier during each iteration, and number of iterations (60) to be performed to learn the classifier. Twenty percent of the training set was used for the evaluation of the run-time classifier to understand during the training whether this phase was performed correctly and whether there was overfitting. The model and its relative weights, after being saved, were used to classify images from the test-set, and based on this classification, an assessment of the model was made by calculating its precision relative to both the training and the test phase.

3.3 Visualization of Classified Cells

The system we designed aims to provide support to doctor who is responsible for assigning final classification and formulation of the diagnosis. As such, it was necessary to design a 'check panel' allowing the doctor to verify classified cells much faster than normal. This panel allows the doctor to select the slide to view and check each class of cells by choosing various enlargements. He can now confirm the correct classification or move elements to different classes, if necessary (see Fig. 2).

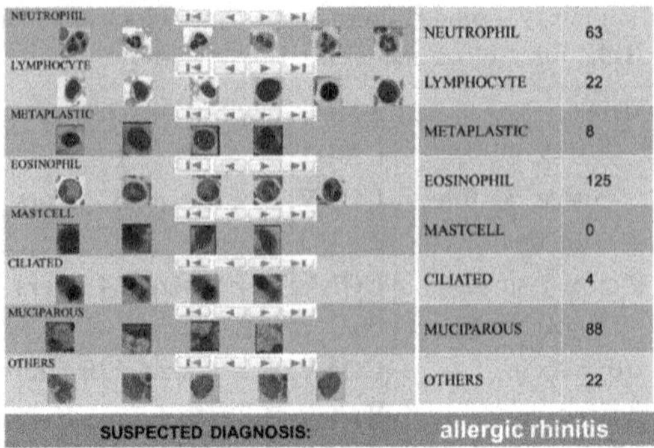

Fig. 2. Supervised classification review

4 Results

4.1 Image Extraction

To estimate the efficacy of the system presented herein to correctly select cells, we chose five tiles from each of the ten slides that were scanned (50 tiles containing 2351 'objects'). Then, we determined how many elements were correctly identified as cells, defined as coloured areas enclosed by an outline, among all other 'objects'.

Figure 3 shows, within the circles, some examples of elements identified by the system as cells. The rounded rectangles, which were subsequently drawn manually, indicate other "objects" present. The red arrow indicates a stain incorrectly classified as a cell. We considered as true positive elements correctly classified as cells, whereas true negative elements, 'not cells', were those that the system correctly discarded. False positive elements were 'not cells' that the system identified as cells, and false negative elements included cells that the system incorrectly discarded (Table 4). Sensitivity = TP/(TP + FN) was equal to 96.4%, specificity = TN/(FP + TN) was 82.5%, and accuracy = (TP + TN)/(TP + TN + FP + FN) was 94.0%. Overall, we think this is a satisfactory result.

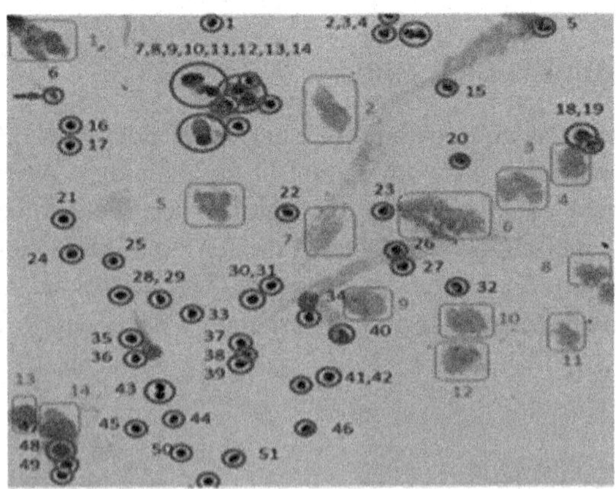

Fig. 3. 'Objects' identified by the system (Color figure online)

Table 4. Confusion matrix of cell image extraction

Hypothesis	Positive	Negative
True	1870	340
False	72	69

Table 5. Results of the test-set cell classification

Predicted class									
		Metaplastic	Neutrophil	Eosinophil	Ciliated	Mast cell	Lymphocyte	Goblet cell	Others
True class	Metaplastic	205	0	0	2	2	1	0	0
	Neutrophil	0	195	0	0	0	6	0	0
	Eosinophil	0	0	204	0	0	0	0	0
	Ciliated	2	0	0	186	3	0	0	0
	Mast cell	0	0	0	0	202	0	0	0
	Lymphocyte	0	1	0	1	0	189	0	0
	Goblet cell	0	0	0	0	0	0	203	0
	Others	5	0	0	1	0	0	1	191

4.2 Cell Classification

As we previously described, the classification algorithm was tested on 1600 cell images of the test-set. The confusion matrix of those results is shown in Table 5. As mentioned above, in this case, only some of the images acquired were considered, according to the characteristics previously specified. Sensitivity, specificity and accuracy were calculated for each cell type and are reported in Table 6. We have considered true positive cells belonging to a class that the classifier has correctly labelled, true negative cells not classified for a specific cytotype and not belonging to it, false positive cells not belonging to a certain cytotype but classified as if they belonged to it, and false negative cells belonging to a cytotype but classified as belonging to others.

Table 6. Sensitivity, specificity and accuracy

	Sensitivity	Specificity	Accuracy
Metaplastic	0,97	0,99	0,99
Neutrophil	0,97	1,00	0,99
Eosinophil	1,00	1,00	1,00
Ciliated	0,97	0,99	0,99
Mast cell	1,00	0,99	0,99
Lymphocytes	0,99	0,99	0,99
Goblet	1,00	1,00	1,00
Others	0,96	1,00	0,99

5 Conclusion and Future Work

Our results are satisfactory, and the system presented herein could be useful in cases where many patients need to be evaluated. One of the most important aspects of rhinocytology is the possibility that it offers the specialist an opportunity to make a correct differential diagnosis by means of a low-cost analysis without having to send their patient to a laboratory for further testing. To further improve our system, we could give the specialist the ability to quickly choose the ROIs from which to extract cells: the system could allow the specialist to photograph a certain number of microscopic fields (typically 50 or 100) that they consider useful and subsequently allow the software to classify and count the cells recognized in the selected fields. With respect to automating the slide scanning, there are at least two goals: the first one involves moving from the current semi-quantitative estimation to a quantitative one, which is more precise and valuable on a scientific level for standardization; second, cataloguing cellular elements will require less time. These changes may help in the dissemination and use of nasal cytology, a diagnostic investigation that is rarely used and about which little is known.

Acknowledgements. We thank the Dr. Alfredo Zito, head of the Department of Pathological Anatomy of I.R.C.C.S., for making the D-Sight available to scan the preparations used.

References

1. Gelardi, M.: Atlas of Nasal Cytology for the Differential Diagnosis of Nasal Diseases. Edi. Ermes, Milano (2012)
2. Piuri, V., Scotti, F.: Morphological classification of blood leucocytes by microscope images. In: 2004 IEEE International Conference on Computational Intelligence for Measurement Systems and Applications, CIMSA 2004, pp. 103–108 (2004)
3. Qiao, G., Zong, G., Sun, M., Wang, J.: Automatic neutrophil nucleus lobe counting based on graph representation of region skeleton. Cytom. Part A **81**(9), 734–742 (2012)
4. Li, Q., Wang, Y., Liu, H., Wang, J., Guo, F.: A combined spatial-spectral method for automated white blood cells segmentation. Opt. Laser Technol. **54**, 225–231 (2013)
5. Bevilacqua, V., Buongiorno, D., Carlucci, P., Giglio, F., Tattoli, G., Guarini, A., Sgherza, N., De Tullio, G., Minoia, C., Scattone, A., Simone, G., Girardi, F., Zito, A., Gesualdo, L.: A supervised CAD to support telemedicine in hematology. In: Proceedings of the International Joint Conference on Neural Networks (2015)
6. Python 3.6.5: https://docs.python.org/3. Accessed 03 May 2018
7. Pycharm: https://www.jetbrains.com/pycharm/documentation/. Accessed 03 May 2018
8. Keras: https://keras.io/. Accessed 03 May 2018
9. Hyperas: https://github.com/maxpumperla/hyperas. Accessed 03 May 2018
10. Scipy: https://www.scipy.org/docs.html. Accessed 03 May 2018
11. van der Walt, S., et al.: Scikit-image: image processing in python. PeerJ **2**, e453 (2014)
12. Agarap, A.F.: An architecture combining convolutional neural network (CNN) and support vector machine (SVM) for image classification. arXiv1712.03541 (2017)

Recovering Segmentation Errors in Handwriting Recognition Systems

Claudio De Stefano[1], Francesco Fontanella[1(✉)], Angelo Marcelli[2], Antonio Parziale[2], and Alessandra Scotto di Freca[1]

[1] Cassino and Southern Lazio University, Via G. Di Biasio 43, 03043 Cassino, FR, Italy
{destefano, fontanella, a.scotto}@unicas.it

[2] Salerno University, Via Giovanni Paolo II 132, Fisciano, SA, Italy
{amarcelli, aparziale}@unicas.it

Abstract. Most handwriting recognition systems need a mechanism for handling classification errors. These errors are typically caused by the large shape variability of the handwriting produced by different writers and by the segmentation errors, which occur when the word recognition process is performed by extracting and classifying single characters. In this paper, in order to reduce the segmentation errors, we propose a hierarchical recognition system composed of two classification modules. The first one discriminates isolated characters from cursive fragments using specifically devised features. The second one is an OCR engine that receives as input only those samples classified as isolated characters in the previous module. The whole system works like a highly reliable OCR that rejects most of the cursive fragments avoiding their incorrect classification. The experimental results confirmed the effectiveness of the proposed system.

Keywords: Handwriting segmentation · Feature extraction
Feature evaluation · Reject option

1 Introduction

Handwritten word recognition is a crucial aspect of any application involving document image processing. In this framework, the approaches proposed in the literature can be subdivided in two classes: *analytical* and *holistic*. The approaches belonging to the former class treat words as a collection of characters and proceed by segmenting the word into characters and then classifying each single characters. On the contrary, holistic approaches treat words as single, indivisible entities, attempting to recognize them based on features characterizing the word as a whole: thus, there are as many classes as the number of words in the considered dictionary. Analytical approaches have been successfully applied to machine-printed and hand-printed words with large size dictionary, but their performance significantly decreases when applied to cursive words: in these cases, in fact, it becomes difficult to correctly segment the ink into parts corresponding to isolated characters [17]. Holistic approaches, instead, have been successfully exploited when the size of the dictionary, i.e. the number of classes, is

D.-S. Huang et al. (Eds.): ICIC 2018, LNCS 10955, pp. 631–642, 2018.
https://doi.org/10.1007/978-3-319-95933-7_72

limited to a few hundreds, but the recognition rate sensibly decreases as the dictionary size increases [17].

Segmentation errors may also be very relevant in applications in which isolated characters are produced, such as for instance in handwritten form processing. In these cases, in fact, character images are extracted from the whole image by a segmentation algorithm that exploits layout information on where the characters should be within the form layout. Unfortunately, also in presence of boxed forms, writers often connect adjacent characters or write outside the boxes: the effect is that the images processed by the OCR engine do not necessarily correspond to isolated characters, reducing classification results. For this reason, many real world OCR systems requires operator's intervention, which are obviously very expensive. To limit the number interventions, these systems usually consider classification schemes allowing the implementation of a reject option: such schemes provide a measure of the classification reliability, generally based on an estimate of the probability that an input sample belongs to the class it has been assigned to.

However, implementing a reject option is not an easy task and many solutions have been proposed in the literature [5–7, 11, 12, 15]. In [2] the author proved that if the conditional probability densities of the classes at hand are known, then it is possible to find the optimal trade-off between the error and rejection rate. However, in real world problems, these densities are unknown and can be only estimated by using the available data [3, 6]. Therefore, the reject option is implemented by using the reliability measures provided by the classifiers, and by heuristically setting a reject threshold. Once the threshold value has been set, then unknown samples are rejected when their class probabilities are below such value.

In this framework, our proposal is that of designing a hierarchical handwriting recognition system consisting of two modules, each using a different set of features. The first module aims at rejecting the samples corresponding to segmentation errors and uses a set of features specifically devised for distinguishing isolated characters from cursive handwriting. As for the second module, it implements an OCR, which receives as input only the samples classified as isolated characters in the first module and performs the final classification. In this case, features able to discriminate patterns corresponding to the uppercase and lowercase letters of the English alphabet have been used. It is worth noticing that our approach does not require the use of any classification reliability measure, or the setting of a threshold value: the reject option is implemented by simply rejecting all the samples classified as cursive handwriting in the first step.

The experimental results, obtained by using data extracted from the RIMES database, confirmed the effectiveness of our approach, showing that it is possible to reject most of the samples corresponding to cursive fragments with a small reduction of the classification rate on samples representing isolated characters.

The remainder of the paper is organized as follows: Sect. 2 illustrates the proposed method, Sect. 3 describes the feature extraction process, while in Sect. 4 the criteria for evaluating the effectiveness of the considered features are critically discussed. Finally, Sect. 5 presents the experimental results, while Sect. 6 discusses the experimental findings.

2 The Proposed System

The proposed system is a hierarchical classifier that takes as input a previously segmented image and provides as output: (i) a character label, if the system recognizes the ink traces in the image as belonging to a single character; (ii) a reject, if the ink traces are classified by the system as a cursive fragment, containing more than a single character. The layout of the proposed system is shown in Fig. 1.

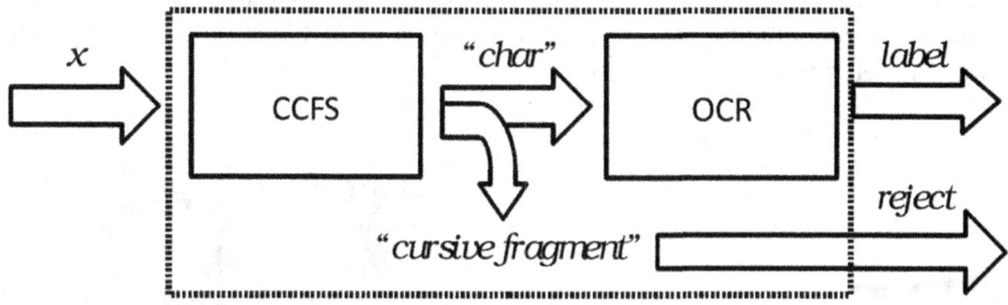

Fig. 1. The architecture of the proposed system.

The first classifier, denoted as *Cursive/Character Filtering System* (CCFS in the following), performs a binary classification, and it is able to distinguish isolated characters from cursive fragments. In the first case, they are sent to the next stage, which implements an OCR module, in the second case they are rejected. The CCFS extracts 18 features, specifically devised for discriminating between cursive fragments and isolated character, as it will be described in Sect. 3. As concerns the classification scheme to be adopted for the CCFS, we have chosen a Multilayer perceptron.

The OCR module has been implemented by adopting the widely used feature set described in [14]. According to this approach, each image is divided into six parts and it is described by a feature vector containing measures associated to the different parts. For each part, 22 features are computed, totaling 132 features: 13 of these features represent concavity measures, while the remaining ones are related to other contour information. As regards the classifier, we opted for the Random Forest (RF) [1], since it has proven to be very performing in dealing with classification problems with a large number of classes [4], as it is the case of our problem.

3 CCFS Module: Feature Extraction

The aim of the feature extraction process in the CCFS module is that of allowing the classification of connected components of ink traces, possibly produced by writers without lifting the pen, in two main classes: isolated characters and cursive. The basic idea behind our approach has been derived from studies on handwriting generation, according to which handwriting results form time superimposition of strokes, each one corresponding to a simple shape. Thus, simpler shapes are generated by simpler motor program, producing smaller quantity of ink in the trace.

Moreover, in order to improve the fluency of handwriting, a writer may introduce extra strokes, or ligatures, to connect the last stroke of a character and the first of the following one, instead of lifting the pen between the final point of the former and the initial point of the latter. Accordingly, we expect that images of isolated characters will contain less ink (and less strokes) than those of cursive and the ink will not span prevalently along the writing direction.

In order to estimate the characteristics of the connected components of ink traces, we proceeded as follows. The word image is processed for extracting the bounding box of each connected component (see Fig. 2a). Then, each component is analyzed by considering its size, the number and the distribution of its black pixels and the size of the word it belongs to (see Fig. 2b).

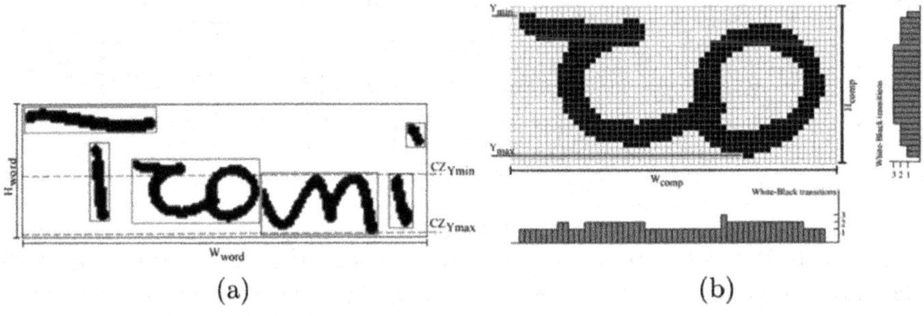

Fig. 2. (a) The image of word "Trani" with the bounding box of each connected component. (b) A connected component extracted from the word image.

In particular, for each component, we consider the coordinates of the top-left and bottom right vertices of the bounding box $(X_{\min}, Y_{\min}, X_{\max}, Y_{\max})$, the width and the height of the bounding box $\left(W_{comp}, H_{comp}\right)$, the total number of pixels and the number of black pixels included in the bounding box $\left(P_{comp}, BP_{comp}\right)$, the width and the height of the bounding box of the word $\left(W_{word}, H_{word}\right)$. Starting from these basic features, we computed an additional set of features, namely the *height ratio HR*, the *aspect ratio AR*, the *proportional aspect ratio PAR* and the *fill factor FF*. These features are summarized in Table 1. Note that the features *HR*, *AR* and *PAR* are meant to capture the space, and hence the temporal, extension of the handwriting, whereas *FF* is meant to capture the spatial density of the ink.

Table 1. Description of additional features

HR	AR	PAR	FF
$HR = \frac{H_{comp}}{H_{word}}$	$AR = \frac{W_{comp}}{H_{comp}}$	$PAR = \frac{W_{comp}}{H_{word}}$	$FF = \frac{P_{comp}}{BP_{comp}}$

Moreover, in order to evaluate the shape complexity of the ink trace, we have considered the number of transitions between white and black pixels along consecutive rows/columns of the component. These values have been arranged in two histograms, namely ink-mark on the horizontal and vertical axis, where each bin represents the above number of transitions along a row or a column, respectively (see Fig. 2b). We have used as features, the maximum number of transitions obtained on each histogram (say IM_x and IM_y, respectively). Note that, transitions corresponding to sequences of black pixels lower than a given threshold θ, are assumed as noise and are discarded. These features can be seen as a measurement of the complexity of the ink: an empty or flat ink-mark on both horizontal and vertical axis suggests that the component presents scattered black pixels and it is likely to be noise, whereas higher values correspond to shapes that are more complex.

Finally, we have estimated the center-zone of the word and we have used this information to discriminate characters or cursive from the other possible classes (like dots, vertical and horizontal dashes, etc.), since characters and cursive are located in the center-zone of a word, even if they can extend beyond this area. $CZ_{Y\min}$ and $CZ_{Y\max}$ are the y-coordinates of the upper and lower side of the center-zone respectively (see Fig. 2a).

4 Features Analysis

In order to identify the set of features having the highest discriminant power, we have used five standard univariate measures that evaluate the goodness of each single feature. The univariate measures evaluate the effectiveness of the features at hand in discriminating samples belonging to different classes. Each univariate measure can be used to rank all the available features. Starting from the ranking, a subset of N features can be selected choosing the first N features of the ranking.

In our study, we have considered the following univariate measures: *Chi-square* [16], *Relief* [13], *Gain ratio*, *Information Gain* and *Symmetrical uncertainty* [10].The Chi-Square measure estimates feature merit by using a discretization algorithm: if a feature can be discretized to a single value, it has not discriminative power and the it can safely be removed from the data. The discretization algorithm, adopts a supervised heuristic method based on the χ^2 statistic. The range of values for each feature is initially discretized by considering a certain number of intervals (heuristically determined). Then, the χ^2 statistic is used to determine if the relative frequencies of the classes in adjacent intervals are similar enough to justify the merging of such intervals. The second considered measure is the Relief, which uses an instance based learning approach to assign a relevance weight to each feature. The assigned weights reflect the feature ability to distinguish among the different classes at hand. The algorithm works by randomly sampling instances from the training data. For each sampled instance, the nearest instance of the same class (nearest hit) and that of the different class (nearest miss) are found. A feature weight is updated considering how well its values distinguish the sample distance from its nearest hit and nearest miss. A feature will receive a high weight if it differentiates between instances from different classes and has the same value for instances of the same class.

The last three considered univariate measures are based on the well-known information-theory concept of entropy *H(X)*, which is an estimate of the uncertainty of the random discrete variable *X*. Entropy can be used to define the conditional entropy *H (X|Y)* of two random discrete variables *X* and *Y* and it represents the amount of randomness of *X* when the value of *Y* is known. These quantities can be used to estimate the usefulness of a feature *X* to predict the class *C* of unknown samples. More specifically, such quantities can be used to define the *information gain* (*I_G*) concept [18]:

$$I_G = H(C) - H(C \mid X) \tag{1}$$

I_G represents the amount by which the entropy of *C* decreases when *X* is given, and reflects the additional information about *C* provided by the feature *X*.

The last three considered univariate measures uses the information gain defined in Eq. (1). The first one is the information Gain itself. The second one, called *Gain Ratio* (I_R), is defined as the ratio between the information gain and the entropy of the feature *X* to be evaluated:

$$I_R = \frac{I_G}{H(X)} \tag{2}$$

Finally, the third univariate measure taken into account, called *Symmetrical Uncertainty* (I_S), compensates for information gain bias toward attributes with more values and normalizes its value to the range [0, 1]:

$$I_S = \frac{I_G}{H(C) + H(X)} \tag{3}$$

5 Experimental Results

In order to ascertain the effectiveness of the proposed approach, we performed two sets of experiments. In the first one, we evaluated the features defined in Sect. 3 by using the univariate measures described in Sect. 4. In the second one, instead, we assessed the classification performances of the proposed system. The experiments have been performed on the results have been obtained by using data extracted from the standard RIMES database, which contains real world handwritten words and is a publicly available.

The goal of our experiments was to evaluate both the ability of our system in rejecting samples corresponding to segmentation errors, and the global effect of such errors on classification results. To this aim, we processed 4047 word images to extract sub-images containing connected components of ink. These words were then shown to six human experts, which classified manually the connected components previously extracted. At the end of this process, 9869 samples were manually classified, 5101 of them as cursive fragments corresponding to segmentation errors, and 4768 as isolated characters. A further label has been added to each sample, depending on its meaning: the first one, provided by the human experts, associates the sample to its corresponding letter in the alphabet, distinguishing in uppercase and lowercase, while the second one specifies whether the sample is an isolated character or a trace of cursive.

This latter labeling was used for setting up the filtering module CCFS of Fig. 1, while the former was used for the OCR module. Notice that, as mentioned in Sect. 2, each sample image is described by two feature vectors: the first one contains the 18 features used by the first module to discriminate between cursive fragments and isolated characters; the second one consists of the 132 features used by the OCR module. The following subsections detail the two sets of experiments performed.

5.1 Feature Evaluation

To validate the devised features, we analyzed them by using the univariate measures detailed in Sect. 4. In order to better understand the results provided by the different rankings, we have also computed a global score for each feature, according to the following equation derived from the Borda count rule:

$$W_i = M \cdot N_F - \sum_{j=1}^{M} pos_{ij} \tag{4}$$

where W_i is the overall score of the i-th feature, while pos_{ij} is the position of the i-th feature in the j-th ranking. M is the total number of rankings to merge, while N_F is the total number of features to evaluate: in our case $M = 5$ and $N_F = 18$. The obtained global scores are shown in the histogram of Fig. 3.

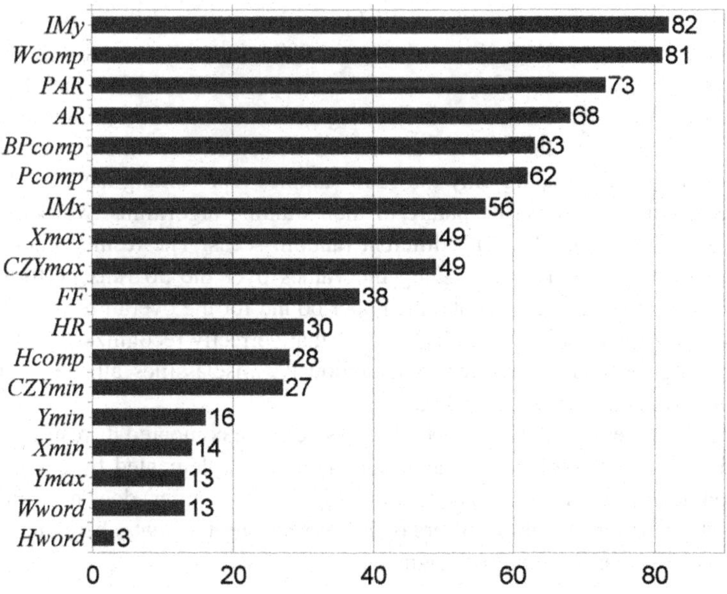

Fig. 3. Features Overall Ranking.

A similar approach has been already used in [8]. Note that to perform this analysis we used the whole database of 9869 samples. Looking at the histogram, it can be noted that the most discriminating feature is IM_y. Such a feature is related to the shape complexity of the ink trace along the vertical axes. This confirms the basic idea that simple shapes are generated by simple motor programs, and that an isolated character typically has a shape simpler than that of a cursive trace. As concerns the next three most discriminating features, they measure the component width $\left(W_{comp}\right)$, and the aspect ratio of the component width with respect to the height of the whole word (*PAR*) and the component height (*AR*). These results confirm the importance of comparing the component size with respect to the word size in discriminating between isolated characters and cursive fragments. In the fifth and in the sixth position we find two features related to pixel distribution within the bounding box of the component. Finally, lower positions of the histogram are occupied by the coordinates of the bounding box of the component (except for $X_{\max}$) and by the height and the width of the whole word bounding box.

5.2 Classification Performance Evaluation

This set of experiments was aimed at evaluating the performance of the whole system shown in Fig. 1. As mentioned above, for the first stage of our system we adopted a Multilayer Perceptron (MLP in the following), which takes as input the features defined in Sect. 3. As concerns the OCR module, we used a Random Forest (RF). We considered the RF classifier because it has shown to be very performing in dealing with classification problems with a large number of classes [4], as is the case of our problem, as the samples have been labeled in such a way to distinguish between uppercase and lowercase letters (52 classes).

In this set of experiments, the available samples were split in two statistically independent sets: a training set consisting of 6878 samples (namely, 3313 isolated characters and 3565 cursive fragments) and a test set made of 2989 samples (namely 1457 isolated characters and 1532 cursive fragments). The MLP classifier in the CCFS module was trained on the whole training set, while the RF classifier in the second module was trained by using only the 3313 samples representing isolated characters.

To deal with the stochastic nature of their training algorithms, we performed 30 runs for both MLP and RF, with different random seeds. The results reported in the following were obtained by averaging the values over the 30 runs performed. It is worth noticing that, with these data, the upper bound for the overall recognition rate is equal to 48,7%, attainable by an ideal system that correctly recognizes all the samples corresponding to isolate characters but, obviously, misclassifies all the samples corresponding to cursive fragments (Fig. 4).

In order to assess the performance of our system, we compared its results with those obtained by two different single stage systems, both implemented by using the same OCR module developed by us. The first one ($System_1$, hereafter) does not consider any filtering stage or reject option, whereas the second one ($System_2$, hereafter) adopts a reject rule for discarding cursive fragments.

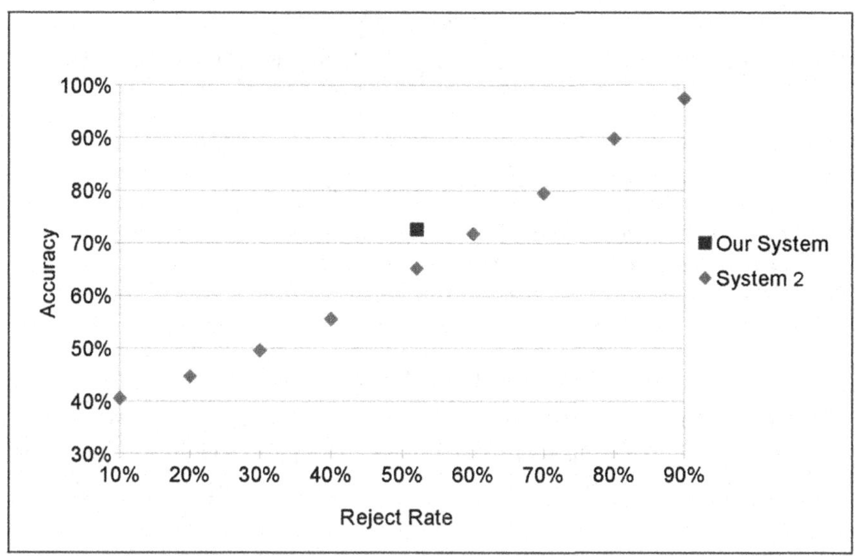

Fig. 4. The accuracy reject curve for *System*$_2$. The single squared point shows the accuracy and the reject rate achieved by our system.

Since, in the first system, the RF classifier receives as input samples representing both isolated characters and cursive fragments, it provides by definition wrong results for all cursive fragments. We considered such system because its results represent both an upper bound for the recognition rate on isolated characters and a lower bound for the accuracy.

As regards the second system, it exploits the information provided by the RF classifier for implementing a reject option and for computing the reject-accuracy curve. In fact, the RF provides as output both the class to which a sample has been assigned to, and the probability of that class given the input sample: this probability value can be used as a reliability assessment of the labeling given by the classifier, and for rejecting those samples having such a value below a given threshold. We used this reliability measure and the Chow's rule for computing the optimal reject-accuracy curve, following the approach discussed in [9]. The rationale of this system is that cursive fragments should be classified with a low reliability by the OCR: thus, it is expected that the majority of cursive fragments could be rejected without reducing too much the recognition rate on isolated characters. Obviously, the higher the threshold value, the higher the accuracy but also the lower the recognition rate on isolated characters.

The results obtained by using our approach are summarized in Tables 2, 3 and 4. Table 2 shows that the first stage is able to reject about 93\% of the cursive fragments, rejecting only about 9% of isolated characters. Table 3 shows that the OCR module correctly recognizes about 78% of isolated characters received by the first stage, but obviously provides wrong results on the cursive fragments also received by the first stage. Finally, Table 4 shows that the whole system allows us to correctly recognize more than 71% of isolated characters, with an accuracy equal to 72.6% and a reject rate

Table 2. Results of CCFS module.

# samples	Overall rec. rate	Char. accepted	Cursive rejected	Overall rej. rate
2989	91.7%	90.6%	92.8%	52.2%

Table 3. Results of the OCR engine in the second stage.

# samples	Overall rec. rate	Char. rec. rate	Char. err. rate	Cursive err. rate
1430	72.6%	78.6%	21.4%	100.0%

Table 4. Results of the whole system proposed by us.

# samples	Accuracy	Rec. rate	Error rate		Reject rate		Overall
		On char	On char	On cursive	On char	On cursive	Reject rate
2989	72.6%	71.2%	19.4%	7.2%	9.4%	92.8%	52.2%

Table 5. Classification results of $System_1$.

Total samples	Accuracy	Rec. rate	Error rate		Reject rate		Overall
		On char	On char	On cursive	On char	On cursive	Reject rate
2989	38.2%	78.4%	21.6%	100.0%	0.0%	0.0%	0.0%

equal to 52.2%. Note that more than 91% of rejected samples are cursive fragments, thus confirming the effectiveness of our approach.

The results provided by $System_1$ are reported in Table 5. The analysis of this data shows that this system correctly recognizes more than 78% of isolated characters but, as expected, the accuracy is very low (about 38%) since it is not able to discard samples representing cursive fragments.

6 Conclusion

We presented a novel approach for recovering character segmentation errors in handwriting recognition systems. The proposed system consists of two modules. The first one implements a two-class classifier, which distinguishes isolated characters from cursive fragments. This distinction is based on novel, specifically devised features, which measures the connected components of the ink traces. The second module, instead, is an OCR-based classifier, which takes as input only the patterns labeled as isolated characters by the first module. The second module performs the final classification assigning the processed samples to one of the 52 classes, taking into account the shape of the characters to be classified.

In the experiments, the performance of the proposed system has been compared with those achieved by using two systems: the first one ($System_1$) does not consider any reject option, while the second one ($System_2$) implements a rejection rule for discarding cursive fragments by using a classification reliability measure. Both systems have been implemented by adopting the same OCR engine developed for our system. The comparison results show that, in terms of accuracy, our system performs much better than $System_1$, with a relatively slight reduction of the recognition rate on isolated characters (less than 7%). The results also show that the recognition rate on isolated characters of our system is higher than that of $System_2$ (about 10%) with the same overall rejection rate.

In conclusion, the experimental results confirmed the effectiveness of both the devised features, and the hierarchical architecture of our system. This approach allowed us to reject most of the input samples corresponding to segmentation errors with a small reduction of the classification rate on the samples representing isolated characters.

References

1. Breiman, L.: Random forests. Mach. Learn. **45**(1), 5–32 (2001)
2. Chow, C.: On optimum recognition error and reject trade of. IEEE Trans. Inf. Theor. **16**(1), 41–46 (2006)
3. Cordella, L.P., De Stefano, C., Fontanella, F., Scotto di Freca, A.: A weighted majority vote strategy using bayesian networks. In: Petrosino, A. (ed.) ICIAP 2013, Part II. LNCS, vol. 8157, pp. 219–228. Springer, Heidelberg (2013). https://doi.org/10.1007/978-3-642-41184-7_23
4. De Stefano, C., Fontanella, F., Marcelli, A., Parziale, A., Scotto di Freca, A.: Rejecting both segmentation and classification errors in handwritten form processing. In: Proceedings of International Conference on Frontiers in Handwriting Recognition, ICFHR, pp. 569–574, September 2014
5. De Stefano, C., Fontanella, F., Marrocco, C., Scotta di Freca, A.: A hybrid evolutionary algorithm for bayesian networks learning: an application to classifier combination. In: Di Chio, C., Cagnoni, S., Cotta, C., Ebner, M., Ekárt, A., Esparcia-Alcazar, A.I., Goh, C.-K., Merelo, J.J., Neri, F., Preuß, M., Togelius, J., Yannakakis, G.N. (eds.) EvoApplications 2010, Part I. LNCS, vol. 6024, pp. 221–230. Springer, Heidelberg (2010). https://doi.org/10.1007/978-3-642-12239-2_23
6. De Stefano, C., Fontanella, F., Scotto di Freca, A.: A novel naive bayes voting strategy for combining classifiers. In: Proceedings of International Workshop on Frontiers in Handwriting Recognition, IWFHR, pp. 467–472 (2012)
7. De Stefano, C., Sansone, C., Vento, M.: To reject or not to reject: that is the question-an answer in case of neural classifiers. IEEE Trans. Syst. Man Cyber. Part C **30**(1), 84–94 (2000)
8. De Stefano, C., Fontanella, F., Maniaci, M., Scotto di Freca, A.: A method for scribe distinction in medieval manuscripts using page layout features. In: Maino, G., Foresti, G.L. (eds.) ICIAP 2011, Part I. LNCS, vol. 6978, pp. 393–402. Springer, Heidelberg (2011). https://doi.org/10.1007/978-3-642-24085-0_41
9. Fumera, G., Roli, F., Giacinto, G.: Reject option with multiple thresholds. Pattern Recognit. **33**, 2099–2101 (2000)

10. Hall, M.: Correlation-based feature selection for machine learning. Ph.D. thesis. University of Waikato (1999)
11. Hanczar, B., Dougherty, E.R.: Classification with reject option in gene expression data. Bioinformatics **24**(17), 1889–1895 (2008)
12. Koerich, A.: Rejection strategies for handwritten word recognitions. In: IWFHR-9, pp. 479–484. IEEE Computer Society Press (2004)
13. Kononenko, I.: Estimating attributes: Analysis and extensions of RELIEF. In: Bergadano, F., De Raedt, L. (eds.) ECML 1994. LNCS, vol. 784, pp. 171–182. Springer, Heidelberg (1994). https://doi.org/10.1007/3-540-57868-4_57
14. Oliveira, L.S., Sabourin, R., Bortolozzi, F., Suen, C.: Automatic recognition of handwritten numerical strings: a recognition and verification strategy. IEEE Trans. Pattern Anal. Mach. Intell. **24**(11), 1438–1454 (2002)
15. Landgrebe, T., Tax, D.M.J., Paclk, P., Duin, R.P.W.: The interaction between classification and reject performance for distance-based reject-option classifiers. Pattern Recognit. Lett. **27** (8), 908–917 (2006)
16. Liu, H., Setiono, R.: Chi2: feature selection and discretization of numeric attributes. In: ICTAI, pp. 88–91. IEEE Computer Society (1995)
17. Plamondon, R., Srihari, S.: On-line and o-line handwriting recognition: a comprehensive survey. IEEE Trans. Pattern Anal. Mach. Intell. **22**(1), 63–84 (2000)
18. Quinlan, J.R.: C4.5: Programs for Machine Learning. Morgan Kaufmann Series in Machine Learning. Morgan Kaufmann, Los Altos (1993)

A Deep Learning Approach for the Automatic Detection and Segmentation in Autosomal Dominant Polycystic Kidney Disease Based on Magnetic Resonance Images

Vitoantonio Bevilacqua[1(✉)], Antonio Brunetti[1], Giacomo Donato Cascarano[1], Flavio Palmieri[1], Andrea Guerriero[1], and Marco Moschetta[2]

[1] Department of Electrical and Information Engineering, Polytechnic University of Bari, Bari, Italy
vitoantonio.bevilacqua@poliba.it

[2] Department of Emergency and Organ Transplants, University of Bari Medical School, Bari, Italy

Abstract. The automatic segmentation of kidneys in medical images is not a trivial task when the subjects undergoing the medical examination are affected by Autosomal Dominant Polycystic Kidney Disease.

In this work, two different approaches based on Deep Learning using Convolutional Neural Networks (CNNs) for the semantic segmentation of images containing polycystic kidneys are described, leading to a fully-automated classification of each pixel of the images without the need to extract hand-crafted features. In details, the first approach performs the automatic segmentation of the kidney considering the whole image as input, without any pre-processing. Conversely, the second approach is based on a two-steps classification procedure constituted by a CNN for the automatic detection of Regions of Interest (ROIs), according to the R-CNN approach, and a subsequent convolutional classifier performing the semantic segmentation on the ROIs previously extracted.

Results for both the approaches are reported considering different metrics evaluated on the test set. Even though the R-CNN shows an overall high number of false positives, the subsequent semantic segmentation on the extracted ROIs allows achieving good performance in terms of mean accuracy. Moreover, the results show that both the investigated approaches seem to be reliable for the automatic segmentation of polycystic kidneys, as in both the cases an accuracy of about 85% was reached.

Keywords: Deep Learning · Convolutional Neural Network
Semantic segmentation · Regions with CNN
Autosomal Dominant Polycystic Kidney Disease

D.-S. Huang et al. (Eds.): ICIC 2018, LNCS 10955, pp. 643–649, 2018.
https://doi.org/10.1007/978-3-319-95933-7_73

1 Introduction

Autosomal Dominant Polycystic Kidney Disease (ADPKD) is a hereditary disease characterised by the onset and growth of renal cysts that lead to a slow, gradual and steady enlargement of the Total Kidney Volume (TKV). As there is no specific therapy, the adoption of some countermeasures can significantly extend patient's life. In fact, since the estimation of the ADPKD severity is performed considering the TKV variations over time, a non-invasive and accurate assessment of renal volume is of fundamental importance.

The traditional methods to perform TKV estimation, based on imaging acquisitions, are stereology and manual segmentation. In these approaches, the segmentation of the kidney is a tricky task, due to both the morphological alterations of the kidney itself, and the similarities of the grey levels in adjacent organs, which may contain cysts too, thus requiring expert training to achieve an accurate assessment of the TKV.

To overcome manual methodologies, several approaches for semi-automatic TKV computation have been investigated, allowing to estimate the kidney volume considering a few slices only. Despite their speed and compliance, these techniques are far from being accurate enough to be used in clinical protocols.

Recently, innovative approaches based on Deep Learning (DL) strategies have been introduced for the classification of images. Thanks to DL architectures, the design of procedures for extraction of hand-crafted features is overcome, as the classifier structure is able to compute the most characteristic features for a specific dataset automatically. These features let DL approaches to be investigated in different fields, including medical imaging [1, 2].

In this work, two different approaches based on DL architectures have been investigated to perform the automatic detection and segmentation of polycystic kidneys starting from Magnetic Resonance (MR) images of subjects affected by ADPKD. Specifically, Convolutional Neural Networks have been designed and evaluated for the semantic segmentation of MR images, discriminating all the pixels in the images. Subsequently, the object detection approach using the Regions with CNN (R-CNN) technique has been investigated to automatically detect ROIs containing parts of the kidneys, with the aim to subsequently perform the semantic segmentation on the extracted ROIs.

This work is organized as follows: Sect. 2 reports the materials, in terms of patients and images' acquisition protocol and characteristics; Sect. 3 describes the classification approaches, including the semantic segmentation, the R-CNN technique and the combination of the two algorithms; Sect. 4 reports the results obtained, whereas Sect. 5 contains the discussion and the conclusion of the work.

2 Materials

From February 2017 to July 2017, 32 patients affected by ADPKD (mean age 33.4; range 19–47) underwent MR examinations for TVK assessment. The examinations were performed on a 1.5 TMR device (Achieva, Philips Medical Systems, Best, The Netherlands) by using a four-channel breast coil. The protocol did not use contrast material intravenous injection and consisted of three different sequences, namely Transverse and coronal short TI inversion recovery (STIR) turbo-spin-echo (TSE) sequences, Transverse and coronal T2-weighted TSE and Three-dimensional (3D) T1-weighted high-resolution isotropic volume (THRIVE) sequence.

In this work, the MR images from coronal T2-weighted TSE sequence have been considered for processing and classification. The acquisitions of four patients (mean age 31.25 $\pm$ 15.52 years) were manually segmented with a digital tool for ROIs contouring properly designed and implemented. The final set of images was obtained extracting from all the slices belonging to the four patients (155 images) those containing at least 1 pixel in the kidney area, thus reaching 57 images. The final dataset used for the classification was constituted by image and label samples, as depicted in Fig. 1.

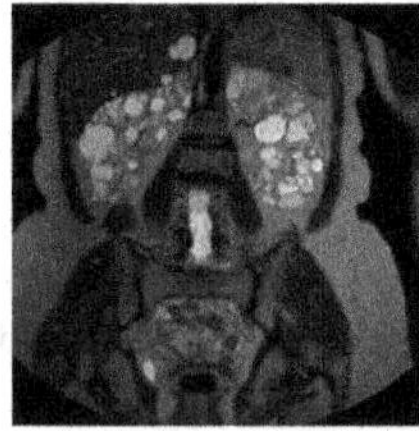
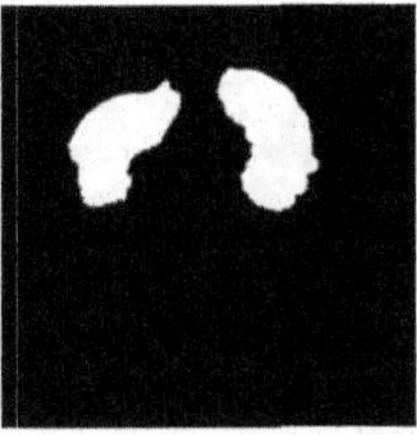

Fig. 1. Example of a segmented MR Image; left: Original DICOM image; right: the mask obtained after the manual contouring of the considered slice.

3 Classification

Two different approaches based on DL techniques have been investigated. Specifically, the first approach, which is reported in Sect. 3.1, consists in the semantic segmentation of the MR images. Then, in Sect. 3.2, the technique of Regions with Convolutional Neural Networks has been investigated and evaluated in the field of medical imaging.

3.1 Semantic Segmentation

To perform semantic segmentation, several CNN topologies have been designed and tested, inspired to the architecture of SegNet [3] and Fully Convolutional Network (FCN) [4], which are constituted by an encoder – decoder design. CNN architectures were designed varying the number of encoders (and decoders), the number of layers for each encoder, the number of convolutional filters for each layer and the learner used for the CNN training.

To discriminate the pixels of kidneys, the semantic segmentation was performed considering two classes: the pixels in the image corresponding to those white in the mask were labelled as “Kidney”, whereas the remaining pixels were labelled as “No-Kidney” (Fig. 1). For all the tests, 3x3 kernels were used for the convolutional layers, with stride [1 1] and padding [1 1 1 1] allowing to keep unchanged the dimension of the input across each encoder, thus performing down-sampling with the max-pooling layers only (with stride [2 2] and dimension 2×2). According to recent works [5], images augmentation was performed to improve the overall classification performance; the transformations taken into account are: horizontal shift in the range [–200; 200] pixels, horizontal flip and scaling with scale factor in the range [0.5; 4]. Table 1 reports the CNN configurations designed and tested for this task, whereas Fig. 2 shows the result obtained considering an image sample.

Table 1. Configurations designed and tested for the semantic segmentation of the full image. Each layer is a sequence of a convolutional layer, a batch normalization layer and a ReLu layer.

Network ID	Number of layers per encoder	Number of convolutional filters per layer	Learner
VGG16	[2 2 3 3 3]	[64 128 256 512 512]	ADAM
S-CNN-1	[3 2 3 3 3]	[64 128 256 512 512]	ADAM
S-CNN-2	[3 2 3 3 3]	[96 128 256 512 512]	ADAM

3.2 Regions with Convolutional Neural Networks

Since the dimensions of the input images negatively affected the computational performance of the semantic classifiers, an approach based on object detection using R-CNN have been investigated to allow the automatic detection of smaller regions to be subsequently classified. To do this, several CNNs with different architectures have been evaluated, according to the Fast R-CNN approach [6].

After the automatic detection of the ROIs, the same architectures designed for the segmentation of the whole image (reported in the previous section), have been considered for semantic segmentation (Table 1). Besides the network architectures, the training parameters were kept unchanged. Images augmentation was performed, as well, considering image horizontal shift in the range [–25; 25] pixels, image vertical shift in the range [–25; 25] pixels, image horizontal flip and scaling with scale factor ranging [0.5; 1.1].

4 Results

The input dataset, which was constituted by 57 images, was divided into training, validation and test sets randomly splitting the images into 60% for the training, 20% for the validation and 20% for the test set. In all the cases, the final evaluation has been performed considering the same permutation of the dataset for all the semantic classifiers, whereas the final R-CNN training was performed considering the union of training and validation sets. In the following, the performance obtained considering the R-CNN approach and those obtained by the classifiers working of the full image and the ROIs detected automatically are reported.

Several metrics have been considered for the evaluation of the different classifiers. In particular, Accuracy (Eq. 1), Boundary F1 Score (BF Score) (Eq. 2) and Jaccard Similarity Coefficient (Eq. 5) have been evaluated on the test set, according to the confusion matrix reported in Table 2, and the Precision (Eq. 3) and Recall (Eq. 4) values.

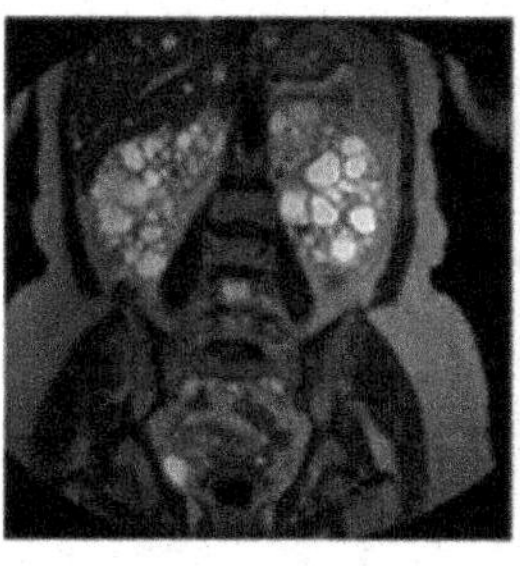
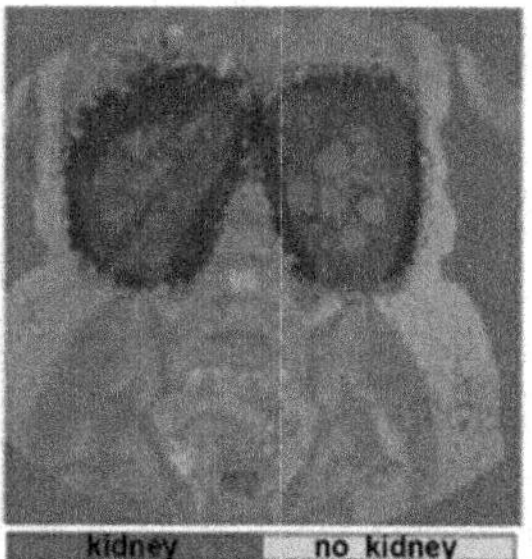

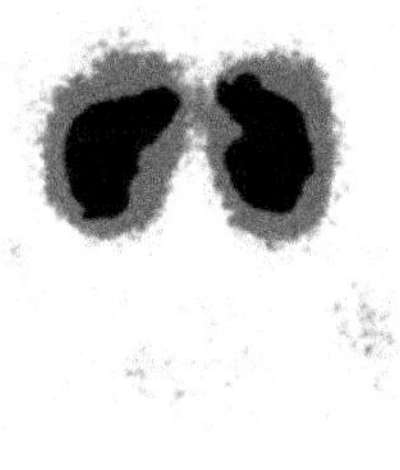

Fig. 2. Semantic segmentation result on the full image. Left: the MR slice; middle: the classification result; right: superimposition of the classification result to the ground-truth mask

Table 2. Confusion matrix for metrics computation

		True condition	
		Positive	Negative
Predicted condition	Positive	True positive (TP)	False positive (FP)
	Negative	False negative (FN)	True negative (TN)

In addition, the average precision and the logAverageMissRate have been evaluated for the R-CNN approach, considering the Miss Rate (MR) as reported in Eq. 6.

Concerning the R-CNN approach, all the implemented classifiers reach good performance, reporting an average precision of 0.78 in all the cases. However, the classifier reaching the higher recall (equals to 0.9, meaning that the 90% of the regions classified as positives are true positives) shows a precision equal to 0.6. Since in this phase the main objective is to correctly detect smaller regions to process with the subsequent semantic classifier, a relatively high number of false positives is still acceptable.

Regarding the semantic segmentation on the full image, all the classifiers reached an Accuracy value higher than 86%, whereas 84% of Accuracy was reached by the ROIs classifier. In both the cases, the S-CNN-1 classifier reached the highest performance.

Generally, the introduction of one additional layer into the first encoder of VGG16 allowed the net to create a more significant set of features respect to a simpler architecture, thus leading to a more accurate classification of the pixels. Conversely, the increment of the number of convolutional filters in the first layer of the first encoder did not improve the overall performance.

$$Accuracy = \frac{TP + TN}{TP + FP + FN + TN} \tag{1}$$

$$Boundary\,F1\,Score = \frac{2 * Precision * Recall}{Precision + Recall} \tag{2}$$

$$Precision = \frac{TP}{TP + FP} \quad (3)$$

$$Recall = \frac{TP}{TP + FN} \quad (4)$$

$$IoU(Jaccard\ Similarity\ Index) = \frac{TP}{TP + FP + FN} \quad (5)$$

$$Miss\ Rate = \frac{FN}{FN + TP} \quad (6)$$

5 Discussion and Conclusion

In this work, two different approaches based on DL architectures have been investigated to perform the automatic segmentation of polycystic kidneys starting from MR images. In a first approach, the convolutional classifiers considered the full image as input. Subsequently, R-CNNs have been investigated to automatically detect ROIs containing parts of the kidneys, with the aim to semantically segment the whole kidney. Considering the results obtained, it is clear that both the approaches are comparable and may be considered as reliable methods to perform a fully-automated segmentation of kidneys affected by ADPKD, when automatic or semi-automatic methodologies, such as feature-, atlas- or model-based techniques, lack in terms of performances. In future works, it will be investigated the combination of the Deep Learning strategies considered in this work with approaches based on image processing techniques, or rather use DL algorithms in pipelines with segmentation algorithms of different nature to perform a refinement on a previous automatic coarse image segmentation.

Acknowledgment. This work has been partially funded from the PON MISE 2014-2020 "HORIZON2020" program - project PRE.MED - Innovative and integrated platform for the predictive diagnosis of the risk of progression of chronic kidney disease, targeted therapy and proactive assistance for patients with autosomal dominant polycystic genetic disease.

References

1. Brunetti, A., Buongiorno, D., Trotta, G.F., Bevilacqua, V.: Computer vision and deep learning techniques for pedestrian detection and tracking: a survey. Neurocomputing (2018). https://doi.org/10.1016/j.neucom.2018.01.092
2. Litjens, G., Kooi, T., Bejnordi, B.E., Setio, A.A.A., Ciompi, F., Ghafoorian, M., van der Laak, J.A.W.M., van Ginneken, B., Sánchez, C.I.: A survey on deep learning in medical image analysis. Med. Image Anal. (2017). https://doi.org/10.1016/j.media.2017.07.005
3. Badrinarayanan, V., Kendall, A., Cipolla, R.: SegNet: a deep convolutional encoder-decoder architecture for image segmentation. IEEE Trans. Pattern Anal. Mach. Intell. **39**, 2481–2495 (2017). https://doi.org/10.1109/tpami.2016.2644615

4. Brostow, G.J., Fauqueur, J., Cipolla, R.: Semantic object classes in video: a high-definition ground truth database. Pattern Recognit. Lett. **30**, 88–97 (2009). https://doi.org/10.1016/j.patrec.2008.04.005
5. Bevilacqua, V., et al.: A supervised breast lesion images classification from tomosynthesis technique. In: Huang, D.-S., Jo, K.-H., Figueroa-García, J.C. (eds.) ICIC 2017. LNCS, vol. 10362, pp. 483–489. Springer, Cham (2017). https://doi.org/10.1007/978-3-319-63312-1_42
6. Girshick, R.: Fast R-CNN. In: Proceedings of the IEEE International Conference on Computer Vision, pp. 1440–1448 (2015). https://doi.org/10.1109/iccv.2015.169

A Model-Free Computer-Assisted Handwriting Analysis Exploiting Optimal Topology ANNs on Biometric Signals in Parkinson's Disease Research

Vitoantonio Bevilacqua[1(✉)], Claudio Loconsole[1], Antonio Brunetti[1], Giacomo Donato Cascarano[1], Antonio Lattarulo[1], Giacomo Losavio[2], and Eugenio Di Sciascio[1]

[1] Department of Electrical and Information Engineering (DEI), Polytechnic University of Bari, Bari, Italy
vitoantonio.bevilacqua@poliba.it
[2] Medica Sud s.r.l., Bari, Italy

Abstract. In this paper, we propose a novel model-free technique for differentiating both Parkinson's Disease (PD) patients from healthy subjects and mild PD patients from moderate ones by using a handwriting analysis tool. The tool is based on the analysis of biometric signals and the application of Artificial Neural Network (ANN)-based classifier. Experimental tests have been carried on with both healthy and PD subjects to identify the most representative features and to assess the accuracy and repeatability of classification performances achieved through optimal topology ANNs. Finally, the obtained results are reported and discussed to infer some important properties on classification approaches and the role of muscular activities on the handwriting analysis applied to neurodegenerative disease research.

Keywords: Handwriting analysis · Model-free · sEMG · Parkinson Disease ANN · MOGA

1 Introduction

The handwriting is a highly overlearned fine and complex manual skill involving an intricate blend of cognitive, sensory and perceptual-motor components [1]. For these reasons, the presence of abnormality in the handwriting process is a well-known and well-recognized manifestation of a wide variety of neuromotor diseases. There are two main difficulties related to the handwriting and affecting Parkinson's Disease (PD) patients: (i) the difficulty in controlling the amplitude of the movement, i.e., decreased letter size (micrographia) and failing in maintaining stroke width of the characters as writing progresses [2], and (ii) the irregular and bradykinetic movements, i.e., increased movement time, decreased velocities and accelerations, and irregular velocity and acceleration trends over time [3]. For these reasons, in literature, there are several works investigating the possibility of a differentiation between PD patients and healthy subjects by means of computer-aided handwriting analysis tools [3, 4].

D.-S. Huang et al. (Eds.): ICIC 2018, LNCS 10955, pp. 650–655, 2018.
https://doi.org/10.1007/978-3-319-95933-7_74

In our previous work [5], we proposed a preliminary approach for differentiating PD patients from healthy subjects using a reduced set of features (4 features) by exploiting computer vision techniques applied on the scan of common paper sheets and surface ElectroMyoGraphy (sEMG) signal processing. We found dynamic features being more representative for the differentiation. In a subsequent work [6], instead, we used a graphic tablet and a sEMG bracelet during handwriting tasks to respectively extract biometric signals related to pen movements and to muscular activity. The resulting performance allowed us to assess both the comparison among different classification approaches and the differentiation between PD patients and healthy age-related subjects.

In this paper, we improved our previous research extending both the number and type of proposed features and the subject dataset. We have also investigated the most representative features to be used in handwriting research applied to PD and the differentiation between mild and moderate PD patients.

2 The Proposed Model-Free Technique

2.1 Handwriting Feature Extraction

The features related to handwriting were extracted from biometric signals acquired during the handwriting tasks. In particular, it is possible to group the proposed features into two categories - sEMG related and pen tip related features:

- *sEMG related features* – these features are related to the muscular activity of the subject and are extracted from the sEMG signals acquired at the subject's forearm:
  - *Root Mean Square (RMS) features* extracted for each sEMG channel. RMS is computed as the square root of the mean of the sample squares.
  - *Zero Crossing (ZC) features,* an index related to the signal sign variation. To normalize the features among the subjects, its value is divided by the length of the signal.
- *Pen tip related features* - these features are extracted from the signals generated by a graphic tablet during the handwriting task:
  - *Cartesian and XY features* are referred to the pen tip position and are extracted starting from the XY axes position: *Cartesian and XY (i) velocity, (ii) acceleration, and (iii) jerk.* This lead to a total of nine signals.
  - *Pen tip pressure feature,* a scalar feature and corresponds to the pressure applied by the pen tip on the surface of the tablet.
  - *Azimuth and altitude feature:* the azimuth feature is the value of the angle between a reference direction (e.g., the Y axes of the tablet) and the pen direction projected on the horizontal plane. The altitude feature is the value of the angle between the pen direction and the horizontal plane.
  - *Pattern specific features* associated to a specific writing pattern (WP). For letter-based WPs, the features are mainly related to the writing size, whereas for the spiral-based WPs, the features are mainly related to the writing precision. For the features extracted from the letter-based WPs, the upper and the lower peaks of the Y coordinate of the pen tip position are computed and, then, used as input

data of a linear regressor. Finally, the angle α between the R_{up} and R_{low} regression line and the coefficient of determination (R^2) are computed and selected as features. For spiral WPs, instead, the feature extracted is an index representative of the variability of the strokes. For each point P of the X-Y pen tip position, the vector $\vec{r}$ with respect to the spiral centroid point C having origin in P is computed. The angle β between $\vec{r}$ and the direction vector $\vec{d}$ tangent to the spiral in P is, then, calculated. The spiral precision index feature is the standard deviation of the β angles computed for each point P.

2.2 Feature Selection and Classification

To reduce the number of features to be classified and to infer which of them are the most representative of the subject's status, we used a classification decision tree technique based on Gini's diversity index. To classify the extracted features, we used Artificial Neural Network (ANN) based classifier. The optimal topology for an ANN classifier was found by exploiting a Multi-Objective Genetic Algorithm (MOGA) and by maximizing the average test accuracy on a certain number of training, validation and test iterations for each ANN topology using different permutations of the dataset [7].

The performance for both the MOGA algorithm and the ANN-based classification were evaluated in terms of accuracy, specificity and sensitivity.

3 Experiments and Results

3.1 Participants

32 participants (21 males, 11 females, age: 71.4 ± 8.3 years old) took part in the experimental tests. In detail, the age-matched control group was composed of 11 healthy subjects (4 males, 7 females, age: 70.2 ± 10.2 years old), whereas the PD group was composed of 21 subjects (17 males, 4 females, age: 72.1 ± 8.3). According to the degree of the disease, the PD group was following divided into two subgroups: mild and moderate. The mild group was composed of 12 patients (9 males, 3 females, age: 70.5 ± 10.0), whereas the moderate one was composed of 9 patients (8 males, 1 female, age: 73.8 ± 6.0).

3.2 System Setup

The system setup for data acquisition is reported in Fig. 1. It includes two main sensors: (i) the Myo™ Gesture Control Armband allowing us to synchronously acquire 8 different sEMG sources at the forearm, and (ii) the WACOM Cintiq 13" HD, a graphics tablet providing visual feedback for acquiring pen tip planar coordinates and pressure, and the tilt of the pen with respect to the writing surface.

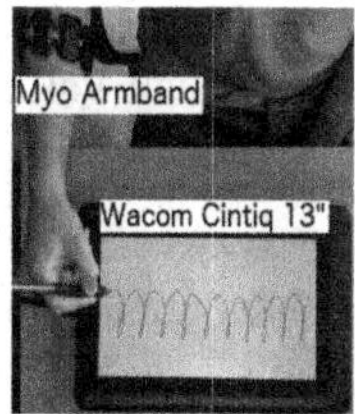

Fig. 1. Example of the system set-up used for data acquisition

3.3 Experimental Description

For the experiments, we used three writing patterns (WPs) leading to as many writing tasks, these are: a five-turn spiral drawn in anticlockwise direction (WP 1), a sequence of 8 Latin letter "l" with a size of 2.5 cm (WP 2) and with a size of 5 cm (WP 3). Since the last two WPs were size-constrained, a visual marker was provided as reference. In the experiment, we asked each subject to perform the three writing tasks four times each for a total of twelve tasks: first for familiarization purposes, whereas the other three were acquired and stored for the subsequent feature extraction and processing. The subject was asked to rest between two subsequent handwriting tasks for at least three seconds. The beginning of the task signal acquisition was triggered by a positive pen pressure applied on the graphic tablet. The processing of the acquired raw signals led to the extraction of 41 features for WP 1 and to 43 features for WP 2 and 3.

3.4 Experimental Data Processing Description

We conducted the experiments under two main objectives: (i) the separation of the PD patients from healthy ones, and (ii) the classification of mild and moderate Parkinson subjects. The following features extracted during the experiments, conducted according to Sect. 3.3, were grouped in three datasets: (i) dataset A with 41 features, (ii) dataset B with 43 features and (iii) dataset C with 43 features extracted from WP 1, WP 2 and WP 3, respectively. Then, a feature selection algorithm was applied on the three datasets to select and reduce the number of the features. This led to the creation of six new different feature datasets: dataset with all features included in set A, B and C (Case 1, 2 and 3, respectively) and dataset with only the features resulting from the feature selection algorithm applied on dataset A, B and C (Case 4, 5 and 6, respectively).

3.5 Results and Discussion

Since we performed 250 iterations of the net training procedure for each case, the performance results have been reported in percentage (standard deviation in brackets).

Objective 1 - Separating PD patients and healthy subjects:

- for dataset A (WP 1–41 features – Case 4), the 6 selected features were: one RMS value, 3 ZC values, the mean cartesian velocity and acceleration on X axes;
- for dataset B (WP 2–43 features – Case 5), the 6 selected features were: the mean jerk on Y axes, 3 ZC values, the mean cartesian acceleration and velocity on X axes;

- for dataset C (WP 3–43 features – Case 6), the 7 selected features were: 2 RMS values, one ZC value, the mean cartesian velocity, the altitude STD, the azimuth RMS and the mean velocity on X axes.

The best accuracy value (96.85%) was achieved in case 6 (classification on the dataset composed of the selection of 7 features from the dataset of 43 features extracted from WP 3, i.e., the sequence of 8 Latin letter "l" with a size of 5 cm). In case 6, three out of seven features were related to sEMG signals (RMS and ZC), whereas the other features were related to pen tilt and velocity.

Objective 2 - Separating mild and moderate PD patients:

- for dataset A (WP 1–41 features – Case 4), the 6 selected features were: 2 RMS values, 2 ZC values, the mean pressure and the mean altitude;
- for dataset B (WP 2–43 features – Case 5), the 5 selected features were: 2 RMS values, 2 ZC values and the mean cartesian velocity;
- for dataset C (WP 3–43 features – Case 6), the 5 selected features were: 2 RMS values, one ZC value, the mean cartesian velocity on X axes and the mean pressure.

The best accuracy value (96.00%) was achieved in Case 4 (dataset A - 6 features selected over 41 features extracted from WP 1, i.e., the spiral WP). In Case 4, four out of six features were related to sEMG signals (RMS and ZC), whereas the other features were related to pen tilt and pressure.

The obtained classification accuracy for all four cases for both objectives, instead, are reported in Table 1. As it can be observed, the obtained accuracy values x for both objectives are high ($86 < x < 97$) and present a limited standard deviation d ($d < 0.09$), thus demonstrating the repeatability of the classification performances and the stability of the optimal topology ANN architectures. It is worth to observe also that the highest values of resulting accuracy have been obtained for both objectives for the classification of the selected features. The obtained results allow us to confirm the relevance of the sEMG signals not only in differentiating PD patients and healthy patients, but also, and especially, in differentiating mild and moderate PD patients. Furthermore, the results confirm the choice of acquiring signals related also to pen tilt, pressure and velocity.

Table 1. Accuracy and standard deviation values obtained for each case.

	Case	Objective 1	Objective 2
All features	1	90.76% (0.0764)	94.34% (0.0626)
	2	92.98% (0.0523)	87.26% (0.0850)
	3	95.95% (0.0479)	91.86% (0.0830)
Selected features	4	93.78% (0.0566)	96.00% (0.0658)
	5	91.58% (0.0526)	86.71% (0.0837)
	6	96.85% (0.0405)	91.66% (0.0858)

4 Conclusion

In this work, we proposed a model-free technique for computer-assisted handwriting analysis. The technique is based on the extraction of different features from biometric signals acquired during the handwriting task and on their classification by means of optimal topology ANNs whose characteristics result from a MOGA processing.

The proposed technique allowed us to tackle two main research objectives: the differentiation between healthy subjects and PD patients (objective 1) and between mild and moderate PD patients (objective 2) both with a high classification accuracy (over 90%). Furthermore, we demonstrated that a limited number of representative feature, selected by means of a classification decision tree technique based on Gini's diversity index, allowed us to obtain a more performing classification for both the objectives of the study (up to 96.85% and 96.00% for objective 1 and 2, respectively).

Acknowledgments. This work has been partially funded from the FutureInResearch program of the Regione Puglia - project n. JTFWZV0 ABIOSAN - Advanced BIOmetric analysiS Against Neuromuscular disease.

Bibliography

1. Carmeli, E., Patish, H., Coleman, R.: The aging hand. J. Gerontol. Ser. A Biol. Sci. Med. Sci. **58**, M146–M152 (2003)
2. Van Gemmert, A.W.A., Teulings, H.-L., Contreras-Vidal, J.L., Stelmach, G.E.: Parkinsons disease and the control of size and speed in handwriting. Neuropsychologia **37**, 685–694 (1999)
3. Drotar, P., Mekyska, J., Smekal, Z., Rektorova, I., Masarova, L., Faundez-Zanuy, M.: Prediction potential of different handwriting tasks for diagnosis of Parkinson's. In: 2013 E-Health and Bioengineering Conference (EHB), pp. 1–4 (2013)
4. Rosenblum, S., Samuel, M., Zlotnik, S., Erikh, I., Schlesinger, I.: Handwriting as an objective tool for Parkinson's disease diagnosis. J. Neurol. **260**, 2357–2361 (2013)
5. Loconsole, C., et al.: Computer vision and EMG-based handwriting analysis for classification in Parkinson's disease. In: Huang, D.-S., Jo, K.-H., Figueroa-García, J.C. (eds.) ICIC 2017. LNCS, vol. 10362, pp. 493–503. Springer, Cham (2017). https://doi.org/10.1007/978-3-319-63312-1_43
6. Loconsole, C., Cascarano, G.D., Brunetti, A., Trotta, G.F., Losavio, G., Bevilacqua, V., Di Sciascio, E.: A model-free technique based on computer vision and sEMG for classification in Parkinson's disease by using computer-assisted handwriting analysis. Pattern Recognit. Lett. (2018, in press). https://doi.org/10.1016/j.patrec.2018.04.006
7. Bevilacqua, V., Brunetti, A., Triggiani, M., Magaletti, D., Telegrafo, M., Moschetta, M.: An optimized feed-forward artificial neural network topology to support radiologists in breast lesions classification. In: Proceedings of the 2016 on Genetic and Evolutionary Computation Conference Companion - GECCO 2016 Companion, pp. 1385–1392 (2016)

Particle Swarm Optimization with Convergence Speed Controller for Sampling-Based Image Matting

Yihui Liang[1], Han Huang[1](✉), Zhaoquan Cai[2], and Liang Lv[1]

[1] South China University of Technology, Guangzhou, China
hhan@scut.edu.cn
[2] Huizhou University, Huizhou, China

Abstract. Image matting is a challenging task and has become the basis of various digital multimedia technologies. The aim of image matting is to extract the foreground from a given image with the user-provided information. This study focuses on sampling-based image matting methods. The key issue in sampling-based image matting methods is to search the best foreground-background (F-B) sample pair for each unknown pixel which is generally known as a large-scale "sample optimization problem". This study explores a new variant particle swarm optimization algorithm based on convergence speed controller, a premature-convergence-prevented strategy, to improve the performance of image matting. Particularly, we embed the convergence speed controller into particle swarm optimization and proposed a efficient variant algorithm of it for the sample optimization problem. We conducted extensive experiments to verify the efficiency of the proposed algorithm. The experimental results show that the proposed algorithm, compared to the existing algorithms, is competitive and can achieve higher-quality matting.

Keywords: Sampling-based image matting · Sample optimization
Particle Swarm Optimization · Convergence Speed Controller

1 Introduction

Image matting technologies originated in [3]. The process of image matting is to extract the foreground, which is specified by the user, from a given image and compose it onto the new background [23], as shown in Fig. 1. This process is mathematically modeled by considering the observed color of a pixel as the linear composite of foreground and background colors [15]:

$$I_k = \alpha_k F_k + (1 - \alpha_k) B_k. \tag{1}$$

where I_k is the observed color of the pixel k, $k = 1, 2, \ldots, N_I$; N_I is the number of pixels in a given image; F_k and B_k are the true foreground color and background color of the pixel k, respectively. The color is a three-dimensional vector representing three color channels (RGB). Each color channel takes value in the range $[0, 255]$. The α_k is the opacity of pixel k and it takes value in the range $[0, 1]$: if $\alpha_k = 1$, it indicates that the

D.-S. Huang et al. (Eds.): ICIC 2018, LNCS 10955, pp. 656–668, 2018.
https://doi.org/10.1007/978-3-319-95933-7_75

Fig. 1. Illustration of the processes of image matting.

pixel k belongs to the foreground; if $\alpha_k = 0$, it indicates that the pixel k belongs to the background. Normally, before starting image matting, a user would provide some extra information to guide the matting. The information is usually provided by the so-called trimap [10, 22], which can be drawn by the user or generated automatically [10, 22]. Specifically, given an image, the corresponding trimap divides this image into three regions: known foreground region, known background region, and the unknown region. The goal of image matting is to determine which unknown pixels belong to the foreground.

In the literature, many methods have been proposed to achieve image matting. According to how to utilize the user-provided information, these methods can be classified into propagation-based image matting methods and sampling-based image matting methods.

The propagation-based image matting methods utilize the correlation among neighboring pixels and propagate the alpha value from the known pixel to the unknown pixel [5, 10, 11, 18]. The definition of correlation is based on some prior assumptions, such as affinity between the pixels [5], the large kernel [10] and the local color line [11]. The propagation-based image matting methods can produce smooth matting result with less input information. However, different assumptions have direct influence on the quality of matting result. Once the assumption is invalid on an image, the quality of matting result will decline. In this sense, the propagation-based methods may suffer from the issue of weak adaptability.

In the sampling-based image matting methods, some representative pixels are sampled from the both known foreground regions and background regions. The process of image matting is carried out by searching the best foreground-background (F-B) sample pair for every unknown pixel. Ideally, the best F-B sample pair for an unknown pixel represents the true foreground and background colors of the pixel. Therefore, if the best foreground and background colors of all unknown pixels are sampled, the sampling-based image matting methods can achieve accurate matting. Earlier sampling-based methods are based on some local strategies [6, 16, 24]. In these strategies, a part of pixels are collected from the known (foreground and background) regions nearby each unknown pixel according to color or spatial characteristics. Usually, the number of foreground and background samples is not huge, and thus the best F-B sample pair of every unknown pixel can be found by the brute-force method [6, 16, 24]. However, it is not guaranteed that the best F-B sample pairs of all unknown pixels can always be obtained due to the locality limitation of these local strategies. Moreover, the quality of matting result will decline if we fail to include the best F-B sample pairs when sampling. Later, to avoid losing the best F-B sample pairs, the global sampling strategy [7] was proposed, which collects all known foreground and background pixels around the boundaries of the known regions to construct the sample sets. The global sampling method can generate sufficient number of samples to alleviate the loss of the best F-B sample pairs, meanwhile it may also incur growing time cost of searching the best F-B sample pair for each unknown pixel. To address this challenge, the brute-force method is usually unacceptable, especially when processing the matting of a high-resolution image. Therefore, how to search the best F-B sample pair for each unknown pixel within a reasonable time cost becomes the key of improving the matting quality. This is referred to as the "sample optimization problem". In this paper, our study focuses on designing an efficient sample optimization algorithm to solve the sample optimization problem.

The random search algorithm [1] was firstly used in [7] to solve the sample optimization problem. The experimental results shown that it can produce better matting result than [6, 16, 24]. However, the random search algorithm has its limitation regarding to the exploration capacity and robustness. Later, to solve the sample optimization problem more efficiently, several meta-heuristic algorithms were introduced, which include the particle swarm optimization (PSO) [13] and cooperative coevolution differential evolution [4]. The experimental results shown that the search performance of meta-heuristic algorithms is better than the random search algorithm. However, premature convergence is the major obstacle of the meta-heuristic algorithms [8, 9], especially when solving a large-scale optimization problem [12]. Once the premature convergence occurs, the meta-heuristic algorithms usually fail to find the optimum in the search space, i.e., fail to find the best F-B sample pairs for image matting.

In this study, we used the convergence speed controller to avoid the premature convergence of the meta-heuristic algorithm in the process of searching the best F-B sample pair for each unknown pixel and improve its search efficiency. The convergence speed controller is a general algorithm strategy [25, 26]. It can be directly embedded into algorithm to control the convergence speed and prevent the premature convergence. We embedded the convergence speed controller (CSC) into particle swarm optimization (PSO) [13] to solve sample optimization problem. The proposed sample optimization algorithm is called as the CSC-PSO-IM algorithm. This paper is a substantial extension of [14] where core contents related to the implementation of CSC-PSO-IM algorithm, including the operator of recovering diversity and the operator of acceleration, are further explained and discussed in this paper. The more comprehensive comparison among existing sample optimization algorithms are shown in this paper. Additionally, a novel operator of acceleration is also proposed in this paper.

2 Sample Optimization Problem

In this section, the mathematical model of the sample optimization problem will be introduced in detail. For a given image, suppose that the number of the unknown pixels is N_u according to the trimap and the numbers of foreground samples and background samples are N_F and N_B according to specific sampling strategy. Once the best F-B sample pair of the kth unknown pixel is found, we will calculate its alpha value (opacity) by Eq. (2).

$$\alpha_k = \frac{(I_k - B_k)(F_k - B_k)}{\|F_k - B_k\|^2} \tag{2}$$

where I_k is the observed color of the kth unknown pixel, $k = 1, 2, \ldots, N_u$, F_k and B_k are the colors of the best F-B sample pair. The color is a three-dimensional vector. The Eq. (2) is based on the Eq. (1). According to the alpha value α_k, we can estimate whether the kth unknown pixel is from the foreground. Obviously, we can get seven unknowns at every unknown pixel. Therefore, the sample optimization problem is an ill-posed problem.

To distinguish the qualities of different F-B sample pairs and find the best F-B sample pair for each unknown pixel, different evaluation criteria considering the photometric, the spatial or the probabilistic characteristics of the F-B sample pair have been proposed [7, 16]. In this study, we used a classic evaluation model [7], which evaluates the quality of candidate F-B sample pair by considering its color and spatial characteristics. Formally, suppose that the candidate F-B sample pair of the kth unknown pixel is (i,j), we firstly calculate the approximate alpha value $\hat{\alpha}_k$ of the kth unknown pixel by Eq. (2). Then, the approximate alpha value $\hat{\alpha}_k$ is used to calculate the color evaluation value of (i,j) by Eq. (3).

$$f_c(i,j) = \left\| I_k - \left(\hat{\alpha} F^i + (1 - \hat{\alpha}) B^j\right) \right\| \tag{3}$$

where i represents the ith foreground sample in the foreground sample set and F^i represents the color of this foreground sample; $i = 1, 2, \ldots, N_F$; N_F is the number of the foreground samples; j represents the jth background sample in the background sample set and B^j represents the color of this background sample; $j = 1, 2, \ldots, N_B$; N_B is the number of the background samples. $f_c(i,j)$ is the color evaluation value of the candidate sample pair (i,j). A smaller color evaluation value indicates that the quality of the candidate sample pair (i,j) is better.

If there is an overlap in the color distribution of the foreground and background. The color evaluation value cannot distinguish the best F-B sample pair from other candidate F-B sample pair for the unknown pixels distributed in the overlap region [21]. A worst F-B sample pair may obtain a very small color evaluation value by the Eq. (3) Therefore, to eliminate the misguide of the single-color evaluation value, the spatial characteristic is also considered. The spatial evaluation values of the candidate foreground sample i and background sample j are calculated by Eq. (4) and Eq. (5), respectively.

$$f_s(i) = \frac{\|S_{F^i} - S_k\|}{min_F} \tag{4}$$

$$f_s(j) = \frac{\|S_{B^j} - S_k\|}{min_B} \tag{5}$$

where S_{F^i} is the spatial coordinate of ith foreground sample; S_{B^j} is the spatial coordinate of jth background sample; S_k is the spatial coordinate of the kth unknown pixel. min_F is the nearest distance of the kth unknown pixel to the foreground sample set. min_B is the nearest distance of the kth unknown pixel to the background sample set. $f_s(i)$ and $f_s(j)$ are the spatial evaluation values of the candidate sample pair (i,j), respectively. A smaller spatial evaluation value indicates that the candidate sample is closer to the kth unknown pixel in Euclidean space.

The final evaluation value of the candidate sample pair (i,j) of the kth unknown pixel is calculated by Eq. (6).

$$f(i,j) = f_c(i,j) + f_s(i) + f_s(j) \tag{6}$$

This evaluation function considers the color and spatial characteristics of the candidate sample pair (i,j). A smaller evaluation value indicates that the quality of (i,j) is better. We need to search the best F-B sample pair with the smallest evaluation value for all unknown pixels. This is the sample optimization problem. For every unknown pixel, the process of solving the sample optimization problem can be regarded as a search process in the two-dimensional F-B search space, as shown in Fig. 2.

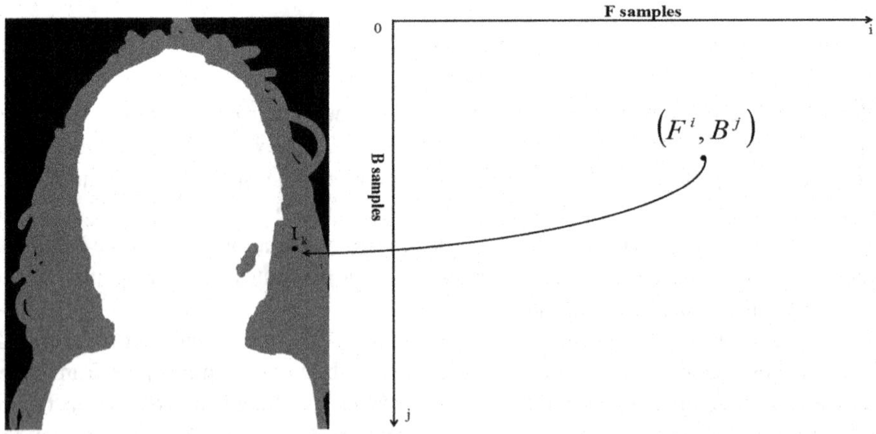

Fig. 2. Sample optimization problem. From left to right: trimap, foreground-background (F-B) sample pair search space.

3 Particle Swarm Optimization with Convergence Speed Controller

Let us suppose that the given image has N_u unknown pixels and the numbers of the foreground samples and the background samples are N_F and N_B, respectively. If we use the brute-force method to consider every possible combination of the foreground-background (F-B) sample pair for each unknown pixel, the worst time cost of searching the best F-B sample pairs for all unknown pixels is $(N_F N_B N_u)$. This time cost grows rapidly with the numbers of unknown pixels and samples, especially when processing the matting of a high-resolution image. To produce high-quality matting result in the acceptable time cost, we designed a sample optimization algorithm, which is based on the convergence speed controller. The convergence speed controller (CSC) is a general algorithm strategy [25, 26], which has two basic operations and the corresponding trigger conditions. The CSC can be embedded into algorithm and detect periodically convergence state. If the trigger condition is met, the corresponding operation will be applied for preventing premature convergence or controlling the convergence speed. The CSC has been embedded successfully into differential evolution [26] and particle swarm optimization [25] to solve numerical optimization problem. In this paper, we embedded the CSC into particle swarm optimization (PSO) [13] to solve the sample optimization problem, with the goals of avoiding the premature convergence of the PSO in the process of searching the best F-B sample pair and improving its search performance. The proposed sample optimization algorithm is called as the CSC-PSO-IM.

3.1 Particle Representation and Evaluation

We modeled the sample optimization problem into a large-scale optimization problem because of every unknown pixel needs to find the best F-B sample pair. In our algorithm, every particle is a $2N_u$ dimensional vector. N_u is the number of the unknown pixels. Every odd dimensional value of a particle represents the index of a foreground sample in the foreground sample set and takes value in the range $[1,..,N_F]$. Every even dimensional value of a particle represents the index of a background sample in the background sample set and takes value in the range $[1,..,N_B]$. N_F and N_B are the numbers of the foreground samples and the background samples. An adjacent pair of odd dimensional value and even dimensional value represents a candidate F-B sample pair for an unknown pixel. Therefore, every particle obtains N_u candidate F-B sample pairs and represents an overall solution for all unknown pixels. The formal representation of a particle is shown in Eq. (7). In addition, we sorted all foreground samples and background samples by their gray values to distinguish the indexes of different samples.

$$X = \left(x^1, \ldots, x^j, \ldots, x^{2N_u}\right), j = 1, \ldots, 2N_u \tag{7}$$

To find the best solution, which obtains the best F-B sample pairs of all unknown pixels, we used an evaluation function to distinguish the qualities of various particles. This evaluation function is based on the Eq. (6). In particular, for the ith particle in the swarm, its evaluation value was calculated by Eq. (8):

$$\varphi(X_i) = \sum_{j=1}^{N_u} f\left(x_i^{2j-1}, x_i^{2j}\right) \tag{8}$$

where $i = 1, 2, \ldots, N_p$, N_p is the size of the swarm. $f()$ is defined by Eq. (6). The Eq. (8) is the sum of the evaluation values of all candidate F-B sample pair in the particle X_i. Therefore, a smaller $\varphi()$ indicates that the quality of current particle is better.

3.2 The Implement of the CSC-PSO-IM

The terminology convergence of meta-heuristic algorithm has been discussed in many theoretical researches [1, 2, 19]. Swarm convergence [20] is one of the main types. It means that the positions of all particles in the swarm are very similar. In the process of searching the best solution, we hope the the swarm keep diversity for exploring more promising areas. If the swarm loses diversity in search process, we think the premature convergence is most likely to occur. In CSC-PSO-IM, we use the Condition 1 to detect whether the swarm loses diversity in the process of searching the best F-B sample pairs for all unknown pixels. The formulation of Condition 1 is shown in Inequality (9).

$$cos(X_a, X_b) > \Gamma_0 \tag{9}$$

where X_a and X_b are two particles selected randomly in the current swarm. The $cos(X_a, X_b)$ is the cosine value of these two particles and calculated by Eq. (10).

$$cos(X_a, X_b) = \frac{X_a X_b}{\|X_a\| \|X_b\|} \tag{10}$$

If this cosine value is greater than the threshold Γ_0, it indicates that the positions of these two particles in the search space are very similar. The current swarm may lose diversity and the premature convergence has occurred. The operation of recovering swarm diversity will be carried out to avoid the premature convergence. The detailed implement of this operation is described as follow.

N_p is the size of the current swarm. X_i is the ith particle in the swarm, $i = 1, \ldots, N_p$. If the Eq. (9) is established, the diversity of swarm will be recovered by regenerating this swarm in search space. In order to help the regenerated swarm find the high-quality solution, the historical best solutions of particles and the historical best solution of the swarm are unchanged. In addition, due to every dimensional value of the particle represents an index of the foreground sample or the background sample, every dimensional value of the particle must be an integer. We use the rounding operation to remain the integer part of every dimensional value as the final value.

We use the Inequality (9) to detect whether the premature convergence of particle swarm optimization has occurred in the process of searching the best F-B sample pairs for all unknown pixels and use the Eq. (10) to recover the swarm diversity. However, if the high-quality solutions are not found in the search process, the convergence speed of particle swarm optimization may be very slow. To improve the search efficiency of particle swarm optimization, we use the Condition 2 to detect whether the convergence speed of particle swarm optimization is too slow. The formulation of Condition 2 is shown in Inequality (11).

$$\frac{\varphi(lgBest) - \varphi(gBest)}{\varphi(lgBest)} < \Gamma_1 \tag{11}$$

where $\varphi(gBest)$ is the evaluation value of the historical best solution of the current swarm, $\varphi(lgBest)$ is the evaluation value of the historical best solution of the swarm before p generations. p is a period parameter. Γ_1 is a threshold which is used to measure the convergence speed of particle swarm optimization within recent p generations. If the Inequality (11) is established, it indicates that the convergence speed of particle swarm optimization is too slow within recent p generations. Then, the operation of acceleration will be carried out to accelerate the convergence speed. The detailed implement of this operator is described as follow.

The $normrand(0, \sigma_1)$ and $normrand(0, \sigma_2)$ are two adaptive normal random numbers. The standard deviation σ_1 and σ_2 are calculated by Eqs. (12) and (13).

$$\sigma_1 = \frac{(N_F - 1)}{\Delta} \tag{12}$$

$$\sigma_2 = \frac{(N_B - 1)}{\Delta} \tag{13}$$

The parameter Δ is a multiple of p and p is not equal to zero. If the particle swarm optimization maintains fast convergence speed for a long time, the parameter Δ will be larger and the operation of acceleration will regenerate the swarm in smaller nearby region of *gBest*. Conversely, the parameter Δ will be smaller and the operation of acceleration will regenerate the swarm in the larger nearby region of *gBest*. Overall, the operation of acceleration will regenerate the swarm around the nearby region of *gBest*. In addition, if a dimensional value of a particle is greater than N_F or N_B or smaller than 1, it will be modified into the original value. This operation is used for preventing particles from reaching infeasible space. The complete procedure of the CSC-PSO-IM algorithm is shown in Algorithm 1.

4 Experiments and Results

The experiments and the corresponding results are presented in this section. The purpose of these experiments is to prove that the convergence speed controller can improve the search performance of particle swarm optimization in solving the sample optimization problem and help particle swarm optimization find higher-quality foreground-background (F-B) sample pairs. we compared the matting results obtained by the proposed CSC-PSO-IM algorithm with the matting results obtained by the existing random search algorithm [7], particle swarm optimization [13] and the cooperative coevolution differential evolution [4]. These algorithms were implemented by ourselves according to [4, 7, 13].

To avoid losing the best F-B sample pairs of all unknown pixels, we used the global sampling method [7] to construct the sample sets. Besides, for a fair comparison, only the sample optimization step was replaced with different sample optimization algorithms, which are the random search algorithm [7], particle swarm optimization [13], cooperative coevolution differential evolution [4], and the CSC-PSO-IM algorithm. Other steps keep the same. Therefore, the differences of matting results are only due to the different sample optimization algorithms. An open image data set was used as the test images [17]. This image data set has twenty-seven test images and has been used in lots of matting researches [6, 16, 24]. In addition, the maximum iteration number of the CSC-PSO-IM algorithm was set to 3×10^5. This number keeps the same with the particle swarm optimization [13] and the cooperative coevolution differential evolution [4]. The maximum iteration number of the random search algorithm keeps the same with [7]. The value of the parameter Γ_0 used in the operation of recovering diversity was set to 0.9. The value of the parameter Γ_1 used in the operation of acceleration was set to 0.001. The value of the period parameter p was set to 150. We found that these setups are conducive to find higher-quality F-B sample pairs.

Algorithm 1. The CSC-PSO-IM Algorithm

Input: The number of the unknown pixels: N_u. The numbers of the foreground samples and the background samples: N_F and N_B. The foreground samples and the background samples are sorted by their gray values, respectively.

for $i = 1$ to N_p **do**
    for $j = 1$ to $2N_u$ **do**
        if j is a odd number **then**
            $X_i^j = \lfloor 1 + rand * (N_F - 1) \rfloor$, $V_i^j = \left\lfloor (rand - 0.5) * (\frac{N_F - 1}{2}) \right\rfloor$
        else
            $X_i^j = \lfloor 1 + rand * (N_B - 1) \rfloor$, $V_i^j = \left\lfloor (rand - 0.5) * (\frac{N_B - 1}{2}) \right\rfloor$
        end if
    end for
    $pBest_i = X_i$.
    if $\varphi(pBest_i) < \varphi(gBest)$ **then**
        $gBest = pBest_i$
    end if
end for
Set Δ to 0, $iter$ to 0
while termination criterion is not met **do**
    Particle swarm optimization is used to update the positions of particles, the $pBest$ and the $gBest$
    Set $\Delta = \Delta + 1$, $iter = iter + 1$
    if $iter$ is a multiple of p **then**
        if condition 1 is met **then**
            Run operation of recovering diversity
        end if
        if condition 2 is met **then**
            Run operation of acceleration
            Set Δ to 0
        end if
    end if
end while

Output: The best solution that was found: $gBest$;

The quantitative comparison of matting results obtained by different sample optimization algorithms are also conducted in the experiment. The mean squared error (MSE) value is used as the quantitative metric for measuring the accuracy of the experimental matting results. The calculation of the MSE value is based on the difference between the opacity of experimental matting result and the opacity of accurate matting result at every pixel. A smaller MSE value indicates that the experimental matting result is closer to the accurate matting result. The MSE values of the best matting results obtained by different sample optimization algorithms are recorded in the Table 1. By comparing the MSE values of different sample optimization algorithms, we found that the MSE values of the particle swarm optimization is smaller than the MSE values of the random search algorithm on twenty-six test images. However, comparing with the MSE values of the cooperative coevolution differential evolution, the MSE values of the particle swarm optimization are only smaller on the *Image_03*, *Image_08*, *Image_09*, *Image_11*, *Image_15*, *Image_19*, *Image_20* and *Image_24*. Next, we further compared the MSE values of the cooperative coevolution differential evolution with the MSE values of the CSC-PSO-IM algorithm. We found that the MSE values of the CSC-PSO-IM algorithm are smaller than the MSE values of the cooperative coevolution differential evolution on twenty-four test images. With the help of the convergence speed controller, the search performance of the particle swarm optimization has a significant improvement in solving the sample optimization problem. Therefore, the results of MSE value verify the above conclusion. The convergence speed controller can improve the search performance of particle swarm optimization and help particle swarm optimization find higher-quality foreground-background sample pairs.

Table 1. The comparison of different algorithms in terms of MSE

NO.	Random search	PSO	CCDE	CSC-PSO-IM
Image_01	13.90×10^{-3}	1.25×10^{-3}	1.18×10^{-3}	0.77×10^{-3}
Image_02	7.02×10^{-3}	8.23×10^{-3}	2.72×10^{-3}	3.52×10^{-3}
Image_03	13.00×10^{-3}	4.21×10^{-3}	10.02×10^{-3}	5.19×10^{-3}
Image_04	100.17×10^{-3}	20.10×10^{-3}	13.16×10^{-3}	8.48×10^{-3}
Image_05	15.80×10^{-3}	2.20×10^{-3}	2.01×10^{-3}	1.05×10^{-3}
Image_06	10.22×10^{-3}	2.67×10^{-3}	1.95×10^{-3}	1.07×10^{-3}
Image_07	15.65×10^{-3}	2.57×10^{-3}	2.06×10^{-3}	0.56×10^{-3}
Image_08	57.26×10^{-3}	12.32×10^{-3}	23.23×10^{-3}	10.39×10^{-3}
Image_09	12.54×10^{-3}	2.50×10^{-3}	4.58×10^{-3}	3.32×10^{-3}
Image_10	8.12×10^{-3}	3.11×10^{-3}	3.08×10^{-3}	1.73×10^{-3}
Image_11	12.97×10^{-3}	4.04×10^{-3}	4.26×10^{-3}	3.78×10^{-3}
Image_12	4.86×10^{-3}	2.82×10^{-3}	2.33×10^{-3}	0.29×10^{-3}
Image_13	57.17×10^{-3}	21.58×10^{-3}	19.74×10^{-3}	6.92×10^{-3}
Image_14	4.60×10^{-3}	2.00×10^{-3}	1.94×10^{-3}	0.74×10^{-3}
Image_15	9.25×10^{-3}	3.88×10^{-3}	4.33×10^{-3}	1.18×10^{-3}
Image_16	80.49×10^{-3}	62.30×10^{-3}	61.85×10^{-3}	23.57×10^{-3}
Image_17	10.92×10^{-3}	2.58×10^{-3}	2.32×10^{-3}	1.11×10^{-3}
Image_18	11.64×10^{-3}	3.81×10^{-3}	3.61×10^{-3}	2.02×10^{-3}
Image_19	6.18×10^{-3}	0.94×10^{-3}	0.97×10^{-3}	0.69×10^{-3}
Image_20	4.45×10^{-3}	1.94×10^{-3}	2.15×10^{-3}	0.41×10^{-3}
Image_21	22.50×10^{-3}	4.84×10^{-3}	4.56×10^{-3}	3.41×10^{-3}
Image_22	11.15×10^{-3}	1.94×10^{-3}	1.84×10^{-3}	0.76×10^{-3}
Image_23	4.49×10^{-3}	3.76×10^{-3}	3.15×10^{-3}	0.62×10^{-3}
Image_24	6.29×10^{-3}	4.45×10^{-3}	4.76×10^{-3}	2.42×10^{-3}
Image_25	24.96×10^{-3}	15.50×10^{-3}	15.20×10^{-3}	15.29×10^{-3}
Image_26	47.48×10^{-3}	25.44×10^{-3}	24.87×10^{-3}	26.72×10^{-3}
Image_27	20.88×10^{-3}	16.33×10^{-3}	15.90×10^{-3}	15.72×10^{-3}

5 Conclusion

The sample optimization problem is the core optimization problem of the sampling-based image matting methods. The difficult of solving this optimization problem is how to search the best foreground-background sample pairs for all unknown pixels within an acceptable time cost. In this study, to improve the search performance of particle swarm optimization in solving the sample optimization problem, we used the convergence speed controller to prevent particle swarm optimization from premature convergence in the process of searching the best foreground-background sample pairs. We compared the matting results of the proposed CSC-PSO-IM algorithm and other sample optimization algorithms. The visual matting results and the quantitative comparison of MSE value show that the CSC-PSO-IM algorithm can produce higher-quality matting result.

Acknowledgement. This work is supported by National Natural Science Foundation of China (61772225), Guangdong Natural Science Funds for Distinguished Young Scholar (2014A03 0306050), the Ministry of Education - China Mobile Research Funds (MCM20160206) and Guangdong High-level personnel of special support program (2014TQ01X664).

References

1. Barnes, C., Shechtman, E., Finkelstein, A., Goldman, D.B.: PatchMatch: a randomized correspondence algorithm for structural image editing. ACM Trans. Graph. (TOG) **28**(3), 24 (2009)
2. van den Bergh, F., Engelbrecht, A.P.: A convergence proof for the particle swarm optimiser. Fundam. Inform. **105**(4), 341–374 (2010)
3. Beyer, W.: Traveling-matte photography and the blue-screen system: a tutorial paper. J. SMPTE **74**(3), 217–239 (1965)
4. Cai, Z.Q., Lv, L., Huang, H., Hu, H., Liang, Y.H.: Improving sampling-based image matting with cooperative coevolution differential evolution algorithm. Soft Comput. **21**(15), 4417–4430 (2017)
5. Chen, Q., Li, D., Tang, C.K.: KNN matting. In: IEEE Conference on Computer Vision and Pattern Recognition CVPR 2012, pp. 869–876 (2012)
6. Gastal, E.S.L., Oliveira, M.M.: Shared sampling for real-time alpha matting. Comput. Graph. Forum **29**(2), 575–584 (2010)
7. He, K., Rhemann, C., Rother, C., Tang, X., Sun, J.: A global sampling method for alpha matting. In: IEEE Conference on Computer Vision and Pattern Recognition (CVPR), pp. 2049–2056, June 2011
8. Joshi, R., Deshpande, B.: Empirical and analytical study of many-objective optimization problems: analysing distribution of nondominated solutions and population size for scalability of randomized heuristics. Memet. Comput. **6**(2), 133–145 (2014)
9. Lalwani, S., Kumar, R., Gupta, N.: A novel two-level particle swarm optimization approach for efficient multiple sequence alignment. Memet. Comput. **7**(2), 1–15 (2015)
10. Lee, P., Wu, Y.: Nonlocal matting. In: IEEE Conference on Computer Vision and Pattern Recognition CVPR 2011, pp. 2193–2200 (2011)
11. Levin, A., Lischinski, D., Weiss, Y.: A closed-form solution to natural image matting. IEEE Trans. Pattern Anal. Mach. Intell. **30**(2), 228–242 (2008)

12. Lin, Y., Yao, X., Zhao, Q., Higuchi, T.: Scaling up fast evolutionary programming with cooperative coevolution. In: Proceedings of the 2001 IEEE Congress on Evolutionary Computation, pp. 1101–1108 (2001)
13. Lv, L., Huang, H., Cai, Z., Hu, H.: Using particle swarm large-scale optimization to improve sampling-based image matting. In: Proceedings of the 2015 Annual Conference on Genetic and Evolutionary Computation, pp. 957–961 (2015)
14. Lv, L., Huang, H., Cai, Z., Liang, Y.: Improving sample optimization with convergence speed controller for sampling-based image matting. In: Gong, M., Pan, L., Song, T., Zhang, G. (eds.) BIC-TA 2016. CCIS, vol. 682, pp. 400–406. Springer, Singapore (2016). https://doi.org/10.1007/978-981-10-3614-9_49
15. Porter, T., Duff, T.: Compositing digital images. ACM Siggraph Comput. Graph. **18**(3), 253–259 (1984)
16. Rhemann, C., Rother, C., Gelautz, M.: Improving color modeling for alpha matting. In: BMVC (2008)
17. Rhemann, C., Rother, C., Wang, J., Gelautz, M., Kohli, P., Rott, P.: A perceptually motivated online benchmark for image matting. In: CVPR, June 2009
18. Ruzon, M.A., Tomasi, C.: Alpha estimation in natural images. In: IEEE Conference on Computer Vision and Pattern Recognition CVPR 2000, pp. 18–25 (2000)
19. Schmitt, B.I.: Convergence Analysis for Particle Swarm Optimization. FAU University Press, Erlangen (2015)
20. Schmitt, M., Wanka, R.: Particle swarm optimization almost surely finds local optima. Theor. Comput. Sci. **561**, 57–72 (2015)
21. Shahrian, E., Rajan, D.: Weighted color and texture sample selection for image matting. In: IEEE Conference on Computer Vision and Pattern Recognition (CVPR), vol. 22, no. 11, pp. 4260–4270 (2012)
22. Wang, J., Cohen, M.F.: An iterative optimization approach for unified image segmentation and matting. In: Tenth IEEE International Conference on Computer Vision (ICCV), vol. 2, pp. 936–943 (2005)
23. Wang, J., Cohen, M.F.: Image and video matting: a survey. Found. Trends Comput. Graph. Vis. **3**(2), 97–175 (2007)
24. Wang, J., Cohen, M.F.: Optimized color sampling for robust matting. In: IEEE Computer Society Conference on Computer Vision and Pattern Recognition (CVPR), pp. 1–8 (2007)
25. Xu, C., Huang, H., Lv, L.: An adaptive convergence speed controller framework for particle swarm optimization variants in single objective optimization problems. In: IEEE International Conference on Systems, Man, and Cybernetics (2014)
26. Ye, S., Huang, H., Xu, C.: Enhancing the differential evolution with convergence speed controller for continuous optimization problems. In: Proceedings of the 2014 Annual Conference on Genetic and Evolutionary Computation, pp. 161–162 (2014)

iCorr-GAA Algorithm for Solving Complex Optimization Problem

Fangyuan Ding, Min Huang(✉), Yongsheng Deng, and Han Huang

School of Software Engineering, South China University of Technology, Guangzhou 510000, Guangdong, China
minh@scut.edu.cn

Abstract. Optimization is widely used to solve problems in many fields. With the development of society, the complexity of optimization problems is also increasing. Genetic algorithm (GA) is one of the most powerful stochastic optimizer. As a well-known GA variant, Correlation-based Genetic Algorithm (Corr-GAA) has been successfully applied to solve these optimization problems. Although highly effective, Corr-GAA tends to converge quickly at early evolution, and may fall into the local optimum in the later evolution stage. Non-uniform mutation operator can effectively improve this situation by adjusting dynamically search step of each iteration. In this paper we present an improved genetic algorithm (iCorr-GAA) that combines Corr-GAA with non-uniform mutation operator to solve complex optimization problems. The performance of the algorithm was evaluated by solving a set of benchmark functions provided for CEC 2014 special session and competition. Experimental results give evidence that iCorr-GAA has good global search capability and fast convergence speed.

Keywords: Complex optimization · Genetic algorithm
Non-uniform mutation

1 Introduction

In practice, optimization is widely used to solve many problems without clear forms or structures [1]. These problems can be found in many fields such as science, engineering and economics [2]. The global optimization problem can be defined as follows [3].

$$\begin{array}{ll} \textit{minimize} & f(x) \\ \textit{subject to} & x \in X \subseteq R^n \end{array}$$

where, $f(x)$ denotes the object function, the vector $\mathrm{x} = (x_1, x_2, \ldots, x_D)^T$ denotes decision variables, $lb_i \leq \mathrm{x_i} \leq ub_i$, where lb_i is the lower bounds and ub_i is the upper bounds, and D denotes the dimensionality of the problem.

When dealing with global optimization problems, we are faced with troubles such as increasing the scale of problems and deepening the complexity. It is not easy to find high-quality solutions to these optimization problems. Traditional optimization methods, such as linear programming, quadratic programming and hybrid planning, are

D.-S. Huang et al. (Eds.): ICIC 2018, LNCS 10955, pp. 669–680, 2018.
https://doi.org/10.1007/978-3-319-95933-7_76

usually used for structuring problems, whose search procedure requires more explicit information and prerequisites for these problems [4]. If the optimization problem is nonlinear and multimodal and its search space is very large, these algorithms may be in trouble. A stochastic optimization technique called evolutionary algorithm (EAs) that mimics the natural evolution process of organisms has shown superior performance in solving such optimization problems. Genetic algorithm is one of representative and widely used evolutionary algorithms [5].

In recent years, scholars have made many researches on GA and its variants in order to solve complex optimization problems. Chuang develops a parallel structure genetic algorithm for numerical optimization [6]. An algorithm that combines GA with the crossover operator in the differential evolution (DE) algorithm for improving the search ability of GA is proposed in [7]. Besides, mutation operator which restores the lost or undeveloped genetic material of an individual can increase the ability of the GA to obtain the global optimal solution [8]. Lu proposes an adaptive genetic algorithm that improves the performance of GA by adaptively adjusting the selection pressure and the mutation parameter [9]. A hybrid island model genetic algorithm is proposed in [10], which uses taboo search and a combination of three random mutation operators to search global best individuals. Hybridization of GA with local search technique in presence of archive can also better search the solution space [11]. González proposes a hybrid approach that combines GA with local search techniques [12]. It uses local search operators to improve the search capability of GA. Kundu proposes a correlation-based genetic algorithm named Corr-GAA that embeds local search and similarity between individuals in GA to solve global optimization problems [13]. This algorithm selects the next generation of individuals by considering the external archives and the correlation between the best individual and other individuals, thereby accelerating the convergence speed of the genetic algorithm. Although highly effective, Corr-GAA tends to converge quickly in the early stage of the search, and may fall into the local optimum in the later stage. So we use non-uniform mutation operator to effectively adjust this situation. In this paper, we propose an improved algorithm called iCorr-GAA that combines Corr-GAA with non-uniform mutation to solve complex optimization problems.

The rest of the paper is organized as follows: Sect. 2 gives research background for this work, where shows a description of the Corr-GAA algorithm and its limitation. Section 3 presents the proposed algorithm and its components. Section 4 gives the experimental results of the proposed algorithm and analysis of those results. Finally, the conclusions are given in Sect. 5.

2 Background

Corr-GAA algorithm shows good performance in improving the convergence speed and avoiding premature convergence [13].

2.1 Corr-GAA Algorithm

Corr-GAA algorithm proposes a new selection operator (named Correlation Selection CS) which takes into account correlation between individuals. In addition, it uses three external archives in selection process for fast convergence, where archive T_b saves best individual of current population, archive T_{xb} saves the predicted best individuals by observing archive T_b, and archive T_e contains individuals with the least correlation with the best individual but higher than average fitness. A pseudo-code of the correlation selection operator is shown in Algorithm 1.

Algorithm 1. Correlation Selection algorithm [13]

```
procedure CSelection (X_best,G, X)
  Compute the mean CC of the population and store it in μ_CC
  CC ← 0.0
  flag ← -1
  for i ← 1 to N_p/8 do
    Randomly select a population member X_rand
    if fit(X_rand) ≥ mean(fit(X)) then
      Calculate the CC between X_best,G and X_rand, and store it in temp_CC
      if temp_CC ≥ μ_CC then
        CC ← 0.0
        Store X_rand as the reference member.
        flag ← 0
    if flag == 0 then
      Uniformly randomly select a member X_rand
    return X_rand
```

The correlation coefficient (CC) in correlation selection operator is calculated as follows: Firstly, the relative weight of each gene of a chromosome is calculated. At the beginning of each run, it is assumed that each gene is equally important in each dimension (i.e. initialize $W_i = 0.5$). After each run, W_i is recalculated according to the followings:

$$\mathrm{F_i} = \frac{\left|X_{i,A} - X_{i,B}\right|}{\left|fit(X_A) - fit(X_{i,B})\right|} \tag{1}$$

$$\mathrm{W_i} = \begin{cases} \mathrm{W_i} + \frac{(1-\mathrm{W_i})}{2^{\mathrm{F_i}}}, & \mathrm{F_i} < 1 \\ \mathrm{W_i} - \frac{\mathrm{W_i}}{\frac{1}{2^{\mathrm{F_i}}}}, & \mathrm{F_i} > 1 \end{cases} \tag{2}$$

Where, $fit(X_A)$ and $fit(X_B)$ are the fitness values of individual X_A and individual X_B respectively, and $\mathrm{F_i}$ is a custom factor in this algorithm. If $fit(X_A)$ and $fit(X_B)$ are equal, W_i is not modified.

If $fit(X_A)$ is greater than $fit(X_B)$, then the FR (fitness ratio) is set to 1, otherwise, the FR is set to the ratio of the absolute values of $fit(X_A)$ and $fit(X_B)$. If X_A and X_B are not equal, WSS and SS of each dimension between individual X_A and individual X_B are calculated as followed.

$$\text{Weighted Square Sum (WSS)} = \sum_{i=1}^{K}\left(W_i * \left|x_{i,A} - x_{i,B}\right|^2\right) \tag{3}$$

$$\text{Square Sum (SS)} = \sum_{i=1}^{K}\left|x_{i,A} - x_{i,B}\right|^2 \tag{4}$$

Finally the correlation coefficient of X_B with respect to X_A $(X_B \neq X_A)$ is given by,

$$CC = \left(1 - \sqrt{\frac{WSS}{SS} * FR}\right) \tag{5}$$

2.2 Limitation

The correlation-based genetic algorithm selects the individuals for the production of the next generation through the correlation between other individuals and the best individual in current population, and also adds three external archives for rapid convergence. This algorithm improves the convergence speed to some extent, and increases the chance of obtaining the optimal solution within a limited time frame. However, Corr-GAA tends to select the candidate individuals in the population that have the highest correlation with the current best individual and whose fitness is higher than the average fitness of the population. It may fall into the local optimum. If an effective mutation operator is introduced in the algorithm, the search space will be explored in a wider range. It can further enrich the solution space and increase the possibility of obtaining a better solution.

3 The Proposed Algorithm

In this section, our iCorr-GAA algorithm and its components for solving complex global optimization problems are described.

3.1 Mutation

At early evolution, correlation selection operator that is too dependent on the optimal individuals in current population tends to reduce the population diversity and produce precocity. If the search space that mutation operator explores is too small, the result will tend to be locally optimal. In the later stages of evolution, too larger search space will make the result random due to the randomness introduced by archive T_e. We need to dynamically adjust the search space that mutation operator explores. Therefore, we combine this algorithm with the non-uniform mutation operator [14] which is adaptive.

The entire population is searched extensively to find the possible area obtained the optimal solution. The range of search space shrinks as the number of generation increases, and only the small area of the current solution is searched at the end of the evolution. In this way, the optimal solution can be accurately located and it will not be far away from the current optimal area. The mutation operates as follows:

$$v_k' = \begin{cases} v_k + \Delta(g, u_k - v_k) & \text{if } \gamma = 0 \\ v_k - \Delta(g, v_k - l_k) & \text{if } \gamma = 1 \end{cases} \tag{6}$$

Where, γ randomly takes 0 or 1, and the function $\Delta(g, y)$ returns a number from $[0, y]$. The probability that $\Delta(g, y)$ approaches 0 increases as generation increases. The $\Delta(g, y)$ is given by,

$$\Delta(g, y) = y * \left(1 - r^{\left(1-\frac{g}{G}\right)^b}\right) \tag{7}$$

Where, r is a randomly chosen number from [0, 1], G denotes the maximum of iterations, and b is the system parameter that determines the non-uniformity, which plays the role of adjusting the local search area.

3.2 Crossover

We use BLX $- \alpha$ crossover operator [15] which is proposed by Eshelman and Schaffer. The BLX $- \alpha$ operator are highly exploratory, which causes diversity in offspring. For two parent individuals: X_A: $\{x_{1,A}, x_{2,A}, x_{3,A}, \ldots, x_{D,A}\}$ and X_B: $\{x_{1,B}, x_{2,B}, x_{3,B}, \ldots, x_{D,B}\}$, the BLX $- \alpha$ operator randomly chooses a solution from the interval $[C_{min} - I\alpha, C_{max} + I\alpha]$, where C_{max}, C_{min} and I are calculated as follows,

$$\begin{aligned} C_{max} &= \max\{x_{i,A}, x_{i,B}\} \\ C_{min} &= \min\{x_{i,A}, x_{i,B}\} \\ I &= C_{max} - C_{min} \end{aligned} \tag{8}$$

3.3 The iCorr-GAA Algorithm

The iCorr-GAA algorithm first initializes the trial population. In each iteration, it calculates the fitness value of each individual in the current population and updates three external archives T_b, T_{xb}, and T_e. It uses the correlation selection method adopted from Corr-GAA algorithm to select individuals to be parents. If external archives are not full, the selection pool contains only individuals from the current population. Otherwise, individuals from archive T_{xb} and T_e are also added to the selection pool along with the current population. After external archives are full, one of the parents is selected from the current population, and the other is selected from archive T_{xb} or archive T_e or from the current population. The selected parent individuals produce offspring using the BLX $- \alpha$ operator with crossover probability P_c, and then offspring perform mutation with probability P_m according to the non-uniform mutation strategy to generate a new population. Finally, the selection, crossover, and mutation operations are repeated for the new population until the termination criteria are met. The flow chart of iCorr-GAA algorithm is presented in Fig. 1.

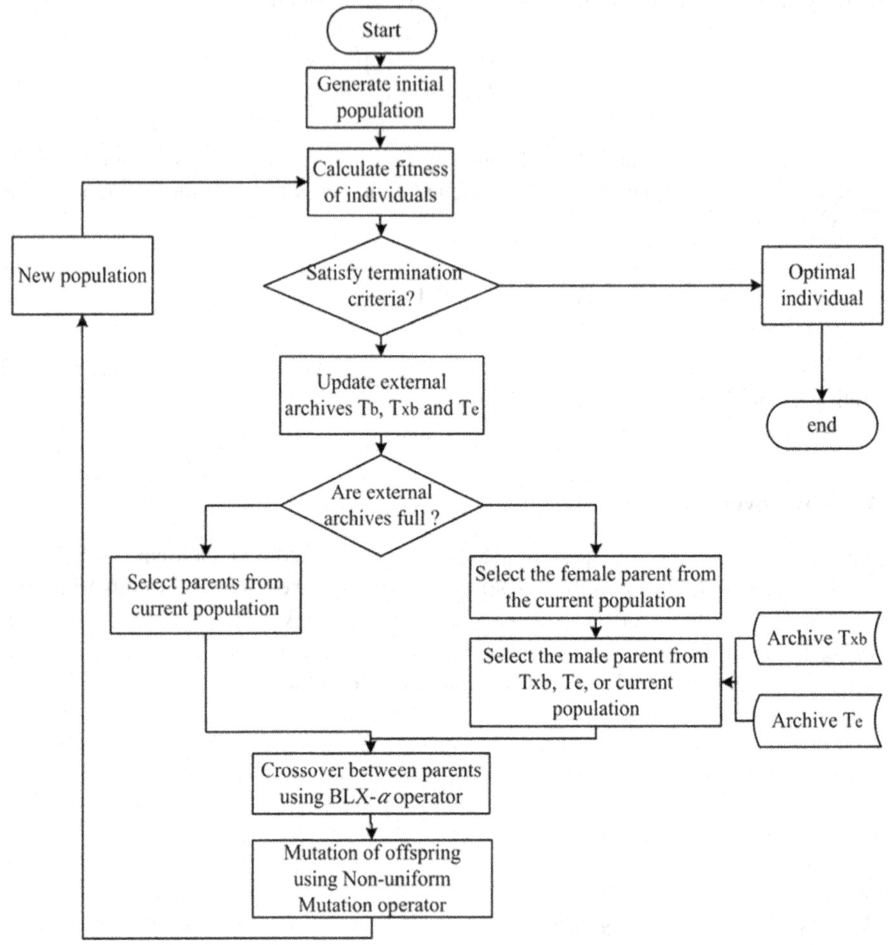

Fig. 1. Flow chart of iCorr-GAA algorithm

4 Experiment Results

In this section, the performance of iCorr-GAA algorithm was evaluated by solving a set of single objective problems introduced in the 2014 IEEE Evolutionary Computing International Conference special session and competition [16]. The algorithm was coded using Matlab R2014a and was run on a PC with a Intel(R) Core(TM) i3-4170 CPU @ 3.70 GHz processor, 4.0 GB RAM, and with Windows 7 Enterprise OS.

The algorithm was tested on a set of 30 benchmark functions, whose dimensions are 10 and 30. And 30 runs of the algorithm were needed for each test problem.

4.1 Initialization of the Parameters

In this paper, the value of T (i.e. the size of T_b and T_{xb}) and R (i.e. the size of T_e) should maintain a good balance between exploitation and exploration. T is set to $max\_iteration \times 0.4$ and R is set to $max\_iteration \times 0.2$. Similarly, P_{xb} which denotes the usage probability of archive T_{xb} is set to 0.5, and P_e which denotes the usage probability of archive T_e is set to 0.2. The system parameter that determines the non-uniformity b is set to 2. Besides, the population size N_p is 100, the crossover probability P_c is set to 0.9, and the mutation probability P_m is set to 0.051. All variable were randomly initialized with $X_{max} = 100$ while $X_{min} = -100$.

4.2 Result for 10D

Table 1 lists the best, worst, median, mean, and standard deviation of the best solution that iCorr-GAA obtains in 30 runs. The maximum iterations is set to 300 for each run.

From this results, it is clear that iCorr-GAA can show good performance on the optimization problem of a simple multi-modal function and can be very close to the global optimum. For the composition function, the proposed algorithm is able to reach near optimal solutions. However, the algorithm is not able to obtain the global optimal solutions for all test problems, indicating that the search capability of the proposed algorithm needs to be further strengthened.

4.3 Result for 30D

Table 2 shows the computational results of the proposed algorithm for 30D test problems, and the best, worst, median, mean, and standard deviation of the best solution are presented in each column in the table. The maximum iterations is set to 500 for each run.

From this table, the error value of the optimization problem calculated by the proposed algorithm increases with the dimension of the problems. But the algorithm still exhibit good performance in solving high dimension multimodal function, and can be closed to the global optimal solution.

4.4 Results for Two Function Instances

Two single objective optimization functions are analyzed in detail as follows:

A. Shifted Rastrigins Function

$$F_8(\mathrm{x}) = f_8\left(\frac{5.12(\mathrm{x} - \mathrm{o}_8)}{100}\right) + F_7^*$$

Table 1. Result for 10D

Func.	Best	Worst	Average	Std.
F01	6.2331E+04	1.0463E+08	1.4892E+07	2.1963E+07
F02	1.6928E+04	2.2738E+05	6.2827E+04	6.2388E+04
F03	4.7729E+02	3.7653E+04	1.0381E+04	1.0483E+04
F04	4.0078E+02	4.7480E+02	4.2529E+02	2.3558E+01
F05	5.2000E+02	5.2025E+02	5.2009E+02	5.6700E−02
F06	6.0053E+02	6.0645E+02	6.0312E+02	1.4470E+00
F07	7.0009E+02	7.0091E+02	7.0052E+02	1.9880E−01
F08	8.0000E+02	8.0011E+02	8.0005E+02	5.4506E−02
F09	9.0597E+02	9.2487E+02	9.1362E+02	4.6832E+00
F10	1.0041E+03	1.2989E+03	1.1040E+03	8.2894E+01
F11	1.2573E+03	2.2690E+03	1.7765E+03	2.5353E+02
F12	1.2001E+03	1.2011E+03	1.2004E+03	1.8555E−01
F13	1.3001E+03	1.3004E+03	1.3003E+03	8.5244E−02
F14	1.4001E+03	1.4008E+03	1.4003E+03	1.7911E−01
F15	1.5014E+03	1.5108E+03	1.5038E+03	2.1559E+00
F16	1.6021E+03	1.6036E+03	1.6031E+03	4.0239E−01
F17	8.8001E+03	3.1356E+06	8.7396E+05	8.5476E+05
F18	1.8155E+03	3.8343E+04	1.1910E+04	1.0430E+04
F19	1.9006E+03	1.9060E+03	1.9026E+03	1.4596E+00
F20	2.0547E+03	4.3939E+04	1.3814E+04	1.4930E+04
F21	3.4763E+03	5.8863E+06	7.0165E+05	1.2673E+06
F22	2.2004E+03	2.4556E+03	2.3097E+03	8.5287E+03
F23	2.6294E+03	2.6397E+03	2.6300E+03	1.9316E+00
F24	2.5128E+03	2.5610E+03	2.5281E+03	1.1520E+01
F25	2.6422E+03	2.7051E+03	2.6920E+03	2.0343E+01
F26	2.7021E+03	2.7220E+03	2.7070E+03	4.5117E+00
F27	2.7045E+03	3.2141E+03	3.0658E+03	1.2448E+02
F28	3.1198E+03	3.2228E+03	3.1698E+03	4.0638E+01
F29	3.1089E+03	3.5488E+03	3.2076E+03	1.6744E+02
F30	3.3370E+03	3.9862E+03	3.5716E+03	2.1301E+02

Where,

$$f_8(x) = \sum_{i=1}^{D} \left(x_i^2 - 10\cos(2\pi x_i) + 10\right),$$

$$F_7^* = 800$$

Table 2. Result for 30D

Func.	Best	Worst	Average	Std.
F01	1.9196E+07	3.2781E+09	5.0793E+08	8.6114E+07
F02	8.5349E+07	1.2943E+08	1.6514E+08	3.2734E+07
F03	1.3375E+03	6.8501E+05	4.3805E+04	1.2059E+05
F04	5.5153E+02	7.3134E+02	5.8668E+02	5.9936E+01
F05	5.2013E+02	5.2044E+02	5.2028E+02	7.8482E−02
F06	6.1472E+02	6.2362E+02	6.1939E+02	1.8363E+00
F07	7.0103E+02	7.0176E+02	7.0123E+02	2.0428E−02
F08	8.0020E+02	8.0416E+02	8.0152E+02	1.0053E+00
F09	9.5119E+02	1.0317E+03	9.9372E+02	2.2093E+01
F10	1.1546E+03	1.6335E+03	1.3998E+03	1.3725E+02
F11	3.0361E+03	5.7032E+03	4.4369E+03	7.0339E+02
F12	1.2003E+03	1.2007E+03	1.2005E+03	1.2296E−01
F13	1.3003E+03	1.3006E+03	1.3004E+03	1.1042E−01
F14	1.4001E+03	1.4006E+03	1.4003E+03	9.4176E−02
F15	1.5181E+03	1.3828E+05	4.3028E+04	4.3028E+04
F16	1.6021E+03	1.6036E+03	1.6030E+03	4.0239E−01
F17	4.6025E+05	4.3365E+05	1.7310E+05	1.2075E+05
F18	4.6322E+03	3.7960E+04	1.4522E+03	1.0120E+03
F19	1.9006E+03	1.9060E+03	1.9026E+03	3.3099E+01
F20	8.9073E+03	8.2246E+04	3.2182E+04	2.4660E+04
F21	3.6799E+04	1.7249E+05	3.7785E+05	3.4124E+05
F22	2.6987E+03	3.8506E+03	3.1095E+03	2.6208E+02
F23	2.6144E+03	2.6756E+03	2.6278E+03	1.4516E+01
F24	2.6265E+03	2.6457E+03	2.6357E+03	6.4316E+00
F25	2.7021E+03	2.7220E+03	2.7070E+03	4.5117E+00
F26	2.7002E+03	2.7035E+03	2.7010E+03	8.6951E−01
F27	3.1525E+03	3.6060E+03	3.4831E+03	1.6920E+02
F28	3.2266E+03	3.4246E+03	3.3025E+03	5.1411E+01
F29	3.1120E+03	3.3626E+03	3.2102E+03	3.1520E+02
F30	3.8637E+03	8.2246E+04	1.2182E+04	1.5213E+04

Table 3. The results of the shifted rastrigins function for D = 10 and D = 30

Func.	Dimension	Best	Worst	Average	Std.
F08	10	8.0000E+02	8.0011E+02	8.0005E+02	5.4506E−02
F08	30	8.0020E+02	8.0416E+02	8.0152E+02	1.0053E+00

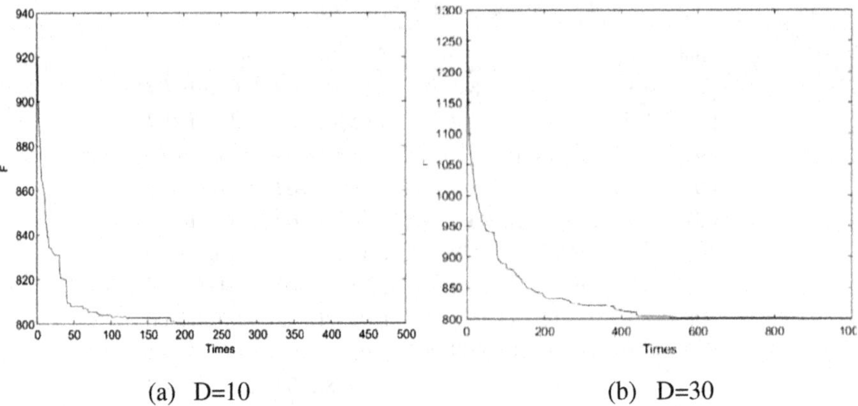

(a) D=10 (b) D=30

Fig. 2. Change curve of optimal fitness of shifted rastrigins function for 10 dimension and 30 dimension

The shifted rastrigins function is a multimodal function with variable dimensions and multiple local optimal solutions. Because these characteristics of the function have strong deceptiveness to ordinary evolutionary algorithms, they are easy to fall into a local optimum and cannot obtain a global optimum. Table 3 shows experimental statistical results of the shifted rastrigins function. The change curve of optimal fitness of this function for 10 dimension and 30 dimension are presented in Fig. 2.

As can be seen from the figure, the 10-dimensional objective function can converge to the global optimal solution after 200 generations, and the 30-dimensional objective function can also tend to the global optimal solution after 600 generations. Therefore, the experiments show that the iCorr-GAA algorithm has a strong ability of searching for optimization.

B. Shifted and Rotated Griewanks Function

$$F_7(x) = f_7\left(\frac{5.12(x - o_8)}{100}\right) + F_6^*$$

Where,

$$f_7 = -\frac{1}{4000}\sum_{i=1}^{n} x_i^2 + \prod_{i=1}^{n}\cos\left(\frac{x_i}{\sqrt{i}}\right) - 1$$

$$F_6^* = 700$$

Table 4. The results of shifted and rotated griewanks function for D = 10 and D = 30

Func	Dimension	Best	Worst	Average	Std.
F07	10	7.0009E+02	7.0091E+02	7.0052E+02	1.9880E−01
F07	30	7.0103E+02	7.0176E+02	7.0123E+02	2.0428E−02

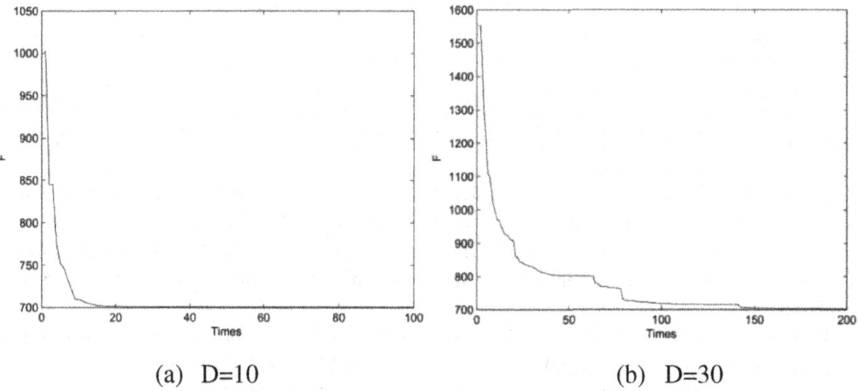

(a) D=10 (b) D=30

Fig. 3. Change curve of optimal fitness of shifted and rotated griewanks for 10 dimension and 30 dimension

The shifted and rotated griewanks function is a non-linear, symmetric, inseparable, multimodal function which has many local optimal solutions but only one global optimal solution. As the dimension of the function increases, the search difficulty increases. Table 4 shows experimental statistical results of the shifted and rotated griewanks function. And the change curve of optimal fitness of this function for 10 dimension and 30 dimension are presented in Fig. 3.

It can be seen from the figure that the 10-dimensional and 30-dimensional objective functions can converge to the global optimal solution after 40 generations and 160 generations respectively. Therefore, it shows that the proposed algorithm can quickly find the best solution within the required time frame, and at the same time, due to the randomness introduced by the non-uniform mutation operator, it also avoids falling into the local optimal solution.

5 Conclusion

In this paper we presented an improved genetic algorithm, called iCorr-GAA, combining the correlation-based genetic algorithm with non-uniform mutation to solve complex optimization problems. The performance of iCorr-GAA algorithm was evaluated on the set of benchmark functions provided for CEC 2014 special session. The experimental results give evidence that iCorr-GAA algorithm has good global search capability and fast convergence speed. But, it may still fall into a local optimal solution for some optimization problems. Therefore, the global search capability of iCorr-GAA algorithm needs to be further strengthened.

Acknowledgement. The project was partly sponsored by Guangdong province science and technology planning projects (Grant: 2016B070704010), and Guangdong province science and technology planning projects (Grant: 2016B010124010).

References

1. Elsayed, S.M., Sarker, R.A., Essam, D.L.: A genetic algorithm for solving the CEC'2013 competition problems on real-parameter optimization. In: Evolutionary Computation, pp. 356–360. IEEE (2013)
2. Segura, C., Coello, C.A.C., Miranda, G., et al.: Using multi-objective evolutionary algorithms for single-objective optimization. 4OR **11**(3), 201–228 (2013)
3. Brest, J., Maučec, M.S., Bošković. B.: iL-SHADE: improved L-SHADE algorithm for single objective real-parameter optimization. In: Evolutionary Computation, pp. 1188–1195. IEEE (2016)
4. Thakur, M., Meghwani, S.S., Jalota, H.: A modified real coded genetic algorithm for constrained optimization. Appl. Math. Comput. **235**(235), 292–317 (2014)
5. Goldberg, D.: Genetic Algorithms in Search, Optimization, and Machine Learning. Addison-Wesley, MA (1989)
6. Chuang, Y.C., Chen, C.T., Hwang, C.: A real-coded genetic algorithm with a direction-based crossover operator. Inf. Sci. **305**, 320–348 (2015)
7. Ali, M.Z., Awad, N.H., Suganthan, P.N., et al.: An improved class of real-coded genetic algorithms for numerical optimization. Neurocomputing **275**, 155–166 (2017)
8. Falco, I.D., Cioppa, A.D., Tarantino, E.: Mutation-based genetic algorithm: performance evaluation. Appl. Soft Comput. J. **1**(4), 285–299 (2002)
9. Lu, H.L., Wen, X.S., Lan, L., et al.: A self-adaptive genetic algorithm to estimate JA model parameters considering minor loops. J. Magn. Magn. Mater. **374**, 502–507 (2015)
10. Kurdi, M.A.: A new hybrid island model genetic algorithm for job shop scheduling problem. Comput. Industr. Eng. **88**(C), 273–283 (2015)
11. Trivedi, A., Srinivasan, D., Biswas, S., Reindl, T.: Hybridizing genetic algorithm with differential evolution for solving the unit commitment scheduling problem. Swarm Evol. Comput. **23**, 50–64 (2015)
12. González, M.A., Vela, C.R., Varela, R.: A new hybrid genetic algorithm for the job shop scheduling problem with setup times. In: Eighteenth International Conference on Automated Planning and Scheduling, ICAPS 2008, Sydney, Australia, pp. 116–123, DBLP, September 2008
13. Kundu, A., Laha, S., Vasilakos, A.V.: Correlation-based genetic algorithm for real-parameter optimization. In: Evolutionary Computation, pp. 4804–4809. IEEE (2016)
14. Michalewicz, Z.: Genetic Algortithms+Data Structure=Programs. Springer, Berlin (1992)
15. Achiche, S., Ahmed-Kristensen, S.: Genetic fuzzy modeling of user perception of three-dimensional shapes. Artif. Intell. Eng. Des. Anal. Manuf. **25**, 101 (2011)
16. Liang, J.J., Qu, B.Y., Suganthan, P.N.: Problem definitions and evaluation criteria for the cec 2014 special session and competition on single objective real-parameter numerical optimization. Computational Intelligence Laboratory, Zhengzhou University, Zhengzhou China and Technical report, Nanyang Technological University, Singapore (2013)

Two Possible Paradoxes in Numerical Comparisons of Optimization Algorithms

Qunfeng Liu[1], Wei Chen[1], Yingying Cao[1], Yun Li[1(✉)], and Ling Wang[2(✉)]

[1] School of Computer Science and Network Security, Dongguan University of Technology, Dongguan, China
Yun.Li@ieee.org
[2] Department of Automation, TsingHua University, Beijing, China
wangling@tsinghua.edu.cn

Abstract. Comparison strategies of benchmarking optimization algorithms are considered. Two strategies, namely "C2" and "C2+", are defined. Existing benchmarking methods can be regarded as different applications of them. Mathematical models are developed for both "C2" and "C2+". Based on these models, two possible paradoxes, namely the cycle ranking and the survival of the non-fittest, are deduced for three optimization algorithms' comparison. The probabilities of these two paradoxes are calculated. It is shown that the value and the parity of the number of test problems affect the probabilities significantly. When there are only dozens of test problems, there is about 75% probability to obtain a normal ranking result for three optimization algorithms' numerical comparison, about 9% for cycle ranking, and 16% for survival of the non-fittest.

Keywords: Optimization algorithm · Benchmarking · Paradox
Survival of the non-fittest · Cycle ranking

1 Introduction

Numerous optimization algorithms have been developed to solve the following minimization problem

$$\mathrm{Min} f(x)\ s.t. x \in \Omega \tag{1}$$

where $f(x)$ is the objective function and Ω is the feasible region. If Ω is countable, then (1) is called as a discrete optimization problem, otherwise, a continuous optimization problem. If Ω comes from some constrain conditions, then problem (1) is constrained, otherwise unconstrained. Any maximization problem can be easily modeled as the above minimization problem through replacing $f(x)$ with $-f(x)$.

When the objective function $f(x)$ is nonconvex, problem (1) is often hard to solve. Therefore, in the mathematical programming community, local optimum $\hat{x}$ satisfied

$$f(\hat{x}) \leq f(x), \quad \forall x \in B_\delta(\hat{x}) \tag{2}$$

is often seeked, where $B_\delta(\hat{x})$ is a neighborhood of $\hat{x}$. The gradient information of $f(x)$ is helpful in algorithm design and mathematical analysis. However, in the evolutionary

D.-S. Huang et al. (Eds.): ICIC 2018, LNCS 10955, pp. 681–692, 2018.
https://doi.org/10.1007/978-3-319-95933-7_77

computation community and the global optimization community, global optimum x^* satisfied

$$f(x^*) \leq f(x), \ \forall x \in \Omega \tag{3}$$

is investigated. Global optimization is often harder than local optimization, one reason is that there is no information which guides to x^* mathematically.

Therefore, it is necessary to compare optimization algorithms' performance numerically. Firstly, there are many optimization algorithms, and which one is the "best" on some specified functions is often unclear. Numerical comparison can bring helpful insight. Secondly, there is no suitable mathematical convergence for global optimization algorithms, and numerical comparison is the only way to show their efficiency.

Extensive studies have been done on how to compare optimization algorithms numerically, especially on the design of test problems and the development of data analysis methods. Numerous test problems including many sets of benchmark functions [1–4] and hundreds of practical test problems [5–7] have been designed or modeled for numerical comparison of optimization algorithms.

Furthermore, many methods for analyzing experimental data are developed. For instance, the popular performance profiles [8, 9] and data profiles [10, 11] for comparing deterministic optimization algorithms. More methods are developed for comparing stochastic optimization algorithms, e.g., calculating means and standard deviations [12, 13], displaying the history of the found best function values [14, 15], applying statistical inferences [16, 17], employing empirical distribution functions [18–20], and visualizing confidence intervals [18, 19].

However, there are few literatures discuss the selection of comparison strategy, which relates to but is different from the analysis method. When analyze empirical data, if there are only two optimization algorithms, then comparison strategy is unnecessary. However, when algorithms exceed two, there are two basic comparison strategies, namely "C2" strategy and "C2+" strategy. In this paper, "C2" strategy means to compare two algorithms at every match and repeats several matches to obtain an aggregated ranking [12–17]. On the contrary, "C2+" strategy means to rank all algorithms through one or few grand matches [2, 8–11, 18, 19]. In other words, the main difference between "C2" and "C2+" is how many algorithms are compared in each match: two for "C2" while more than two for "C2+".

In this paper, we dedicate to answer the following questions: What is the difference between the ranking results when employing the "C2" or "C2+" strategy? are the ranking results compatible? These questions are well known in the community of political elections and some other social science [21–23]. However, they are unfamiliar in the numerical optimization community, especially the evolutionary computation community.

Through considering the properties and conditions of numerical comparison of optimization algorithms, we will show that the results of "C2" and "C2+" strategy may be different and even incompatible. Specifically, two paradoxes are shown to be possible and their probabilities are calculated to determine the extent of incompatibility.

The rest of this paper is organized as follows. In next Section, the “C2” strategy and the “C2+” strategy are modeled mathematically for convenient analysis. Based on the model, possible paradoxes are deduced in Sect. 3, and probabilities of paradoxes are calculated in Sect. 4. Finally, some conclusions are summarized in Sect. 5.

2 Model the Comparison Strategies

In this section, we describe and model mathematically both the “C2” and the “C2+” strategies.

2.1 The “C2” Strategy

Under this strategy, the whole comparison is divided into several matches (sub-comparisons), and only two algorithms are considered in each match. There are two popular applications of the “C2” strategy, namely the one-play-all comparison and the all-play-all comparison.

One-Play-All Comparison. One-play-all comparison means to compare one special algorithm with all other algorithms, one by one. It is often applied when a new algorithm (including the improvement of an existing algorithm) is proposed. In this case, whether the proposed algorithm performs better than existing similar algorithms is often concerned, and therefore, popular choice is to compare the proposed algorithm with some popular existing algorithms [12, 15, 17, 18, 24, 25]. Different data analysis methods are allowable for applying the one-play-all comparison. For example, the statistical test methods [12, 15, 17, 24, 25], the cumulative distribution function methods [2, 18, 26], and the visualizing confidence intervals method [19].

All-Play-All Comparison. All-play-all comparison means to compare each algorithm with all other algorithms, one by one, and is often called as the Round-robin comparison. It is often applied in algorithms competition [3]. All-play-all comparison can be regarded as a repeated version of one-play-all comparison.

Suppose there are k algorithms, then $k-1$ matches are needed to finish a one-play-all comparison, while $\frac{k(k-1)}{2}$ matches are needed to finish an all-play-all comparison. In other words, k one-play-all comparisons are executed. To aggregate several one-play-all comparisons' ranking results, it is popular to sum up each algorithm's ranking number.

Mathematical Model of “C2” Strategy. Although different data analysis methods are allowable for applying the “C2” strategy, only two ranking results are possible in any match of two algorithms A_1, A_2: A_1 performs better than A_2, or A_1 does not perform better than A_2. In numerical optimization, there are several different standards to judge which algorithm performs better, e.g., convergence, robustness or efficiency. Any single standard is allowable in this paper. For convenience of later discussion, we select one of the most popular standards in global optimization. Specifically, given computational budget, the found best objective function values are employed to determine which algorithm performs better.

Definition 1. Given a fixed computational cost, suppose that f^i_{min} is the found minimal objective function value by the algorithm A_i, $i = 1, 2$. Then $A_1 \succsim A_2$ if and only if $f^1_{min} \leq f^2_{min}$. Moreover, $A_1 > A_2$ if and only if $f^1_{min} < f^2_{min}$, and $A_1 = A_2$ if and only if $f^1_{min} = f^2_{min}$.

Based on Definition 1, there are two possible ranking results of the match of algorithms A_1 and A_2 on each test problem: $A_1 \succsim A_2$ or $A_2 \succsim A_1$. Therefore, if m test problems are tested, then there are 2^m possible ranking combinations. This can be regarded as a random sampling with size m from the population with a binomial distribution, which is described in Table 1.

Table 1. Ranking distribution when comparing algorithms A_1, A_2

Ranking	$A_1 \succsim A_2$	$A_2 \succsim A_1$
Probability	p	$1 - p$

In Table 1, p measures the occurrence probability of the event "$A_1 \succsim A_2$", and it is problem dependant. If the test problem biases A_1, then p is close to 1. On the contrary, p is close to 0 if the test problem biases A_2. More details about the parameter p will be discussed in Sect. 4.

Given some test problems, the sampling can be described as a matrix, e.g.,

$$\begin{bmatrix} A_1 & A_2 & A_1 & \ldots & A_2 \\ A_2 & A_1 & A_2 & \ldots & A_1 \end{bmatrix}_{2 \times m} \tag{4}$$

one column for each problem. In this matrix, the first column $[A_1, A_2]^T$ means A_1 performs better than A_2, and the rest is similar.

2.2 The "C2+" Strategy

This strategy compares all the algorithms at a single or few matches, and it is often used to determine the winner(s) in algorithms competitions [2, 27] or new algorithm proposing [9, 10, 28].

Difference Between "C2+" and "C2". An obvious difference is that "C2+" allows to compare more than two algorithms in a single match while "C2" always compare two algorithms in each match. This brings another difference that "C2" often needs much more matches than "C2+" to finish the whole comparison.

The third but maybe the most important difference between "C2" and "C2+" is that "C2+" adopts statistical aggregation method to obtain all algorithms' ranking results directly. On the contrary, "C2" has to obtain ranking in each match firstly and then aggregate them to obtain a final ranking.

Mathematical Model of "C2+". Suppose there are k algorithms $A_i, i = 1, \ldots, k$. After testing these k algorithms on a test problem, k found best function values f^i_{min} are obtained for any fixed computational budget, where f^i_{min} is the best function value found by $A_i, i = 1, \ldots, k$. Through comparing these function values, a ranking

$$A_{i1} \succsim A_{i2} \succsim \ldots \succsim A_{ik} \tag{5}$$

is obtained, where $(i1, i2, \ldots, ik)$ is a permutation of $(1, 2, \ldots, k)$, and the relationship $\succsim$ is defined in Definition 1. Obviously, different test problem often brings different ranking.

Since there are totally $k!$ possible ranking series, we obtain a multinomial distribution. When $k = 3$, the distribution is summarized in Table 2, where the parameter $p_i, i = 1, \ldots, 6$ and satisfied $\sum_{i=1}^{6} p_i = 1$. When $k > 3$, the distribution is similar as but more complex than that in Table 2.

Table 2. Ranking distribution when comparing algorithm $A_i, i = 1, 2, 3$.

Ranking	$A_1 \succsim A_2 \succsim A_3$	......	$A_3 \succsim A_2 \succsim A_1$
Probability	p_1	......	p_6

If m problems are tested, then it can be regarded as a random sampling with size m from the distribution. For convenience, denote (5) as the following column vector

$$[A_{i1}, A_{i2}, \ldots, A_{ik}]^T.$$

Then a matrix

$$M_1 = \begin{bmatrix} A_{j1} & \cdots & A_{i1} \\ A_{j2} & \cdots & A_{i2} \\ \cdots & \cdots & \cdots \\ A_{jk} & \cdots & A_{ik} \end{bmatrix} \tag{6}$$

can be used to represent the random sampling with size m, each column corresponds the ranking on a test problem. Denote X_i as the number of the i-th ranking, $i = 1, \ldots, k$, then $\sum_{i=1}^{k!} X_i = m$ and the random vector $X = [X_1, X_2, \ldots, X_k]^T$ follows the multinomial distribution with parameter m and $p = [p_1, p_2, \ldots, p_{k!}]$.

Since the sampling matrix (6) of "C2+" includes the ranking information of any pair of algorithms, it contains the sample matrix (4) of "C2". Therefore, it can be adopted to analyze the relationship of ranking results from both "C2+" and "C2". Based on the matrix (6), two paradoxes are presented in Sect. 3, and their probabilities are calculated in Sect. 4 by the help of the multinomial distribution in Table 2.

3 Two Paradoxes

In this section, we adopt the majority rule, which is very popular in numerical comparisons of optimization algorithms [2, 12, 15, 17, 19], to deduce two paradoxes.

Assumption 1 (Majority rule). An algorithm performs better than other algorithms if it can perform better on more test problems than the others do.

For simplicity, only 3 algorithms (A_1, A_2, A_3) are considered in this and next sections, and it is enough for our purpose. In this case, there are 6 possible ranking series, and its ranking distribution is the multinomial distribution listed in Table 2.

Given any problem and the computational budget, test these 3 algorithms on it, and we can obtain a ranking result. According to the discussions in Sect. 2, it can be regarded as a random sampling from Table 2. Specifically, testing on a problem is regarded as a random sampling from the distribution in Table 2. Repeat such process until a desired number of problems are tested, then we obtain a sampling matrix, e.g.,

$$M_1 = \begin{bmatrix} A_2 & A_3 & A_2 & A_3 & A_1 \\ A_1 & A_1 & A_1 & A_2 & A_3 \\ A_3 & A_2 & A_3 & A_1 & A_2 \end{bmatrix}, \quad M_2 = \begin{bmatrix} A_2 & A_3 & A_2 & A_3 & A_1 \\ A_1 & A_1 & A_1 & A_1 & A_2 \\ A_3 & A_2 & A_3 & A_2 & A_3 \end{bmatrix}. \tag{7}$$

Both M_1 and M_2 have 3 rows and 5 columns, indicating that 3 algorithms have been tested on 5 problems. The first column means $A_2 \succsim A_1 \succsim A_3$, i.e., A_2 performs better than or similarly as A_1 on this problem, and A_1 performs better than or similarly as A_3 on this problem. The rest is similar.

Paradox from "C2": Cycle Ranking. Suppose that there are totally 5 test problems, and the ranking results are given by the matrix M_1 in (7). If we adopt the "C2" strategy to compare these 3 algorithms, then $A_2 \succsim A_1$ since 3 problems bias A_2 and 2 bias A_1. Similarly, $A_1 \succsim A_3$ since 3 problems bias A_1 and 2 bias A_3, $A_3 \succsim A_2$ since 3 problems bias A_3 and 2 bias A_2. As a result, we obtain a cycle ranking $A_2 \succsim A_1 \succsim A_3 \succsim A_2$, and we cannot tell which algorithm performs the best.

The cycle ranking paradox is also called as Condorcet paradox [29, 30], which is very popular in voting theory and was found firstly by Marquis de Condorcet in the 18th century when he investigated a voting system. In next section, we will discuss the occurrence probability of cycle ranking, and how the number of test problems affect the probability.

Paradox from "C2+": Survival of the Non-fittest. Suppose that the ranking results are given by the matrix M_2 in (7). If we adopt the "C2+" strategy to compare these 3 algorithms, then A_2 or A_3 is the winner since both perform the best on 2 problems while A_1 only performs the best on the fifth problem.

However, if we compare A_1 and A_2 alone, then A_1 performs better than A_2 on 3 problems (2nd, 4th and 5th) while worse only on 2 problems (1st and 3rd). Therefore, A_1 performs better than A_2 on the whole test set. Similarly, A_1 performs better than A_3 on the whole test set, too. Therefore, A_1 is the winner of the "C2" strategy.

In other words, the winner of the "C2+" strategy do not perform well in "C2" comparisons. Such phenomenon is called as the survival of the non-fittest in this paper, which is also called as the Borda paradox, it was also found in the 18th century [30]. We will discuss its occurrence probabilities in next section and show how the number of test problems affect the probability.

4 Probability Analysis

To calculate the occurrence probabilities of cycle ranking and survival of the non-fittest, the parameters in Table 2 should be determined firstly. In this paper, we adopt the following No Free Lunch (NFL) assumption.

Assumption 2 (The NFL assumption). For any given 3 optimization algorithms and any given test problem, all 6 possible rankings of these algorithms on this problem are equally likely, i.e., $p_i = \frac{1}{6}$, $i = 1, \ldots, 6$ in Table 2.

The NFL assumption is a direct application of the No Free Lunch theorem in optimization [31]. According to the NFL theorem, if the test problems are selected randomly from all possible problems, then the average performance of any algorithm is equal. In other words, any ranking in Table 2 has the same occurrence probability, and therefore $p_i = \frac{1}{6}$.

If these 3 algorithms have been tested on m problems, and the i-th ranking in Table 2 has appeared X_i times, $i = 1, \ldots, 6$, then the random vector $X = [X_1, X_2, \ldots, X_6]^T$ satisfies the following multinomial distribution.

$$P(X_1 = x_1, \ldots, X_6 = x_6) = \frac{m!}{x_1! \ldots x_6!} \frac{1}{6^m}, \tag{8}$$

where $x_i \in [0, m], i = 1, \ldots, 6$ and satisfy $\sum_{i=1}^{6} x_i = m$. In this paper, $P(A)$ is denoted as the probability of a random event A.

Then we calculate the occurrence probabilities of cycle ranking and survival of the non-fittest based on the NFL assumption.

4.1 Division of the Sample Space

Firstly, we give some definitions below, which define possible random events when benchmarking optimization algorithms.

Definition 2. When we adopt the "C2" strategy, if the ranking of these 3 algorithms form a cycle, i.e., $A_1 \succsim A_2 \succsim A_3 \succsim A_1$ or $A_3 \succsim A_2 \succsim A_1 \succsim A_3$, then we say that the random event of cycle ranking happens, or random event C happens for simplicity.

Definition 3. If the final winner of the "C2+" strategy is not the final winner of the "C2" strategy, then we say that the random event of survival of the non-fittest happens, or random event S happens for simplicity.

Definition 4. The final winner of the "C2+" strategy is exactly the final winner of the "C2" strategy, then we say that the random event N happens.

Then we have the following theorem, whose proof is not presented in this paper due to the limitation of space.

Theorem 1. Only three random events C, S, N are possible in the numerical comparisons of three optimization algorithms.

Theorem 1 implies that

$$\mathrm{P}(C) + P(S) + P(N) = 1. \tag{9}$$

Therefore, in later subsections, we will calculate the probabilities of $P(C)$ and $P(S)$, and then calculate the probability of $P(N)$ indirectly.

4.2 Probabilities of the Random Event C

Theorem 2. When comparing 3 optimization algorithms on m test problems, the probability of random event C is

$$\mathrm{P}(C) = \frac{1}{6^m}\left(2\sum\nolimits_{\{x_i\}\in C_1}\frac{m!}{x_1!\ldots x_6!} - \sum\nolimits_{\{x_i\}\in C_2}\frac{m!}{x_1!\ldots x_6!}\right), \tag{10}$$

where C_1, C_2 are determined as follows.

$$C_1 : \begin{cases} x_1 + \ldots + x_6 = m \\ x_1 + x_2 + x_5 \geq \frac{m}{2} \\ x_1 + x_3 + x_4 \geq \frac{m}{2} \\ x_4 + x_5 + x_6 \geq \frac{m}{2} \\ x_i = 0, 1, \ldots, m, i = 1, \ldots, 6, \end{cases} \qquad C_2 : \begin{cases} x_1 + \ldots + x_6 = m \\ x_1 + x_2 + x_5 = \frac{m}{2} \\ x_1 + x_3 + x_4 = \frac{m}{2} \\ x_4 + x_5 + x_6 = \frac{m}{2} \\ x_i = 0, 1, \ldots, m, i = 1, \ldots, 6. \end{cases} \tag{11}$$

Theorem 2's proof is omitted in this paper partly due to the limitation of space.

It is clear from (11) that C_2 is the border of C_1, and it is empty when m is odd. Furthermore, $\log_{m\to\infty} P(C_2) = 0$. Therefore, in the literatures of calculating Condorcet paradox's probabilities, only odd m are often considered [30].

Given the number of test problems m, we can calculate the probability $P(C)$ through formula (10) in Theorem 2. Figure 1 shows the numerical results of $P(C)$ for $m = 1, 2, \ldots, 100$. From Fig. 1 we found that $P(C) = 0$ when $m = 1$ and $P(C) = 0.5$ when $m = 2$. As m increases, $P(C)$ changes zigzagged, and there are two opposite trends of $P(C)$. When m is even, $P(C)$ decreases from 0.5 to near 0.13 as m increases. On the contrary, when m is odd, $P(C)$ increases from 0 to near 0.9 as m increases. These results are the same as those reported in [30] when m is odd. However, we provide the probabilities when m is even, which bring helpful insight.

Based on these calculations, we conclude that odd number of test problems is a good choice for numerical comparisons of optimization algorithms, since it decreases the occurrence probability of cycle ranking. Under this choice, the occurrence probability of cycle ranking is less than 9%. In other words, cycle ranking is only occasionally happened.

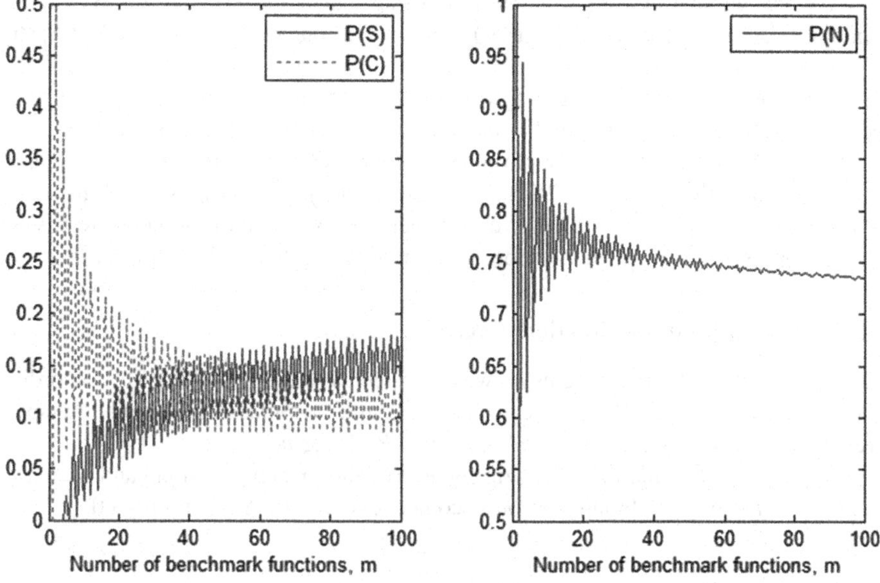

Fig. 1. Probabilities of $P(C), P(S)$ and $P(N)$.

4.3 Probabilities of the Random Event S

Although there are several probability calculations [30] of the Condorcet paradox (i.e.,$P(C)$), to our knowledge, there is no published results of $P(S)$. In [32, 33], the occurrence probabilities of the strict Borda paradox and the strong Borda paradox were analyzed, however, which are significantly different from $P(S)$.

We present the theoretical formula of $P(S)$ as the following theorem, whose proof is not included here due to the limitation of space.

Theorem 3. When comparing 3 optimization algorithms on m test problems, the probability of random event S is given by

$$\mathrm{P}(S) = \frac{3}{6^m}\left(\sum\nolimits_{\{x_i\}\in S_1} \frac{m!}{x_1!\ldots x_6!} - \sum\nolimits_{\{x_i\}\in S_2} \frac{m!}{x_1!\ldots x_6!}\right), \tag{12}$$

where the dominant S_1, S_2 are defined as follows.

$$S_1 : \begin{cases} x_1 + \ldots + x_6 = m \\ x_1 + x_2 + x_5 > \frac{m}{2} \\ x_1 + x_2 + x_3 > \frac{m}{2} \\ \max(x_3 + x_4, x_5 + x_6) > x_1 + x_2 \\ x_i = 0, 1, \ldots, m, i = 1, \ldots, 6, \end{cases} \quad S_2 : \begin{cases} x_1 + \ldots + x_6 = m \\ x_1 + x_2 + x_5 = \frac{m}{2} \\ x_1 + x_2 + x_3 > \frac{m}{2} \\ x_1 + x_3 + x_4 > \frac{m}{2} \\ x_5 + x_6 > x_1 + x_2 \\ x_5 + x_6 > x_3 + x_4 \\ x_i = 0, 1, \ldots, m, i = 1, \ldots, 6. \end{cases} \tag{13}$$

Given the number of test problems m, we can calculate the probability $P(S)$ through formula (12) in Theorem 3. Figure 1 shows the numerical results of $P(S)$ for $m = 1, 2, \ldots, 100$.

From Fig. 1 we found that $P(S) = 0$ until $m \geq 5$, and changes zigzagged as m increases. Roughly speaking, $P(S)$ increases from almost 0 to about 0.18 as $m = 5, 7, 9, \ldots$ increases, and increases from 0 to about 0.13 as $m = 6, 8, 10, \ldots$ increases.

To conclude, the probability $P(S)$ is larger than $P(C)$. What is more, $P(S)$ increases as m increases, whatever m is odd or even. Therefore, survival of the non-fittest is not too rare, and should be taken it seriously when adopting the "C2+" strategy.

4.4 Probabilities of the Random Event N

Given the number of test problems m, we can calculate the probability $P(N)$ according to (9), where $P(C)$ and $P(S)$ are calculated through formulas (10) and (12), respectively. Figure 1 shows the numerical results of $P(N)$ for $m = 1, 2, \ldots, 100$.

From Fig. 1 we found that $P(N)$ zigzagged violently when m is small and decreases roughly as m increases. Finally, $P(N)$ becomes less than 0.75 when $m > 80$.

5 Conclusions and Future Work

Numerical comparisons of optimization algorithms are analyzed through considering is as a selection, where optimization algorithms are regarded as candidates while test problems are regarded as voters. Two popular comparison strategies of benchmarking optimization algorithms are discussed, namely the "C2" strategy and the "C2+" strategies.

It was shown that two paradoxes, cycle ranking and survival of the non-fittest, are possible. Their probabilities are calculated when only three optimization algorithms are compared. It was shown that the value and the parity of the number of test problems m affect the probabilities significantly.

To decrease the probability of paradox, we suggest adopting an odd m test problems to implement a "C2" comparison, while an even m for a "C2+" comparison. However, our calculations show that it is impossible to eliminating both paradoxes except $m = 1$, which is impractical.

Roughly speaking, there is about 9% probability to find a cycle ranking when adopting the "C2" strategy, about 16% to find a survival of the non-fittest when adopting "C2+", and about 75% to obtain a normal ranking result for three optimization algorithms' numerical comparison. Therefore, "C2" is more suitable than "C2+" from the view of bringing less probability of paradox.

Although only three optimization algorithms are considered in this paper, the paradoxes happen in more general cases, and the probability calculation are ongoing. Several relevant issues are necessary to investigate.

Acknowledgment. This work was supported by National Key R&D Program of China (No. 2016YFD0400206), NSF of China (No. 61773119) and NSF of Guangdong Province (No. 2015A030313648).

References

1. Gaviano, M., Kvasov, D., Lera, D., Sergeyev, Y.D.: Algorithm 829: software for generation of classes of test functions with known local and global minima for global optimization. ACM Trans. Math. Softw. **9**, 469–480 (2003)
2. Hansen, N., Auger, A., Ros, R., Finck, S. and Pošík P.: Comparing results of 31 algorithms from the black-box optimization benchmarking bbob-2009. In: Proceedings of the 12th annual conference companion on genetic and evolutionary computation, pp. 1689–1696 (2010)
3. Awad, N.H., Ali, M.Z., Liang, J.J., Qu, B.Y., Suganthan, P.N.: Problem definitions and evaluation criteria for the CEC2017 special session and competition on single objective bound constrained real parameter numerical optimization. Nanyang Technological University, Singapore, Technical report, November 2016
4. Hansen, N., Auger, A., Mersmann, O., Tušar, T., Brockhoff, D.: Coco: A platform for comparing continuous optimizers in a black-box setting. ArXiv e-prints arXiv:1603.08785 (2016)
5. Gong, M., Wang, Z., Zhu, Z., Jiao, L.: A similarity-based multiobjective evolutionary algorithm for deployment optimization of near space communication system. IEEE Trans. Evol. Comput. **21**, 878–897 (2017)
6. Valle, Y., Venayagamoorthy, G.K., Mohagheghi, S., Hernandez, J.-C., Harley, R.G.: Particle swarm optimization: Basic concepts, variants and applications in power systems. Inf. Sci. **12**, 171–195 (2008)
7. Wang, Y., Xu, B., Sun, G., Yang, S.: A two-phase differential evolution for uniform designs in constrained experimental domains. IEEE Trans. Evol. Comput. **21**, 665–680 (2017)
8. Dolan, E.D., Moŕe, J.J.: Benchmarking optimization software with performance profiles. Math. Program. **91**, 201–213 (2002)
9. Liu, Q., Zeng, J.: Global optimization by multilevel partition. J. Glob. Optim. **61**, 47–69 (2015)
10. Liu, Q., Zeng, J., Yang, G.: MrDIRECT: a multilevel robust DIRECT algorithm for global optimization problems. J. Glob. Optim. **62**, 205–227 (2015)
11. Moŕe, J., Wild, S.: Benchmarking derivative-free optimization algorithms. SIAM J. Optim. **20**, 172–191 (2009)
12. Omidvar, M.N., Yang, M., Mei, Y., Li, X., Yao, X.: Dg2: A faster and more accurate differential grouping for large-scale black-box optimization. IEEE Trans. Evol. Comput. **21**, 929–942 (2017)
13. Yang, M., Omidvar, M.N., Li, C., Li, X., Cai, Z., Kazimipour, B., Yao, X.: Efficient resource allocation in cooperative co-evolution for large-scale global optimization. IEEE Trans. Cybern. **21**, 493–505 (2017)
14. Li, X., Yao, X.: Cooperatively coevolving particle swarms for large scale optimization. IEEE Trans. Evol. Comput. **16**, 210–224 (2012)
15. Qin, Q., Cheng, S., Zhang, Q., Li, L., Shi, Y.: Particle swarm optimization with interswarm interactive learning strategy. IEEE Trans. Cybern. **46**, 2238–2251 (2015)
16. Gong, Y.-J., Li, J.-J., Zhou, Y., Li, Y., Chung, H.S.-H., Shi, Y.-H., Zhang, J.: Genetic learning particle swarm optimization. IEEE Trans. Cybern. **46**, 2277–2290 (2016)
17. Yang, Q., Chen, W.-N., Gu, T., Zhang, H., Deng, J.D., Li, Y., Zhang, J.: Segment-based predominant learning swarm optimizer for large-scale optimization. IEEE Trans. Cybern. **47**, 2896–2910 (2017)
18. Liu, Q.: Order-2 stability analysis of particle swarm optimization. Evol. Comput. **23**, 187–216 (2015)

19. Liu, Q., Chen, W.-N., Deng, J.D., Gu, T., Zhang, H., Yu, Z., Zhang, J.: Benchmarking stochastic algorithms for global optimization problems by visualizing confidence intervals. IEEE Trans. Cybern. **47**, 2924–2937 (2017)
20. Hansen N., Auger A., Brockhoff D., Tušar D., and Tušar T.: Coco: Performance assessment. ArXiv e-prints arXiv:1605.03560 (2016)
21. Maassen, H., Bezembinder, T.: Generating random weak orders and the probability of a Condorcet winner. Soc. Choice Welf. **19**, 517–532 (2002)
22. Dwork C., Kumar R., Naor M., and Sivakumar D.: Rank aggregation methods for the web. In: Proceedings of the 10th International Conference on World Wide Web, pp. 613–622. ACM (2001)
23. Cucuringu, M.: Sync-rank: Robust ranking, constrained ranking and rank aggregation via eigenvector and SDP synchronization. IEEE Trans. Netw. Sci. Eng. **3**, 58–79 (2016)
24. Li, Y.H., Zhan, Z.-H., Lin, S.J., Zhang, J., Luo, X.N.: Competitive and cooperative particle swarm optimization with information sharing mechanism for global optimization problems. Inf. Sci. **293**, 370–382 (2015)
25. Chen, W.-N., Zhang, J., Lin, Y., Chen, N., Zhan, Z.-H., Chung, H.S.-H., Li, Y., Shi, Y.-H.: Particle swarm optimization with an aging leader and challengers. IEEE Trans. Evol. Comput. **17**, 241–258 (2013)
26. Liu, Q., Wei, W., Yuan, H., Zhan, Z.-H., Li, Y.: Topology selection for particle swarm optimization. Inf. Sci. **363**, 154–173 (2016)
27. Rios, L.M., Sahinidis, N.V.: Derivative-free optimization: a review of algorithms and comparison of software implementations. J. Glob. Optim. **56**, 1247–1293 (2013)
28. Paulavičius, R., Sergeyev, Y.D., Kvasov, D.E., Žlinskas, J.: Globally-biased DISIMPL algorithm for expensive global optimization. J. Glob. Optim. **59**, 545–567 (2014)
29. Deemen, A.V.: On the empirical relevance of condorcet's paradox. Pub. Choice **158**, 311–330 (2014)
30. Gehrlein, W.V.: Condorcet's Paradox. Springer, Berlin (2006)
31. Wolpert, D.H., Macready, W.G.: No free lunch theorems for optimization. IEEE Trans. Evol. Comput. **1**, 67–82 (1997)
32. Diss, M., Gehrlein, W.V.: Borda's Paradox and weighted scoring rules. Soc. Choice Welf. **38**, 121–136 (2012)
33. Gehrlein, W.V., Lepelley, D.: On the probability of observing Borda's paradox. Soc. Choice Welf. **35**, 1–23 (2015)

Mosquito Host-Seeking Algorithm Based on Random Walk and Game of Life

Yunxin Zhu[1], Xiang Feng[1,2(✉)], and Huiqun Yu[1]

[1] Department of Computer Science and Technology, East China University of Science and Technology, Shanghai 200237, China
xfeng@ecust.edu.cn
[2] Smart City Collaborative Innovation Center, Shanghai Jiao Tong University, Shanghai 200240, China

Abstract. Mosquito Host-seeking Algorithm (MHSA) is a novel bionic algorithm. It simulates the behavior of mosquito seeking host. MHSA can find near-optimum solutions for the traveling salesman problem (TSP), however there are two drawbacks. First, it may be trapped into local optimum. Second, the solution exists several circles sometimes. In this paper, we adopt the Random Walk and the Game of Life strategies to improve MHSA, and propose a Random Walk and Game of Life Host-seeking Algorithm (RGHSA). RGHSA model is proposed to solve these two drawbacks. We use set theory and probability theory to prove the validity of the model. TSPlib is a benchmark for TSP. In the simulation, we choose server datasets from TSPlib, and compare the simulation result of RGHSA with original MHSA, Simulated Annealing Algorithm (SA) and Ant Colony Optimization Algorithm (ACO). The result shows that RGHSA have a good performance in TSP.

Keywords: Mosquito host-seeking algorithm · Traveling salesman problem
Random walk · Game of life

1 Introduction

1.1 Background

TSP is easy to describe but difficult to solve. Since it was proposed by Menger in 1932, no scholar proposed effective method to solve large-scale TSP exactly up to now. It has been confirmed that the large-scale TSP can not be solved by exact algorithm. Therefore, some methods to find the near-optimum solution has been proposed.

Mosquito Host-seeking Algorithm (MHSA) [1] is a parallel algorithm. Each artificial mosquito moves according to the energy mechanism. The artificial mosquitoes gradually consume energy when they move. After all the energy is exhausted, the mosquitoes stop moving. At the same time, the optimal solution of TSP can be found. Due to the parallel structure, MHSA can find optimal very fast. But it may be trapped into local optimum and the result exist several circles.

D.-S. Huang et al. (Eds.): ICIC 2018, LNCS 10955, pp. 693–704, 2018.
https://doi.org/10.1007/978-3-319-95933-7_78

1.2 Related Work

In order to solve the TSP, the scholars proposed lots of evolutionary algorithms. These algorithms are inspired by the biological behavior, physical or social phenomena.

It's difficult to find the exactly result for a large-scale TSP with classical approach such as exhaustive method. Compared with classical approach, evolutionary algorithms can be more adaptive. Such as Genetic Algorithm (GA) [2], Ant Colony Optimization (ACO) [3, 4], Particle Swarm Optimization (PSO) [5], Group Search Algorithm (GS) [6], Crystal Energy Optimization Algorithm [7], Simulated Annealing Algorithm (SA) [8] and Elastic Network (EN) [9].

In addition, lots of improved algorithms have been proposed based on these evolutionary algorithms. In order to reduce the execution time of the algorithms, Some scholars improve the ACO and GA with parallel structure [10, 11]. Some improved algorithms proposed had been considered to get higher accuracy such as ACO algorithm modified by particle swarm optimization algorithm (PS-ACO) [12].

1.3 Motivation

All these algorithms have their drawbacks. The elastic net has good parallelism exist local optimum and parameter adjustment problems [13]. The computation time of ACO and GA is long. PSO is easy to trap into the local optimum. SA has better optimization ability, but the efficiency is low.

MHSA has the fast convergence. If we can solve the following two drawbacks of MHSA, it can have a good performance on solving TSP.

- Easy to trapped in the local optimum and can't find the global optimum.
- In the solutions, several loops may exist. This is not the except result.

In order to handle these two problems, we extend MHSA with Random Walk [14] and Game of Life [15]. Therefore, the RGHSA is proposed in this paper.

1.4 Contribution

The main contribution of this paper is as follows.

- Extend the Mosquito Host-seeking Algorithm with Random Walk in order to jump out from the local optimum.
- Combined with the Game of Life and Random Walk in matrix, we proposed a novel matrix transformation.
- The theory of probability is used to analyze the onvergence of Game of Life.
- The simulation results validate the effectiveness of RGHSA.

1.5 Organization

The rest of this paper is organized as follows. In Sect. 2, we formalize the problem model of TSP and introduce MHSA. The biological model and mathematical model of RGHSA are proposed in Sect. 3. In Sect. 4, we presented the RGHSA steps. In Sect. 5, we discuss the convergence of RGHSA and Game of Life and an analysis on the efficiency of Game of Life. The simulation results present in the Sect. 6. Finally, we draw the conclusions in Sect. 7.

2 TSP and Mosquito Host-Seeking Algorithm

2.1 Tsp

TSP is a famous problem. The description of problem: a traveling salesman wants to visit n cities, each city should be visited one and only once. Finally return to the start city and the path is the shortest. Main variables in TSP are listed in Table 1.

Table 1. Main variable in TSP

Variable	Meaning	About
n	Total number of TSP cities	Problem
C_i	The i-th city $(i = \overline{1,n})$	Problem
(x_i,y_i)	Coordinates of city C_i	Problem
d_{ij}	Distance between city pair $(C_i,\ C_j)$ $(i,j = \overline{1,n})$ $d_{ij} = \sqrt{(x_i - x_j)^2 + (y_i - y_j)^2}$	Problem
p_{ij}	Path between C_i and C_j	Problem
x_{ij}	If $x_{ij} = 1$, then p_{ij} can be passed; or else $x_{ij} = 0$, then $d_{ij} = \infty$, p_{ij} can't be passed	Problem
Z	The shortest path through n cities	Solution
r_{ij}	(1) if $r_{ij} = 1$, then Z pass p_{ij}; or else $r_{ij} = 0$, then Z not pass p_{ij}. (2) when $x_{ij} = 0$, then $r_{ij} = 0$.	Solution

TSP can be described as:

$$Minimize : Z = \sum_{i,j} d_{ij} \cdot r_{ij}, \tag{1}$$

2.2 Mosquito Host-Seeking Algorithm

MHSA can find near-optimal solutions to TSP. In Fig. 1(a), a group of artificial mosquito distribute around the host, each artificial mosquito is attracted by host and fly to it. After all artificial stop, the solution to TSP can be found.

MHSA set an artificial mosquito between each city-pair, each artificial mosquito has three attributes r, c, x.

- r_{ij}: With the movement of artificial mosquito, this value changed between 1 and 0. This value will be converge to 0 or 1 when the artificial mosquito stop. The value equal to 1 means the artificial mosquito catch the host.
- c_{ij}: Represent the strength of each mosquito's trace ability.

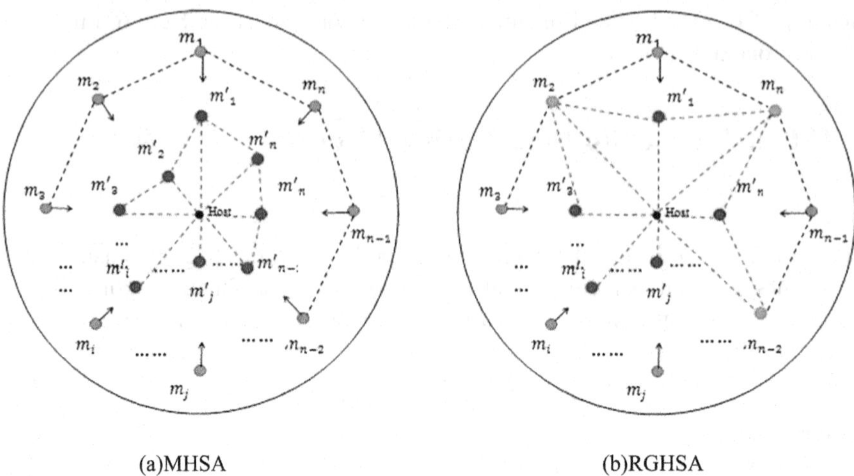

(a)MHSA (b)RGHSA

Fig. 1. MHSA and RGHSA model

- x_{ij}: $x_{ij} = 1$ means female mosquito, 0 means male. Only females have the ability to seeking host. Males do not have trace ability, that means there is no route between this city-pair in TSP.

The structure of TSP matrixes is represented in Table 2. MHSA iteration matrix R, C and X to find solutions. Finally, the elements of matrix R will convergence to 0 or 1.

3 RGHSA Model

3.1 Biological Model

In nature, mosquito live on nectar. However, female mosquito need suck blood for multiply. By this biological characteristics, we set part of mosquito's sex be male randomly. In Fig. 1(b), the green points in periphery of host are male mosquito. They won't seek host and always stay still. But blue points will be attracted by host and fly to it. When they stopped, we will get a near-optimum solution. This solution is related to the sex of mosquito.

3.2 Mathematics Model

The matrix X represents the sex of artificial mosquito. The element in the matrix is either 0 or 1. The number 0 represents male and 1 represents female.

A simple method is that use the random matrix X to iterate. We named this method as Random Host-Seeking Algorithm (RHSA). This paper proposed a more complex and effective transformation for matrix X, named Random Walk and Game of Life Host-seeking Algorithm (RGHSA). In RGMHA, the Game of Life and Random Walk is simulated. We use two methods to generate a new matrix X.

- A new matrix X is generated randomly
- The Game of Life is used to evolve current matrix X, and then a new matrix is generated.

3.3 Random Walk Model

The main idea of Random Walk model is from one or a series of nodes to traverse a graph. In each node, traverser has a probability a teleport to any node of the graph randomly. a is called teleport probability. Meanwhile, traverser can walk to the neighbor node by probability $1 - a$. After several iterations, each node's probability of visited converges to a constant.

In Fig. 2, node A is the start node, node B and C are the neighbors of start node, the straight lines are the edges of the graph. We can find that there is no edge between node D and node A, therefore, the node D is not the start node's neighbor. But in Random Walk model, traverser can walk to node D directly by teleport.

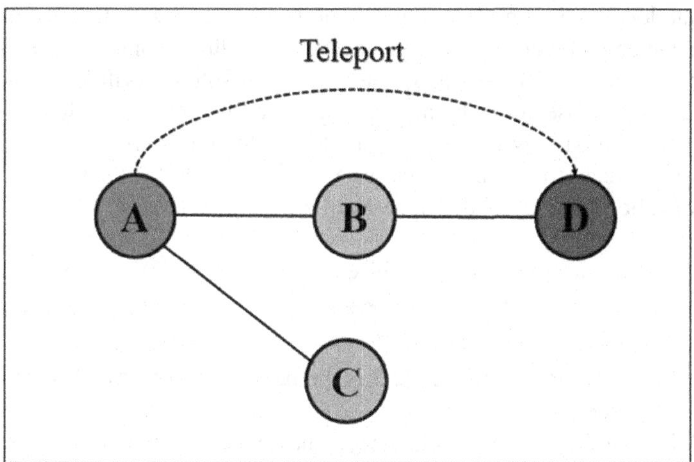

Fig. 2. Random Walk model

In RGHSA, teleport generates a new matrix X with probability a, and the Game of Life is used to evolve current matrix X to a new matrix with probability $1 - a$.

3.4 Game of Life Model

Game of Life is a kind of cellular automata. It simulates the phenomenon of life evolve. At the beginning, only a few cells alive, then the number increase gradually and bloom, finally after a threshold, the number of cells will decrease till wither. This process is similar to life evolution. According to this model diverse pattern can be formed.

The Game of Life considered as a matrix only has element 0 or 1. Each element in the matrix represents a cell. The number 1 means alive while the number 0 means dead.

Thus, we can deal with this matrix as matrix X because that all elements of them are either 0 or 1. A element's neighbors represent 8 elements around it. The evolution rule of Game of Life as below:

- If a cell has three alive neighbors, then it will alive.
- If a cell has two alive neighbors, then it's state will not be change.
- In other condition, this cell will die.

After initialize the matrix X, the evolution can be start. By using these three rules, matrix X can transform into many different patterns, it represents the sex of mosquito.

4 RGHSA

4.1 Optimization for MHSA

In MHSA, we set all city pairs to connected with each other, but the solution is unsatisfactory. Consider some edges have great influence on the solution, this may lead to obtain the local optimum. Therefore, in the beginning of search, we remove some edges from the graph randomly, then use the MHSA to find a solution. If the solution is better than current best solution, and then the current solution will be replaced.

In RGHSA, we use different matrix X to start the search, finally a solution is obtained. The solution may be different due to the different start matrix X. We repeat this method for many times and set the best solution as the final solution.

The evaluation function is the same as MHSA.

$$Z = \sum_{i=1}^{n} \sum_{j=1}^{n} d_{ij} r_{ij} x_{ij}, \tag{2}$$

The geometric meaning of formula 2 is the path length of TSP. The value of Z is the smaller the better.

MHSA has another drawback, the results may exist several circles. This is not the solution of TSP. RGHSA repeats on a problem for many times, each time get a solution. The algorithm judges whether the solution is a connected graph or not. If not, the solution will be discard and find another solution.

4.2 Algorithm Steps

We use formulas 3 and 4 to initialize the Matrix C and R.

$$c_{ij} = \max_{i,j} d_{ij} - d_{ij}, \tag{3}$$

$$r_{ij} = \frac{1}{n}, \tag{4}$$

```
Algorithm 1 RGHSA
Input: search times k, cities` location and numbers n.
Output: minimum length of route Z and adjacency matrix of shortest route.
Begin:
For i = 1: k
 X = round (1 * rand(n)) or X = GameOfLife(X);
 X` = 1 - X`;
 Matrix C initialized by formula 3;
 Matrix R initialized by formula 4;
 (R, Z)=MHSA(X`, R, C);
 If R is connected graph and Z < minZ
     minZ = Z;
     minR = R;
 End if
End for
Output minZ, minR;
End.
```

5 Model Analysis

5.1 Convergence of RGHSA

Lemma 1: As show in Fig. 3.

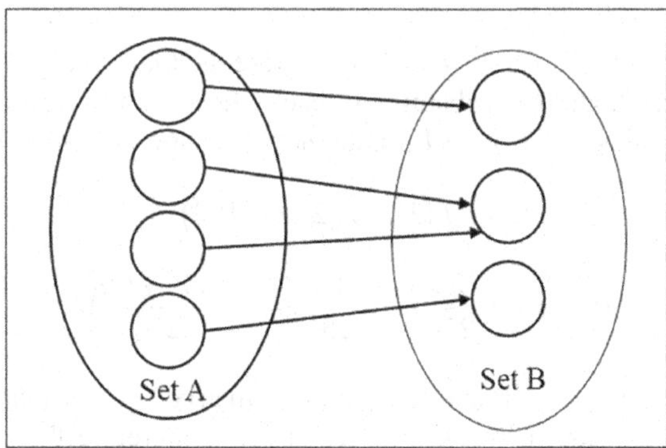

Fig. 3. Epimorphism model

$$A = \{M = (m_{ij}) \in R^{n \times n} | m_{ij} = 0\, or\, m_{ij} = 1\}$$
$$B = \{M | M \in A \text{ and } \sum_{i=1}^{n} m_{ij} = 2 \text{ and } \sum_{j=1}^{n} m_{ij} = 2\}$$
$$R = MHSA(R, C, X)$$

$X \in A, R \in B$, C and R were initialized by formulas 3 and 4.
Prove that: $f : A \rightarrow B$ is a epimorphism.

Prove: For $\forall R \in B$ and $\exists X \in A$, let $X = R$. According to TSP, if matrix X is already a loop and other cities are all not connected. MHSA can find the solution R which must be the same as matrix X.

$$R = MHSA(R, C, R), \tag{5}$$

Formula 5 is an identity, thus the lemma has been proved. Because of the epimorphism, the result is based on matrix X. The change of matrix X must lead to the solution convergence to a shorter route to TSP.

5.2 Convergence Analysis on Game of Life

Game of Life usually has a process with start, bloom and fading. The convergence analysis of Game of Life is based on probability theory.

Assume in the whole matrix, the ratio of all the number of 1 and all the number of elements in matrix is λ, the matrix is a square matrix of order n. The total number of 1 is m.

$$\lambda = \frac{m}{n^2}, \tag{6}$$

Then, we use $i(i \in Z, 0 <= i <= 8)$ represents the number of 1 around one element, we can calculate the probability of each i from 0 to 8. According to Formal 6 and the evolution rules of Game of Life, the next time the ratio of 1 and all element is

$$P(next) = [P(3) + P(2)]\lambda + P(3)(1 - \lambda), \tag{7}$$

$$P(next) = \frac{28}{256}\lambda^2 + \frac{56}{256}\lambda^2 = (\frac{28\lambda + 56}{256})\lambda, \tag{8}$$

Because of $0 <= \lambda <= 1$, and from the start matrix X's λ is approximately equals 0.5. Obviously, the value in brackets is less than 1, therefore the number of 1 will decrease during the evolution. Finally, λ will converge to 0. Most of matrix X will converge to 0 except some special conditions. These conditions will form a stable state before converge to 0.

6 Simulations

6.1 10-Cities TSP

In this section, a random 10-cities TSP is chosen, RGHSA is used to find the solution, and the simulation results are compared with MHSA and RHSA. The RHSA and RGHSA need search for several times and choose the best solution. In this simulation, the value of k is set to 2000, it means that RHSA search for 2000 times. RGHSA search 200 times and each time with 10 steps of Game of Life evolution.

The optimal result of three algorithms are shown in Fig. 4. RGHSA has the best performance.

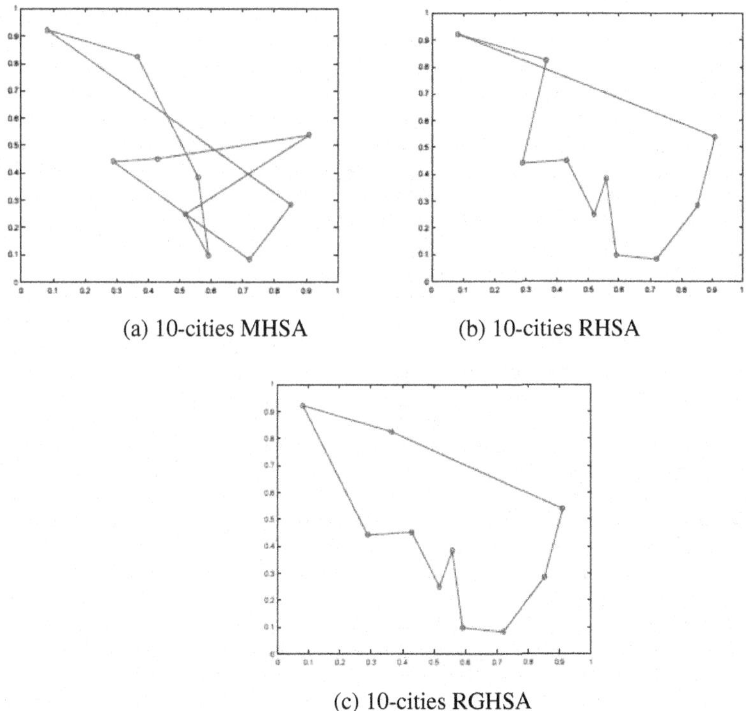

(a) 10-cities MHSA

(b) 10-cities RHSA

(c) 10-cities RGHSA

Fig. 4. Optimal result of RHSA, MHSA and RGHSA

6.2 TSPLIB Simulations

TPSLIB is a library of sample instances for TSP, it is the benchmark for TSP. We can find different scale of datasets from it. We choose ch150, ts225, u2319 and pcb3038, the numbers in name represent the number of cities. The comparison between RGHSA and MHSA, ACO and SA is shown in Fig. 5. The solution results of RGHSA on different scale data sets is shown in Fig. 6.

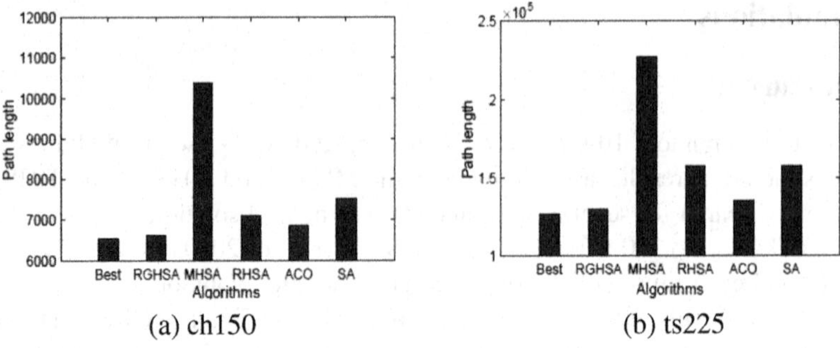

(a) ch150 (b) ts225

Fig. 5. Optimal value comparison between RGHSA and other algorithms

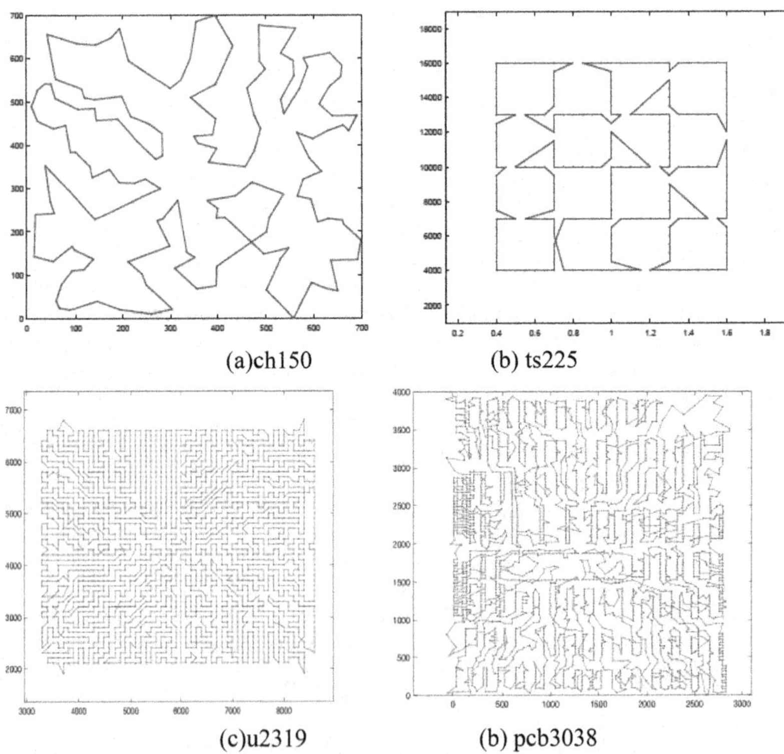

(a)ch150 (b) ts225

(c)u2319 (b) pcb3038

Fig. 6. Solution results of RGHSA on different scale data sets

7 Conclusion

MHSA is a fast-convergence algorithm but there are two drawbacks, one is easy to trapped into local optimum, another is the results exist several circles sometimes. By extend MHSA with Random Walk and Game of Life, we proposed RGHSA. This algorithm can find a better near-optimum solution than MHSA for TSP. Random Walk makes the search jump out from local optimum. New matrix is generated by Game of Life or randomly, these rules make the search well-distributed. In addition, the results with several loops will be discard, therefore the final solution must have only one loop. Next step, we will continue to improve the RGHSA, let it can solve TSP quickly.

Acknowledgement. This work was supported by the National Natural Science Foundation of China (Grant No. 61472293). Research Project of Hubei Provincial Department of Education (Grant No. 2016238).

References

1. Hamerly, G., Elkan, C.: Alternatives to the k-means algorithm that find better clusterings. In: Proceedings of the Eleventh International Conference on Information and Knowledge Management, pp. 600–607. ACM (2002)
2. Matas, J., Kittler, J.: Spatial and feature space clustering: Applications in image analysis. In: Hlaváč, V., Šára, R. (eds.) CAIP 1995. LNCS, vol. 970, pp. 162–173. Springer, Heidelberg (1995). https://doi.org/10.1007/3-540-60268-2_293
3. Natali, A., Toschi, E., Baldeweg, S., et al.: Clustering of insulin resistance with vascular dysfunction and low-grade inflammation in type 2 diabetes. Diabetes **55**(4), 1133–1140 (2006)
4. Ben-Dor, A., Shamir, R., Yakhini, Z.: Clustering gene expression patterns. J. Comput. Biol. **6**(3–4), 281–297 (1999)
5. Steinbach, M., Karypis, G., Kumar, V.: A comparison of document clustering techniques In: KDD Workshop on Text Mining, vol. 400(1), pp. 525–526 (2000)
6. Hu, T., Liu, C., Tang, Y., et al.: High-dimensional clustering: a clique-based hypergraph partitioning framework. Knowl. Inf. Syst. **39**(1), 61–88 (2014)
7. Bouveyron, C., Brunet-Saumard, C.: Model-based clustering of high-dimensional data: a review. Comput. Stat. Data Anal. **71**, 52–78 (2014)
8. Roweis, S.T., Saul, L.K.: Nonlinear dimensionality reduction by locally linear embedding. Science **290**(5500), 2323–2326 (2000)
9. Zhu, X., Huang, Z., Yang, Y., et al.: Self-taught dimensionality reduction on the high-dimensional small-sized data. Pattern Recogn. **46**(1), 215–229 (2013)
10. Song, Q., Ni, J., Wang, G.: A fast clustering-based feature subset selection algorithm for high-dimensional data. IEEE Trans. Knowl. Data Eng. **25**(1), 1–14 (2013)
11. Soltanolkotabi, M., Elhamifar, E., Candes, E.J.: Robust subspace clustering. Ann. Stat. **42** (2), 669–699 (2014)
12. Bouveyron, C.: Model-based clustering of high-dimensional data in Astrophysics. EAS Publ. Ser. **77**, 91–119 (2016)
13. Han, E.H., Karypis, G., Kumar, V., et al.: Hypergraph based clustering in high-dimensional data sets: a summary of results. IEEE Data Eng. Bull. **21**(1), 15–22 (1998)
14. Sun, L., Ji, S., Ye, J.: Hypergraph spectral learning for multi-label classification. In: Proceedings of the 14th ACM SIGKDD International Conference on Knowledge Discovery and Data Mining, pp. 668–676. ACM (2008)
15. Huang, Y., Liu, Q., Zhang, S., et al.: Image retrieval via probabilistic hypergraph ranking. In: 2010 IEEE Conference on Computer Vision and Pattern Recognition (CVPR), pp. 3376–3383. IEEE (2010)

16. Wang, M., Liu, X., Wu, X.: Visual classification by $\ell 1$-hypergraph modeling. IEEE Trans. Knowl. Data Eng. **27**(9), 2564–2574 (2015)
17. Fiduccia, C.M., Mattheyses, R.M.: A linear-time heuristic for improving network partitions. In: Papers on Twenty-Five Years of Electronic Design Automation, pp. 241–247. ACM (1988)
18. Huang, D.J.H., Kahng, A.B.: When clusters meet partitions: new density-based methods for circuit decomposition. In: Proceedings of the 1995 European Conference on Design and Test. IEEE Computer Society (1995)
19. Karypis, G., Aggarwal, R., Kumar, V., et al.: Multilevel hypergraph partitioning: applications in VLSI domain. IEEE Trans. Very Large Scale Integr. VLSI Syst. **7**(1), 69–79 (1999)
20. Cai, W., Young, E.F.Y.: A fast hypergraph bipartitioning algorithm. In: 2014 IEEE Computer Society Annual Symposium on VLSI (ISVLSI), pp. 607–612. IEEE (2014)
21. Lotfifar, F., Johnson, M.: A Serial Multilevel Hypergraph Partitioning Algorithm. arXiv preprint arXiv:1601.01336 (2016)
22. Henne, V., Meyerhenke, H., Sanders, P., et al.: n-Level Hypergraph Partitioning. arXiv preprint arXiv:1505.00693 (2015)
23. Liu, H., Latecki, L.J., Yan, S.: Dense subgraph partition of positive hypergraphs. IEEE Trans. Pattern Anal. Mach. Intell. **37**(3), 541–554 (2015)
24. Jagannathan, J., Sherajdheen, A., Deepak, R.M.V., et al.: License plate character segmentation using horizontal and vertical projection with dynamic thresholding. In: 2013 International Conference on Emerging Trends in Computing, Communication and Nanotechnology (ICE-CCN), pp. 700–705. IEEE (2013)
25. Tuba, E., Bacanin, N.: An algorithm for handwritten digit recognition using projection histograms and SVM classifier. In: 2015 23rd Telecommunications Forum Telfor (TELFOR), pp. 464–467. IEEE (2015)
26. Hinton, G., Roweis, S.: Stochastic neighbor embedding. In: NIPS. 15, pp. 833–840 (2002)
27. Maaten, L., Hinton, G.: Visualizing data using t-SNE. J. Mach. Learn. Res. **9**, 2579–2605 (2008)
28. Eppstein, D., Löffler, M., Strash, D.: Listing all maximal cliques in sparse graphs in near-optimal time. In: Cheong, O., Chwa, K.-Y., Park, K. (eds.) ISAAC 2010. LNCS, vol. 6506, pp. 403–414. Springer, Heidelberg (2010). https://doi.org/10.1007/978-3-642-17517-6_36
29. The Semeion dataset. https://archive.ics.uci.edu/ml/datasets/Semeion+Handwritten+digit
30. Fowlkes, E.B., Mallows, C.L.: A method for comparing two hierarchical clusterings. J. Am. Stat. Assoc. **78**(383), 553–569 (1983)
31. The MNIST dataset. http://yann.lecun.com/exdb/mnist/index.html
32. The USPS dataset. http://www.cs.nyu.edu/~roweis/data/html
33. The Binaryalphadigs dataset. http://www.cs.toronto.edu/~roweis/data/binaryalphadigs.mat
34. Van der Maaten, L.: A new benchmark dataset for handwritten character recognition, pp. 2–5. Tilburg Universit (2009)
35. Kaufman, L., Rousseeuw, P.: Clustering by Means of Medoids. North-Holland, Amsterdam (1987)
36. Sun, X., Tian, S., Lu, Y.: High dimensional data clustering by partitioning the hypergraphs using dense subgraph partition. In: Ninth International Symposium on Multispectral Image Processing and Pattern Recognition (MIPPR2015). International Society for Optics and Photonics (2015)

A Novel Energetic Ant Optimization Algorithm for Routing Network Analysis

Xiang Feng[1,2](✉) and Hanyu Xu[1]

[1] Department of Computer Science and Engineering, East China University of Science and Technology, Shanghai, China
xfeng@ecust.edu.cn

[2] Smart City Collaborative Innovation Center, Shanghai Jiao Tong University, Shanghai, China

Abstract. The latest biological research results show that it is natural to see that ants at different age group play roles and responsibilities differently. As inspired by the same, the concept of age and intra-groups is thus introduced into traditional Ant Colony Optimization (ACO) algorithm. A new intelligent parallel algorithm, Energetic Ant Optimization model (EAO), is put forward and applied for energy-aware routing network analysis. The proposed algorithm is designed to calculate the routing probability and phenomenon increment by taking the remaining energy of node as a heuristic factor. By EAO, the age of ant corresponds to the energy of the Ad Hoc network. Not only was mathematical model built for the EAO theoretically, but also its application was described detailedly. Finally, the proposed algorithm is simulated and analyzed in different scenarios, and the experimental results are compared with the results of Ad hoc on-demand distance vector routing (AODV). The simulation results show that EAO routing algorithm (EAORA) performs much better in packet delivery ratio, the average end-to-end delay and lifetime of network. Besides, the EAORA has better performance in balancing the energy consuming between nodes.

Keywords: Energetic ant optimization model · Routing network analysis
Energy-aware

1 Introduction

Nowadays, with the growing demanding for wireless communication, the applications of Ad Hoc network technology have been focusing on the normal communication of pubic areas in life more than military and emergency situations. However, it is of great challenge to achieve the reliable and efficient communication in Ad Hoc network with unstable features causing by the highly dynamic topology and limited energy. Therefore, the researches on the Ad Hoc network routing algorithms are of profound theoretical and practical significance to improve the quality of service in wireless communication network [1–3].

D.-S. Huang et al. (Eds.): ICIC 2018, LNCS 10955, pp. 705–716, 2018.
https://doi.org/10.1007/978-3-319-95933-7_79

2 Related Work

Traditional algorithms and mechanism have been applied into route protocols for Ad Hoc networks. Some latest work focus on Geographic Routing, which is either Self-Adaptive On-Demand [4] or Edge-Constrained Localized [5]. They can avoid incurring unnecessary control overhead, as well as have very low forwarding overhead and transmission delay. Adaptiveness is such an important and necessary signature of Ad Hoc networks that routing algorithms are mean to achieve [6]. Termed Progress Face uses an additional traversal step to decide the direction of perimeter forwarding [7]. Moreover, cooperation and coordination between nodes in wireless network is another effective method to improve the routing algorithms [8].

Besides the traditional route approaches, there're also some swarm intelligent algorithms [9], such as the ant colony algorithm (ACO) [10], the greedy algorithm [11] and the genetic algorithm [12], applied to the Ad Hoc network route problems. With the advantages of high efficiency and ability for complex problems, intelligent algorithms usually perform better performance than linear algorithms on path-finding problems. Especially the ACO, inspired by the ants' behavior of searching food, is believed to be good at building routing tables. [13] applied the ACO to the energy-aware Ad Hoc routing protocol and had a good performance in the minimal energy broadcast problems. However, they only consider the optimization of overall consumption of energy, rather than the residual energy of each node. Moreover, the ACO is so easy to be premature that the routing table may not be the optimal solution. Therefore, in [14], the ACO is improved focusing on the global searching with the elitist strategy, however, there's no consideration of energy of the Ad Hoc network. [15] also surveyed and proposed a few ways to improve the ACO, for instance, introducing the aging in which lesser and lesser pheromone as it moves from node to node.

It is pointed out in Mersch's paper on Science that roles and responsibilities of ants in the ant group change along with their ages [16]. Ants are divided into three social groups based on different ages with different roles and responsibilities. All the three groups differ from each other in their distribution density which increases with age to allow them move from one group to the next in sequence. Expected interaction rates in one group and among groups are figured out based on information accurately distributed in the space-time. Statistics indicate that change to ants in the groups with age serves as the key regulator in ant group interaction [14].

The main contribution of this paper is to propose Energetic Ant Optimizer Routing Algorithm (EAORA) based on improved ant colony algorithm inspired the age and intra-groups for routing network analysis. It combined Ad Hoc network routing algorithm and energy-saving strategy, and improved ant colony algorithm routing rules and pheromone update rules. By these means, the reliability of the network and quality of service are increased. The specific mathematical model of the proposed algorithm was described theoretically and its application was verified experimentally. The simulations of improved routing protocol are taken in NS2. And our algorithm performs much better in packet delivery ratio, the average end-to-end delay, lifetime of network, and the balance of the energy consuming between nodes.

3 Energetic Ant Optimization Routing Algorithm

3.1 Theory of the EAORA

The ACO is famous for its ability of solving "path" problems, such as the traveler sales man, vehicle route problems, while the premature and low efficiency are also serious shortcomings. From the perspective of its basic model, ant colony, age and intra-groups are introduced as two major factors to improve its performance fundamentally.

Each energetic ant has its own age since it is born and hence will die when its age is old enough. During the evolution, age is detected first before ants go for the food. Searching strategy of ants with different age varies. As shown in Fig. 1, ants from 70 to 80 days have fewer paths and from 80 to 90 days have barely paths. Ants over 97 days, which are at end of their life, have no job to do. Therefore, by recording and calculating the age of ants, i.e. their ability on work, division of work can be more efficiency. Moreover, adjustments for the "retired" ants could be made in advance.

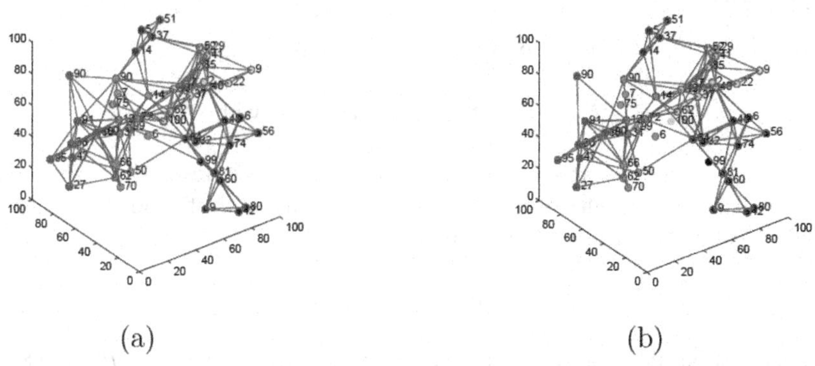

(a) (b)

Fig. 1. Groups and division of work of EAO (a) Intra-Groups, (b) Division of work by age

Furthermore, in large space or network, ants don't search the place blindly. Instead, they are going to separate into a few intra-groups in charge of different regions. Interactions inside group are much frequent than those among groups. Multi-objects are more easily to be found in parallel. As shown in Fig. 1, five groups are divided in a colony according to their relative positions. Enhancing the communication among ants in the same group avoid the unnecessary search in long distance.

3.2 Model Definitions and Mechanisms

In EAORA, a network structure diagram $G = (V, E)$ is defined, where V is set of nodes and E is set of edges between nodes. Each node saves some information tables and the path information represented by the pheromone. Node information table contains pheromone matrix and neighbor information table.

Calculation of Residual Energy of Neighbor Nodes

The following energy model is used in simulation to calculate the size of the energy consumption. Initialize energy of each node in the beginning. The initial energy can be different. Because the energy consumption of the source node and destination node is relative large, the initial energy of them is set to be 10 units, and others are 5 units. In simulation, we use a simple energy consumption calculation formula,

$$\text{Energy} = \text{Power} * \text{Time} \tag{1}$$

According to the Eq. 1, the energy consumption of a node is determined by the transmission power or received power. Processing time of a data packet is calculated as follows.

$$\text{Time} = 8 * \text{PacketSize}/\text{Bandwith} \tag{2}$$

Thus, we can obtain the equations of sending energy consumption and receiving energy consumption.

$$E_{tx} = P_{tx} * 8 * \text{PacketSize}/\text{Bandwith} \tag{3}$$

$$E_{rx} = P_{rx} * 8 * \text{PacketSize}/\text{Bandwith} \tag{4}$$

Where, P_{tx} and P_{rx} represent sending and receiving power, respective. When a node forwards data package, the energy consumption is the sum of sending energy consumption and receiving energy consumption.

$$E_f = E_{tx} + E_{rx} \tag{5}$$

In simulation, the ratio of sending power P_{tx} and receiving power P_{rx} is 0.8:1.

Router Selecting Rules

When the EAORA is used in Ad Hoc network, we use a probability routing table to represent routing table for each node. It contains the pheromone and probability. Probability is obtained by pheromone and other heuristic factors according to Eq. 6. In EAORA, when forward ants passed by node i, next hop follows the probability as follows.

$$P_{ij}^{k} = \begin{cases} \frac{\tau_{ij}^{\alpha}(t)\eta_{ij}^{\beta}(t)f(t)}{\sum_{s\in N_i}\tau_{is}^{\alpha}(t)\eta_{is}^{\beta}(t)} & if\, j \in N_i \\ 0 & otherwise \end{cases} \tag{6}$$

$$\mathrm{f(t)} = -(t - C_1)^2 + C_2$$

Where f(t) is the ability of ant, where t is the ant's age. P_{ij}^{k} is the probability that ant k transfer from node i to node j at time t, j $\in N_i$, N_i is the set of node i's neighbor nodes. $\tau_{ij}(t)$ is the strength of pheromone on path (i, j), $\eta_{ij}(t)$ is the heuristic value to represent the residual energy function. α is the heuristic factor of pheromone and β is the heuristic factor of energy.

Rules of Updating Pheromone
When forward ants reach the destination node, they turn into backward ants and go back to the source node with information. Backward ants will update the pheromone on the way back. For node i, backward ants will calculate the pheromone from node i + 1 to node i to update the pheromone value $\tau_{ij}(t)$. Updating equation is as follows.

$$\tau_{ij}(t+\Delta t) = (1-\rho)\tau_{ij}(t) + \tau_{ij}(t, t+\Delta t) \tag{7}$$

Where, ρ is the pheromone evaporation rate, $(1-\rho)$ is the pheromone strength residual factor, and $\tau_{ij}(t, t+\Delta t)$ is the period between t and $t+\Delta t$.

The length of path or hop is taken as the measurement standard of the increment of pheromone, such as $\Delta\tau_{ij}^k = \frac{Q}{L_k}$, where Q is the constant of pheromone and L_k is the total length of path. However, this paper will introduce the residual energy to the update of pheromone.

$$\Delta\tau_{ij}^k = Q * \left(\frac{\sum_{i=1}^{hop} ER_i}{hop}\right) \tag{8}$$

Where hop is the number of hops that forward ants passed by, and ER_i is the residual energy of path that forward ants passed by.

Process Rules of Link Interruption
If a node doesn't receive a responds from neighbor node, this neighbor node is out of path and leads to link interrupt. Therefore, nodes in Ad Hoc network send Hello packages to check the link interruption periodically. Node i broadcasts Hello packages to neighbor node j periodically. If it gets responds Hello package from node j, node j is its neighbor node. And if node j isn't inside neighbor node i's table, add node j's information, including pheromone and probability, into both neighbor node table and probability table. Otherwise, if node i doesn't get responds from node j, it means that path (i, j) is interrupted, so delete relevant information of node j.

4 Algorithm Application

4.1 Exploration Process of Router

Path exploration is used to generate the path from source node to destination node. The establishment of routing table is based on on-demand routing protocol. When source node needs to communicate with the destination node, it will check its own probability table first to find out if there is any relevant entry. If not, exploration process is established. It is completed by both forward ants and backward ants, which are responsible for looking router from source to destination and update pheromone, respectively.

Source node sends ants to search and build routing table on demand. Ants choose next hop according to probability. Source address, destination address, number of hops passed by, list of nodes passed by and information of residual energy is carried by forward ants. When forward ants pass by a node outside routing table, it will be added. Otherwise, choose next node according to the information table until reach destination. Routing process of forward ants is as shown in Fig. 2.

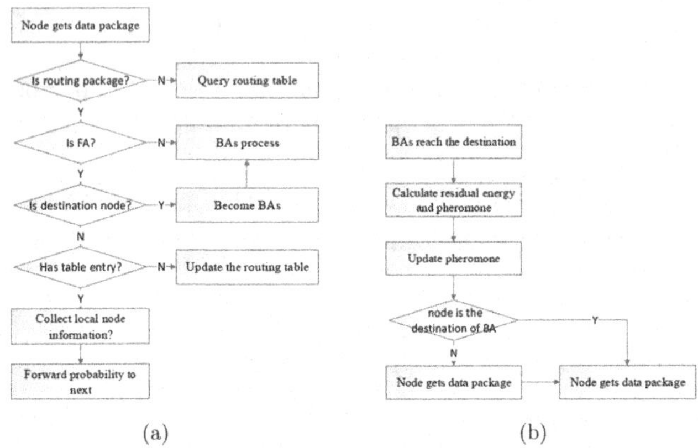

Fig. 2. Routing process of ants (a) forward ants, (b) backward ants

When Forward ants reach the destination, they become backward ants. Backward ants calculate the total energy of path passed by. Their information includes the ant number, source address, destination address, hops, and list of nodes on path and increment of pheromone. The increment of pheromone is calculated by Eq. 8. Path with higher residual energy and less hops has higher pheromone. Backward ants return to source node according to its own path list and update the pheromone on path. Updating equation is obtained by Eq. 7. Only the optimal path is remained after each loop, therefore, only the ants have the most residual energy have the right to alter pheromone. Routing process of backward ants is as shown in Fig. 2.

4.2 Routing Updating

Each node in ad Hoc network performs the same routing algorithm. The source node broadcasts the forward ants to destination node periodically to find the optimal path.

1. When the source node S needs to send data and communicate with the destination node, firstly, S checks the local routing table if there is a route to the destination node. If so, S will choose a high probability of pheromone from its neighbors as the next hop, otherwise, go to 2.
2. S node stores the packets in cache and broadcasts a forward ant (Fant) to all neighbors of ants. Select the next hop randomly based on initialized information.

3. If intermediate node j receives the forward ants from node i, checks the destination address, if the address matches the local address, generates the backward ant, then goes to 5. If it is not the destination, the forward ant will be destroyed when the current node belongs to the list of routing path. Otherwise, go to 4.
4. Intermediate node j interprets the information in the Forward ant and uses Eq. 7 to calculate the probability of all the neighbor nodes according to information stored in the information table. Then update the phenomenon value from previous node and current node, the 3rd and 4th process are repeated and the Fant is relayed by intermediate nodes till it reaches the destination.
5. When the Fant reaches the destination, its information is extracted and it is destroyed. Backward ant which has the same structure with Forward ant is created in the destination and sent towards the source. Backward ant updates the pheromone values according to Eq. 6 when returns to the source. Once the source node receives Backward ant from destination, Backward ant is destroyed and the path is generated.

5 Simulation and Analysis

5.1 Convergence Analysis

In this section, the convergence property of EAO was verified and analyzed according to four basic benchmark functions, Sphere f1, Rosenbrock f2, Sum of Squares f3, and Rastrigin f4, respectively, and their dimensions are all 30. We compare EAO with three classical nature-based algorithms, the Ant Colony Optimization (SSO) algorithm, the Particle Swarm Optimization (PSO) method, and the Artificial Bee Colony (ABC) algorithm, respectively. In all experiments, the maximum iteration is set as 1000 times. The initial population size is 50. For ACO algorithm, $\alpha = 1$, $\beta = 2$, $\rho = 0.6$, $Q = 100$ For PSO method, $c1 = 1.5$ and $c2 = 2$, and the weight factor decreases linearly from 0.9 to 0.2. For ABC approach, the parameter limit = 100. For the proposed algorithm EAO, $Q = 100$, $\alpha = 1$, $\beta = 2$, $\rho = 0.3, \tau_0 = 10$. Each algorithm was tested 30 times for every optimization function to get the Average Best-so-far (AB) solution and the Standard Deviation (SD) of best-so-far solution, and the results are reported in Table 1. The best results of the four algorithms are shown in bold. According to this table, EAO performed better than the other algorithms for all functions. Figure 3 described the convergent curve of EAO, ACO, ABC and PSO for four functions. Among them, the rate of convergence curve of EAO is the fastest for each function.

5.2 Network Analysis

The performance of algorithm is evaluated based on the following four indicators:

1. The ratio of the number of received data packet and the number of delivered data packet to the destination

$$\text{Packagedeliveryratio} = \frac{\sum_{i=0}^{N} numR_i}{\sum_{i=0}^{N} numS_i}$$

Where, $numR_i$ is the number of package receive, $numS_i$ is the number of package send.

2. End-to-end average delay. rt_i is the receiving time and st_i is the sending time.

$$\text{averagedelaytime} = \frac{1}{M}\sum_{i=0}^{M}(rt_i - st_i)$$

3. The network lifetime. The survival time of network in simulation is defined by the time from beginning to the first node failure.
4. Energy consumption. The total consumption of energy during the process of building route table and deliver message.

Table 1. Minimization results of benchmark functions

		EAO	ABC	PSO	ACO
f1	AB ± SB	**3.29E−06** ± 1.10E−06	1.94E−04 ± 4.25E−04	2.71E−04 ± 1.41E−04	1.25E−03 ± 3.14E−04
f2	AB ± SB	**1.24E+01** ± 3.11E+01	1.04E+02 ± 3.54E+01	1.18E+02 ± 2.55E+03	3.14E+03 ± 3.18E+02
f3	AB ± SB	**6.15E−07** ± 3.25E−07	4.98E−05 ± 1.97E−05	9.17E−02 ± 1.78E−02	6.87E+00 ± 3.48E+00
f4	AB ± SB	**2.74E−02** ± 4.25E−02	2.59E+01 ± 1.13E+00	2.24E+00 ± 1.46E+00	1.05E+02 ± 3.24E+01

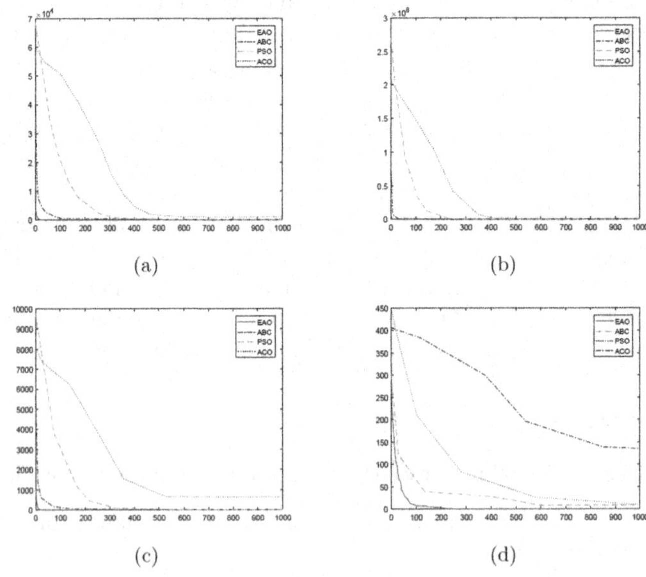

Fig. 3. Evolution curves of four algorithms considering for functions (a) f1, (b) f2, (c) f3 and (d) f4.

In order to verify the running efficiency of EAORA algorithm, we compare the performance of NS2 with the classical protocol AODV in simulation. Some basic parameters of the EAORA algorithm are the same in Sect. 5.1.

Energy consumption under two algorithms is shown in Fig. 4. As we can see from the figure, energy consumption of EAORA algorithm is relatively small, and energy consumption gap between the nodes is also smaller, i.e. the curve is relatively smooth. Therefore, energy consumption is relatively balanced with EAORA. It is due to the consideration of residual energy by EAORA, while AODV considers shortest hop, which will lead to the excessive use of node and reduce the life time of the node.

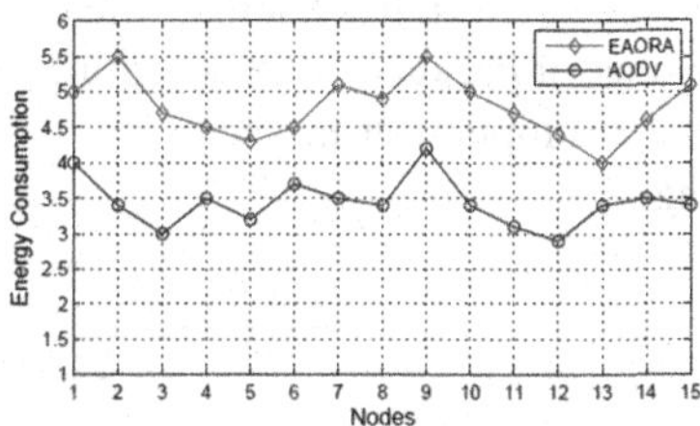

Fig. 4. Energy balance degree of network nodes

Pause time (PT) of different nodes is used to evaluate the performance of algorithm under three indicators.

As shown in Fig. 5, the packet delivery rate indicates the throughput of the algorithm. While with the consideration of node energy, path won't be disabled because of running out of energy with EAORA algorithm. Therefore, success rate of sending a data packet is higher.

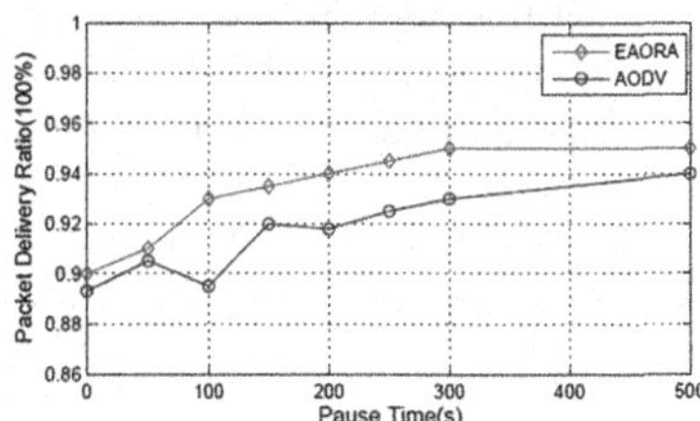

Fig. 5. Packet delivery ratio and Pause Time

Figure 6 shows the relationship between the average end-to-end delay and node pause time. EAORA algorithm considers information of node hops and residual energy and selects the path with less remaining energy according to the probability, which achieves the goal of energy balance of the network nodes and avoids node death caused by routing failure.

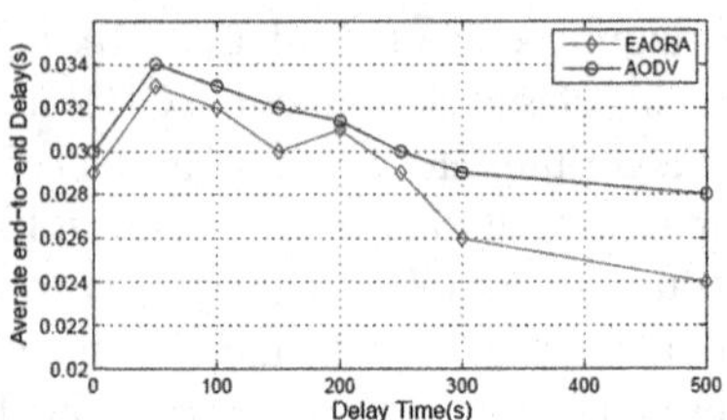

Fig. 6. Average end-to-end delay and Pause time

Figure 7 shows the relationship between network survival time and pause time. When the pause time is 0 s, the network has a longer life cycle. Because in the case of frequent movement, the number of packets nodes accepting and forwarding is less than the average. So the energy consumption of the node is relatively average, and the life cycle of the network is longer.

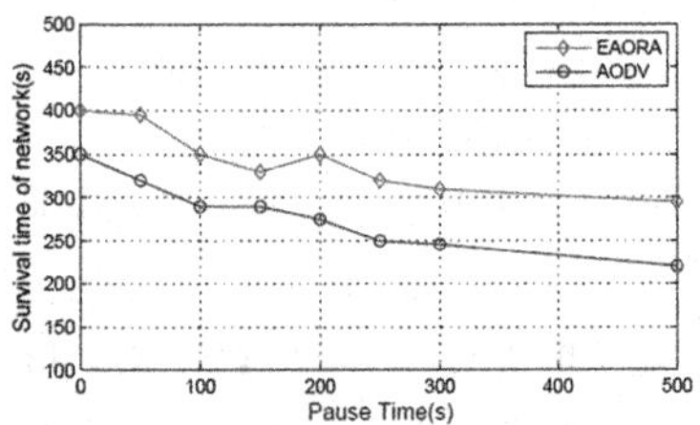

Fig. 7. Survival time and Pause time of network

Moving speed of different nodes is used to evaluate the performance of algorithm under three indicators.

Figure 8 shows the relationship between delivery rate under and node moving speed. The greater the strength of the node mobility leads to the greater chance of re-explore the route. But due to the adaptability of EAORA, re-routing can be very fast. Packet delivery rate will not make a huge change in node speed.

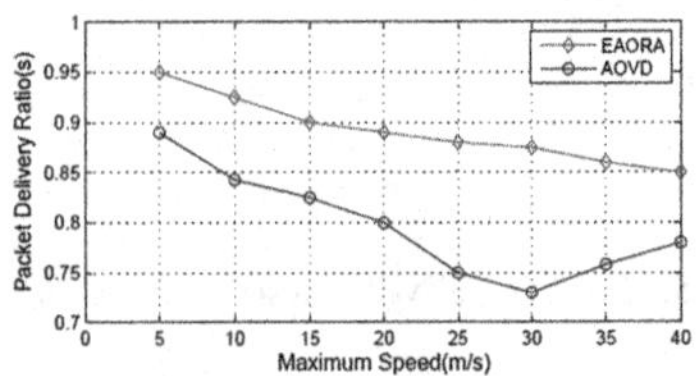

Fig. 8. Packet delivery ratio and Maximum speed

As shown in Fig. 9, the time delay of EAORA has great progress, which is due to the adaptive routing algorithm. Take the residual energy information into account and balance the node energy of network, thus extend the time of the nodes death. So that the link uses relatively less time to find the router, and node end-to-end transmission delay increases slowly with the rapid changes in the network topology.

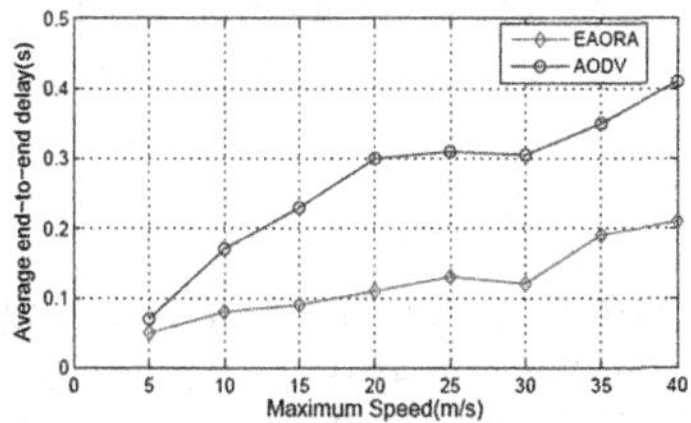

Fig. 9. Average end-to-end delay and Maximum speed

Figure 10 shows the relationship between survival time and node movement speed in the Ad Hoc network with the first node disabled. Because nodes moving fast will cause the increasing energy consumption, and involving in the routing communication will make the energy decrease quickly. In the same node moving speed, EAORA algorithm survival time is longer than AODV.

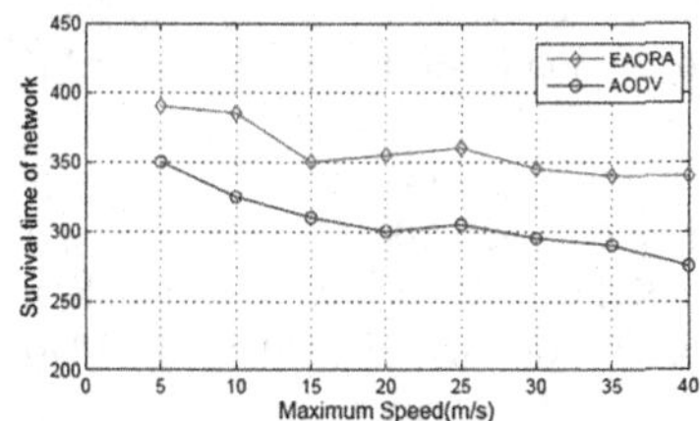

Fig. 10. Survival time of network and Maximum speed

6 Conclusion

Ad Hoc Networks are self-organized and dynamic networks with broad applications in reality. When a node runs out of energy, the connection between nodes will lose, which will result in reducing life of the entire network. To overcome the unbalanced energy of nodes, EAORA is proposed, and age and intra-group factors are introduced to overcome the premature and improve the performance. In order to balance the energy of nodes and extend the lifetime of the whole network, the EAO uses the remaining energy of nodes as a heuristic factor to calculate the routing probability and update phenomenon.

Furthermore, there are also some difficulties and limitations in the proposed algorithm. Firstly, the parameters setting of our algorithm needs to be optimized. Secondly, dynamic changes of network topology, the limitations of network wireless channel bandwidth, and the limited energy of network nodes are huge challenges for designing Ad Hoc network routing. Therefore, deeper theory research of the proposed algorithm is our important work in future.

Acknowledgement. This work was supported in part by the National Natural Science Foundation of China under Grant Nos. 61472139 and 61462073, the Information Development Special Funds of Shanghai Economic and Information Commission under Grant No. 201602008, the Open Funds of Shanghai Smart City Collaborative Innovation Center.

References

1. Cavalcanti, E.R., Spohn, M.A.: On improving temporal and spatial mobility metrics for wireless ad hoc networks. Inf. Sci. **188**(4), 182–197 (2012)
2. de Moraes, R.M., Kim, H., Sadjadpour, H.R., Garcia-Luna-Aceves, J.J.: A new distributed cooperative MIMO scheme for mobile ad hoc networks. Inf. Sci. **232**(5), 88–103 (2013)
3. Wang, W., Wang, H., Wang, B., Wang, Y., Wang, J.: Energy-aware and self-adaptive anomaly detection scheme based on network tomography in mobile ad hoc networks. Inf. Sci. **220**(1), 580–602 (2013)
4. Sun, Y., Jiang, Q., Singhal, M.: An edge-constrained localized delaunay graph for geographic routing in mobile ad hoc and sensor networks. IEEE Trans. Mob. Comput. **9**(4), 479–490 (2010)
5. Xiang, X., Wang, X., Zhou, Z.: Self-adaptive on-demand geographic routing for mobile Ad Hoc networks. IEEE Trans. Mob. Computing **11**(9), 1572–1586 (2012)
6. Zhang, X., Wang, E., Xia, J., et al.: A neighbor coverage based probabilistic rebroadcast for reducing routing overhead in mobile Ad hoc networks. IEEE Trans. Mob. Comput. **12**(3), 424–433 (2013)
7. Zhu, J., Wang, X.: Model and protocol for energy-efficient routing over mobile ad hoc networks. IEEE Trans. Mob. Comput. **10**(11), 1546–1557 (2011)
8. Mersch, D.P., Crespi, A., Keller, L.: Tracking individuals shows spatial fidelity is a key regulator of ant social organization. Science **340**(6136), 1090–1093 (2013)
9. Ren, F., Zhang, J., Wu, Y., et al.: Attribute-aware data aggregation using potential-based dynamic routing in wireless sensor networks. IEEE Trans. Parallel Distrib. Syst. **24**(5), 881–892 (2013)
10. Tan, G., Kermarrec, A.M.: Greedy geographic routing in large-scale sensor networks: a minimum network decomposition approach. IEEE/ACM Trans. Netw. (TON) **20**(3), 864–877 (2012)
11. Lorenzo, B., Glisic, S.: Optimal routing and traffic scheduling for multihop cellular networks using genetic algorithm. IEEE Trans. Mob. Comput. **12**(11), 2274–2288 (2013)
12. Qaed, A.S.M., Devi, T.: Ant colony optimization based delay and energy conscious routing protocol for mobile Adhoc networks. Int. J. Comput. Appl. **41**, 1–5 (2012)
13. Hernández, H., Blum, C., Francès, G.: Ant colony optimization for energy-efficient broadcasting in ad-hoc networks. In: Dorigo, M., et al. (eds.) ANTS 2008. LNCS, vol. 5217, pp. 25–36. Springer, Heidelberg (2008). https://doi.org/10.1007/978-3-540-87527-7_3
14. Ho, S.L., Yang, S., Wong, H.C., et al.: An improved ant colony optimization algorithm and its application to electromagnetic devices designs. IEEE Trans. Magn. **41**(5), 1764–1767 (2005)
15. Sim, K.M., Sun, W.H.: Ant colony optimization for routing and load-balancing: survey and new directions. IEEE Trans. Syst. Man Cybern. Part A Syst. Hum. **33**(5), 560–572 (2003)
16. Mersch, D.P., Crespi, A., Keller, L.: Tracking individuals shows spatial fidelity is a key regulator of ant social organization. Science **340**(6136), 1090–1093 (2013)

A Comparison Study of Surrogate Model Based Preselection in Evolutionary Optimization

Hao Hao, Jinyuan Zhang, and Aimin Zhou(✉)

Shanghai Key Laboratory of Multidimensional Information Processing, Department of Computer Science and Technology, East China Normal University, Shanghai, China
{51174506005, jyzhang}@stu.ecnu.edu.cn, amzhou@cs.ecnu.edu.cn

Abstract. In evolutionary optimization, the purpose of preselection is to identify some promising solutions in a set of candidate offspring solutions. The surrogate model is a popular method employed in preselection. A surrogate model is built to approximate the original objective function and to estimate the fitness values of the candidate solutions. Based on the estimated fitness values, the promising solutions can be identified. This paper aims to study and compare the surrogate model based preselection strategies in evolutionary algorithms. Systematic experiments are conducted to study the performance of four surrogate models. The experimental results suggest the surrogate model based preselection can significantly improve the performance of evolutionary algorithms.

Keywords: Surrogate model · Preselection · Evolutionary algorithm

1 Introduction

In scientific and engineering areas, there are many kinds of optimization problems. This paper considers the following box-constrained global optimization problem.

$$min_{x\in\Omega}\mathrm{f}(\mathrm{x}) \tag{1}$$

Where $\mathrm{x} = (x_1, x_2, \ldots, x_n)^T$ is a decision variable vector, $\Omega = \sum_{i=1}^{n} [a_i, b_i]$ defines the feasible region of the search space, and $\mathrm{f} = R^n \rightarrow R$ is the objective function.

In order to solve continuous optimization problem, a variety of methods have been proposed. Most of them are mathematical methods [1]. Apart from these mathematical methods, derivative-free or heuristic optimization methods, which do not need strong assumptions, are of interest [2, 3]. Among them, the evolutionary algorithm (EA) [4, 5] is a promising one due to global convergence and weak assumptions about the problems to solve. In EAs, there are two main components: reproduction and selection. The reproduction operator generates new trail solutions and the selection operator chooses the promising solutions. Actually, in EAs, the selection includes three different

D.-S. Huang et al. (Eds.): ICIC 2018, LNCS 10955, pp. 717–728, 2018.
https://doi.org/10.1007/978-3-319-95933-7_80

selection procedure, the mating selection, the preselection and the environmental selection. This paper focuses on preselection.

Preselection [6, 7] generally refers to determine promising solutions in a set of generated candidate offspring solutions. Since it is not a necessary step, preselection has not been paid enough attention in the traditional EAs. However, the preselection may play an important role in improving the performance of the algorithms when the computation resources are very limited. In which case, by using the preselection operator, the generated unpromising candidate solutions will be discarded directly without evaluating. A main issue in preselection is to measure the quality of the candidate offspring solutions, such as whether a solution is promising or not. According to this idea, the strategies in preselection can be classified into three categories:

- Fitness based strategy [8, 9]: In this case, all of the generated candidate offspring solutions are evaluated with the real fitness function, and the one with the best value is chosen out.
- Surrogate model based strategy [10, 11]: In this category, a surrogate model is built to approximate the objective function, then use the model to estimate the fitness values of the candidate solutions, and a solution will be chosen based on the approximated value.
- Classification based strategy [12, 13]: This category is a special case of surrogate model based strategy. It builds a classification model to measure the quality of the candidate offspring solutions. And the ones with 'good' classification labels are chosen out.

Undoubtedly fitness based strategy is an accurate method to measure the quality of candidate offspring solutions. However, as many evaluated solutions are used only once for choosing the promising solutions, the computational resource is largely wasted. This is vital especially when the function evaluation is computationally expensive. In order to solve this problem, researchers resort to surrogate model based approaches, where a surrogate model is built to estimate the objective function values and to replace the original function evaluation [10, 11]. Some popularly used surrogate models include the Kriging [14] which is also known as Gaussian process [15], radial basis function [16], artificial neural networks [17], and polynomial response surfaces [18]. This strategy is usually applied to expensive optimization problems [19].

This paper focuses on studying and comparison of surrogate model based strategies in EAs. The rest of the paper is organized as follows. Section 2 introduces the basic idea *of surrogate model based preselection (SPS)*, and presents the details of implementing SPS in EAs. The employed surrogate models and EAs used in this study are introduced as well. Section 3 dedicates to study SPS on 4 test instances. Finally, the paper is concluded with some marks for future work in Sect. 4.

2 A General Evolutionary Algorithm with Surrogate Model Based Preselection

This section introduces a general EA framework with SPS. Let $Y = \{y^1, \ldots, y^M\}$ be a set of M candidate offspring solutions of current solution, x. The target of a general preselection method is to select $k \leq M$ promising candidate solutions. Therefore, how to measure the quality of the candidate solutions in preselection is a key issue. A natural way is to find an approximated fitness function to replace the original one and to evaluate the candidate solutions. Fortunately, in the community of statistical and machine learning, the regression model, or the surrogate model, are designed to do so. In EAs, the current solutions can be used as training data points to build a surrogate model to approximate the original problem. When the new candidate solutions are generated, the surrogate model can be used to estimate the approximated fitness values. And then the approximated values instead of the real values are used to do preselection.

A general framework of evolutionary algorithm with surrogate model based preselection, denoted as SPS is presented in Fig. 1. The SPS procedure works as follows: at first, a surrogate model is built with the solutions in current population and their evaluated fitness values; then, the model is used to estimate the approximated fitness

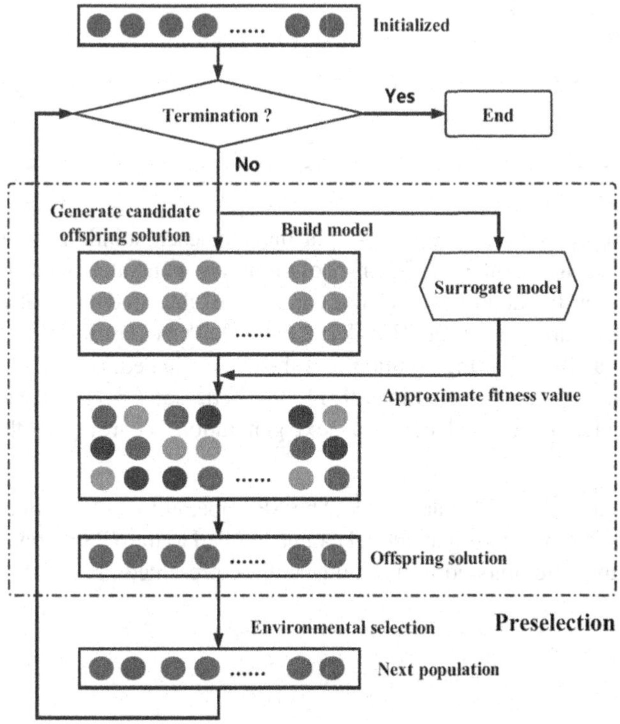

Fig. 1. Illustration of the SPS-EA procedure

values of the newly generated candidate offspring solutions; finally, according to the estimated fitness values, choose a promising offspring solution out. The corresponding algorithm framework is shown in Algorithm 1. Some components in framework are explained as follows.

- Population initialization: In Line 1, population P with N solutions are uniformly and randomly sampled from Ω, and all of the solutions are evaluated.
- Stopping condition: In Line 2, the algorithm stops when the number of function evaluations exceeds the given maximum number FES.
- Model building: In Line 3, a surrogate model is built based on the solutions and their fitness values.

Algorithm 1: Frameworks of SPS-EA

1 Initialize the population $P = \{x^1, x^2, \ldots, x^N\}$, and evaluate them;
2 **while** *termination condition is not satisfied* **do**
3   Build a surrogate model $f' = SA(x)$ based on the data set $\{< x^i, f(x^i) >, i = 1, 2, \ldots, N\}$;
4   **foreach** $x \in P$ **do**
5   Generate K candidate offspring solutions $Z = \{z^1, \ldots, z^K\}$;
6   For each $z^i \in Z$, estimate its fitness value $f'(z^i), i = 1, \ldots, K$ by the surrogate model;
7   Select $u = argmin_{z \in Z} f'(z)$ as the offspring solution of x;
8   Evaluate u;
9   **if** $f(u) < f(x)$ **then**
10   Set $x = u$;
11   **end**
12   **end**
13 **end**

- Offspring reproduction: In Lines 5–8, at first, for each solution in current population, a set of K candidate offspring solutions are generated with a reproduction operator. Then the approximated fitness values of the candidate solutions are estimated by the surrogate model. Finally, the one with best estimated fitness value is chosen out as the offspring solution and then is evaluated.
- Environmental selection: In Lines 9–11, the better one between current and offspring solution is selected out into next generation according to the real fitness values.

SPS-EA is a general EA framework. The SPS operator can be applied to a variety of EAs, and most of the surrogate model can be applied to the SPS operator. The details of surrogated model employed in this study will be presented in the following section.

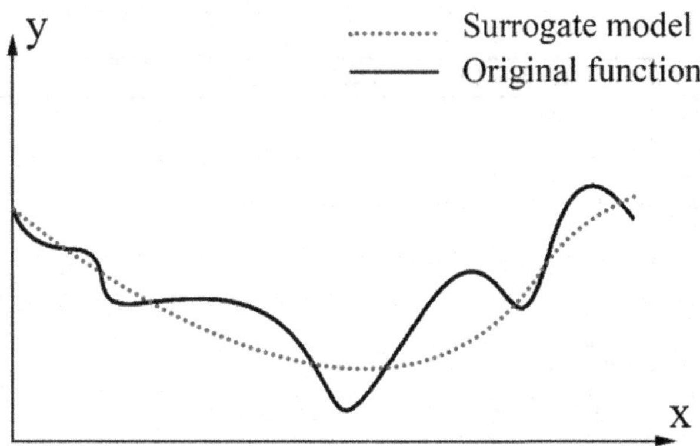

Fig. 2. Illustration of the surrogate model

2.1 Surrogate Model Building

Surrogate model is a kind of regression method. It takes less computational resource to approximate the original problem. By using the model, the approximate fitness value of a test solution can be obtained. An illustrating of the surrogate model is present in Fig. 2. The black curve indicates a complex real problem and red curve is a simple surrogate model.

In this paper, 4 surrogate models are employed to do preselection: Gaussian processing (GP) [15], radial basis function (RBF) [15], support vector machine (SVM) [20] and binary regression decision tree (Btree) [21]. GP is also known as the kriging model, and it can be seen as a combination of a global model plus a localized deviation. RBF is a kind of neural networks (NN) and it has shown to be effective tools for function approximation. SVM is mainly inspired from statistical learning theory, and the main advantages are that there is no local minima during learning and the generalization error does not depend on the dimension of the space. Btree is a kind of white box model, and if the given situation is observable in a model, then the condition is easy to be explained by boolean logic.

2.2 Offspring Reproduction

To study the efficiency of SPS, a reproduction operator from CoDE [8] is chosen. The details of the employed operator are presented in Algorithm 2.

Algorithm 2: CoDE based reproduction operator

1 Randomly select the control parameter $< F, CR >$ from $\{< 1.0, 0.1 >, < 1.0, 0.9 >, < 0.8, 0.2 >\}$;

2 Randomly select three parents $x^{r1}, x^{r2}, x^{r3}, x^{r4}$, and x^{r5} from P that are different from each other;

3 Generate a trail individual $y^1 = (y_1^1, \cdots, y_n^1)$ by

$$y_j^1 = \begin{cases} x_j^{r1} + F(x_j^{r2} - x_j^{r3}) & \text{if } rand() < CR \text{ or } j = j^{rnd} \\ x_j & \text{otherwise} \end{cases}$$

for $j = 1, \cdots, n$;

4 Generate a trail individual $y^2 = (y_1^2, \cdots, y_n^2)$ by

$$y_j^2 = \begin{cases} x_j^{r1} + F(x_j^{r2} - x_j^{r3}) + F(x_j^{r4} - x_j^{r5}) & \text{if } rand() < CR \text{ or } j = j^{rnd} \\ x_j & \text{otherwise} \end{cases}$$

for $j = 1, \cdots, n$;

5 Generate a trail individual $y^3 = (y_1^3, \cdots, y_n^3)$ by

$$y_j^3 = x_j + rand()(x_j^{r1} - x_j) + F(x_j^{r2} - x_j^{r3}).$$

for $j = 1, \cdots, n$;

6 Return a trial individual randomly selected from $\{y^1, y^2, y^3\}$.

Basically, CoDE employees a fitness based preselection, where three candidate offspring solutions are generated by using three operators, then all of them are evaluated and the best value is chosen out. Different from the original one, in SPS based CoDE, it employees the same operators to generate three candidate offspring solutions, then all of them get estimated fitness values through the surrogate model and the one with approximated fitness value is chosen out.

3 Experimental Results

This section aims to study the performance of SPS strategies. The 4 benchmark functions named as LZG01–LZG04 from [22] are chosen as the test instance. These test instances have different characteristics: LZG01 is simple unimodal; LZG02 is multimodal and with a narrow valley near the global optimum; Both LZG03 and LZG04 are multimodal.

In the experiments, the maximal number of function evolutions for all the algorithms is set to 60000 on all instances. The population size is N = 30 for all algorithms. Each algorithm is executed on each test instance for 10 independent runs. All the algorithms are implemented in Matlab, and the four surrogate models use the default settings as in the Matlab toolbox.

In the following subsections, we firstly study the performance of different surrogate model on different dimensions, then study the efficiency of SPS, and finally investigate the influence of the number of generated candidate offspring solutions.

The Wilcoxon rank sum test is applied to compare the experimental results. The "+", "-", "$\sim$" in the following tables indicate the value obtained by an algorithm is

Table 1. Statistical results for SPS-CoDE with 4 surrogate models on 4 instances after 60,000 FES over 10 runs.

Instance	n	SVM	GP	Btree	RBF
		mean_{std} $[rank]$	mean_{std} $[rank]$	mean_{std} $[rank]$	mean_{std} $[rank]$
LZG01	5	$4.32e{-}99_{1.16e-98}$ [3]	$3.95e{-}94_{8.33e-94}$ [4]	$1.39e{-}159_{1.86e-159}$ [1]	$5.61e{-}101_{6.79e-101}$ [2]
	10	$1.24e{-}42_{1.38e-42}$ [3]	$2.16e{-}42_{1.45e-42}$ [4]	$7.60e{-}60_{1.44e-59}$ [1]	$1.37e{-}46_{1.06e-46}$ [2]
	20	$1.84e{-}19_{1.46e-19}$ [4]	$5.62e{-}22_{2.54e-22}$ [3]	$1.65e{-}28_{4.19e-28}$ [1]	$1.68e{-}23_{1.76e-23}$ [2]
LZG02	5	$5.11e{-}25_{9.06e-25}$ [4]	$0.00e{+}00_{0.00e+00}$ [1]	$0.00e{+}00_{0.00e+00}$ [2]	$1.45e{-}30_{1.83e-30}$ [3]
	10	$1.28e{-}13_{6.94e-14}$ [4]	$3.39e{-}19_{4.87e-19}$ [2]	$2.35e{-}30_{3.99e-30}$ [1]	$2.23e{-}18_{1.87e-18}$ [3]
	20	$4.56e{-}05_{7.91e-05}$ [1]	$7.31e{-}01_{2.60e-01}$ [4]	$9.50e{-}03_{8.22e-03}$ [3]	$1.94e{-}04_{3.93e-04}$ [2]
LZG03	5	$8.88e{-}16_{0.00e+00}$ [1]	$8.88e{-}16_{0.00e+00}$ [2]	$8.88e{-}16_{0.00e+00}$ [3]	$8.88e{-}16_{0.00e+00}$ [4]
	10	$4.44e{-}15_{0.00e+00}$ [1]	$4.44e{-}15_{0.00e+00}$ [2]	$4.44e{-}15_{0.00e+00}$ [3]	$4.44e{-}15_{0.00e+00}$ [4]
	20	$1.12e{-}11_{6.65e-12}$ [2]	$1.25e{-}11_{8.79e-12}$ [3]	$8.22e{-}14_{4.63e-14}$ [1]	$2.17e{-}11_{4.13e-11}$ [4]
LZG04	5	$0.00e{+}00_{0.00e+00}$ [1]	$0.00e{+}00_{0.00e+00}$ [2]	$0.00e{+}00_{0.00e+00}$ [3]	$6.12e{-}04_{1.93e-03}$ [4]
	10	$4.91e{-}03_{1.04e-02}$ [3]	$5.41e{-}03_{1.47e-02}$ [4]	$1.72e{-}03_{5.45e-03}$ [2]	$0.00e{+}00_{0.00e+00}$ [1]
	20	$1.11e{-}17_{3.51e-17}$ [1]	$1.73e{-}03_{3.68e-03}$ [3]	$7.40e{-}04_{2.34e-03}$ [2]	$1.06e{-}02_{1.96e-02}$ [4]
Mean rank	5	2.25	2.25	2.25	3.25
	10	2.75	3.00	1.75	2.50
	20	2.00	3.25	1.75	3.00

smaller than, greater than, or similar to that obtained by its OCPS based version at 95% significance level.

3.1 Influence of Surrogate Models

This section studies the influence of the 4 surrogate models, SVM, GP, Btree and RBF, which are introduced in Sect. 2.1. To be simple, the name of these employed models are used to denote these SPS based CoDE algorithms. And the number of generated candidate offspring solutions is K = 3.

The statistical results for SPS-CoDE with 4 surrogate models on 4 instances after 60,000 FES over 10 runs are shown in Table 1. The mean and standard deviation values of the final obtained best solutions are calculated. The rank of SPS-CoDE with each model is presented as well.

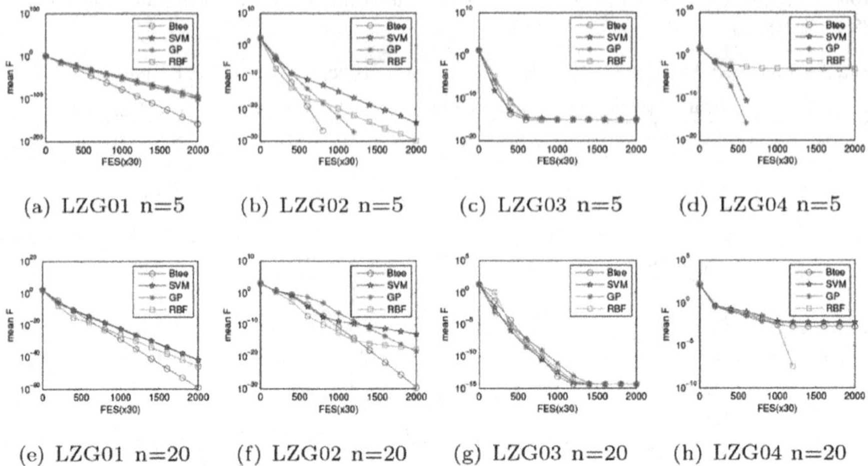

Fig. 3. The mean function values of the best solutions obtained by SPS-CoDE with 4 surrogate models versus FES on 4 instances and 2 dimensions over 10 runs

From Table 1, we can see that from the angle of the number of best results, SPS-CoDE with 4 surrogate models obtains 2, 1, 1, 0 best results on n = 5, and 1, 0, 2, 1 best results on n = 10, as well as 2, 0, 2, 0 best results on n = 20. It shows that the SVM-CoDE and Btree-CoDE work better than other two approaches. From the angle of the mean rank values, Table 1 suggests that Btree-CoDE achieves the best results and works significantly better than the other approaches.

Figure 3 plots the mean function values of the best solutions obtained by SPS-CoDE with 4 surrogate models versus FES on 4 instances and 2 dimensions over 10 runs. The figure suggests that on LZG01 and LZG02 with n = 5, 20, the Btree-CoDE convergence faster than the other approaches and obtains the best results. And on LZG03–LZG04, the 4 approaches have similar performance.

Considering the performance on 4 instances and 3 dimensions from Table 1 and Fig. 3, we can conclude that the CoDE with Btree works better than CoDE with other surrogate models.

3.2 Efficiency of SPS

This section studies the efficiency of SPS based CoDE. The results obtained by SPS-CoDE and CoDE are compared. The parameters are as follows: the Btree model is used, and the number of generated candidate offspring solutions is K = 3. The other control parameters are the same as in the previous section. The statistical results are shown in Table 2.

Table 2 suggests when n = 5, 10, the performance of median and mean values are same. Btree-CoDE works better, worse or similar than CoDE on 1, 0, 3 instances on n = 5, and 3, 0, 1instances on n = 10. When n = 20, 30, for median values, Btree-CoDE works

Table 2. Statistical results for Btree-CoDE and CoDE with 4 dimensions on 4 instances after 60,000 FES over 10 runs.

Instance	n	Btree-CoDE		CoDE	
		Median	$mean_{std}$	Median	$mean_{std}$
LZG01	5	2.83e–160(+)	3.54e–159$_{9.83e-159}$ (+)	5.32e–75	1.72e–74$_{2.40e-74}$
	10	2.65e–61(+)	3.94e–60$_{8.90e-60}$ (+)	3.74e–33	6.50e–33$_{7.53e-33}$
	20	3.17e–29(+)	4.84e–29$_{8.90e-60}$ (+)	8.01e–17	7.61e–17$_{7.53e-33}$
	30	5.95e–20(+)	9.46e–20$_{1.42e-19}$ (+)	9.20e–12	8.57e–12$_{5.18e-12}$
LZG02	5	0.00e+00(∼)	0.00e+00$_{0.00e+00}$ (∼)	0.00e+00	0.00e+00$_{0.00e+00}$
	10	0.00e+00(+)	7.68e–29$_{2.36e-28}$ (+)	9.17e–15	6.61e–14$_{1.71e-13}$
	20	4.63e–03(+)	6.38e–03$_{7.19e-03}$ (+)	5.16e+00	5.07e+00$_{6.24e-01}$
	30	1.32e+01(+)	1.32e+01$_{1.21e+00}$ (+)	2.00e+01	1.99e+01$_{5.31e-01}$
LZG03	5	8.88e–16(∼)	8.88e–16$_{0.00e+00}$ (∼)	8.88e–16	8.88e–16$_{0.00e+00}$
	10	4.44e–15(∼)	4.09e–15$_{1.12e-15}$ (∼)	4.44e–15	4.44e–15$_{0.00e+00}$
	20	6.13e–14(+)	7.37e–14$_{5.71e-14}$ (+)	2.73e–08	2.87e–08$_{9.38e-09}$
	30	4.90e–10(+)	5.73e–10$_{2.82e-10}$ (+)	3.74e–06	4.27e–06$_{1.54e-06}$
LZG04	5	0.00e+00(∼)	0.00e+00$_{0.00e+00}$ (∼)	0.00e+00	0.00e+00$_{0.00e+00}$
	10	0.00e+00(+)	0.00e+00$_{0.00e+00}$ (+)	5.12e–10	3.83e–07$_{1.20e-06}$
	20	0.00e+00(+)	1.23e–03$_{3.89e-03}$ (∼)	4.24e–13	1.64e–10$_{2.51e-10}$
	30	0.00e+00(+)	9.86e–04$_{3.12e-03}$ (∼)	8.07e–10	2.00e–09$_{2.56e-09}$
+/–/∼	5	1/0/3	1/0/3		
	10	3/0/1	3/0/1		
	20	4/0/0	3/0/1		
	30	4/0/0	3/0/1		

better than CoDE on all instances. And for mean values, Btree-CoDE works better than CoDE on 3 instances and similar on 1 instance.

The results clearly show that the SPS strategy is able to improve the performance of the original algorithms on LZG test instances. The reason might be that in SPS-CoDE, there are more candidate offspring solutions generated in one generation. This situation provides a high opportunity to obtain better offspring solutions in one generation without evaluate them. Thus the computational resource is saved.

3.3 Influence of Candidate Offspring Number

This section studies the influence of the number of candidate offspring solutions. In the experiment, the Btree surrogate model is used. The number of candidate offspring solutions K, is set to 2–5 and the other algorithm parameters are the same as in the previous sections.

Figure 4 plots the mean function values of the best solutions obtained by Btree-CoDE with different K values versus FES on 4 instances and 2 dimensions over 10 runs. The run time performance suggests the algorithm with different K value perform similar.

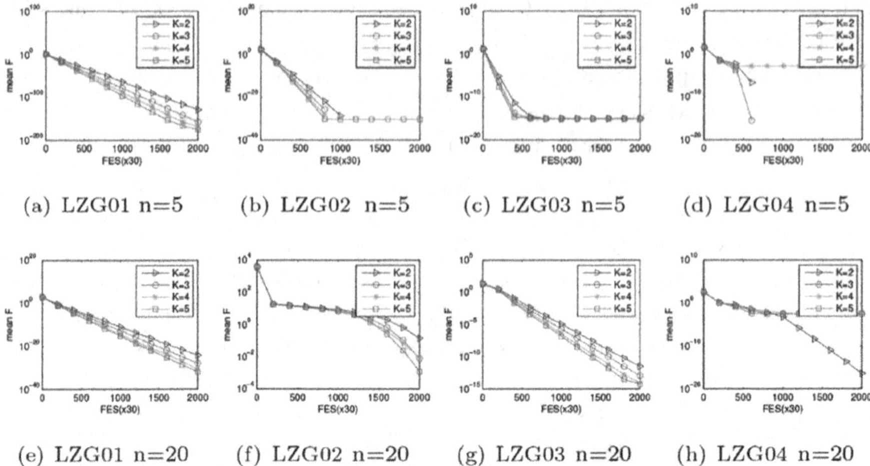

(a) LZG01 n=5 (b) LZG02 n=5 (c) LZG03 n=5 (d) LZG04 n=5

(e) LZG01 n=20 (f) LZG02 n=20 (g) LZG03 n=20 (h) LZG04 n=20

Fig. 4. The mean function values of the best solutions obtained by Btree-CoDE with different K values versus FES on 4 instances with 2 dimensions over 10 runs

The experimental results suggest that the number of candidate offspring solutions does not influence the results very much. However, if the number is big, it may need more cost to generate solutions and estimate the fitness values. Therefor a small number might be suitable.

4 Conclusion

This paper studies the surrogate model based preselection (SPS) in EAs. To apply SPS in EA, the solutions in current population and their fitness values are used to build a surrogate model. Then when new candidate offspring solutions are generated, the model is used to estimate their fitness values and the one with approximate fitness value is chosen out as the offspring solution. By using SPS, the low quality candidate solutions can be discarded without evaluation.

We firstly study the influence of 4 different surrogate models on the 4 LZG test instances with 3 different dimensions, the experimental results suggest that the Btree based algorithm has best performance. Then, the efficiency of the SPS is studied with 4 dimensions, and the results show that the using of SPS can improve the performance of original algorithm. Finally, the number of generated candidate offspring solutions are studied with 2 dimensions, the results shows the SPS is not really sensitive to the number of candidates.

Through the experiments, we can conclude that SPS is easy to implement and use. And the SPS can improve the performance of EAs on different dimensions.

Some further work about SPS in the further could be in the follows: (a) study the SPS on other EAs or multi-objective evolutionary algorithms (MOEAs), (b) analysis the efficiency of SPS more properly, (c) improve the efficiency of the approach.

Acknowledgement. This work is supported by the National Natural Science Foundation of China under Grant No. 61731009, 61673180, and 61703382.

References

1. Polak, E.: Optimization: Algorithms and Consistent Approximations. Springer, New York (1997). https://doi.org/10.1007/978-1-4612-0663-7
2. Lu, X.-F., Tang, K., Sendhoff, B., Yao, X.: A new self-adaptation scheme for differential evolution. Neurocomputing **146**(C), 2–16 (2014)
3. Mallipeddi, R., Suganthan, P.N.: Unit commitment - a survey and comparison of conventional and nature inspired algorithms. Int. J. Bio-Inspir. Comput. **6**(2), 71–90 (2014)
4. Back, T., Schwefel, H.-P.: Evolutionary computation: an overview. In: 1996 IEEE International Congress on Evolutionary Computation (CEC), pp. 20–29 (1996)
5. Back, T.: Evolutionary Algorithms in Theory and Practice: Evolution Strategies, Evolutionary Programming. Genetic Algorithms. Oxford University Press, New York (1996)
6. Cavicchio, D.J.: Adaptive search using simulated evolution. Unpublished doctoral dissertation, University of Michigan, Ann Arbor (1970)
7. Mahfoud, S.W.: Crowding and preselection revisited. In: Parallel Problem Solving from Nature (PPSN), pp. 27–36. Amsterdam Press, North-Holland (1992)
8. Wang, Y., Cai, Z., Zhang, Q.: Differential evolution with composite trial vector generation strategies and control parameters. IEEE Trans. Evol. Comput. **15**, 55–66 (2011)
9. Li, Y., Zhou, A., Zhang, G.: An MOEA/D with multiple differential evolution mutation operators. In: 2014 IEEE Congress on Evolutionary Computation (CEC), pp. 397–404 (2014)
10. Jin, Y.: A comprehensive survey of fitness approximation in evolutionary computation. Soft. Comput. **9**(1), 3–12 (2003)
11. Jin, Y.: Surrogate-assisted evolutionary computation: recent advances and future challenges. Swarm Evol. Comput. **1**(2), 61–70 (2011)
12. Zhang, J., Zhou, A., Zhang, G.: Preselection via classification: a case study on global optimization. Int. J. Bio-Inspir. Comput. (2018, accepted)
13. Lu, X., Tang, K., Yao, X.: Classification-assisted differential evolution for computationally expensive problems. In: 2011 IEEE Congress on Evolutionary Computation (CEC), pp. 1986–1993 (2011)
14. Emmerich, M., Giotis, A., Özdemir, M., Bäck, T., Giannakoglou, K.: Metamodel—assisted evolution strategies. In: Guervós, J.J.M., Adamidis, P., Beyer, H.-G., Schwefel, H.-P., Fernández-Villacañas, J.-L. (eds.) PPSN 2002. LNCS, vol. 2439, pp. 361–370. Springer, Heidelberg (2002). https://doi.org/10.1007/3-540-45712-7_35
15. El-beltagy, M.A., Keane, A.J.: Evolutionary optimization for computationally expensive problems using Gaussian processes. In: Arabnia, H. (ed.) Proceedings of International Conference on Artificial Intelligence IC-AI'2001. CSREA Press (2001)
16. Sun, C., Ding, J., Zeng, V., Jin, Y.: A fitness approximation assisted competitive swarm optimizer for large scale expensive optimization problems. Memetic Comput. **10**(2), 123–134 (2018)
17. Jin, Y., Olhofer, M., Sendhoff, B.: A framework for evolutionary optimization with approximate fitness functions. IEEE Trans. Evol. Comput. **6**(5), 481–494 (2002)
18. Tenne, Y., Armfield, S.W.: A framework for memetic optimization using variable global and local surrogate models. Soft. Comput. **13**(8), 781–793 (2009)

19. Tabatabaei, M., Hakanen, J., Hartikainen, M., Miettinen, K., Sindhya, K.: A survey on handling computationally expensive multiobjective optimization problems using surrogates: non-nature inspired methods. Struct. Multidiscip. Opt. **52**(1), 1–24 (2015)
20. Cortes, C., Vapnik, V.: Support-vector networks. Mach. Learn. **20**(3), 273–297 (1995)
21. Breiman, L., Friedman, J., Olshen, R., Stone, C.J.: Classification and regression trees. Biometrics **40**(3), 17–23 (1984)
22. Liu, B., Zhang, Q., Gielen, G.G.E.: A Gaussian process surrogate model assisted evolutionary algorithm for medium scale expensive optimization problems. IEEE Trans. Evol. Comput. **18**(2), 180–192 (2014)

A Grouping Genetic Algorithm Based on the GES Local Search for Pickup and Delivery Problem with Time Windows and LIFO Loading

Feng Zhang, Bin Li(✉), and Kun Qian

School of Information Science and Technology,
University of Science and Technology of China, Hefei 230027, China
{zfeng331,qk92}@mail.ustc.edu.cn, binli@ustc.edu.cn

Abstract. This paper investigates the pickup and delivery problem with time windows and last-in-first-out (LIFO) loading (PDPTWL). In this problem, the compartment of a vehicle is modeled as a linear LIFO stack. The last picked up goods are placed on the top of the stack, and the goods can be delivered only when they are on the top of the stack. The LIFO constraint makes the feasible solution space more tightly constrained and the design of an effective algorithm more difficult. A grouping genetic algorithm combined with the guided ejection search is proposed to solve the PDPTWL problem of large-size, in which, an evaluation function is defined to guide the selection of genes for crossover and mutation, and a local search based on the guided ejection search is embedded into the genetic algorithm to improve the quality of the solutions. Then, a population-based metaheuristic is ready for the PDPTWL problem. It can solve instances with 50–300 requests in the Li and Lim's benchmarks. Compared with the existing state-of-the-art algorithms, the experimental results confirm that the proposed algorithm works more efficiently. It improves 164 best-known solutions out of 236 instances and reduces 424 vehicles.

Keywords: Vehicle routing problem with pickup and delivery
Last-in- first-out loading · Grouping genetic algorithm · Guided ejection search

1 Introduction

The pickup and delivery problem with time windows and last-in-first-out (LIFO) loading (PDPTWL) has one more LIFO loading constraint than the PDPTW problem. Load and unload operations must be performed in the last-in-first-out order. The compartment of a vehicle is modeled as a linear LIFO stack. The loaded goods are placed on the top of the stack. The goods can be delivered only when they are on the top of the stack. In this paper, the vehicle with only one stack is considered. A set of requests need to be served by several vehicles. A depot is used as the start and end points of each vehicle. There are some other constraints that need to be observed:

D.-S. Huang et al. (Eds.): ICIC 2018, LNCS 10955, pp. 729–741, 2018.
https://doi.org/10.1007/978-3-319-95933-7_81

- The pickup and delivery constraint. Each request contains a pickup node and a delivery node. The goods must be picked up first and then delivered. Each node can only be visited once.
- The capacity constraint. Each vehicle cannot exceed the capacity limit at any time.
- The time window constraint. There is a time window limit for each pickup node and delivery node. If the vehicle arrives at a node before the earliest time window, it must wait. It cannot arrive later than the latest time window.

Valid routes are constructed by the least vehicles to complete all requests with the minimum cost. This problem can be regarded as a combination of two interdependent subproblems. The first one is the grouping problem (clustering problem), that is, clustering and distributing requests to each vehicle. The second one is the routing problem, which is, planning a reasonable route for each vehicle. Two goals are considered. The first goal is to minimize the number of vehicles, and the second goal is to reduce the total driving distance. People usually use the two stage method to optimize the two targets individually. In this paper, the first goal is the focus of our algorithm.

The LIFO constraint of PDPTWL is common in practice, such as the transport of heavy goods, dangerous goods, or livestock, and so on, to reduce the handing cost. In recent years, more and more attention has been paid to the PDPTWL problem. There are two main ways to deal with this LIFO constraint. The first is to treat it as a hard constraint that cannot be violated. The second is to treat it as a soft constraint, and violation will increase the additional handling cost.

Carrabs et al. [3] introduced a variable neighborhood search (VNS) heuristic for the Pickup and Delivery Traveling Salesman Problem with LIFO Loading (TSPPDL or PDTSPL). Cordeau et al. [16] developed a branch-and-cut algorithm for TSPPDL. Liabbabc [7] developed a novel variable neighborhood search heuristic for the TSPPDL with a tree representation. Carrabs et al. [4] proposed an additive branch-and-bound algorithm for TSPPD with LIFO or FIFO loading (TSPPDF). Wei et al. [12] investigated perturbation operators for variable neighborhood search approaches for TSPPDL and FIFO loading. Gao et al. [6] extended the TSPPDL with multiple vehicles and distance constraints called multiple pickup and delivery traveling salesman problem with LIFO and distance constraints (MTSPPD-LD). They proposed a variable neighborhood search heuristic which includes a dynamic programming component for optimally solving TSPPDL instances. Cheang et al. [8] proposed six new neighborhood operators and a two-stage approach for solving the MTSPPD-LD. Côté et al. [9] described a large neighborhood search heuristic for PDTSP with multiple stacks. Two branch-price-and-cut algorithms were implemented to solve the pickup and delivery problem with time windows and multiple stacks by Cherkesly et al. [13], and each stack is rear-loaded with LIFO fashion. The pickup and delivery traveling salesman problem with handling cost (PDTSPH), which is a generalization of PDTSP and PDTSPL, was introduced by Veenstra et al. [15]. For PDTSPH, a large neighborhood search heuristic with new removal operators was proposed. Branch-price-and-cut algorithms with ad hoc dominance criteria were presented by Veenstra et al. [14] to solving the pickup and delivery problem with time windows and handling operations (PDPTWH) with two handling policies including compulsory and preventive handling.

The size of the solution space for PDTSP and PDTSPL with several requests was compared by Veenstra et al. [15]. PDTSPL has one more LIFO loading constraint than PDTSP, and its solution space is only a subset of PDTSP as shown in Table 1. The existence of the last-in-first-out constraint makes the feasible solution space tightly constrained. The column n indicates the problem size. The column $P(\%)$ indicates the proportion of the number of feasible solutions of PDTSPL to PDTSP.

Due to strict constraints, the number of valid routing arrangements for a given set of requests is limited. The first sub-problem (grouping problem) is proved to be more dominant than the second sub-problem (routing problem) in improving the quality of the solution [1]. The main object in this paper is to optimize the first goal, reducing the number of vehicles.

PDPTWL was firstly raised by Cherkesly et al. [11], and three related branch-price-and-cut algorithms were proposed to solve instances with up to 75 requests. Cherkesly et al. [10] proposed a population-based metaheuristic and gave the computation results for instances with 30–300 requests. The population is managed in order to maintain high quality solutions in terms of total cost and population diversity.

This paper develops a grouping genetic algorithm (GGA) based on the guided ejection search (GES) for solving large-size instances of PDPTWL. Two major heuristic search principles form the basis of this paper. Firstly, the proposed genetic algorithm employs an adaptation of the GGA frame which has been applied to solve the PDPTW by Pankratz [2]. It uses a group-oriented genetic encoding introduced by Falkenauer [21] in which each gene represents a group of requests rather than a single request. In this paper, the new crossover and mutation operators are proposed and they are denoted as the guided crossover and mutation operators. An evaluation function of each gene in a chromosome is defined. Two criteria are designed to guide the selection of genes for crossover and mutation. Secondly, the proposed GES-based local search is embedded in genetic algorithm. The GES for PDPTW was proposed by Nagata and Kobayashi [5], and successfully applied to solve a wide variety of combinational optimization problems, such as the vehicle routing problem with time windows (VRPTW) [18], the job shop scheduling problem [17]. The main goal of GES is to reduce the number of vehicles. In this paper, it is slightly modified to suit the PDPTWL. After each iteration, if a solution with fewer vehicles can't be found, the one with the poor performance can be accepted. The GES principle is performed on the local search process. The experimental results confirm that our algorithm works efficiently.

2 Problem Description

A problem instance of PDPTWL contains n pair of requests which will be severed by a fleet of vehicles. The set of n pair of requests is presented by $I = \{1, \ldots, n, n+1, \ldots, 2n\}$, where $P = \{1, \ldots, n\}$ denotes the set of pickup nodes, and $D = \{n+1, \ldots, 2n\}$ denotes the set of delivery nodes. Each request $i \in I$ consists a pickup node $i \in P$ and a delivery node $n+i \in D$. K denotes a set of vehicles with the same capacity Q. The PDPTWL problem described in this paper considers only one depot, which is denoted by $\{0, 2n+1\}$

Table 1. The number of feasible solutions with n requests [16].

n	PDTSP	PDTSPL	$P(\%)$
1	1	1	100.00
2	6	4	66.67
3	90	30	33.33
4	2520	336	13.33
5	113,400	5040	4.44
6	7,484,400	95040	1.27
7	681,080,400	2,162,160	0.32
8	81,729,648,000	57,657,600	0.07

with the start depot 0 and the end depot $2n+1$. Each node can only be accessed once except the depot, and the goods at each node can not be split.

The problem can be described on a graph $G(N, A)$, where $N = P \cup D \cup \{0, 2n+1\}$ is the set of the nodes, and A is the set of the arcs. For each node $i \in N$, let q_i represent the weight of the goods. For $i \in P$, $q_i > 0$ and $q_{n+i} < 0$ for $n+i \in D$, where $q_i = -q_{n+i}$. Let $q_i = 0$ for $i \in \{0, 2n+1\}$. The quantity of goods loaded by a vehicle cannot exceed its capacity Q at any time. For each node $i \in N$, let nonnegative variable s_i represent the service time, where $s_i > 0$ for $i \in P \cup D$ and $s_i = 0$ for $i \in \{0, 2n+1\}$. Each node $i \in N$ has a time window $[e_i, l_i]$ with the earliest time e_i and the latest time l_i. The node $i \in N$ only can be accessed between e_i and l_i. If the vehicle $k \in K$ arrives the node $i \in N$ before its earliest time e_i, it must wait until e_i. It can not arrived later than l_i. For each arc $(i,j) \in A$, $d_{ij} \geq 0$ denotes the distance, where $d_{ij} > 0$ for $i \neq j$ and $d_{ij} = 0$ for $i = j$.

Each arc $(i,j) \in A$ must satisfy the constraints mentioned above. In consideration of the pickup and delivery constraint, for $i \in P$, the arc $(i+n, i)$ is infeasible. The arc (i,j) is unviable for $i = 0$ and $j \in D$ because the delivery node can't be accessed before its pickup node. In the same way, the arc (i,j) is infeasible for $i \in P$ and $j = 2n+1$ because the pickup good in node i is not delivered. According to the last-in-first-out constraint, the arc (i,j) is infeasible for $i \in P$ and $j \in D/\{n+i\}$ because the goods of pickup node i are placed on the top of the stack, and the goods can be delivered only when they are on the top of the stack.

3 Description of the Algorithm

A grouping genetic algorithm based on GES local search (GGA-GES) is presented to solve the large-size instances of PDPTWL. In this section, the algorithm is introduced in two parts. The first part is based on the GGA framework proposed by Pankratz [2] for PDPTW. The guided crossover and mutation operations proposed in this paper play an important role in GGA-GES. An evaluation function which is defined to judge the quality of the gene is used to specify the crossover genes and the mutation genes. The second part is a local search based on GES. Its role is to reduce the number of vehicles. The general search scheme of the proposed GGA-GES for PDPTWL can be described in Algorithm 1.

Algorithm 1. The frame of GGA-GES

Initialize population P;
Repeat
  select parents x, y: choosing a pair of individuals x, y randomly from P as parents;
  generate two children x', y': assessing the performance of each gene using (1) and applying the guided crossover operator to x, y with probability p^{cross};
  generate two modified children x'', y'': evaluating each gene using (1) and applying the guided mutation operator with the GES local search to x', y' respectively with probability p^{mut};
  update population P: insert x'', y'' into P and in turn remove the two worst individuals from P ;
Until termination criterion is met
Return best individual from P as solution.

The population size n^{pop} is a user-defined parameter. At the beginning, each pair of requests is assigned to one vehicle. Then the removing and inserting heuristic described in 3.2 is implemented to each individual for reducing the number of vehicles. Removing a route randomly and inserting its requests into other routes repeat until the route can't be removed successfully. Because of the random execution of the removing and inserting process, the initial solutions are different. The crossover and mutation operations are a set of sequential execution processes. The parameters p^{cross} and p^{mut} are the crossover probability and mutation probability. Termination condition is determined by the maximum number of iterations *iter*. If the program runs for more than three hours, the program is ended even if *iter* iterations are not completed.

3.1 Grouping Genetic Operator

In our paper, the group-oriented genetic encoding proposed by Falkenauer [21] is used, where each gene in a chromosome represents the route of each vehicle with several requests. The number of vehicles is the length of a chromosome, i.e. the number of genes. We also propose two new crossover and mutation operators with the well-designed evaluation method for each gene.

The Guided Crossover Operator. Due to capacity, pickup and delivery, time window and last-in-first-out constraints, most of the modified solutions would be invalid. It is very important to design an effective crossover operator. The guided crossover operator takes the following four steps. After the crossover, a feasible solution will be generated.

a. Select two individuals randomly as father 1 and father 2.
b. Specify a crossover section, i.e. several incoherent genes. Assume that *crossN* crossover genes are selected. The maximum and minimum ratios of the number of the genes in a chromosome are indicated by $[lnum, unum]$. In our experiment, we

choose 5%–15% of all genes as crossover genes. Two selection criteria are defined. The first one is to select the genes with the best performance with *gcross* probability of all crossover genes. The second is to select the genes randomly with $1 - gcross$ probability. The evaluation function f_V for each gene is defined as the following Formula (1), where V is set of the requests in the gene. The numerator of the fraction is the sum of the total waiting time *twait* and travel time *distance* in that gene. In this paper, the vehicle speed is set to one. The denominator is the number of requests. It can be interpreted as the average time spent on each request in that gene, including waiting time and driving time, excluding service time. The smaller the value, the better the gene performance.

$$f_V = \frac{\sum_{v \in V} twait + \sum_{v \in V} distance}{|V|} \tag{1}$$

c. Insert the genes in crossover section of father 1 at corresponding crossover points of father 2, which are selected randomly.
d. Eliminate some conflicts. Clear all duplicate nodes in father 2 that have been assigned by the crossover section. The corresponding route in the crossover gene imported from the parent 1 remains unchanged. Then get the complete offspring. Exchange the parent's role to generate the second offspring.

The Guided Mutation Operator. The number of genes may increase after crossover operation. In order to improve the quality of the solution, the guided mutation operation is carried out. The main idea is to remove some genes and then reinsert their requests into other genes. It can be described as the following steps.

a. Select a gene in the chromosome. Two choose schemes are designed. The first scheme is based on the evaluation function of Formula (1) for each gene. Here we choose the gene with the worst performance with *bmut* probability. Intuitively, poor genetic performance means that it serves fewer requests, but its cost is very higher. The second scheme is to choose the gene randomly with $1 - bmut$ probability.
b. Remove the selected gene from the chromosome and insert its requests into the other genes by means of the local search described in 3.2.
c. Repeat the above process *mutN* times, where $mutN = j^{mua} \times crossN$ and j^{mua} is a user-defined parameter.

3.2 The GES-Based Local Search

In this paper, GES [5] is modified to solve the more complicated PDPTWL problem. The GES-based local search has made the most contribution to reduce the number of vehicles in the process of constructing initial solutions and the mutation operator. The overall idea is described below.

Select and eliminate a route randomly from the current solution. The ejection pool *EP* is initialized with the requests in the removed route. It is used to restore the temporarily unassigned requests. A request is removed from the *EP* randomly and inserted into other routes. If a request couldn't be inserted elsewhere, the ejection-insertion procedure works. For example, if a request i removed from EP cannot be inserted into the route r, some requests from r are ejected and restored into EP in order to make room for inserting i feasibly. $kmax$ requests can be ejected at most. Other parameters refer to [5]. If the termination condition is met and several requests are unassigned, where $EP \neq 0$, the modified solution is restored to its previous state in the original GES. This means that some operations are wasted, for example the successful insertion operations for the pickup and delivery nodes which have the new positions.

In our algorithm, we allocate a new vehicle if possible to route the unassigned requests in EP. After this operation, although the number of vehicles does not decrease and the total travel distance may increase, a new solution will be obtained. It will improve individual diversity and the search efficiency. The new route is created by a simple removing and inserting heuristic. A request is randomly removed from EP to create a new route. Then other requests in EP are removed one by one to insert in a randomly selected feasible position in the route. After the process, if $EP \neq 0$, the solution will be restored to the last feasible state. Otherwise a new solution is obtained.

Considering the LIFO constraint in PDPTWL, more judgment and pruning conditions can be added to the lexicographic search in GES in order to find the best insertion-ejection combination more efficiently. Given a request h_{in} (including the pickup node p_{in} and the delivery node d_{in}) to be inserted, the lexicographic search is used to find the insertion positions for p_{in} and d_{in} and the ejection requests which make room for inserting p_{in} and d_{in}. For each p_{in}, the combination of possible insertion position for d_{in} and ejections of at most $kmax$ requests ($2kmax$ nodes) can be denoted as $\{i(1), \ldots i(jd), \ldots, i(j)\}(1 \leq i(1) < \ldots < i(jd) < \ldots < i(j) \leq n, 1 \leq jd \leq j \leq 2kmax)$. Here, $i(jd)$ denotes that d_{in} is inserted just after $i(jd)$. The requests to be ejected are denoted as a set of the corresponding nodes $\{i(1), \ldots i(jd-1), i(jd+1) \ldots, i(j)\}$, where j is the number of temporarily ejected nodes plus one. Let $H(h_{in})$ denote a partial route starting at the node p_{in} and ending at the node d_{in}. If $H(h_{in})$ violates the LIFO constraint, the lower-level lexicographic search $\{i(1), \ldots i(jd), *, *, \ldots\}$ can be pruned because the LIFO constraint is never satisfied.

In original GES, for inserting a request h_{in}, all feasible insertion-ejection combinations are obtained in all routes, and only one is selected. In order to accelerate the computation time, we randomly select a route to calculate until some suitable combinations have been achieved.

4 Experiments and Analysis

The algorithm[1] is implemented in C++ and tested on Li and Lim's data set [20], which is available at http://www.sintef.no/projectweb/top. There are 10 different problem sizes, including 50 requests, 100 requests, 200 requests, 300 requests, 400 requests and 500 requests, and for each instance size, six groups are tested including LC1, LC2, LR1, LR2, LRC1, and LRC2. Compared with the population-based metaheuristic proposed by Cherkesly et al. [10], we do experiments on the first four problem sizes.

[1] https://github.com/fengzxl/GGA-GES-PDPTWL.git.

All tests are performed on a computer equipped with an Intel (R) Xeon(R) E5-2690V4 processor (2.6 GHz). In this section, we report the computational results and analyze the impact of the guided crossover and mutation operation and the GES-based local search. The user-defined parameters ($n^{pop}, p^{cross}, p^{mut}, lnum, unum, gcross, bmut, j^{mua}$, $iter$) are set to (6, 0.8, 1, 0.05, 0.15, 0.6, 0.5, 3,100). These parameter values are first set to reasonable values, then sequentially modified individually to find the most appropriate values.

4.1 Experiment Results

Table 2 shows the summary of the experiment results. Let n represent the problem size. For each problem size, let $\#new\ best$ denote the number of instances whose solutions obtained by our algorithm are better than the best known solutions obtained by Cherkesly et al. [10]. Their experiment environment is Intel (R) Xeon(R) X5675 processor (3.07 GHz). According to the SPEC's CPU2006 benchmark [19], their computer is faster than ours. Let $P(\%)$ denote the proportion of instances that get better results, that is the new-best solutions, by use our method in all instances. Let $V(\%)$ denote the average deviation on the minimum number of vehicles after 10 runs. Let *Seconds* denote the average time to get the best found solutions after 10 runs.

Table 2. The summary of the experiment results.

n	50	100	200	300
$\#new\ best$	11	45	52	56
$P(\%)$	19.64	75.00	86.67	93.33
$V(\%)$	0.07	1.56	1.40	1.08
$Seconds(s)$	7.10	160.59	1223.59	3935.52

For the instances with 50 requests, 11 new best solutions are found with an average deviation of 0.07%, accounting for 19.64%, compared with [10]. For the instances with 300 requests, 56 new best solutions are found with an average deviation of 1.08%, accounting for 93.33%. From Table 2, we can see that as the problem size increases, more new best solutions are found by our algorithm.

Table 3 presents the best found results obtained by GGA-GES proposed in this paper and the comparison with best known solution values obtained by the population-based metaheuristic [10]. The first column n represents the problem size. The first line indicates different types of instances. For each instance, we report the number of vehicles ($Veh.$) and the total travel distance ($Dist.$) with respect to the best found solution after 10 runs by our GGA-GES. For each instance, the column Δ. ows the reduced number of vehicles compared with the best know solutions in [10]. The last line *total* counts the sum of the instances for each type. It is clear that our method has achieved better results in 164 instances out of 236 instances with respect to the number of vehicles and reduced 424 vehicles. For LC1, our solutions reduced 72 vehicles than

Table 3. The best found solutions obtained by the proposed GGA-GES and the comparison with best known solutions obtained by Cherkesly et al. [10].

n		LC1			LC2			LR1			LR2			LRC1			LRC2		
		Veh.	*Dist.*	Δ	*Veh.*	*Dist.*	Δ	*Veh.*	*Dist.*	Δ	*Veh.*	*Dist.*	Δ	*Veh.*	*Dist.*	Δ	*Veh.*	*Dist.*	Δ
50	1	15	4707.2	0	7	11261.4	0	30	4265.3	0	6	2115.7	0	18	2072.5	0	5	2377.7	0
	2	13	3581.0	−1	5	6249.3	0	21	2929.9	0	4	2286.7	−1	16	2049.6	0	4	2368.4	−1
	3	10	2304.6	0	4	3820.5	0	14	1965.7	0	3	1784.1	0	12	1542.3	0	4	1910.3	0
	4	9	1684.6	0	4	3383.2	0	10	1215.6	0	3	1681.6	0	11	1376.7	0	3	1552.1	0
	5	13	3167.6	0	4	3183.4	−1	17	2084.7	0	4	1832.3	0	18	2102.1	0	5	2459.7	0
	6	13	3451.2	0	4	3344.4	0	15	1965.5	0	3	1841.7	−1	13	1781.9	−1	4	2117.3	0
	7	12	2394.7	0	4	2805.9	0	12	1571.9	0	3	1676.4	0	12	1585.7	−1	4	1952.5	0
	8	11	1938.9	0	4	2746.8	0	11	1371.8	0	3	1429.3	0	11	1397.6	0	3	1657.1	−1
	9	10	1529.1	0				13	1639.3	0	3	1703.8	−1						
	10							11	1342.1	0	3	1831.9	−1						
	11							12	1540.0	−1	3	1435.7	0						
	12							11	1376.3	0									
100	1	31	13168.9	−2	13	23158.8	−2	32	14544.1	−2	8	8508.7	−1	25	5722.5	−2	10	7196.6	−1
	2	24	9263.4	−2	9	11774.8	−1	20	9179.6	−2	7	9158.7	−1	19	5646.5	−1	7	6951.5	−1
	3	19	6228.6	−1	7	6575.0	−1	15	6661.5	−2	5	7877.8	0	14	4928.9	−1	5	7314.9	−1
	4	18	4900.7	+1	7	6494.5	0	11	4831.8	−1	4	5553.5	0	10	3686.4	0	4	5695.3	0
	5	25	7758.6	−2	8	8213.9	−2	23	10257.3	−1	6	8500.9	−1	18	5010.3	−1	6	6215.8	−1
	6	24	7462.3	−2	8	7783.0	−1	17	7420.7	−1	5	8042.4	−1	19	4769.8	−1	6	6384.8	−1
	7	23	6319.2	−1	8	7153.3	0	13	5477.6	−1	4	6346.2	0	16	4458.2	−1	5	5732.7	−1
	8	21	4782.8	−1	7	5293.8	−1	10	4050.5	0	3	4767.5	0	14	4084.6	−1	5	5800.8	0
	9	20	5017.7	−1	7	5609.1	0	17	7127.8	−1	5	7479.8	−1	15	4225.1	0	4	5536.6	−1
	10	18	4698.7	−1	7	5457.1	0	13	5121.6	−1	4	6396.5	−1	13	3956.8	−1	4	4914.3	0

(continued)

Table 3. *(continued)*

n		LC1			LC2			LR1			LR2			LRC1			LRC2		
		Veh.	*Dist.*	Δ	*Veh.*	*Dist.*	Δ	*Veh.*	*Dist.*	Δ	*Veh.*	*Dist.*	Δ	*Veh.*	*Dist.*	Δ	*Veh.*	*Dist.*	Δ
200	1	58	27102.0	−5	21	32397.4	−4	56	32479.1	−4	16	22638.5	−3	47	13900.6	−2	19	17818.7	−4
	2	46	21035.4	−5	17	21771.7	−3	37	21772.2	−4	14	23944.1	−2	39	14544.2	−4	13	18806.7	−3
	3	36	16627.0	−4	15	16768.2	0	27	16510.1	−5	8	19400.0	−1	29	13588.2	−3	11	21129.5	−2
	4	33	12570.5	+1	14	14043.3	1	18	9688.9	−2	6	12464.5	0	19	9202.7	−2	6	13585.6	−1
	5	52	20782.1	−2	17	19641.7	−3	41	23694.7	−6	11	20193.3	−2	38	12211.6	−3	14	16460.8	−2
	6	48	16542.6	−2	15	14083.0	−1	32	18292.7	−4	9	18800.6	−1	35	12351.3	−4	12	14864.6	−3
	7	46	15177.7	−1	15	14791.7	−1	24	13988.9	−3	7	15060.3	0	33	11161.4	−2	11	15748.2	−2
	8	44	13738.6	−1	14	12175.0	−1	16	8513.2	−1	5	11815.2	0	29	10801.4	−3	10	14020.1	−1
	9	40	12562.6	−1	14	12523.3	−1	33	18171.4	−4	10	19205.6	−2	29	10737.7	−3	9	13591.4	−2
	10	38	12317.9	0	14	12477.2	0	24	12240.7	−2	8	17253.1	−2	26	9857.6	−3	8	12900.4	−1
300	1	100	57858.1	−7	35	63874.1	−8	70	64784.6	−7	27	53363.1	−4	68	30080.9	−7	29	39929.7	−6
	2	73	40457.4	−7	26	39598.0	−5	50	48337.0	−7	19	53333.8	−3	55	32560.1	−4	21	43142.4	−5
	3	56	27625.0	−4	22	26991.4	−1	37	36197.2	−4	13	46435.0	−2	40	30323.5	−3	14	45655.4	−3
	4	52	22611.3	0	20	21475.6	+1	28	25584.6	0	7	26693.7	−1	25	21071.3	−2	9	34765.7	−1
	5	79	37432.2	−7	27	38227.3	−5	58	52969.6	−8	18	48169.7	−4	51	26036.5	−6	22	36621.9	−3
	6	73	32175.2	−4	23	26399.9	−3	43	40876.3	−8	14	46707.8	−3	52	27308.1	−8	19	37873.2	−4
	7	71	30064.9	−5	22	24239.9	−3	29	28172.1	−6	10	38862.6	−2	44	23742.8	−5	17	36165.6	−3
	8	67	26411.8	−1	21	21596.6	−2	21	18452.7	−2	7	27040.6	0	39	22732.9	−6	14	33581.4	−3
	9	61	23945.2	−2	22	24554.6	−1	49	44192.1	−7	16	45282.6	−3	39	22331.0	−4	14	33035.5	−2
	10	57	23812.5	−2	20	19286.3	−1	34	29830.7	−6	13	39323.7	−3	34	20596.7	−4	11	31303.1	−2
total		1459	–	−72	511	–	−50	1075	–	−103	327	–	−48	1045	–	−89	371	–	−62

theirs, for LC2 reducing 50 vehicles, for LR1 reducing 103 vehicles, for LR2 reducing 48 vehicles, for LRC1 reducing 89 vehicles, and for LRC2 reducing 62 vehicles.

4.2 Impact of the Guided Crossover and the Mutation Operations

After crossover operation, the number of genes (vehicles) increases and it can be effectively reduced by mutation operation. The quality of the solution is closely related to the effectiveness of the crossover and mutation operations.

In mutation operation, the program of removing a gene by the evaluation function or randomly selection are sequentially executed with the same number of executions. For the first experiment where $gcross = 0.66$ and $bmut = 0.5$, the success times of removing the genes by the evaluation function and randomly selection are counted separately. The average ratio is 1.80. The success mentioned here means that after the process of removing a gene, this gene can be successfully deleted or the requests in ejection pool can form a new route (gene) in the mutation operator. In other words, a new solution can be obtained.

For the second experiment where $gcross = 0$ and $bmut = 0$, all the crossover genes and mutation genes are selected randomly and the evaluation function doesn't work. We report Δ the number of instances whose solutions are worse than the results in the first experiment, and $\Delta P(\%)$ the corresponding proportion for each problem size in Table 4. In the first experiment with the proposed guided crossover and mutation operators, 22 better results have been achieved compared with the second experiment with the original operators.

4.3 Impact of the GES-Based Local Search

We investigate the effect of the creating a new route operator in the GES-based local search. Two experiments are carried out. The first one is the GGA algorithm with creating a new route operator, and the second one is the GGA algorithm without creating a new route operator. For each instance, a hash table is used to count all the searched solutions without repetition. The diversity of the search can be reflected in the size of the hash table. Experimental results show that the average ratio of the size of the hash table of the first experiment to the second experiment is 1.23. In Table 5, we report Δ the number of instances whose solutions in the second experiment that are worse than those in the first experiment, and $\Delta P(\%)$ the corresponding proportion of the instances in each problem size. When the original GES is performed in the second experiment, there are 20 instances that have achieved worse solutions with more vehicles than the first experiment.

Table 4. Results on the guided crossover and mutation operations.

n	50	100	200	300
Δ	0	4	6	12
$\Delta P(\%)$	0	6.67	10.00	20.00

Table 5. Results on the GES-based local search.

n	50	100	200	300
Δ	0	4	6	10
$\Delta P(\%)$	0	6.67	10.00	16.67

5 Conclusions

A new algorithm is presented for the pickup and delivery problem with time window and last-in-first-out loading, which combines the grouping genetic algorithm (GGA) and the GES-based local search. The selection strategy based on the quality of genes in crossover and mutation operations has played a great role in improving the quality of the solution. In order to adapt to the strict constraints of PDPTWL, we modify the GES to accelerate the exploration of the solution space. The experimental results on the Li and Lim's benchmark show that our algorithm improves 164 best known solutions out of 236 instances and reduces 424 vehicles compared with the existing algorithm. The result shows that the proposed algorithm is more effective.

References

1. Savelsbergh, M., Sol, M.: Drive: dynamic routing of independent vehicles. Oper. Res. **46**(4), 474–490 (1998)
2. Pankratz, G.: A grouping genetic algorithm for the pickup and delivery problem with time windows. OR Spectr. **27**(1), 21–41 (2005)
3. Carrabs, F., Cordeau, J.F., Laporte, G.: Variable neighborhood search for the pickup and delivery traveling salesman problem with LIFO loading. INFORMS J. Comput. **19**(19), 618–632 (2007)
4. Carrabs, F., Cerulli, R., Cordeau, J.F.: An additive branch-and-bound algorithm for the pickup and delivery traveling salesman problem with LIFO or FIFO loading. INFOR Inf. Syst. Oper. Res. **45**(4), 2007 (2008)
5. Nagata, Y., Kobayashi, S.: Guided ejection search for the pickup and delivery problem with time windows. In: Cowling, P., Merz, P. (eds.) EvoCOP 2010. LNCS, vol. 6022, pp. 202–213. Springer, Heidelberg (2010). https://doi.org/10.1007/978-3-642-12139-5_18
6. Gao, X., Lim, A., Qin, H., Zhu, W.: Multiple pickup and delivery TSP with LIFO and distance constraints: a VNS approach. In: Mehrotra, K.G., Mohan, C.K., Oh, J.C., Varshney, P.K., Ali, M. (eds.) IEA/AIE 2011. LNCS (LNAI), vol. 6704, pp. 193–202. Springer, Heidelberg (2011). https://doi.org/10.1007/978-3-642-21827-9_20
7. Liabbabc, Y.: The tree representation for the pickup and delivery traveling salesman problem with LIFO loading. Eur. J. Oper. Res. **212**(3), 482–496 (2011)
8. Cheang, B., Gao, X., Lim, A., Qin, H., Zhu, W.: Multiple pickup and delivery traveling salesman problem with last-in-first-out loading and distance constraints. Eur. J. Oper. Res. **223**(1), 60–75 (2012)
9. Côté, J.F., Archetti, C., Speranza, M.G., Gendreau, M., Potvin, J.Y.: A branch-and-cut algorithm for the pickup and delivery traveling salesman problem with multiple stacks. Networks **60**(4), 212–226 (2012)

10. Cherkesly, M., Desaulniers, G., Laporte, G.: A population-based metaheuristic for the pickup and delivery problem with time windows and LIFO loading. Comput. Oper. Res. **62**, 23–35 (2015)
11. Cherkesly, M., Desaulniers, G., Laporte, G.: Branch-price-and-cut algorithms for the pickup and delivery problem with time windows and last-in-first-out loading. Transp. Sci. **49**, 752–766 (2015)
12. Wei, L., Qin, H., Zhu, W., Wan, L.: A study of perturbation operators for the pickup and delivery traveling salesman problem with LIFO or FIFO loading. J. Heuristics **21**(5), 617–639 (2015)
13. Cherkesly, M., Desaulniers, G., Irnich, S., Laporte, G.: Branch-price-and-cut algorithms for the pickup and delivery problem with time windows and multiple stacks. Eur. J. Oper. Res. **250**(3), 782–793 (2016)
14. Veenstra, M., Cherkesly, M., Desaulniers, G., Laporte, G.: The pickup and delivery problem with time windows and handling operations. Transp. Res. Part B Methodol. **77**(7), 127–140 (2017)
15. Veenstra, M., Roodbergen, K.J., Vis, I.F.A., Coelho, L.C.: The pickup and delivery traveling salesman problem with handling costs. Eur. J. Oper. Res. **257**(1), 118–132 (2017)
16. Cordeau, J.F., Iori, M., Laporte, G., González, J.J.S.: A branch-and-cut algorithm for the pickup and delivery traveling salesman problem with LIFO loading. Networks **55**(1), 46–59 (2010)
17. Nagata, Y., Tojo, S.: Guided ejection search for the job shop scheduling problem. In: Cotta, C., Cowling, P. (eds.) EvoCOP 2009. LNCS, vol. 5482, pp. 168–179. Springer, Heidelberg (2009). https://doi.org/10.1007/978-3-642-01009-5_15
18. Nagata, Y., Ysy, O.: A powerful route minimization heuristic for the vehicle routing problem with time windows. Oper. Res. Lett. **37**(5), 333–338 (2009)
19. SPEC, CPU 2006 results. http://www.spec.org/cpu2006/results/cpu2006.html. Accessed 29 Mar 2018
20. Li, H., Lim, A.: A metaheuristic for the pickup and delivery problem with time windows. In: International Conference on TOOLS with Artificial Intelligence, pp. 160–167 (2001)
21. Falkenauer, E.: Genetic Algorithms and Grouping Problems. Wiley, New York (1998)

Evolutionary Structure Optimization of Convolutional Neural Networks for Deployment on Resource Limited Systems

Qianyu Zhang, Bin Li(✉), and Yi Wu

School of Information Science and Technology,
University of Science and Technology of China, Hefei 230027, China
binli@ustc.edu.cn

Abstract. Convolutional neural networks (CNNs) have achieved great success in various computer vision tasks. However, trial and error are still the most adopted way to design the structure of neural networks, which are time consuming. In this work, a population based evolutionary structure optimization approach of CNNs for deployment on resource limited systems is proposed. The method evolved several different kinds of individuals together in a way called natural selection. Evolutionary operators in conventional genetic algorithms are well defined based on structure design problem. Objectives of minimizing space cost or time cost of one deep neural network is considered in optimization process. Experiments on MNIST datasets show that the proposed method can evolve networks with state-of-the-art accuracy and have low storage or time cost, which give inspiration to hand-made structure by the obtained network structure.

Keywords: Convolutional Neural Networks · Structure design
Neuroevolution · Evolutionary Algorithms

1 Introductions

As one of the most popular methods in machine learning problems over the last decade, deep neural networks (DNNs) have achieved good performance on various computer vision tasks [1–4]. There's a trend of designing DNNs with multiple layers, which can extract advanced semantic features from raw data, updating results in image classification problems from real world. DNNs which obtained impressive results always have tens and hundreds of thousands of parameters, which are easy to become overfitting in one learning system. Moreover, they are always computationally intensive, which are difficult to deploy on mobile systems with limited hardware resources. ResNet [5] has shown that with carefully design of network structure, small amount of parameters are enough to reach or even surpass state-of-the-art accuracy on image classification tasks. However, structure design of DNNs needs expertise and rich experience, yet is a waste of time by trial and error.

Genetic algorithm (GA) [6] is inspired by Darwin's theory of evolution, which is effective for solving optimization problem with non-continuous, non-differentiable, and

D.-S. Huang et al. (Eds.): ICIC 2018, LNCS 10955, pp. 742–753, 2018.
https://doi.org/10.1007/978-3-319-95933-7_82

non-convex conditions. During one generation in the iteration procedure, individuals perform small change called mutation and are valued on the optimization function to get evaluation scores called fitness. Individuals with higher fitness have larger opportunity to be selected to appear in the next generation. In structure design problem of CNNs, CNN structure is encoded as genome and performs genetic operators like conventional GA does. There are two ways of encoding scheme: Direct ways [7] and indirect ways [8]. To take good use of human expertise in architecture design and evolve the structure from a high abstractly level, we use an indirect way to form the structure which seems like a stack of highly functional module, which will be introduced in details in Sect. 3.

The evolutionary structure optimization algorithm to design CNN structure proposed in this work takes advantage of population based strategies like many other applications of evolutionary algorithms (EAs). They can perform non-gradient optimization and be benefit from natural selection rules. Genetic operators such as mutation and crossover are carefully designed according to encoding scheme. As a necessary part to be considered in deployment of networks on mobile systems, storage cost or time cost of CNN structure is involved in fitness evaluation, in which way the algorithm tends to find structure with high accuracy during test-time evaluation while low storage or time cost. The whole process is conducted on one GPU, which makes it easy to use.

The rest of the work is organized as follows: Sect. 2 talks about hyperparameter optimization and structure design in neuroevolution field, which declares the related works of designing CNN structure. Section 3 displays the population based approach for designing convolutional neural networks in details. Section 4 shows the experiment parts and analysis. Section 5 draws conclusions.

2 Related Works

2.1 Hyperparameter Optimization

It can be taken as model selection or hyperparameter optimization problem [9] for structure design of CNNs to achieve high accuracy in test-time evaluation since structure parameters such as layer size, kernel numbers, need to be specified. There are some methods for optimization parameters: Hand selection, random search [10], grid search [11], etc.

Gradient search [12] is a methodology based on the computation of the gradient of a model selection criterion. [10] shows theoretically that random chosen is efficient compared to grid search and manual search. Bayesian optimization [13] maintains a surrogate model and is used to tune the parameters of Markov chain Monte Carlo algorithms. [14] considers optimization on performance for the learning algorithm as a sample from Gaussian process and automatically conducts the procedures.

Evolutionary algorithms are naturally used for hyperparameter tuning [15]. Since deep learning community defaults to use back-propagation for weight leaning, it follows an alternation of back-propagation and evolution in which only the architecture is evolved [16]. A Gender-based genetic algorithm (GGA) [17, 18] studies non-parametric models and involves offspring reproduction and sexual selection to improve configure algorithms. Population based training (PBT) [19] jointly optimizes hyperparameters of a group of individuals during training time, such as learning rate and unroll length of reinforcement learning models. With small changes to the variables by disturbing and resampling, PBT gradually performs smoothly change to find profit parameters in training phase. In hyperparameter tuning perspective of network training, weights can sustained in the network architecture during training, which takes good advantage of computing resources. However, above works usually tune a pre-defined network which lacks flexibility compared to structure design approaches.

2.2 Neuroevolution for Structure Design of Convolutional Neural Networks

At first, optimization process can only evolve weights in a fixed matrix [20]. Research works on evolving both network structure and weights together by EAs starts from the method called NeuroEvolution of Augmenting Topologies (NEAT) [7] and is well known with highly improved computing resources [21, 22]. From network training perspective, there are three ways of combining EAs in structure design of deep learning field: Structure design, hyperparameter tuning, and weight learning.

NEAT [7] encodes nodes and connections in genome to gain an advantage from evolving neural network topologies along with weights using direct encoding ways. NEAT has three kinds of ways to perform mutation: Modify a weight, add a connection between existing nodes, and insert a node while splitting an existing connection. Hypercube-based NEAT [23] employs connective Compositional Pattern Producing Networks (connective CPPNs) to produce connectivity patterns within a hypercube in a lower-dimensional space. CoDeepNEAT [24] involves a chain of blueprints to store information of trained architectures, which includes topology, components and hyperparameters. Those works started from trivial start to evolve network structures along with weight training, which are able to evolve shallow networks compared to todays' state-of-the-art network structure. However, there isn't clear layered structure, resulting in disorderly and uncomprehensive in the obtained models.

[21] uses novel and intuitive genetic operators to perform learning models for CIFAR-10 and CIFAR-100 datasets, starting from a trivial conditions and yielding an unprecedented scale involving 250 workers in parallel. [9] directly encodes network structures as highly functional modules using Cartesian genetic programming (CGP) to build state-of-the-art network structure. [25] views the networks as a variable-length string and uses recurrent neural networks to train them. These works take in the manuscript of hand designed structure and recombine the modules from a macro perspective, which yield the state-of-the-art results in image classification challenges.

EAs can be viewed as a small neighborhood sample of the individual in non-gradient direction, which can perform weight updating instead of back-propagation theoretically. [22] points out that evolutionary strategies (ES) can be used as a scalable alternative in reinforcement learning (RL), as they are easy to be parallelized. Moreover, although gradient descent direction is a popular optimization method, it is not always effective to train a high quality network. GAs can find solutions in non-continues feasible domain, as is illustrated in [26, 27].

3 Evolutionary Structure Optimization of Convolutional Neural Networks

In this section the whole process of evolutionary structure optimization approach for designing CNNs in this work is introduced. A population of individuals denoting different kinds of network structures are evolved together. After generating new individuals, accuracy on test dataset is valued and allocated as their fitness. The iteration performs a way like exploration process to find better performed structures in unknown search space. The network structure represented by the best individual after iterations has high accuracy on test data and low storage or time cost, which shows in Fig. 1.

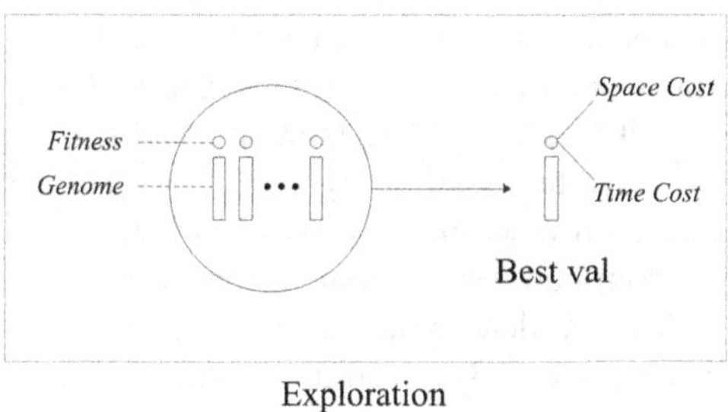

Fig. 1. The whole process of the evolutionary structure optimization approach to design CNNs performs like exploration.

3.1 Encoding Scheme

Networks are encoded in the way like a cascade of highly functional modules. Layer types, number of convolutional channels in the structures and other structure parameters are all stored in a genome instead of weight matrix in neural networks, which saves the storage during the whole process. Several kinds of layers which exist in traditional CNNs are used as highly functional modules in the proposed approach. They are: Convolutional layer, pooling layer, and fully connected layer.

A convolutional layer [28] has four dimensions, which denote number, height, width and channel. Height and width are the kernel size of one convolution layer, and they always have same size. Channel or number denotes the number of dimensions in input or output feature maps. The evolved parameters are constrained in bound: [1, 10] for kernel size, [1,100] for output channels [29]. Paddings are set surrounding input images to ensure that output images have same size as input maps [30].

A pooling layer has one tunable parameter, which is the kernel size. The evolved parameters are constrained in bound: [1, 10] for kernel size. A stride of 2 is used as fixed setting of a pooling layer. This kind of layer cut down feature maps to 1/4 of its size, which saves storage and time cost, acting as fusion of semantic features. Max pooling and average pooling are frequently types of pooling layers in use [1], in which the former is used in this work.

A fully connected layer performs as neuron connections in traditional feedforward networks. Number of nodes denotes the size of this layer. The trainable parameters are constrained in bound: [10,100] for nodes.

$$y = \sigma(wx) \tag{1}$$

Trainable weights denotes by weight matrix of this layer depend on types of last layer. If it is the first fully connected layer, input map is a 4 dimension tensor, which you have to resize it to a vector as an acceptable input. If the last layer is also a fully connected layer, the weight matrix of this layer calculated as:

$$size(w) = node\_of\_f1 \times node\_of\_f2 \tag{2}$$

Dropout [31] performs a variety kinds of network structure, which used to be take in training phase. In this work we exploit much test and find they are involved usually in early iteration times, but are disappeared in middle and late stages. So in the release encoding scheme we remove this kind of layer (Tables 1 and 3).

Individuals generate from random initialization. When it is beyond the maximum length limitations, a random number is sampled from uniform sampling to control the length of the layer, which also determines the scale of trainable parameters. The network structure is set up layer by layer according to the genome.

Table 1. Parameter bounds for kinds of layer [29].

Layer	Bound
conv	[1, 10] for kernel size, [1,100] for channel
pool	[1, 10] for kernel size
fc	[10,100] for nodes

Table 2. Hyper-parameter in exploration stage.

Popsize	Max_gen	Genetic operators	Epochs
50	10	Mutation, crossover	3

Table 3. Hyper-parameter in network training after iterations.

Learning rate	Momentum	Validation frequency	Epochs
0.01	0.9	200	20

3.2 Genetic Operations

Genetic algorithms always perform different kinds of operators to explore the search space. Mutation adds a small change on original genome so as to bring a new gene to the population [29]. Many works have exploit on this operations [9, 29]. In this work a similar way is performed.

Crossover operation takes advantages of well-explored space and mix the different kinds of genomes together. In a limitation of computation cost, it can find better combination of past generations. We perform in a way like single-point crossover. The individual of the spliced part is carefully chosen to avoid exceeding of maximum length. Figure 2 shows a difference of the genetic operations.

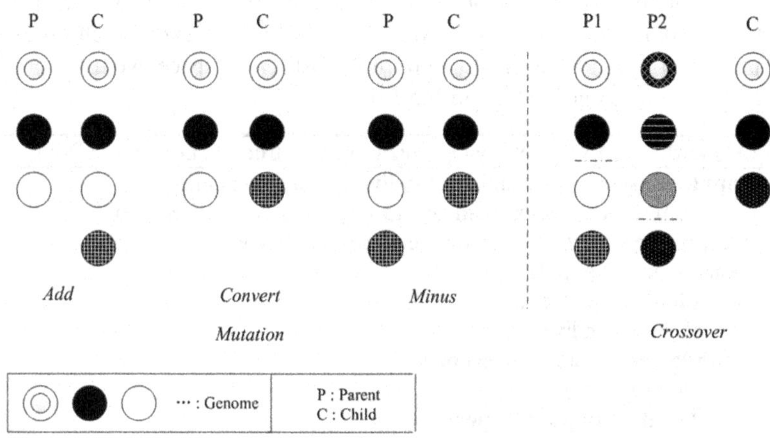

Fig. 2. Genetic operations.

3.3 Fitness Evaluation

When networks are generated and undergo training process, test error and validation error are obtained per evaluation frequency. Validation error is kept as fitness value.

The optimization problem can be defined as follows

$$min\,fitness' = min\,(1 - accuracy) \tag{3}$$

When network of different structures have same accuracy on test time evaluation, they are of different storage and time cost. Storage cost leads to whether it is easy to deploy on mobile systems. It is added as penalty on the fitness. Evolutionary structure optimization prefers to keep network structures with same accuracy and has low storage cost.

$$fitness\_Np = -accuracy - 1/Np \tag{4}$$

Time cost determines whether a network can give instant actions or not in test-time evaluation, which is another optimal target added on original fitness. Evolution process prefers to keep network structures with same accuracy and has low time cost.

$$fitness\_Tp = -accuracy - 1/Tp \tag{5}$$

3.4 Iteration Process

In the evolutionary optimization for structure design of CNNs, a population of different structures of individuals representing different kinds of neural networks are born. Mutation and crossover operators are well-designed and performed to exploit much more uncovered search space. Networks are set up according to the genomes and undergo training phase. Test accuracy is allocated as fitness of an individual. Storage or time cost of that network is calculated according to number of trainable weights of that network. Procedures are in details in Algorithm 1. When in exploration process, it is important to find a well behaved structure from vast search space, we use a population based method same as genetic algorithm [6].

Algorithm 1. Evolutionary Optimization for Structure Design of CNNs

```
Input: popsize – The number of individual in a population
       max_gen – Maximum generations in the whole iteration
Output: bestindiv – best structure of neural network
generate initial population();
for indiv in pop: Repeat
  eval = train(indiv)
  while gen < max_gen: Repeat
    choose_pop();
    for indiv in pop: Repeat
      do_reproduction(genetic operators);
      indiv = chose_best();
    bestindiv = min(indiv)
Return bestindiv
```

3.5 Network Training After Iterations

When structure of a network is set up, weights are initialized by 0.1. A training process is undergo with Stochastic Gradient Descent (SGD) [32] to learn weights better performed on specific data. Validation error is collected by a frequency of 200 steps. Minimum validation error is set as fitness evaluation. After changing the network structure, it is re-initialized and do another training process.

4 Experiments

Evolutionary structure optimization of CNNs proposed in this work performed on MNIST [28] dataset. It is a classical image classification problem with 60000 training data and 10000 testing data, which has 10 classes of 0 to 9 numbers. Each image is a hand-writing binary image. In this work we split out 5000 data as validation set. During iteration the newly formed neural networks performs forward testing on validation set and the validation error is stored as fitness of that individual. The test data is invisible in the evolutionary process. Test accuracy is used as an index after the stop criteria of the whole iterations. The experiments are conducted on one single NVIDIA GPU and the neural network architectures are set up on tensorflow [30].

4.1 Experiments and Results

Experiments are first conducted on MNIST datasets using Eq. (4) as fitness evaluation (called as experiments considered in storage cost). The left subfigure of Fig. 3 illustrates best individuals of each generation during the whole iteration. Hyperparameters of the optimization process in experiments considered in storage cost are set according to Table 2. To obtain a confidence conclusion, experiments are repeated for 3 times using different random seed. After one iteration, a group of individuals representing specific structure of CNNs are obtained with their error rates on validation dataset.

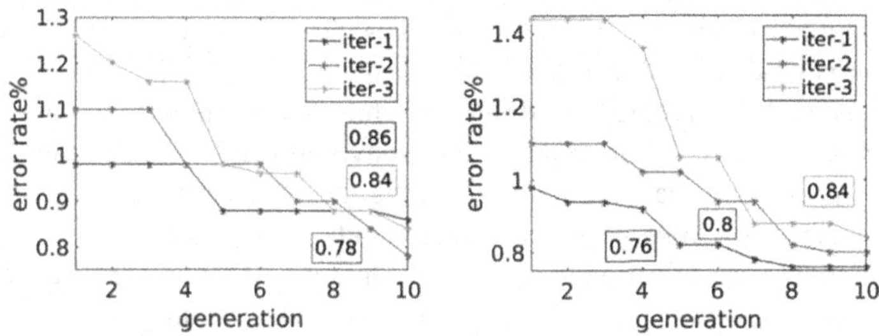

Fig. 3. (a) The best individual of each generation from experiments considered in storage cost. (b) The best individual of each generation from experiments considered in time cost.

To well understand the effective of the structure design approach, we choose the best child of every iteration and form the model it represents to undergo a training process, in which epoch number of 20 is used. Some points are not always well-initialized and cannot achieve the results collected in iterations, then a group of sub-optimal individuals is used as additional. Table 4 shows the results compared with LeNet-5 like neural network, which has same structure with Ref. [28] but the feature maps after convolutional layers have same size with input feature maps, like the usually adopted way in tensorflow [30]. It shows that using the approach proposed in this work we can get a network structure better performed on image classification tasks and has lower storage cost.

Experiments conducted on MNIST datasets using Eq. (5) as fitness evaluation (called as experiments considered in time cost). The right subfigure of Fig. 3 illustrates best individuals of each generation during the whole iteration. Table 5 shows the results of best child and suboptimal children obtained by the optimal solution after training process compared with LeNet-5 like neural network, which achieved a result that the method proposed in this article can find network structure better performed on image classification tasks and has lower time consuming in test-time evaluation.

4.2 Structure Optimized by Evolutionary Approach

In the process, network structures are encoded by layers, which like a stack of highly functional modules. After a whole iteration, the last generation of individuals denote different kinds of network structures that are survived. Channel numbers of convolutional layers in survived individuals are always lower or equal to 5. It is reasonable that in MNIST datasets the size of input images is 28, so small kernel size is enough to extract continues feature from raw data. Meanwhile, convolutional layers have a trend to perform a lower channel number after a higher channel number, supposing a structure of alternating thickness which sames like sandwiches.

Figures 4 and 5 demonstrate structures obtained by the structure design method proposed in this work. When the fitness is calculated by Eq. (4), in which storage cost is considered, as illustrated in Fig. 4, networks are tend to form a stack of pooling layers before fully-connected layers. The structure design method automatically finds that connections between convolution layers and fully-connected layers are highly parameterization-consuming. When the fitness is calculated by Eq. (5), in which time computation is crucial, as illustrated in Fig. 5, there has been many pooling layers in the former layer of convolutional structure, while few of them are stacked. This is reasonable that the size of feature maps after convolutional layers determine time cost, which are expectedly to be cut down by pooling layers as early as possible. However, in former layers of CNNs, features inside input images are essential to get high accuracy on image classification dataset. Thus structure of a pooling layer followed by a convolutional layer and its repetition can get better results with state-of-the-art accuracy considered in time consuming.

Table 4. Best individual of each iteration, using Eq. (4).

Iteration	Validation		Test		Np
	AVG	MIN	AVG	MIN	
1	0.63	0.62	**0.60**	**0.59**	**168 K**
2	0.92	0.62	**0.64**	**0.61**	**71 K**
3	0.78	0.74	0.75	0.74	700 K
LeNet-5 like	*0.68*	*0.66*	*0.73*	*0.71*	*1.3 M*

Table 5. Best individual of each iteration, using Eq. (5).

Iteration	Validation		Test		Tp
	AVG	MIN	AVG	MIN	
1	0.67	0.66	**0.65**	**0.62**	**10 M**
2	0.64	0.58	**0.64**	**0.59**	**7 M**
3	0.59	0.56	**0.64**	**0.62**	–
LeNet-5 like	*0.68*	*0.66*	*0.73*	*0.71*	*13 M*

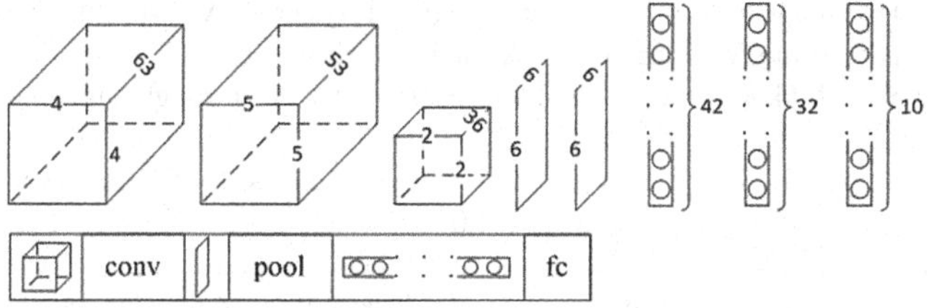

Fig. 4. Example network structure of best individual considered in storage cost.

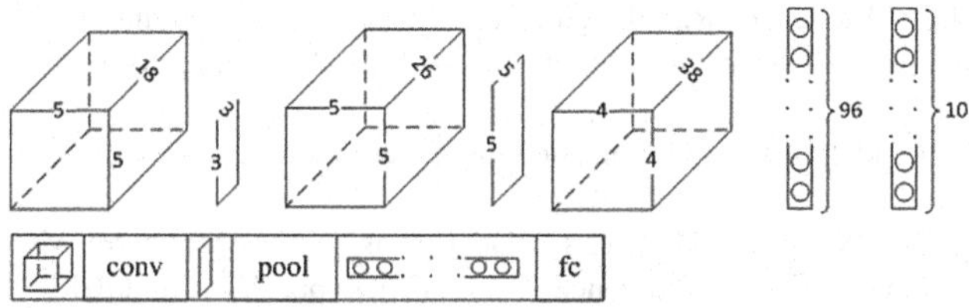

Fig. 5. Example network structure of best individual considered in time cost.

5 Conclusions

This work provides an approach of evolutionary structure optimization for CNN structure design which views the model selection problem in machine learning as a structure optimization problem to use evolutionary algorithms to solve the problem. The structure design approach is capable to evolve deep neural networks with high accuracy while considering storage cost or time consuming. Crossover strategies are introduced in structure design problem to better exploit the search space of variables within parameter boundaries. The whole process can be conducted on a single GPU in 2–4 days. Future works will better research parallelization approach of evolutionary process and apply the structure design evolutionary optimization to other complex datasets.

References

1. Krizhevsky, A., Sutskever, I., Hinton, G.E.: Imagenet classification with deep convolutional neural networks. In: Advances in Neural Information Processing Systems, pp. 1097–1105 (2012)
2. Simonyan, K., Zisserman, A.: Very deep convolutional networks for large-scale image recognition. arXiv preprint arXiv:1409.1556 (2014)
3. Szegedy, C., Liu, W., Jia, Y., Sermanet, P., Reed, S., Anguelov, D., Erhan, D., Vanhoucke, V., Rabinovich, A.: Going deeper with convolutions. In: CVPR (2015)
4. Girshick, R.: Fast R-CNN. arXiv preprint arXiv:1504.08083 (2015)
5. He, K., Zhang, X., Ren, S., Sun, J.: Deep residual learning for image recognition. In: Proceedings of the IEEE Conference on Computer Vision and Pattern Recognition (CVPR), pp. 770–778 (2016)

6. Holland, J.H.: Adaptation in Natural and Artificial Systems: An Introductory Analysis with Applications to Biology, Control, and Artificial Intelligence. MIT Press, Cambridge (1992)
7. Stanley, K.O., Miikkulainen, R.: Evolving neural networks through augmenting topologies. Evol. Comput. **10**, 99–127 (2002)
8. Kim, M., Rigazio, L.: Deep clustered convolutional kernels. In: Feature Extraction: Modern Questions and Challenges, pp. 160–172 (2015)
9. Suganuma, M., Shirakawa, S., Nagao, T.: A genetic programming approach to designing convolutional neural network architectures. In: Proceedings of the Genetic and Evolutionary Computation Conference, pp. 497–504. ACM (2017)
10. Bergstra, J., Bengio, Y.: Random search for hyper-parameter optimization. J. Mach. Learn. Res. **13**, 281–305 (2012)
11. Bergstra, J.S., Bardenet, R., Bengio, Y., Kégl, B.: Algorithms for hyper-parameter optimization. In: Advances in Neural Information Processing Systems, pp. 2546–2554 (2011)
12. Bengio, Y.: Gradient-based optimization of hyperparameters. Neural Comput. **12**, 1889–1900 (2000)
13. Mahendran, N., Wang, Z., Hamze, F., De Freitas, N.: Adaptive MCMC with Bayesian optimization. In: Artificial Intelligence and Statistics, pp. 751–760 (2012)
14. Snoek, J., Larochelle, H., Adams, R.P.: Practical Bayesian optimization of machine learning algorithms. In: Advances in Neural Information Processing Systems, pp. 2951–2959 (2012)
15. Loshchilov, I., Hutter, F.: CMA-ES for hyperparameter optimization of deep neural networks. arXiv preprint arXiv:1604.07269 (2016)
16. Breuel, T., Shafait, F.: AutoMLP: simple, effective, fully automated learning rate and size adjustment. In: The Learning Workshop, p. 51. Utah (2010)
17. Ansótegui, C., Sellmann, M., Tierney, K.: A gender-based genetic algorithm for the automatic configuration of algorithms. In: Gent, I.P. (ed.) CP 2009. LNCS, vol. 5732, pp. 142–157. Springer, Heidelberg (2009). https://doi.org/10.1007/978-3-642-04244-7_14
18. Ansótegui, C., Malitsky, Y., Samulowitz, H., Sellmann, M., Tierney, K.: Model-based genetic algorithms for algorithm configuration. In: IJCAI, pp. 733–739 (2015)
19. Jaderberg, M., Dalibard, V., Osindero, S., Czarnecki, W.M., Donahue, J., Razavi, A., Vinyals, O., Green, T., Dunning, I., Simonyan, K.: Population Based Training of Neural Networks (2017)
20. Miller, G.F., Todd, P.M., Hegde, S.U.: Designing neural networks using genetic algorithms. In: ICGA, pp. 379–384 (1989)
21. Real, E., Moore, S., Selle, A., Saxena, S., Suematsu, Y.L., Tan, J., Le, Q., Kurakin, A.: Large-scale evolution of image classifiers. arXiv preprint arXiv:1703.01041 (2017)
22. Salimans, T., Ho, J., Chen, X., Sidor, S., Sutskever, I.: Evolution strategies as a scalable alternative to reinforcement learning. arXiv preprint arXiv:1703.03864 (2017)
23. Stanley, K.O., D'Ambrosio, D.B., Gauci, J.: A hypercube-based encoding for evolving large-scale neural networks. Artif. Life **15**, 185–212 (2009)
24. Miikkulainen, R., Liang, J., Meyerson, E., Rawal, A., Fink, D., Francon, O., Raju, B., Shahrzad, H., Navruzyan, A., Duffy, N.: Evolving deep neural networks. arXiv preprint arXiv:1703.00548 (2017)
25. Zoph, B., Le, Q.V.: Neural architecture search with reinforcement learning. arXiv preprint arXiv:1611.01578 (2016)
26. Such, F.P., Madhavan, V., Conti, E., Lehman, J., Stanley, K.O., Clune, J.: Deep Neuroevolution: Genetic Algorithms Are a Competitive Alternative for Training Deep Neural Networks for Reinforcement Learning. arXiv preprint arXiv:1712.06567 (2017)
27. Lehman, J., Chen, J., Clune, J., Stanley, K.O.: ES Is More Than Just a Traditional Finite-Difference Approximator. arXiv preprint arXiv:1712.06568 (2017)

28. LeCun, Y., Bottou, L., Bengio, Y., Haffner, P.: Gradient-based learning applied to document recognition. Proc. IEEE **86**, 2278–2324 (1998)
29. Dufourq, E., Bassett, B.A.: EDEN: Evolutionary Deep Networks for Efficient Machine Learning. arXiv preprint arXiv:1709.09161 (2017)
30. Abadi, M., Barham, P., Chen, J., Chen, Z., Davis, A., Dean, J., Devin, M., Ghemawat, S., Irving, G., Isard, M.: TensorFlow: a system for large-scale machine learning. In: OSDI, pp. 265–283 (2016)
31. Srivastava, N., Hinton, G., Krizhevsky, A., Sutskever, I., Salakhutdinov, R.: Dropout: a simple way to prevent neural networks from overfitting. J. Mach. Learn. Res. **15**, 1929–1958 (2014)
32. Bottou, L.: Large-scale machine learning with stochastic gradient descent. In: Lechevallier, Y., Saporta, G. (eds.) Proceedings of COMPSTAT 2010, pp. 177–186. Physica-Verlag, Heidelberg (2010). https://doi.org/10.1007/978-3-7908-2604-3_16

Research on Vehicle Routing Problem and Its Optimization Algorithm Based on Assembled Building

Kun Jiang[1], Jun-qing Li[1,2,3,4](✉), Ben Niu[3], Yongqin Jiang[3], Xiaoping Lin[3], and Pei-yong Duan[1](✉)

[1] School of Information, Shandong Normal University, Jinan 250014, China
lijunqing.cn@gmail.com, duanpeiyong@sdnu.edu.cn
[2] School of Computer, Liaocheng University, Liaocheng 252059, China
[3] China Key Laboratory of Computer Network and Information Integration, Southeast University, Ministry of Education, Nanjing 211189, People's Republic of China
[4] State Key Laboratory of Synthetical Automation for Process Industries, Northeastern University, Shenyang 110819, China

Abstract. Assembled building flow distribution has become the main problem facing the industry. The vehicle routing problem (VRP) is the key link in the distribution system. This article systematically summarizes the classification of common and the basic algorithm of VRP problems. It fully understands the commonly used and efficient heuristic algorithms for solving VRPs and the corresponding research status. Finally, it summarizes the problems existing in the research. The future research and prospective solution methods of VRPs are discussed.

Keywords: Assembly building · VRP · Heuristic algorithm

1 Introduction

A prefabricated building is mainly made of prefabricated concrete elements, and the concrete structure of the building is constructed by means of on-site assembly. The main components of a prefabricated building need to be prefabricated at the factory and transported to the site for assembly and irrigation to complete the building of the structure. Owing to the characteristics of the components of the prefabricated building, the transportation costs represent a high proportion of the entire project cost, and the transportation problem has become a major issue for the development of prefabricated buildings. Distribution is a key link in logistics. Among them, the vehicle routing problem (VRP) in the distribution process is a key issue in the planning and scheduling decisions of distribution systems. The VRP is one of the core issues in logistics management and transportation organization optimization. It refers to a series of receiving points and shipments under certain constraints (such as time limits, vehicle capacity restrictions, and traffic restrictions). The objective of the problem is to reasonably arrange driving routes, under the premise that the customer's needs are met, to

D.-S. Huang et al. (Eds.): ICIC 2018, LNCS 10955, pp. 754–762, 2018.
https://doi.org/10.1007/978-3-319-95933-7_83

achieve the goals of the fewest delivery vehicles, shortest delivery time, lowest delivery costs, and shortest delivery distance. The VRP was introduced in 1959 [1], and has aroused widespread interest from researchers in operational research. It has now become a classic combinatorial optimization problem in operations research and is a typical NP-conundrum.

2 Classification of Vehicle Routing Problems

The typical components of the VRP are vehicles, cargo, customer sites, distribution centers, road networks, constraints, and objective functions. According to different focuses, VRPs are divided into different types. Table 1 presents the different types of VRP.

Table 1. Different types of vehicle routing problem

Classification basis	Type	Problem description
Task characteristics	Unidirectional Logistics	Vehicle purchase or delivery problem
	Two-way Logistics	Can pick up and ship at the same time
Distribution centers	Single distribution center	Only one distribution center serves customers
	Multiple distribution centers	Multiple distribution center service customers
Car shipment status	Fully loaded	The amount of goods is not less than the capacity of the car
	Not full	The amount of goods is less than the Capacity of the car
Vehicle type	Single model	Same for all vehicle types and maximum load
	Multiple model	Vehicle type and maximum load vary
Vehicle travel constraints	With travel constraints	All vehicles cannot exceed maximum travel
	No constraint	All vehicles have no maximum travel limit
By time	With time window constraints	Serve within the specified time period
	Without time window constraints	No time limit
Optimize target classification	Single optimization goal	Single target optimization
	Multiple optimization goals	Multiple objective optimization

Owing to the different constraint conditions of VRP, it has various classifications, and the model construction and the solution algorithm of different types of VRP vary significantly.

3 VRP Solution Algorithm and Research Status

The methods to solve the VRP can be roughly divided into two types [2]: exact algorithms and heuristic algorithms. Because the computational difficulty and computational complexity of exact algorithms increase exponentially with the increase in customer points, their application range is limited in practice. Moreover, heuristic algorithms have the characteristics of strong global search capability and high solution efficiency, and the solution obtained also has the characteristics of better reference; Therefore, most of the current researchers mainly focus on how to construct high-quality heuristic algorithms. This article also discusses some of the more recent heuristic optimization algorithms. Many heuristic algorithms have been proposed for the VRP. The most studied of these include the following algorithms:

3.1 Genetic Algorithm

A genetic algorithm (GA) is a computational model that simulates the natural selection and genetic mechanism of Darwinian biological evolution. It is a method for searching for optimal solutions by simulating the natural evolutionary process. This method is used to select, cross, and mutate populations. The essential steps of a GA are to generate a population that represents a new solution set, select individuals based on the individual fitness, and gradually evolve the population to an approximately optimal solution state through iterations. However, the algorithm coding cannot be expressed in terms of the constraints of the optimization problem. One way to consider the constraint is to use a threshold for the infeasible solution. This time will inevitably increase. GA have no effective quantitative analysis methods for the accuracy, feasibility, computational complexity, etc. The efficiency is also lower than that of other traditional methods.

The research status of using GA to study VRP issues includes: Baker and Ayechew [3] presented calculation results of a pure GA. The further result is the use of this hybrid GA and a neighborhood search method, which shows that this method is competitive with tabu search and simulated annealing (SA) in terms of the solution time and quality; Prins [4] proposed a relatively simple but effective hybrid GA. In terms of the average solution cost, this algorithm outperforms most published travelling salesman (TS) heuristics on the 14 classical Christofides instances and becomes the best solution method for the 20 large-scale instances generated by Golden et al.; Zhang and Yan [5] established a hybrid model for multiple distribution center and multi-vehicle VRPs and used a new fuzzy GA to solve the problem; Hwang [6] improved a GA operator and initial population to develop a GA-TSP model. The calculation results show that this method is very effective for a set of standard testing problems, it can solve vehicle routing problems, and is potentially useful.

3.2 Simulated Annealing

SA is a stochastic optimization algorithm based on the Monte-Carlo iterative solution strategy. The starting point is based on the similarity between the solid material annealing process and general combinatorial optimization problems. Similar to the tabu search algorithm, it is a local search algorithm, but the SA algorithm can give a time-varying and eventually zero probability jump to the search process, which can effectively avoid falling into a local minimum and eventually tend to the global maximum. SA is an optimal serial structure optimization algorithm.

The research status of using simulated annealing algorithm to study VRP problems includes the following. Pei and Jia [7] built on the traditional SA algorithm by the introduction of memory functions, combined with GIS methods, and used SPSS cluster analysis to determine the initial state population. A variety of parallel mechanisms and the emergence of new states were observed, based on the size of the population with different generation algorithms to improve the calculations. Wang et al. [8]. constructed a two-objective mathematical model for VRPs. Based on this, a hybrid SA algorithm was proposed. The SA algorithm and the 2-opt optimization algorithm were organically combined such that the mixed algorithm not only had the advantages of these two algorithms, but also overcome their corresponding shortcomings. Mu et al. [9]. proposed a parallel SA algorithm, and the application of the algorithm is extended to other VRPs and combinatorial optimization problems. Mirabi et al. [10]. proposed a three-step heuristic algorithm based on SA to solve the multi-distribution center VRP model that minimizes delivery time.

3.3 Ant Colony Optimization

The Ant colony algorithm is inspired by the natural phenomenon that people can quickly find food near ants. The mechanisms established by the ant colony algorithm mainly include three aspects: the memory of ants, the interactive communication of ants using pheromones, and the cluster activity of ants. An ant's walking path is used to represent a feasible solution of the problem to be optimized, and all the paths of the entire ant population constitute the solution space of the problem. Ants with shorter paths release more pheromones. As time advances, the concentration of pheromone accumulated in shorter paths gradually increases, and the number of ants that choose the path increases. In the end, the entire ant population will concentrate on the best path under the action of positive feedback, and the corresponding solution is the optimal solution to the optimization problem. The ant colony algorithm has the advantages of easy implementation, fast convergence, and high precision. It has achieved good results on various optimization problems.

The research status of using the ant colony algorithm to study VRP issues is as follows: Chen et al. [11] improved the pheromone update strategy and heuristic factors for the shortcomings of the ant colony algorithm, and introduced a search hotspot mechanism to effectively solve the deficiencies of the ant colony algorithm. Zhang et al. [12] introduced genetic operators such as replication, crossover, and mutation to the ant colony algorithm to improve the algorithm's convergence speed and global search ability. Yue et al. [13] designed an improved ant colony algorithm to solve this problem, introduced improved operators such as selecting operators, interpolation operations, and dynamically changing algorithm parameters. Liu and Shen [14] modified the pheromone update rules and conversion rules. Through the introduction of solution uniformity and the choice of window improvement, ants are attracted to the arc in order to reduce the computation time, to avoid the basic behavior of the ant colony algorithm.

3.4 Particle Swarm Optimization

The particle swarm optimization (PSO) algorithm, also called the bird swarming algorithm, is a group parallel optimization algorithm based on research on the foraging behavior of flock birds. It starts from random solution and finds the optimal solution through iteration. It also evaluates the quality of solution through fitness, but it is simpler than the rules of the GA. It does not have the crossover and mutation operations of the GA. It seeks the global optimum by following the current searched optimal value. This type of algorithm has attracted much attention from academia because of its advantages of easy implementation, high precision, and fast convergence. It also shows its superiority in solving practical problems. However, because the PSO algorithm finds the optimal solution through the interaction between particles and lacks the mutation mechanism of the GA, the PSO algorithm is easily trapped in a local optimum.

The current research status of using PSO to study VRP issues includes: Qin et al. [15] proposed an improved mutation operator for the PSO algorithm. By adding a mutation operator to the algorithm, not only is the attraction of the local minima in the convergence period avoided, but also the early convergence speed is maintained; Mohemmed et al. [16] proposed an improved priority-based heuristic coding method to reduce the possibility of loop generation during path construction; Wu and Zhang [17] proposed a local neighbor PSO algorithm with a self-adjustment mechanism to solve the VRP; Wei et al. [18] used a new polar path-based PSO algorithm for path planning. The result shows that the method is more efficient and adaptive to the environment than the traditional PSO algorithm and GA for obstacle avoidance.

3.5 Bat Algorithm

The bat algorithm (BA) is a new type of swarm intelligence evolution algorithm proposed by Yang [19] of Cambridge University in 2010. It simulates the biological characteristics of bats in the natural world through ultrasonic search and prey and is a population-based random optimization algorithm. The algorithm is an iterative optimization technique that is initialized as a set of random solutions, iteratively searches for optimal solutions, and generates local new solutions around the optimal solution by random flight, which strengthens the local search. Compared with other algorithms, the BA is far superior in terms of accuracy and effectiveness, and there are not many parameters to adjust. As of now, the BA is mainly used to solve function optimization problems in continuous domains, and only a few scholars use it to solve discrete problems, which holds great potential for research.

The research status of using the BA to study VRP issues includes: Ma et al. [20] applied the BA to solve the VRP, introduced inertia weights in the bat speed updating formula, improved the basic BA, and overcame the inadequacies of the basic BA; Wang et al. [21] proposed an improved BA combined with differential evolution (DE) to optimize the three-dimensional path planning problem for the first unmanned combat aircraft; the BA is a new intelligent optimization algorithm, which has broad application prospects. However, it cannot be directly used to solve discrete problems, and like most intelligent optimization algorithms, it is quite susceptible to fall into local optima, and the late convergence rate is slower than that of Sun and Zhang [22] redefines the bat coding method and uses the GRASP heuristic algorithm to generate the initial population to improve the BA, and then apply it to solve the VRP.

3.6 Dijkstra Algorithm

The Dijkstra algorithm is a well-known algorithm for finding the optimal path in the shortest path search problem. It is the shortest path algorithm from one vertex to the other vertices. It solves the shortest path problem in directed graphs. The Dijkstra algorithm was used to establish the single source shortest path solution model, and the shortest rectangular road array between the distribution center and distribution point i, and distribution points i and j was obtained. The optimal distribution path solution was solved by the model.

The status of research on the VRP problem using Dijkstra's algorithm includes: Deng et al. [23] solved the problem of the shortest path under uncertain conditions through Dijkstra's algorithm. Noto and Sato [24] proposed a method to obtain the path in a short time, using the Dijkstra method to get the path as close as possible to optimal (path optimization). The new method extends the conventional Dijkstra method so as to obtain a solution to a problem given within a specified time, such as a path search in a vehicle navigation system; Yuan et al. [25] presented the problem of time complexity in large-scale network computation for the classic Dijkstra path optimization algorithm. It improved the traversal process of unlabeled nodes and directly sought to reach the target node in the range and direction of traversal, performing the search process. It is not necessary to traverse all or untraversed nodes. The time complexity is reduced from $O(n^2)$ to $O(n)$ to improve the efficiency of the algorithm; Wang et al. [26] based on the classical Dijkstra algorithm, introduced the decision mechanism in the AI domain into the path search and proposed a heuristic optimal path search algorithm. This algorithm introduces a cost function in the path-finding process. The function determines the path-finding strategy (i.e., which intermediate nodes are searched first) in order to reduce the number of search nodes.

4 Summary and Outlook

Owing to the different constraint conditions, the VRP leads to a variety of classifications and different emphasis focuses on the mathematical model and the solution algorithm. By summarizing the research on the VRP and the results of some related scholars in recent years, I learned a lot of excellent optimization algorithms. By systematically reviewing the VRP, this paper summarizes the following problems in the current research and the future possible research directions:

- The research goal is too idealistic. At present, the study of the VRP by scholars focuses on the least cost and the shortest path. Most of them are single-objective optimization problems. However, in practical applications, the existence of various sudden conditions cannot be as simple as single-objective optimization. For example, drivers may delay planned trips for many reasons. Customers' needs may be different or even conflicting. There is a contradiction between customer satisfaction and the goal of minimizing costs. Future research can combine cost, distance, driver's rest, customer satisfaction, time window, and other goals to study, and can be integrated by a linear weighted approach and a penalty function.

- The single-objective optimization VRP has been researched for a long time, and its application limitations are significant. It is necessary to consider combining multiple constraints to establish a vehicle path problem that meets the actual multi-constraint conditions to better solve the company's distribution optimization.
- Although the heuristic algorithm has the advantages of strong global search capability and convenient operation, there are also problems such as poor local search capability, long convergence time, and easy local optimization. Using a single swarm intelligence algorithm is not the most effective method for solving the VRP. Combining two and more swarm intelligence algorithms to study VRPs and complement each other is an approach that should be considered in the future. At the same time, more intelligent optimization algorithms should be considered for solving the VRP.

Acknowledgement. This research is partially supported by the National Science Foundation of China (61773192, 61503170, 61603169, 61773246), Shandong Province Higher Educational Science and Technology Program (J17KZ005, J14LN28), Natural Science Foundation of Shandong Province (ZR2016FL13, ZR2017BF039), Key Laboratory of Computer Network and Information Integration (Southeast University), Ministry of Education (K93-9-2017-02), and State Key Laboratory of Synthetical Automation for Process Industries (PAL-N201602).

References

1. Dantzig, G.B., Ramser, J.K.: The truck dispatching problem. Manag. Sci. **6**, 80–91 (1959)
2. Bi, G.-t.: Business School, Henan University, Kaifeng 475000, China
3. Baker, B.M., Ayechew, M.A.: A genetic algorithm for the vehicle routing problem. Comput. Oper. Res. **30**, 787–800 (2003)
4. Prins, C.: A simple and effective evolutionary algorithm for the vehicle routing problem. Comput. Oper. Res. **31**, 1985–2002 (2004)
5. Zhang, Q., Yan, R.: School of Economics and Management, University of Science & Technology Beijing, Beijing 100083, China
6. Hwang, H.S.: An improved model for vehicle routing problem with time constraint based on genetic algorithm. Comput. Ind. Eng. **42**, 361–369 (2002)
7. Pei, X.-b., Jia, D.-f.: School of Management, Tianjin University of Technology, Tianjin 300384, China
8. Wang, B., Shang, X.-c., Li, H.-f.: Applied Institute, University of Science and Technology Beijing, Beijing 100083, China; Transport Planning and Research Institute Ministry of Communications, Beijing 100028, China
9. Mu, D., Wang, C., Wang, S.-c., Zhou, S.-c.: .School of Economics and Management, Beijing Jiaotong University, Beijing 100044, China; School of Economics and Management, Beijing University of Technology, Beijing 100124; School of Electronic and Information Engineering, Beijing Jiaotong University, Beijing 100044, China; Qingdao Geotechnical Investigation and Surveying Research Institute, Qingdao 266032, China

10. Mirabi, M., Ghomi, S.F., Jolai, F.: Efficent stochastic hybrid heuristics for the multi-depot vehicle routing problem. Rob. Cim.-Int. Manufac. **26**, 564–569 (2010)
11. Chen, Y.-x.: School of Economics & Management, Harbin Engineering University, Harbin 150001, China
12. Zhang, W.-z., Lin, J.-b., Wu, H.-s., Tong, R.-f., Dong, J.-x.: Institute of Artificial Intelligence, Zhejiang University, Hangzhou 310027, China; Zhejiang Jinji Electronic Co. Ltd. Hangzhou 310013, China; Basic Study, Zhejiang Police College, Hangzhou 310053, China
13. Yue, Y.-x., Zhhou, L.-s., Yue, q.-x., Sun, Q.: Sch. of Traffic & Transportation, Beijing Jiaotong Univ., Beijing 100044, China; Sch. of Economics & Management, Beihang Univ., Beijing 100083, China
14. Liu, Z., Shen, J.: An adaptive ant colony algorithm for vehicle routing problem based on the evenness of solution. Acta Simulata Systematica Sinica. **5**, 016 (2002)
15. Qin, Y.Q., Sun, D.B., Li, N., et al.: Path planning for mobile robot using the particle swarm optimization with mutation operator. In: Proceedings of 2004 IEEE International Conference on Machine Learning and Cybernetics 2004, vol. 4, pp. 2473–2478 (2004)
16. Mohemmed, A.W., Sahoo, N.C., Geok, T.K.: Solving shortest path problem using particle swarm optimization. Appl. Soft Comput. **8**, 1643–1653 (2008)
17. Wu, Y.-H., Zhang, N.-Z.: Modified particle swarm optimization algorithm for vehicle routing problem with time windows. Comput. Eng. Appl. **46**(15), 230–234 (2010)
18. Wei, Z.U., Gang, L.I., Zhengxia, Q.I.: Study on a path planning method based on improved particle swarm optimization. J. Projectiles Rockets Missiles Guidance (2008)
19. Yang, X.S.: A new metaheuristic bat-inspired algorithm. In: González, J.R., Pelta, D.A., Cruz, C., Terrazas, G., Krasnogor, N. (eds.) Nature Inspired Cooperative Strategies for Optimization (NICSO 2010). Studies in Computational Intelligence, vol. 65. Springer, Heidelberg (2010). https://doi.org/10.1007/978-3-642-12538-6_6
20. Ma, X.-I., Zhang, H.-z., Ma, L.: School of Management, University of Shanghai for Science and Technology, Shanghai 200093, China
21. Wang, G.G., Chu, H.C.E., Mirjalili, S.: Three-dimensional path planning for UCAV using an improved bat algorithm. Aerosp. Sci. Technol. **49**, 231–238 (2008)
22. Sun, Q., Zhang, H.: Business School, University of Shanghai for Science and Technology, Shanghai 200093, China
23. Deng, Y., Chen, Y., Zhang, Y., et al.: Fuzzy Dijkstra algorithm for shortest path problem under uncertain environment. Appl. Soft Comput. **12**, 1231–1237 (2012)
24. Noto, M., Sato, H.: A method for the shortest path search by extended Dijkstra algorithm. In: 2000 IEEE International Conference on Systems, Man, and Cybernetics.vol. 3, pp. 2316–2320. IEEE (2000)
25. Yuan, B., Liu, J.-s., Qian, D., Luo, D.-h.: School of Institute of mechanical and electrical engineering, Nanchang University, Nanchang 330031, China
26. Wang, J., Zhang, X., Chen, B., Chen, H.: Information Engineering School, University of Science and Technology Beijing, Beijing 100083, China; Informat ion Engineering School, Wuhan University of S cience and Technology, Wuhan 430081, China
27. Zhou, Y., Luo, Q., Xie, J., Zheng, H.: A hybrid bat algorithm with path relinking for the capacitated vehicle routing problem. In: Yang, X.-S., Bekdaş, G., Nigdeli, S.M. (eds.) Metaheuristics and Optimization in Civil Engineering. MOST, vol. 7, pp. 255–276. Springer, Cham (2016). https://doi.org/10.1007/978-3-319-26245-1_12
28. Dorigo, M., Gambardella, L.M.: Ant colony system: a cooperative learning approach to the traveling salesman problem. IEEE Trans. Evol. Comput. **1**, 53–66 (1997)
29. Dorigo, M., Caro, G.D., Gambardella, L.M.: Ant algorithms for discrete optimization. Artif. Life. **5**, 137–172 (1999)

30. Roberge, V., Tarbouchi, M., Labonté, G.: Comparison of parallel genetic algorithm and particle swarm optimization for real-time UAV path planning. IEEE Trans. Ind. Inform. **9**, 132–141 (2012)
31. Li, G., Shi, H.: Path planning for mobile robot based on particle swarm optimization. Robotica, 3290–3294(2004)
32. Wang, G., Guo, L., Duan, H., et al.: A bat algorithm with mutation for UCAV path planning. Sci. World J. **2012**, 15 (2012)
33. Liberatore, F., Ortuño, M.T., Tirado, G., et al.: A hierarchical compromise model for the joint optimization of recovery operations and distribution of emergency goods in Humanitarian Logistics. Comput. Oper. Res. **42**, 3–13 (2014)
34. Sheu, J.B.: A hybrid fuzzy-optimization approach to customer grouping-based logistics distribution operations. Appl. Math. Model. **31**, 1048–1066 (2007)
35. Gu, Q.I.N.: Logistics distribution center allocation based on ant colony optimization. Syst. Eng. Theor. Pract. **4**, 120–124 (2006)
36. Caramia, M., Dell'Olmo, P.: Multi-Objective Management in Freight Logistics: Increasing Capacity, Service Level and Safety with Optimization Algorithm. Springer, London (2008). https://doi.org/10.1007/978-1-84800-382-8
37. Zhang, J., Zhou, Q.: Study on the optimization of logistics distribution VRP based on immune clone algorithm. J. Hunan Univ. (Natural Science) **5**, 013 (2004)
38. Luo, Y., Chen, Z.Y.: Path optimization of logistics distribution based on improved genetic algorithm. Syst. Eng. **30**, 118–122 (2012)
39. Li, R., Yuan, J.: Research on the optimization of logistics distribution routing based on improved genetic algorithm. J. Wuhan Univ. Technol. **12**, 028 (2004)
40. Marinakis, Y., Marinaki, M.: A particle swarm optimization algorithm with path relinking for the location routing problem. J. Math. Model. Alg. **7**, 59–78 (2008)
41. Wang, X., Li, Y.: Research on optimization of logistics distribution routing under electronic commerce. Jisuanji Gongcheng/ Comput. Eng. **33**, 202–204 (2007)
42. Wang, Y., Ma, X., Xu, M., et al.: Two-echelon logistics distribution region partitioning problem based on a hybrid particle swarm optimization–genetic algorithm. Exper. Syst. Appl. **42**, 5019–5031 (2015)
43. Bell, J.E., Griffis, S.E.: Swarm intelligence: application of the ant colony optimization algorithm to logistics-oriented vehicle routing problems. J. Bus. Logistics **31**, 157–175 (2010)
44. Wang, H., Li, W.: Study on logistics distribution route optimization by improved particle swarm optimization. ACM Trans. Model Comput. Simul. **5**, 243–246 (2012)
45. Jiang, Z., Wang, D.: Model and algorithm of location optimization of distribution centers for B2C E-commerce. Control Decis. **20**, 1125 (2005)
46. Jie-ming, W.U.: Vehicle routing optimization problem of logistics distribution. ACM Trans. Model Comput. Simul. **7**, 357–360 (2011)
47. Jiang, Z., Wang, D.: Model and algorithm for logistics distribution routing of B2C e-commerce. Inf. Control-Shenyang **34**, 481 (2005)
48. Jianya, Y.Y.G.: An efficient implementation of shortest path algorithm based on dijkstra algorithm. J. Wuhan Tech. Univ. Surv. Mapping. **24**(3), 208–212 (1999)
49. Kang, H.I., Lee, B., Kim, K.: Path planning algorithm using the particle swarm optimization and the improved Dijkstra algorithm. In: 2008 Pacific-Asia Workshop on Computational Intelligence and Industrial Application, PACIIA 2008, vol. 2, pp. 1002–1004. IEEE (2008)

Research on Vehicle Routing Problem with Time Windows Restrictions

Yun-Qi Han[1], Jun-Qing Li[1,2,3,4(✉)], Yong-Qin Jiang[3], Xing-Rui Chen[3], Kun Jiang[3], Xiao-Ping Lin[3], and Pei-Yong Duan[1(✉)]

[1] School of Information, Shandong Normal University, Jinan 250014, China
lijunqing.cn@gmail.com, duanpeiyong@sdnu.edu.cn
[2] School of Computer, Liaocheng University, Liaocheng 252059, China
[3] China Key Laboratory of Computer Network and Information Integration, Southeast University, Ministry of Education, Nanjing 211189, People's Republic of China
[4] State Key Laboratory of Synthetical Automation for Process Industries, Northeastern University, Shenyang 110819, China

Abstract. The multi-objective optimization problem of vehicle routing is a hot issue in many industries in recent years, because it is more fully considered for real-world constraints and has always been a hot issue in shipping transport. Vehicle path planning issues with time window constraints are also the most concerned. Therefore, this paper presents a large number of papers on vehicle routing problem with time window limit, and details the application of various algorithms in this problem. In this paper, several novel and widely applied algorithms are introduced and compared.

Keywords: Vehicle routing · Multi-Objective optimization
Time window restrictions · Algorithm

1 Introduction

The multi-objective optimization problem of vehicle routing is a hot issue in many industries in recent years, because it is more fully considered for real-world constraints and has always been a hot issue in shipping transport. Vehicle path planning issues with time window constraints are also the most concerned. Over the years, all kinds of problems have been studied at home and abroad.

For vehicle routing optimization problems with time window constraints, a large degree of research staff chose to use genetic algorithm or improved genetic algorithm to solve the problem, because the adaptive and random searching characteristics of genetic algorithm. Enables it to have a good global search capability in multi-objective optimization. For example, a bat algorithm, a bat algorithm is an intelligent heuristic algorithm that is based on the echo location system of the bat. It is characterized by simple and robust model, but also has the disadvantage of premature convergence and slow convergence. Combining it with genetic algorithm has good convergence characteristics, and the resulting hybrid bat algorithm has strong advantage. Think of bat as

D.-S. Huang et al. (Eds.): ICIC 2018, LNCS 10955, pp. 763–770, 2018.
https://doi.org/10.1007/978-3-319-95933-7_84

a vehicle. The bat algorithm simulates the echo positioning function of the bat in nature by simulating the frequency of sound waves, loudness of sound, pulse rate (changing the speed of vehicle, vehicle body capacity, etc.), locating the target prey, locating (finding the optimal path between vehicle and destination), finishing predator (complete distribution), and updating the population position (using genetic algorithm's local search, elite parent cross-operator, gene transposition, etc.). Iterate to the next predator. There are also a number of researchers who choose to use particle swarm optimization, ant colony algorithm, simulated annealing method, tabu search, and novel glowworm algorithm in recent years to solve this problem. Because the constraints of vehicle path optimization problem with time window limit vary widely, it is almost impossible to solve the central idea of solving, i.e., using different search strategies of various algorithms, under constraints, find out the optimal path to satisfy the current demand. Then iterate. For example, tabu search algorithm is a search process that uses memory to guide the algorithm, and takes advantage of the short-term memory boot algorithm to jump out locally optimal algorithm, memory the most recently searched solution, and disallow the move of the algorithm back to the previous solution. The simulated annealing method is a local search algorithm that enhances the diversity of the algorithm by introducing a probability mechanism to prevent the algorithm from becoming locally optimal. In this paper, several novel and widely applied algorithms are introduced and compared.

The first part introduces the research significance and present situation of vehicle path planning with time window limitation. The second part introduces what is the problem of vehicle path planning with time window limitation. The third part introduces four algorithms about VRPTW: Firefly algorithm, genetic algorithm, tabu search algorithm and bat algorithm. The last part, that is, the fourth part, sums up the advantages and limitations of these algorithms.

With the development of computer science in recent years, researchers in the world have made great contributions to the field of algorithms, and put forward a variety of intelligent optimization algorithms. Efficient intelligent group optimization algorithm.

2 VRPTW

VRPTW is a hot issue based on VRP. Due to the continuous development of VRP, the time window is added to the vehicle routing problem under the requirement that the demand point is required for the arrival time of the vehicle, and becomes the vehicle path problem with time window(VRP with Time Windows, VRPTW). A time window vehicle path problem(VRPTW) is a time window constraint that adds a customer's access on the VRP. In VRPTW issues, the cost function also includes the latency and customer-required service time due to the early arrival of a customer, in addition to the driving costs.

In VRPTW, in addition to that limitation of the VRP problem, the vehicle must meet the time window limit of the demand point, while the time window limit of the demand point can be divided into two, one is hard time window, the hard window requires that the vehicle must arrive within the time window, while being late is rejected; Another is Soft Time Window, which does not necessarily arrive within the time window, but must be punished outside the time window to punish alternative wait and rejection as the maximum difference between the window and the hard time window.

The vehicle transit problem (VRPTW), which is limited by the time window, relative to the vehicle transit problem (VRP), must additionally take into account the shipping time and time window, which is mainly due to the customer's deadline for service time and the earliest starting service time limit. Therefore, under this limitation, the original VRP problem must be taken into account in addition to the spatial path (Routing) considerations. At the same time, because the field station also has the limitation of time window and indirectly causes the limitation of the path length, it can be seen that the total patrol cost of VRPTW not only contains the transportation cost, but also considers the time cost, and the penalty cost which is not delivered within the time window limit. Therefore, it is very important to find a good solution, time and space.

3 Important Algorithms

Intelligent algorithm refers to the fact that in engineering practice, a number of comparisons are often brought into contact "New" Algorithms or theories such as simulated annealing, genetic algorithm, taboo search, neural network, etc. These algorithms or theories have some common characteristics (such as simulation of natural processes). They are useful in solving complex engineering problems.

In general, the intelligent optimization algorithm is to solve the problem. The optimization problem can be divided into (1) solving a function, optimizing the function optimization problem of the argument value with the smallest function value and (2) searching the optimal solution in a solution space, and minimizing the combination optimization problem of the target function value. Typical combinatorial optimization problems are: Travel Salesman Problem, TSP, Scheduling Problem, 0-1 knapsack problem, and Bin Packing Problem.

The optimization algorithms are many, and the classical algorithms include: linear programming, dynamic programming, etc.; The improved local search algorithm includes the climbing method, the steepest descent method and the like, and the simulated annealing, the genetic algorithm and the taboo search described herein are referred to as instructive searching methods. The neural network and chaos search belong to the dynamic evolution of the system.

In the optimization idea, a neighborhood function is often referred to, and its function is to indicate how to get a new solution from the current solution. The specific implementation method is to be determined according to the specific problem.

In general, local search is based on the greedy idea to use the neighborhood function to search, if finding a solution better than the existing value, discard the former and take the latter. However, it can only be obtained "local minimal solution" That means, maybe it's just a rabbit. "Mount Tai and Little World" But he didn't find it. Simulated annealing, genetic algorithm, tabu search, neural network and so on have been improved from different angles and strategies. Global Minimum Solution And the like.

3.1 Firefly Algorithm

Firefly Algorithm are derived from the natural phenomenon of simulating natural phenomena in nature in the evening, and in the crowd-gathering activity of the lies, each of the lies is exchanged with the companion through the distribution of fluorescein and the companion for information exchange. In general, the brighter glow of the fluoresce, the stronger its number, will eventually appear to gather a lot of lies around the brighter Fireflies. The artificial lies algorithm is a novel intelligent optimization algorithm based on this phenomenon. In the artificial worm swarm optimization algorithm, each of the lies is regarded as a solution of the solution space, which is distributed in the search space as the initial solution, and then the movement of every lies in the space is carried out according to the movement mode of the nature lies. Through every generation of movement, the final made lies gather around the better lies, that is, to find a plurality of extreme points, thus achieving the aim of population optimization. The traditional Firefly algorithm is suitable for solving the problem of continuous optimization and can not solve the problem of discrete optimization of VRPTW. Therefore, it is necessary to improve the traditional Firefly algorithm.

Osaba et al. [1] put forward a Firefly algorithm. An evolutionary discrete firefly algorithm has a time window routing problem for a novel operator of a vehicle. This novel uses some new route optimization operators who have targeted extraction attempts to minimize the number of nodes in the current solution. Use the random path size, path size, the distance from the center of gravity. In other words, try deleting any path and then re-insert the extracted node. In order to transform continuous optimization problems to discrete problems, this paper proposed EDFA, each firefly in the swarm is a possible and feasible solution for the VRPTW [1].

Osaba et al. [2] put forward another Firefly algorithm. In this paper, they proposed DFA. Each Firefly represents a viable solution for the ACVRP- SPDVCFP. All lies are initialized randomly.

3.2 Genetic Algorithm

Genetic algorithm is a computational model to simulate the biological evolution process of Darwin's theory of natural selection and genetics, which is a method of searching optimal solution by simulating the natural evolution process. Genetic algorithms are beginning with a population of potential solutions that represent problems, while one population is composed of a number of individuals encoded by the gene. Each individual is actually a chromosome bearing entity. Chromosome is the main vector of genetic material, i.e. the collection of multiple genes, whose internal representation is a gene combination, which determines the external representation of the individual's shape, such as black hair, which is determined by a combination of genes that control this feature in the chromosome. Thus, a mapping from phenotype to genotype, i.e. coding, is required at the outset. Because of the complexity of gene coding, we tend to simplify, such as binary coding, generation of early generations, the principle of survival and survival of the fittest, and generation of more and more approximate solutions from generation to generation. In each generation, An individual is selected according to the fitness size of an individual in the problem domain, and cross and variation are combined by means of genetic operators of natural genetics to produce a population representing a new solution set. This process will result in the population, like the natural evolution, more adaptive to the environment than the previous generation, and the optimal individual in the last generation population is decoded and can be used as the approximate optimal solution for the problem.

The main feature of genetic algorithm is to directly operate the structural object without the limitation of derivation and function continuity. has inherent hidden parallelism and better global optimization ability; By adopting the method of probability optimization, the optimized search space can be automatically acquired and optimized, the search direction can be adjusted adaptively, and the determined rule is not needed.

Ombuki et al. [3] propose a paper. In this paper, VRPTW is represented as multi-objective optimization problem, and a genetic algorithm solution is proposed using Pareto ranking technique. The VRPTW is interpreted directly as a multi-target problem, where two target dimensions are the number of vehicles and the total cost (distance). One advantage of this approach is that the weights of the weighting and scoring formulas need not be derived. The result of our research is that the multi-objective optimization genetic algorithm returns a group of solutions that take into account the two dimensions. Our approach is quite effective and provides solutions that compete with the best solutions known in the literature, and are not biased towards the number of vehicles. A set of well-known reference data is used to compare the effectiveness of the proposed approach to VRPTW.

3.3 Tabu Search Heuristic

Taboo search heuristic is a meta-heuristic random search algorithm which selects a series of specific search directions as heuristic from an initial feasible solution. In order to avoid the local optimal solution, the TS search is flexible "Memory" The technology is to record and select the optimization process that has been carried out, and guide the search direction of the next step, which is the establishment of Tabu table.

In order to find the global optimal solution, it should not be devoted to a particular area. The disadvantage of local search is that it is too greedy to search for a local area and its neighborhood, resulting in results that are not optimal. Tabu search is a part of the local optimal solution found, consciously avoiding it but not completely isolating it, thus obtaining more search intervals.

Taillard et al. [4] proposed a paper. In this paper, a taboo search algorithm is proposed to solve the problem of vehicle path with soft time window. In this problem, once the penalty has occurred, it is added to the target value allowing reordering at the customer location. The problem of vehicle routing with a hard time window can be simultaneously solved by adding a larger penalty value. In a taboo search, the neighborhood of the current solution is created by an exchange process where the sequence of successive customers (or fragments) is exchanged between the two paths. The Tabu search also leverages an adaptive memory that contains the routing of the best solution previously accessed. A new starting point for a taboo search is generated by a combination of routes taken from different solutions found in the memory. A number of known solutions are reported on classic test issues.

3.4 Bat Algorithm

Bats are the only mammals with wings, and they have advanced echo location capabilities. Most microbats are carnivorous animals. Mini bats use acoustic echo to locate, detect prey, avoid obstacles, and find themselves in the dark to find their habitat in the crack. The bats emit a loud voice and hear echoes reflected from the surrounding objects. For different bats, their impulses are related to hunting strategies. Most bats use a short and high-frequency signal scan around a filter, while others often use fixed-frequency signals for echo localization. The variation of its signal bandwidth depends on the species of bats and often increases by using more harmonics.

The algorithm of bats algorithm is to simulate bats in nature to detect prey using a random search algorithm, that is, a random search algorithm to avoid obstacles, that is, to simulate bats' most basic detection, positioning and connection with optimized target functions by using ultrasound to detect obstacles or prey. The bionic principle of the BA algorithm maps the population of bats as NP feasible solutions in the D dimension problem space, and the optimization process and the search simulation are simulated into a population bat individual moving process and a fitness function value for searching the prey utilization solving problem to measure the position of the bats at the position of the bat, Analogy is an iterative process for optimizing and searching an individual's superiority in the process of optimization and search.

Yongquan Zhou, Jian Xie, and Hongqing Zheng proposed a paper. In this paper, since the standard BA is a continuous optimization algorithm, the coding scheme of the BA cannot be directly used to solve the CVRP problem. In order to apply the BA to solve the CVRP problem, a suitable representation of the candidate solution scheme is a hybrid BAT algorithm designed for a specific problem. Each individual is a sequence with an integer to visit these customers' orders.

4 Conclusion

Genetic algorithm, which is the earliest multi-objective optimization algorithm, has very deep research significance, but its limitation is obvious. Genetic algorithm has the inaccuracy of coding and coding. Moreover, a single genetic algorithm coding cannot fully express the constraints of optimization problems. One way to consider constraints is to apply a threshold to the unfeasible solution, so that the calculated time must be increased. Genetic algorithms generally have a lower efficiency than other conventional optimization methods. Genetic algorithms tend to converge prematurely. Genetic algorithm has no effective quantitative analysis on the accuracy, feasibility and computational complexity of the algorithm.

It can be seen from the optimization principle that the homing ability of bats is mainly dependent on the interaction and influence among bats, but the individual lacks variation mechanism, and once it is bound by a certain local extreme value, it is difficult to get rid of; and in the course of evolution, Super bats in the population may attract other individuals to gather rapidly around them, so that the diversity of the population is greatly decreased, while the convergence rate is greatly reduced and even evolution has stagnated due to the increasing approaching of the individual species, and the population loses the ability to further evolve. In many cases, especially for the optimization space with high dimension, multimodal and complex terrain, the algorithm does not converge to the global extreme value, so it is difficult to find the global optimal distribution in the local optimal neighborhood. Therefore, the improvement of the basic bat optimization algorithm should be put on improving the diversity of the population, so that the population can maintain the ability of continuous optimization during the iterative process. In contrast, the glow-worm algorithm simulates the light-emitting characteristics of the nature worm, and achieves the purpose of exchanging information by comparing the size of the fluorescein value so as to realize the optimization of the problem. The algorithm has the advantages of less parameters, simple operation, good stability and the like. In addition, the improved lies can autonomously look for the superior individual within the sensing range, reduce the degree of dependence on the excellent individuals, and additionally, by comparing the size of the neighborhood average distance, the individual movement step size can be appropriately adjusted within the sensing range, Thereby reducing the oscillation phenomenon and improving the solution precision.

Acknowledgement. This research is partially supported by the National Science Foundation of China (61773192, 61503170, 61603169, 61773246), Shandong Province Higher Educational Science and Technology Program (J17KZ005, J14LN28), Natural Science Foundation of Shandong Province (ZR2016FL13, ZR2017BF039), Key Laboratory of Computer Network and Information Integration (Southeast University), Ministry of Education (K93-9-2017-02), and State Key Laboratory of Synthetical Automation for Process Industries (PAL-N201602).

References

1. Osaba, E., Carballedo, R., Yang, X.-S., Diaz, F.: An evolutionary discrete firefly algorithm with novel operators for solving the vehicle routing problem with time windows. In: Yang, X.-S. (ed.) Nature-Inspired Computation in Engineering. SCI, vol. 637, pp. 21–41. Springer, Cham (2016). https://doi.org/10.1007/978-3-319-30235-5_2
2. Osaba, E., Yang, X.S., Diaz, F., Onieva, E., Masegosa, A.D., Perallos, A.: A discrete firefly algorithm to solve a rich vehicle routing problem modelling a newspaper distribution system with recycling policy. Soft. Comput. **21**(18), 5295–5308 (2017)
3. Ombuki, B., Ross, B.J., Hanshar, F.: Multi-objective genetic algorithms for vehicle routing problem with time windows. Appl. Intell. **24**(1), 17–30 (2006)
4. Taillard, É., Badeau, P., Gendreau, M., Guertin, F., Potvin, J.Y.: A tabu search heuristic for the vehicle routing problem with soft time windows. Transp. Sci. **31**(2), 170–186 (1997)
5. Solomon, M.M.: Algorithms for the vehicle routing and scheduling problems with time window constraints. Oper. Res. **35**(2), 254–265 (1987)
6. Desrochers, M., Desrosiers, J., Solomon, M.: A new optimization algorithm for the vehicle routing problem with time windows. Oper. Res. **40**(2), 342–354 (1992)
7. Solomon, M.M., Desrosiers, J.: Survey paper—time window constrained routing and scheduling problems. Transp. Sci. **22**(1), 1–13 (1988)
8. Gambardella, L.M., Taillard, É., Agazzi, G.: MACS-VRPTW: a multiple colony system for vehicle routing problems with time windows. In: New Ideas in Optimization (1999)

9. Govindan, K., Jafarian, A., Khodaverdi, R., Devika, K.: Two-echelon multiple-vehicle location–routing problem with time windows for optimization of sustainable supply chain network of perishable food. Int. J. Prod. Econ. **152**, 9–28 (2014)
10. Schneider, M., Stenger, A., Goeke, D.: The electric vehicle-routing problem with time windows and recharging stations. Transp. Sci. **48**(4), 500–520 (2014)
11. Dalmeijer, K., Spliet, R.: A branch-and-cut algorithm for the time window assignment vehicle routing problem. Comput. Oper. Res. **89**, 140–152 (2018)
12. Golden, B.L., Assad, A.A.: OR forum—perspectives on vehicle routing: exciting new developments. Oper. Res. **34**(5), 803–810 (1986)
13. Feillet, D., Dejax, P., Gendreau, M., Gueguen, C.: An exact algorithm for the elementary shortest path problem with resource constraints: application to some vehicle routing problems. Networks **44**(3), 216–229 (2004)
14. Coelho, L.C., Laporte, G.: A branch-and-cut algorithm for the multi-product multi-vehicle inventory-routing problem. Int. J. Prod. Res. **51**(23–24), 7156–7169 (2013)
15. Lysgaard, J., Letchford, A.N., Eglese, R.W.: A new branch-and-cut algorithm for the capacitated vehicle routing problem. Math. Program. **100**(2), 423–445 (2004)
16. Raff, S.: Routing and scheduling of vehicles and crews: the state of the art. Comput. Oper. Res. **10**(2), 63–211 (1983)

Research on Swarm Intelligence Algorithm Based on Prefabricated Construction Vehicle Routing Problem

Xingrui Chen[1], Jun-Qing Li[1,2,3,4(✉)], Yongqin Jiang[3], Yunqi Han[3], Kun Jiang[3], Xiaoping Lin[3], and Pei-Yong Duan[2(✉)]

[1] School of Information, Shandong Normal University, Jinan 250014, China
lijunqing.cn@gmail.com
[2] School of Computer, Liaocheng University, Liaocheng 252059, China
duanpeiyong@sdnu.edu.cn
[3] China Key Laboratory of Computer Network and Information Integration, Southeast University, Ministry of Education, Nanjing 211189, People's Republic of China
[4] State Key Laboratory of Synthetical Automation for Process Industries, Northeastern University, Shenyang 110819, China

Abstract. Prefabricated buildings are becoming increasingly popular in China. Logistics distribution is an important aspect of their deployment. At present, there are few logistics management issues, and the logistics and distribution problems are gradually increasing. The planning of path problems is also one of the issues that many scholars are concerned about. With the attention of many experts, a single intelligent optimization algorithm fails to achieve the optimal path and does not apply to large-scale and complicated path planning. Of the existing swarm intelligence algorithms, the ant colony algorithm is the most widely studied one, whereas other swarm intelligence algorithms or hybrid swarm algorithms are relatively less studied. This study combines the research of swarm intelligence algorithms at home and abroad, and thus presents a comprehensive review and analysis of the swarm intelligence algorithms proposed by scholars, which is of significant theoretical importance for the solution of realistic path optimization problems.

Keywords: Assembly building · Swarm intelligence · Ant colony algorithm Logistics and distribution

1 Introduction

With the gradual development of the logistics industry, the problem of vehicle routing [1] has drawn increasing attention. The vehicle routing problem (VRP) was first proposed by Dantzig and Ramser in 1959. It refers to a certain number of customers, each with different cargo requirements. The distribution center provides goods to customers and a fleet of vehicles distributes the goods. The purpose of organizing appropriate driving routes is to satisfy the customers' needs and to achieve goals such as the shortest route, lowest cost, and least time spent under certain constraints.

D.-S. Huang et al. (Eds.): ICIC 2018, LNCS 10955, pp. 771–779, 2018.
https://doi.org/10.1007/978-3-319-95933-7_85

The community intelligence emerging from the group behavior of social groups is gaining considerable attention. In 1957, Professor John Holland [2] of the University of Michigan in the United States published his groundbreaking book Adaptation in Natural and Artificial Systems. Therein, Professor John Holland conducted an adaptive change mechanism in intelligent systems and nature, explained it in detail and proposed an adaptive change mechanism for computer programs. The publication of this work is considered to be the starting point of swarm intelligence [3] algorithms. Certain algorithms designed to mimic the behaviors of colonies, such as searching for food and cleaning lairs, have successfully solved problems such as combinatorial optimization and vehicle routing. Many experts believe that swarm intelligence is an algorithm and a distributed problem-solving device designed to inspire collective behavior of the colony of insects and other animal groups. The characteristics of swarm intelligence are that the smallest intelligent but autonomous individuals use direct inter-individual interactions and indirect interactions with the environment to achieve complete distributed control and are self-organizing, extensible, and robust.

After a long period of evolution and natural selection, the biological community has created many wonderful group phenomena. It has been amazing and also offered us endless inspiration. This laid the foundation for our swarm intelligence algorithm. In recent years, swarm intelligence optimization algorithms have attracted increasing attention. For example, the ant colony algorithm (ACA) [4] was a new swarm intelligence algorithm recently proposed by the Italian scholar M. Dorigo. It is a swarm intelligence algorithm that simulates the foraging behavior of real ant colonies in nature. It uses artificial ants with memory to find the shortest path to food sources in caves through information exchange and collaboration among individuals. Currently, the ant colony algorithm has been used to solve optimization problems such as VRP, assignments, and job-shop scheduling. There are other swarm intelligence algorithms, such as the wolves algorithm [5], cat swarm algorithm [6–9], fish school algorithm [10], and particle swarm algorithm [11], which can all be applied to vehicle path problems.

2 Model Classification of Vehicle Routing Problem

The term VRP was first proposed by the famous scholar Dantzig and Ramser in 1959. Since then, it has attracted the attention of experts in transportation research, combinatorial mathematics, graphic and network analysis, and computer applications, and has found wide applications such as postal delivery problems, vehicle scheduling issues, management paving issues, and network topology issues. The VRP is an extension of the traveling salesman problem (TSP) [12], adding unequal restrictions to the TSP.

Here, we present a brief description of the VRP. There are N customers who purchase different quantities of goods. The distribution centers that provide the goods can be one or many. Each distribution center has a group of vehicles that can complete the delivery of the goods. At this moment, it needs to be selected. A suitable truck route allows trucks to pass them in order, and under certain constraints, such as vehicle load capacity, customer demand, or deadlines, to achieve certain indicators, such as routes, the shortest total distance, the highest degree of customer satisfaction, the least time consumption, or the minimum number of vehicles.

After the VRP was proposed, many scholars such as Linus (1981), Bodin and Golden (1981), Assad (1988), and Desrochers (1990) classified the problem according to different standards from different perspectives [13]. For example, according to the type of vehicle, it can be divided into single vehicle model problem and multimodel problem; according to the number of distribution centers (carriage yards), it can be divided into single distribution center (carriage) problems and multiple distribution centers (carriage) issues; according to the task characteristics, it can be divided into pure delivery (fetch) goods problems and loading and unloading problems; according to whether there are time constraints, it can be divided into no time window and time window problems in addition to the vehicle loading situation, according to the number of optimization goals, and according to the vehicle pairs. The affiliation of parking lots is classified according to different classification criteria, such as the certainty of known information.

2.1 Vehicle Routing Problem with Time Windows

The vehicle routing problem with time windows (VRPTWs) is generally described as starting from a logistics distribution center, using multiple vehicles to deliver to multiple customers. The vehicles return to the distribution center after completing the delivery task. Knowing the position and demand of each customer, the capacity of each car is certain. The delivery of goods to the customer needs to meet certain time constraints, and it is required to rationally arrange the car route so that the objective function can be optimized. The study of this issue has attracted the attention of many scholars. It is an extension of the capacity-constrained VRP, and is an NP problem. The solution algorithm can be divided into an exact algorithm and a heuristic algorithm. When the number of customer points is large, it is difficult to obtain a global optimal solution within an acceptable time using precise algorithms. Therefore, the heuristic algorithm is an important technique for the study to obtain a satisfactory solution to the problem within an acceptable time.

2.2 VRPSPD

The VRPSPD is also an extension of the VRP. In the VRPSPD problem, the customers' needs are twofold; that is, there are both pickup and delivery requirements. It is not possible to separate the two services independently. Only one service is acceptable.

The VRPSPD was first proposed by Min in 1989, which solved the problem of book sending and returning between a central library and 22 local libraries when the vehicle number was determined, and the vehicle load capacity was limited. In recent years, certain scholars have begun to pay attention to and continue to study this issue. For example, Tang and Galvao first proposed the mathematical model of the VRPSPD with the maximum travel constraints of the vehicle, and used the tabu search algorithm [14, 15] and a hybrid local optimization algorithm to solve the problem; Angelelli and Mansini used the exact algorithm of the branch demarcation method and the branch price method to solve the VRPSPD with time window constraints.

3 Population Optimization Algorithm

The swarm intelligence algorithm realizes the solution to the problem by learning from certain life phenomena or natural phenomena in nature. This class of algorithm includes the self-organization, self-learning, and self-adaptive characteristics of the natural life phenomena. In the calculation process, the population is searched for the solution space through the obtained calculation information. During the search process, the population evolves in accordance with the fitness function values set in advance and the survival of the fittest. Therefore, the algorithm has certain intelligence.

Owing to its advantages, when the swarm intelligence algorithm is used to solve a problem, it is not necessary to deal with the solution problem in advance to obtain a detailed solution. It is thus possible to efficiently solve some high-complexity problems.

3.1 Group (Cluster) Intelligence

Typical swarm intelligence algorithms include the ACA, fish swarm algorithm, swarm algorithm, and cat swarm algorithm. Inspired by the behavior of social insects, researchers have produced a series of new solutions to traditional problems through the simulation of social insects. These studies represent cluster intelligence research. Swarm intelligence algorithms use some type of evolutionary mechanism to guide the population to search the solution space through the simulation of intelligent collaborative evolutionary phenomena from nature or other things in nature. In cluster intelligence, a group refers to a group of subjects that can communicate directly or indirectly with each other (by changing the local environment), and this group of subjects can cooperate to solve the distribution problem. The so-called cluster intelligence refers to the phenomenon of nonintellectual agents showing the characteristics of intelligent behavior through cooperation. Cluster intelligence provides a basis for finding solutions to complex distributed problems without centralized control and without providing a global model.

The characteristics and advantages of cluster intelligence are that the individuals who cooperate with each other in a group are distributed, and thus can better adapt to the current working conditions in the network environment. Without central control and data, such a system is more robust and the solution of the entire problem in not affected by the failure of one or several individuals. It is possible to cooperate without direct communication between individuals but through indirect communication, and such a system has better scalability. The increased communication overhead of the system due to the increased number of individuals in the system is very small here. The ability of each individual in the system is very simple, such that the execution time of each individual is relatively short, and the implementation is relatively simple. Because of these advantages, although cluster intelligence research is still in its infancy and there are many difficulties, it can be predicted that research on cluster intelligence represents an important direction for future computer research and development.

3.2 Wolves Algorithm and Vehicle Routing Problem

Owing to the relatively large amount of literature on the ACA for VRPs, this study uses the wolves algorithm as an example to illustrate its feasibility to solve VRPs. As a new swarm intelligence algorithm, the wolf colony algorithm is similar to the ant colony algorithm, particle swarm algorithm, bee swarm algorithm, and fish swarm algorithm. It simulates certain special behaviors of biological groups and uses information exchanged between individuals to achieve the purpose of collaborative search. Ants form a positive feedback mechanism through the sharing of confidence between individuals; individuals in particle groups rely on their own and their peers' experience to interact with each other to achieve optimization goals; fish individuals realize information exchange by observing the actual situation of their peers in order to make the best choice. In contrast, the wolf colony algorithm and the bee colony algorithm have similarities and differences, and the use of individual division of labor and information exchange between groups to achieve collaborative operations.

In [16], a wolf population algorithm was proposed to solve the VRP. That study used the wolves algorithm to establish a mathematical model of a multi-distribution center vehicle path. The data in the MATLAB program example in [17] were referenced. A verification was performed and the results confirmed the performance of the wolves algorithm. Based on this, it can be seen that the wolves algorithm has a good effect in solving this problem.

Combining the literature [18], the wolves algorithm is summarized as follows:

Wolves algorithm has good robustness and global convergence for complex functions with different features, which can effectively avoid the premature convergence commonly exhibited by intelligent algorithms. Especially for multipeak and high-dimensional complex functions, the optimization effect is better, and it can provide new ideas and solutions for a large number of nonlinear and multipeak complex optimization problems.

3.3 Artificial Fish School Algorithm

The artificial fish swarm algorithm (AFSA) [19] is a new optimization algorithm proposed by Li Xiaolei et al. in 2002 based on the intelligent behavior of animal groups. The algorithm is based on the fact that the region in the water with the most abundant fish is the region with the most abundant nutrients. In accordance with this feature, we imitate the behaviors of fish, such as feeding, gathering, and rear-ending, so as to achieve a global optimum.

ASFA has the characteristics of a simple concept, easy implementation, high flexibility, good robustness, high versatility, and high searching speed [20]. At present, it has been applied to continuous optimization, combination optimization, and on-line identification of time-varying systems. Furthermore, robust PID parameter settings, optimization of forward neural network, power system reactive power optimization, multiuser detectors, information retrieval, and multistage station positioning of oilfields, have been addressed, achieving good results [21].

AFSA is a newly developed intelligent optimization algorithm; it is a continuous algorithm and cannot be directly used to solve discrete optimization problems. The authors of Ref. [22] used the AFSA to combine optimization for the first time by redefining concepts such as distance and neighborhood of AFSA.

3.4 Comparison of Swarm Intelligence Algorithms

With the continuous development of the logistics industry, the scope of problems encountered in logistics transportation and distribution has been continuously expanding, and the scale and complexity of practical problems have also been continuously increasing. The effect of optimizing and solving problems through a single algorithm has become increasingly unsatisfactory. Each algorithm has its own deficiencies and faces challenges such as time performance and optimization performance. There are currently several advantages and disadvantages of intelligent optimization algorithms, as summarized in the following table. A variety of single algorithms can be used to complement one another, and the advantages of each algorithm can be fused, to construct a new hybrid algorithm with higher optimization efficiency.

Intelligent optimization algorithm	The basic idea	Advantages	Shortcomings
Genetic algorithm	Evolutionary mechanisms of simulating the genetics and variations of the survival of the fittest in the biological world	Ability to search globally in parallel, ability of self-adaptation, can handle a large amount of information, focus on local search with high performance, not easy to fall into local minimum	Convergence is poor, it is easy to prematurely converge, and it is more difficult to deal with optimization problems
Ant colony algorithm	Imitate ants in the natural world. During the process of foraging, ants transfer pheromones to each other to find food in the shortest time	Distributed computing, parallelism, global search capability, self-organization, positive feedback	Convergence is slow, easy to precocious, easy to fall into local optimal solution
Cat population algorithm	In recent years, a bionic swarm intelligence algorithm has been proposed. This algorithm simulates the daily behavior patterns of cats, namely search mode and tracking mode	High optimization efficiency, simple operation, fast convergence, suitable for real-value processing	Without jumping out of the local optimal mechanism, premature convergence may easily occur. At the same time, this algorithm will take longer CPU time

(continued)

(*continued*)

Intelligent optimization algorithm	The basic idea	Advantages	Shortcomings
Wolves algorithm	The main body of the wolf colony algorithm is composed of three kinds of intelligent behavior: wolf exploration, wolf calling, and wolf siege. The "winner is king" wolf competition rules and the "survival of the fittest" wolves renewal rules are operative	In the solving strategy, the global search and local development of the solution space can be well balanced with excellent search performance	Easy to fall into local optimal solution
Particle swarm optimization	Simulate behaviors of predation by birds, artificially establish a dynamic system with a directed graph as a topological structure and perform information processing on the status of consecutive or intermittent input states	Simple operation, fast convergence, suitable for real-value processing	Easy to fall into local optimal solution

4 Conclusion

Swarm intelligence algorithms are still at the initial stage. This paper listed several swarm intelligence algorithms. Through the characteristics of the algorithms, the advantages and disadvantages of several algorithms were compared. The ACA still has greater potential in the VRP. In the process of solving the vehicle scheduling problem, we fully considered how to combine the objective functions of vehicle scheduling with the optimization features of the swarm intelligence algorithm, constructed an appropriate coding method, a path segmentation, and a combination strategy, such that the swarm algorithm can use its own search. The ability to find solutions, and to introduce appropriate operators based on this to improve the performance of the algorithm remain open issues to be considered later. As an optimization algorithm, the swarm intelligence optimization algorithm is inspired by the phenomena of biological groups living in the natural world, but there is no mature guiding mathematical theory, including analysis and verification of the convergence of the algorithm. There is no universal parameter setting to the initial value. The methods of searching the solution space lack strict mathematical theory and other issues. In spite of this, the field has broken the shackles of the traditional algorithm model and designed a new intelligent algorithm, which has made significant contributions to the development of artificial intelligence.

Acknowledgement. This research is partially supported by the National Science Foundation of China (61773192, 61503170, 61603169, 61773246), Shandong Province Higher Educational Science and Technology Program (J17KZ005, J14LN28), Natural Science Foundation of Shandong Province (ZR2016FL13, ZR2017BF039), Key Laboratory of Computer Network and Information Integration (Southeast University), Ministry of Education (K93-9-2017-02), and State Key Laboratory of Synthetical Automation for Process Industries (PAL-N201602).

References

1. Fisher, M.L.: Vehicle routing problem. Oper. Res. Manag. Sci. **8**, P1–P33 (1995)
2. Holland, J.H.: Outline for a logical theory of adaptive systems. J. Assoc. Comput. Mach. **9** (3), 297–314 (1962)
3. Kennedy, J., Eberhart, R.C., Shi, Y.: Swarm Intelligence. Morgan Kaufman Publisher, San Francisco (2001)
4. Dorigo, M., Maniezzo, V., Colorni, A.: Ant system: optimization by a colony of cooperating agents. IEEE Trans. Syst. Man Cybern. Part B **26**(1), 29–41 (1996)
5. Liu, C.A., Yan, X.H., Liu, C.Y., et al.: The wolf colony algorithm and applications. Chin. J. Electron. **20**(2), 212–216 (2011)
6. Tsai, P.W., Pan, J.S., Chen, S.M., et al.: Parallel cat swarm optimization. In: International Conference on Machine Learning and Cybernetics, vol. 6, pp. 3328–3333. IEEE (2008)
7. Santosa, B., Ningrum, M.K.: Cat swarm optimization for clustering. In: International Conference of Soft Computing and Pattern Recognition, SOCPAR 2009, pp. 54–59. IEEE (2009)
8. Chittineni, S., Abhilash, K., Mounica, V., et al.: Cat swarm optimization based neural network and particle swarm optimization based neural network in stock rates prediction. In: Proceedings of the 3rd International Conferences on Machine Learning and Computing, pp. 292–296 (2011)
9. Ganapati, P., Pyari, M.P., Babita, M.H.: System identification using cat swarm optimization. Expert Syst. Appl. **38**(10), 12671–12683 (2011)
10. Carmelo, J.A., Filho, B., Fernando, B., Lins, J.C.C.: A novel search algorithm based on fish school behavior. In: IEEE International Conference on Systems, pp. 2645–2651 (2008)
11. Ayed, S., Imtiaz, S., Sabah, A.M.: Particle swarm optimization for task assignment problem. Microprocess. Mincrosyst. **26**, 363–371 (2002)
12. Hoffman, K.L., Padberg, M., Rinaldi, G.: Traveling salesman problem. In: Gass, S.I., Fu, M. C. (eds.) Encyclopedia of Operations Research and Management Science. Springer, Boston (2013)
13. Fisher, M.L.: Vehicle routing problem. Oper. Res. Manag. Sci. **8**, 1–3 (1995)
14. Liu, R., Jiang, Z., Geng, N.: A hybrid genetic algorithm for the multi-depot open vehicle routing problem. OR Spectr. **36**(2), 401–421 (2014)
15. Zou, T., Li, N., Sun, D.: Genetic algorithm for multiple-depot vehicle routing problem. Comput. Eng. Appl. **40**(21), 82–83 (2004)

16. Korayem, L., Khorsid, M., Kassem, S.S.: Using grey wolf algorithm to solve the capacitated vehicle routing problem. In: IOP Conference Series Materials Science and Engineering, May 2015
17. Zhi, Y., Ye, C.: Hierarchical algorithm model for vehicle delivery scheduling problem in multiple distribution centers. J. Syst. Manag. **23**(4), 602–606 (2014)
18. Wu, H., Zhang, F.: A uncultivated wolf pack algorithm for high-dimensional functions and its application in parameters optimization of PID controller. In: IEEE Congress on Evolutionary Computation, pp. 1477–1482. IEEE (2014)
19. Li, X.L., Lu, F.: Applications of artificial fish school algorithm in combinatorial optimization problems (2004)
20. Fang, J., Zhang, Q.: Distribution center decision-making problem and fish school algorithm. Comput. Appl. **34**(5), 1652–1655 (2011)
21. He, S., Belacel, N., Hamam, H., Bouslimani, Y.: Fuzzy clustering with improved artificial fish swarm algorithm. Comput. Sci. Optim. (CSO) **2**(1), 317–321 (2009)
22. Li, X., Lu, F., Tian, G.: Application of artificial fish swarm algorithm for combinatorial optimization. J. Shandong Univ. Eng. Edn. **34**(5), 64–67 (2004)

Improved Digital Password Authentication Method for Android System

Bo Geng[1,2], Lina Ge[1,2(✉)], Qiuyue Wang[1,2], and Lijuan Wang[1,2]

[1] College of Information Science and Engineering, Guangxi University for Nationalities, Nanning 530006, Guangxi, China
472076256@qq.com

[2] ASEAN Research Center (Guangxi Science Experimental Center), Guangxi University for Nationalities, Nanning 530006, Guangxi, China

Abstract. Proposes a simple digital password authentication method for mobile phones. Improves the authentication program by introducing random numbers and increasing the number of constant numbers. The input number of the six passwords increases correspondingly with location. The authentication order of the current authentication password is determined according to the cyclic order of the cyclic constants and the dynamic nature of the authentication cipher is achieved. The improved authentication process: this paper calculates the user's input code and the random number, preserves the final number of the results. Authentication is accomplished by verifying the correspondence of the last digits to the stored passwords. If the result is consistent the authentication is passed otherwise the authentication fails. The randomness of the cipher is improved by adding random numbers. The randomness of the password is increased by increasing the random number. At the same time, the upper limit of authentication failure is set and the password is automatically updated when the upper limit is reached. Experiments show that this improved Android system digital password authentication method can effectively resist attacks such as shoulder peeping attacks, stain attacks and guessing attacks, thus increasing the security of mobile authentication.

Keywords: Digital password · Random number · Dynamic · Certification Attack

1 Introduction

Due to the rapid development of smartphones, these have become a necessity in people's daily lives. At this stage of increasing computing power of smart phones, users are enabled to access the Internet anytime and anywhere [1]. Smartphones however are small and portable devices that are consequently easily stolen and often lost. Since these are personal devices smartphones store a significant amount of personal private and otherwise sensitive data, such as contacts short messages and text files that record sensitive information [2]. If such a phone is stolen or lost the privacy data will cause additional loss for the user. Therefore, after loss the internally stored data of a smartphone denies access for the illegal user, to minimize the security risk.

D.-S. Huang et al. (Eds.): ICIC 2018, LNCS 10955, pp. 780–791, 2018.
https://doi.org/10.1007/978-3-319-95933-7_86

The authentication mechanism of smartphones is the most important barrier to prevent illegal users from accessing the internal data of smartphones [3]. The authentication mechanism of smartphones is mainly verified by unlocking the phone screen. At present, this authentication mechanism mainly includes a PIN code a digital password a gesture password fingerprint identification and face recognition [4–6]. Digital passwords are most common; however, passwords are also vulnerable to attacks such as shoulder prying attacks, stain attacks, simple guessing attacks [7].

Currently, smartphones equipped with an android operating system account for more than 79% of all smart phones [8] and digital passwords are most widely used. Therefore, based on the android system this paper designs dynamic digital passwords, thus increasing the diversity of passwords. Adding a random number increases the unpredictability of the password increases the difficulty to crack the password and improves the overall security of the phone.

2 Digital Password Authentication Method

The digital password authentication method is currently the highest authentication method. The digital password authentication method verifies whether the password is identical to the password stored internally and the smartphone is unlocked when passwords match. The number of passwords for this authentication method is generally six; therefore, the number of unlocking passwords of this authentication method is 531,441 [9].

The method of digital authentication is only one resulting in a single password, which is vulnerable to attacks [6, 10, 11] such as stain attack, shoulder attack and guessing attack as previously outlined and this method is therefore easily breached. Aviv et al. [12] reported about the stains attack and Andriotis et al. [13] described the intelligent mobile phone screen stains attack using residual oil or grease with a tool to clearly visualize traces of oil and using this method of attack to obtain password lock patterns. Another security threat of the initial digital password authentication method is the shoulder peeping attack, i.e., users frequently enter their unlock password at public places, where prospective attackers can easily see or otherwise obtain the user's unlock password. The initial digital password authentication method unlocks a single password and is easily conjectured. In 2012, Bonneau et al. [14] conducted relevant experiments and concluded that the use of digital passwords is a preferred method. According to the frequency of ciphers, the authentication method can be guessed.

Due to these problems with digital cryptography, a solution is required that can resist stains shoulder peep attacks and brute force. To solve the existing problems with the initial digital password authentication method, the authentication method of digital password was improved in this study.

3 Methodology

To increase the security of the digital cryptographic authentication method, the authentication process was improved. The six initially stored passwords are placed in a different ordered array List<Integer> respectively. The six passwords in the digital password authentication method are changed to six one bit ciphers increasing the number of three random numbers and displaying the random number on the unlocked interface. The user inputs the remaining three numbers, adds three random numbers to three user entered passwords where the last number is identical to the reserved number; this unlocks the phone. The pre-stored six passwords are unlocked in turn, each unlocking success is updated to the next password and three failures automatically update to the next password.

3.1 The Principle of the Improved Digital Cryptographic Authentication Method

A random class was added to generate a random number program. Adding a number of constants, these corresponded to six times the unlock code and a cyclic order constant determines the password to unlock the order thus unlocking the password cycle. Then, a dynamic change of the password is realized; the improved unlocking flow chart is shown in Fig. 1.

According to the principle of the unlocking screen program designed by the Android system, the number of ordered array List<Integer> is added to store six digital ciphers. The six passwords collected by SettingActivity are stored in six ordered array List<Integer> to complete the storage of six passwords. The password storage flowchart is shown in Fig. 2.

When the LockActivity collected user passwords authentication digital touch, three touch passwords will be collected with three random numbers and will be computed retaining the last figure. Querying the specific value of the current Cishu variable, the final retained number is compared to the number stored in the PassList corresponding to the current Cishu variable which is successful. The successful flowchart of the improved authentication mechanism of the digital password authentication mechanism is shown in Fig. 3.

If the authentication fails, then the variable cishu2+1 is introduced. When the cishu2 value is 3, cishu+1, i.e., when authentication fails three times, the digital authentication password is updated to the next digital authentication password. At the same time, after the first error, the authentication will use related hints of the first few authentication passwords. If authentication failed once, the variable cishu3+1 is used; when cishu3=2, the device automatically sends the specific location to a specified mailbox and the device automatically turns off. The variable Cishu is replaced between 1 to 6, and circular authentication of the digital authentication password is realized. The specific flow chart of the improved authentication failure of the digital password authentication mechanism is shown in Fig. 4.

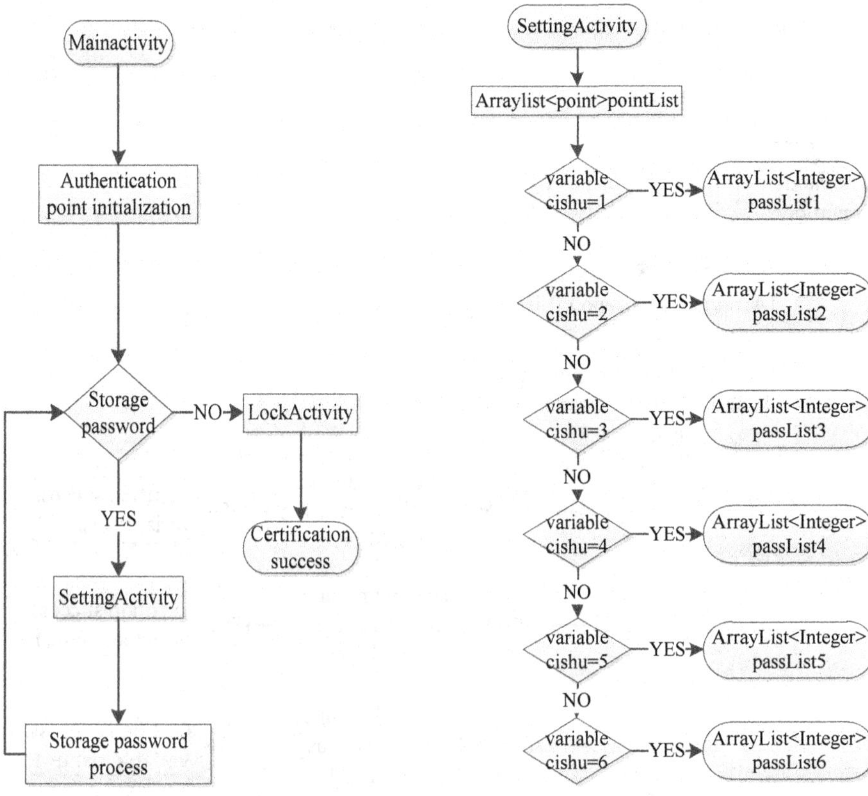

Fig. 1. Improved authentication flow

Fig. 2. Password storage flow chart

3.2 Improved Digital Password Authentication Process Instance

As an example, the password of the user registration number is 256189, as shown in Fig. 5.

After completion of the order cycle on the basis of digital certification registration password authentication and when the interface displayed a three bit random number the registered user must enter three digits. The three-digit random number and user input are operated only at the end, after the digital authentication password authentication is successful. As shown above, the first certification number is 2. The authentication interface is shown in Fig. 6.

The number of authentication is 2 and the sum of the random numbers 5, 7, and 9 is 1 and the number of the remaining digit is 1. The number of users who enter the three digits with the remaining number of 1 can be certified successfully. After the first certification of authentication fails no information is prompted. If authentication fails again the last successful certification number is 1, indicating the last successful certification number, as shown in Fig. 7.

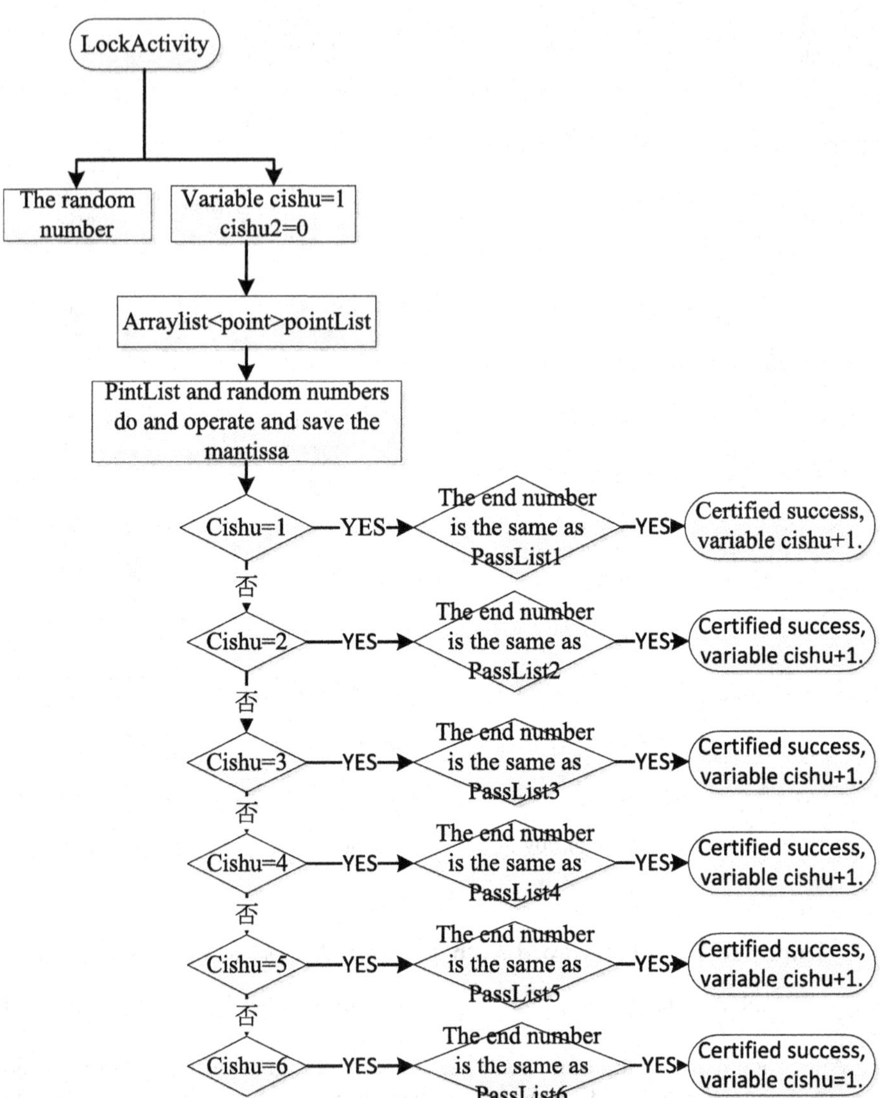

Fig. 3. The improved flow chart for the success of the authentication mechanism of digital cryptography

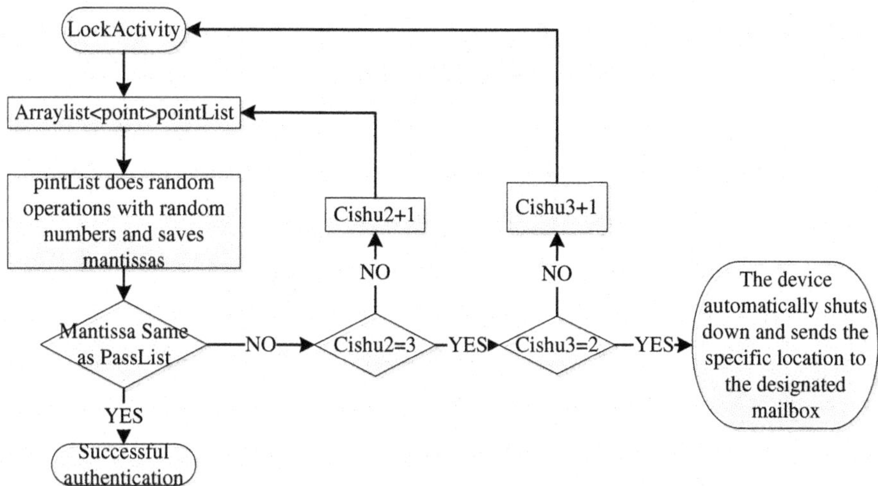

Fig. 4. Flow chart of improved authentication failure of digital password authentication mechanism

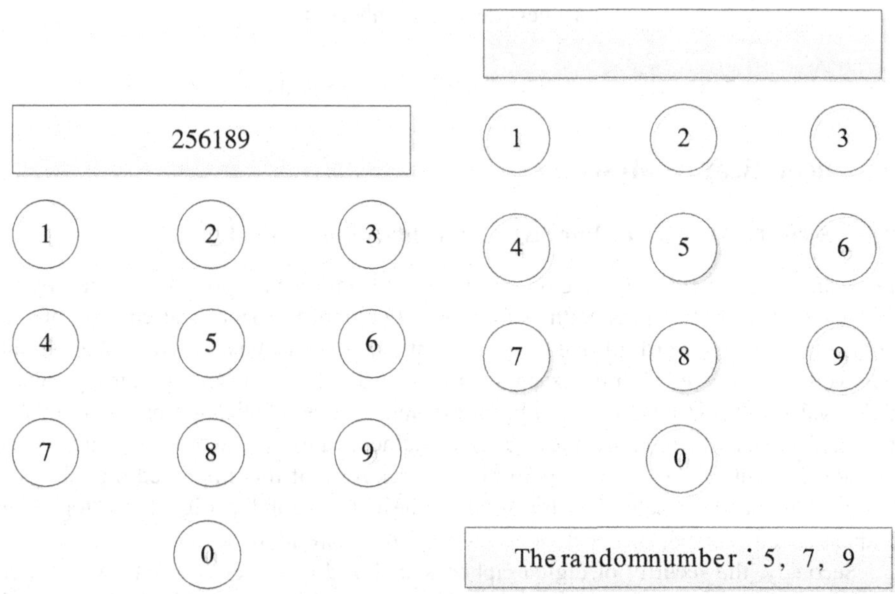

Fig. 5. Set the digital authentication cipher interface

Fig. 6. Authentication interface for improved digital password

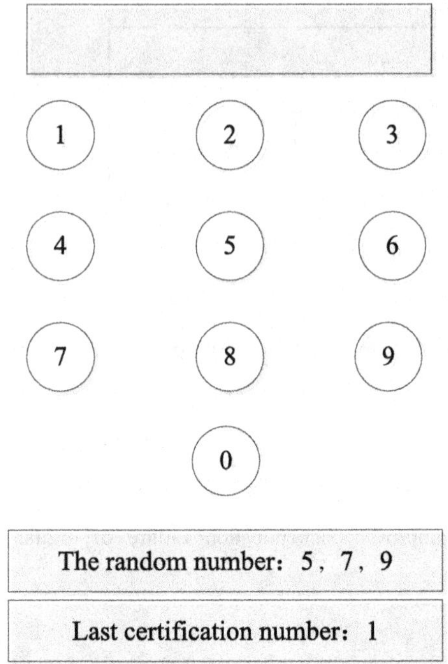

Fig. 7. Authentication failure interface for improved digital password

4 Theoretical Analysis

4.1 Security Analysis of Improved Authentication Method

First, the security of the digital cipher is analyzed from the viewpoint of the entropy of the improved digital cipher both before and after improvement. The entropy of the authentication password of the user authentication mechanism before the digital password is improved. Using six passwords as example, the resulting entropy calculation value is 19.93 bits. The entropy of the improved authentication password of the user authentication mechanism also uses six authentication passwords as example; its entropy computation value is 119.59 bits. The entropy of this improved digital password authentication method is higher than the entropy of the digital cryptography authentication method, the intelligent device is therefore more secure.

Secondly, the security of digital cipher is analyzed to achieve an improvement of the cryptographic space of digital cipher both before and after the process. The utilized digital password user authentication method used to improve before the password space is 1000000 and the digital password authentication mechanism improved the password space for 6000000; consequently, the digital password user authentication method improved the password more than prior to the improvement and the digital password strength achieved better improvement of the digital password user authentication method, making it more difficult to break by others.

4.2 Analysis of the Availability of the Improved Authentication Method

The improved digital password authentication method has very high availability. First, the security has been improved. The improved digital password authentication method realizes the dynamic change of the digital cipher and thus, enlarges the cryptographic space of the digital cipher authentication mechanism. Secondly, the convenience of the method has been improved. This digital password authentication mechanism could be improved with user input of only three password for authentication, thus reducing the user input password digits, while improving the convenience of user the authentication mechanism.

5 Experiment and Analysis

5.1 Experimental Environment

The mobile phone system Android 6.0.1 was used as platform, with the mobile phone model: millet 3, equipped with a mobile RAM capacity of 2 GB and a cell phone ROM capacity of 16 GB. The computer operating system was Windows 10.

5.2 Experimental Design

Usability Experimental Design. To detect the usability of the improved method, a single-factor repeated experiment was designed within the group. The independent variable is the unlocked password. By entering correct and incorrect passwords multiple times, it was detected whether the improved method can correctly differentiate between a legitimate user and an illegal user; Furthermore, whether the random number can be correctly displayed on the screen of the mobile phone was tested. This was done to verify whether the improved method is feasible.

Safety Experiment Design. To evaluate the safety of this improved method, a single factor repeated experiment was designed. The independent variable still remained unlocked. Providing a single unlocked password to simulate a shoulder attack, the mobile phone received a number of unlocked high-definition photos, thus simulating a stains attack. No hints were provided and the number of times within a specific time to guess the phone to unlock the password was not limited, thus simulating a guess attack. The security of the improved method was evaluated by simulating various attacks.

5.3 Experimental Steps

Testing the Availability of Improved Methods. One hundred participants received a mobile phone with an Android system that ran the improved authentication method and participants were told the unlocking method. Then, participants were allowed to set their own password and repeat authentication. It was tested whether it was possible to differentiate between a legitimate user and an illegal user. Each participant was required to experiment 10 times. After the experiment, record the corresponding data.

Testing the Security of the Improved Method

(1) One hundred participants received a single correct password each to unlock. It was ensured that participants received different passwords, to simulate shoulder surfing attacks. Each participant evaluated 10 times whether the digital password authentication method improved the security. At the end of the experiment, two types of mobile phone numbers were recorded as well as the number of failed cracking attempts.
(2) One hundred participants received a mobile phone with a number of unlocked high-definition photos, clarity of the photos was ensured. Participants could freely work on the photos and each participant could try whether the mobile phone authentication method improved security compared to the original digital password authentication method 10 times each. At the end of the experiment, two types of mobile phone numbers were recorded as well as the number of failed cracking attempts.
(3) One hundred participants were not equipped with a digital password authentication method of the improved mobile phone. Each participant had 30 min to try whether the number of password guesses and run mobile phone screen to unlock the original digital password authentication method of digital password authentication method had improved the operation. The end of the experiment was reached after the mobile phone number was unlocked, recording the number of cracking attempts.

5.4 Experimental Results and Experimental Analysis

Results and Analysis of the Availability of the Improved Method. In this experiment, 100 participants were tested 10 times per person. The results of the experiment were that for the five correct password inputs, the success rate of authentication was therefore 100%. For the five erroneous password attempts of 100 participants, all smartphones refused to unlock. The authentication time before and after the mobile operation improved as shown in Table 1.

Table 1. The initial authentication method and the authentication time of the improved authentication method (unit: Second)

	Minimum unlocking time	Maximum unlocking time	Unlocking time average
Initial authentication method	0.953	2.022	1.451
Improved authentication method	0.834	1.591	1.082

We can conclude from Table 1 that it can be seen that the minimum, maximum, and average value of the improved user authentication time after the improved authentication mechanism of digital cryptography are less than that before the improvement, the convenience of the improved user authentication mechanism are better than that before the improvement.

The experimental results show that the availability of the improved digital password authentication mechanism is better than that before the improvement.

The Security Result and Analysis of the Improved Method. The number of mobile phones that ran both the initial authentication method and the improved authentication method are shown in Table 2.

Table 2. The number of cell phones (unit: Unit) that runs the initial authentication method and the improved authentication method is cracked.

	Guessing attack	Stain attack	Shoulder glimpse
Initial authentication method	10	30	30
Improved authentication method	0	0	0

The result of the security experiment was analyzed. Table 2 indicates that the success rate of the digital password authentication mechanism is higher than that of the improved digital password authentication mechanism.

First, this improved digital cryptographic authentication method can resist guessing attacks well. Guessing attacks fail, because the authentication password of the improved digital password authentication mechanism is a single password so that the success rate of cracking is very high. However, the digital password authentication mechanism was improved by adding a random number of elements, realizing dynamic changes of the improved authentication password. Consequently, the password authentication password space has increased, and the password guessing attack authentication mechanism increased the success rate of 0.

Furthermore, the improved digital password authentication method can be very good against stain attacks. In this experiment, the stains and photos of participants' smartphone screens were very clear. The following drawbacks of the authentication password of the digital password authentication mechanism could be detected: It was easy to input the correct number password according to the stain marks on the photos. Due to the addition of random numbers, the improved digital password authentication mechanism does not use a fixed authentication password, resulting in stain and chaos in the photos, thus improving the ability to resist attacks. Therefore, the success rate of the improved digital password authentication mechanism is 0.

Finally, the improved digital password authentication method can resist shoulder attacks very well. Having added a random number, password authentication increases uncertainty and there is no fixed password authentication. Therefore, the improved digital password authentication mechanism can perfectly resist shoulder attacks. According to the above experiments, the improved digital password authentication method is more secure on the basis of guaranteeing availability.

6 Conclusions and Future Work

To improve the safety of smartphone authentication methods for the purpose of digital password authentication for the Android system, a random number was introduced in the digital password authentication method, realizing the automatic updating of password authentication. This realizes the dynamic phone unlock function and improves the safety of the smartphone.

Several aspects of the improved digital password authentication method can be further studied and improved.

(1) The memory optimization problem. While this paper improves the user authentication method, it does not optimize the method significantly. A notable power consumption problem was present in the test as well as the problem of smartphone Carton. Optimizing memory will be a further focus for future research work.
(2) The generation of random numbers. The random number used in this paper was generated according to the class of Random; however, there is still small probability with which the same sequence is produced. Optimizing the generation of random numbers is a central point for future research.

Acknowledgements. This work is supported by the National Science Foundation of China under Grant No. 61462009, Science Research Project 2014 of the China-ASEAN Study Center (Guangxi Science Experiment Center) of Guangxi University for Nationalities, No. TD201404 and Innovation Project of Guangxi University for Nationalities Graduate Education, No. gxunchxps201766. The authors would like to thank Prof. Lina Ge for their valuable comments which greatly improved the presentation of the paper.

References

1. Ahmad, H.M., Abdulkareem, B.J.: Biometric authentication system based on iris patterns. J. Comput. Commun. **04**(1), 23–32 (2016)
2. Shah, A., Thapa, P., Dwivedi, M., et al.: Method and system to provide customizable and configurable lock screen in a user device. US20150254464 (2015)
3. Samangouei, P., Patel, V.M., Chellappa, R., et al.: Attribute-based continuous user authentication on mobile devices. US20170026836 (2017)
4. Ali, M.L., Monaco, J.V., Tappert, C.C., et al.: Keystroke biometric systems for user authentication. J. Signal Process. Syst. **86**(2–3), 175–190 (2016)
5. Faruki, P., Bharmal, A., Laxmi, V., et al.: Android security: a survey of issues, malware penetration, and defenses. IEEE Commun. Surv. Tutor. **17**(2), 998–1022 (2015)
6. Lee, S., Park, J., Hong, S., et al.: Study on the improvement about user authentication of android third party application through the vulnerability in Google voice. J. KIISE **42**(1), 23–32 (2015)
7. Pathangay, V., Rath, S.P.: Biometric user authentication system and a method therefor. US20160132669 (2016)
8. Wójtowicz, A., Joachimiak, K.: Model for adaptable context-based biometric authentication for mobile devices. Pers. Ubiquit. Comput. **20**(2), 195–207 (2016)

9. Wu, J., Cao, T., Zhai, J.: BlindLock: a pattern lock system that effectively prevents stains attack. Comput. Sci. (b11), 364–367 (2015)
10. Todeschini, E.: User authentication system and method. US20160188861 (2016)
11. Kim, G.L., Lim, J.D., Kim, J.N.: Secure user authentication based on the trusted platform for mobile devices. EURASIP J. Wirel. Commun. Netw. **2016**(1), 233 (2016)
12. Aviv, A.J., Gibson, K., Mossop, E., et al.: Smudge attacks on smartphone touch screens. In: Proceedings of the 4th USENIX Conference on Offensive Technologies, pp. 1–7. USENIX Association (2010)
13. Andriotis, P., Tryfonas, T., Oikonomou, G.: Complexity metrics and user strength perceptions of the pattern-lock graphical authentication method. In: Tryfonas, T., Askoxylakis, I. (eds.) HAS 2014. LNCS, vol. 8533, pp. 115–126. Springer, Cham (2014). https://doi.org/10.1007/978-3-319-07620-1_11
14. Bonneau, J., Preibusch, S., Anderson, R.: A birthday present every eleven wallets? The security of customer-chosen banking PINs. In: Keromytis, A.D. (ed.) FC 2012. LNCS, vol. 7397, pp. 25–40. Springer, Heidelberg (2012). https://doi.org/10.1007/978-3-642-32946-3_3

Steepest Descent Bat Algorithm for Solving Systems of Non-linear Equations

Gengyu Ge, Jiaxian Song, Juan Wang, and Aijia Ouyang(✉)

School of Information Engineering, Zunyi Normal University, Zunyi 563006, Guizhou, China
ouyangaijia@163.com

Abstract. Bat algorithm (BA) is a kind of heuristic algorithm imitating the echolocation behavior of bats. In consideration of BA shortcomings such as that it could easily fall into traps like local optimum, low accuracy and premature convergence, a new algorithm is proposed by combining steepest descent (SD) algorithm and bat algorithm based on their respective advantages and disadvantages so as to achieve the goal of solving systems of non-linear equations effectively. The results of simulation experiments show that this proposed algorithm (SD-BA) can help improve the accuracy of problem solving and make the optimization results more accurate, and therefore, it is a very efficient and reliable algorithm for solving systems of non-linear equations.

Keywords: Bat Algorithm · Steepest Descent algorithm · Hybrid algorithm Systems of non-linear equations · Searching

1 Introduction

With the rapid development of modern science and technology as well as the wide application of electronic information technology, solving systems of non-linear equations has been receiving considerable attention. Although there has been a lot of discussions in academic circles about systems of nonlinear equations both theoretically and numerically, solving system of non-linear equations is still a tough problem. Therefore, it is necessary to explore an efficient and reliable algorithm. Bat algorithm (BA) [1] is a kind of heuristic algorithm imitating the echolocation behavior of bats, which is first put forward by Yang X.S in 2010. This algorithm is a random searching mechanism based on population, efforts are made to seek the optimal solution by changing frequency, loudness and impulse transmission rate. This method is characterized by a simple structure with fewer parameters, and it can be easily understood and implemented, with its satisfying searching performance and good robustness, etc. At present [2], BA has been widely used in the field of natural science and engineering science as a new swarm intelligence optimization algorithm, such as multi-objective optimization [3], engineering optimization [4], K-means clustering optimization [5], permutation flow-shop scheduling problem (PFSP) [6], large scale optimization problems [7], etc. However, its application range is greatly limited due to the fact that this algorithm has its drawbacks such as that it could easily fall into the trap of local optimal and its accuracy is not high, etc. [3].

D.-S. Huang et al. (Eds.): ICIC 2018, LNCS 10955, pp. 792–801, 2018.
https://doi.org/10.1007/978-3-319-95933-7_87

Wang et al. [8] proposed an improved bat algorithm called an adaptive bat algorithm(ABA). Pan et al. [9] proposed a communication strategy of Mixed Particle Swarm Optimization (PSO) with Bat Algorithm (BA) for Solving Numerical Optimization Problems. Luo et al. [10] proposed a discrete bat algorithm (DBA) for optimal permutation flow shop scheduling problem (PFSP). Yang et al. [11] proposes an optimal design method for passive power filters (PPF) in order to suppress critical harmonics and improve power factor. Yuvaraj et al. [12] proposes a new method of scheduling for optimal placement and sizing of Distribution Static Compensator in the radial distribution networks to minimize the power loss. Sun et al. [13] proposes a modified fast ensemble empirical model decomposition (FEEMD)-bat algorithm (BA)-least support vector machines model combined with input selected by deep quantitative analysis.

The steepest descent method (also called as the gradient method) is an adaptive method based on gradient [14]. This method has been applied in many fields [15]. It has the advantages such as stable convergence and easy implementation [16] but the speed of convergence is often slow at the extreme point.

Parlos et al. [17] proposed an accelerated learning algorithm (ABP adaptive back propagation) for the supervised training of multilayer perception networks. Nguyen et al. [18] proposed efficient block circular precondition for solving the Tikhonov regularized super resolution problem by the conjugate gradient method. Li and Santosa [19] proposed algorithm minimizes a piece wise linear l/sub 1/function (a measure of total variation) subject to a single 2-norm inequality constraint (a measure of data fit) Murota and Tanaka [20] propose a steepest descent algorithm for minimizing an M-convex function on a constant-parity jump system. Burgs et al. [21] propose a simple probabilistic cost function, and we introduce Rank Net, an implementation of these ideas using a neural network to model the underlying ranking function.

Therefore, an effective and hybrid algorithm SD-BA algorithm is proposed in this paper based on BA global searching ability and fast convergence rate as well as SD fast convergence with the initial value. SD-BA algorithm is introduced in five parts. First of all, systems of non-linear equations, BA and SD are briefly introduced and SD-BA algorithm is put forward in the introduction part. Secondly, these two basic algorithms are further introduced. Then, the fitness function model and SD-BA algorithm are introduced in details. After that, the experimental simulation is carried out for SD-BA algorithm and the data is recorded. Finally, the data of experimental simulation is analyzed so as to evaluate the effect of SD-BA algorithm by comparison.

2 Mathematical Model

The general form of systems of nonlinear equations is:

$$\begin{cases} f_1(x_1, x_2, \cdots, x_n) = 0 \\ f_2(x_1, x_2, \cdots, x_n) = 0 \\ \cdots \\ f_n(x_1, x_2, \cdots, x_n) = 0 \end{cases} \tag{1}$$

Where $f_i(x_1, x_2, \cdots, x_n)(i = 1, 2, \cdots, n)$ is defined as the real function of n variables at $D \subset R^n$ and there is at least one non-linear function.

$$x = (x_1, x_2, \cdots, x_n)^T \tag{2}$$

$$F(x) = (f_1(x), f_2(x), \cdots, f_n(x))^T \tag{3}$$

Then, this system of non-linear equations can be briefly recorded as $F(x) = 0$. If there is the vector quantity $x^* \in D \subset R^n$ make $F(x) = 0$ and then x^* can be regarded as the solution of the system of Eq. (1). The fitness function is constructed.

$$F(x) = \sum_{i=1}^{n} |f_i(x)| \tag{4}$$

Therefore, the problem of finding the roots of this system of non-linear Eq. (1) is converted to be an unconstrained optimization problem by working out the minimum value of this fitness function $F(x)$.

3 Basic Algorithm

3.1 Bat Algorithm

The echolocation behavior of bats is idealized and then the following conclusions can be made: Each virtual bat randomly flies at a speed of vi at xi (the solution to the problem), and meanwhile different bats have different frequency or wave length, loudness A0 and impulse transmission rate. When the bat is searching for prey and find it, it changes frequency, loudness and impulse transmission rate and selects the optimal solution until the objective stops or the conditions are met. In essence, tuning techniques are utilized to control the dynamic behaviors of a swarm of bats, balance and adjust the corresponding parameters of this algorithm, hoping thereby to achieve the optimum of BA. The basic bat algorithm is a kind of swarm intelligence searching algorithm. Steps of the bat algorithm may be referred to Literature [22] for more details: A bat is regarded as a solution position in the searching space in the process of initialization, and this position corresponds to a fitness value of the problem; in the process of iterative search, the moving process of solution position is regarded as the process that the bat is searching for prey. The bat adjusts the wavelength (or frequency) of impulse transmitted according to the distance between itself and the prey, and also adjusts the frequency $r \in [0, 1]$ of impulse transmitted when it gets closer to the prey. The bat flies randomly at a speed of v_i at x_i with a fixed frequency $f_{\min}$ the variable-wavelength λ and loudness A_0 in order to search for prey. In a D-dimensional space, the updating formula of speed v_i^t and position x_i^t at the time of t is:

$$f_i = (f_{\max} - f_{\min}) \cdot \beta \tag{5}$$

$$v_i^t = v_i^{t-1} + (x_i^{t-1} - x^*) \cdot f_i \tag{6}$$

$$x_i^t = x_i^{t-1} + v_i^t \tag{7}$$

Where $\beta \in [0, 1]$ is the evenly distributed random value, $f_i \in [f_{\min}, f_{\max}]$ is the frequency of impulse transmitted by the bat, x^* is the global optimal solution among n bats. For the part of local searching, if the randomly generated value is greater than the current impulse transmission rate, a new solution is randomly generated from the existing set of optimal solutions. The new solution is generated according to Formula (4):

$$x_{new} = x_{old} + \varepsilon A^t \tag{8}$$

Where $\varepsilon \in [-1, 1]$ is a random value subject to even distribution, A^t is the average loudness of all the bats at the time of t. According to the echolocation characteristics of bats, the loudness value of impulse transmitted by the bat in the initial phase is large and the rate is low, which facilitates the search for prey in a large space; once the prey is found, the loudness of impulse would decrease gradually and the rate increases, in order to accurately know about the spatial position of target.

The updating formulas of loudness and impulse transmission rate are:

$$A_i^{t+1} = \alpha A_i^t \tag{9}$$

$$r_i^{t+1} = r_i^0[1 - \exp(-\gamma t)] \tag{10}$$

Where $0 < \alpha < 1$ and $\gamma > 1$, both of them are constants.

3.2 Steepest Descent Algorithm

The searching direction with the method of steepest descent algorithm is the negative gradient direction of objective function. It keeps going forward along with the negative gradient direction of objective function until it reaches the minimum value of function. It is known that the gradient of objective function at $X_{(k)}$ is:

$$\nabla f\left(X_{(k)}\right) = \left[\frac{\partial f(X_{(k)})}{\partial x_1} \quad \frac{\partial f\left(X_{(k)}\right)}{\partial x_2} \quad \cdots \quad \frac{\partial f\left(X_{(k)}\right)}{\partial x_n}\right]^T \tag{11}$$

When efforts are made to find the minimum value of objective function, because the function values descend most steeply along the negative gradient direction, and therefore, the searching direction at $X_{(k)}$ should be the negative gradient direction at this point, that is to say:

$$S_{(k)} = -\frac{\nabla f(X_{(k)})}{\left\|\nabla f(X_{(k)})\right\|} \tag{12}$$

Obviously, $S_{(k)}$ is the unit vector, and then the new point computed at the $k + 1$th iteration is:

$$X_{(k)} = X_{(k)} + \alpha_{(k)} S_{(k)} = X_{(k)} - \frac{\alpha_{(k)} \nabla f\left(X_{(k)}\right)}{\left\|\nabla f\left(X_{(k)}\right)\right\|} \tag{13}$$

The negative gradient direction only implies the optimal direction but not the step size, and therefore, there may be various kinds of processes of steepest descent, which depend on the size of $\alpha_{(k)}/\left\|\nabla f\left(X_{(k)}\right)\right\|$: two methods are available to define the step size $\alpha_{(k)}$: One is to give an initial step size randomly but make it meet the following conditions:

$$f(X_{(k)} + \alpha_{(k)} S_{(k)}) < f(X_{(k)}) \tag{14}$$

The other is to make one-dimensional search along the negative gradient direction so as to find the optimal step size of one-dimensional optimization problem, and in other words, to get the optimal step size by minimizing the value of objective function:

$$\min_{\alpha > 0} f(X_{(k)} + \alpha S_{(k)}) = f(X_{(k)} + \alpha_{(k)} S_{(k)}) \tag{15}$$

This optimal step size is regarded as the step size $\alpha_{(k)}$ in the process of searching starting from the point $X(k)$ in the negative gradient direction of this point. The convergence property of iterative computation with this method can be determined with one or two of the following three formulas as the criterion:

$$\begin{cases} \left\|\nabla f(X_{(k)})\right\| \le \varepsilon_1 \\ \frac{\left\|f(X_{(k)}) - f(X_{(k-1)})\right\|}{\left\|f(X_{(k)})\right\|} \le \varepsilon_2 \\ \left\|X_{(k)} - X_{(k-1)}\right\| \le \varepsilon_3 \end{cases} \tag{16}$$

4 SD-BA Algorithm

4.1 Hybrid Algorithm

(1) BA has global searching ability and fast convergence rate can be combined with SD fast convergence with the initial value.

(2) BA has the characteristics such as simple structure with fewer adjustable parameters, easy-to-understand and implement, with its excellent searching performance and good robustness; SD has the advantages such as stable convergence property and low computational complexity. The combination of both algorithms can help improve the convergence property and stability effectively.

(3) For BA, the values of fitness function are used to determine the evolutionary direction, but for SD, first-order derivatives are used to determine the evolutionary direction, and relatively speaking, it would be more convenient with the combination of both.

4.2 Hybrid Algorithm

Step 1: Initialize the number of individual bats n at the time of t, the location of the bat x_i^t, the speed of the bat v_i^t, the loudness of impulse transmitted by the bat A_i^t and impulse transmission rate r_i randomly, the randomized location of the bats regarded as the initial point, accuracy $\varepsilon > 0$.

Step 2: Work out the searching direction $v_{(k)} = -\nabla f(x_{(k)})$ and make a comparison between the calculated searching direction $v_{(k)}$ and accuracy $\|v_{(k)}\| > \varepsilon$, make one-dimensional search along $v_{(k)}$ from $x_{(k)}$, in order to work out λ_k, make $f(x_{(k)} + \lambda_k v_{(k)}) = \min_{\lambda \geq 0} f(x_{(k)} + \lambda v_{(k)})$, and then make $x_{(k+1)} = x_{(k)} + \lambda_k v_{(k)}, k = k+1$, continue searching until a series of new points $x_j (j = 1, 2, 3, \cdots, n)$ is acquired.

Step 3: Work out the values of fitness function for each bat according to algorithm model with the new points $x_j (j = 1, 2, 3, \cdots, n)$ and find the optimal value, record the frequency of impulse transmitted by the bat $fit_{\min}$ as well as the location of the bat x^* at the optimal value.

Step 4: Update the parameters according to Formulas (5) and (7), the random number rand_1 is generated, select a solution from the set of the optimal solutions if $\text{rand}_1 > r_i$ and generate a new solution X_{new} near this solution according to Formula (8), which should be dealt with in case of out of range.

Step 5: Generate the random number rand_2, accept the new solution x_{new} if $\text{rand}_2 < A_i$ and $fit(\text{x}_{\text{new}}) < fit(x^*)$, and then update loudness A_i and rate r_i according to Formulas (9) and (10).

Step 6: Sort the bats based on the values of fitness function, find and save the current optimal value x^*.

Step 7: Output the globally optimal location x_{best} if the maximum number of iterations is achieved; otherwise go back to Step 4.

5 Experiment and Simulation

In order to check the performance of SD-BA algorithm in solving systems of non-linear equations, some systems of non-linear equations are selected as the numerical examples for test in this paper. In order to facilitate comparison and analysis, several typical systems of non-linear equations are selected as the examples for explanation. In addition, the experimental platform needed for this experiment in this paper as well as the parameter-setting of algorithm are provided as the reference. Specifically, CPU model is Inter i5-5200U, frequency is 2.20 GHz, memory is 4.00 GB, Win8 64-bit operating system is used, the software tools such as MATLAB R2012a, VISIO and MATHTYPE are used as well; for BA, the size of population is set to be 40, the number of iterations is 1000, the maximum frequency is set to be 2 Hz and the minimum frequency is 0 Hz, loudness A_i^t is **, impulse transmission rate r_i is **; for SD-BA algorithm, the size of population is set to be 40, the number of iterations is 1000. The experimental results show that SD-BA algorithm is characterized by high-accuracy computation, with a very high success rate.

In this paper, SD-BA is used to solve these three systems of non-linear equations mentioned above, and for each system of non-linear equations, BA and SD-BA run 30 times respectively, average values and standard deviations are obtained. Table 1 shows the experimental results of Examples 1–3. Example 1 shows that SD-BA performs better than BA; in Examples 2 and 3, the effects of both algorithms are almost the same; while according to the convergence curves, SD-BA performs better than BA in the aspect of optimization on the whole.

Table 1. Comparison of computation between SD-BA and BA

No.	Algorithm	Best	Worst	Mean	STD.
F1	BA	1.56E–06	5.24E–05	2.12E–05	1.28E–05
	SD-BA	6.99E–07	3.62E–05	1.17E–05	7.13E–06
F2	BA	3.30E–05	1.39E+00	1.25E–01	3.08E–01
	SD-BA	4.62E–05	8.60E–01	2.88E–02	1.57E–01
F3	BA	1.13E–06	4.88E–05	2.53E–05	1.42E–05
	SD-BA	4.56E–06	4.03E–05	1.58E–05	9.96E–06

Convergence comparison for example 1, example 2, example 3 are shown in Fig. 1, Fig. 2, Fig. 3 respectively.

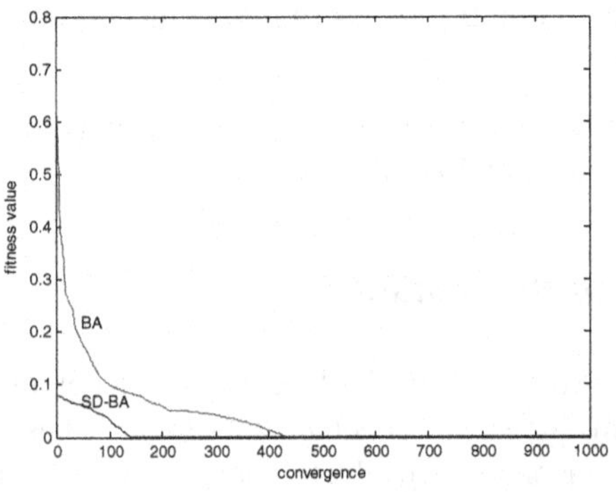

Fig. 1. Convergence comparison for example 1

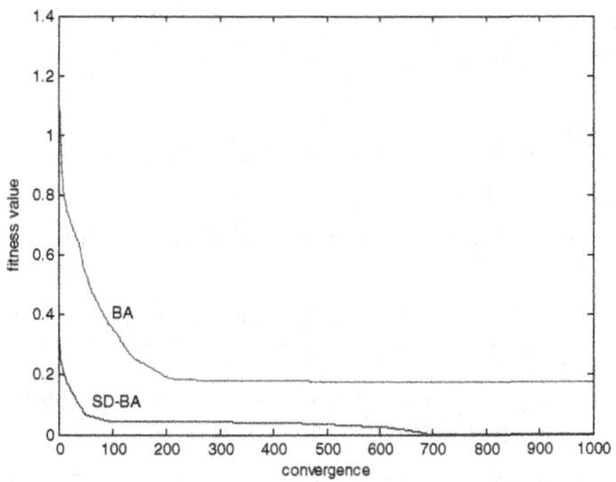

Fig. 2. Convergence comparison for example 2

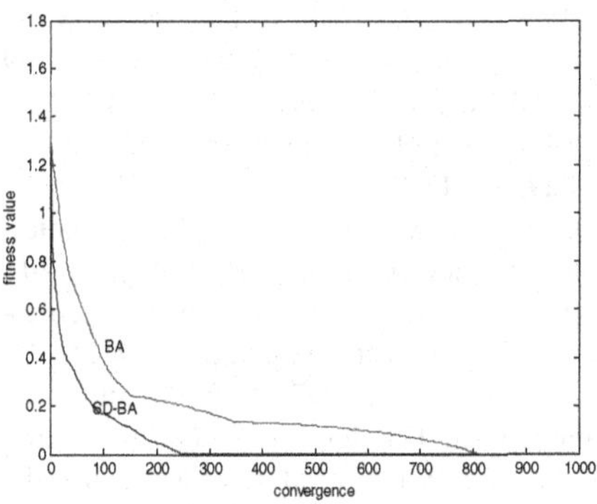

Fig. 3. Convergence comparison for example 3

6 Conclusion

In this paper, a hybrid algorithm based on BA and SD is proposed, which is SD-BA algorithm. Their respective advantages and disadvantages are taken into consideration so as to achieve the goal of solving systems of non-linear equations effectively. This combined method integrates SD with BA tactfully so as to give full play to the advantages of both algorithms. This hybrid algorithm has been applied to solve systems of non-linear equations in this paper and achieved a good effect.

Acknowledgements. The research was partially funded by the science and technology project of Guizhou ([2017]1207), the training program of high level innovative talents of Guizhou ([2017]3), the Guizhou province natural science foundation in China (KY[2016]018), the Science and Technology Research Foundation of Hunan Province (13C333).

References

1. Yang, X.S.: Nature-Inspired Meta Heuristic Algorithms. Luniver Press, Bristol (2010)
2. Xiao, H.H., Duan, Y.M.: Research and application of improved bat algorithm based on DE algorithm. Comput. Simul. **31**(1), 272–301 (2014)
3. Yang, X.S.: Bat algorithm for multi-objective optimization. Int. J. Bio-Inspir. Comput. **3**(5), 267–274 (2011)
4. Yang, X.S., Hossein, G.A.: Bat algorithm: a novel approach for global engineering optimization. Eng. Comput. **29**(5), 464–483 (2012)
5. Komarasarmy, G., Wahi, A.: An optimized k-means clustering technique using bat algorithm. Eur. J. Sci. Res. **84**(2), 263–273 (2012)
6. Sheng, X.H., Ye, C.M.: Application of bat algorithm to permutation flow-shop scheduling problem. Ind. Eng. J. **16**(1), 119–124 (2013)
7. Huang, G.Q., Zhao, W.J., Lu, Q.Q.: Bat algorithm with global convergence for solving large-scale optimization problem. Appl. Res. Comput. **30**(5), 1323–1328 (2013)
8. Wang, X., Wang, W., Wang, Y.: An adaptive bat algorithm. In: Huang, D.-S., Jo, K.-H., Zhou, Y.-Q., Han, K. (eds.) ICIC 2013. LNCS (LNAI), vol. 7996, pp. 216–223. Springer, Heidelberg (2013). https://doi.org/10.1007/978-3-642-39482-9_25
9. Pan, T.S., Dao, T.K., Chu, S.C.: Hybrid particle swarm optimization with bat algorithm. Genet. Evol. Comput. **329**, 37–47 (2015)
10. Luo, Q.F., Zhou, Y.Q., Xie, J., Ma, M.Z., Li, L.L.: Discrete bat algorithm for optimal problem of permutation flow shop scheduling. Sci. World J. **2014**, 15 (2014)
11. Yang, N.C., Le, M.D.: Optimal design of passive power filters based on multi-objective bat algorithm and pare to front. Appl. Soft Comput. **35**, 257–266 (2015)
12. Yuvaraj, T., Ravi, K., Devabalaji, K.: DSTATCOM allocation in distribution networks considering load variations using bat algorithm. Ain Shams Eng. J. **8**(3), 391–403 (2017)
13. Sun, W., Liu, M., Liang, Y.: Wind speed forecasting based on FEEMD and LSSVM optimized by the bat algorithm. Energies **8**(7), 6585–6607 (2015)
14. Zhang, X.W., Xing, Z.D., Dong, J.M.: A hybrid genetic algorithm to seeking the optimum solution of a class of unconstrained optimization. J. Northwest Univ. (Nat. Sci. Edit.) **35**(2), 130–132 (2005)
15. Haario, H., Saksman, E.: Simulated annealing process in general state space. Adv. Appl. Probab. **23**(4), 866–893 (1991)
16. Zhao, X.P.: Convergence on the steeped decent method using difference quotient. J. East China Inst. Chem. Technol. **18**(6), 807–812 (1992)
17. Parlos, A.G., Fernandez, B., Atiya, A.F., Muthusami, J., Tsai, W.K.: An accelerated learning algorithm for multilayer perception networks. IEEE Trans. Neural Netw. **5**(3), 493–497 (1994)
18. Nguyen, N., Milanfar, P., Golub, G.: A computationally efficient super resolution image reconstruction algorithm. IEEE Trans. Image Process. **10**(4), 573–583 (2001)
19. Li, Y., Santosa, F.: A computational algorithm for minimizing total variation in image restoration. IEEE Trans. Image Process. **5**(6), 987–995 (1996)
20. Murota, K., Tanaka, K.: A steepest descent algorithm for M-convex functions on jump systems. IEICE Trans. Fundam. Electron. Commun. Comput. Sci. **E89A**(5), 1160–1165 (2006)

21. Burges, C., Shaked, T., Renshaw, E., Lazier, A., Deeds, M., Hamilton, N., Hullender, G.: Learning to rank using gradient descent. In: Proceeding of the 22nd International Conference on Machine Learning, pp. 89–96 (2005)
22. Yang, X.S.: A new meta-heuristic bat-inspired algorithm. In: González, J.R., Pelta, D.A., Cruz, C., Terrazas, G., Krasnogor, N. (eds.) Nature Inspired Cooperative Strategies for Optimization, vol. 284, pp. 65–74. Springer, Heidelberg (2010). https://doi.org/10.1007/978-3-642-12538-6_6

User Engagement Prediction Using Tweets

Ameesha Mittal[1], Geetika Arora[1], Kamlesh Tiwari[1(✉)], Vandana Dixit Kaushik[2], and Phalguni Gupta[3]

[1] Birla Institute of Technology and Science Pilani, Jhunjhunu 333031, Rajasthan, India
{f2014107, p2016406, kamlesh.tiwari}@pilani.bits-pilani.ac.in

[2] Harcourt Butler Technical University, Kanpur, Utter Pradesh, India
vandanadixitk@yahoo.com

[3] National Institute of Technical Teachers' Training & Research Kolkata, Kolkata, India
pg@cse.iitk.ac.in

Abstract. People are spending huge amount of time on social media platforms these days. Through this they get engage in various real-world activities, either for awareness or just for participation. Twitter has 330 million active users, who generates around 6,000 tweet per second. This forms a huge corpus of data that is widely available for analysis, monitoring and research. Different forms of user behavior can be studied with this data. Analysis in this paper shows how simple machine learning and natural language processing techniques can be used to predict user interests based on his/her past tweets. The paper proposes to use a keyword extraction and semantic clustering based approach to do the analysis. The proposed approach has been tested on a dataset of 1,69,000 tweets and has achieved an accuracy of 80%.

Keywords: Tweet · Feature · K-means · Clustering · Lin-similarity

1 Introduction

Online networking is turning out to be crucial in our day-to-day lives. The way in which data is created, viewed, and shared has significantly changed individuals' engagement in real life events [8]. Social media platforms like Twitter, Facebook, Instagram, Google+ *etc.* have become the most effective means of communication. People communicate over social network about different real-life events like sports, music concerts, presidential elections, environmental pollution, terrorism *etc.* However, there are multiple factors affecting a user's engagement on social media like demographics, past activities, friends and so on. As a result, it is worth to create tools for event identification and detection on social media. User engagement prediction is useful to identify right set of audience for specific events. For instance, if a conference organizer knows who are the persons interested to join the conference, he may directly send the advertisement to them. A real-time interaction of people in events like earthquakes on Twitter has been studied in [12]. An algorithm using probabilistic spatiotemporal model to investigate available tweets and detect a target event for users'

D.-S. Huang et al. (Eds.): ICIC 2018, LNCS 10955, pp. 802–808, 2018.
https://doi.org/10.1007/978-3-319-95933-7_88

interaction has been discussed in [11]. Techniques for analyzing a stream of tweets to distinguish them between real-world events and non-events have been explored in [2]. Identifying real time events on Twitter data is quite a challenging problem, because the data is highly heterogeneous and huge in size. The problem of predicting user engagement on Twitter using real world events has been addressed in [4]. It has attempted to solve the problem by first analyzing a person's engagement on twitter in real-world events through activities like posting tweets, retweeting, or replying to tweets about such events. Statistical models to examine different predictive factors have been designed.

It is important to address the problem in hand because this may provide some new insights about the factors capable of influencing the presence. It may also predict the degree of engagement of Twitter users in real-world events and the extent of their influence. Several researchers have tried to understand user behavior through tweets about specific real world events. For example, the tweets related to political events, local events and natural hazards have been considered in [5, 6, 14].

This problem has many important and diverse applications like marketing, political campaigns, and citizen journalism. It can be used to improve recommendation engines. Event organizers can take advantage of such a framework to identify potential audiences for their events and organize things accordingly. But analysis of twitter data has many challenges because of three reasons. First one is due to use of limited number of characters. The other one is due to usage of slang words and incorrect spellings. Such words do not exist in any English language corpus and are thus treated as unknown words during similarity measurements. The third one is agglomeration of multiple words into one. People commonly try to combine several words into one such as in hashtags. The problem of extracting key phrases as a way to summarize Twitter content has been studied in [16]. It has followed three steps for key phrase extraction: keyword ranking, candidate keyphrase generation and keyphrase ranking. A work to generate automatically personalized tags for Twitter users has been presented in [15].

This paper attempts to determine the effectiveness of social media in analyzing and predicting real-world user behavior, and to explore the events or topics where social media can be a reliable predictor for real-world behavior. It has used a machine learning based approach for predicting user interests. While most of the previous research work in this area focuses on building sophisticated methodologies for studying user behavior using social media data, we propose to use simple established methods for the same task. The proposed method has achieved a more accurate model for predicting twitter user's engagement in real-life events. It consists of 4 sections. Next section discusses some of the well-known techniques which are used in designing the proposed method. Section 3 presents the proposed method along with experimental results. Conclusion is presented in the last section.

2 Proposed Approach

To perform our analysis, it is required to get data that has correlation between twitter feeds and real life events. To get that, one has to either scrape a list of events from newspapers and then extract tweets about it or first detect real world events from Twitter streams, and then infer their geolocations later. The second approach is more suitable because not all real-life events go trending on social media. However, we may face a problem of scarcity while taking the first approach. In [2] a framework for this that first clusters tweets into multiple clusters and then for each cluster, classifies events as real-world events and non-events has been proposed. Geolocation of the event is a tricky thing to find and thus the work till date relies on annotators for that. Twitter streaming or rest API can be used to obtain as many tweets as required. However not all the tweets are about real-life events so only a fraction of them would actually be used for analysis. This paper uses *lin* similarity method [9] for calculating semantic similarity between the words. The relatedness value returned by the lin measure is given by Eq. 1.

$$2 \times \frac{IC\,(lcs)}{IC\,(synset1)} + IC\,(synset2) \quad (1)$$

In this, IC(x) is the information content of x. LCS is least common subsequence (LCS). The information content of x is defined as $-\log(p(x))$ where p(x) refer to the probability of the concept occurrence in the Brown corpus. One can observe that the relatedness value will be greater-than or equal-to zero and less-than or equal-to one. In this paper, three different clustering techniques have been considered, namely k-means, spectral clustering and agglomerative clustering. K-means [7, 10] is one of the simplest unsupervised learning clustering algorithms. It follows an easy and straightforward method to distribute any given dataset into a certain given number of clusters. The central idea is to calculate k centroids, one for each cluster. These centroids should be placed as much far away from each other as possible. The next step is to pick each point from the dataset and find the centroid closest to this point and associate that point to the respective cluster. When no point is pending, the first cycle gets over. Now k new centroids are re-calculated for the clusters resulting from the previous step. Once we have these k new centroids, the points are redistributed. A loop is thus generated. As a result of this loop, k centroids change their locations in each step. The loop is iterated over again and again until no more changes are done. The aim is to minimize:

$$J(V) = \sum_{i=1}^{c} \sum_{j=1}^{ci} \left\|x_i - v_j\right\|^2 \quad (2)$$

where, $\|x_i - v_j\|$ is the Euclidean distance between x_i and v_j, c_i is the number of data points in i^{th} cluster, c is the number of cluster centers. Then points are assigned to the closest centroid. Then the new centroids are calculated.

In spectral clustering methods [11], top eigen vectors of a matrix are derived from the distance between points. It first performs dimensionality reduction before clustering in fewer dimensions. The similarity matrix is given as an input. It consists of a quantitative value of the relative similarity of each pair of words in the dataset. The aim of spectral clustering is to cluster data that is connected but not always compact or clustered within convex boundaries. Agglomerative clustering [3] is a hierarchical clustering method. It is a bottom up clustering technique. In this technique, the clusters have sub-clusters, which in turn have sub-clusters and so on. One example of this type of clustering is the species taxonomy. Agglomerative clustering starts with every single object (word in this case) represented as a single cluster. Then, in each successive cycle, it agglomerates or merges the closest pair of clusters using the similarity matrix, till the whole data becomes one cluster.

This paper considers tweets belonging to UK only. We have collected required tweets using the static twitter API and have extracted a subset of most frequently occurring keywords from these tweets and found semantic similarities between each pair of keywords using lin similarity [9]. We have then used the similarity matrix to form clusters of these words using k-means clustering. This formed our trained corpus. User tweets are divided into predicting and testing data. For each of these datasets, we have again extracted the most frequently used keywords and checked to which cluster do most of them lie in. The block diagram depicting the proposed approach is given in Fig. 1. Our hypothesis is that most of the keywords from testing dataset should lie in the same cluster as those from predicting datasets.

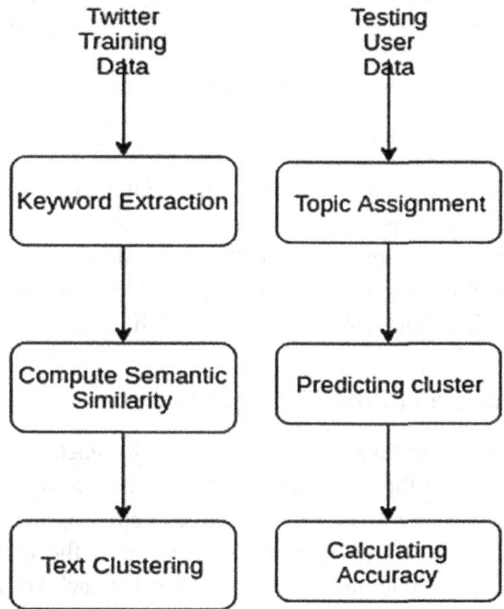

Fig. 1. Control flow of the prediction system

2.1 Keyword Extraction

In the proposed model, we used python package, 'pattern' to do Parts of Speech (POS) tagging/Phrase extraction of the tweets. Pattern is a web mining module for the Python programming language. It bundles tools for data mining (Google + Twitter + Wikipedia API, web crawler, HTML DOM parser), natural language processing (part-of-speech taggers, n-gram search, sentiment analysis, WordNet), machine learning (vector space model, k-means clustering, Naive Bayes + k-NN + SVM classifiers) and network analysis (graph centrality and visualization).

The standard POS tagging/chunking tools do not function well for free form texts like tweets, so we had to utilize a tool that is designed and trained for twitter data. From Pattern tool output, we extract required keywords. One can choose to use on NP (Noun Phrase), but our default is to use NP (Noun Phrase) and ADJP (Adjective Phrase). With the created module, one can also extract urls, usernames, hashtags from the tweet.

2.2 Compute Semantic Similarity

Semantic Similarity refers to the closeness between concepts which are not lexicographically similar. The similarity is based on the semantic meaning of the words. This is an crucial problem in Natural Language Processing and has received intense attention in research. Several algorithms have been proposed for computing semantic similarity [13]. This paper uses Sematch framework. Sematch is an integrated framework for the development, evaluation, and application of semantic similarity for Knowledge Graphs (KGs). We have used lin similarity to form the similarity matrix for the chosen keywords. We have extracted the top 1000 keywords from the Keyword Extraction step and formed a 1000 by 1000 similarity matrix using the lin similarity method.

2.3 Clustering

Out of various clustering options, we have found out that the results yielded by k-means clustering are better and are therefore, used for further analysis. Out of the clusters formed, one of the cluster represents the words that are either unavailable in the wordnet corpus or could not be fit with the remaining clusters. Each of the cluster formed through clustering represents an event in which some user may be interested in.

2.4 Topic Assignment and Predicting User Interest

For the last part of the project, we divided the tweets for each user into predicting and testing datasets. For each of these datasets we have extracted the major keywords. For the most frequently occurring keywords from the predicting set, we have found the average similarity of each of the keyword with each of the clusters formed in the previous step. By taking the maximum average value, for each keyword we have found the cluster it should belong to amongst those formed in the last step. The most commonly stated clusters identified as the interest cluster of the user. As per the hypothesis the testing dataset should yield the same result as the predicting when the same set of steps is performed on it. We have repeated this process for multiple users to get the percentage accuracy of the hypothesis.

3 Experimental Results

The system has been tested on the tweets of 100 users and the results are calculated in terms of accuracy of the prediction.

Database. The dataset is partially obtained from UK geolocated tweets [1] and partially from the static twitter API. Standard twitter API along with python library Tweepy provides a convenient way to store twitter stream. However, it provides tweets not more than a week old and needs to be run for the whole month if one needs the tweets for any particular month. We have collected 1,69,000 tweets from UK for April 2016. These tweets came from 24,264 different users. The accuracy of the system is calculated by dividing the number of users for with the calculated cluster is same for the training and the testing datasets by the total number of users under study i.e. 100.

Experimental Settings. The model has been tested on a HP Pavilion 15-AU621TX Notebook using Linux 16.04 Operating System. The system consists of a 7th Generation Core i5 2.50 GHz processor by Intel, 8 GB DDR4 RAM and 1 TB memory. We have used python libraries and have tried to utilize them for predicting any user's interests based on his past tweets. During the training, top 500 keywords have been used of clustering. The keywords have been clustered using k-means with $k = 11$. For each of the 100 users, the tweets have been divided in a ratio of 20:80 as predicting and testing datasets. Top 70 keywords from the predicting datasets are used for topic extraction while top 50 keywords from the testing dataset are considered used to validate the prediction.

Observations. Confusion matrix showing how activity of an user is get classified by the proposed system is shown in Fig. 2. The model have achieved an accuracy of 80%.

		Actual Cluster										
		0	1	2	3	4	5	6	7	8	9	10
Predicted Cluster	0	2	0	0	1	0	0	0	0	0	0	0
	1	0	0	0	0	0	0	0	0	0	0	0
	2	0	0	2	0	0	0	0	0	0	0	0
	3	0	0	0	0	0	0	0	0	0	0	0
	4	0	0	0	0	1	0	0	0	0	0	0
	5	0	0	0	0	0	3	0	0	4	0	0
	6	0	0	0	0	0	0	6	0	0	0	0
	7	0	0	0	0	0	0	0	1	0	0	0
	8	0	0	0	0	0	0	0	0	0	0	0
	9	0	0	0	0	0	0	0	0	0	0	0
	10	0	0	0	0	0	0	0	0	0	0	0

Fig. 2. Confusion matrix on a small representative database of 10 users

4 Conclusion

The paper proposes an approach to design a model to predicting users' interests based on semantic clustering of keywords extracted from his tweets. Such an analysis has a variety of applications like targeted events planning and advertisements delivery. Proposed method analyses Twitter feed data by applying a k-means clustering algorithm along with semantic similarity measure. We have utilized a modified lin similarity measure to get closeness of phrases based upon the words they contain. The method was able to achieve 80% of correct predictions (accuracy). Our analysis does not include factors like, the interests of the followers of the user, personal details like his age, gender *etc.* that could be included in future work.

References

1. Twitter dataset of UK geolocated tweets. http://www.followthehashtag.com/datasets/170000-uk-geolocated-tweets-free-twitter-dataset/
2. Becker, H., Naaman, M., Gravano, L.: Beyond trending topics: real-world event identification on twitter. In: ICWSM 2011, pp. 438–441 (2011)
3. Dubien, J.L., Warde, W.D.: A mathematical comparison of the members of an infinite family of agglomerative clustering algorithms. Can. J. Stat. **7**(1), 29–38 (1979)
4. Hu, Y., Farnham, S., Talamadupula, K.: Predicting user engagement on twitter with real-world events. In: ICWSM, pp. 168–178 (2015)
5. Hu, Y., Farnham, S.D., Monroy-Hernández, A.: Whoo.ly: facilitating information seeking for hyperlocal communities using social media. In: Proceedings of the SIGCHI Conference on Human Factors in Computing Systems, pp. 3481–3490. ACM (2013)
6. Hu, Y., John, A., Seligmann, D.D., Wang, F.: What were the tweets about? Topical associations between public events and twitter feeds. In: ICWSM (2012)
7. Jain, A.K.: Data clustering: 50 years beyond k-means. Pattern Recogn. Lett. **31**(8), 651–666 (2010)
8. Kwak, H., Lee, C., Park, H., Moon, S.: What is twitter, a social network or a news media? In: Proceedings of the 19th International Conference on World Wide Web, pp. 591–600. ACM (2010)
9. Li, Y., Bandar, Z.A., McLean, D.: An approach for measuring semantic similarity between words using multiple information sources. IEEE Trans. Knowl. Data Eng. **15**(4), 871–882 (2003)
10. MacQueen, J., et al.: Some methods for classification and analysis of multivariate observations (1967)
11. Ng, A.Y., Jordan, M.I., Weiss, Y.: On spectral clustering: analysis and an algorithm. In: Advances in Neural Information Processing Systems, pp. 849–856 (2002)
12. Sakaki, T., Okazaki, M., Matsuo, Y.: Earthquake shakes twitter users real-time event detection by social sensors. In: Proceedings of the 19th International Conference on World Wide Web, pp. 851–860. ACM (2010)
13. Varelas, G., Voutsakis, E., Raftopoulou, P., Petrakis, E.G., Milios, E.E.: Semantic similarity methods in wordnet and their application to information retrieval on the web. In: Proceedings of the 7th Annual ACM International Workshop on Web Information and Data Management, pp. 10–16. ACM (2005)
14. Vieweg, S., Hughes, A.L., Starbird, K., Palen, L.: Microblogging during two natural hazards events: what twitter may contribute to situational awareness. In: Proceedings of the SIGCHI Conference on Human Factors in Computing Systems, pp. 1079–1088. ACM (2010)
15. Zhao, W.X., Jiang, J., He, J., Song, Y., Achananuparp, P., Lim, E.P., Li, X.: Topical keyphrase extraction from twitter. In: Proceedings of the 49th Annual Meeting of the Association for Computational Linguistics: Human Language Technologies, vol. 1, pp. 379–388. Association for Computational Linguistics (2011)
16. Zhu, G., Iglesias, C.A.: Sematch: semantic similarity framework for knowledge graphs. Knowl. Based Syst. **130**, 30–32 (2017)

Discrimination and Prediction of Protein-Protein Binding Affinity Using Deep Learning Approach

Rahul Nikam, K. Yugandhar, and M. Michael Gromiha(✉)

Department of Biotechnology, Bhupat and Jyoti Metha School of Biosciences, Indian Institute of Technology Madras, Chennai 600036, Tamilnadu, India
gromiha@iitm.ac.in

Abstract. Protein-protein interactions (PPIs) mediate myriad biological functions. Estimating the binding affinity for PPIs help us to understand the underlying molecular recognition mechanism. In this work, we utilized deep learning approach to discriminate protein-protein complexes based on their binding affinity. We setup a database of 464 protein-protein complexes along with their experimental binding affinities and developed a deep learning based binary classification model, which showed an accuracy of 81.75% using 5-fold cross-validation. Furthermore, we refined the method for predicting the binding affinity of protein-protein complexes using a large set of complexes (PPA-Pred2). It could predict the binding affinity of the training set and blind test set with a mean absolute error of 1.24 kcal/mol and 1.31 kcal/mol, respectively. We suggest that our methods could serve as efficient tools to study PPIs and provide crucial insights about the underlying mechanism for the molecular recognition process.

Keywords: Protein-protein interaction · Binding affinity · Deep learning
Machine learning

1 Introduction

PPIs play essential roles in several biomolecular processes such as DNA replication, signalling pathways, regulation of metabolic pathways, immunologic recognition, cell cycle progression etc. [1]. The efficacy of a PPI in carrying out its normal function is governed by the underlying recognition mechanism. Protein-protein binding affinity is one of the key factors that regulate binding specificity thereby affecting the eventual function. Furthermore, PPI affinity information can analyse complex interaction networks to reveal interesting patterns that could potentially unveil the mechanism of interaction [2]. Hence, estimating protein-protein binding affinity has been a very important research problem.

R. Nikam and K. Yugandhar—Contributed equally to this work.

D.-S. Huang et al. (Eds.): ICIC 2018, LNCS 10955, pp. 809–815, 2018.
https://doi.org/10.1007/978-3-319-95933-7_89

Protein-protein binding affinity is typically estimated using experimental techniques such as Surface Plasmon resonance (SPR) and Isothermal titration calorimetry (ITC). However, complex and expensive experimental setup hinder their utilization in a high-throughput approach for binding affinity estimation for PPIs. Hence, several computational methods have been developed to predict the binding affinity of protein-protein complexes using features derived using structural [3] and sequence information [4]. However, structure-based methods have limitations to large-scale proteome-wide studies and the performance is moderate in sequence-based methods. Hence, there is ample scope for improvement and need for the development of new prediction methods that could efficiently capture the key descriptors and employ them inefficiently predicting binding affinities of PPIs with a diverse set of functions.

In this work, we systematically evaluated the potential of various important sequence-based features of PPIs in discriminating and estimating the binding affinity and devised a deep learning based binary classification method to discriminate the PPIs into high and low affinity interactions with an accuracy of 81.75% using 5-fold cross-validation. Furthermore, we refined PPA-Pred, our previously published method for binding affinity prediction with training it on more number of complexes. The refined method (PPA-Pred2) could predict binding affinity for PPIs with MAE 1.31 kcal/mol on a blind test set. We suggest that our models could be effectively utilized for large-scale classification and prediction of protein-protein binding affinities with potential applications in diverse areas of study.

2 Materials and Methods

2.1 Dataset

We collected and compiled a set of 741 binary protein-protein complexes with experimental binding affinity data from PDBbind database [5] and literature [6–10]. By employing PISCES method [11] to remove redundant complexes with sequence identity greater than 25%, we obtained a set of 464 non-redundant complexes.

2.2 Deep Learning Approach

Deep learning is a branch of machine learning, mainly based on a set of algorithms that attempt to model high-level abstraction in data by using model architectures. In deep learning method, one can stack enough layers that consist of several neurons to represent a complex relationship, by the composition of many non-linear functions. Hence, deep learning has the potential and capability to efficiently discriminate protein-protein complexes based on their binding affinities. We used Keras sequential model [12] with three hidden layers and 8 neurons in each layer. The model was trained on 500 epochs using Adam optimizer.

2.3 Feature Selection

For reducing redundancy in the feature set and to identify the most informative features for the binary classification problem, we employed a systematic approach that iteratively verifies the importance of each feature towards discriminating the complexes based on their binding affinities. We evaluated features by estimating their individual contribution for the model's performance by examining the drop in accuracy each time a feature was dropped from the feature set.

2.4 Validation Procedures

We used n-fold cross-validation approach for evaluating the performance of the generated binary classification models. In this approach, n − 1 data are utilized to develop a model and the remaining data are used to test the method. Additionally, the performance of the method has been assessed using the following measures:

$$\text{Accuracy} = (\text{TP} + \text{TN})/(\text{TP} + \text{TN} + \text{FP} + \text{FN}) \tag{1}$$

$$\text{Specificity} = \text{TN}/(\text{TN} + \text{FP}) \tag{2}$$

$$\text{Sensitivity} = \text{TP}/(\text{TP} + \text{FN}) \tag{3}$$

In the above equations, TP, TN, FP, and FN, represent, true positives, true negatives, false positives and false negatives, respectively. In addition, AUC (Area under the ROC curve) has been estimated to verify the correspondence between true positive rate and false positive rate.

We employed multiple validation techniques at various stages of model development to ensure the reliability and robustness of the method. We used Akaike Information Criterion (AIC) for evaluating the quality of the generated models as an initial filter. Further, we applied inner Leave-One-Out Cross-Validation (LOOCV) loop and an external blind test set, evaluated based on Pearson's Correlation Coefficient (PCC) and Mean Absolute Error (MAE). Apart from this systematic validation procedure, we compared the MAE obtained using our method with that of Average Assignment Method (AAM) [13] to ensure its reliable predictive ability. The statistical significance of PCC at various validation steps has been verified with the help of p-values.

3 Results and Discussion

3.1 Discrimination of High and Low Affinity Complexes

We have classified the dataset into training and test set with 400 and 64 complexes respectively. We selected a set 17 features that are important for discriminating protein-protein complexes based on their binding affinities. The combination of these selected features showed an accuracy of 81.75% with 73.68% sensitivity and 85.76% specificity in 5-fold cross-validation. These accuracy levels are better than other methods reported in the literature. The tradeoff between true positive rate and false positive rate for training and test data showed an AUC of 0.71 (Fig. 1).

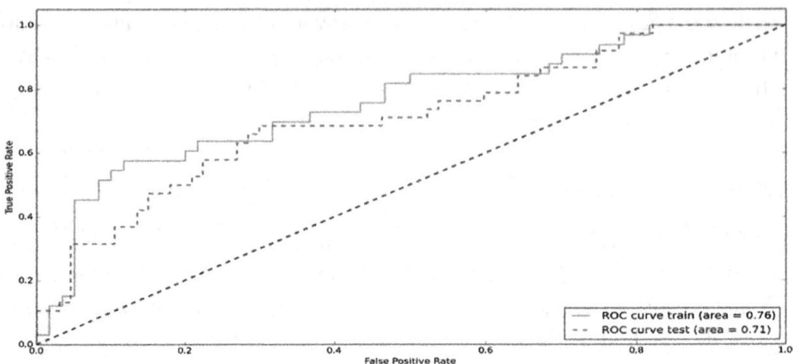

Fig. 1. Receiver operating characteristic (ROC) for training and test dataset

3.2 Prediction of Binding Affinity

From the initial non-redundant set of 464 complexes, we removed nine complexes that had an affinity in the form of range instead of absolute value. This resulted in a set of 453 non-redundant complexes along with their experimental binding affinity information. Further, these complexes were grouped into eight classes based on their biological functions viz. Antigen-Antibody (Ag-Ab) with 15 complexes, Enzyme-Inhibitor (EI) with 42 complexes, Other enzymes (OE) with 107complexes, G-Protein containing (GC) with 43 complexes, Receptor containing (RC) with 71 complexes along with Miscellaneous M1 (56 complexes), M2 (60 complexes) and M3 (59 complexes) classes with sub-grouped complexes, which does not belong to any of the aforementioned six functional classes based on the criteria described in our previous work [14]. As a next step, we randomly split the complexes in each class into training and test datasets. Furthermore, we derived a set of 271 important sequence-based features representing physicochemical, biophysical and thermodynamic properties of protein-protein interactions from AAindex database [15], predicted binding sites [16] and literature [17]. Additionally, we employed multiple validation techniques at various stages of multiple regression based feature selection and model development as described in Sect. 2.4. This enabled us to identify the most important and informative features ranging from three to fourteen across different functional classes. The prediction results are presented in Table 1.

With Closer examination of selected features, we observed that our approach identified several features that were previously reported to potentially play important roles in protein-protein interactions. Identified features across different functional classes include key features such as (i) the number of aromatic and positively charged residues in EI, which was previously shown to be vital in protein-protein complexes [18], (ii) sequence composition of the interacting proteins has been selected across most of the classes showing its importance determining binding principles of a functional

Table 1. Performance of PPA-Pred2 across different functional classes

Class	No. of features	No. of complexes		Training set LOOCV (Jack-knife test)			Test set		
		LOOCV	Test set	Correlation r-value	MAE kcal/mol	p-value	Correlation (r)	MAE (kcal/mol)	p-value
Ag-Ab	03	12	03	0.802	0.77	0.002	NA	0.89	NA
EI	05	32	10	0.705	1.30	7×10^{-6}	0.908	1.09	0.0003
OE	14	91	16	0.639	1.45	$<1 \times 10^{-7}$	0.737	1.28	0.001
GC	07	36	07	0.647	1.09	2×10^{-5}	0.861	1.34	0.013
RC	12	61	10	0.650	1.42	$<1 \times 10^{-7}$	0.714	1.42	0.020
M1	06	49	07	0.701	1.09	$<1 \times 10^{-7}$	0.777	1.31	0.040
M2	08	51	09	0.714	1.05	$<1 \times 10^{-7}$	0.800	1.20	0.010
M3	06	50	09	0.728	1.18	$<1 \times 10^{-7}$	0.825	1.58	0.006

Ag-Ab: Antigen-Antibody; EI: Enzyme-Inhibitor; OE: Other Enzymes; GC: G-protein containing; RC: Receptor containing; M: Miscellaneous.

protein, (iii) features representing propensity of secondary structural elements have been selected across multiple classes such as EI, OE, GC, M1and M3, thus reiterating their importance reported by previous studies [14, 19, 20], and (iv) identification of features accounting for thermodynamic properties for RC class is in agreement with the previous study by Maenaka et al. [21], which suggested that the binding affinity of specific complexes from RC class are attributed with thermodynamic properties such as enthalpy and entropy.

We observed that the correlation lies in the range of 0.639 to 0.802 using LOOCV and 0.714 to 0.908 using test datasets. The average MAE obtained for the eight functional classes by using LOOCV experiment and test set is 1.24 kcal/mol and 1.31 kcal/mol, respectively. The performance of the model on training and test datasets is shown in Fig. 2.

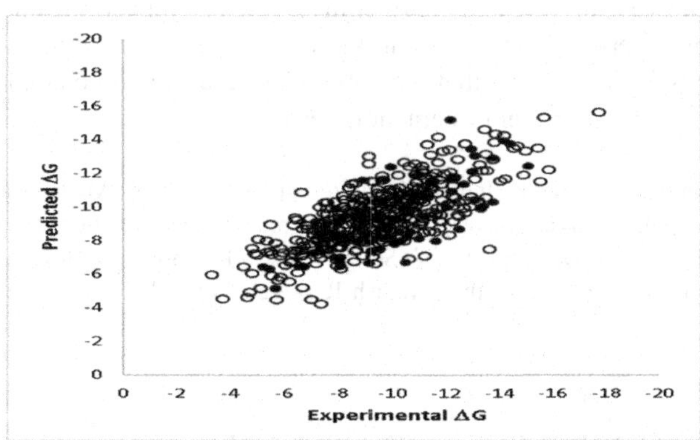

Fig. 2. Scatter plot for experimental and predicted ΔG. Training and test set data are shown in empty and filled circles, respectively

We compared these results with those obtained using average assignment method [13] to check predictive power of our method, which showed an MAE of 1.70 kcal/mol (r = 0.411), whereas MAE obtained using our method is 1.24 kcal/mol (r = 0.737). We have also compared the MAE in individual classes and observed that the current method performs significantly better than average assignment method (p-value: 0.001).

Furthermore, our method does not depend on the range of ΔG in each class. EI and RC have the ΔΔG of 7.65 and 4.44 kcal/mol, respectively and the MAEs are 1.09 kcal/mol and 1.42 kcal/mol. Although the number of complexes and the number of selected features in each class showed a linear relationship between them, this trend is not observed in all the classes. Typically, GC class with 36 complexes utilizes seven features whereas M1 class with 49 complexes utilizes only six features, with similar MAE.

We observed that 75% of all complexes used in the study are predicted within a deviation of 1.84 kcal/mol. Considering the remaining as outliers, we assume that factors such as (i) different binding poses during interaction and (ii) concentration and pH dependency of binding affinity [22] might be some of the limiting aspects, which could not be accounted by our method.

3.2.1 Web Server

The revised prediction method has been implemented as a web tool viz. PPA-Pred2, and it is freely accessible at https://www.iitm.ac.in/bioinfo/PPA_Pred/. It accepts amino acid sequences of the two interacting proteins in fasta format as input with additional option choose functional class if known, and return affinity result as ΔG and K_d values.

4 Conclusion

We developed a binary classification model for discriminating protein-protein complexes based on their binding affinities using deep learning approach, with 81.75% accuracy with 80% sensitivity and 86% specificity on the blind test set. Furthermore, we refined our previously published binding affinity prediction method (PPA-Pred) using more than two times the number of complexes. The refined method (PPA-Pred2) showed consistent performance across all the functional classes with an overall MAE 1.31 on the test set. We suggest that the current methods could be employed in large-scale analyses of protein-protein interaction networks.

Acknowledgements. We thank Indian Institute of Technology Madras and the High-Performance Computing Environment (HPCE) for computational facilities. The work was partially supported by the Department of Science and Technology, Government of India (DST/INT/SWD/P-05/2016) and of the Swedish Research Council.

References

1. Bahadur, R.P., Chakrabarti, P., Rodier, F., Janin, J.: A dissection of specific and non-specific protein-protein interfaces. J. Mol. Biol. **336**, 943–955 (2004)
2. Yugandhar, K., Gromiha, M.M.: Analysis of protein-protein interaction networks based on binding affinity. Curr. Prot. Pept. Sci. **17**, 72–81 (2016)

3. Moal, I.H., Agius, R., Bates, P.A.: Protein-protein binding affinity prediction on a diverse set of structures. Bioinformatics **27**, 3002–3009 (2011)
4. Yugandhar, K., Michael Gromiha, M.: Feature selection and classification of protein-protein complexes based on their binding affinities using machine learning approaches. Proteins **82**(9), 2088–2096 (2014)
5. Liu, Z., Li, Y., Han, L., Li, J., Liu, J., Zhao, Z., Nie, W., Liu, Y., Wang, R.: PDB-wide collection of binding data: current status of the PDBbind database. Bioinformatics **31**, 405–412 (2015)
6. Kastritis, P.L., Bonvin, A.M.J.J.: On the binding affinity of macromolecular interactions: daring to ask why proteins interact. J. R. Soc. Interface **10**, 20120835 (2013)
7. Chen, J., Sawyer, N., Regan, L.: Protein-protein interactions: general trends in the relationship between binding affinity and interfacial buried surface area. Protein Sci. **22**, 510–515 (2013)
8. Lomax, J.E., Christopher, M.B., Chang, A., George, N.P.J.: Functional evolution of ribonuclease inhibitor: insights from birds and reptiles. J. Mol. Biol. **426**, 3041–3056 (2014)
9. Spencer, A.L., Bagai, I., Becker, D.F., Zuiderweg, E.R., Ragsdale, S.W.: Protein-protein interactions in the mammalian heme degradation pathway heme oxygenase-2, cytochrome p450 reductase, and biliverdin reductase. J. Biol. Chem. **289**, 29836–29858 (2014)
10. Vreven, T., Iain, H.M., Anna, V., Brian, G.P.: Updates to the integrated protein-protein interaction benchmarks: docking benchmark version 5 and affinity benchmark version 2. J. Mol. Biol. **427**, 3031–3041 (2015)
11. Wang, G., Dunbrack, R.L.J.: PISCES: a protein sequence culling server. Bioinformatics **19**, 1589–1591 (2003)
12. Chollet, F.: "Keras" (2015)
13. Saraboji, K., Gromiha, M.M., Ponnuswamy, M.N.: Average assignment method for predicting the stability of protein mutants. Biopolymers **82**, 80–92 (2006)
14. Yugandhar, K., Michael Gromiha, M.: Protein-protein binding affinity prediction from the amino acid sequence. Bioinformatics **30**, 3583–3589 (2014)
15. Kawashima, S., Pokarowski, P., Pokarowska, M., Kolinski, A., Katayama, T., Kanehisa, M.: AAindex: amino acid index database, progress report 2008. Nucleic Acids Res. **36**, D202–D205 (2008)
16. Ofran, Y., Rost, B.: Interaction sites identified from sequence. Bioinformatics **23**, e13–e16 (2007)
17. Gromiha, M.M.: A statistical model for predicting protein folding rates from amino acid sequence with structural class information. J. Chem. Inf. Model. **45**, 494–501 (2005)
18. Gromiha, M.M., Saranya, N., Selvaraj, S., Jayaram, B., Fukui, K.: Sequence and structural features of binding site residues in protein-protein complexes: comparison with protein-nucleic acid complexes. Proteome Sci. **9**, S13 (2011)
19. Li, W., Hamill, S.J., Hemmings, A.M., Moore, G.R., James, R., Kleanthous, C.: Dual recognition and the role of specificity-determining residues in colicin E9 DNase-immunity protein interactions. Biochemistry **37**, 11771–11779 (1998)
20. Eathiraj, S., Pan, X., Ritacco, C., Lambright, D.G.: Structural basis of family-wide Rab GTPase recognition by rabenosyn-5. Nature **436**, 415–419 (2005)
21. Maenaka, K., van der Merwe, P.A., Stuart, D.I., Jones, E.Y., Sondermann, P.: The human low affinity Fc receptors IIa, IIb, and IIIbind IgG with fast kinetics and distinct thermodynamic properties. J. Biol. Chem. **276**, 44898–44904 (2001)
22. Blanco, M.A., Tatiana, P., Vincenzo, M., Mauro, M., Christopher, J.R.: Protein-protein interactions in dilute to concentrated solutions: α-Chymotrypsinogen in acidic conditions. J. Phys. Chem. B **118**, 5817–5831 (2014)

Principal Component Analysis-Based Unsupervised Feature Extraction Applied to Single-Cell Gene Expression Analysis

Y-h. Taguchi(✉)

Department of Physics, Chou University, Tokyo 112-8551, Japan
tag@granular.com

Abstract. Due to missed sample labeling, unsupervised feature selection during single-cell (sc) RNA-seq can identify critical genes under the experimental conditions considered. In this paper, we applied principal component analysis (PCA)-based unsupervised feature extraction (FE) to identify biologically relevant genes from mouse and human embryonic brain development expression profiles retrieved by scRNA-seq. When evaluating the biological relevance of selected genes by various enrichment analyses, the PCA-based unsupervised FE outperformed conventional unsupervised approaches that select highly variable genes as well as bimodal genes in addition to the recently proposed dpFeature.

Keywords: Principal component analysis · Feature selection
Embryonic brain development

1 Introduction

Single-cell analysis is a newly developed high-throughput technology that enables us to identify gene expression profiles of individual genes. There is a critical difference between single-cell analysis and conventional tissue-specific analysis; tissue samples are labeled distinctively (e.g., patients vs healthy controls) while single-cell samples are not always. Inevitably, we need an unsupervised methodology, such as highly variable genes [1, 2] and bimodal genes [3] or recently proposed dpFeature [4]. Highly variable genes are able to select genes that can discriminate the underlying cluster structure depicted by unsupervised clustering, namely tSNE [5]. In contrast, bimodal genes are selected because unimodal genes are unlikely to be expressed distinctly between multiple classes, e.g., healthy controls and patients. While the combination of tSNE and highly variable genes or bimodal genes approach is often employed and empirically successful, biological validation of selected genes is rarely addressed. Generally, very few studies have evaluated multiple gene selection procedures for single-cell RNA-seq. The purpose of this paper is to compare multiple gene selection procedures and to identify the best method.

Principal component analysis (PCA)-based unsupervised feature extraction (FE) has been previously shown to be an effective method to investigate tissue-specific gene expression profiles [6–27]. In this paper, PCA-based unsupervised FE was applied to single-cell gene expression analysis. In addition, its effectiveness was evaluated from

D.-S. Huang et al. (Eds.): ICIC 2018, LNCS 10955, pp. 816–826, 2018.
https://doi.org/10.1007/978-3-319-95933-7_90

the biological point of view and compared to conventional approaches, including the highly variable genes approach as well as the bimodal genes approach.

2 Materials and Methods

2.1 Gene Expression

Gene expression profiles used in this study were downloaded from the Gene Expression Omnibus (GEO) database under the GEO ID GSE76381. Specifically, the files named "GSE76381_EmbryoMoleculeCounts.cef.txt.gz" (for human) and "GSE76381_MouseEmbryoMoleculeCounts.cef.txt.gz" (for mouse) were downloaded. Gene expression profiles were standardized such that each sample had zero mean and unit standard deviation. That is, when x_{ij} represented the expression of ith gene in jth sample, $\sum_i x_{ij} = 0$ and $\sum_i x_{ij}^2 = N$ where N represented the number of genes. These two gene expression profiles were generated from single-cell RNA-seq datasets that represented the following: human embryo ventral midbrain cells between 6 and 11 weeks of gestation, mouse ventral midbrain cells at six developmental stages between E11.5 to E18.5, Th+ neurons at P19–P27, and FACS-sorted putative dopaminergic neurons at P28–P56 from Slc6a3-Cre/tdTomato mice.

2.2 PCA-Based Unsupervised FE

Suppose that matrix X has element x_{ij} representing the gene expression of ith gene of jth sample, then the kth PC score attributed to ith gene u_{ki} can be computed as ith element of kth eigen vector of Gram matrix XX^T as

$$XX^T \boldsymbol{u}_k = \lambda_k \boldsymbol{u}_k$$

kth PC loading attributed to jth sample, v_{kj}, can be obtained by $\boldsymbol{v}_k = X^T \boldsymbol{u}_k$ since

$$X^T X \boldsymbol{v}_k = X^T X X^T \boldsymbol{u}_k = X^T \lambda_k \boldsymbol{u}_k = \lambda_k \boldsymbol{v}_k$$

Initially, PC loading attributed to samples were identified, which were coincident with distinction between considered class labels attributed to the samples. Since single-cell RNA-seq (scRNA-seq) lacks sample labeling, the first k PCs were employed. Subsequently, assuming multiple Gaussian distribution to PC scores, P-values were attributed to gene i using χ^2 distribution,

$$P_i = P_{\chi^2}\left[> \sum_{k=1}^{K} \left(\frac{u_{ki}}{\sigma_k} \right)^2 \right]$$

where σ_k represented standard deviation and $P_{\chi^2}[>x]$ represented the cumulative probability of χ^2 distribution that the argument was larger than x. The summation was taken over the selected first K PC scores. Obtained P-values were adjusted using

Benjamini and Hochberg (BH) criterion [28] and genes associated with adjusted *P*-values less than 0.01 were selected.

2.3 Enrichment Analysis

In order to perform enrichment analyses, selected genes were uploaded to Enrichr [29], which included various enrichment analysis.

2.4 Highly Variable Genes

The procedure was performed as previously described [1], and a brief description is provided below. Suppose that x_{ij} represented gene expression of *i*th gene of *j*th sample, then the mean expression of *i*th gene was defined as

$$\mu_i = \frac{\sum_i x_{ij}}{M}$$

where M represented the number of samples. The standard deviation σ_i was defined as

$$\sigma_i^2 = \sum_i \frac{\left(x_{ij} - \mu_i\right)^2}{M}$$

Following, the regression relation between σ_i and μ_i was assumed as

$$\log_{10}\left(\frac{\sigma_i}{\mu_i}\right) = \frac{1}{2}\log_{10}\left(\frac{\beta}{\mu_i} + \alpha\right) + \epsilon_i$$

where α and β were the regression coefficients. Subsequently, *P*-value P_i' was attributed to ϵ_i assuming χ^2 distribution as

$$P_i' = P_{\chi^2}\left[> \left(\frac{\epsilon_i}{\sigma'}\right)^2\right]$$

where σ' was the standard deviation. Finally, genes *i* with an adjusted *P*-value less than 0.01 by BH criterion were selected as highly variable genes.

2.5 Bimodal Genes

P-values attributed to genes that reject the null hypothesis (unimodal genes) were computed by the dip.test function in R using the default setting. Obtained *P*-values were adjusted by BH criterion, and genes associated with adjusted *P*-values less than 0.01 were selected.

2.6 dpFeature

dpFeature was performed using monocle package in R. More detailed instruction was in Supplementary Document.

3 Results

Applying PCA-based unsupervised FE to human and mouse embryonic brain developmental gene expression profiles, 116 genes for human ($K = 2$, i.e., the first two PC scores were used for gene selection) and 118 genes for mouse ($K = 3$, i.e., the first three PC scores were used for gene selection) were selected, respectively. Interestingly, 53 genes of the selected genes were common both in human and mouse samples. The large overlap between the two genes sets with highly restricted numbers of genes is not plausible to occur by chance; therefore, it is very likely that these selected genes play critical roles in embryonic midbrain development.

To validate the biological relevance of the selected genes, various enrichment analyses were applied using Enrichr. Table 1 shows an Enrichment analysis by Enrichr, "MGI Mammalian Phenotype 2017", of the 118 genes selected in mice. Among the top five ranked terms, four were brain-related terms. As all terms described abnormal morphology, this is an expected result since fetal gene expression is often distinct from adults that lack fetal-specific gene expression. Tables 2 and 3 show another Enrichment analysis by Enrichr, "Allen Brain Atlas down", of the 116 genes selected in humans and 118 genes selected in mice, respectively. All genes were downregulated in brain regions. This is again reasonable since fetal genes expressed in the embryo is unlikely to be expressed in adult tissue. The lack of fetal brain-specific gene expression in the selected genes can also be seen in Tables 4 and 5.

Table 1. Enrichment analysis by Enrichr, "MGI Mammalian Phenotype 2017", of 118 selected genes in mice (Top 5 ranked terms)

Term	Overlap	*P*-value	Adjusted *P*-value
MP:0000788_abnormal_cerebral_cortex_morphology	7/145	2.45×10^{-5}	4.55×10^{-5}
MP:0003651_abnormal_axon_extension	5/48	9.18×10^{-6}	4.55×10^{-3}
MP:0000812_abnormal_dentate_gyrus_morphology	5/58	2.34×10^{-5}	4.55×10^{-3}
MP:0000807_abnormal_hippocampus_morphology	5/86	1.56×10^{-4}	2.04×10^{-2}
MP:0000819_abnormal_olfactory_bulb_morphology	4/48	1.83×10^{-4}	2.04×10^{-2}

While these above results are highly significant, they are also negative results. As such, we speculated whether these results could provide enough support and confidence for the selected genes. Therefore, in order to show positive results, gene expression in the embryonic brain was assessed. Subsequently, it was found that these genes were enriched in "Jensen TISSUES" by Enrichr. Table 6 shows that selected genes are enriched in the embryonic brains of humans and mice, respectively. Thus, confidence of the selected genes is supported by both negative and positive selection.

Table 2. Enrichment analysis by Enrichr, "Allen Brain Atlas down", of 116 selected genes in humans (Top 5 ranked terms)

Term	Overlap	*P*-value	Adjusted *P*-value
Periventricular stratum of cerebellar vermis	18/300	1.33×10^{-13}	3.72×10^{-11}
Simple lobule	18/300	1.33×10^{-13}	3.72×10^{-11}
Simple lobule, molecular layer	18/300	1.33×10^{-13}	3.72×10^{-11}
Simple lobule, granular layer	18/300	1.33×10^{-13}	3.72×10^{-11}
White matter of cerebellar vermis	18/300	1.33×10^{-13}	3.72×10^{-11}

Table 3. Enrichment analysis by Enrichr, "Allen Brain Atlas down", of 118 selected genes in mice (Top 5 ranked terms)

Term	Overlap	*P*-value	Adjusted *P*-value
Pyramus (VIII), granular layer	18/300	1.81×10^{-13}	4.66×10^{-11}
Pyramus (VIII)	18/300	1.81×10^{-13}	4.66×10^{-11}
Pyramus (VIII), molecular layer	18/300	1.81×10^{-13}	4.66×10^{-11}
Paraflocculus, molecular layer	18/300	1.81×10^{-13}	4.66×10^{-11}
Cerebellar cortex	18/300	1.81×10^{-13}	4.66×10^{-11}

Table 4. Enrichment analysis by Enrichr, "GTEx Tissue Sample Gene Expression Profiles down", of 116 selected genes in humans (Top 5 ranked terms)

Term	Overlap	*P*-value	Adjusted *P*-value
GTEX-Q2AG-0011-R10A-SM-2HMLA_brain_female_40–49_years	51/1467	1.47×10^{-27}	3.29×10^{-24}
GTEX-TSE9-3026-SM-3DB76_brain_female_60–69_years	49/1384	1.06×10^{-26}	1.19×10^{-23}
GTEX-S7SE-0011-R10A-SM-2XCDF_brain_male_50–59_years	44/1278	3.20×10^{-23}	1.43×10^{-20}
GTEX-QMR6-1426-SM-32PLA_brain_male_50–59_years	41/1066	2.57×10^{-23}	1.43×10^{-20}
GTEX-RNOR-2326-SM-2TF4I_brain_female_50–59_years	47/1484	2.02×10^{-23}	1.43×10^{-20}

We further sought the regulatory elements that can regulate the selected genes, as there are likely common regulatory elements if the selected genes are truly co-expressed. In order to perform this, we investigated "ENCODE and ChEA Consensus TFs from ChIP-X" by Enrichr for both human and mouse, respectively. Specifically, 42 TFs for humans and 23 TFs for mice were associated with adjusted *P*-values less than 0.01 (Table 7). Thus, they are likely co-regulated by these TFs. Moreover, most mouse TFs were also identified in humans (bold faces in Table 7). Therefore, it is very likely that we successfully identified common (species non-specific or conserved) TFs that regulate genes expression during embryonic brain development.

Table 5. Enrichment analysis by Enrichr, "GTEx Tissue Sample Gene Expression Profiles down", of 118 selected genes in mice (Top 5 ranked terms)

Term	Overlap	*P*-value	Adjusted *P*-value
GTEX-U8XE-0126-SM-4E3I3_testis_male_30–39_years	15/376	6.13×10^{-9}	3.45×10^{-6}
GTEX-X4XX-0011-R10B-SM-46MWO_brain_male_60–69_years	23/938	5.25×10^{-9}	3.45×10^{-6}
GTEX-U4B1-1526-SM-4DXSL_testis_male_40–49_years	13/282	1.23×10^{-8}	3.71×10^{-6}
GTEX-Q2AG-0011-R10A-SM-2HMLA_brain_female_40–49_years	29/1467	5.11×10^{-9}	3.45×10^{-6}
GTEX-RNOR-2326-SM-2TF4I_brain_female_50–59_years	29/1484	6.62×10^{-9}	3.45×10^{-6}

Table 6. Selected gene enrichment in the embryonic brain of "Jensen TISSUES" by Enrichr

Term	Overlap	P-value	Adjusted P-value
Human			
Embryonic_brain	71/4936	2.52×10^{-16}	4.07×10^{-15}
Mouse			
Embryonic_brain	75/4936	8.90×10^{-20}	1.06×10^{-18}

Table 7. TF enrichment in "ENCODE and ChEA Consensus TFs from ChIP-X" by Enrichr for human and mouse (Bold TFs are common)

Human	**ATF2**, **BCL3**, BCLAF1, BHLHE40, **BRCA1**, **CEBPB**, **CEBPD**, **CHD1**, **CREB1**, CTCF, **E2F1**, E2F4, **EGR1**, ELF1, ETS1, FLI1, GABPA, **KAT2A**, KLF4, **MAX**, **MYC**, NANOG, **NELFE**, NFYA, NFYB, NR2C2, **PBX3**, **PML**, **RELA**, SALL4, **SIN3A**, SIX5, SOX2, SP1, SPI1, **TAF1**, **TAF7**, **TCF3**, USF2, **YY1**, ZBTB33, **ZMIZ1**
Mouse	**ATF2**, **BCL3**, **BRCA1**, **CEBPB**, **CEBPD**, **CHD1**, **CREB1**, **E2F1**, **EGR1**, **KAT2A**, KLF, **MAX**, **MYC**, **NELFE**, **PBX3**, **PML**, **RELA**, **SIN3A**, **TAF1**, **TAF7**, **TCF3**, **YY1**, **ZMIZ1**

Additionally, we assessed whether the TF constructs functioned cooperatively. These TFs were uploaded to regnetwork server [30], and TF networks were identified as shown in Fig. 1. It is evident, even partially, that these TFs interact with each other.

We also investigated whether the identified TFs were related to fetal brain development. TAF7 was reported to play a critical role in embryonic development [31]. KAT2A, ATF2 and TAF1 were also suggested to be included in brain development [32]. BRCA1 has been shown to play critical roles in brain development [33]. Additionally, CEBPD and CREB were reported to be related to brain disease [34, 35]. E2F1 was reported to be related to postnatal brain development [36] while functional EGR1 was found in the embryonic rat brain [37]. PML and SIN3A were also reported to be

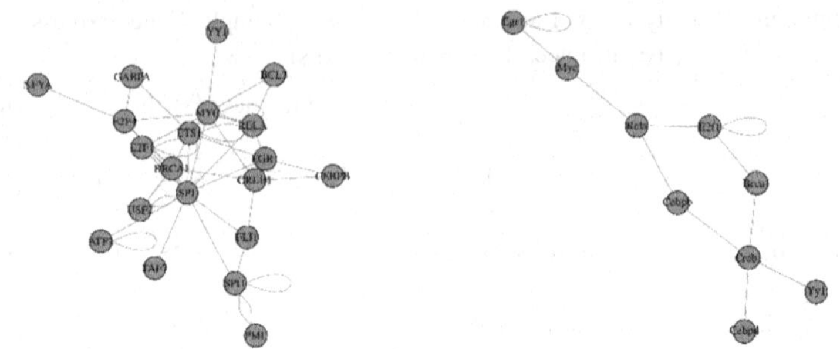

Fig. 1. TF network identified by regnetworkweb for TFs in Table 7 (Left: human, right: mouse)

involved in brain development [38, 39]. TCF3 has been shown to play a role in zebrafish brain development [40]. YY1 was also reported in brain development [41]. In conclusion, most of the selected genes in common (bold faces in Table 7) between humans and mice are related to brain development. Thus, the selection of genes is possibly reasonable.

In addition to the unconventional PCA-based unsupervised FE, we applied the widely used highly variable genes approach to the present data set and determined which strategy was more consistent with the enrichment analysis [1]. After applying the highly variable genes strategy, we obtained 168 genes for human and 171 genes for mouse. The numbers of selected genes were similar to those selected by PCA-based unsupervised FE. Additionally, there were 44 commonly selected genes between human and mouse. Thus, the highly variable genes method shows some efficacy.

However, after a detailed investigation, there were very few overlaps between genes selected by PCA-based unsupervised FE and highly variable genes (only four genes were commonly selected by PCA-based unsupervised FE and highly variable genes for both human and mouse). This result did not provide much confidence since genes selected by PCA-based unsupervised FE were demonstrated to be biologically reliable. However, there was still a possibility that highly variable genes were coincident to the enrichment analysis without significant overlap with genes selected by PCA-based unsupervised FE.

For confirmation, genes were uploaded to Enrichr. Accordingly, the results were substantially distinct from that given by PCA-based unsupervised FE. Specifically, the top five ranked enriched terms in “MGI Mammalian Phenotype 2017” for mouse did not include anything related to the brain, which is inferior to the results in Table 1. In contrast, no biological terms were significantly enriched in “Allen Brain Atlas down” for human, yet substantially large numbers of terms were enriched in mouse. This is inconsistent since the highly variable genes were substantially common between mouse and human; therefore, the inconsistency between human and mouse suggests that the selection of highly variable genes might be abiological. Subsequently, the “GTEx Tissue Sample Gene Expression Profiles down” was considered, and the top five ranked terms did not include anything related to brain, instead relations to skin and blood were found. This suggests that highly variable genes are not likely more biologically reliable than genes selected by PCA-based unsupervised FE. Finally, “Jensen

TISSUES" was considered as was completed above (Table 6). Similarly, no brain-related terms were found to be significantly enriched in highly variable genes. TFs were also investigated to determine whether they could regulate highly variable genes. Nevertheless, "ENCODE and ChEA Consensus TFs from ChIP-X" only identified one TF whose target genes were significantly enriched in highly variable genes in human or mouse. As such, one TF identified is significantly less than that in Table 7. All of these suggest that highly variable genes are unlikely to be biologically more reliable than genes selected by PCA-based unsupervised FE.

In addition, we investigated bimodal genes as an alternative strategy (see Supplementary document). Genes associated with adjusted *P*-values less than 0.01 were 11344 and 10849 for human and mouse, respectively. Due to the large volume, it suggests that the bimodal genes approach has no ability to select a reasonable (restricted) number of genes. Nevertheless, in order to evaluate bimodal genes further, 200 top ranked (i.e., associated with smaller *P*-values) were intentionally selected. Specifically, of the 200 selected genes, only 21 genes were commonly selected between human and mouse, as compared to 53 and 44 commonly selected genes between human and mouse by PCA-based unsupervised FE and highly variable genes, respectively. Furthermore, there were no commonly selected genes between bimodal genes and PCA-based unsupervised FE. Enrichment analyses by Enricher for bimodal genes were also inferior to that of PCA-based unsupervised FE. "MGI Mammalian Phenotype 2017" of 200 bimodal genes selected for mouse included no terms associated with adjusted *P*-values less than 0.01 (Table S1). Top five ranked terms by "Allen Brain Atlas down" (Tables S2 and S3) were less significant than PCA-based unsupervised FE (Tables 2 and 3) as *P*-values were generally larger for bimodal genes. In addition, top five ranked terms by "GTEx Tissue Sample Gene Expression Profiles down" included no brain-related terms (Tables S4 and S5). Taken together, all of these suggests that bimodal genes are unlikely to be biologically more reliable than genes selected by PCA-based unsupervised FE.

Nevertheless, bimodal genes are slightly better than highly variable genes. Specifically, "Jensen TISSUES" by Enrichr included Embryonic_brain as a significant term (Table S6). In addition, "GTEx Tissue Sample Gene Expression Profiles down" identified 40 TFs associated with adjusted P-values less than 0.01 for human and mouse, among which as many as 30 TFs were commonly selected (Table S7). These TFs were also highly connected in the regnetworkweb (Figure S1).

Interestingly, among the 30 commonly selected TFs between human and mouse in Table S7, 11 TFs were also commonly selected between human and mouse in Table 7. When considering that no genes were commonly selected between top ranked 200 bimodal genes and genes selected by PCA-based unsupervised FE, the high number of commonly selected TFs between PCA-based unsupervised FE and bimodal genes suggests the robustness of TFs selected. Thus, as an overall evaluation PCA-based unsupervised FE is better than the other two.

Although we have also compared with the newly proposed approach, dpFeature, because of lack of space, it was included in Supplementary document. dpFeature could not select biologically more reliable genes than PCA based unsupervised FE, either.

While PCA is not a new technology and highly variable genes and bimodal genes are also not new concepts, the application to scRNA-seq is innovative and valuable.

Therefore, even if methods themselves applied are not new, their application to new technology, scRNA-seq, can be innovative, especially if they have never been applied to scRNA-seq or were successful. Especially if PCA is more successful than even newly proposed approach, e.g., dpFeature.

4 Conclusions

We applied PCA-based unsupervised FE to gene expression profiles retrieved by scRNA-seq analysis. Since scRNA-seq primarily lacks the labeling of sample (each cell), an unsupervised approach is necessary. Genes selected by PCA-based unsupervised FE for human and mouse embryonic brain development were not only associated with numerous significant biological terms enrichment but also highly coincident between mice and humans. In contrast, the frequently employed highly variable genes approach as well as the bimodal genes approach or recently proposed dpFeature could not identify as many genes associated with significant biological terms enrichment as the PCA-based unsupervised FE achieved. Thus, PCA-based unsupervised FE is more favorable than the highly variable genes approach, the bimodal genes approach or dpFeature from the biological point of view.

5 Supplementary Materials

Supplementary materials are available at https://github.com/tagtag/SC.

Full list of genes, enrichment analyses for genes selected by PCA-based unsupervised FE, bimodal genes and dpFeature, supplementary document that includes supplementary tables and figure for bimodal gene analyses and dpFeature, R codes that identify genes selected by PCA-based unsupervised FE, highly variable genes, bimodal genes and dpFeature.

References

1. Chen, H.-I.H., Jin, Y., Huang, Y., Chen, Y.: Detection of high variability in gene expression from single-cell RNA-seq profiling. BMC Genom. **17**, 508 (2016)
2. Costa-Silva, J., Domingues, D., Lopes, F.M.: RNA-Seq differential expression analysis: an extended review and a software tool. PLoS one **12**(12), e0190152 (2017)
3. DeTomaso, D., Yosef, N.: FastProject: A tool for low-dimensional analysis of single-cell RNA-Seq data. BMC Bioinform. **17**, 315 (2016)
4. Qiu, X., Mao, Q., Tang, Y., Wang, L., Chawla, R., Pliner, H.A., Trapnell, C.: Reversed graph embedding resolves complex single-cell trajectories. Nat. Methods **14**, 979–982 (2017)
5. Van Der Maaten, L., G, H.: Visualizing Data using t-SNE. J. Mach. Learn. Res. **1**(620), 267–284 (2008)
6. Ishida, S., Umeyama, H., Iwadate, M., Taguchi, Y.H.: Bioinformatic screening of autoimmune disease genes and protein structure prediction with FAMS for drug discovery. Protein Pept. Lett. **21**, 828–839 (2014)

7. Taguchi, Y.-H.: microRNA-mRNA interaction identification in Wilms tumor using principal component analysis based unsupervised feature extraction. In: 2016 IEEE 16th International Conference on Bioinformatics and Bioengineering (BIBE), pp. 71–78 (2016)
8. Murakami, Y., et al.: Comprehensive analysis of transcriptome and metabolome analysis in Intrahepatic Cholangiocarcinoma and Hepatocellular Carcinoma. Sci. Rep. **5**, 16294 (2015)
9. Taguchi, Y.-H.: Identification of More Feasible MicroRNA-mRNA Interactions within Multiple Cancers Using Principal Component Analysis Based Unsupervised Feature Extraction. Int. J. Mol. Sci. **17**, 696 (2016)
10. Murakami, Y., Toyoda, H., Tanahashi, T., Tanaka, J., Kumada, T., Yoshioka, Y., Kosaka, N., Ochiya, T., Taguchi, Y.h: Comprehensive miRNA expression analysis in peripheral blood can diagnose liver disease. PLoS one **7**, e48366 (2012)
11. Taguchi, Y.-H.: Identification of candidate drugs using tensor-decomposition-based unsupervised feature extraction in integrated analysis of gene expression between diseases and DrugMatrix datasets. Sci. Rep. **7**, 13733 (2017)
12. Tamori, A. et al.: MicroRNA expression in hepatocellular carcinoma after the eradication of chronic hepatitis virus C infection using interferon therapy. Hepatol. Res. 46 (2016)
13. Taguchi, Y.-H., Iwadate, M., Umeyama, H., Murakami, Y.: Principal component analysis based unsupervised feature extraction applied to bioinformatics analysis. Comput. Methods Appl. Bioinforma. Anal. 153–182 (2017)
14. Taguchi, Y.H.: Principal components analysis based unsupervised feature extraction applied to gene expression analysis of blood from dengue haemorrhagic fever patients. Sci. Rep. **7**, 44016 (2017)
15. Taguchi, Y.-H., Wang, H.: Exploring microRNA biomarker for amyotrophic lateral sclerosis. Int. J. Mol. Sci. **19**, 1318 (2018)
16. Taguchi, Y.-H.: Principal component analysis based unsupervised feature extraction applied to publicly available gene expression profiles provides new insights into the mechanisms of action of histone deacetylase inhibitors. Neuroepigenetics **8**, 1–18 (2016)
17. Taguchi, Y.-H., Murakami, Y.: Universal disease biomarker: can a fixed set of blood microRNAs diagnose multiple diseases? BMC Res. Notes. **7**, 581 (2014)
18. Taguchi, Y.-H.: Principal component analysis based unsupervised feature extraction applied to budding yeast temporally periodic gene expression. BioData Min. **9**, 22 (2016)
19. Umeyama, H., Iwadate, M., Taguchi, Y.-H.: TINAGL1 and B3GALNT1 are potential therapy target genes to suppress metastasis in non-small cell lung cancer. BMC Genom. **15**, S2 (2014)
20. Taguchi, Y.H., Murakami, Y.: Principal component analysis based feature extraction approach to identify circulating microRNA biomarkers. PLoS one **8**, e66714 (2013)
21. Taguchi, Y.-H., Wang, H.: Genetic association between amyotrophic lateral sclerosis and cancer. Genes (Basel) **8**, 243 (2017)
22. Taguchi, Y.-H., Iwadate, M., Umeyama, H.: SFRP1 is a possible candidate for epigenetic therapy in non-small cell lung cancer. BMC Med. Genomics **9**, 28 (2016)
23. Taguchi, Y.-H., Iwadate, M., Umeyama, H.: Principal component analysis-based unsupervised feature extraction applied to in silico drug discovery for posttraumatic stress disorder-mediated heart disease. BMC Bioinform. **16**, 139 (2015)
24. Taguchi, Y.-H., Iwadate, M., Umeyama, H.: Heuristic principal component analysis-based unsupervised feature extraction and its application to gene expression analysis of amyotrophic lateral sclerosis data sets. In: IEEE Conference on Computational Intelligence in Bioinformatics and Computational Biology (CIBCB), pp. 1–10 (2015)

25. Taguchi, Y-H., Umeyama, H., Iwadate, M., Murakami, Y., Okamoto, A.: Heuristic principal component analysis-based unsupervised feature extraction and its application to bioinformatics. In: Wang, B., Li, R., Perrizo, W. (eds.): Big Data Analytics in Bioinformatics and Healthcare, pp. 138–162. IGI global (2015)
26. Murakami, Y., Tanahashi, T., Okada, R., Toyoda, H., Kumada, T., Enomoto, M., Tamori, A., Kawada, N., Taguchi, Y.H., Azuma, T.: Comparison of hepatocellular carcinoma miRNA expression profiling as evaluated by next generation sequencing and microarray. PLoS one **9**, e106314 (2014)
27. Taguchi, Y.-H.: Integrative analysis of gene expression and promoter methylation during reprogramming of a non-small-cell lung cancer cell line using principal component analysis-based unsupervised feature extraction. In: ICIC 2014, pp. 445–455 (2014)
28. Benjamini, Y., Hochberg, Y.: Controlling the false discovery rate: a practical and powerful approach to multiple testing. J. R. Stat. Soc. **B57**, 289–300 (1995)
29. Kuleshov, M.V., Jones, M.R., Rouillard, A.D., Fernandez, N.F., Duan, Q., Wang, Z., Koplev, S., Jenkins, S.L., Jagodnik, K.M., Lachmann, A., McDermott, M.G., Monteiro, C. D., Gundersen, G.W., Ma'ayan, A.: Enrichr: a comprehensive gene set enrichment analysis web server 2016 update. Nucleic Acids Res. **44**, W90–W97 (2016)
30. Liu, Z.-P., Wu, C., Miao, H., Wu, H.: RegNetwork: an integrated database of transcriptional and post-transcriptional regulatory networks in human and mouse. Database 2015 (2015). bav095
31. Gegonne, A., et al.: The general transcription factor TAF7 is essential for embryonic development but not essential for the survival or differentiation of mature T cells. Mol. Cell. Biol. **32**, 1984–1997 (2012)
32. Tapias, A., Wang, Z.Q.: Lysine acetylation and deacetylation in brain development and neuropathies. Genomics, Proteomics Bioinform. **15**, 19–36 (2017)
33. Pao, G.M., Zhu, Q., Perez-Garcia, C.G., Chou, S.-J., Suh, H., Gage, F.H., O'Leary, D.D.M., Verma, I.M.: Role of BRCA1 in brain development. Proc. Natl. Acad. Sci. **111**, E1240–E1248 (2014)
34. Sun, Y., et al.: Temporal gene expression profiling reveals CEBPD as a candidate regulator of brain disease in prosaposin deficient mice. BMC Neurosci. **9**, 1–20 (2008)
35. Mantamadiotis, T., et al.: Disruption of CREB function in brain leads to neurodegeneration. Nat. Genet. **31**, 47–54 (2002)
36. Suzuki, D.E., Ariza, C.B., Porcionatto, M.A., Okamoto, O.K.: Upregulation of E2F1 in cerebellar neuroprogenitor cells and cell cycle arrest during postnatal brain development. Vitr. Cell. Dev. Biol. - Anim. **47**, 492–499 (2011)
37. Wells, T., Rough, K., Carter, D.A.: Transcription mapping of embryonic rat brain reveals EGR-1 induction in SOX2 + neural progenitor cells. Front. Mol. Neurosci. **4**, 1–12 (2011)
38. Korb, E., Finkbeiner, S.: PML in the brain: from development to degeneration. Front. Oncol. **3**, 1–5 (2013)
39. Witteveen, J.S., et al.: Haploinsufficiency of MeCP2-interacting transcriptional co-repressor SIN3A causes mild intellectual disability by affecting the development of cortical integrity. Nat. Genet. **48**, 877–887 (2016)
40. Dorsky, R.I.: Two tcf3 genes cooperate to pattern the zebrafish brain. Development **130**, 1937–1947 (2003)
41. Beagan, J.A., et al.: YY1 and CTCF orchestrate a 3D chromatin looping switch during early neural lineage commitment. Genome Res. **27**, 1139–1152 (2017)

Finding Protein-Binding Nucleic Acid Sequences Using a Long Short-Term Memory Neural Network

Jinho Im, Byungkyu Park, and Kyungsook Han(✉)

Department of Computer Engineering, Inha University, Incheon, South Korea
22162012@inha.edu, {bpark, khan}@inha.ac.kr

Abstract. With an increasing amount of data of protein-nucleic acid interactions, several machine learning-based methods have been developed to predict protein-nucleic acid interactions. However, most of these methods are classification models either for finding binding sites within a sequence or for determining whether a pair of sequences interacts. In this paper we propose a generative model for constructing nucleic acids binding to a target protein using a long short-term memory (LSTM) neural network. Nucleic acid sequences generated by the model showed high affinity for several target proteins. The generative model will be useful for constructing an initial library of nucleic acid sequences for *in vitro* selection of nucleic acid sequences that bind to a target protein with high affinity and specificity.

Keywords: Neural network · LSTM · Protein-binding nucleic acids

1 Introduction

Motivated by a huge amount of interaction data generated by high-throughput technologies, several learning-based computational methods have been developed to predict protein-nucleic acid interactions. Most of these methods are classifiers for finding binding sites in a sequence [1–3] or for determining whether a pair of sequences interacts or not [4]. DeepBind [4], for instance, contains hundreds of convolutional neural network (CNN) models, each for different target proteins, to provide a predictive binding score between protein and nucleic acid sequences. A more recent model known as DeeperBind [5] predicts the protein-binding specificity of DNA sequences using a long short-term recurrent convolutional network. However, both DeepBind and DeeperBind are classification models rather generative models, so cannot be used to construct nucleic acid sequences that potentially bind to a target protein.

In this paper we propose a generative model for constructing single-stranded nucleic acids binding to a target protein using a long short-term memory (LSTM) neural network. The preliminary results showed that the model can generate nucleic acid sequences that bind to a target protein with high affinity and specificity. To the best of our knowledge, this is the first attempt to generate protein-binding nucleic acid sequences using a recurrent neural network.

D.-S. Huang et al. (Eds.): ICIC 2018, LNCS 10955, pp. 827–830, 2018.
https://doi.org/10.1007/978-3-319-95933-7_91

2 Materials and Methods

2.1 Sequence Generator

A sequence generator model was implemented using char-rnn (https://github.com/karpathy/char-rnn), which is a long short-term memory (LSTM) neural network for text data. The generator model is composed of two layers of LSTM with 128 hidden neurons in each layer. During training, the generator reads one nucleotide at a time from a given nucleic acid sequence and predicts the next nucleotide in the sequence. After reading a given sequence, the hidden neurons are updated based on the loss function to reflect prediction accuracy.

We use two representations, matrix representation and vector representation, to encode a nucleic acid sequence N of length l. In the matrix representation, the sequence N is represented by a $l \times 4$ matrix x, in which the t-th row x_t represents the t-th nucleotide in N. Only one element of x_t is 1 and the other elements are all zeros. If t-th nucleotide is A, the first element of the t-th row is 1, and the remaining three elements are 0. In the vector representation, a nucleic acid sequence N of length l is encoded into a vector of length l, which is constructed by mapping A in the sequence to 1, C to 2, G to 3, and T (or U) to 4.

Given the matrix representation of a nucleic acid sequence, LSTM calculates z_t, and softmax transforms z_t to a vector which can be interpreted as a probability. Each value of the vector is between 0 and 1, and the sum of the values is 1. In our model, the loss is defined by the mean of the negative log-likelihood of the prediction (loss $= -\sum_{t=1}^{|x|} ln \left(\text{softmax}_{y_t}(z_t)\right)/|x|$). One nucleotide is sampled from the multinomial distribution of the nucleotides and the vector representation of the nucleotide is fed back to the model as the next x_t. The generator repeats this process until it generates a sequence of pre-determined length.

For protein-binding DNA sequences, the model was trained on a set of DNA sequences obtained by HT-SELEX experiments as binding sequences to proteins [6]. Since the training set consists of DNA sequences of 20 nucleotides, the length of DNA sequences generated by the model was also set to 20 nucleotides in this study.

2.2 Binding Affinity of the Generated Sequences

To evaluate the binding affinity of generated sequences to their target protein, we defined the binding affinity using the predictive binding score of DeepBind [4] (hereafter called DeepBind score). The binding affinity AF in Eq. 1 is the normalized DeepBind score. Since DeepBind scores are arbitrary in scale, it is not possible to directly compare the binding scores from different DeepBind models. Thus, we normalized the DeepBind scores to compare the scores from different DeepBind models.

To define the binding affinity function AF_p for protein p, we computed the background distribution of the scores of DeepBind model m by running the model on 200,000 random nucleic acid sequences of 20 nucleotides. We also derived an empirical cumulative distribution function F_m from the background distribution. The F_m is a step function that increases by $1/n$ at each score of the background distribution.

In this study, n = 200,000 since we used 200,000 random sequences. $F_m(s)$ gives the portion of the scores smaller than or equal to $score_m(s)$. The binding affinity AF is always between 0 and 1.

$$\begin{array}{c} \mathrm{AF_p}(s) = F_m(score_m(s)) = \frac{1}{n}\sum_{i=1}^{n} \delta(x_i \leq score_m(s)), \\ \text{where } \delta(\mathrm{A}) = 1 \text{ if } A \text{ is true; } \delta(A) = 0 \text{ otherwise.} \end{array} \quad (1)$$

3 Results and Discussion

We built sequence generators for several target proteins shown in Table 1, and evaluated the binding affinity of the DNA sequences for their target proteins. For each target protein, Table 1 shows the AUC of the DeepBind model and the median binding affinity of the DNA sequences for their target proteins. For comparative purposes, both the median binding affinity of the DNA sequences generated by our model and that of random DNA sequences are shown. The reason that the median binding affinity was used instead of the mean binding affinity because the mean affinity was largely distorted by outliers in several DeepBind models. Overall, the DNA sequences generated by our model showed much higher binding affinity than random DNA sequences.

Table 1. The AUC of DeepBind model and the median binding affinity (AF) of the generated DNA sequences and random DNA sequences for eight target proteins.

	DRGX	GCM1	OLIG1	RXRB	SOX2	BHLHE23	MTF1	FOXP3
AUC	0.897	0.841	0.733	0.720	0.605	0.557	0.538	0.499
Median AF of generated sequences	0.999	0.995	0.763	0.644	0.473	0.513	0.557	0.502
Median AF of random sequences	0.497	0.500	0.504	0.494	0.504	0.501	0.502	0.501

In Table 1, the difference in median binding affinities between the generated sequences and random sequences is much larger for proteins with a large AUC of the DeepBind model. For example, binding affinities of the generated sequences for DRGX, GCM1, OLIG1 and RXRB are larger than those of random sequences. Figure 1 shows the distribution of binding affinities of the generated sequences and random sequences for the four target proteins. Unlike the generated sequences by our model, the distributions of binding affinities of the random sequences for the four proteins are all centered around 0, and very similar to each other.

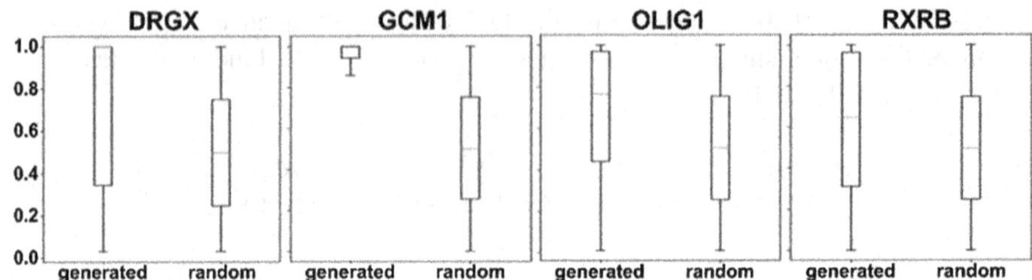

Fig. 1. The binding affinity (AF) of the generated and random DNA sequences for 4 proteins (DRGX, GCM1, OLIG1 and RXRB).

4 Conclusion

In this paper, we presented a generator model using LSTM to construct single-stranded nucleic acid sequences binding to a protein. After we trained the generator model on data of interactions between protein and DNA sequences obtained from high-throughput experiments, we constructed DNA sequences for several target proteins and evaluated them. DNA sequences generated by the model showed high binding affinity for several proteins. Although preliminary, our approach will help design efficient biochemical experiments by constructing an initial pool of nucleic acid sequences for a target protein.

Acknowledgments. This work was supported by the National Research Foundation of Korea (NRF) grant funded by the Ministry of Science and ICT (2015R1A1A3A04001243, 2017R1E1 A1A03069921) and the Ministry of Education (2016R1A6A3A11931497).

References

1. Walia, R.R., Xue, L.C., Wilkins, K., El-Manzalawy, Y., Dobbs, D., Honavar, V.: RNABindRPlus: a predictor that combines machine learning and sequence homology-based methods to improve the reliability of predicted RNA-binding residues in proteins. PLoS ONE **9**, e97725 (2014)
2. Tuvshinjargal, N., Lee, W., Park, B., Han, K.: PRIdictor: protein-RNA interaction predictor. BioSystems **139**, 17–22 (2016)
3. Choi, D., Park, B., Chae, H., Lee, W., Han, K.: Predicting protein-binding regions in RNA using nucleotide profiles and compositions. BMC Syst. Biol. **11**(Suppl. 2), 16 (2017)
4. Alipanahi, B., Delong, A., Weirauch, M.T., Frey, B.J.: Predicting the sequence specificities of DNA- and RNA-binding proteins by deep learning. Nat. Biotechnol. **33**, 831–838 (2015)
5. Hassanzadeh, H.R., Wang, M.D.: DeeperBind: enhancing prediction of sequence specificities of DNA binding proteins. In: Proceedings of IEEE International Conference on Bioinformatics and Biomedicine (BIBM), pp. 178–183 (2016)
6. Jolma, A., Yan, J., Whitington, T., et al.: DNA-Binding specificities of human transcription factors. Cell **152**, 327–339 (2013)

Constructing Gene Co-expression Networks for Prognosis of Lung Adenocarcinoma

Byungkyu Park, Jinho Im, and Kyungsook Han(✉)

Department of Computer Engineering, Inha University, Incheon, South Korea
{bpark, khan}@inha.ac.kr, 22162012@inha.edu

Abstract. Many studies of prognostic genes for cancer have focused on comparative analysis of gene expressions in cancer cells and normal cells. However, prognosis of cancer patients can be done more accurately by comparative analysis of patients with different conditions. In this study we partitioned the patients with lung adenocarcinoma into two groups, one with a wide-type TP53 gene and the other with somatic mutations in the TP53 gene, and constructed gene co-expression networks for the two groups. From the comparative analysis of the two GCNs we obtained several gene pairs with significantly different co-expression patterns in the two groups. The GCNs constructed in our study are more informative than other GCNs in the sense that ours provide the specific type of correlation between genes, the concordance and prognostic type of a gene. The GCNs will be informative for prognosis of lung adenocarcinoma, which is the most common type of lung cancer.

Keywords: Gene co-expression network · Differential expression analysis Cancer · Prognostic gene

1 Introduction

Lung cancer is the leading cause of cancer death in both men and women worldwide [1]. For therapeutic purposes, lung cancer is classified into two major types: non-small-cell lung carcinoma and small-cell lung carcinoma. This distinction is useful since treatment for non-small-cell lung cancers is quite different from that for small-cell lung carcinoma cancers. Non-small-cell lung carcinomas are the most common form, accounting for about 85–90% of total lung cancers [2].

Recent progress has significantly advanced the understanding of the types of lung cancer and their subtypes. The three main subtypes of non-small-cell lung carcinoma are adenocarcinoma, squamous cell carcinoma and large-cell carcinoma, which are classified according to the types of cells involved. Among the three subtypes, adenocarcinoma is the most common type of non-small-cell lung cancer, which accounts for over 40% of all lung cancer cases [3, 4]. It is commonly found in current or former smokers, but is also the most common type of lung cancer found in people who have never smoked.

The primary focus of this study is to construct gene co-expression networks prognosis of lung adenocarcinoma. So far, many studies for identifying prognostic factors in cancer patients have focused on differentially expressed genes in cancer cells

D.-S. Huang et al. (Eds.): ICIC 2018, LNCS 10955, pp. 831–839, 2018.
https://doi.org/10.1007/978-3-319-95933-7_92

and normal cells. For example, Gov and Arga [5] identified a prognostic gene module by comparing gene expression levels in epithelial cells from ovarian tumor and healthy samples. However, tumor heterogeneity in conjunction with genetic profiling and histopathologic examination is one of the most promising prognostic factors in the prediction of patient survival [6]. In particular, the presence of TP53 gene mutations is known to be closely associated with poor prognosis in patients with lung cancer [7].

In this study we partitioned the patients with lung adenocarcinoma into two groups, one with a wide-type TP53 gene (hereafter called wtTP53) and the other with somatic mutations in the TP53 gene (mTP53), and constructed gene co-expression networks for the two groups. The reason that we selected the TP53 gene for classifying gene expression samples is because somatic mutations in the TP53 gene occur frequently in all lung cancer subtypes. The rest of this paper presents our approach to constructing gene co-expression networks (GCNs) for two groups of patients and comparative analysis of the networks for prognosis of lung adenocarcinoma.

2 Materials and Methods

2.1 Gene Expression Data of Lung Cancer

We obtained gene expression data in primary tissue samples of lung adenocarcinoma (LUAD) patients using the Genomic Data Commons (GDC) data transfer tool of The Cancer Genome Atlas (TCGA) [8]. We used the gene expression and clinical data from the TCGA LUAD project. The data includes the expression levels of 60,483 genes in 513 LUAD patients. Subtypes of LUAD based on gene expression include terminal respiratory unit (TRU), proximal proliferative (PP), and proximal inflammatory (PI), which are also known as the bronchioid, magnoid, and squamoid subtypes, respectively [1, 4, 9]. Table 1 shows the number of LUAD patients and their subtypes used in our study.

Table 1. The number of lung adenocarcinoma (LUAD) patients with their subtypes, obtained from TCGA. mTP53: LUAD patients with somatic mutations in the TP53 gene. wtTP53: LUAD patients with a wild-type TP53 gene.

LUAD subtype	mTP53	wtTP53	Total
Terminal respiratory unit (TRU, bronchioid)	30	59	89
Proximal proliferative (PP, magnoid)	22	41	63
Proximal inflammatory (PI, squamoid)	51	27	78
Unknown	160	123	283
Total	263	250	513

2.2 Gene Filtering

From the initial 60,483 genes, we filtered out 39,097 genes with very low counts across all samples using edgeR [10] and obtained 21,386 genes for further analysis. We divided the LUAD patients into two groups according to the existence of somatic mutations in the TP53 gene: one group of LUAD patients with a wide-type TP53 gene (wtTP53) and another of LUAD patients with somatic mutations in the TP53 gene (mTP53). We performed differentially expressed gene (DEG) analysis between wtTP53 and mTP53 with the limma R package [11]. From the DEG analysis, we obtained a total of 357 genes with an adjusted p-value < 0.001.

2.3 Concordance Index and Prognostic Type

To evaluate the prognostic significance of genes, we built Cox proportional hazard models for each gene in wtTP53 and mTP53. A pair of patients is called concordant if the risk of the event predicted by a model is lower for the patient who experiences the event at a later time point. The concordance index (C-index) is the frequency of concordant pairs among all pairs of subjects. It can be used to measure and compare the predictive power of a model. For each Cox regression model, we calculated the concordance index [12] by Eq. 1. In the equation, T is the recorded survival time and $\widehat{T}$ is the survival time estimated by the Cox model. Patients with no follow-up data were excluded when computing the concordance index.

$$\text{Concordance Index} = \frac{1}{N}\sum\nolimits_{T_i \text{uncensored}} \sum\nolimits_{T_j > T_i} \delta\left(\widehat{T}_i > \widehat{T}_i\right), \tag{1}$$
where $\delta(A)$is 1 if $A$is true, 0 otherwise.

In addition to the concordance index, we identified the prognostic type of genes based on the study by Uhlen *et al.* [13]. Depending on clinical outcome, genes are classified into two prognostic types: (1) prognostically favorable genes are those with a correlation between a higher expression and a longer survival time and (2) prognostically unfavorable genes are those with a correlation between a higher expression and a shorter survival time.

2.4 Gene Co-expression Network

For every pair of core genes in wtTP53 and mTP53, we computed the Pearson correlation coefficients (PCC) between their expression levels by Eq. 2. In the equation, N is the number of patients and $\bar{x}$ is the mean of x.

$$\mathrm{PCC}\left(\mathrm{x_i}, x_j\right) = \frac{\sum_{k=1}^{N}\left(x_{ik} - \overline{x_i}\right)\left(x_{jk} - \overline{x_j}\right)}{\sqrt{\sum_{k=1}^{N}\left(x_{ik} - \overline{x_i}\right)^2 \sum_{k=1}^{N}\left(x_{jk} - \overline{x_j}\right)^2}} \tag{2}$$

We constructed two separate gene co-expression networks (GCNs) for wtTP53 and mTP53, in which the weight of an edge represents a PCC between two genes. For comparison of GCNs for wtTP53 and mTP53, we selected gene pairs that satisfy two conditions: (1) the sign of PCC between the gene pairs is opposite in the two GCNs and (2) the absolute value of PCC > 0.3 in at least one GCN.

3 Results and Discussion

3.1 Differentially Expressed Genes in Two Groups

Figure 1 shows a volcano plot for the fold change of expression levels of 21,386 genes between wrTP53 and mTP53 on a log2 scale. Genes with positive fold changes indicate those with higher expression levels in mTP53 than in wtTP53 and genes with negative fold changes indicate the opposite.

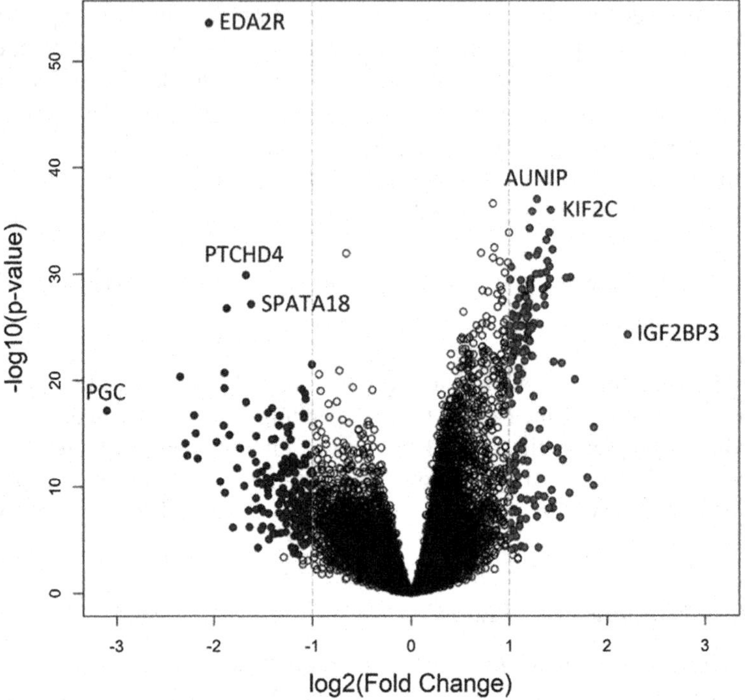

Fig. 1. (A) Volcano plot for the fold change (FC) of gene expression levels between two groups, wtTP53 and mTP53. The horizontal axis represents the fold change (FC) between the two groups on a log2 scale, and the vertical axis shows the negative logarithm of p-values from the t-test to the base 10. A gene with a higher expression level in mTP53 than in wtTP53 has a positive FC and shown as a red dot. A gene with a lower expression level in mTP53 than in wtTP53 has a negative FC and shown as a blue dot. Genes with significantly different expressions in two groups are marked by their names. (Color figure online)

Using an adjusted p-value < 0.001, we selected 357 genes with significant fold changes. Among the 357 genes, the fold change of 194 genes was smaller than −1 and the fold change of 163 genes was larger than 1. Since the fold change was computed on a log2 scale, the expression levels of the 194 genes were higher more than two-fold in wtTP53 than in mTP53, whereas the expression levels of the 163 genes were higher more than two-fold in mTP53 than in wtTP53. Table 2 shows the top 20 genes with significant fold changes.

Table 2. The concordance index (C-index) computed for the top 20 genes with significantly different expressions in mTP53 and wtTP53. The prognostic type of the genes was obtained from [13]. The 20 genes were selected from the analysis of fold changes shown in Fig. 1.

Gene	Log2(FC)	−log10(p-value)	C-index		Prognostic type
			mTP53	wtTP53	
EDA2R	−2.046	53.564	0.530	0.563	Favorable
AUNIP	1.287	37.011	0.514	0.573	Unfavorable
CDCA8	1.209	34.253	0.502	0.605	Unfavorable
RAD54L	1.281	31.748	0.528	0.420	Unfavorable
XRCC2	1.199	31.639	0.503	0.608	Unfavorable
KIF4A	1.391	31.152	0.587	0.609	Unfavorable
TPX2	1.416	30.637	0.552	0.626	Unfavorable
MCM2	1.023	30.610	0.508	0.588	Unfavorable
HJURP	1.378	30.105	0.550	0.646	Unfavorable
PTCHD4	−1.672	29.870	0.518	0.556	Favorable
MYBL2	1.626	29.641	0.496	0.607	Unfavorable
DEPDC1	1.419	29.504	0.547	0.611	Unfavorable
KIFC1	1.215	28.861	0.516	0.626	Unfavorable
MELK	1.360	28.655	0.527	0.608	Unfavorable
KIF15	1.130	28.565	0.508	0.602	Unfavorable
KIF23	1.165	27.978	0.537	0.627	Unfavorable
CENPI	1.161	27.852	0.540	0.419	Unfavorable
GINS1	1.099	27.690	0.487	0.591	Unfavorable
SPATA18	−1.620	27.140	0.517	0.562	Favorable
TICRR	1.175	26.614	0.530	0.619	Unfavorable

For some genes, very low p-values or significantly different expressions between wtTP53 and mTP53 were observed. For instance, the expression level of IGF2BP3 was four-fold higher in mTP53 than in wtTP53, whereas the expression levels of eight genes, EDA2R, SFTPB, PGC, GKN2, ERN2, SCHB3A2, SFTPA1B, SFTPA1, were higher more than four-fold in wtTP53 than in mTP53.

3.2 Gene Co-expression Network

From the correlation of genes, concordance index, and prognostic type of genes, we constructed gene co-expression networks (GCNs). Figure 2 shows the GCN for wtTP53, which contains 175 correlations between 171 genes. In the network, a red node represents a prognostically favorable gene, for which higher expression of the gene is correlated with a longer patient survival outcome. A blue node represents a prognostically unfavorable gene, for which higher expression of the gene is correlated with a poor patient survival outcome. The node size is proportional to the concordance index of the gene. The larger the concordance index of a gene, the better it predicts survival time. Solid edges represent positive correlations between gene expressions and dashed lines represent negative correlations between gene expressions.

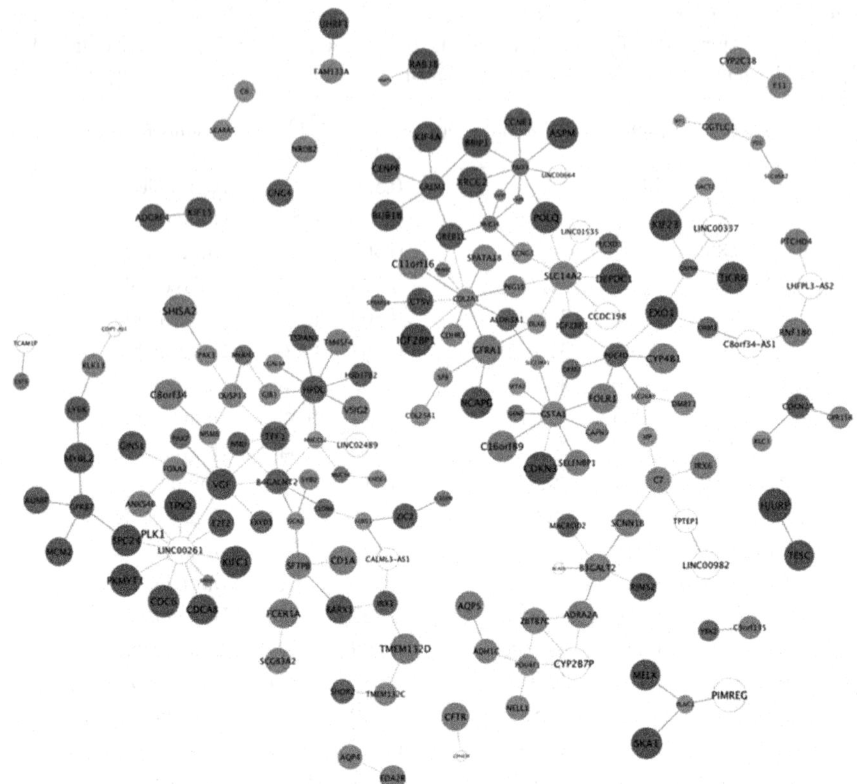

Fig. 2. Gene co-expression network of 175 correlations between 171 genes for wtTP53. A red node represents a prognostically favorable gene, for which higher expression of the gene is correlated with a longer patient survival outcome. A blue node represents a prognostically unfavorable gene, for which higher expression of the gene is correlated with a poor patient survival outcome. The node size is proportional to the concordance index. The larger the concordance index of a gene, the better it predicts survival time. Solid edges represent positive correlations and dashed lines represent negative correlations. (Color figure online)

Figure 3 displays the GCN for mTP53, which is composed of 175 correlations between 171 genes. The GCNs constructed in our work are more informative for prognosis analysis than typical GCNs in the following sense: (1) the edge type (solid line vs dashed line) indicates whether two genes are positively or negatively correlated with other, (2) the prognostic type of a gene (prognostically favorable vs unfavorable) is represented by the color of a node (red vs blue), and (3) the size and color brightness of a node indicates the concordance of a gene.

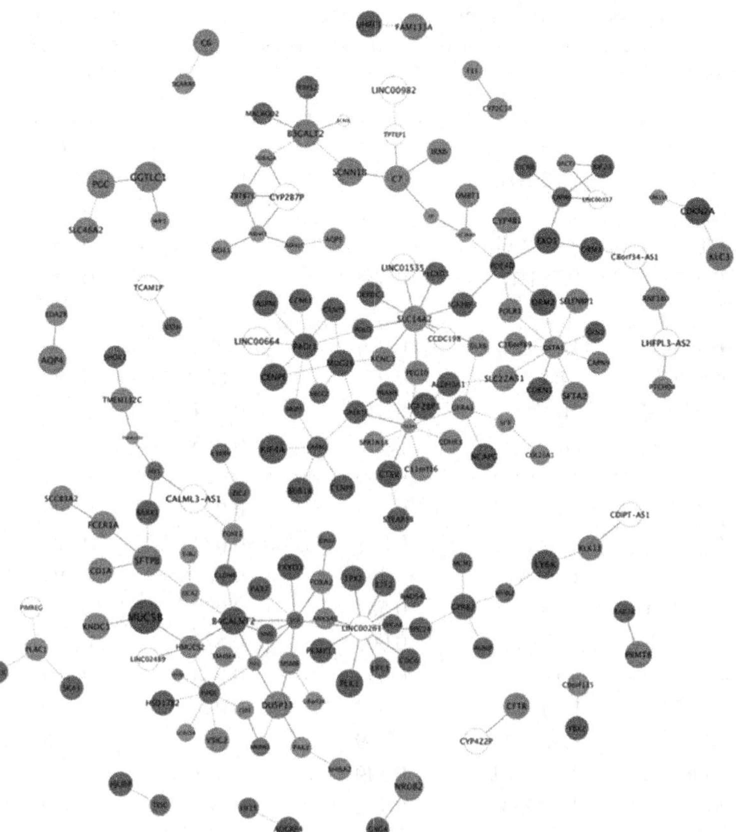

Fig. 3. Gene co-expression network of 175 correlations between 171 genes for mTP53. A red node represents a prognostically favorable gene, for which higher expression of the gene is correlated with a longer patient survival outcome. A blue node represents a prognostically unfavorable gene, for which higher expression of the gene is correlated with a poor patient survival outcome. The node size is proportional to the concordance index. The larger the concordance index of a gene, the better it predicts survival time. Solid edges represent positive correlations and dashed lines represent negative correlations. (Color figure online)

3.3 Comparative Analysis of GCNs in Two Groups

The two GCNs for wtTP53 and mTP53 are different in several respects. First, many genes showed different expression patterns in wtTP53 and mTP53. Second, but more importantly, several gene pairs showed different co-expression patterns. In particular, we found 175 gene pairs that satisfy two criteria: (1) PCCs have opposite signs in the two GCNs and (2) the absolute value of PCC > 0.3 in at least one GCN. Table 3 shows the top 20 gene pairs with significantly different PCCs in mTP53 and wtTP53.

Table 3. Gene pairs that are differentially co-expressed in wtTP53 and mTP53. PCC: Pearson correlation coefficient. Difference in PCC: absolute value of the difference in PCCs between mTP53 wtTP53.

Gene pairs		PCC in mTP53	PCC in wtTP53	Difference in PCC
COL2A1	GREB1L	0.782	−0.059	0.841
C7	HP	0.700	−0.066	0.766
CTSV	COL2A1	0.732	−0.028	0.761
POU4F1	ADH1C	0.665	−0.080	0.745
VGF	MSMB	0.683	−0.028	0.711
HMGCS2	HPDL	−0.020	0.681	0.701
GFRA1	SP8	−0.073	0.627	0.699
LGALS4	HPDL	−0.046	0.604	0.650
BARX1	IRX1	0.549	−0.086	0.635
LHFPL3-AS2	PTCHD4	0.589	−0.014	0.603
C7	IRX6	0.578	−0.021	0.599
COL2A1	PRAME	0.571	−0.024	0.596
NCAPG	GFRA1	−0.267	0.327	0.595
SLC14A2	LINC01535	0.552	−0.034	0.586
HSD17B2	HPDL	−0.009	0.575	0.585
COL2A1	GFRA1	−0.052	0.528	0.580
RNF180	LHFPL3-AS2	0.543	−0.036	0.579
IRX1	CALML3-AS1	0.469	−0.107	0.576
COL2A1	IGF2BP1	0.529	−0.028	0.557
SLC14A2	KCNG3	0.442	−0.108	0.550

4 Conclusion

Lung cancer is a heterogeneous cancer with many subtypes, and different subtypes require different treatments. The most common subtype of lung cancer is adenocarcinoma, which accounts for over 40% of all lung cancer cases.

In this study we clustered the 513 patients with lung adenocarcinoma into two groups: one with a wild-type TP53 gene (wtTP53) and the other with somatic mutations in the TP53 gene (mTP53). From the analysis of differentially expressed genes in wtTP53 and mTP53, we selected 357 selected genes and constructed gene co-expression networks (GCNs) enriched with prognostic information. From the comparative analysis of the two GCNs we obtained several gene pairs with opposite correlations in the two groups. The GCNs constructed in our study are more informative than other GCNs in the sense that ours provide the specific type of correlation between genes, the concordance and prognostic type of a gene. The enriched GCNs will be informative for prognosis of survival and for selecting a drug target for lung cancer.

Acknowledgments. This work was supported by the National Research Foundation of Korea (NRF) grant funded by the Ministry of Science and ICT (2017R1E1A1A03069921) and the Ministry of Education (2016R1A6A3A11931497).

References

1. Collisson, E., Campbell, J., Brooks, A., et al.: Comprehensive molecular profiling of lung adenocarcinoma. Nature **511**, 543–550 (2014)
2. American Cancer Society. http://www.cancer.org/cancer/lungcancer-non-smallcell/detailedguide/non-small-cell-lung-cancer-what-is-non-small-cell-lung-cancer
3. Meza, R., Meernik, C., Jeon, J., Cote, M.L.: Lung cancer incidence trends by gender, race and histology in the United States. PLoS ONE **10**, e0121323 (2015)
4. Faruki, H., Mayhew, G., Serody, J., Hayes, D., Perou, C., Lai-Goldman, M.: Lung adenocarcinoma and squamous cell carcinoma gene expression subtypes demonstrate significant differences in tumor immune landscape. J. Thorac. Oncol. **12**, 943–953 (2017)
5. Gov, E., Arga, K.Y.: Differential co-expression analysis reveals a novel prognostic gene module in ovarian cancer. Sci. Rep. **7**(1), 4996 (2017)
6. Yoon, S., Park, C., Park, S., Yoon, J., Hahn, S., Goo, J.: Tumor heterogeneity in lung cancer: assessment with dynamic contrast-enhanced MR imaging. Radiology **280**, 940–948 (2016)
7. Gu, J., Zhou, Y., Huang, L., Ou, W., Wu, J., Li, S., Xu, J., Feng, J., Liu, B.: TP53 mutation is associated with a poor clinical outcome for non-small cell lung cancer: evidence from a meta-analysis. Mol. Clin. Oncol. **5**, 705–713 (2016)
8. Weinstein, J., Collisson, E., Mills, G., Shaw, K., Ozenberger, B., Ellrott, K., Shmulevich, I., Sander, C., Stuart, J., Cancer Genome Atlas Research Network: The cancer genome atlas pan-cancer analysis project. Nat. Genet. **45**, 1113–1120 (2013)
9. Wilkerson, M., Yin, X., Walter, V., Zhao, N., Cabanski, C., Hayward, M., Miller, C., Socinski, M., Parsons, A., Thorne, L., Haithcock, B., Veeramachaneni, N., Funkhouser, W., Randell, S., Bernard, P., Perou, C., Hayes, D.: Differential pathogenesis of lung adenocarcinoma subtypes involving sequence mutations, copy number, chromosomal instability, and methylation. PLoS ONE **7**, e36530 (2012)
10. Robinson, M., McCarthy, D., Smyth, G.: edgeR: a Bioconductor package for differential expression analysis of digital gene expression data. Bioinformatics **26**, 139–140 (2010)
11. Ritchie, M., Phipson, B., Wu, D., Hu, Y., Law, C., Shi, W., Smyth, G.: limma powers differential expression analyses for RNA-sequencing and microarray studies. Nucl. Acids Res. **43**, e47 (2015)
12. Raykar, V.C., Steck H., Krishnapuram, B., Dehing-Oberije, C., Lambin, P.: On ranking in survival analysis: bounds on the concordance index. Advances in Neural Information Processing Systems 20 (2007)
13. Uhlen, M., Zhang, C., Lee, S., et al.: A pathology atlas of the human cancer transcriptome. Science **357**, 660 (2017)

RMCL-ESA: A Novel Method to Detect Co-regulatory Functional Modules in Cancer

Jiawei Luo(✉) and Ying Yin(✉)

College of Computer Science and Electronic Engineering of Hunan University, Collaboration and Innovation Center for Digital Chinese Medicine in Hunan Province, Changsha 410082, Hunan, China
{luojiawei,yinying}@hnu.edu.cn

Abstract. Considering the increasingly large scale of gene expression data, common module identification algorithms exist many problems, such as large search space and long running time. A novel co-regulatory modules identification algorithm RMCL-ESA (Regularized Markov Cluster & Explosion Search Algorithm) based on improved Markov cluster and explosion search strategy has been proposed. Improved Markov cluster is adapted to preprocess gene expression profiles through three subprocedure: expansion, inflation, prune, which filter redundant genes and save computational cost. Then, two-stage explosion search strategy has been explored for identifying co-regulatory modules. Comparing with existing methods on breast cancer and ovary cancer datasets from TCGA, CRMs (Co-regulatory Functional Modules) of RMCL-ESA include more significant biological function GO-terms and regulation pathways with high enrichment score.

Keywords: Co-regulatory modules · Markov cluster · Explosion search

1 Introduction

With the extension and development of high-throughput sequencing technology, a large number of biomolecular interactions data have been accumulated [1]. These data display important theoretical and practical value on gene expression, gene mutations, gene identification and cancer diagnosis [2]. The research of cancer co-regulatory network helps understand the generation process of oncogenes inside cancer tissues.

miRNAs, TFs, and target genes (mRNAs) are the basic factor types of co-regulatory functional modules (CRMs) [3]. Overlapping CRMs is also important for co-regulatory network. Yu et al. put forward a method called Regularized Markov Cluster (R-MCL) based on iteratively executing regularized Markov cluster [4]. Li et al. put forward a algorithm Misynergy, which use overlapping neighbourhood expansion to detect synergistic miRNA regulatory modules by two stages [5].

In this work, we propose a novel method called RMCL-ESA (Regularized Markov Cluster & Explosion Search Algorithm) to detect overlapping CRMs based on improved Markov cluster and explosion search. Compared with traditional modules detection algorithms, our algorithm framework includes two modified aspects:

D.-S. Huang et al. (Eds.): ICIC 2018, LNCS 10955, pp. 840–846, 2018.
https://doi.org/10.1007/978-3-319-95933-7_93

- In order to reduce running time of algorithm, improved Markov cluster is adapted to preprocess co-regulatory network combined which make the subsequent processing more quickly and save storage space.
- Considering special regulate patterns among regulators and target genes, RMCL-ESA researches a two-stage explosion search greedy algorithm to identify overlapping CRMs.

To validate the validity of the algorithm, breast cancer (BRCA) and ovarian cancer (OVCA) datasets are applied to RMCL-ESA. Besides, we also select two contrast algorithms: the classical spectral clustering algorithm (NJW) [6] and non-negative matrix factorization (SNMNMF) [7].

2 Method

2.1 Improved Markov Cluster

Markov cluster algorithm aims to finish iterative processes of random walk continuously in the complex network. $G(V, E, W)$ expresses the co-regulatory network which is built by OSC [8]. Markov cluster algorithm defined three matrix operations: expansion, inflation and prune. The realization processes are as follows:

Expansion. Expansion process means adjacency matrix multiplication essentially. Standardizing the new adjacency matrix:

$$M_G(i,j) = \frac{A(i,j)}{\sum_{x=1}^{n} A(x,j)} \tag{1}$$

Where A is the adjacency matrix of co-regulatory network and $M_G(i,j)$ is the standardized migration matrix. Considering the unbalanced distribution in the results from classic R-MCL algorithm, the penalty factor is introduced to punish the nodes with high attribute expression values. So the standardized migration matrix is regenerated as:

$$M_R = normaliza(M_G * (diag(M'_G * \sum_j M(i,j)))^b) \tag{2}$$

Where b is the penalty factor. Therefore, continue to finish the expansion operation:

$$Expand(M_{Ex}) = M_G * M_R \tag{3}$$

Expansion aims to connect different areas within the complex network. With the increasing of multiplication times, the tightness of same clusters would be weaken.

Inflation. Inflation process designs to enhance the clustering probability of internal migration, and weaken the migration probability of external migration. Firstly, apply point multiplication and normalization processing for M_{Ex}:

$$Inflate(M_{In}) = \frac{M_{Ex}(i,j)^r}{\sum_{k=1}^{n} M_{Ex}(k,j)^r}, \quad r = 2 \tag{4}$$

Considering the slow convergence speed of classic R-MCL algorithm, the mutation factor is introduced to avoid restriction from initial transition probability. Select the attractive nodes in co-regulatory network:

$$Attract(X_{At}) = \left\{ \sum\nolimits_{k}^{n} A(i,k) > \frac{1}{n} \sum\nolimits_{i}^{m} \sum\nolimits_{j}^{n} A(i,j) | x_i \in G \right\} \tag{5}$$

The punitive migration process can be represented as:

$$PenaltyInflate(M_{Pe}) = \frac{M_{In}(i,j)^{Pe}}{\sum_{k=1}^{n} M_{In}(k,j)^{Pe}}, \quad x_i, x_j \in Attract(X_{At}) \tag{6}$$

Where Pe is the mutation factor for punitive migration.

Prune. Prune process aims to delete the unobvious nodes and edges with low transition probability. We use the threshold pruning method, take the fixed threshold d, and retain the probability value which is greater than d, which is represented as:

$$Prune(M_{Pr}) = \begin{cases} M_{Pe}(i,j), & M_{Pe}(i,j) > d \\ 0 & M_{Pe}(i,j) \leq d \end{cases} \tag{7}$$

As to cancer co-regulatory network, there are more than 10,000 nodes and edges. Therefore, the RMCL only run one time for cancer network. The rest process of nodes gathered is completed by two-stage explosion search algorithm.

2.2 Two-Stage Explosion Search Algorithm

Based on fireworks algorithm [9] and fireworks explosion optimization [10], two-stage explosion search algorithm is proposed to identify overlapping CRMs. Due to the number of biological network nodes is too much, different start nodes would be generate different results. So we start search operation from nodes with high betweenness. The calculation of betweenness is as follows:

$$Center(i) = \sum_{j,k \in v} \frac{n_{jk}(i)}{n_{jk}} \tag{8}$$

Where, v expresses the node set, $n_{jk}(i)$ is the number of shortest paths through i between j and k. n_{jk} means the number of shortest paths between j and k. And we use *Dijkstra* algorithm to calculate the shortest path between nodes.

Stage 1. Regulators are sorted by their betweenness from large to small, and then start clustering from the largest regulator by maximizing fitness function $CenterModule(V_c)$ in turn. Fitness function $CenterModule(V_c)$ comes from two criterions: the minimum external linkage and the maximal internal accumulation, which describes in OSC [8]. $CenterModule(V_c)$ is calculated as:

$$CenterModule(V_c) = \frac{\sum W_{in}(v)}{\sum E_{max}(v) + \sum_{i,j=1}^{m,n} M(i,j)} + C_{in}(v) \tag{9}$$

$$C_{in}(v) = \frac{2 \times n_{in}}{Ne_{in} \times (Ne_{in} - 1)} \tag{10}$$

Where, $\sum W_{in}(v)$ shows the weight of internal connections for node v in module V_c. $\sum E_{max}(v)$ shows the weight of connections to external modules. $\sum M(i,j)$ expresses the whole weights of node v in module V_c. $C_{in}(v)$ is the internal clustering coefficient of node v in module V_c, Ne_{in} shows the number of neighbors for node v in module V_c, n_{in} expresses the number of linkages among these neighbors. The entire process terminates until all of the miRNAs and TFs are considered.

Stage 2. We apply the similar clustering procedure as in stage 1 for mRNAs, which means add mRNAs to previous regulator clusters through maximizing fitness function $FiresModule(V_c)$. Different from stage 1, considering the center regulator clusters from stage 1 have strong cohesiveness, so expansion factor α and β have been added into the new fitness function $FiresModule(V_c)$:

$$FiresModule(V_c) = (\sum W_{in}(v) * \alpha) / \sum E_{max}(v) + C_{in}(v) * \beta \tag{11}$$

3 Results

3.1 Data Set Collection

The miRNA/mRNA/TF expression profiles of OVCA and BRCA are downloaded from the Cancer Genome Atlas (TCGA, http://cancergenome.nih.gov/). The total information of datasets are showed in Table 1.

Table 1. Summary of two kinds of genomic datasets

Cancers	Sample (T/N)	miRNA	mRNA	TF
Breast	541 (461/80)	358	16344	112
Ovary	292 (292/0)	351	11980	100

3.2 Comparison of Module Characteristics

Gene ontology terms (GO) and Kyoto Encyclopedia of Genes and Genomes pathways (KEGG) are the important enrichment analysis methods in bioinformatics. Li et al. proposed the GO enrichment score (GOES) to measure GO-BP enrichment degree for each module [5]. Similarly, we put forward the KEGG enrichment score (KEGGES) to measure KEGG pathway enrichment degree for each module, which is defined as:

$$KEGG\ enrichment\ score = \frac{1}{P} \times \sum_{g}^{P} - \log_{10}(FDR_g) \tag{12}$$

Where P represents the number of KEGG pathways that the genes enriched in. Finally, the detailed results can be seen in Table 2.

Table 2. Performance summary of RMCL-ESA, NJW and SNMNMF

Algorithm	Cancer	M#	miRNA	TF	GOBP	GOES	KEGG	KEGGES	Time
RMCL-ESA	BRCA	24	**6.54**	1.13	**18535**	**3.2391**	**526**	**5.2737**	**6 h**
NJW		27	2.07	1.26	6567	1.7860	103	1.7735	24 h
SNM		50	17.86	-	8771	1.8068	124	1.7244	12 h
RMCL-ESA	OVCA	28	**4.64**	**2.68**	**12026**	**1.9529**	**209**	**2.0345**	**5 h**
NJW		27	1.67	1.19	6164	1.8043	141	1.7365	24 h
SNM		50	16.4	-	7432	1.8007	146	1.6698	12 h

Comparing with NJW and SNMNMF, RMCL-ESA gets the best enrichment effect. As to breast cancer, RMCL-ESA achieves 18535 GO-BP terms, and each term has significant biological significance with 3.2391 average. Our method also achieves 526 KEGGE pathways, and each pathway has significant biological significance with 5.2737 average, which is three times higher than NJW and SNMNMF. So overall, RMCL-ESA attains the most biological enrichment information in CRMs.

3.3 Evaluating Modules by Functional Enrichments

After exploring basic information of CRMs, we compared functional enrichments of RMCL-ESA with NJW and SNMNMF specially. Therefore, we contrasted the empirical cumulative distribution function of GOES and KEGGES in Fig. 1.

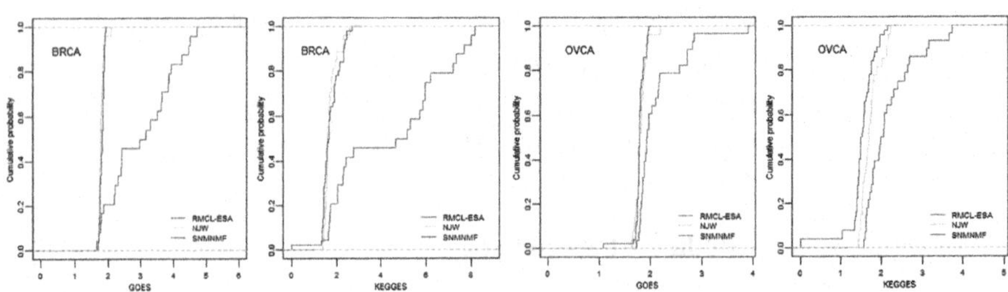

Fig. 1. The empirical cumulative distribution analysis of GOES and KEGGES in two cancer datasets.

It can be seen from the comparison results among three algorithms, RMCL-ESA can get more GO-BP terms and KEGG pathways with high enrichment score. However, the enrichment score of NJW-CRMS and SNMNMF-CRMS only distributes in (1,2). But GOES and KEGGES of many CRMs from RMCL-ESA can achieve (4,5), which is three time higher than other two methods. From the above analysis, CRMs from RMCL-ESA have the best performance and possess significant biological significance.

3.4 Survival Analysis

Survival analysis is the method to research relationships between survival time and many influence factors, which analyzes and judges from multifactor performance in clinical datasets. Therefore, based on the clinical data of OVCA from TCGA, the associations of patient survival time from diagnosis to death are explored on CRMs of OVCA. As a result, we found four CRMs (CRM-9, CRM-10, CRM-21, CRM-24) with FDR < 0.05 for OVCA, and the detailed analysis results have been showed in Fig. 2.

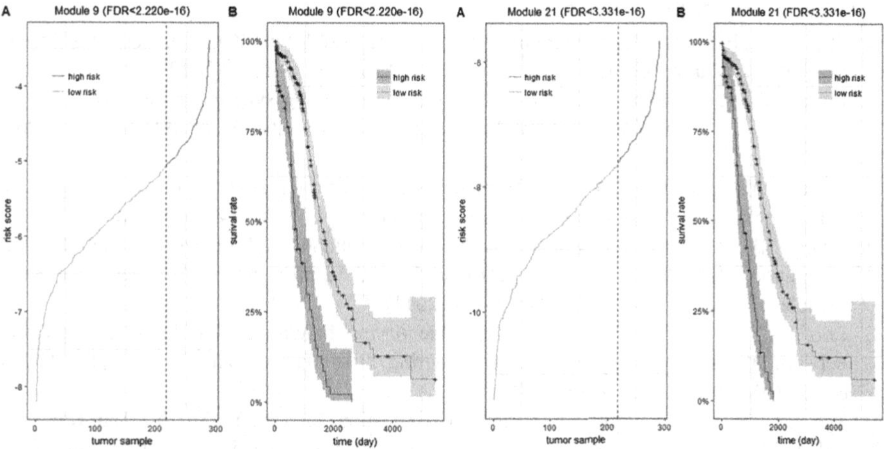

Fig. 2. KM survival analysis using CRMs in ovarian cancer. Survival rates between initial diagnosis and death were plotted for high-risk group (pink area) and low-risk group (turquoise area). And the significant separation of these two curves are assessed by log-rank test. CRM-9 and CRM-21 with significant FDRs were displayed.

4 Conclusion

In this article, we developed a novel and effective method RMCL-ESA to identifying overlapping CRMs based on improved markov cluster and explosion search strategy. Firstly, considering the large scale of miRNA/mRNA/TF expression profiles, markov cluster algorithm is adapted to reduce cancer co-regulatory network through random walk. Then, after removing redundant nodes and edges in RMCL, two-stage explosion search strategy is studied to identifying modules. In order to demonstrate the utility of our method, RMCL-ESA was compared with NJW and SNMNMF on breast cancer and ovarian cancer. We analyzed that CRMs identified by RMCL-ESA are more functionally enriched. Moreover, through the study of accumulated experience distribution analysis on GOES and KEGGES, RMCL-ESA can achieve co-regulatory modules with high enrichment degree. And CRMs of RMCL-ESA can separate patients samples and normal samples through survival analysis, which reflect significant biological significance.

References

1. Ding, P.J., Luo, J.W., Xiao, Q., Chen, X.T.: A path-based measurement for human miRNA functional similarities using miRNA-disease associations. Sci. Rep. **6**, 32533 (2016)
2. Xiao, Q., Luo, J.W., Liang, C., Cai, J., Ding, P.J.: A graph regularized non-negative matrix factorization method for identifying microRNA-disease associations. Bioinformatics **34**(2), 239–248 (2018)
3. Luo, J.W., Xiang, G., Pan, C.: Discovery of microRNAs and transcription factors co-regulatory modules by integrating multiple types of genomic data. IEEE Trans. Nanobiosci. **16**(1), 51–59 (2017)
4. Shih, Y.K., Parthasarathy, S.: Identifying functional modules in interaction networks through overlapping Markov clustering. Bioinformatics **28**(18), i473–i479 (2012)
5. Li, Y., Liang, C., Wong, K.C., et al.: Mirsynergy: detecting synergistic miRNA regulatory modules by overlapping neighbourhood expansion. Bioinformatics **30**(18), 2627–2635 (2014)
6. Andrew, Y.N., Michael, I.J., Yair, W.: On spectral clustering: analysis and an algorithm. In: Advance in Neural Information Processing Systems, vol. 2, pp. 849–856 (2002)
7. Zhang, S.H., Li, Q., Liu, J., Zhou, X.J.: A novel computational framework for simultaneous integration of multiple types of genomic data to identify microRNA gene regulatory modules. Bioinformatics **27**(13), 401–409 (2011)
8. Luo, J., Yin, Y., Chu Pan, G.X., et al.: Identifying functional modules in co-regulatory networks through overlapping spectral clustering. IEEE Trans. Nanobiosci. (2018). https://doi.org/10.1109/TNB.2018.2805846
9. Tan, Y., Zhu, Y.: Fireworks algorithm for optimization. In: Tan, Y., Shi, Y., Tan, Kay Chen (eds.) ICSI 2010. LNCS, vol. 6145, pp. 355–364. Springer, Heidelberg (2010). https://doi.org/10.1007/978-3-642-13495-1_44
10. Zhang, Q., Liu, H., Dai, C.: Fireworks explosion optimization algorithm for parameter identification of PV model. In: 2016 IEEE 8th International Power Electronics and Motion Control Conference (IPEMC-ECCE Asia), pp. 1587–1591. IEEE (2016)

Author Index

Printed in the United States
By Bookmasters